Lecture Notes in Physics

Springer
Berlin
Heidelberg
New York
Barcelona
Hong Kong
London
Milan
Paris
Tokyo

Physics and Astronomy ONLINE LIBRARY

http://www.springer.de/phys/

Editorial Policy

The series *Lecture Notes in Physics* (LNP), founded in 1969, reports new developments in physics research and teaching -- quickly, informally but with a high quality. Manuscripts to be considered for publication are topical volumes consisting of a limited number of contributions, carefully edited and closely related to each other. Each contribution should contain at least partly original and previously unpublished material, be written in a clear, pedagogical style and aimed at a broader readership, especially graduate students and nonspecialist researchers wishing to familiarize themselves with the topic concerned. For this reason, traditional proceedings cannot be considered for this series though volumes to appear in this series are often based on material presented at conferences, workshops and schools (in exceptional cases the original papers and/or those not included in the printed book may be added on an accompanying CD ROM, together with the abstracts of posters and other material suitable for publication, e.g. large tables, colour pictures, program codes, etc.).

Acceptance

A project can only be accepted tentatively for publication, by both the editorial board and the publisher, following thorough examination of the material submitted. The book proposal sent to the publisher should consist at least of a preliminary table of contents outlining the structure of the book together with abstracts of all contributions to be included.
Final acceptance is issued by the series editor in charge, in consultation with the publisher, only after receiving the complete manuscript. Final acceptance, possibly requiring minor corrections, usually follows the tentative acceptance unless the final manuscript differs significantly from expectations (project outline). In particular, the series editors are entitled to reject individual contributions if they do not meet the high quality standards of this series. The final manuscript must be camera-ready, and should include both an informative introduction and a sufficiently detailed subject index.

Contractual Aspects

Publication in LNP is free of charge. There is no formal contract, no royalties are paid, and no bulk orders are required, although special discounts are offered in this case. The volume editors receive jointly 30 free copies for their personal use and are entitled, as are the contributing authors, to purchase Springer books at a reduced rate. The publisher secures the copyright for each volume. As a rule, no reprints of individual contributions can be supplied.

Manuscript Submission

The manuscript in its final and approved version must be submitted in camera-ready form. The corresponding electronic source files are also required for the production process, in particular the online version. Technical assistance in compiling the final manuscript can be provided by the publisher's production editor(s), especially with regard to the publisher's own Latex macro package which has been specially designed for this series.

Online Version/ LNP Homepage

LNP homepage (list of available titles, aims and scope, editorial contacts etc.):
http://www.springer.de/phys/books/lnpp/

LNP online (abstracts, full-texts, subscriptions etc.):
http://link.springer.de/series/lnpp/

P. Garbaczewski R. Olkiewicz (Eds.)

Dynamics of Dissipation

Springer

Editors

Piotr Garbaczewski
University of Zielona Góra
Institute of Physics
pl. Słowiański 6
65069 Zielona Góra, Poland

Robert Olkiewicz
University of Wrocław
Institute of Theoretical Physics
pl. M. Borna 9
50205 Wrocław, Poland

Cover Picture: (see contribution by D. Cohen in this volume. The Brownian motion is induced due to the interaction with chaotic degrees of freedom.)

Cataloging-in-Publication Data applied for

Bibliograhpic information published by Die Deutsche Bibliothek
Die Deutsche Bibliothek lists this publication in the Deutsche Nationalbibliografie; detailed bibliographic data is available in the Internet at http://dnb.ddb.de

ISSN 0075-8450
ISBN 3-540-44111-5 Springer-Verlag Berlin Heidelberg New York

Springer-Verlag Berlin Heidelberg New York
a member of BertelsmannSpringer Science+Business Media GmbH

http://www.springer.de

Printed in Germany

Typesetting: Camera-ready by the authors/editor
Camera-data conversion by Steingraeber Satztechnik GmbH Heidelberg
Cover design: *design & production*, Heidelberg

Printed on acid-free paper
SPIN: 10887917 54/3141/du - 5 4 3 2 1 0

Preface

The present volume contains expanded and substantially reworked records of invited lectures delivered during the 38th Karpacz Winter School of Theoretical Physics on "Dynamical Semigroups: Dissipation, Chaos, Quanta", which took place in Lądek Zdrój, (Poland) in the period 6–15 February 2002.

The main purpose of the school was to create a platform for the confrontation of viewpoints and research methodologies represented by two groups of experts actually working in the *very same* area of theoretical physics. This situation is quite distinct in non-equilibrium statistical physics of open systems, where classical and quantum aspects are addressed separately by means of very different and even incompatible formal tools.

The school topics selection by the Lecturers reads: dissipative dynamics and chaotic behaviour, models of environment–system coupling and models of thermostats; non-equilibrium statistical mechanics and far from equilibrium phenomena; quantum open systems, decoherence and links to quantum chaos; quantum and classical applications of Markov semigroups and the validity of Markovian approximations.

The organizing principle for the whole endeavour was the issue of the *dynamics of open systems* and more specifically – the *dynamics of dissipation*. Since this research area is extremely broad and varied, no single book can cover all important developments. Therefore, links with dynamical chaos were chosen to represent a supplementary constraint.

The programme of the school and its final outcome in the form of the present volume has been shaped with the help of the scientific committee comprising: R. Alicki, Ph. Blanchard, J. R. Dorfman, G. Gallavotti, P. Gaspard, I. Guarneri, F. Haake, M. Kuś, A. Lasota, B. Zegarliński and K. Życzkowski. Some of the committee members took charge of lecturing too. We convey our thanks to all of them.

We would like to express words of gratitude to members of the local organizing committee, W. Cegła and P. Ługiewicz, for their help. Special thanks must be extended to Mrs Anna Jadczyk for her help at various stages of the school organization and the competent editorial assistance.

The school was financially supported by the University of Wrocław, University of Zielona Góra, Polish Ministry of Education, Polish Academy of Sciences, Foundation for the Karpacz Winter School of Theoretical Physics and the donation from the Dr Wilhelm Heinrich Heraeus und Else Heraeus Stiftung.

Wrocław and Zielona Góra, Poland
June 2002

Piotr Garbaczewski
Robert Olkiewicz

Table of Contents

List of Contributors

R. Alicki
Institute of Theoretical Physics and Astrophysics, University of Gdańsk, Wita Stwosza 57, 80-952 Gdańsk, Poland
fizra@univ.gda.pl

J. Bellissard
Université Paul-Sabatier, 118 route de Narbonne, 31062 Toulouse, France
jeanbel@irsamc2.ups-tlse.fr

F. Benatti
Dipartimento di Fisica Teorica, Università di Trieste
Strada Costiera 11, 34014 Trieste, Italy
benatti@ts.infn.it

A. Buchleitner
Max-Planck-Institut für Physik komplexer Systeme, Nöthnitzer Str. 38, 01187 Dresden, Germany
abu@mpipks-dresden.mpg.de

D. Cohen
Department of Physics, Ben-Gurion University, Beer-Sheva, Israel
dcohen@bgumail.bgu.ac.il

E.G.D. Cohen
The Rockefeller University
1230 York Avenue,
New York, NY 10021-6399, USA
egdc@rockefeller.edu

J.R. Dorfman
Department of Physics,
University of Maryland, College Park, Maryland, 20910, USA
jrd@glue.umd.edu

L. Erdős
School of Mathematics, Georgia Institute of Technology, Atlanta, GA 30332, USA
lerdos@math.gatech.edu

M. Fannes
Instituut voor Theoretische Fysica,
K.U. Leuven,
Celestijnenlaan 200D,
3001 Heverlee,
Belgium
Mark.Fannes@fys.kuleuven.ac.be

S. Fishman
Department of Physics, Technion, Haifa 32000, Israel
fishman@physics.technion.ac.il

P. Gaspard
Center for Nonlinear Phenomena and Complex Systems,
Université Libre de Bruxelles,
Campus Plaine, Code Postal 231,
1050 Brussels, Belgium
gaspard@ulb.ac.be

C. Jarzynski
Theoretical Division, Los Alamos National Laboratory, Los Alamos, NM 87545, USA
chrisj@lanl.gov

D. Kusnezov
Center for Theoretical Physics, Sloane Physics Lab, Yale University, New Haven, CT 06520-8120, USA
dimitri@mirage.physics.yale.edu

W.A. Majewski
Institute of Theoretical Physics, Gdańsk University
ul. Wita Stwosza 57, 80-592 Gdańsk, Poland
fizwam@univ.gda.pl

L. Rondoni
Dipartimento di Matematica, Politecnico di Torino, Corso Duca degli Abruzzi 24, 10129 Torino, Italy
rondoni@polito.it

R. Rudnicki
Silesian University, Bankowa 14, PL-40-007 Katowice, Poland
rudnicki@us.edu.pl

W.T. Strunz
Theoretische Quantendynamik, Fakultät für Physik, Universität Freiburg, Hermann-Herder-Str. 3, 79104 Freiburg, Germany
walter.strunz@physik.uni-freiburg.de

S. Tasaki
Department of Applied Physics and Advanced Institute for Complex Systems, Waseda University, 3-4-1 Okubo, Shinjuku-ku, Tokyo 169-8555, Japan
stasaki@waseda.jp

Introduction

Lecture notes are organized around the key word **dissipation**, while focusing on a presentation of modern theoretical developments in the study of irreversible phenomena. A broad cross-disciplinary perspective towards non-equilibrium statistical mechanics is backed by the general theory of nonlinear and complex dynamical systems. The classical-quantum intertwine and semiclassical dissipative borderline issue (decoherence, "classical out of quantum") are here included.

Special emphasis is put on links between the theory of classical and quantum dynamical systems (temporal disorder, dynamical chaos and transport processes) with central problems of non-equilibrium statistical mechanics like e.g. the connection between dynamics and thermodynamics, relaxation towards equilibrium states and mechanisms capable to drive and next maintain the physical system far from equilibrium, in a non-equilibrium steady (stationary) state.

The notion of an equilibrium state - towards which a system naturally evolves if left undisturbed - is a fundamental concept of equilibrium statistical mechanics. Taken as a primitive point of reference that allows to give an unambiguous status to near equilibrium and far from equilibrium systems, together with the dynamical notion of a relaxation (decay) towards a prescribed asymptotic invariant measure or probability distribution (properties of ergodicity and mixing are implicit). A related issue is to keep under control the process of driving a physical system away from an initial state of equilibrium and either keeping it in another (non-equilibrium) steady state or allowing to restore the initial data (return back, relax). To this end various models of environment (heat bath, reservoir, thermostat, measuring instrument etc.), and the environment - system coupling are analyzed.

The central theme of the book is the **dynamics of dissipation** and various mechanisms responsible for the irreversible behaviour (transport properties) of open systems on classical and quantum levels of description.

A distinguishing feature of these lecture notes is that microscopic foundations of irreversibility are investigated basically in terms of "small" systems, when the "system" and/or "environment" may have a finite (and small) number of degrees of freedom and may be bounded. This is to be contrasted with the casual understanding of statistical mechanics which is regarded to refer to systems with a very large number of degrees of freedom. In fact, it is commonly accepted that the accumulation of effects due to many (range of the Avogadro number) particles is required for statistical mechanics reasoning. Albeit those large numbers are not at all sufficient for transport properties.

A helpful hint towards this conceptual turnover comes from the observation that for chaotic dynamical systems the random time evolution proves to be compatible with the underlying purely deterministic laws of motion. Chaotic features of the classical dynamics already appear in systems with two degrees of freedom and such systems need to be described in statistical terms, if we wish to quantify the dynamics of relaxation towards an invariant ergodic measure. The relaxation towards equilibrium finds a statistical description through an analysis of statistical ensembles. This entails an extension of the range of validity of statistical mechanics to small classical systems. On the other hand, the dynamics of fluctuations in macroscopic dissipative systems (due to their molecular composition and thermal mobility) may render a characterization of such systems as being chaotic.

That motivates attempts of understanding the role of microscopic chaos and various "chaotic hypotheses" - dynamical systems approach is being pushed down to the level of atoms, molecules and complex matter constituents, whose natural substitute are low-dimensional model subsystems (encompassing as well the mesoscopic "quantum chaos") - in non-equilibrium transport phenomena.

On the way a number of questions is addressed like e.g.: is there, or what is the nature of a connection between chaos (modern theory of dynamical systems) and irreversible thermodynamics; can really quantum chaos explain some peculiar features of quantum transport? The answer in both cases is positive, modulo a careful discrimination between viewing the dynamical chaos as a necessary or sufficient basis for irreversibility.

In those dynamical contexts, another key term **dynamical semigroups** refers to major technical tools appropriate for the "dissipative mathematics", modelling irreversible behaviour on the classical and quantum levels of description. Dynamical systems theory and "quantum chaos" research involve both a high level of mathematical sophistication and heavy computer "experimentation". One of the present volume specific flavors is a tutorial access to quite advanced mathematical tools. They gradually penetrate the classical and quantum dynamical semigroup description, while culminating in the noncommutative Brillouin zone construction as a prerequisite to understand transport in aperiodic solids.

Lecture notes are structured into chapters to give a better insight into major conceptual streamlines.

Chapter I is devoted to a discussion of non-equilibrium steady states and, through so-called chaotic hypothesis combined with suitable fluctuation theorems, elucidates the role of Sinai-Ruelle-Bowen distribution in both equilibrium and non-equilibrium statistical physics frameworks (**E. G. D. Cohen**). Links between dynamics and statistics (Boltzmann versus Tsallis) are also discussed. Fluctuation relations and a survey of deterministic thermostats are given in the context of non-equilibrium steady states of fluids (**L. Rondoni**). Response of systems driven far from equilibrium is analyzed on the basis of a central assertion about the existence of the statistical representation in terms of an ensemble of dynamical realizations of the driving process. Non-equilibrium work relation is deduced for irreversible processes (**C. Jarzynski**). The survey of non-equilibrium steady states in statistical mechanics of classical and quantum systems employs heat bath models and the random matrix theory input. The quantum heat bath analysis and derivation of fluctuation-dissipation theorems is performed by means of the influence functional technique adopted to solve quantum master equations (**D. Kusnezov**).

Chapter II deals with an issue of relaxation and its dynamical theory in both classical and quantum contexts. Pollicott-Ruelle resonance background for the exponential decay scenario is discussed for irreversible processes of diffusion in the Lorentz gas and multibaker models (**P. Gaspard**). The Pollicott-Ruelle theory reappears as a major inspiration in the survey of the behaviour of ensembles of chaotic systems, with a focus on model systems for which *no* rigorous results concerning the exponential decay of correlations in time is available (**S. Fishman**). The observation, that non-equilibrium transport processes in simple classical chaotic systems can be described in terms of fractal structures developing in the system phase space, links their formation and properties with the entropy production in the course of diffusion processes displaying a low dimensional deterministic (chaotic) origin (**J. R. Dorfman**).

Chapter III offers an introduction to the theory of dynamical semigroups. Asymptotic properties of Markov operators and Markov semigroups acting in the set of probability densities (statistical ensemble notion is implicit) are analyzed. Ergodicity, mixing, strong (complete) mixing and sweeping are discussed in the familiar setting of "noise, chaos and fractals" (**R. Rudnicki**). The next step comprises a passage to quantum dynamical semigroups and completely positive dynamical maps, with an ultimate goal to introduce a consistent framework for the analysis of irreversible phenomena in open quantum systems, where dissipation and decoherence are crucial concepts (**R. Alicki**). Friction and damping in classical and quantum mechanics of finite dissipative systems is analyzed by means of Markovian quantum semigroups with special emphasis on the issue of complete positivity (**M. Fannes**). Specific two-level model systems of elementary particle physics (kaons) and rudiments of neutron interferometry are employed to elucidate a distinction between positivity and complete positivity (**F. Benatti**). Quantization of dynamics of stochastic models related to equilibrium Gibbs states results in dynamical maps which form quantum stochastic dynamical semigroups (**W. A. Majewski**).

Chapter IV addresses diverse but deeply interrelated features of driven chaotic (mesoscopic) classical and quantum systems, their dissipative properties, notions of quantum irreversibility, entanglement, dephasing and decoherence. A survey of non-perturbative quantum effects for open quantum systems is concluded by outlining the discrepancies between random matrix theory and non-perturbative semiclassical predictions (**D. Cohen**). As a useful supplement to the subject of bounded open systems, methods of quantum state control in a cavity (coherent versus incoherent dynamics and dissipation) are described for low dimensional quantum systems (**A. Buchleitner**). The dynamics of open quantum systems can be alternatively described by means of non-Markovian stochastic Schrödinger equation, jointly for an open system and its environment, which moves us beyond the Linblad evolution scenario of Markovian dynamical semigroups. The quantum Brownian motion is considered (**W. Strunz**).

Chapter V enforces a conceptual transition 'from "small" to "large" systems with emphasis on irreversible thermodynamics of quantum transport. Typical features of the statistical mechanics of infinitely extended systems and the dynamical (small) systems approach are described by means of representative examples of relaxation towards asymptotic steady states: quantum one-dimensional lattice conductor and an open multibaker map (**S. Tasaki**). Dissipative transport in aperiodic solids is reviewed by invoking methods on noncommutative geometry. The anomalous Drude formula is derived. The occurence of quantum chaos is discussed together with its main consequences (**J. Bellissard**). The chapter is concluded by a survey of scaling limits of the N-body Schrödinger quantum dynamics, where classical evolution equations of irreversible statistical mechanics (linear Boltzmann, Hartree, Vlasov) emerge "out of quantum". In particular, a scaling limit of one body quantum dynamics with impurities (static random potential) and that of quantum dynamics with weakly coupled phonons are shown to yield the linear Boltzmann equation (**L. Erdös**).

Various interrelations between chapters and individual lectures, plus a detailed fine-tuned information about the subject matter coverage of the volume, can be recovered by examining an extensive **index**.

Chapter I

Nonequilibrium Dynamics

Some Recent Advances in Classical Statistical Mechanics

E.G.D. Cohen

The Rockefeller University, 1230 York Avenue New York, NY 10021, USA

Abstract. Two recent developments, each introducing a new, not previously available basic aspect of classical statistical mechanics, are discussed. The presentations concern a new distribution function describing a system in a nonequilibrium stationary state, which leads in turn to a Fluctuation Theorem for large deviations, possibly far away from equilibrium. Furthermore an attempt is made to place the statistics introduced by Tsallis some time ago, in the general context of statistical mechanics. This is based on a remark by Einstein and the consistency of this statistics with some recent experiments on fully developed turbulence.

Introduction

In the last decade or so a great number of new developments in classical Statistical Mechanics have come to the foreground, which all, in some way, extend the scope of the field, using new approaches. In addition, of all nineteenth century physics one could argue that perhaps mainly optics (through the laser) and Statistical Mechanics have still sufficiently new and unknown aspects, that they survive, at least at some Universities, into the 21st century. In the case of Statistical Mechanics, this is mainly due to the enormous difficulty and variety of the questions it wants to understand. It is impossible to survey all new developments within the framework of this contribution, even if one confines oneself to those treated at this meeting. Therefore, important new developments such as the fully worked out theory of the microcanonical ensemble with application to phase transitions in small systems as developed by Gross[1] or the new approach to phase transitions, as developed by Pettini et al[2], connecting phase transitions with topological transitions in phase space (the "Topological Hypothesis") and many others had to be left out. I will restrict myself to two new developments which all concern the foundations of Statistical Mechanics.

Many new developments were inspired or at least influenced by the emergence of dynamical systems theory (modern mechanics), where, however, the emphasis has been in general on systems with a few degrees of freedom, while traditional Statistical Mechanics referred to macroscopic systems with a very large number of degrees of freedom. While on the one hand, one of the new developments is the extension of equilibrium statistical mechanics to small systems[1], on the other hand the relevance of dynamical systems theory of small systems for large systems still needs assessment as to the degree this is so, in view of the enormously more complicated (and unknown) structure of the phase space for macroscopic or very large systems.

I. My first contribution will concern a generalization of the usual phase space distribution function for systems in equilibrium to nonequilibrium stationary states (ness). A brief account will be given of this new distribution function, the Sinai-Ruelle-Bowen (SRB) distribution. A possible physical application of this distribution function, based on a 'Chaotic Hypothesis", to nonequilibrium stationary states, which could be far from equilibrium, will be considered in particular. Then as a consequence of the SRB distribution, a Fluctuation Theorem in a nonequilibrium stationary state and some of its applications will be discussed. Since this Theorem has been studied in many previous publications, I will restrict myself here to an outline of, what I will call, the Fluctuation Theorem (FT). I will also discuss a wide-spread confusion of this FT with another, identically looking, but physically completely different Theorem. Finally, I will very briefly mention some recent new developments, which try to push statistical mechanics deeper into the largely unknown realm of far from equilibrium and large deviation phenomena.

II. My second topic will be a new classical statistics, introduced in 1988 by Tsallis, which has spawned an enormous amount of publications. This mainly because this generalization of the classical Boltzmann-Gibbs (BG) statistics, which had dominated classical statistical mechanics for a century and a half, seems to describe a very large number of entirely different experiments very well, by the adjustment of a single parameter q, absent in BG statistics (which corresponds to the fixed value $q = 1$). To the best of my knowledge, no physical prediction has yet emerged from this statistics and subsequently been confirmed by experiment. Also, a foundation and justification from first principles, i.e. from dynamics, is lacking. Yet, the stunning agreement with very many experiments seems to suggest that it may well contain something new, which at present can only be classified as either profound or accidental. I will discuss a few physical applications of this new statistics and justify perhaps, in principle, its validity on the basis of an old and consistently ignored criticism of Einstein's of Boltzmann's principle: $S = k_B \log W$.

A hopefully helpful survey of what I consider as the present situation here, composes then the second part of my contributions to these Proceedings.

1 Dynamical Systems and Statistical Mechanics

In this chapter I will discuss the application of a distribution function, the so-called Sinai-Ruelle-Bowen (SRB) distribution, first introduced in the mathematical theory of dynamical systems, to nonequilibrium statistical mechanics. Contrary to the distribution functions describing systems in equilibrium, this distribution function is based on dynamical rather than static weights (e.g. Boltzmann factors) in phase space.

Then I will discuss a Fluctuation Theorem (FT) for a system in a nonequilibrium stationary state[3], which can be proved with this SRB distribution function[4].

Finally I will try to clarify the connection of this FT with another, also called FT, introduced by Evans and Searles[5], which in a special case looks identical to the first FT, but is physically entirely different.

1.1 The SRB Distribution

As mentioned above, one of the recent advances in Statistical Mechanics is the application of a distribution function, the Sinai-Ruelle-Bowen (SRB) distribution, originally introduced in the context of the mathematical theory of dynamical systems[6–8] to the Statistical Mechanics of a nonequilibrium stationary state. A consequence of this SRB distribution is a Fluctuation Theorem (FT) for a nonequilibrium stationary state that can be proved with this distribution function[4]. Historically this FT was first found numerically, stimulated by theoretical arguments, by Evans, Morriss and myself for a viscous system in a nonequilibrium stationary state in 1993[3]. A rigorous proof of the theorem and the connection between the SRB distribution (mathematics) and the nonequilibrium stationary state (physics) was given by Gallavotti and myself in 1995[4], establishing a new connection between mathematics and Statistical Mechanics. In 1994, Evans and Searles[5] proved an identically looking, but physically entirely different Theorem, which they also called a FT, which has caused much confusion and which will be discussed later in this article.

The SRB distribution concerns a generalization of the well-known classical theory, due to Boltzmann and Gibbs for the Statistical Mechanics of macroscopic systems in thermal equilibrium[9] to that of non-equilibrium stationary states. Equilibrium systems are usually described by the Gibbs ensembles with measures $\mu : \mu_{\mathrm{ME}}$ for the microcanonical ensemble for a system at constant energy E or μ_{CE} for the canonical ensemble for a system at constant temperature T.

This classical situation can be summarized as follows:

1. The Existence of an Equilibrium State = zeroth law of Thermodynamics, which can be formulated as:

"An isolated (conservative) dynamical system, consisting of many particles, which is initially not in an equilibrium state, will approach equilibrium, where all macroscopic properties have reached steady values, independent of the initial state of the system".

2. These steady values can be computed with the ergodic hypothesis (EH) of Boltzmann (1884)[11], which says that the time averages, that *define* macroscopic quantities, can be obtained from phase space averages computed with e.g. the microcanonical ensemble for a system at constant energy:

$$
\begin{aligned}
F_{eq}^{\mathrm{macr}} &\equiv \lim_{T\to\infty} \frac{1}{T} \int_0^T dt F(x(t) \\
&= \int_C F(x) \mu_{\mathrm{ME}}(x) dx \,. \qquad (1.1)
\end{aligned}
$$

Here T and t are times, $x = \{q, p\}$ represents the phase, i.e. the coordinates ($\boldsymbol{q}_1, ..., \boldsymbol{q}_N$ and the momenta ($\boldsymbol{p}_1, ..., \boldsymbol{p}_N$) all the N particles of the system, F is the microscopic (i.e. in terms of q's and p's) representation of the macroscopic quantity of interest and is supposed to be a smooth function of x, while C represents the phase space.

The question that I am interested in here is:
Is there a probability distribution function (measure) μ for a non-conservative, dissipative system in a nonequilibrium stationary state?

Such systems can be stationary states generated by a stationary heat flow $\boldsymbol{J}_h^{\text{macr}}$ through a system, which is coupled to two reservoirs held at different temperatures T, or a viscous flow $\boldsymbol{J}_v^{\text{macr}}$ due to a velocity gradient of the fluid layers in a sheared fluid, or an electric flow (current) in a conductor J_e^{macr}, due to the presence of a (constant) electric field. All these currents, which are due to what I will call "external fields", vanish in equilibrium.[1] Systems in a nonequilibrium stationary state are dissipative and produce heat (entropy): e.g. viscous or Ohmic heating etc.

The situation can be summarized as follows:
1. The existence of a non-equilibrium Stationary State = extended zeroth law of thermodynamics.[2] A sufficiently chaotic dissipative dynamical system, consisting of many particles, subject to an external field and initially not in a nonequilibrium stationary state, will approach a nonequilibrium stationary state, where all macroscopic properties have reached steady values, independent of the initial state.
2. These steady values can be computed by using a Chaotic Hypothesis(CH) (Gallavotti and Cohen, 1995[4]). The Chaotic Hypothesis reads:

"A reversible dissipative macroscopic system in a nonequilibrium stationary state can be regarded as a (transitive)[3] Anosov (i.e. sufficiently chaotic) system for the calculation of its macroscopic properties".

Sinai has proved that for such a system a unique distribution μ exists, which is called the Sinai-Ruelle-Bowen (SRB) distribution μ_{SRB}, and can be used, in principle, to compute the macroscopic properties of the system in a nonequilibrium stationary state.

The chaoticity of a macroscopic system is physically due to the very many collisions between the very many particles in a many particle (macroscopic)

[1] The external fields do work on the system, thus heating it up indefinitely. In order to obtain stationary states a "thermostatting" mechanism is introduced usually in the equations of motion. The systems discussed in this paper and the Fluctuation Theorem as well, all refer to thermostatted systems. Although they are bona fide dynamical systems, the question to what extent the way of thermostatting corresponds physically to that occurring in Nature, continues to be discussed. [See contributions of Rondoni and Klages in these Proceedings]. It does not seem to be a serious physical question to me though.

[2] While the equilibrium state is "absolutely" stable, non-equilibrium stationary states are not necessarily stable, but this has not been considered in the present context as far as I know.

[3] Transitivity implies the x-independence of the statistics.

system. The identification of the SRB distribution as the appropriate distribution function for a sufficiently chaotic nonequilibrium stationary state is really the main new step taken here. This implies that the time averages, which define the macroscopic properties, can be obtained for sufficiently chaotic systems from phase space averages, computed with the SRB distribution function μ_{SRB}:

$$\begin{aligned} F_{\mathrm{ness}}^{\mathrm{macr}} &\equiv \lim_{T\to\infty} \int_0^T F(x(t)) \\ &= \int_C F(x)\mu_{\mathrm{SRB}}(dx)\,. \end{aligned} \tag{1.2}$$

Here $\mu_{\mathrm{SRB}}(dx)$ allows the absence of a phase space density in the SRB distribution for a nonequilibrium stationary state, i.e. the absence of absolute continuity of μ_{SRB}, when $\mu_{\mathrm{SRB}}(dx) = \rho(x)dx$, with $\rho(x)$ a smooth density function and $\mu_{\mathrm{SRB}} \sim dx$. $F(x)$ is again supposed to be a smooth function of x and I note that the CH implies the much weaker EH, which therefore allows Eq. (1.2).

1.2 Construction of the SRB Distribution[4]

The SRB distribution is constructed by making a special (Markov) partition of the phase space of the system in cells, which Sinai has proved is possible for transitive Anosov systems (cf. Fig. (1a)). To each cell E_j, with center x_j, is then

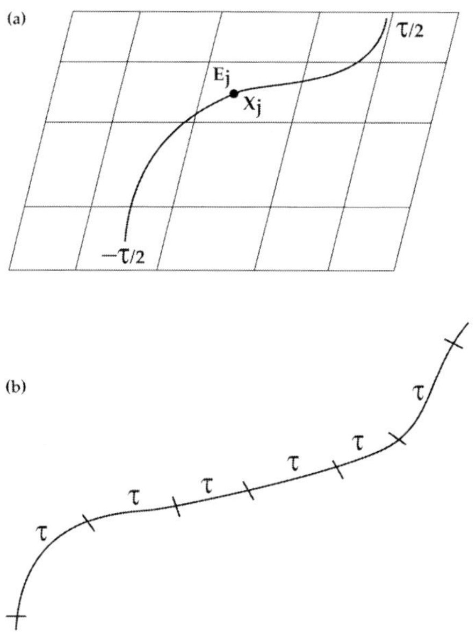

Fig. 1. Trajectory segments of duration τ: (a) in phase space (for simplicity a two dimensional phase space is plotted with parallelograms as (Markov) partition cells); (b) along a trajectory in a nonequilibrium stationary state.

associated a weight which indicates the frequency with which the dynamics of the system allows the system to visit this cell. This crucial weight is based on a "characteristic" change in the phase space volume of a small element of phase space at $-\tau/2$ along a trajectory of the system of duration τ from $-\tau/2$ to $+\tau/2$, through the center x_j of each cell E_j (cf. Fig. 1a). The larger/smaller this characteristic change in phase space volume along this trajectory is the less/more the system "wants" to be in the cell considered. In other words, physically, this phase space volume change over a time τ characterizes the instability of trajectories near x_j and probes the probability that the system will be far or near, respectively, from the cell E_j under consideration. This probability will be inversely proportional to the volume change over τ and determines therefore also the frequency of finding the system in the cell E_j. For further details of this rather crude description of the generation of the SRB distribution, I refer to the literature[4].

From the above, it will be clear that the expansion or contraction of the small volume element in phase space at $-\tau/2$, when it moves along a phase space trajectory to $+\tau/2$, will be needed to obtain the appropriate weight in phase space for a system in a nonequilibrium stationary state to be in the corresponding region of phase space. This implies that the equations of motion of the system have to be solved explicitly in order to compute these phase space volume changes.

Contrary, therefore, to the μ_{ME} and μ_{CE} in equilibrium, which contain static Boltzmann factors, the μ_{SRB} for a nonequilibrium state explicitly involves the *dynamics* of the system and can be expressed over the entire phase space in terms of the above mentioned phase space volume changes associated with each cell E_j of the (Markov) partition of the phase space. This then leads to the following expression for the average value of a macroscopic quantity analogous to (1.1) and (1.2):

$$\begin{aligned} F_{\mathrm{ness}}^{\mathrm{macr}} &\equiv \lim_{T\to\infty} \frac{1}{T}\int_0^T F(x(t)dt \\ &= \int_C F(x)\mu_{SRB}(dx) \\ &= \lim_{\tau\to\infty}\lim_{E_j\to 0} \frac{\sum_j F(x_j)\Lambda_{\tau,u}^{-1}(x_j)}{\sum_j \Lambda_{\tau,u}^{-1}(x_j)}\,. \end{aligned} \tag{1.3}$$

Here for mathematical precision, the trajectory segments through the centers x_j of the partition cells E_j of phase space are taken in the large time limit $(\tau \to \infty)$, while the size of the cells $E_j \to 0$ first. The characteristic dynamical weights associated with the phase space volume change over the time τ (from $-\tau/2$ to $\tau/2$) are indicated by the symbols $\Lambda_{\tau,u}(x_j)$, which can be expressed in terms of products of the absolute values of Jacobian determinants (i.e. volume changes) along the trajectory from $-\tau/2$ to $\tau/2$. The actual weights in (1.3), i.e.

the frequency of visits of the phase point to the cells E_j, are then proportional to the inverse of the Λ's, i.e. to the stability of the trajectories near E_j.[4]

1.3 The Fluctuation Theorem

As mentioned above the FT was discovered numerically in a viscous flow in 1993. In the light of later developments it was the first indication that the - for physicists - rather abstruse SRB distribution could be relevant for a new class of physical systems, viz, those in a nonequilibrium stationary state. In particular, the FT concerns a property of (possibly large) fluctuations, in a nonequilibrium stationary state, which itself could be far from equilibrium. The theorem can be rigorously proved with the SRB distribution and presents therefore a (first) test for the possible physical relevance of the SRB distribution and the CH, on which it is based. To formulate the FT we proceed as follows.

As above, one considers a system in the presence of external fields, which induce currents J (heat, viscous, electric) in the system, and of thermostats, which maintain the system in a (non-equilibrium) stationary state. One considers a single trajectory in this stationary state, which one partitions in trajectory segments, on which the system spends times of equal duration τ (cf. Fig. (1b). Then one asks the question: what is the ratio of the probability $\pi_\tau(a)$ to find on a trajectory segment of duration τ, (an average) current J_τ of magnitude a, to the probability $\pi_\tau(-a)$ to find there (an average) current of magnitude $-a$. Here a refers, in general, to a fluctuation of the current in the direction of the external field, while $-a$ to a fluctuation of the current of equal magnitude opposite to the external field. The FT then states, that for a system in a nonequilibrium stationary state[5]

$$\frac{\pi_\tau(a)}{\pi_\tau(-a)} = \exp dN\tau a\,, \tag{1.4}$$

where d is the dimension of space and N the number of particles.

This FT is supposed to hold for sufficiently large τ, so that a mathematically better presentation is:

$$lim_{\tau\to\infty}\frac{1}{dN\tau}\log\frac{\pi_\tau(a)}{\pi_\tau(-a)} = a\,, \tag{1.5}$$

where no corrections of higher order in a ($\sim a^3$ etc.), or in general no a and τ dependent corrections, occur even for large a. As such it is a "large deviation"

[4] The extra subscript u on the Λ's indicates the precise nature of the "characteristic" dynamical weights, i.e. taken along the so-called unstable manifold u. For a further discussion of this point we refer again to the literature[4].

[5] This theorem is often formulated in terms of entropy production rates or phase space contraction rates, which differ from (1.4) and (1.5) in that, instead of the values of the currents, those of the entropy production rate σ or the phase space contraction rate - due to the dissipative nature of the system - appear, as the observable a. For constant gradients, $\sigma = JX$, where X represents a constant gradient in the system. In the electric case, $J = \boldsymbol{J}$ the electric current and $X =$ the external electric field $\boldsymbol{E}$.

theorem, not restricted to the Central Limit dominated regime of a Gaussian. In fact, the $\pi_\tau(a)$ are not Gaussian. The fluctuation theorem can be formulated more precisely as follows:

"If a dynamical system is sufficiently chaotic, reversible as well as dissipative then in a nonequilibrium stationary state the probability for a certain value a or $-a$, respectively, of a current in this system, averaged over a sufficiently long time τ, satisfies Eqs. (1.4) or (1.5)".

The actual proof of the FT uses all three conditions mentioned in its formulation: chaoticity, time reversibility and dissipation.[6] Many numerical experiments of the FT have been performed, necessarily, however, for systems with a small number of particles[10]. They all confirm the FT as formulated above and constitute parameter free tests of the SRB distribution and the CH. In addition one laboratory experiment has been performed[12], which appears to confirm the FT, although its precise interpretation is not entirely clear.

The important feature of the FT, that it is not restricted to small fluctuations near equilibrium, but also valid for large deviations far from equilibrium, seems directly related to its dynamical nature[7]

Gallavotti and Gallavotti and Ruelle[13] have shown that near equilibrium the FT agrees with Irreversible Thermodynamics (IT) and that, in particular, the Onsager reciprocal relations between the (linear) transport coefficients as well as the Green-Kubo expressions for them, hold. This implies that the (full) FT must contain information about nonequilibrium stationary states beyond the linear regime of IT.

As mentioned before, this FT for nonequilibrium stationary states was discovered in 1993. The connection with the SRB distribution, was done in 1995. In between, Evans and Searles discovered in 1994 another FT, identical in form, but completely different in physical content than the FT (1.4). That Theorem did not refer to a single trajectory in a nonequilibrium stationary state, but to an ensemble of trajectories emanating from an initial state with a Lebesgue or Liouville measure. One then follows this ensemble of trajectories subject to an external field, and proves that after a time τ the equation (1.4) is valid for these trajectories for all τ. The physical content was that the ratio of ensemble trajectories moving in the direction of the external field, with current value a, and those moving against the external field, with current value $-a$, would be given be given by (1.4) as a function of time. Like the first Fluctuation Theorem, the second FT can be formulated more precisely as follows:

"If a dynamical system is reversible as well as dissipative then for a set of initial conditions with Liouville measure,the phase space trajectories over any time τ satisfy the relation (1.4)".

[6] It may seem contradictory that a time reversible system can be dissipative. The contradiction disappears if - as we do here - one only consider the time evolution of the system in positive times (as in the Eqs. (1.1)-(1.3)).

[7] There is a restriction on the external field (e.g. $\boldsymbol{E}$) that is not so large that it causes ordering in the system.

We note that this Theorem is more general than the first, in that no chaoticity of the system is necessary, but that there is a special condition for the initial state, which is absent in the first.

Therefore, an obvious difference between the two FT's is that while the first FT referred to the statistics of trajectory *segments* along any *single* trajectory in a nonequilibrium stationary state probed at equal time intervals τ, the second FT always referred to (an ensemble of) *many* trajectory *"histories"*, emanating from a special (Lebesgue or Liouville) initial state. This Theorem depends on the choice of the initial state, so that for non-Liouville initial states this second FT is different, while the first FT is independent of which initial state one chooses in the stationary state.

Like the first FT also the second FT has been confirmed by a computer experiment[14], as it has to be, since it is a mathematical identity[16]. It has also been tested by one laboratory experiment[15], the precise meaning of which, however, as a proof of the second FT, is not entirely clear, since this experiment concerns not a usual dynamical system with a precise number of particles and a well defined associated phase space. For further details and examples I refer to the literature quoted.

4. Finally, there are, of course, very many open points, of which I will only mention three:

1. Which class of dynamical systems, consisting of many particles and subject to an external field, satisfies the Chaotic Hypothesis, i.e, are Anosov-like? The Anosov-like assumption is almost certainly (much) too strong, but the proper assumption is unclear at present.

2. Is there a local FT? The FT discussed here always referred to a macroscopic system as a whole, where any fluctuation is bound to be small and hardly detectable, if at all, because of the large number of particles involved. There are indications, theoretically as well as experimentally, that a local FT, of the same or a very similar form, as the global FT discussed here, is valid[cf.17]. However, this remains still an important open question to be resolved.

3. As mentioned above, the more detailed exploitation of the FT far from equilibrium and the derivation from it of observable consequences for macroscopic systems, remains one of the most intriguing challenges for the future. In this respect the recent work of Bertini et al[18], Derrida et al[19] and Gallavotti[20] may well be relevant.

2 Statistics and Dynamics

For more than a century and a half the statistics of Boltzmann and Gibbs (BG) has dominated classical statistical mechanics[9]. These statistics have been very successful in understanding most of the macroscopic properties of the matter surrounding us, including their phase transitions[21].

In 1988 a new classical statistics was proposed by Tsallis[22] on a rather ad hoc basis by introducing a new real number q in the statistics, which would classify a class of systems with similar statistics. These statistics would differ from those of BG, but reduce to them for $q = 1$. In that sense it was a generalization of the BG statistics, but the motivation to introduce an unknown parameter q in the statistics as well as the particular form chosen for the new q statistics is still lacking. Although the Tsallis statistics has been able to describe a stunningly wide variety of phenomena by a proper choice of q, few predictions other than probability densities have as yet been made and later verified on the basis of it.

When I will discuss here the Tsallis statistics and a few of its applications, I do this because the empirical success of this new statistics is, in my opinion, too good to be thrown away out of hand, in spite of the lack of a deeper understanding of it, as mentioned above. In addition, a criticism of Einstein already made in 1904 against Boltzmann's statistics[23], but never taken seriously, may provide a conceptual frame work in which Tsallis' and possibly other, not yet discovered, statistics might play a crucial role[24].

In order to place Tsallis' statistics and Einstein's remark in context, I will start in Section 1 with a brief survey of how Boltzmann derived his statistics and in doing so point out even there some as yet not quite understood problems. In Section 2 I will discuss a dynamical version of Boltzmann's statistics, due to Boltzmann and Einstein. In Section 3 I give my interpretation of the significance of Einstein's criticism of Boltzmann's statistics[24]. In Section 4 I will introduce the basic probabilistic assumption of Tsallis' q-statistics, show that it reduces to Boltzmann's statistics for $q = 1$ and discuss its non-extensive nature[3]. In Section 5 Tsallis' assumption will be used to obtain a new physical statistics, which generalizes Boltzmann's statistics and classical thermodynamics. A number of possible forms of Tsallis statistics will be mentioned[27]. In Section 6 I will discuss the possible relevance of Tsallis statistics to understand in the context of classical mechanics, phenomena in Nature, to which Boltzmann's statistics is not applicable. They all involve a dynamics of a kind which is much more complicated than those encountered in those systems which can be understood on the basis of classical statistics. First I discuss (under A) the recent experiment of Lewis and Swinney[28] on the statistics of longitudinal velocity differences in a fully developed turbulent Couette-Taylor flow; next (under B) the statistics of the Lagrangian acceleration of a particle in a fully developed turbulent flow as measured by La Porta et al[29] is presented. Finally (under C) a theory by Chr. Beck[30], which derives under a number of assumptions the Tsallis statistics as suitable for the description of the observations in these two experiments will be surveyed (a simple version under a. and an extended version under b.). A conclusion and a further discussion of some of the many open points will be given in Section 7.

2.1 Boltzmann's Statistics

It is good to keep in mind that Boltzmann's statistics, as derived by him, only applied to an ideal gas in thermal equilibrium[23,24]. He considered N non-interacting particles with a total energy E in a volume V. Boltzmann proceeded in four steps.

a. In order to be able to apply statistics he partitioned the six dimensional phase $(\boldsymbol{r}, \boldsymbol{v})$ space (μ-space) of a single particle into M physically infinitesimal cells, where $M \ll N$ (cf. Fig. 2). Physically infinitesimal meant that the cells of the partitions were large enough to contain many particles so that statistics could be done, but small enough so that the variation of external fields acting on the particles over the cells is negligible. I will not consider external fields here, so only sketch the derivation of Maxwell's velocity distribution function and not the full Maxwell-Boltzmann distribution function, which includes external fields.

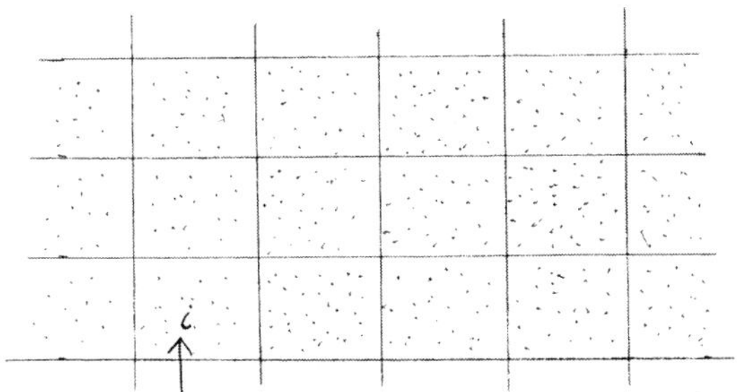

Fig. 2. Coarse grained μ-space.

b. The six dimensions of μ-space are spanned by the position $\boldsymbol{r}$ and the velocity $\boldsymbol{v}$ of a molecule. One can then characterize the state of the N particle gas, in a "coarse-grained" way, as given by the set of occupation numbers $\{n_i\}$ of the M cells $i = 1, 2, ..., M$ in μ-space (cf. Fig. 2). I note that only for an ideal gas the energy ϵ_i associated with any cell i can be determined, since if there are interparticle interactions, the energy of the particles in cell i will depend on the particles in neighboring cells. Furthermore, because of the coarse-graining and finiteness of the cells, the energy of the particles in each cell is only approximately known and we shall identify it for convenience, with the kinetic energy $\frac{1}{2}mv_i^2$, where $\boldsymbol{v}_i$ is the velocity associated with the center of the cell i. In fact, there is a range of velocities along the $\boldsymbol{v}$ axis associated with each cell so that in reality, the coarse-grained description gives only a range of energies of the gas on an energy shell, between energies E and $E + \Delta E$ for the entire system, where $\Delta E/E \ll 1$ and ΔE depends on the degree of coarse-graining. As shown in Fig. 2 each cell i is characterized by its particle occupation number n_i, i.e. how many of all the N particles in the gas have coordinates and velocities, which fall inside the cell i, their energy $\epsilon_i = \frac{1}{2}mv_i^2$ and their volume ω_i, which I will often consider to be a constant independent of i.

c. One now makes a transition from the coarse-grained description of the gas in the 6-dimensional μ-space (the phase space of one particle) to one in Γ-space, the $6N$-dimensional phase space of the entire gas of N particles.

In fact, the $\{n_i\}$- distribution in μ-space corresponds to a volume on the energy shell $W(\{n_i\})$ in Γ-space, given by:

$$W(\{n_i\}) = \frac{N!}{\Pi_{i=1}^{M} n_i!} \cdot \Pi_{i=1}^{M} \omega_i^{n_i} , \tag{2.1}$$

which Boltzmann called the "thermodynamic probability"for the gas to be in the state $\{n_i\}$.[8] The result (2.1) for the volume in Γ-space follows from the facts that a) each of the n_i particles in cell i can move over the entire cell i and generate then together a corresponding total volume $\omega_i^{n_i}$ in Γ-space. The preceding combinatorial factor in (2.1) weeds out of all possible $N!$ permutations of the N particles over the cells, those that would not lead to a different number of them in each cell, but only affect how many are put in. Thus if cell $i = 1$ has the particles 1,2 and 3 in it, the coarse-grained description of the gas only knows that there are 3 particles in cell i, not which particles and in which of the 3! possible ways they have been put in there.

Eq. (2.1) assumes - not unreasonably for an ideal gas - that there is an equal a priori probability for a particle to be in any cell i so that there is no bias for any particle to prefer any cell and that this probability is only proportional to the size ω_i of this cell. I mention the following properties of the thermodynamic probability $W(\{n_i\})$:

1. From a probabilistic mathematical point of view $\log W(\{n_i\})$ is the unique solution of four reasonable mathematical assumptions of an information measure for $\log W(\{n_i\})$, as Khinchin showed[25].

2. Log $W(\{n_i\}$ is an extensive quantity, i.e. for two non-interacting systems A and B,

$$W(\{n_{i,j}^{A+B}\}) = W(\{n_i^A\}) \cdot W(\{n_j^B\}) , \tag{2.2}$$

where $W(\{n_{i,j}^{A+B}\})$ is the thermodynamic probability to find n_i particles of system A and n_j particles of system B in a cell of a partition of the μ-space of the combined system $A + B$; therefore $\log W(\{n_i\})$ is extensive.

3. Also:

$$\begin{aligned} \log W(\{n_i\}) &= N \log N - \sum_{i=1}^{M} n_i \log \frac{n_i}{\omega_i} \\ &= - \sum_{i=1}^{M} p_i \log p_i + \text{ constant} , \end{aligned}$$

if $p_i = \frac{n_i}{\omega_i} = n_i$, when $\omega_i = 1$ for all i.

d. The next step Boltzmann took was to look for those $\{n_i\}$ that would give the maximum volume in Γ-space, i.e. would give the maximum thermodynamic probability, under two constraints:

[8] This thermodynamic probability can easily be transformed into a real probability, if one replaces each ω_i by ω_i/Ω_μ, where Ω_μ is the total (finite) volume of all the M cells in μ-space.

a) the total number of particles is fixed to N:

$$\sum_{i=1}^{M} n_i = N\,, \tag{2.3a}$$

b) the total energy of the particles is fixed to E:

$$\sum_{i=1}^{M} n\epsilon_i = \sum_{i=1}^{M} n_i \frac{1}{2} m v_i^2 = E\,. \tag{2.3b}$$

I note, for later, that this and the following steps have not been taken by authors proposing other thermodynamic probabilities than Boltzmann did, like e.g. Rényi's etc., with the exception of Tsallis. Without this step, the results remain in the domain of mathematics, but may give interesting mathematical geometrical information about the behavior of the system in Γ-space, but do not reach the level of the physics of the system as expressed, for example, by its thermodynamic behavior.

Maximizing $W(\{n_i\})$ of (2.1) with respect to the $\{n_i\}$, while taking into account the constraints (2.3a-2.3b) by two Lagrange multipliers, one obtains for the $\{\bar{n}_i\}$ maximizing $W(\{n_i\})$, the Maxwell velocity distribution function:

$$\bar{n}_i = A\omega_i e^{-\beta\epsilon_i}\,, \tag{2.4}$$

where the two constants A and β are determined from the two constraints (2.3a-2.3b). Here $\epsilon_i = \frac{1}{2} m v_i^2$ and $\beta = 1/k_B T$ follows from the ideal gas law.[9] For large N the MB distribution leads to the laws of equilibrium thermodynamics for an ideal gas. We note for later, that equilibrium thermodynamics is extensive, which is based on the relation (2.2) and says that the thermodynamic quantities like the entropy S, the energy E etc. are all proportional to N.

In particular one can derive from the above formulae that, in equilibrium, apart from a constant:

$$S^{eq} = k_B \log W^{eq}(\{\bar{n}_i\}) = -k_B \sum_{i=1}^{M} \bar{p}_i \log \bar{p}_i\,. \tag{2.5}$$

I note that on Boltzmann's grave, instead of Eq. (2.5), one finds the much more general relation $S = k \log W$, where W is the "probability" of the macroscopic state with given N, E, say, but the range of validity and the proper interpretation of this relation are unclear to me.

2.2 Boltzmann-Einstein Statistics[23,24]

While Boltzmann's derivation of Boltzmann statistics in the previous section is a statistical derivation, Boltzmann and later Einstein proposed an alternative

[9] In the presence of an outside field $V(\boldsymbol{r}_i)$, Eq. (2.4) generalizes to the Maxwell-Boltzmann (MB) distribution function given by (2.4) with $\epsilon_i = \frac{1}{2} m v_i^2 + V(\boldsymbol{r}_i)$.

dynamical derivation of the probability $W(R)$ to find the system in a region R on the energy shell in Γ-space. They set:

$$W(R) = \lim_{T\to\infty} \frac{t(R)}{T} = \frac{\text{time spent by system in } R}{\text{total time}} . \tag{2.6a}$$

Using ergodicity, (6a) can also be written:

$$W(R) = \frac{V(R)}{V} = \frac{\text{volume of } R \text{ on energy shell}}{\text{total volume of energy shell}} . \tag{2.6b}$$

2.3 Einstein's Criticism of $S = k_B \log W$ [23,24]

Einstein criticized what he called "Boltzmann's principle": $S = k_B \log W$ on the ground that probabilities (W) cannot be assigned at will, but have to follow from the dynamics of the system. For, the probability to find the system in a certain region R in phase space (or on the energy shell) is determined by the *frequency*, the *dynamics*, i.e. the equations of motion, which move the phase point, representing the system, through phase space [23], allow the system to visit the region R. Two quotations of Einstein in Pais' translation follow here:

Einstein (1905)

"The word *probability*" is used in a sense that does not conform to its definition as given in the theory of probability. In particular, "cases of equal probability" are often hypothetically defined in instances where the theoretical pictures used are sufficiently definite to give a deduction rather than a hypothetical assertion".

Einstein (1910)

"Usually W equals the number of complexions. To compute W one needs a complete (molecular-mechanical) theory of the system. It is therefore dubious whether the Boltzmann principle has any meaning without a complete molecular-mechanical theory or some other theory which describes the elementary processes. In equilibrium, the expression, $S^{eq} = k \log W^{eq}+$ constant, seems without content from a phenomenological point of view, without giving in addition such an elementary Theory".

In my opinion, Einstein's criticism implies that there is nothing sacred about Boltzmann's - as well as Gibbs' later - statistics and that, in principle, other classical statistics, than what is usually called Boltzmann-Gibbs statistics, are possible. From the enormous success of the BG statistics for very many physical systems, one would surmise that unusual dynamics would be a precondition for different statistics to be necessary. Bose-Einstein and Fermi-Dirac statistics for quantum dynamical system are a point in case.

2.4 Tsallis "Entropy" (1988)

a. In the realm of classical statistical mechanics, based on Newton's equations of motion, Tsallis proposed an extension of the MB or BG statistics in the spirit of Einstein, using a new, additional (real) parameter q. The use of such a parameter to obtain and characterize statistics was not new and had been used by many authors before, such as Khinchin, Rényi, Procaccia and Grassberger, etc.[25]. New was that unlike these authors Tsallis imitated all four of Boltzmann's steps and did not stop at and exploit exclusively what corresponded with the above mentioned authors, only to Boltzmann's first steps: the introduction of a thermodynamic probability. On the other hand, by using information theoretic methods, they generalized, it seems, Boltzmann's coarse graining and also solved in a way his difficulty to generalize his thermodynamic probability to interacting (non-ideal) particle systems.[10]

b. Thus Tsallis posed - seemingly "out of the blue" - the following - what I will call the analogue of the logarithm of a q-thermodynamic probability:

$$S_q(\{p_i\}) = k_B \frac{1 - \sum_{i=1}^{W} p_i^q}{q-1} \quad (q \epsilon R) , \tag{2.7}$$

where $S_q(\{p_i\})$ corresponds to Boltzmann's $\log W(\{n_i\})$. Eq. (2.7) implies that for $q = 1$:

$$S_1(\{p_i\}) = k_B \log W(\{p_i\}) = -k_B \sum_{i=1}^{W} p_i \log p_i , \tag{2.8}$$

which is of the same form as Boltzmann's $k_B \log W(\{n_i\})$, apart from a constant. This can be seen as follows. Since for $q = 1$ both numerator and denominator on the right hand side (rhs) of Eq. (2.7) vanish, application of L'Hôpital's rule and setting $q = 1$ immediately gives Eq. (2.8).

c. However, for $q \neq 1$, different thermodynamic probabilities are obtained than Boltzmann's and it seems reasonable to suppose that the equal a-priori probability assumption for a particle to be in cell i with energy ϵ_i in μ-space, as used in Boltzmann's $W(\{n_i\})$, would no longer hold. While this assumption seemed reasonable for an ideal gas and appears to hold also for systems with short range inter-particle forces, there is no a-priori reason to assume this to be universally valid for all dynamical systems, as Einstein might say. A more precise interpretation of the ad-hoc parameter q will be given in the example treated in Section 6a.

d. To be sure Tsallis' "entropy" S_q, which I regard here as the logarithm of a thermodynamic probability, appears to be an inspired and to all appear-

[10] Gibbs solved this in a different way by considering only Γ-space and defining the entropy for a system in equilibrium by the expression:

$$S = \int \rho(\boldsymbol{r}_1, ..., \boldsymbol{v}_N) \log \rho(\boldsymbol{r}_1, ..., \boldsymbol{v}_N) d\boldsymbol{r}_1, ..., d\boldsymbol{v}_N$$

where ρ is assumed to be an appropriate equilibrium distribution function.

ances, non trivial choice, albeit devoid of any physical foundation, as Boltzmann's $\bar{W}(\{n_i\})$ had.

e. The most striking property of Tsallis' $S_q(\{p_i\})$ seems to be its non-extensivity contrary to Boltzmann's $\log W(\{n_i\})$[22]. For two non-interacting systems A and B, one has, contrary to (2.2):

$$\begin{aligned} S_q(A+B)/k_B &= S_q(A)/k_B + S_q(B)/k_B \\ &\quad + (1-q)[(S_q(A)/k_B) \cdot (S_q(B)/k_B] \,. \end{aligned} \tag{2.9}$$

Here $S_q(A) \equiv S_q(\{p_i^A\})$, where the p_i^A are the probabilities to find the system A in a state with energy ϵ_i etc. Extensivity only obtains for $q = 1$, when S_q reduces to Boltzmann's expression. Recently Abe showed[26] that Tsallis' S_q is the unique solution of four information measure axioms, which, of course, differ from those of Khinchin's for Boltzmann's $\log W(\{n_i\})$.

2.5 q-Thermodynamics

a. So far no physics has entered Tsallis' considerations. For that, maximization under certain physical constraints are necessary to give a maximum likely distribution $\{\bar{p}_i\}$ and a corresponding thermodynamics.

b. Therefore we now maximize $S_q(\{p_i\})$ with respect to the $\{p_i\}$ under the constraints:

$$\sum_{i=1}^{W} p_i = 1 \quad \text{and} \quad \sum_{i=1}^{W} p_i \epsilon_i = U \,, \tag{2.10}$$

where the ϵ_i are the energy states of the system. This leads to the generalized canonical distribution:

$$\bar{p}_i = \frac{1}{Z_q}[1-(q-1)\beta\epsilon_i]^{\frac{1}{q-1}} \,, \tag{2.11}$$

where Z_q follows from the constraint $\sum_{i=1}^{W} \bar{p}_i = 1$. For $q = 1$, the $\bar{p}_i$ reduce to the (exponential) MB distribution. This follows by writing

$$\bar{p}_i \sim \exp\{\log[1-(q-1)\beta\epsilon_i]^{\frac{1}{q-1}}\},$$

and using that for $q \to 1, \log(1+\alpha)$ can be replaced by α, giving $\log \bar{p}_i \sim e^{-\beta\epsilon_i}$. We note that for $q \neq 1, \bar{p}_i$ is given by a power law, where for $(q-1)\beta\epsilon_i \gg 1, \bar{p}_i \sim (\beta\epsilon_i)^{\frac{1}{q-1}}$.

c. Now that the maximum distribution function is known, one can derive from the $\{\bar{p}_i\}$ a non-extensive q-thermodynamics, since the q-thermodynamic $S_q(\{\bar{p}_i\})$ still satisfies the relation (2.9). This q-thermodynamics has for all q the usual Legendre transformation structure as conventional ($q = 1$) thermodynamics has[22].

d. There seems to be no necessity to choose the second Eq. (2.10) as the energy constraint[27]. In fact, the proper choice may depend on the nature of the system or the quantity we are interested in, which, since its dynamics for $q \neq 1$

will be complicated, may not be obvious. In particular, for certain applications a different choice of the energy U condition has been used, such as:

$$\sum_{i=1}^{W} P_i(q)\epsilon_i = U_q\,, \tag{2.12}$$

where $P_i(q)$ could be: p_i (as in (10)), p_i^q or $p_i^q/\sum_{i=1}^{W} p_i^q$, respectively. Here only the first and the last choices are normalized, yet the unnormalized choice p_i^q has led to interesting results in agreement with certain experiments as will be briefly discussed in Section 7 sub 1.

From Einstein's point of view - at least in my interpretation - there seems to be no a priori reason why there could not be complicated dynamical system for which $P_i(q) \neq p_i$ in (2.12) and a corresponding nonconventional q-thermodynamics could hold for their macroscopic properties.

I note that in the literature Tsallis' original q, as it appears in S_q and the formulae derived from it, is usually replaced by $\tilde{q} = 2-q$, so that in the expression for $\bar{p}_i, q-1 \rightarrow \tilde{q}-1$, leading instead of to (2.11) to a generalized canonical distribution of the form:

$$\bar{p}_i = \frac{1}{Z_q}[1-(1-q)\beta\epsilon_i]^{\frac{1}{1-q}}\,, \tag{2.13}$$

where I have used again q for $\tilde{q}$, to agree with the literature.

The Tsallis statistics combines in a way, a new information theoretic based probabilistic Ansatz with the usual procedure of looking for the maximum probability distribution. In other words, it combines Rényi, Shannon and many others with Boltzmann,and in doing so, generalizes the latter.

2.6 Relevance of Tsallis Statistics

In all this, the crucial question, of course, is whether there are classical systems in Nature, exhibiting, at least for some of their properties Tsallis rather than BG statistics. Since this would be - in my interpretation of Einstein's criticism of Boltzmann - due to the system's dynamics, one would expect a consistent non-BG behavior for all its macroscopic properties, not just a few.

As mentioned in the Introduction, I will restrict my discussion of the application of Tsallis' statistics to properties of two systems mainly, which can both indeed be said to possess complicated dynamics: two experiments for fully developed turbulent systems.

A) One is the statistics of the (Eulerian) longitudinal velocity difference measured in a Couette-Taylor system by Lewis and Swinney (1999)[28].

B) The other is the statistics of the Lagrangian acceleration of a particle by La Porta et al[29] using a Couette-Taylor system (2000).

A theory for both these experiments has been developed by Chr. Beck[30] and in doing so he derives - admittedly under a number of assumptions about the effect of turbulence on the dynamics of fluid particles - the Tsallis distribution for the statistics relevant in both experiments. I will have to restrict myself to the bare minimum in the exposition of these experiments and the theory and refer to the literature for details.

A. The Lewis-Swinney Experiment

The Lewis-Swinney experiment is an Eulerian experiment where the statistics of the longitudinal velocity difference u between two points in the fluid separated by a vector $\boldsymbol{r}$, along their separation line, is determined. This longitudinal velocity difference is defined by:

$$u(r) = [\boldsymbol{v}(\boldsymbol{x}+\boldsymbol{r}) - \boldsymbol{v}(\boldsymbol{x})] \cdot \hat{r}\,. \tag{2.14}$$

Here $\boldsymbol{v}(\boldsymbol{x}+\boldsymbol{r})$ and $\boldsymbol{v}(\boldsymbol{x})$ are the fluid velocities at the points $\boldsymbol{x}+\boldsymbol{r}$ and $\boldsymbol{x}$, respectively and $\hat{r}$ is a unit vector in the direction $\boldsymbol{r}$.

The statistics, i.e. the probability distribution function $p(u)$, is measured as a function of u for various distances $r = |\boldsymbol{r}|$ between the two points as well as for a variety of Reynolds numbers; in fact, apart from other dependencies, $u = u(r, Re)$ is determined. In this experiment, done in the fully developed turbulent flow between two cylinders, where only the inner one rotates, the Reynolds numbers varied from $2000 < Re < 1,000.000$, while variations of r can be found in Table 1. I will only show the results for $p(u)$ for Re= 540.000.

Table 1. $q(r/\eta)$

$\frac{r}{\eta}$	11.6	23.1	46.2	92.5	208	399	830	14,400
q	1.168	1.159	1.124	1.105	1.084	1.065	1.055	1.038

Considering the wide range of values of r used and in order to see more clearly the behavior of the curves $p(u)$ as a function of u for various r, two separate figures have been made: one plot of log $p(u)$ vs u to highlight the behavior of the tails of the distribution $p(u)$ (see Fig. 3a) and one linear $p(u)$ vs u plot, to highlight the behavior of the central part of the distribution $p(u)$ (see Fig. 3b). Eight values of r are considered where $12 < r/\eta < 14,4000$ with η a characteristic (Kolmogorov) length, which is here of the order of 100-1000 μm.

The main result is that all 8 curves, both in Fig. 3a and Fig. 3b can be very well represented by Tsallis' distribution $\bar{p}_i$ of Eq. (2.13) using one value of q for each r curve. These values $q(r)$ are monotonically decreasing as a function of increasing r and approach one for the largest r (cf. Fig. 3 and Table 1). We note that the curves approach Gaussians for large r/η and power laws for small r/η. In Fig. 4 $q(r/\eta)$ is plotted as a function of r/η for 4 values of Re.

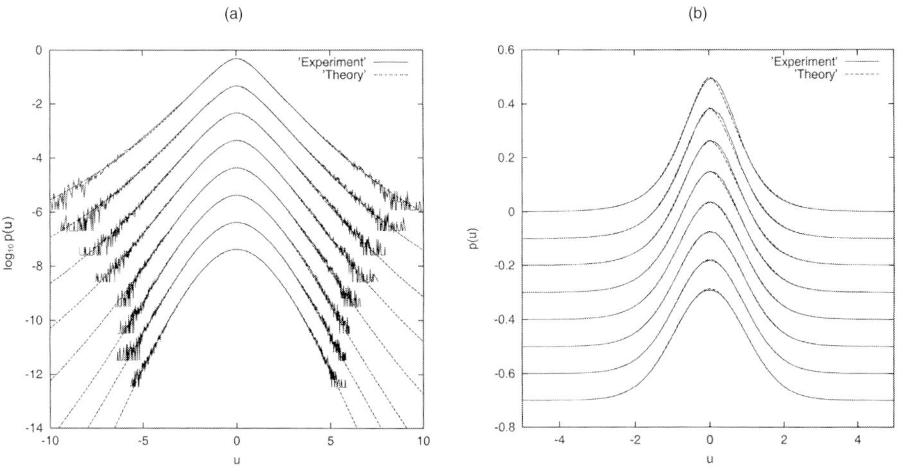

Fig. 3. Experimental results for $p(u)$ of Lewis and Swinney for $Re = 540.000$ compared with the full Beck theory[30] for increasing r/η from top to bottom.
(a) $\log_{10} p(u)$ vs u; (b)$p(u)$ vs u (——— experiment; - - - - - theory). Each curve in (a) is shifted by -1 unit in the y-direction, while each curve in
(b) is shifted by -0.1 in the y-direction.

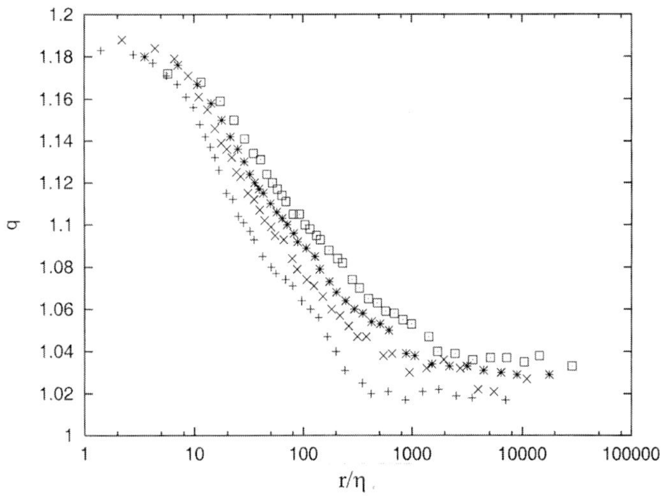

Fig. 4. q versus reduced distance r/η, where η is the Kolmogorov length, for $Re =$ 69000 (+), 133000 (x), 266000 (*), 540000 (□).

B. The La Porta et al Experiment[29]

This is a Lagrangian experiment where the statistics (of a component) of the acceleration a of an inserted polystyrene particle in the fluid in a co-moving (i.e. Lagrangian) frame is measured. The experiment is done in two coaxial counter rotating disks (a Couette-Taylor system) using a strip detector for three Reynolds numbers $2{,}600 \leq Re \leq 63{,}000$, where the fluid exhibits fully developed turbulence. La Porta et al can represent their data for the probability distribution $p(a)$ for a measured acceleration component a of the particle for all Re by a phenomenological fit:

$$p(a) = C \exp -[\frac{a^2}{1 + |\frac{a\beta}{\sigma}|^{\gamma}\sigma^2}] , \tag{2.15}$$

with four parameters C, β, σ and γ. This empirical fit leads to a stretched exponential behavior of the tails $\sim e^{-a^{0.41}}$ for $Re = 63{,}000$, where $\gamma = 1.588$[29].

C. Beck's Theory

I will try to sketch the essentials of Beck's theory, emphasizing mainly his derivation of the generalized canonical distribution Eq. (2.13) for the $p(u)$ and $p(a)$, thus showing the relevance of Tsallis' statistics for the complicated dynamics in these fully developed turbulent flows to which they refer. I will first give Beck's theory in a simplified form, which agrees already well with the above mentioned experiments, especially the La Porta et al experiment. Then I will just indicate how further refinements lead to a very good agreement for the distribution functions measured in the Lewis-Swinney experiment as well. The theory also clarifies the meaning of the "bias" parameter q, but is only able to compute its value for the La Porta et al experiment for $p(a)$. It does allow, however, to represent each of the 8 Lewis-Swinney curves for $p(u)$ for different r with a single parameter q, which decreases with increasing r (cf. Table 1 and Fig. 4), but whose values are not given by the theory[31].

a. Simple Beck Theory[30]

1. One assumes that due to the turbulent fluid motion the fluctuating longitudinal velocity difference u satisfies a Langevin type of equation of the form:

$$\dot{u} = -\gamma u + \sigma L(t) . \tag{2.16}$$

In this equation the friction γ and the fluctuation strength σ both fluctuate (in addition to $(L(t))$, as does $\beta = \gamma/\sigma^2$. The latter relation follows from the (white noise assumed) correlation of the fluctuating term $\sigma L(t)$ on the rhs of Eq. (2.16):

$$\sigma^2 < L(t)L(t') >= 2\frac{\gamma}{\beta}\delta(t - t').$$

From this, the above mentioned relation between the coefficients in this equation follows. The crucial quantity which embodies the effect of the turbulence-induced fluctuations of u is considered to be β, which can be taken to be a generalized temperature. See later for more details.

2. Now the key assumption of the origin of the u-fluctuations can be characterized by those of β and that these fluctuations are due to some random kind of changes of the turbulent environment of a "Brownian"-like particle. In other words, the u-fluctuations are treated as if they were the fluctuations of a "generalized" Brownian particle (with mass one).

3. Next, the fluctuating environment is represented by χ^2 - distributed sum of n independent Gaussian variables $X_i (i = 1, ..., n)$, all with average zero.

If $\beta = \sum_{i=1}^{n} X_i^2$, then a standard result for the distribution of β is[32]:

$$f(\beta) = \frac{1}{\Gamma(\frac{n}{2})}(\frac{n}{2\beta_o})^{\frac{n}{2}}\beta^{\frac{n}{2}-1}\, e^{-\frac{n\beta}{2\beta_0}} . \tag{2.17}$$

Here the average of β is: $<\beta>= \beta_0$, while the variance of $f(\beta)$ is given by: $<\beta^2> - \beta_0^2 = \frac{2}{n}\beta_0$.

4. The next assumption is that there is yet some order in the turbulent motion in that the time scale on which β fluctuates is much longer than that on which the Langevin equation gives the decay of u, i.e. than γ^{-1}. Under this condition, one can consider a "Brownian" particle (or a u) in an environment with a definite value of β, so that the *conditional* probability for u, given β is:

$$p(u|\beta) = \sqrt{\frac{\beta}{2\pi}} e^{-\frac{1}{2}\beta u^2} . \tag{2.18}$$

Then the *joint* probability for u and β is:

$$p(u, \beta) = f(\beta)p(u|\beta) , \tag{2.19}$$

so that the marginal probability for u, no matter what β is, reads:

$$\begin{aligned} p(u) &= \int \beta f(\beta)p(u|\beta) \\ &= \frac{\Gamma(n+\frac{1}{2})}{\Gamma(\frac{n}{2})}(\frac{\beta_0}{2\pi})^{\frac{1}{2}}[1 + 2\frac{\beta_0}{n}\frac{u^2}{2}]^{-\frac{n+1}{2}} . \end{aligned} \tag{2.20}$$

5. It is now interesting to compare Eq. (2.20) with Tsallis' generalized canonical distribution Eq. (2.13):

$$\bar{p}_i = [1 - (1-q)\tilde{\beta}\epsilon_i]^{\frac{1}{1-q}} , \tag{2.21}$$

where we have replaced β in Eq. (2.13) by $\tilde{\beta}$.

Comparing (2.20) and (2.21), we see that they are both power laws and that:
a) The fluctuating "Brownian particle" (or the longitudinal velocity difference u) has effectively an energy $\epsilon_i = \frac{u^2}{2}$ associated with it.
b) By comparing the exponents in the Eqs. (2.20) and (2.21), we have that the "bias" parameter q is given by:

$$q = 1 + \frac{2}{1+n} . \tag{2.22}$$

c) Furthermore there is an effective temperature

$$\tilde{\beta} = \frac{2}{3-q}\beta_0 = (1+\frac{1}{n})\beta_0 \,. \tag{2.23}$$

d) The asymptotic behavior of $p(u)$ in Eq. (2.20) is $\sim u^{-(n+1)}$, i.e. like a power law (not a stretched exponential).

6. La Porta et al's experiment, where the acceleration of a single particle is determined, can be considered as a limiting case of Lewis and Swinney's longitudinal velocity difference, where $\frac{r}{\eta} \approx O(1)$ and the Taylor hypothesis[33] would allow conversion of what is really a spatial gradient into a time derivative. Accepting all this, it seems reasonable to take $n = 3$, i.e that there are for the La Porta et al particle only three independent Gaussian X_i, namely in the x, y and z directions. This leads then for La Porta et al's experiments to a value of $q = \frac{3}{2}$ and a power law behavior of the tail of the $p(u)$ distribution of u^{-4}. Looking at Fig. 5 there seems to be no way to distinguish the representation of La Porta et al's data by the phenomenological formula (2.15) (with a stretched exponential) or by the theoretically based expression (2.21) (with a power law).

7. The same simple Beck theory applied to the Lewis and Swinney experiments still leads to a good agreement with their data, clearly assigning a single q to each of their $p(u)$-curves (cf. Fig. 6).

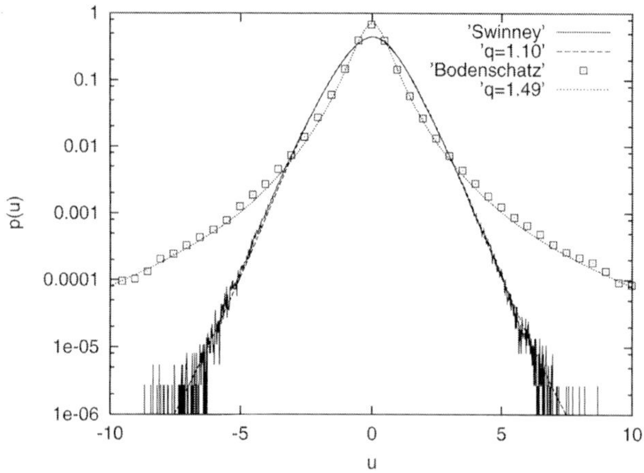

Fig. 5. Comparison of $\log_{10} p(u)$ vs u of Beck theory (- - - - - -) with experiments at $Re = 2600$ of Bodenschatz et al (i.e. La Porta et al): ($\boxdot$) and of Swinney (i.e. Lewis and Swinney) at $Re = 540.000$ (———). In the first case u stands for the acceleration a. For Bodenschatz et al (i.e. La Porta et al): good agreement for q=1.49 and $\alpha = 0.92$ (with the simple Beck Theory: $q = 1.50$ and $\alpha = 1$ cf. Fig. 6). For Lewis and Swinney: good agreement for $q = 1.10$ and α =0.90 (= 2 - q) (with the partly extended Beck theory[30c] (cf. Fig. 7); $q = 1.10$, $\alpha = 0.90 = 2$ - q, $8c\sqrt{\gamma\tau} = q - 1$ (gives very good agreement with the full Beck theory (cf. Fig. 3).

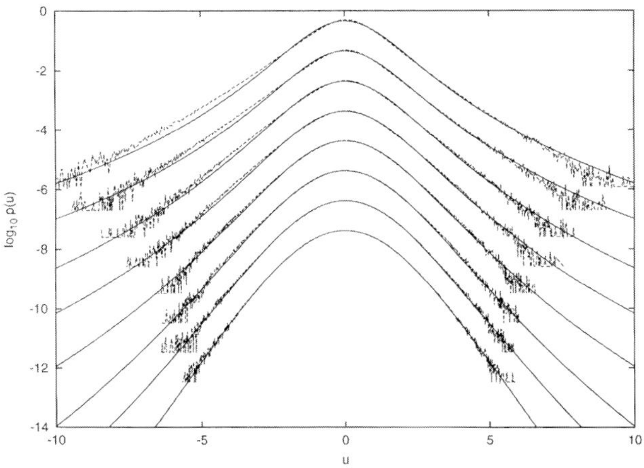

Fig. 6. $\log_{10} p(u)$ vs u with $\alpha = 1$. Each curve is characterized by one value of q; - - - - - experiment, ——— simple Beck theory (shift as in Fig. 3a).

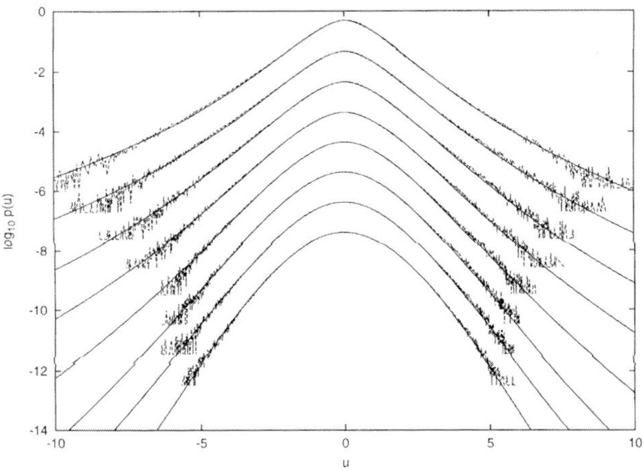

Fig. 7. $\log_{10} p(u)$ vs u with variable α. Each curve is characterized by a value of q and of $\alpha = 2 - q$, - - - - - experiment, ——— partly extended Beck theory[30c] (shift as in Fig. 3a).

b. Extended Beck Theory

In order to account more fully for the Lewis-Swinney experiments a more elaborate theory than the simple one I represented is required. Beck introduces the following modifications:

1) a fluctuating non-linear friction in the Langevin equation (2.16)[30c], so that Eq. (2.16) becomes:

$$\dot{u} = -\gamma \frac{\partial}{\partial u} |u|^{2\alpha} + \sigma L(t) . \tag{2.24}$$

In that case $\epsilon_i = \frac{|u|^{2\alpha}}{2}$ is the "Brownian particle" (or longitudinal velocity difference u) energy and

$$q = 1 + \frac{2\alpha}{1 + n\alpha}, \tag{2.25}$$

which reduces to Eq. (2.22) for the case of a linear friction $\alpha = 1$. In Fig. 7 this improved Beck theory is compared with the Lewis-Swinney data.

2) The $p(u)$ curve exhibits, on closer inspection, a skewness[30,31]. To account for this Beck considers a class of dynamical systems, which mimic in a more microscopic dynamical fashion the changes of the turbulent environment on the "Brownian" particle (or on u)[30b].

Combining the non-linear friction with the skewness correction as derived from the above mentioned dynamical systems, the full Beck theory gives the expression[30,31]:

$$p(u) = \frac{1}{Z(q)}[1 - \tilde{\beta}(1-q)\epsilon(u;\alpha)]^{\frac{1}{1-q}}. \tag{2.26}$$

Here $\tilde{\beta}$ is a formal inverse temperature, representing in reality the effect of the energy production flow in the turbulent system[30a], i.e. the energy cascade in the fluid. It is determined by the condition that the $p(u)$ of Eq. (2.26) has a variance of 1. The energy $\epsilon(u;\alpha)$ is then given by:

$$\epsilon(u;\alpha) = \frac{1}{2}|u|^{2\alpha} + c\sqrt{\gamma\tau}[|u|^{\alpha} - \frac{1}{3}|u|^{3\alpha}], \tag{2.27}$$

where τ is a characteristic time associated with the local kicks on the particle, with $\sqrt{\gamma\tau} \approx 0.1$ as determined from the experimentally observed skewness of the turbulent distributions $p(u)$.

The exponent α and the coefficient $c\sqrt{\gamma\tau}$ are determined from the experimental data, for which $\alpha = 2 - q$ and $c\sqrt{\gamma\tau} \approx \frac{q-1}{8}$ hold. Using these two relations, one can represent virtually perfectly all 8 curves of $p(u)$ by a single value of q each, by requiring that the relative mean squared deviation between the theoretical and experimental $p(u)$ integrated over u is a minimum (cf. Fig. 3). Although two coefficients have been determined here using experimental data, the almost perfect agreement seems to be mainly due to the form of the $p(u)$ curve as a generalized Tsallis canonical distribution as can be seen from Fig. 6. As mentioned before the tails do seem to behave here as power laws, which is contrary to La Porta et al's stretched exponential. We also note the increasing improvement of Beck's theory with the Lewis and Swinney experiments as is shown in Figs. 6,7 and 3, respectively: by including α in Fig. 7 and by including α as well as the skewness correction $\sim \sqrt{\gamma\tau}$, in Fig. 3, respectively, while still each r/η curve corresponds to only one q. Clearly though, a more complete theory should be able to derive the above relations between q and α and $c\sqrt{\gamma\tau}$, as well as the actual values of q needed for each r/η curves, something that has so far been only possible for the La Porta et al experiment.

3 Conclusions

1. Another interesting application of Tsallis' statistics has been made by Boghosian in discussing the drift turbulence observed in a (two-dimensional) pure electron plasma column[34a]. In fact, much of the philosophy espoused in this paper can be found in Boghosian's paper. The most striking point here is that by choosing the energy constraint as $\sum_{i=1}^{W} p_i^q \epsilon_i = U_q$ (cf. Eq. (2.10)), and by using the condition that $p_i = 0$, if $1 - (1-q)\beta\epsilon_i < 0$ (so that p_i remains real and ≥ 0), a natural explanation can be given for the relaxation of such a plasma to a stationary state. Essential is here that the condition on the p_i automatically enforces a vanishing plasma density at the boundary of the column, something that had to be put in by hand by a previous theory[35]. Both theories lead to the same profiles in good agreement with the experiments. Furthermore, as Beck did in the case of the La Porta et al experiment, Boghosian is also able to determine a value for q ($q = \frac{1}{2}$) to obtain agreement with experiment. In addition, Boghosian develops what I called in this paper a q-thermodynamics for the pure electron plasma drift turbulence. These results have been extended [22] and critisized[34b]. We only note that within the context of Tsallis' statistics, the q values found for this system are < 1, while those for La Porta et al and Lewis and Swinney systems are > 1. This would imply higher probabilities for rare/frequent events in the first/second system, respectively. For further details I have to refer to the literature.

2. Recently, Mordant et al[36] have obtained results for the motion of a Lagrangian test particle in a fully developed turbulent flow between two counter rotating disks, using ultra sound techniques. Their results for small r/η appear to be compatible with those of La Porta et al, in particular with $q = 3/2$, as well as for large r/η with those of Lewis and Swinney.

3. As indicated already in the Introduction, this survey is meant to draw attention to a new extension of classical Statistical Mechanics and Thermodynamics, mainly on the basis of Einstein's (never accepted) criticism of Boltzmann's principle. The Tsallis statistics is an example of a new classical statistics in phase space due to "unusual" dynamics, in my interpretation of Einstein's critique of Boltzmann's principle. If there is any real physics in this development - which the examples in the previous sections seem to indicate - it would constitute a real extension of Statistical Mechanics and Thermodynamics to a class of systems whose dynamics is so complicated that they have escaped a simple understanding so far.

4. In all cases considered in this paper the statistics corresponding to a given dynamics are guessed, not derived, as Einstein required. Clearly for many particle systems this is a very difficult, if not a hopeless task. So although Einstein's criticism of Boltzmann's principle, opened the possibilities for other classical statistics, its actual application as to prove the validity of these statistics from the basic dynamics of the system has not been fulfilled as yet, i.e essentially the proofs of ergodic theorems.

5. Concluding[37], one could perhaps say that while in the 19th century Statistical Mechanics was dominated by the dynamical kinetic theories of Maxwell and

Boltzmann - which were mainly concerned with the approach to equilibrium - in the 20th century it was dominated by the static considerations of Gibbs, mainly concerned with problems in equilibrium - in particular with phase transitions[21]. The 21st century seems to shape up as one where again dynamical considerations come to the foreground - in particular to understand systems far from equilibrium - thus alternating the nature of Statistical Mechanics as the pendulum swings.

Acknowledgement

I am very much indebted to Dr. Chr. Beck, Professor L. Rondoni and Professor H. L. Swinney, for reading parts of this paper and for providing me with original figures. In addition, I am particularly obligated to Dr. Beck for his many very helpful comments. I have also appreciated the very generous support of the Karpacz Winterschool, and in particular that of Professors P. Garbaczewski and R. Olkiewicz. Finally I acknowledge financial support from the Office of Basic Energy Sciences of the U.S. Department of Energy under contract No. DE-FG02-88-ER-13847.

References

1. D.H.E. Gross, "Microcanonical Thermodynamics, Phase Transitions in "Small" Systems", World Scientific Lecture Notes in Physics, Vol.66, (World Scientific, Singapore, 2001); see also this issue.
2. L. Casetti, M. Pettini and E.G.D. Cohen, Phys. Rep. **337**, 237 (2000);
   R. Franzosi, M. Pettini, L. Spinelli, cond-mat/0104110;
   M. Casetti, E.G.D. Cohen, and M. Pettini, Phys. Rev. **E65**, 036112 (2002);
   ibid., J. Stat. Phys. submitted (2002).
3. D. Evans, E.G.D. Cohen and G.P. Morriss, Phys. Rev. Lett. **71**, 2401 (1993).
4. G. Gallavotti and E.G.D. Cohen, Phys. Rev. Lett. **74**, 2694 (1995);
   J. Stat. Phys. **80**, 931 (1995);
   G. Gallovotti, Mathematical Physics Electronic Journal, MPEJ, **1**, 1 (1995);
   for a popular account see: E.G.D. Cohen, Physica **A240**, 43 (1997).
5. D. Evans and D. Searles, Phys. Rev. **E50**, 1645 (1994).
6. Ya. Sinai, Lectures in Ergodic Theory, Princeton University, Princeton (1977);
   Funct. Ann. Appl. **2**, 61 (1968); ibid., **2**, 245 (1968).
7. R. Bowen, Am. J. Math. **92**, 725 (1970).
8. D. Ruelle, Am. J. Math. **98**, 619 (1976);
   Ann. N. Y. Acad. Sci., **357**, 1 (1980).
9. See., e.g., J. Kestin and J.R. Dorfman, "A Course in Statistical Thermodynamics", (Academic Press, New York (1971)).
10. See, e.g. (a) F. Bonetto, G. Gallavotti and G. Garrido, Physica **D105**, 226 1997);
    (b) F. Bonetto, N. Chernov and J.L. Lebowitz, Chaos **8**, 823 (1998);
    (c) S. Lepri, R. Livi and P. Politi, Physica **D119**, 140 (1998);
    (d) N. Chernov, G. Eyink, J.L. Lebowitz and Ya. Sinai, Phys. Rev. Lett. **70**, 2209 (1993);
    (e) G. Benettin and L. Rondoni, Math. Phys. Elect. J. (2001).

11. L. Boltzmann, Zeitschr. f. Ru. Angew. Math. (Crelles Journal) **98**, 68 (1884); ibid., Wissensch. Abhandl. Band III, 122, (Chelsea Publ., New York (1968).
12. S. Ciliberto and C. Laroche, J. Phys. IV France, **8**, 215 (1998).
13. G. Gallavotti, Phys. Rev. Lett. **77**, 4334 (1996); J. Stat. Phys. **84**, 899 (1996); G. Gallavotti and D. Ruelle, Comm. Math. Phys. **190**, 279 (1997).
14. D.J. Evans and D. Searles, Advances in Physics (2002).
15. W.M. Wang, E. Sevick, E. Mittag, D. Searles and D.J. Evans, submitted (2002).
16. E.G.D. Cohen and G. Gallavotti, J. Stat. Phys. **96**, 1343 (1999).
17. G. Gallavotti, Physica **A263**, 39 (1999).
18. L. Bertini, A. De Sole, D. Gabrielli, G. Jona-Lasinio and C. Landim, Phys. Rev. Lett. **87**, 040601 (2001); J. Stat. Phys., in print (2002).
19. B. Derrida, J.L. Lebowitz and E.R. Speer, Phys. Rev. Lett. **87**, 150601 (2001); J. Stat. Phys., in print (2002).
20. G. Gallavotti, Nonlinearity, to be published (2002).
21. See, e.g., N. Goldenfeld, *Lectures on Phase Transitions and the Renormalization Group*, (Addison-Wesley, New York (1992)).
22. C. Tsallis, J. Stat. Phys. **52**, 479 (1988; ibid., "Nonextensive Statistical Mechanics and Thermodynamics: Historical Background and Present Status", in: *Nonextensive Statistical Mechanics and its Applications*, S. Abe and Y. Okamoto (eds.), (Lecture Notes in Physics, Springer, New York (2000)).
23. A. Pais, *Subtle is the Lord...*, Chapter II and references therein, (Oxford University Press, New York (1982)).
24. E.G.D. Cohen, Physica **A305**, 19 (2002) and references therein, especially references 2 and 3.
25. C. Beck and F. Schlögl, *Thermodynamics of Chaotic Systems*, Cambridge University Press, Cambridge, UK (1992).
26. S. Abe, Phys. Lett. **A271**, 74 (2000).
27. C. Tsallis, R.S. Mendes and A.R. Plastino, Physica **A262**, 534 (1998).
28. G.S. Lewis and H.L. Swinney, Phys. Rev. **E59**, 5457 (1999).
29. A. La Porta, G.A. Voth, A.M. Crawford, J. Alexander and E.Bodenschatz, Nature **409**, 1017 (2001);
    G.A.Voth, A. La Porta, A.M. Crawford, E. Bodenschantz and J. Alexander, Los Alamos Physics 0110027 (2001).
30. (a) C. Beck, Phys. Rev. Lett. **87**, 180601 (2001);
    (b) ibid., Physica **A277**, 115 (2000);
    (c) ibid., Physica **A286**, 164 (2000).
31. C. Beck, G.S. Lewis and H.L. Swinney, Phys. Rev. **E63**, 035303(R) (2001).
32. H.O. Lancaster, "The Chi-Squared Distribution", John Wiley & Sons, New York (1969).
33. U. Frisch, *Turbulence*, Cambridge University Press, Cambridge, UK 1995).
34. (a) B.M. Boghosian, Phys. Rev. **E53**, 4754 (1996);
    (b) H. Brands, P.H. Chavanis, R. Pasmanter and J. Sommeria, Phys. Fluids **11**, 3465 (1999).
35. X.P. Huang and C.F. Driscoll, Phys. Rev. Lett. **72**, 2187 (1994).
36. N. Mordant, P. Metz, O. Michel and J-F. Pinton, Phys. Rev. Lett. **21**, 4501 (2001).
37. E.G.D. Cohen, "Boltzmann and Statistical Mechanics", in: *Dynamics: Models and Kinetic Methods for Nonequilibrium Many-Body Systems*, J. Karkheck, ed., NATO Science Series E, vol. 371 (Kluwer, Dordrecht (2000)); Pinton, J-F.[36].

Deterministic Thermostats and Fluctuation Relations

L. Rondoni

Dipartimento di Matematica, Politecnico di Torino, Corso Duca degli Abruzzi 24, I-10129 Torino, Italy

Abstract. Deterministic "thermostats" are mathematical tools used to model nonequilibrium steady states of fluids. The resulting dynamical systems correctly represent the transport properties of these fluids, and are easily simulated on modern computers. The purpose and limitations of deterministic thermostats in the context of Irreversible Thermodynamics is discussed, along with some examples of thermostatted fluids, to familiarize the reader with the subject. It is also shown that thermostatted systems ought not to be Hamiltonian, and several fluctuation relations are illustrated.

1 Introduction

The goal of statistical mechanics is to understand the macroscopic properties of physical systems from the dynamics of their microscopic constituents, by means of suitable averaging procedures. For instance, consider a particle system constituted by N classical particles in d dimensions, and let its dynamics be described by:

$$\dot{x} = G(x) \; ; \quad x = (\mathbf{q}, \mathbf{p}) \in \Omega \subset I\!R^{2dN} , \tag{1}$$

where Ω is the phase space, and the vector field G is determined by the forces acting on the system and by the particles interactions. Denote by $S^t x$, $t \in I\!R$, the solution of Eq. (1) with initial condition x. The macroscopic quantity associated with an *observable*, *i.e.* with a function of phase $\Phi : \Omega \to I\!R$, is defined by:

$$\bar{\Phi}(x) = \lim_{T\to\infty} \frac{1}{T} \int_0^T \Phi(S^t x) dt . \tag{2}$$

This represents the fact that macroscopic observations occur on time scales which are long compared to the time scales of the microscopic dynamics (1), and that one observation amounts to a time average of the chosen observable Φ. The functions Φ individuate a function space $\mathcal{F}(\Omega)$ in which the limit (2) exists; but computing that limit may not be a trivial task at all.

For systems in (or evolving towards) equilibrium,[1] the problem is commonly solved by invoking the Ergodic Hypothesis, which states that

$$\bar{\Phi}(x) = \frac{1}{\mu(\Omega)} \int_\Omega \Phi(y) \, d\mu(y) = \langle \Phi \rangle_\mu , \tag{3}$$

[1] A physical system is in an *equilibrium* state if all currents –of heat, momentum, etc.– vanish, and the system is uniquely described by a set of *state variables*, which do not change with time.

for a suitable measure μ (called physical measure), and for μ-almost all $x \in \Omega$. In principle, ergodic theory should identify the cases verifying the Ergodic Hypothesis, and the physical measures μ but, in practice, this is too hard to do in the cases of physical interest. Therefore, for an isolated system whose energy H remains within a thin shell $[E, E + \Delta E]$, one usually *postulates* that μ is the uniform probability distribution, or *microcanonical ensemble*; for a closed system in contact with a heat bath at given temperature, one *postulates* that μ is the *canonical ensemble*; etc. A posteriori one checks whether these assumption are valid or not, and finds that the classical ensembles describe very well most equilibrium situations, despite the fact that the mathematical notion of ergodicity is violated, for instance, by the ideal gas in a cubic box, by chains of harmonic oscillators, and by the blackbody radiation.

The fact is that the Ergodic Hypothesis can be adopted *for practical purposes*, *i.e.* for computing the averages of the *physically relevant* quantities, considering that the detailed differences among different probability distributions are meaningless (undetectable) as long as they yield the same averages for the observables. In other words, the set of observables is often too small, and the observation times too short, to probe the fine differences among certain classes of measures, for the high dimensional systems of interest in Statistical Mechanics (cf. chapter I of [1]). Also, in some circumstances the different classical ensembles become "equivalent" in the thermodynamic limit. This means, for instance, that the microcanonical and canonical averages of the local observables (those that depend only on the particles confined in a fixed, finite box) tend to the same values if the volume of the system grows without bounds while its density and energy density tend to a constant. Obviously, the ensembles do not agree on global quantities, such as the total energy (cf. chapter II of [1]).

The situation is more complex in the case of nonequilibrium systems in a steady state.[2] In particular, the form of a measure μ to be used in Eq. (3) is generally unknown; which generates the need for simple models, on which the various hypothesis could be tested.

The purpose of this paper is to review some of the ideas recently developed in this context, mostly referring to the models of nonequilibrium molecular dynamics (NEMD).

1.1 Molecular Dynamics

In the past decades, a large number of computer algorithms have been developed to simulate in an efficient way various features of equilibrium and nonequilibrium fluids and solids, viewed as systems made of many interacting particles. These algorithms constitute Molecular Dynamics (MD) codes. There is a vast literature on both equilibrium MD and NEMD, including textbooks such as Refs. [2–4]. Reference [5] is the first report on a MD simulation; its purpose was to investigate the phase diagram of a hard sphere system. Since then, MD has been successfully

[2] These are systems whose state parameters remain constant in time, but such that there is transport of mass, energy, entropy, etc.

applied to problems like the calculation of the rheological properties of fluids, or the simulation of polymers in porous media, of defects in crystals, of friction between surfaces, of atomic clusters, of biological macromolecules, etc.

At present, the validity of MD results is limited by: a) quantum mechanical effects play an important role at times;[3] b) the interatomic forces are sometimes too complicated to be correctly modeled; c) the number of particles doesn't get close to Avogadro's number; d) the simulation times are limited by the size of the system. Nevertheless, these problems often do not constitute a hindrance and, in fact, MD is commonly used to better understand the results of real experiments, or even in place of experiments which are either too costly or practically impossible to perform (like the observation of fracture fronts *inside* a solid).

Some NEMD have also inspired G. Gallavotti and E.G.D. Cohen's *Chaotic Hypothesis* (CH) as a new paradigm which unifies equilibrium and nonequilibrium statistical mechanics [7].

However, it is important to keep in mind that the original purpose of NEMD models is to compute transport coefficients. In this respect they are successful, although, under other respects, their behaviour may look inconsistent with the behaviour of the real molecules of a fluid. Indeed, NEMD models usually contain fictitious forces which do not exist in Nature. In some instances, there is a well developed theory which guarantees the correctness of the transport coefficients obtained via NEMD, as explained in, e.g. [3,4,8]. In other cases, as for the extremely simple model of [9] for the calculation of the viscosity of a fluid near its triple point, such a theory is not available. Either way, NEMD models are only supposed to compute transport coefficients, and the presence of fictitious forces makes dubious that they afford the full thermodynamic description of a physical system [10].

1.2 Hamiltonian versus Non-Hamiltonian Dynamics

Typical nonequilibrium systems are the following:

1. A conducting wire subject to a constant electromotive force which produces a drift of the conducting electrons, superimposed on their thermal motion. In a steady state, the work done by the force is converted into heat which flows into the outer environment.

2. A rod whose ends are in contact with heat reservoirs at different temperatures T_1 and T_2. In a steady state, heat flows from the hottest to the coldest end of the rod.

3. A viscous fluid at constant temperature between two moving parallel plates. The fluid sticks to the plates and, because of the friction between adjacent fluid layers, momentum flows from one plate to the other. In a steady state, the energy given by the plates to the fluid particles flows into the outer environment.

[3] For attempts to produce quantum mechanical MD models, see [6] and references therein

Mechanical models of these systems require infinitely many degrees of freedom: a finite number for the system under consideration (the conducting wire or rod, or the fluid), and infinitely many for the outer environment (the reservoirs). From the mathematical viewpoint, the analysis of these models poses formidable problems; see, for instance, Ref. [11] which proves the existence and uniqueness of the steady state of a finite chain of nonlinear oscillators coupled to two reservoirs at different temperatures. From a more practical standpoint, like the calculation of the viscosity of liquid benzene, say, numerical simulations are essential, but impossible to perform for an infinite system. At the same time, it is numerically difficult and inefficient to handle boundary conditions imposed by moving walls or by walls at different temperatures, since they produce inhomogeneous particle distributions, making the interpretation of numerical results cumbersome (cf. chapter 6 of [3]).

Luckily, susceptibilities of thermal processes were found to be similar to susceptibilities of mechanical processes, and a number of *ad hoc* models could be devised as efficient, *equivalent*, mechanical representations of thermal transport processes [3]. In these models, the boundary driven systems are replaced by purely mechanical systems subjected to fictitious external forces and to periodic boundary conditions. But a driving force pumps energy into the system, and a steady state cannot be reached unless this energy is removed: i.e. unless a *thermostatting* term is added to the equations of motion of the model, in order to dissipate the mechanical energy. A viscous term proportional to the velocity of the particles would do, but would also defeat the purpose of computing the viscosity coefficient, for instance. Alternatively, one may resort to a variety of constrained dynamics, to dissipate the excess energy.[4] The resulting systems, obviously, are not Hamiltonian, which is the case of many NEMD models.

The non Hamiltonian nature of NEMD models sometimes leads to consider even simpler models[5] as physically more appealing than the (apparently "tampered") equations of motion of the NEMD driven, interacting particles systems. But the fact that real systems are Hamiltonian does not imply that the models of thermostatted systems should also be Hamiltonian. In fact, faithful representations of thermostatted systems can hardly be described by Hamiltonian dynamics. In the first place, it is plausible (although not obvious) that the effect of a friction term, and that of real heat exchange at the boundaries of a system must be *equivalent* in some sense. For instance, if the heating rate of the system is slower than the heat exchange at the walls, energy is lost practically instantaneously from the bulk, as in dissipative dynamics. This is the case if the particles cooled at the walls reenter the bulk of the fluid fast enough. But, more importantly, a model of a system in contact with a heat reservoir is ideally obtained projecting out the coordinates of the reservoir particles from the equations of

[4] Popular constraints with these properties are those of constant internal energy H_0 or constant kinetic energy K, which prevent the system from indefinitely "heating up".

[5] For example, Hamiltonian systems of non-interacting particles, or area preserving 2-dimensional maps.

motion of very many interacting particles, part of which constitute the system and part the reservoir (see [12] for one example). In other words, the ideal model of a thermostatted system, $\{x_s(t)\}$ say, is made of the first $2dN_s$ components of the vector $\{S^t x\}$ determined by the (Hamiltonian) equations (1), written in the form

$$\dot{x} = \begin{pmatrix} \dot{x}_s \\ \dot{x}_r \end{pmatrix} = G(x) = \begin{pmatrix} G_s(x_s, x_r) \\ G_r(x_s, x_r) \end{pmatrix} , \quad \begin{array}{l} x_s = (\mathbf{q}_i, \mathbf{p}_i)_{i=1}^{N_s} , \\ x_r = (\mathbf{q}_i, \mathbf{p}_i)_{i=N_s+1}^{N} , \end{array} \tag{4}$$

where the subscript s refers to the N_s particles of the thermostatted system, and the subscript r refers to the N_r particles of the reservoir, with $N = N_s + N_r$.

Remark: *To construct the projected dynamics $\{x_s(t)\}$ may be too difficult, in practice, but assuming it can be done, the corresponding $2dN_s$-dimensional dynamics are time reversal invariant, and do not preserve the volumes of the projections of the phase space subsets.*[6]

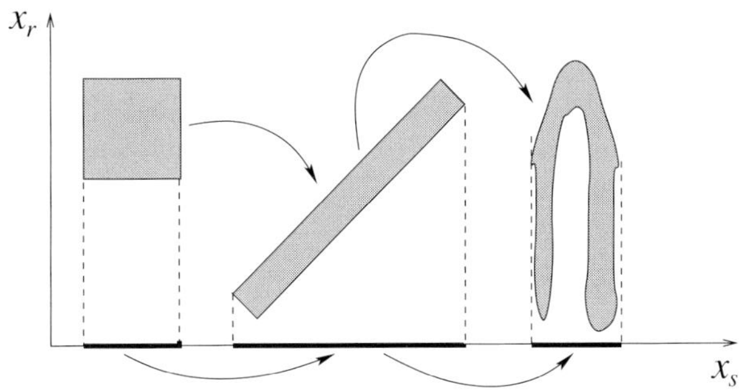

Fig. 1. The two-dimensional dynamics represented by the two top arrows is area preserving, while the projected one-dimensional dynamics, represented by the two bottom arrows, alternates between expanding and contracting states. If the backward evolution occurs in full phase space, so it does in the projected space.

It suffices to look at Fig. 1 to be convinced of that. There, the dynamics in (2-dimensional) phase space is area preserving, while the projected dynamics on the horizontal axis at times expands, other times contracts the size of the projections of the phase space sets. Moreover, if the reverse evolution (from right to left, in Fig. 1) is allowed in the phase space, it is also allowed in the horizontal axis. Similarly, the dynamics of a given particle system is time reversal invariant but not volume preserving, if the degrees of freedom of its reservoir have been

[6] Hamiltonian systems made of N non-interacting particles violate this condition. In this case, the presence or absence of the N_r reservoir particles has no influence on the dynamics of the other N_s particles, and projecting out the coordinates of N_r particles leaves the system Hamiltonian. However, in this case one cannot say that the remaining N_s particles constitute a thermostatted system: no reservoir ever acted on them.

projected out. In particular, a reservoir removes energy from a driven system, on average, hence the corresponding projected dynamics must be dissipative, as is the case of the NEMD models. Unfortunately, NEMD models are not obtained through the ideal procedure described above; hence their dissipative terms are not to be considered as real thermostats. The NEMD models are to be intended "equivalent" to real systems subject to real thermostats only in a limited sense concerning the calculation of specific quantities. Which quantities and under which conditions, will have to be determined by further research; but various works have already provided some examples of this possible equivalence, cf. Refs. [8,13–15].

2 Thermostats

Typically, dissipation is introduced in NEMD models through constraints which either fix some observable in time, or fix its average. The constrained observables depend on the transport process under study, and the constraint terms in the equations of motion are usually called *thermostats*, for brevity, as they are sinks and sources of energy, but must not be confused with real thermostats. Here, certain features of popular *thermostats* are briefly reviewed, along with references to the literature, where more details can be found. A few new thermostats are proposed, to stimulate further research.

2.1 Gaussian Thermostats

Arguably, the most elegant way of *thermostatting* a particle system subjected to external driving is given by Gauss' principle of least constraint [16]:

Gauss Principle (1829):*Consider N point particles of mass m_i, subjected to frictionless bilateral constraints $\mathbf{\Phi}_i$ and to external forces $\mathbf{F}_i$. Among all motions allowed by the constraints, the* **natural** *one minimizes the quantity*

$$C = \sum_{i=1}^{N} m_i \left(\ddot{\mathbf{q}}_i - \frac{\mathbf{F}_i}{m_i} \right)^2 = \sum_{i=1}^{N} \frac{1}{m_i} \mathbf{\Phi}_i^2 \ .$$

According to Gauss, the "Curvature" C is minimized by the accelerations of real motions or, equivalently, real motions minimize the action of the constraints. In the case of holonomic constraints, Gauss' principle is consistent with the principle of least action, and produces Hamiltonian equations of motion. For instance, consider a diatomic molecule whose interatomic distance D must be kept constant in time: $\mathbf{q}_{12}^2 - D^2 = 0$, with $\mathbf{q}_{12} = \mathbf{q}_1 - \mathbf{q}_2$, $\mathbf{q}_i$ = position of atom i. Gauss' principle yields:

$$\begin{cases} m_1 \ddot{\mathbf{q}}_1 = \mathbf{F}_1 + \lambda \mathbf{q}_{12} \ ; \quad m_2 \ddot{\mathbf{q}}_2 = \mathbf{F}_2 - \lambda \mathbf{q}_{12} \, , \\ \lambda = - \left(\mathbf{q}_{12} \cdot (m_1 \mathbf{F}_2 - m_2 \mathbf{F}_1) + m_1 m_2 \dot{\mathbf{q}}_{12}^2 \right) / \, (m_1 + m_2) \, \mathbf{q}_{12}^2 \, . \end{cases}$$

Differently, non-holonomic constraints lead to non-Hamiltonian equations of motion. In particular, consider the isokinetic (IK) constraint, which fixes the kinetic energy of the system, $K = \sum_i \mathbf{p}_i^2/2m$, and the isoenergetic ($IE$) constraint, which fixes the internal energy $H_0 = K + \Phi^{int}$, Φ^{int} = particles interaction potential. For an N-particle system subjected to an external field, these constraints yield:

$$\begin{cases} \dot{\mathbf{q}}_i = \mathbf{p}_i/m\,, \\ \dot{\mathbf{p}}_i = \mathbf{F}_i^{int}(\mathbf{q}) + \mathbf{F}_i^{ext}(\mathbf{q}) - \alpha(x)\mathbf{p}_i\,, \end{cases} \qquad i = 1, \ldots, N\,, \tag{5}$$

where $\mathbf{F}_i^{int}(\mathbf{q})$ and $\mathbf{F}_i^{ext}(\mathbf{q})$ respectively denote the interactions among the particles and the external force doing work on the system. The term $-\alpha(x)\mathbf{p}_i$ makes the dynamics dissipative, allowing the system to reach a steady state in the long time limit. If $\mathbf{F}_i^{ext}(\mathbf{q}) = C_i\mathbf{F}_e$, where C_i is the coupling constant of the ith particle, one then has:

$$\alpha_{IK}(x) = \frac{1}{2K}\left(\mathbf{J}\cdot\mathbf{F}_e + \sum_{i=1}^{N}\frac{\mathbf{p_i}}{m}\cdot\mathbf{F}_i^{int}\right), \quad \mathbf{J} = \sum_{i=1}^{N} C_i\dot{\mathbf{q}}_i\,, \tag{6}$$

$$\alpha_{IE}(x) = \frac{1}{2K}\mathbf{J}\cdot\mathbf{F}_e\,. \tag{7}$$

A popular NEMD model for shear flows, called SLLOD, is the following:

$$\begin{cases} \dot{\mathbf{q}}_i = \mathbf{p}_i/m + \mathbf{i}\gamma\, y_i\,, \\ \dot{\mathbf{p}}_i = \mathbf{F}_i(\mathbf{q}) + \mathbf{i}\gamma\, p_{yi} - \alpha(x)\mathbf{p}_i, \quad i = 1, \ldots, N\,, \end{cases} \tag{8}$$

$$\alpha_{IK}(x) = \frac{\sum_{i=1}^{N}(\mathbf{F}_i\cdot\mathbf{p}_i - \gamma p_{xi}p_{yi})}{\sum_{i=1}^{N}\mathbf{p}_i^2}\,, \tag{9}$$

where γ is the shear rate in the y direction. Neglecting the *thermostat* term, differentiating the first of Eqs. (8) and substituting in the second, one obtains

$$m\ddot{\mathbf{q}}_i = \mathbf{F}_i + \mathbf{i}m\dot{\gamma}\, y_i\,. \tag{10}$$

Usually, γ is taken to vanish for all times up to $t = 0$, and is constant afterwards, so that $\dot{\gamma} = \gamma\delta(t)$. Therefore, the unthermostatted SLLOD equations represent a system kicked at time $t = 0$, which evolves under the sole action of interparticle forces at all other times.

The IK and IE constraints are only two possible options in NEMD. Depending on the physical property to be described, a wide range of constraints is available, including isobaric, isochoric, isoenthalpic, constant stress constraints, etc.

2.2 Nosé-Hoover Thermostat

S. Nosé and W.G. Hoover [17] devised a popular *thermostat*, similar to the Gaussian *thermostat*, defined by the following transformations of the momenta and of the time variable:

$$\tilde{\mathbf{p}}_i = \frac{\mathbf{p}_i}{s}\,; \quad \tilde{t} = \int_0^t \frac{d\tau}{s}\,, \tag{11}$$

and by the following equations of motion:

$$\frac{d\mathbf{q}_i}{d\tilde{t}} = \frac{\tilde{\mathbf{p}}_i}{m} \; ; \quad \frac{d\tilde{\mathbf{p}}_i}{d\tilde{t}} = \mathbf{F}_i - \zeta \tilde{\mathbf{p}}_i \; ; \quad \frac{d\zeta}{d\tilde{t}} = \frac{1}{\tau^2}\left(\frac{K(\tilde{\mathbf{p}})}{K_0} - 1\right) \; ; \quad \frac{ds}{dt} = \zeta s \; . \tag{12}$$

Here, K_0 is the value chosen for the time average of the kinetic energy $K(\tilde{\mathbf{p}})$, and τ is a relaxation time. In the small τ limit, Nosè-Hoover dynamics approximate Gaussian IK dynamics.

2.3 Dirac Brackets

One method for constraining mechanical systems, which is well known in quantum field theory, but does not seem to be used in NEMD, is Dirac's method for Hamiltonian systems [18]. An exhaustive review of Dirac's method goes beyond the purpose of this paper, but it is interesting to see how it could be used in the physical situations of NEMD.

Let H be the Hamiltonian of a particle system with Poisson brackets $\{\cdot,\cdot\}$, and assume the two phase variables ϕ_1 and ϕ_2 ought to be constants of motion. This can be achieved replacing the dynamics given by $\{\cdot,\cdot\}$, with the dynamics given by the following (Dirac) brackets $\{\cdot,\cdot\}_*$:

$$\{A,B\}_* = \{A,B\} - (\{A,\phi_1\},\{A,\phi_2\}) \begin{pmatrix} 0 & \dfrac{1}{\{\phi_2,\phi_1\}} \\ \dfrac{1}{\{\phi_1,\phi_2\}} & 0 \end{pmatrix} \begin{pmatrix} \{\phi_1,B\} \\ \\ \{\phi_2,B\} \end{pmatrix} , \tag{13}$$

where A, B are any two phase variables. In fact, one immediately realizes that:

$$\dot{\phi}_i = \{\phi_i, H\}_* = 0 \; , \quad \text{for} i = 1,2 \; . \tag{14}$$

As an example, consider one particle subject to a conservative field Ψ and to a constant driving force $\mathbf{E}= (E_x, E_y)$, in two spatial dimensions. Its Hamiltonian is given by:

$$H(\mathbf{q},\mathbf{p}) = \frac{\mathbf{p}^2}{2m} + \Psi(\mathbf{q}) - \mathbf{E}\cdot\mathbf{q} \; . \tag{15}$$

Suppose that, as in NEMD, one wants to fix the "internal" energy:

$$H_0 = \phi_1(\mathbf{q},\mathbf{p}) = \frac{\mathbf{p}^2}{2m} + \Psi(\mathbf{q}) \; . \tag{16}$$

Dirac's method requires that a second quantity ϕ_2 be constrained and, usually, one takes $\phi_2 = \{\phi_1, H\}$, i.e. $\phi_2(\mathbf{q},\mathbf{p}) = \mathbf{E}\cdot\mathbf{p}/m$, in our case. Then, simple calculations lead to:

$$\dot{\mathbf{q}} = \{\mathbf{q},H\}_* = \begin{pmatrix} E_y^2 & -E_xE_y \\ -E_xE_y & E_x^2 \end{pmatrix} \frac{\mathbf{p}/m}{\mathbf{E}\cdot\nabla\Psi} \; ; \quad \dot{\mathbf{p}} = \{\mathbf{p},H\}_* = \mathbf{E} - \frac{\mathbf{E}^2\nabla\Psi}{\mathbf{E}\cdot\nabla\Psi} \; , \tag{17}$$

This model is easy to simulate on a computer, but Eqs. (17) do not reduce to the equilibrium equations $\dot{\mathbf{q}} = \mathbf{p}/m$, $\dot{\mathbf{p}} = -\nabla\Psi$ as $\mathbf{E}$ tends to zero, and cannot be used to study near equilibrium systems. Thus, the symplectic structure of Dirac's method, absent in Gaussian dynamics, seems to alter more profoundly the nature of a driven system like ours, than Gaussian thermostats do. Are there variables (ϕ_1, ϕ_2) for which Dirac's equations of motion reduce to the equilibrium equations, in the small field limit?

2.4 Pump-Billiard Thermostat

A new thermostat has been introduced in [19], to model a (rarefied) gas in a pump. The system consists of a periodic chain of billiard cells as in Fig. 2.

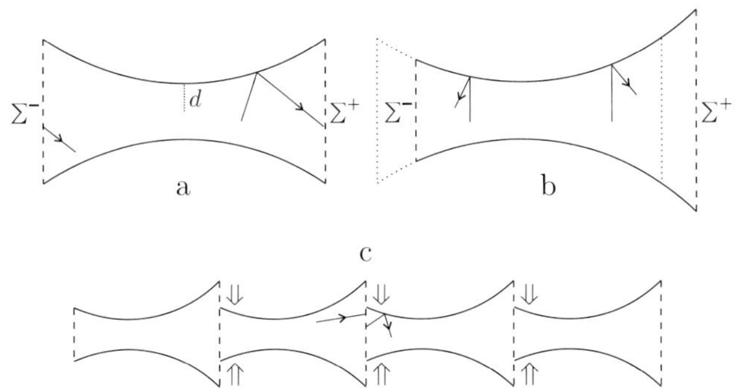

Fig. 2. Pump-billiard equilibrium cell (a); nonequilibrium cell (b); chain of cells (c).

The equilibrium periodic cell is delimited by two equal vertical lines, Σ^- and Σ^+, and by two arcs of a circle of radius r, amplitude $\phi = 2 \arcsin r^{-1}$, and minimum distance $2d$ (cf. Fig. 2a). A moving particle collides elastically with the arcs of circle, while it exits the cell from $\Sigma^\pm$. If a particle hits Σ^-, it renters from Σ^+, and vice versa. In this case there is no transport of particles. The system is not in equilibrium, and a net current is produced, if the periodic cell is asymmetric as in Fig. 2b, so that the ratio of the sizes of Σ^+ and Σ^-, $\alpha = |\Sigma^+|/|\Sigma^-|$ is larger than one. In this case, the current depends on the reinjection rule, Fig. 2c, and we chose the following law:[7]

$$x' = x \mp \ell \quad y' = \alpha^{\mp 1} y \ , \quad v'_x = c\, v_x \quad v'_y = c\, \beta^{\pm 1} v_y \quad c^2 = \frac{v_x^2 + v_y^2}{v_x^2 + \beta^{\pm 2} v_y^2} \tag{18}$$

where ℓ is the length of the cell, and (x, y, v_x, v_y) is a phase point. This represents in an idealized way the effect of a compressing mechanism and of a "thermostat", which fixes the kinetic energy. The dynamics are time reversible, and the Jacobian $J^\pm$ of the map for $v_x^2 + v_y^2 = 1$ is given by:

$$J^\pm = (\beta/\alpha)^{\pm 1} [1 + (\beta^{\pm 2} - 1)\, v_y^2]^{-3/2} \ . \tag{19}$$

[7] Analogous laws can be found in [20].

To ensure dissipation for the forward motion, i.e. $J^+ < 1$, it suffices to take $\beta < \alpha$. For $\beta\alpha > 1$, one obtains an average current $\langle v_x \rangle$ proportional to $\varepsilon = \log \alpha$ for small ε, and this proportionality is lost if ε grows [19].

How can this definitely simple mechanism for the transport of particles be generalized to systems of interacting particles?

3 Properties of Thermostatted Dynamics

Some properties of thermostatted dynamics are now analyzed, with the purpose of introducing the reader to the relevant techniques.

3.1 Equivalence of Isokinetic and Isoenergetic Thermostats

The term α in Eqs. (5) is just the Lagrange multiplier which implements Gauss' principle of least constraint, and there is no reason, in principle, to believe that it correctly represents a real heat reservoir. Similarly, the phase space contraction rate defined by

$$\chi \equiv - \sum_{i=1}^{N} \left[\frac{d}{d\mathbf{q}_i} \cdot \dot{\mathbf{q}}_i + \frac{d}{d\mathbf{p}_i} \cdot \dot{\mathbf{p}}_i \right] , \tag{20}$$

does not necessarily have any physical meaning (see, e.g. [21]). Nevertheless, it is interesting to note that in d dimensions the time average of χ obeys

$$\overline{\chi} = (dN - 1)\, \overline{\alpha} = - \sum_{i=1}^{2dN} \lambda_i \,, \tag{21}$$

where $\{\lambda_i\}_1^{2dN}$ are the Lyapunov exponents of the system, and to compare the values of $\overline{\chi}$ for the IK and IE systems, as in [22]. This sheds some light on the equivalence of nonequilibrium ensembles mentioned in Sect.1, and on the link between the macroscopic properties and the microscopic dynamics of NEMD models. Equation (6) yields

$$\overline{\chi}_{IK} = \langle \chi_{IK} \rangle = (dN - 1) \left[\left\langle \frac{\sum_{i=1}^{N} \frac{\mathbf{p}_i}{m} \cdot \mathbf{F}_i^{int}}{\sum_{i=1}^{N} \frac{\mathbf{p}_i^2}{m}} \right\rangle + \left\langle \frac{\sum_{i=1}^{N} \frac{\mathbf{p}_i}{m} \cdot \mathbf{F}^{ext}}{\sum_{i=1}^{N} \frac{\mathbf{p}_i^2}{m}} \right\rangle \right] , \tag{22}$$

for an IK ergodic system, where the brackets $\langle \cdot \rangle$ represent the average with respect to the physical measure. Because of the conservation of the kinetic energy, the temperature T can (close to equilibrium) be defined by:

$$\sum_{i=1}^{N} \frac{\mathbf{p}_i^2}{m} = \left\langle \sum_{i=1}^{N} \frac{\mathbf{p}_i^2}{m} \right\rangle = (dN - 1) k_{_B} T \,. \tag{23}$$

Considering that

$$\sum_{i=1}^{N} \frac{\mathbf{p}_i}{m} \cdot \mathbf{F}_i^{int}(\mathbf{q}) = - \frac{d}{dt} \Phi^{int}(\mathbf{q}) \,, \quad \text{so that} - \left\langle \frac{d}{dt} \Phi^{int} \right\rangle = - \frac{d}{dt} \langle \Phi^{int} \rangle = 0, \tag{24}$$

and dividing by the volume V of the system, to compare dynamical averages with macroscopic quantities, one obtains:

$$\hat{\chi}_{IK} \equiv \frac{\langle \chi_{IK} \rangle}{V} = \frac{\left\langle \sum_{i=1}^{N} \frac{\mathbf{p}_i}{mV} \cdot \mathbf{F}^{ext} \right\rangle}{k_B T} \ . \tag{25}$$

Noting that $\mathbf{I} = \left\langle \sum_{i=1}^{N} \frac{\mathbf{p}_i}{m} \right\rangle / V$ is the particle current density, one gets:

$$\hat{\chi}_{IK} = \frac{\mathbf{I} \cdot \mathbf{F}^{ext}}{k_B T} \ , \tag{26}$$

where the right hand side of Eq. (26) is *formally* the expression for the entropy production rate in Irreversible Thermodynamics (IT).

In the IE case there is no constraint on the kinetic energy, hence the temperature must be defined by:

$$\left\langle \sum_{i=1}^{N} \frac{\mathbf{p}_i^2}{m} \right\rangle = dNk_B T \ , \tag{27}$$

while Eq. (7) yields:

$$\hat{\chi}_{IE} \equiv \frac{\langle \chi_{IE} \rangle}{V} = (dN - 1) \left\langle \frac{\sum_{i=1}^{N} \frac{\mathbf{p}_i}{mV} \cdot \mathbf{F}^{ext}}{\sum_{i=1}^{N} \frac{\mathbf{p}_i^2}{m}} \right\rangle \ . \tag{28}$$

For large N, one may argue that the average of the ratio on the right hand side of (28) can be replaced by the ratio of the averages. Then, *neglecting* terms of order $O(1/N)$, one obtains

$$\hat{\chi}_{IE} = \frac{\mathbf{I} \cdot \mathbf{F}^{ext}}{k_B T} \ , \tag{29}$$

showing that entropy production and phase space contraction rates can only be equal for large N.

In reality, a large N is required to identify entropy production and phase space contraction rates for other reasons too [22,23]. In the first place, IT is based on local thermodynamic equilibrium, in which the extensive properties are proportional to N and depend further only on the temperature and on the number density $n = N/V$. Without *many interacting* particles that is impossible. Furthermore, the relaxation times for the values of the observables turn out to be unphysically long for systems consisting of a small number of particles. Then, only for large N, and small driving (cf. Eqs. (23,27)), do the average phase space contraction rates of IK and IE systems agree with the entropy production rates of physical systems.

In that case, one also has $\hat{\chi}_{IE} = \hat{\chi}_{IK} + O(1/N)$. This constitutes one example of equivalence of nonequilibrium ensembles in the thermodynamic limit, i.e. for $N, V \to \infty$ in such a way that the density and the energy density tend to a constant. This shows that different equations of motion, reaching different steady states, may describe the same physical situation, at least in part, in the thermodynamic limit.

3.2 Equivalence of Turbulence Models

An equivalence principle has been formulated in the context of turbulence in Ref. [15], where a modified version of the forced Navier-Stokes equation, for a newtonian incompressible fluid, was introduced. Consider the equation:

$$\dot{\mathbf{u}} + (\mathbf{u} \cdot \mathrm{d})\mathbf{u} = -\frac{1}{\rho}\nabla p + \mathbf{g} + \alpha \Delta \mathbf{u} , \quad \nabla \cdot \mathbf{u} = 0 . \tag{30}$$

Here, $\mathbf{u}$ is the velocity field, ρ is the fluid density, p the pressure, $\mathbf{g}$ is a constant forcing term and α can either be the constant viscosity ν, in which case (30) is the Navier-Stokes equation, or can be defined by:

$$\alpha(\mathbf{u}, \omega, \mathbf{f}) = \beta(\mathbf{u}, \omega, \mathbf{f}) = \frac{\int \left[\omega \cdot \mathbf{f} + \omega \cdot (\omega \cdot \nabla)\mathbf{u}\right] \, d\mathbf{x}}{\int (\nabla \times \omega)^2 \, d\mathbf{x}} , \tag{31}$$

with $\omega = \nabla \times \mathbf{u}$ and $\mathbf{f} = \nabla \times \mathbf{g}$. In this case (30) is called Gauss-Navier-Stokes equation, is time reversible, and the enstrophy $Q = \int \omega^2 d\mathbf{x}$ is a constant of motion.

With periodic boundary conditions, one can expand $\mathbf{u}$ in Fourier modes, obtaining an infinite system of ODEs from (30). Truncating the expansion of $\mathbf{u}$ at a finite number of modes yields a dynamical system in a finite dimensional phase space, characterized by a given phase space contraction. In this context, Gallavotti formulated the following equivalence principle [15]:

Equivalence Principle. *The stationary probability distributions of the Navier-Stokes and of the Gauss-Navier-Stokes equations are* equivalent *in the limit of large Reynolds number, provided the enstrophy Q and the viscosity ν are so related that the constant phase space contraction rate of the Navier-Stokes equation and the average of the fluctuating phase space contraction rate of the Gauss-Navier-Stokes equation are equal.*

Similarly to the equivalence of equilibrium ensembles, this principle is supposed to hold for local variables, while the large Reynolds number is invoked for the fluctuations of β to be fast on the observation time scales. In that case, if the average of β equals ν, the macroscopic behaviour of the Navier-Stokes equation and of the Gauss-Navier-Stokes equation should be the same.

3.3 Onsager Relations for Gaussian Isoenergetic Dynamics

In Ref. [24] a derivation of the Onsager reciprocal relations was first given for IE systems. We give a derivation based on periodic orbit theory, not restricted to the smooth dynamics of [25]. Consider an N-particle IE system, in d dimensions with periodic boundary conditions. Assume the forces $\mathbf{F}_i^{ext}(\mathbf{H})$, $i = 1, \ldots N$, depend smoothly on $\mathbf{H} = (H_1, \ldots, H_\nu) \in I\!R^\nu$, representing ν thermodynamic forces, so that the thermostat multiplier of Eqs. (5) can be written as:

$$\alpha(x; \mathbf{H}) = \mathbf{C}(x) \cdot \left[\mathbf{H} + O(H^2)\right] , \qquad \mathbf{C} = (C_1, \ldots, C_\nu) . \tag{32}$$

Because α can be interpreted as the entropy production rate in IE systems, for large N (cf. Sect.3.1), the thermodynamic fluxes associated with the thermodynamic forces take the form:

$$\mathbf{I}(x;\mathbf{H}) = \nabla_{\mathbf{H}}\alpha = (I_1, \ldots; I_\nu)(x, \mathbf{H}) . \tag{33}$$

Denote by $P_n(\mathbf{H})$ the set of unstable periodic orbits w of period $\tau_w \in [T_n, T_n+\epsilon]$, where $\epsilon > 0$ and $T_n \to \infty$ for $n \to \infty$, and let $\boldsymbol{\beta}_w = \int_0^{\tau_w} \mathbf{I}(S^t x_w) dt$, with x_w a point of w. Assume that the current $\mathbf{I}(\mathbf{H})$ at forcing $\mathbf{H}$, admits a periodic orbit expansion [26], i.e. that one can write:

$$\mathbf{I}(\mathbf{H}) = \lim_{n\to\infty} \frac{\sum_{w\in P_n(\mathbf{H})} \Lambda_{w,u}^{-1} \boldsymbol{\beta}_w}{\sum_{w\in P_n(\mathbf{H})} \Lambda_{w,u}^{-1} \tau_w} \equiv \lim_{n\to\infty} \mathbf{I}^{(n)}(\mathbf{H}) . \tag{34}$$

Subdivide $P_n(\mathbf{H})$ as: $P_n(\mathbf{H}) = P_{n,0}(\mathbf{H}) \cup P_{n,R}(\mathbf{H})$, where:

i. $P_{n,0}(\mathbf{H})$ = set of orbits in $P_n(\mathbf{H})$ which are deformations of orbits in $P_n(\mathbf{0})$,

ii. $P_{n,R}(\mathbf{H}) = P_n(\mathbf{H}) \setminus P_{n,0}(\mathbf{H})$.

This subdivision is useful for non smooth dynamics, as in hard-core systems, where any small parameters change "prunes" part of the periodic orbits, and deforms the others. If an orbit is pruned, its contribution to Eq. (34) vanishes. That may happen in a smooth way, as argued in [25], or in a discontinuous way, motivating the present derivation. Writing

$$\mathbf{I}^{(n)}(\mathbf{H}) = \frac{\sum_{w\in P_{n,0}(\mathbf{H})} \Lambda_{w,u}^{-1} \boldsymbol{\beta}_w}{\sum_{w\in P_{n,0}(\mathbf{H})} \Lambda_{w,u}^{-1} \tau_w} + \mathbf{e}^{(n)}(\mathbf{H}) \equiv \mathbf{I}^{(n,0)} + \mathbf{e}^{(n)}(\mathbf{H}) , \tag{35}$$

where $\mathbf{e}^{(n)} = (e_1^{(n)}, \ldots, e_\nu^{(n)})$, Assumption 2 of [25] can be weakened as follows:

Assumption: *There exists $\rho > 0$ such that for $\| \mathbf{H} \| < \rho$ one has:*

a) for every $w \in P_{n,0}(\mathbf{H})$ there are constants $\mathbf{a}_w$, $\mathbf{b}_w \in I\!R^\nu$ such that

$$\begin{cases} \tau_w = \tau_{w(0)} + \mathbf{a}_w \cdot \mathbf{H} + O(H^2) , \\ \Lambda_{w,u}^{-1} = \Lambda_{w(0),u}^{-1} + \mathbf{b}_w \cdot \mathbf{H} + O(H^2) , \end{cases} \quad \textit{for } w(0) \in P_n(\mathbf{0}) , \tag{36}$$

b) the support of the physical measure $\mu_{\mathbf{H}}$ is the whole phase space Ω;

c) $\lim_{H_l\to 0} \lim_{n\to\infty} e_k^{(n)}(0, \ldots, 0, H_l, 0 \ldots, 0)/H_l = 0$, for $k, l = 1, \ldots, \nu$.

Indeed, for sufficiently small $\mathbf{H}$, Anosov systems have $\mathbf{e}^{(n)}(\mathbf{H}) = 0$ for all n. The derivation of Onsager relations now proceeds this way. Denote by H_l the vector $(0, \ldots, 0, H_l, 0 \ldots, 0)$ and let

$$J_w = e^{\tau_w \sum_j \lambda_{w,j}} = \exp\left[-(dN-1)\int_0^{\tau_w} \alpha(S^t x_w)\, dt\right] , \tag{37}$$

be the Jacobian of the dynamics over the orbit w. Introduce $P^+_{n,0}(H_l)$ as the subset of $P_{n,0}(H_l)$ whose orbits have $\beta_{w(0),k} \geq 0$. Then, Eq. (35) and the Assumption above yield:

$$L_{kl} = \lim_{H_l \to 0} \frac{I_k(H_l)}{H_l} = \lim_{H_l \to 0} \lim_{n \to \infty} \frac{I_k^{(n,0)}(H_l) + e_k^{(n)}(H_l)}{H_l}$$
$$= \lim_{H_l \to 0} \lim_{n \to \infty} \frac{I_k^{(n,0)}(H_l)}{H_l} .$$

The k-th component of the numerator of the fraction in Eq. (35) can be written as

$$\sum_{w \in P_{n,0}(H_l)} \Lambda^{-1}_{w,u} \beta_{w,k} = \sum_{w \in P^+_{n,0}(H_l)} \Lambda^{-1}_{w,u}(1 - J_w)\beta_{w,k}$$
$$= \sum_{w \in P^+_{n,0}(H_l)} (dN - 1)\beta_{w,k} \left[\Lambda^{-1}_{w(0),u} \mathbf{c}_w \cdot \mathbf{H} + O(H^2) \right] ,$$

where $\mathbf{c}_w = \int_0^{\tau_w} \mathbf{C}(S^t x_w) dt$. Furthermore, one has

$$\beta_{w,k} = \int_0^{\tau_w} \partial_{H_k} \alpha(S^t x_w; \mathbf{H})\, dt = \int_0^{\tau_w} \left[C_k(S^t x_w) + O(H) \right] dt = c_{w,k} + O(H) , \tag{38}$$

Denoting by $I_k^{(n,0)}$ the k-th component of $\mathbf{I}^{(n,0)}$, this leads to

$$I_k^{(n,0)}(H_l) = (dN - 1) \left[\frac{\sum_{w \in P^+_{n,0}(H_l)} \Lambda^{-1}_{w(0),u} c_{w,k} c_{w,l} (dN - 1)}{\sum_{w \in P_{n,0}(H_l)} \Lambda^{-1}_{w(0),u} \tau_{w(0)}} H_l + O(H^2) \right] \tag{39}$$
$$\equiv A_{kl}^{(n)}(H) H_l + O(H^2) .$$

But $A_{kl}^{(n)}(H_l) = A_{lk}^{(n)}(H_k)$ for all n, if $H_l = H_k$, therefore

$$L_{kl} = \lim_{H_l \to 0} \lim_{n \to \infty} A_{kl}^{(n)}(H_l) = \lim_{H_k \to 0} \lim_{n \to \infty} A_{lk}^{(n)}(H_k) = L_{lk}, \quad l, k = 1, \ldots, \nu , \tag{40}$$

which are the Onsager reciprocal relations. For $k = l$ Eq. (39) also implies the non-negativity of the transport coefficients: $\eta_k = L_{kk} \geq 0$.

3.4 Pump-Billiard Nonnegative Current

The assumptions of Sect.3.4 can be considerably relaxed to prove that the transport coefficients of NEMD models are nonnegative. This is shown here for the pump billiard with $\beta = 1$, and finite horizon, cf. Fig. 3.

Assume that the current $I = \langle v_x \rangle$ can be expressed as:

$$I = \lim_{\tau \to \infty} \frac{\sum_{w \in P_{\tau,\delta}} \Lambda^{-1}_{w,u} \int_w v_x(S^t x_w) dt}{\sum_{w \in P_{\tau,\delta}} \tau_w \Lambda^{-1}_{w,u}} , \quad \Lambda^{-1}_{w,u} = \exp(-\tau_w \lambda_w) , \tag{41}$$

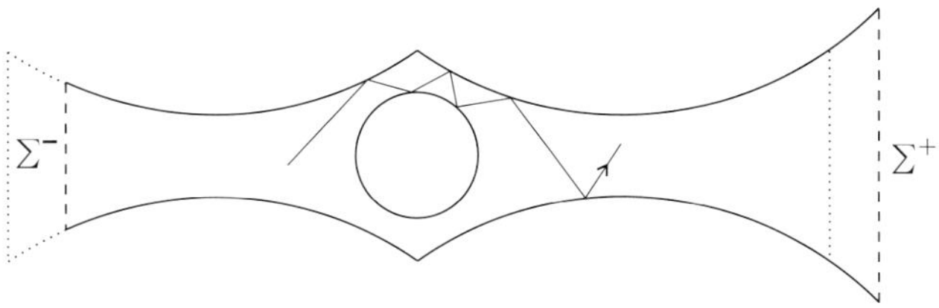

Fig. 3. Elementary cell of the nonequilibrium pump-billiard with finite horizon.

where $P_{\tau,\delta}$ is the set of orbits with a period $\tau_w \in [\tau, \tau+\delta)$, $\delta > 0$;, λ_w is the positive Lyapunov exponent of w; $x_w \in w$; and S^t is the flow. For $\tau > 0$, the numerator of Eq. (41) gives:

$$\sum_{w \in P_{\tau,\delta}} \Lambda_{w,u}^{-1} \int_w v_x(S^t x_w) dt = \sum_{w \in P_{\tau,\delta}} \Lambda_{w,u}^{-1} \triangle x_w \, , \tag{42}$$

where $\triangle x_\omega$, the distance travelled in one period of w, is a multiple of the length ℓ of the periodic cell. If the support of the physical measure is the whole phase space, the time reversibility implies that any orbit w with $\triangle x_w > 0$ has a reverse counterpart w^- such that

$$\triangle x_{w^-} = -\triangle x_w \quad \text{and} \quad \Lambda_{w^-,u} = \frac{\Lambda_{w,u}}{J_w} \, , \tag{43}$$

where J_w is the Jacobian determinant of w. One can then write:

$$\sum_{w \in P_{\tau,\delta}} \Lambda_{w,u}^{-1} \triangle x_w = \sum_{w \in P_{\tau,\delta}^+} \Lambda_{w,u}^{-1} \triangle x_w \left[1 - J_w \right] \, , \tag{44}$$

where $P_{\tau,\delta}^+$ is the subset of $P_{\tau,\delta}$ whose orbits have $\triangle x_w > 0$.

Take $w \in P_{\tau,\delta}^+$ with n crossings of Σ^+ and $N - n < n$ crossings of Σ^-. Using the coordinates (y, v_y) on $\Sigma^\pm$, the corresponding Jacobian determinant is a product of N terms:

$$J_w = J_{w,N} J_{w,N-1} \cdots J_{w,1} \, , \tag{45}$$

with

$$J_{w,i} = \det \begin{pmatrix} 1/\alpha & 0 \\ 0 & \beta/\gamma_i \end{pmatrix} \, , \quad \gamma_i = \left[1 + (\beta^2 - 1) v_{i,y}^2 \right]^{3/2} \, , \tag{46}$$

for crossings of Σ^+, where $v_{i,y}$ is v_y at the i-th crossing, and

$$J_{w,i} = \det \begin{pmatrix} \alpha & 0 \\ 0 & \gamma_i/\beta \end{pmatrix} \, , \tag{47}$$

for crossings of Σ^-. The Jacobian determinants of the motion in the interior of the pump do not play any role, since they equal 1. The result is:

$$J_w = \frac{\prod_{i=1}^{N-n} \gamma_{k_i}}{\prod_{i=1}^{n} \gamma_{l_i}} \left(\frac{\beta}{\alpha} \right)^{2n-N} \, , \tag{48}$$

where k_i represents a crossing of Σ^+ and l_i a crossing of Σ^-. For $\beta = 1$, all γ_i equal 1, hence:

$$J_w = \frac{1}{\alpha^{2n-N}} < 1 \ . \tag{49}$$

Recalling that $\alpha > 1$, the quantity in Eq. (42) is nonnegative for all $\tau, \delta > 0$ and, consequently, the limit of Eq. (41) is also nonnegative. The only assumptions made here are that Eq. (41) holds, and that the physical measure exists and is supported on the whole phase space.

3.5 Temperature for Gaussian Isokinetic Dynamics

A coordinate transformation in (5) leads to a Hamiltonian version of the IK dynamics [27]. Consider for simplicity just one particle, and replace the variable $\mathbf{p}$ in Eqs. (5) with $\boldsymbol{\pi}= \mathbf{p} \exp\left(-\phi/2R\right)$, where $\phi = \phi^{ext}+\phi^{int}$ is the potential of the driving and internal forces, and R is parameter. Then, consider the Hamiltonian:

$$\mathcal{H}(\mathbf{q}, \boldsymbol{\pi}) = \frac{e^{\phi/2R}}{2m} \boldsymbol{\pi}^2 - R\, e^{-\phi/2R} \ , \tag{50}$$

which leads to the following equations of motion:

$$\dot{\mathbf{q}} = \frac{e^{\phi/2R} \boldsymbol{\pi}}{m} \ , \quad \dot{\boldsymbol{\pi}} = \frac{(\mathbf{F}^{ext} + \mathbf{F}^{int}) e^{-\phi/2R}}{2R} \left(\frac{e^{\phi/R} \boldsymbol{\pi}^2}{2m} + R \right) \ . \tag{51}$$

Taking $R = K$, the kinetic energy, and an initial condition $(\mathbf{q}(0),\boldsymbol{\pi}(0))$ such that $\mathcal{H} = 0$, one gets:

$$\dot{\mathbf{q}} = \frac{\mathbf{p}}{m} \ , \quad \dot{\mathbf{p}} = \frac{d\mathbf{p}}{d\boldsymbol{\pi}} \frac{d\boldsymbol{\pi}}{dt} = \mathbf{F}^{ext} + \mathbf{F}^{int} - \frac{(\mathbf{F}^{ext} + \mathbf{F}^{int}) \cdot \mathbf{p}}{\mathbf{p}^2} \mathbf{p} \ , \tag{52}$$

which are the IK equations for a 1-particle system.

One property of Eqs. (50,51) has been investigated in [28], exploiting the dynamical definition of temperature T given by Rugh in [29], for Hamiltonian (equilibrium) particle systems:

$$\frac{1}{T} \equiv \lim_{t \to \infty} \frac{1}{t} \int_0^t \Phi(S^\tau x) d\tau \ , \tag{53}$$

under some ergodic assumption. Here, $S^\tau x$ is the state of the system at time τ, and $\Phi = \nabla \cdot (\nabla \mathcal{H} / \|\nabla \mathcal{H}\|^2)$, ∇ being the gradient with respect to the phase variable x. Thus, T is obtained from both kinetic and configurational degrees of freedom.

In [28] the relation between this T and other kinetic "temperatures" was investigated, using the Hamiltonian dynamics (51). In particular, for IK systems made of N soft particles subjected to a constant external field, it was found that T approaches $T_\perp = \sum_{i=1}^N \mathbf{p}_{i,\perp}^2 / dNm$, in the large N limit, where $\mathbf{p}_{i,\perp}$ is the component of $\mathbf{p}_i$ orthogonal to $\mathbf{F}^{ext}$. This suggests that $T_\perp$ may be a good dynamical definition of temperature for simple liquids. Ref. [28] also shows, that purely configurational definitions of temperature [30] are viable when no kinetic temperature makes sense as, e.g., in the case of nonequilibirum fluids constituted of large, flexible polymers.

4 Fluctuation Relations

Results on the fluctuations of the phase space contraction rate are reviewed, complementing with some technical details the discussion given in E.G.D. Cohen's contribution to this volume.

4.1 Chaotic Hypothesis

The set of properties of thermostatted systems do not constitute a theory of nonequilibrium steady states: a principle is needed to organize them, as the Ergodic Hypothesis does to equilibrium systems. Thus, Gallavotti and Cohen, observing the chaotic properties of NEMD models, and inspired by the results of [31], proposed the following [7]:

Chaotic Hypothesis (CH): *A reversible N-particle system in a stationary state can be regarded as a transitive Anosov system, for the calculation of its macroscopic properties.*

Transitive Anosov flows (Ω, S^t) are the *paradigm* of chaotic systems. They are smooth, and their phase space Ω is hyperbolic, that is to say, for all $x \in \Omega$ the following holds:

i. the tangent space $T_x\Omega$ splits as $T_x\Omega = E_x^u \oplus E_x^s \oplus E_x^0$ continuously with x, and E_x^0 is a one-dimensional linear subspace.
ii. The linear subspaces E_x^s and E_x^u are invariant under the tangent map $D(S^t)$: for any $x \in \Omega$, $D(S^t)(x)E_x^i \subset E_{S^t(x)}^i$ with $i = s, u$.
iii. There exist $C > 0$ and $\theta \in (0,1)$ such that, for all $t \geq 0$, $\| D(S^t)\xi \| \leq C\,\theta^t \| \xi \|$, for all $\xi \in E_x^s$ and $\| D(S^{-t})\eta \| \leq C\,\theta^t \| \eta \|$, for all $\eta \in E_x^u$.

Then, stable and unstable manifolds W_x^s, W_x^u are dense in Ω for every $x \in \Omega$ (transitivity), the periodic orbits are dense in Ω, and the physical measure has a density along the unstable directions. Such a measure is called SRB measure, after Sinai, Ruelle and Bowen [32]. Moreover, the following holds:

Theorem (Sinai, 1968): *Every transitive Anosov system admits Markov partitions.*

Attributing certain dynamical weights to the cells of a Markov partition of Ω, and taking the limit of finer and finer partitions, leads to a representation of the SRB measure. The weight associated with a cell E is $\Lambda_{w,u,\tau}^{-1}$, the inverse absolute value of the Jacobian determinant of the dynamics restricted to the unstable manifold over a trajectory segment $w_\tau = \{S^t x_w\}_{t=-\tau/2}^{\tau/2}$ for $x_w \in E$ and for large τ. Because of the density of periodic orbits in Ω, all w's could be periodic with period in a given interval $[\tau, \tau+\delta)$. In that case, the weights take the form $\Lambda_{w,u}^{-1} = \exp(-\tau \sum^+ \lambda_{w,i})$, where the symbol $\sum^+$ indicates summation over the positive Lyapunov exponents $\lambda_{w,i}$ of w, and can be used to derive Eq. (34), where $\mathbf{I}$ is any smooth observable.

A consequence of the CH is ergodicity at equilibrium. Hence the CH is a unifying principle for both equilibrium and nonequilibrium statistical mechanics.

But there is a difficulty: physical models cannot be expected to be Anosov. Hence, the physical models for which the CH is valid have been called "Anosov-like", an expression which needs to be properly interpreted, as discussed in Sects. 5,7.

The following subsections introduce the reader to the fluctuation theorem, and its modifications.

4.2 Gallavotti-Cohen Fluctuation Theorem

Consider the equations of motion for a time reversible, dissipative and chaotic system $\dot{x} = G(x)$ on Ω, and let $\chi(x) = -(\mathrm{div}G)(x)$. Take a long trajectory w, and break it in consecutive segments $w_{\tau,k}$, $k = 1, 2, ...$, each of duration τ, and possibly separated by a decorrelation time τ_d. The Gallavotti-Cohen fluctuation theorem (GCFT) concerns the probability distribution (the frequency of observation along one trajectory) of the quantity defined by:

$$p_k = \frac{1}{\tau\langle\chi\rangle} \int_{-\tau/2}^{\tau/2} \chi(S^t x_k) dt \,, \tag{54}$$

where $\langle\cdot\rangle$ is the steady state average, and x_k is the center of $w_{\tau,k}$. We call p_k the "phase space contraction" of segment $w_{\tau,k}$. *If the system is Anosov*, the frequency with which p_k falls in $B_{p,\epsilon} = (p-\epsilon, p+\epsilon)$ coincides, in the limit of fine Markov partitions and of long τ's with the weight $\Lambda_{w,u,\tau}^{-1}$ attributed to the cell containing x_k. Let π_τ be the probability distribution of the values of p_k. Then, except for errors which can be made arbitrarily small refining the partition and taking longer τ, one can write

$$\pi_\tau(p \in B_{p,\epsilon}) = \frac{1}{M} \sum_{E_p, p \in B_{p,\epsilon}} \Lambda_{w,u,\tau}^{-1} \,, \tag{55}$$

where M is a normalization constant, and the sum is carried over the cells E_p containing the middle point of a segment with phase space contraction in $B_{p,\epsilon}$. *If the support of the physical measure is* Ω, as in the presence of weak dissipation, time reversibility guarantees that the support of π_τ is symmetric around 0, and one can consider the ratio

$$\frac{\pi_\tau(p \in B_{p,\epsilon})}{\pi_\tau(p \in B_{-p,\epsilon})} = \sum_{E_p, p \in B_{p,\epsilon}} \Lambda_{w,u,\tau}^{-1} \Big/ \sum_{E_p, p \in B_{-p,\epsilon}} \Lambda_{w,u,\tau}^{-1} \,. \tag{56}$$

Denoting by i the involution which replaces the initial condition of one trajectory with the initial condition of the reversed trajectory[8] (which obeys $i^2 = Id$ and $iS^t = S^{-t}i$) time reversibility yields $\chi(x) = -\chi(ix)$, $\Lambda_{iw,u,\tau} = \Lambda_{w,u,\tau}^{-1}$, and $\Lambda_{iw,u,\tau}/\Lambda_{w,u,\tau} = \exp(\tau\langle\chi\rangle p)$, if the phase space contraction of w is p, cf. [7]. Then, taking small ϵ in $B_{p,\epsilon}$, all terms in the numerator of the right hand side of (56) have a counterpart in the denominator, and the division of each term in

[8] For instance, i is the reversal of momenta for Eqs. (5), but is more complicated for SLLOD.

the numerator by its counterpart approximately equals $\exp(\tau\langle\chi\rangle p)$. This value, then, equals the ratio in (56), and in the limit of small ϵ one obtains the GCFT:

Theorem (Gallavotti-Cohen, 1995): *Let (Ω, S^t) be transitive, Anosov and time reversal invariant, with $\langle\chi\rangle > 0$. Then*

$$\lim_{\tau\to\infty} \frac{1}{\tau\langle\chi\rangle p} \log \frac{\pi_\tau(p)}{\pi_\tau(-p)} = 1 \ . \tag{57}$$

The GCFT concerns the large deviations of χ, and for large τ states that positive fluctuations are exponentially more probable than negative ones. The right hand side of Eq. (57) is 1 as long as dissipation is not too high, so that the unstable manifold W^u is dense in Ω.

When χ can be identified with the entropy production rate, the GCFT makes a statement on the physics of nonequilibrium systems which has various consequences, among which the validity of the Onsager reciprocal relations [24,25,33].

4.3 Evans-Searles Theorem

A theorem formally similar to the GCFT, but different in content, was proposed in [34]. We give an independent derivation of this theorem, called Evans-Searles theorem (EST), to clearly distinguish it from the GCFT.[9]

Consider a microcanonical ensemble of (smooth) time reversal invariant, dissipative systems as in Sect.4.2; but do not require any chaos. Denote by E_p the set of all $x \in \Omega$, for which the phase space contraction

$$p(x) = \frac{1}{\tau\langle\chi\rangle} \int_0^\tau \chi(S^t x) dt \,, \tag{58}$$

lies in $B_{p,\epsilon}$, and denote by $\mu_L(E_p)$ its volume. The definition of p implies $\mu_L(S^\tau E_p) = \exp\left(-\tau\langle\chi\rangle p\right) \mu_L(E_p)$, apart from small errors which can be made arbitrarily small taking ϵ correspondingly small. Consider the time reversal operation i introduced above, and set $A_p = iS^\tau E_p$, so that $y \in A_p$ means that there is $x \in E_p$ for which $y = iS^\tau x$. Then,

$$\mu_L(A_p) = \int_{A_p} dy = \int_{E_p} J(iS^\tau x) dx = \left[e^{-\tau\langle\chi\rangle p} + O(\epsilon)\right] \mu_L(E_p) \ , \tag{59}$$

because the Jacobian $J(iS^\tau x)$ is approximately constant in E_p. In fact, taking a point $iS^\tau x$ with $x \in E_p$, and computing its phase space contraction in a time τ, one obtains

$$\begin{aligned} \int_0^\tau \chi\left(S^t i S^\tau x\right) dt &= \int_0^\tau \chi\left(i S^{\tau-t} x\right) dt = \int_0^\tau \chi\left(S^{\tau-t} i x\right) dt \\ &= -\int_0^\tau \chi\left(S^s x\right) ds \in -\tau\langle\chi\rangle B_{p,\epsilon} \,, \end{aligned} \tag{60}$$

[9] We thank C. Liverani for suggesting the present derivation.

where the fact that i is volume preserving has been used together with the substitution $s = \tau - t$. One also realizes that all points with phase space contraction in B_{-p} belong to A_p. Indeed, consider any $x \in \Omega$ with $\int_0^\tau \chi(S^t x)dt = -\tau\langle\chi\rangle p$, and take $y = S^{-\tau} ix$. Then, using (60), the phase space contraction of y in a time τ yields:

$$\int_0^\tau \chi\left(S^{t-\tau} ix\right) dt = \int_0^\tau \chi\left(iS^{\tau-t} x\right) dt = \tau\langle\chi\rangle p \ . \tag{61}$$

Hence, y lies in E_p, and $iS^\tau y = x$, i.e. x is in A_p. This leads to:

$$\frac{\mu_L(E_p)}{\mu_L(E_{-p})} = \frac{\mu_L(E_p)}{\mu_L(A_p)} = e^{-\tau\langle\chi\rangle p} \ , \tag{62}$$

except for corrections of order $O(\epsilon)$. Taking the logarithm of both sides of Eq. (62) a relation formally similar to Eq. (57) emerges. However, Eq. (62), concerns the Liouville measure of the sets of initial conditions of trajectories E_p and E_{-p}, and states that the ratio of their volumes decreases exponentially with τ. If the phase space contraction can be identified with the entropy production, the EST states that the trajectories which produce entropy over a time τ exponentially dominate those which absorb entropy, which is a nice quantitative characterization of how the thermodynamic behaviour may emerge from microscopic reversible dynamics. At the same time, the EST says nothing about the fluctuations of the phase space contraction along *one* trajectory, which are the object of the GCFT. Table 1 summarizes these facts.

Table 1. Main ingredients of the GCFT (left) and of the EST (right). T.R.I. means time reversal invariance. The last row indicates the object of the theorems.

GCFT	EST
T.R.I.	T.R.I.
$\tau \to \infty$	any τ
any absolutely continuous initial distribution	Liouville initial distribution
Chaos needed	Chaos not needed
Weak dissipation (high dissipation changes 1 to $c < 1$ in (57), cf. Sect.4.4)	any dissipation (Eq. (62) is not affected by amount of dissipation)
probability of observing phase space contraction $p\tau$ in a time τ along one steady state evolution	(microcanonical) probability of initial condition which in the time τ will produce a phase space contraction $p\tau$

4.4 Modified Fluctuation Relations

If the dissipation of an NEMD model is made strong, as for high driving fields, the support of the physical measure collapses to a set of zero volume, on a lower dimensional submanifold of the phase space Ω. In this case, the GCFT does not hold, but Bonetto, Gallavotti and Garrido [35,36,15] suggested that Eq. (57) still holds if its right hand side is replaced by some $c \in (0,1)$. To obtain c, one should exclude the vanishing Lyapunov exponents, arrange the remaining ones in decreasing order, $\lambda_1 \geq \lambda_2 \geq ...\lambda_{2K}$, and compute the ratio:

$$c = \frac{\sum^{+}(\lambda_k + \lambda_{2K+1-k})}{\sum_{k=1}^{K}(\lambda_k + \lambda_{2K+1-k})} , \tag{63}$$

where $\sum^{+}$ means summation over the pairs $(\lambda_k, \lambda_{2K+1-k})$ with $\lambda_k > 0$. Dynamical systems for which this hypothesis is verified have been called *axiom-C* systems in [36].

The GCFT and, in particular, its axiom-C extension, concern large deviations of a global (hence non measurable) quantity. However, local fluctuations are observable, and a local version of the fluctuation theorem has been proposed in [37]. There, V_0 denotes a finite region of a chain of coupled maps, $T_0 > 0$ is a fixed time length, and η_+ and p are defined by:

$$\eta_+ = \lim_{V_0, T_0 \to \infty} \frac{1}{|V_0| T_0} \sum_{j=0}^{T_0 - 1} \eta_{V_0}(S^j x) , \quad p = \frac{1}{\eta_+ |V|} \sum_{j=-T_0/2}^{T_0/2} \eta_{V_0}(S^j x) , \tag{64}$$

where η_{V_0} is a properly chosen function, linked to the dissipation rate in V_0. Setting $V = V_0 \times T_0$, the following relation was obtained for the probability distribution of p:

$$\pi_V(p) = e^{\zeta(p)|V| + O(|\partial V|)} , \quad \text{with} \frac{\zeta(p) - \zeta(-p)}{p\eta_+} = 1 \quad \text{and} |p| < p^* . \tag{65}$$

Here, $|\partial V|$ is the size of the boundary of V, $p^* \geq 1$ and $\zeta(\cdot)$ is analytic. This is similar to the GCFT, except for the (space-time) boundary term $|\partial V|$ expected to be negligible for large V.

The problem of the observability of the large fluctuations of the phase space contraction has been addressed also in Ref. [38], where it was empirically observed that the distributions π_τ of Eq. (57) for highly sheared SLLOD systems, verify:

$$\sigma_\tau \, \pi_\tau(\sigma_\tau x + 1) = \sigma_{\tau'} \, \pi_{\tau'}(\sigma_{\tau'} x + 1) + \text{correction term} , \tag{66}$$

where σ_i is the standard deviation of π_i, and the correction term seems to vanish in the large τ, τ' limit. Equation (66) holds for distributions which tend to a Gaussian for growing τ, τ', but this is not the case of our distribution (see also Ref. [39]). Assuming the validity of the modified GCFT with $c \in (0,1)$, and of (66), simple algebra leads to this new fluctuation relation:

$$\log \left[\frac{\pi_\tau(2\hat{p} + y)}{\pi_\tau(2\hat{p} - y)} \right] = c\langle\chi\rangle \left(2\tau' \frac{\sigma_{\tau'}}{\sigma_\tau} - \tau \right) y + \text{correction term} , \tag{67}$$

where $\hat{p} = 1 - (\sigma_\tau / \sigma_{\tau'})$, and the correction term appears to vanish as τ, τ' grow. Unlike Eq. (57), Eq. (67) concerns fluctuations around $2\hat{p}$, rather than around 0.

For large τ, the value 0 is in the tail of π_τ, making impossible to observe the fluctuations around that value. Differently, $2\hat{p} \in [0,2)$, since $\hat{p} = 0$ for $\tau' = \tau$ and $\hat{p} \to 1$ for $\tau'/\tau \to \infty$. Then, for τ'/τ such that $\hat{p} \approx 1/2$, Eq. (67) concerns fluctuations around the peak of π_τ, i.e. *observable* fluctuations [38].

5 Numerical Results

Various papers verify that Eq. (57) holds for NEMD systems. The first is Ref. [31] for the SLLOD system, which inspired [7]. Reference [35] verified the validity of Eq. (57) for hard disks moving within a periodic array of fixed hard circular scatterers. In [35] the distributions π_τ appeared indistinguishable from Gaussian distributions, which is expected for the small fluctuations around the mean of π_τ, but not for the large fluctuations. Non Gaussian distributions verifying (57), instead, were found in [40], for heat transport along chains of anharmonic oscillators; in [38], for high shear SLLOD equations with soft particles, Fig. 4, where also the axiom-C hypothesis and the validity of (67) was supported; and in [19], for the pump-billiard with infinite horizon.

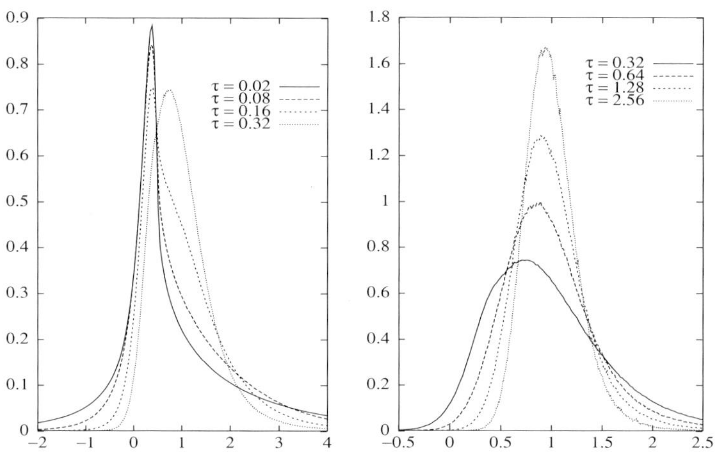

Fig. 4. π_τ for SLLOD dynamics at $\gamma = 2.5$, for different τ's. The left panel shows a transition in the shape of the distributions at small τ's, while the right panel shows a unique shape, which makes all π_τ's with larger τ overlap under the scaling (66). None of the distributions is Gaussian.

The numerical simulations of [41] verified the identity of the viscosities of Gauss-Navier-Stokes and Navier-Stokes fluids, along with the validity of (57), with $c < 1$ in the right hand side, for Gauss-Navier-Stokes steady states. Again, the distributions were not Gaussian. Results on the equality of the Lyapunov spectra of Navier-Stokes and Gauss-Navier-Stokes dynamics, as well as on the validity of the axiom-C hypothesis and of the local fluctuation theorem, are the object of a future paper [42].

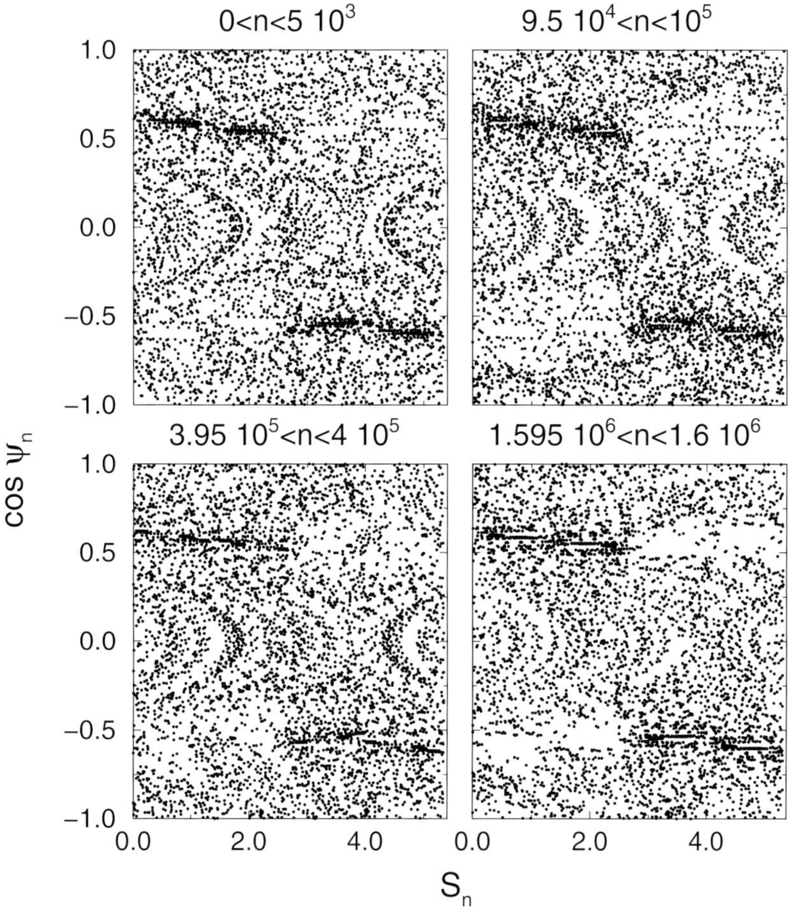

Fig. 5. Phase space portraits for the flat billiard of [43]. S_n and $\cos \Psi_n$ are the usual billiard coordinates representing, respectively, the position and the outgoing direction of a particle undergoing the n-th collision with a scatterer. Each panel shows 5000 consecutive collision points, beginning at different times along one trajectory. The four snapshots are very similar to each other, showing that the system is practically in a steady state. This quasi-steady state lasts for times several orders of magnitude longer than the time needed to compute the average current (the observation time).

In Refs. [19,43], flat billiards with periodic steady states (hence totally different from Anosov systems) have been considered, to compare their average behaviour with those of chaotic systems. It was observed that flat billiards evolve through long *quasi-stationary*, erratic transients, which get longer as the forcing decreases. In these transients, the statistical properties of the dynamics can be studied, practically as in the steady states of chaotic billiards, cf. Fig. 5. The corresponding transport of particles looks similar to that of chaotic systems, and

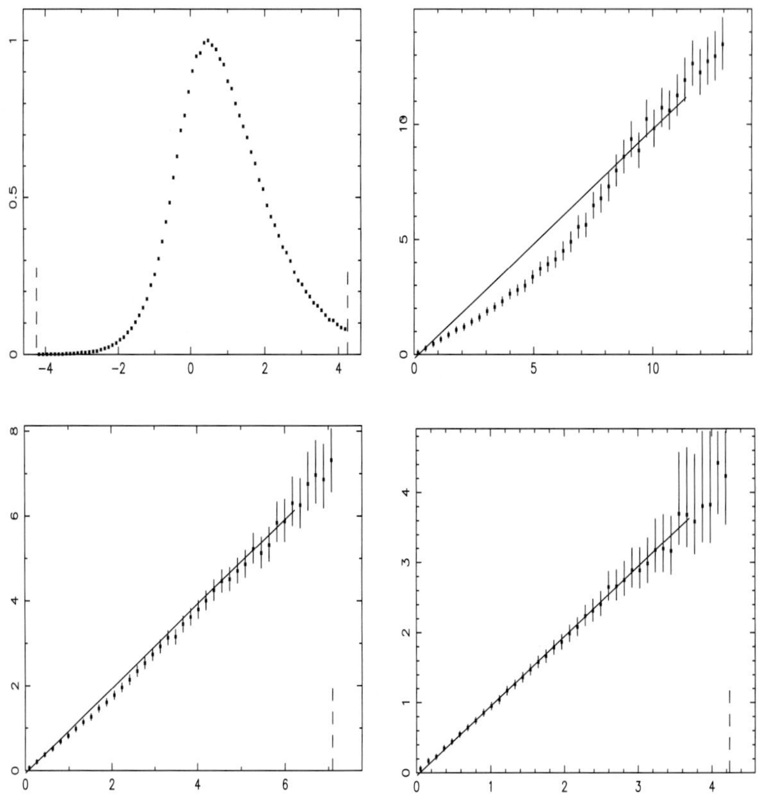

Fig. 6. Distribution $\pi_\tau(p)$ (top left panel) for the flat billiard of [19] obtained from a trajectory of 10^9 collisions, broken in trajectory segments of length $\tau = 2000$ collision times, separated by segments of 200 collisions. In such a long, quasi-steady state, the fluctuation formula can be verified, as shown by the remaining three panels. In these three panels, the logarithm of the left hand side of (56), divided by $\tau\langle\chi\rangle$, is shown as a function of p, for trajectory segments of length $\tau = 200$ collisions (top right panel), $\tau = 700$ collisions (bottom left panel) and $\tau = 2000$ collisions (bottom right panel). If the fluctuation formula holds, the data must converge to a straight line of slope 1, as is the case here.

even Eq. (57) appears to hold, cf. Fig. 6. Therefore, these flat billiards verify the CH, and show that "Anosov-Like" does not necessarily mean chaotic.

6 Experimental Results

The validity of Eq. (57) is supported by two experiments. The first is Ref. [44], and concerns fluctuations of the temperature field in Rayleigh-Benard convection. The second is Ref. [45], and concerns the fluctuations of the dissipated power in far from equilibrium liquid crystal electroconvection. Neither of the two experiments constitutes a severe test of the GCFT and, in particular, it is not clear whether [44] concerns local fluctuations or global ones.

There is also one experiment supporting the validity of the EST [46]. The interest of this experiment lies in the fact that shows negative entropy productions for short times in small systems. This is relevant for the behaviour of nanomachines. In the second place, [46] shows that subtle properties of NEMD dynamics, as the entropy production, are consistent with the behaviour of real systems. This extends our confidence on the NEMD models, well beyond the calculation of transport coefficients.

7 Discussion

The present paper intended to familiarize the reader with some models and techniques currently used in nonequilibrium statistical mechanics, and to stimulate further research. Besides the technical details, we discussed the following points.

1. A realistic thermostatted system typically should be time reversal invariant, but not Hamiltonian. This does not mean that popular toy models, such as Hamiltonian systems of non interacting particles, or volume preserving chains of two-dimensional maps should not be used to gain insight in the physics of certain physical systems. It only says that some care must be used in identifying the systems and the phenomena actually described by toy models.

2. NEMD models are not Hamiltonian, but their primary purpose is to compute in an efficient way the transport coefficients of fluids. Trying to extract more than that from these systems, is not always justified. In particular, the fact that in some circumstances the phase space contraction of NEMD models can be identified with the entropy production, does not imply that the complete entropy balance can be derived for NEMD models. If this is the situation of the NEMD models, whose validity has been demonstrated in a wide range of physical applications, greater care should be used in interpreting the results produced by simpler models.

3. The principle of equivalence of nonequilibrium ensembles, states that different microscopic descriptions of the same physical situation are legitimate if indistinguishable at a macroscopic level, *i.e.* in the thermodynamic limit and for a certain class of observables. This justifies the NEMD dynamics as the simplest mechanical *equivalent* of systems subject to a real thermostat. The meaning of the thermodynamic limit in nonequilibrium statistical mechanics is not fully understood. Sections 3 and 5 propose possible approaches to this problem.

5. The Chaotic Hypothesis, similarly to the Ergodic Hypothesis, assumes that the microscopic dynamics of macroscopic systems are rather "disordered", to the point that the deviations from the Anosov property do not appear on the scale of observations. Chaos is not a necessary condition for that to be the case, and the most important ingredient of statistical mechanics remains the large number of interacting particles.

Chaos and Irreversible Thermodynamics

At the end of the panel discussion, entitled "Chaos and Irreversible Thermodynamics: Is there a connection?" D. Cohen asked: "Is there a unique answer to the question posed by the title of the panel discussion?"

I answered affirmatively: Irreversible Thermodynamics needs local thermodynamic equilibrium, which can only be established in the presence of many interacting particles [23]. Chaos is a secondary issue, while local thermodynamic equilibrium is essential.

Likely, the dynamics which produce local thermodynamic equilibrium are chaotic. But is it necessarily so? The flat (non chaotic) billiards of Refs. [43,19,47] don't speak of local thermodynamic equilibrium, but show that the same transport of particles observed in chaotic billiards can be obtained without chaos, within the proper time scales. Hence, as in Boltzmann's arguments against Zermelo's paradox, large N and time scales seem to play a more fundamental role in IT than chaos might do.

Acknowledgements

The author is grateful to Professors P. Garbaczewski and R. Olkiewicz for the invitation to the 38th Winter School of Theoretical Physics, where this work was initiated, and for the warm hospitality. Thanks are in order to R. Klages and D. Wojcik for insightful remarks. Support from GNFM-INDAM, and from INFM is gratefully acknowledged.

References

1. G. Gallavotti: *Statistical Mechanics: a short treatise*, Springer Verlag, Berlin (2000).
2. M.P. Allen and D.J. Tildesley, *Computer simulation of liquids*, Oxford (1987).
3. D.J. Evans and G.P. Morriss: *Statistical Mechanics of Nonequilibrium Liquids*, Academic Press, New York 1990.
4. W.G. Hoover, *Computational statistical mechanics*, Elsevier (1991).
5. B.J. Alder and T.E. Wainwright, J. Chem. Phys. **27**, 1208 (1957).
6. J. Schnack, Physica, **A259**, 49 (1998); M. Ferrario, D. Laria, G. Ciccotti and R. Kapral, J. Mol. Liq., **61**, 37, (1994).
7. G. Gallavotti and E.G.D. Cohen, J. Stat. Phys. **80**, 931 (1995).
8. S. Sarman, D.J. Evans and P.T. Cummings, Physics Reports, **305**(1-2), 1 (1998).
9. F. Müller-Plathe, Phys. Rev. **E59**(5), 4894 (1999).
10. L. Rondoni and E.G.D. Cohen, Nonlinearity, **13**, 1905 (2000).
11. J.P. Eckmann, C.A. Pillet and L. Rey-Bellet, J. Stat. Phys. **95**, 305 (1999).
12. K. Rateitschak, R. Klages and G. Nicolis, J. Stat. Phys. **99**, 1339 (2000).
13. D.J. Evans and S. Sarman, Phys. Rev. **E48**, 65 (1993).
14. T. Tel, J. Vollmer and W. Breymann, Europhys. Lett. **35**, 659 (1996).
15. G. Gallavotti, Physica **D105**, 163 (1997).
16. K.F. Gauss, J. Reine Angewandte Math. **4**, 232 (1829); C. Lanczos: *The Variational Principles of Mechanics*, Dover, New York (1970).

17. S. Nosé, J. Chem. Phys. **81**, 511 (1984); Mol. Phys. **52**, 255 (1984); W. G. Hoover, Phys. Rev. **A31**, 1695 (1985).
18. P.A.M. Dirac, Can. J. Math. **2**, 129 (1950); Proc. Roy. Soc. **A246**, 326 (1958).
19. G. Benettin and L. Rondoni, Math. Phys. Electronic J. (2001).
20. N.I. Chernov and J.L. Lebowitz, J. Stat. Phys. **86**, 953 (1997).
21. C. Wagner, R. Klages and G. Nicolis, Phys. Rev. **E60**, 1401 (1999).
22. E. G. D. Cohen and L. Rondoni, Chaos, **8**, 357 (1998).
23. E.G.D. Cohen and L. Rondoni, Physica A (to appear).
24. G. Gallavotti, Phys. Rev. Lett. **789**, 434 (1996).
25. L. Rondoni and E.G.D. Cohen, Nonlinearity, **11**, 1395 (1998)
26. W. Parry, Commun. Math. Phys. **106**, 267 (1986).
27. C.P. Dettmann and G.P. Morriss, Phys. Rev. **E54**, 2495 (1996).
28. G.P. Morriss and L. Rondoni, Phys. Rev. **E59**, R5-R8 (1999).
29. H.H. Rugh, Phys. Rev. Lett. **78**, 772 (1997).
30. B.D. Butler, G. Ayton, O.G. Jepps and D.J. Evans, J. Chem. Phys. **109**, 6519 (1998).
31. D.J. Evans, E.G.D. Cohen and G.P. Morriss, Phys. Rev. Lett. **71**, 2401 (1993).
32. Ya. G. Sinai, Russ. Math. Surv. **27**, 21 (1972). R. Bowen, *Equilibrium states and the erogidc theory of Anosov diffeomorphisms*, Lecture Notes in Mathematics 470, Sprimger, Berlin (1975);
D. Ruelle: Am. J. Math. **98**, 619 (1976).
33. G. Gallavotti and D. Ruelle, Commun. Math. Phys. **190**, 279 (1997).
34. D.J. Evans and D.J. Searles, Phys. Rev. **E52**, 5839 (1995)
35. F. Bonetto, G. Gallavotti and P. Garrido, Physica **D105**, 163 (1997).
36. F. Bonetto and G. Gallavotti, Comm. Math. Phys. **189**, 263 (1997).
37. G. Gallavotti, Physica **A263**, 39 (1999).
38. L. Rondoni and G.P. Morriss, *submitted to Nonlinearity.*
39. S.T. Bramwell, P.C.W. Holdsworth and J.-F. Pinton, Nature, **396**, 552, 10 Dec. (1998).
40. S. Lepri, R. Livi and A. Politi, Physica **D119**, 140 (1998).
41. L. Rondoni and E. Segre, Nonlinearity **12**, 1471 (1999).
42. G. Gallavotti, L. Rondoni and E. Segre (in preparation).
43. S. Lepri, L. Rondoni and G. Benettin, J. Stat. Phys. **99**, 857 (2000).
44. S. Ciliberto and C. Laroche, J. Physique IV, **8**, 215 (1998).
45. W.I. Goldburg, Y.Y. Goldschmidt and H. Kellay, University of Pittsburgh preprint (2002).
46. G.M. Wang, E.M. Sevick, E. Mittag and D.J. Evans, submitted to Phys. Rev. Lett.
47. C.P. Dettmann, E.G.D. Cohen and H. van Beijeren, Nature, **401**, 875 (1999).

What Is the Microscopic Response of a System Driven Far From Equilibrium?

C. Jarzynski

Theoretical Division, Los Alamos National Laboratory, Los Alamos, NM 87545, USA

Abstract. The central theme of this lecture is that there exists a surprisingly simple and general answer to the question posed in the title, provided that we frame the question statistically. I will present this result along with several derivations, and will discuss some of its implications and generalizations.

Introduction

It is widely acknowledged that we understand systems in thermal equilibrium far better than those which are out of equilibrium. The statistical foundations laid over a century ago by (above all) Boltzmann and Gibbs, have proven remarkably effective at explaining phenomena as diverse as phase transitions, superconductivity, and blackbody radiation. By contrast, no such universal and elegant framework exists that would simultaneously encompass the mechanics of protein folding, the onset of turbulence, and Fourier's law of heat conduction. In part, the difficulty arises simply because "nonequilibrium phenomena" covers a lot of territory. If the term is understood to include any situation involving a system not in a state of thermal equilibrium, then it is hard to imagine a useful point of reference upon which to build a general theory. By contrast, the very notion of an equilibrium state – toward which a system naturally evolves if left undisturbed – provides the organizing concept at the heart of equilibrium statistical mechanics.

While the field of nonequilibrium physics as a whole might indeed be so broad in scope as to escape a sweeping microscopic foundation (akin to the Boltzmann-Gibbs formalism in the equilibrium case), much progress has been made by considering sub-classes of nonequilibrium phenomena distinguished by a common feature. For instance, the study of systems *near equilibrium* has given us the linear response formalism, whose early successes include the fluctuation-dissipation theorem [1,2] and Onsager's reciprocity relations [3]. It also seems reasonable to expect that *nonequilibrium stationary states* will eventually succumb to an elegant statistical formulation with predictive power.

In this lecture I will focus on *systems driven away from an initial state of thermal equilibrium*, where this is accomplished by the variation of an external parameter. An example would be a container filled with gas (the system), closed off at one end by a piston (the parameter), initially prepared in equilibrium. If we rapidly push the piston into the container by some distance, the gas is driven out of equilibrium. At the microscopic level, the response of the system during this process is described by a single trajectory evolving in the many-body phase space of the gas. This response, however, is typically complicated and non-universal, e.g. shock waves form, propagate, scatter, etc. Therefore we adopt a statistical attitude, and imagine an *ensemble* of realizations, pictured as a swarm of independent trajectories evolving in the many-body phase space. Each trajectory represents one *realization* (possible microscopic outcome) of the piston experiment. The central assertion of this lecture is that *there exists a statistical representation of this ensemble* (defined by Eq. 18 below) *in which the response of the system becomes both simple and universal* (Eq. 20). By "universal", I mean that the validity of this result depends neither on the details of the system, nor – perhaps more surprisingly – on how far it is driven out of equilibrium.

The above-mentioned result is, on its own, somewhat abstract. However, by a single integration it leads to another result, the *nonequilibrium work relation* (Eq. 46), which relates the work performed during an irreversible process, to an equilibrium free energy difference. This result is not only potentially useful, but has recently been tested experimentally, and is closely related to the Second Law of Thermodynamics.

This lecture is organized as follows. I will first present a very brief review of relevant facts from macroscopic thermodynamics (Section 1). Then I will move to the microscopic level of statistical mechanics, where I will again present a brief review, and will establish notation (Section 2). I will then present the central result of this lecture in Section 3, and work through three derivations of this result in Section 4. Section 5 will cover the nonequilibrium work relation, and Section 6 will relate this result to the Second Law. Finally, in Section 7 I will briefly discuss the generalization of these results to nonequilibrium stationary states.

Classical physics is assumed throughout the lecture.

1 Thermodynamics – Brief Review of Relevant Facts

Most of this talk will take place at the microscopic level of statistical mechanics, where one is interested in the behavior of atoms, molecules, and so forth. It is therefore useful at the outset to briefly focus on the bulk behavior of *macroscopic* systems, and to review a few relevant facts of thermodynamics.

Consider a macroscopic system which depends on an external parameter λ, and which is in thermal contact with a heat bath at temperature T. A simple example is a rubber band held stretched between two stiff toothpicks. The rubber band is the system; the distance between the toothpicks is our parameter λ, which we can manipulate as we like; and the surrounding air plays the role of the heat bath. A starting assumption of equilibrium thermodynamics is that for any set of allowable[1] values (λ, T), there exists a unique *equilibrium state* to

[1] By "allowable", I mean in this case a range of distances λ such that the band is neither limp nor stretched beyond its breaking point, and temperatures T for which the band is not frozen solid, nor melting, burning, or otherwise losing its structural integrity.

which the system will relax if λ and T are held fixed. Furthermore, there exist state functions such as the internal energy E, entropy S, and (most importantly for our purposes) the *free energy*

$$F = E - ST , \tag{1}$$

which take on unique values at equilibrium states, e.g. $F = F(\lambda, T)$.

Suppose that we now prepare our system by holding the distance between toothpicks fixed at a value λ_A, and allowing the rubber band to relax to the corresponding equilibrium state, A. We then very slowly pull the toothpicks apart, until the distance between them is λ_B. For sufficiently slow pulling the rubber band will progress through a continuous sequence of equilibrium states, from $A = (\lambda_A, T)$ to $B = (\lambda_B, T)$. In carrying out such a *reversible* process, the external work which we perform on the system will be equal to the free energy difference between the initial and final equilibrium states A and B:

$$W = \Delta F \equiv F_B - F_A \qquad \text{(reversible)} . \tag{2}$$

Now consider an *irreversible* process from A to B: after preparing the system in equilibrium state A we pull the toothpicks apart at a rapid rate, and when the distance reaches λ_B we hold the toothpicks fixed and allow the system to relax to state B. We thus have a process in which the system begins and ends in equililbrium states, but at intermediate times, as we rapidly stretch the rubber band, it is driven out of equilibrium. Physically, this excursion away from equilibrium manifests itself in two ways: (1) the rubber band heats up, and (2) its tension increases relative to the room-temperature value. As a result, more work is required to stretch the rubber band than in the case of slow pulling, i.e.

$$W > \Delta F \qquad \text{(irreversible)} . \tag{3}$$

This is a good point to insert a comment about nomenclature. Throughout this lecture I will be concerned primarily with processes (as the one describe above) in which a system is driven out of equilibrium *in the presence of a single heat reservoir* prepared at a given temperature. I will refer to these as *irreversible-isothermal* processes. It is important to stress that the word "isothermal" here is *not* meant to imply that the temperature of the system remains fixed (if even well-defined!) during the process, but rather to indicate the presence of only one reservoir. Eq. 3 applies universally to irreversible-isothermal processes connecting two equilibrium states.

During a more general thermodynamic process from one equilibrium state to another, the system might be brought into contact with a sequence of reservoirs at different temperatures. In that case the *Clausius inequality* applies:

$$\int_A^B \frac{dQ}{T} < \Delta S . \tag{4}$$

The left side is the integrated heat absorbed by the system over the course of the process, inversely weighted by the temperature of the reservoir from which that heat is absorbed, and the right side is the entropy difference between the initial and final equilibrium states. As with Eqs. 2 and 3, the equality holds if the process is reversible, whereas for irreversible processes we get a strict inequality. Eqs. 2 and 3 represent a special case (one reservoir) of the Clausius inequality.

2 Statistical Mechanics – Notation and Review

Moving now from the macroscopic level of thermodynamics to the microscopic level of statistical mechanics, consider the following experimental set-up described in a recent paper by Liphardt *et al.* [4]. In room-temperature water, one end of a strand of RNA is attached to a small polystyrene bead, the other to a micromechanical cantilever, and a laser trap is used to capture the bead. Using piezo-electric techniques to move the cantilever back and forth, the experimentalists are able to cause the RNA strand to unravel as it is stretched, or coil up as the end-to-end distance is made smaller; moreover, they can measure the work performed on the system as this micro-manipulation is carried out! This experimental arrangement is a molecular-level analogue of the rubber band between the toothpicks, and serves as a nice example to illustrate the theoretical analysis that will make up the main part of this lecture.

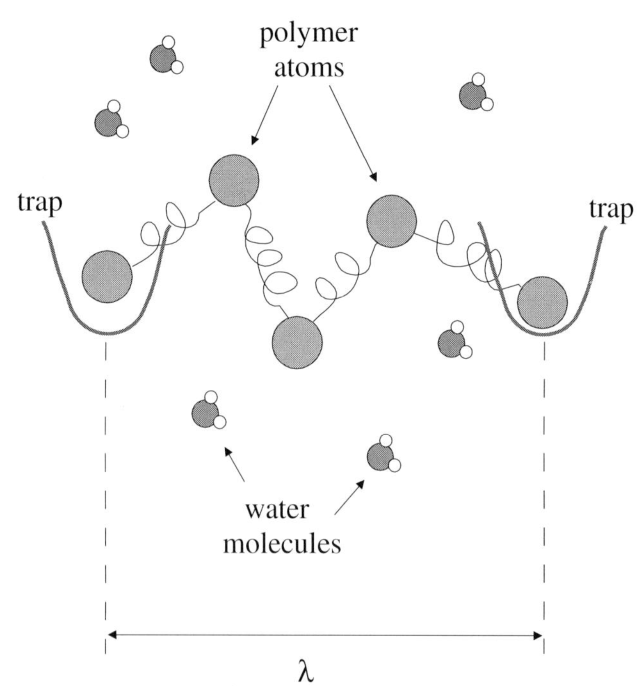

Fig. 1. A toy model of a system (polymer) in contact with a heat reservoir (water). See text for details.

Fig. 1 depicts a cartoon version of the above-mentioned experiment, where for simplicity we consider a chain of atoms (rather than nucleotides) and no beads. The N atoms, of equal mass m, are linked by covalent bonds, and we imagine that the two ends of this chain are trapped by confining harmonic potentials. These N atoms constitute the system in this set-up – we will refer to this simply as the "polymer" – the distance between the traps represents an external parameter λ, and surrounding water molecules provide a heat bath at a temperature T. We will use the variable Γ to denote a *microstate* of the system, in this case a point in the $6N$-dimensional phase space specifying the position and momentum of each of the N atoms:

$$\Gamma = (\mathbf{r}_1, \cdots, \mathbf{r}_N; \mathbf{p}_1, \cdots, \mathbf{p}_N) . \tag{5}$$

Now let $H(\Gamma, \lambda)$ be the parameter-dependent Hamiltonian which gives the internal energy of the system as a function of its microstate, for a given value of λ. According to the central assertion of equilibrium statistical mechanics, if we hold λ fixed and allow our polymer to come to thermal equilibrium then the probability distribution for finding the system in a particular microstate Γ is given by the Boltzmann-Gibbs distribution:

$$p_\lambda^{BG}(\Gamma) = \frac{1}{Z_\lambda} e^{-\beta H(\Gamma,\lambda)} , \tag{6}$$

where $\beta \equiv (k_B T)^{-1}$, and the partition function Z is given by

$$Z_\lambda = \int d\Gamma \, e^{-\beta H(\Gamma,\lambda)} . \tag{7}$$

The Boltzmann-Gibbs distribution can be viewed as the microscopic counterpart of the equilibrium state of macroscopic thermodynamics. For later reference, it is useful to note here that the free energy associated with such an equilibrium state is given by the relation

$$F_\lambda = -\beta^{-1} \ln Z_\lambda , \tag{8}$$

which (under the appropriate microscopic definition of E and S) can easily be shown to be equivalent to Eq. 1.

Since the focus of this lecture is the behavior of systems driven away from an initial state of equilibrium, we will generally be interested in the evolution of our system over some interval of time, with the system assumed to be in equilibrium at the start of that interval. This evolution is described by a trajectory Γ_t in the $6N$-dimensional phase space, which can be viewed as a "microscopic history" chronicling the time-dependence of every degree of freedom of the system over the specified interval. From such a trajectory we can compute various quantities of interest. One which will play a central role in the analysis to follow is the *work* performed on the system. Suppose that we externally vary the parameter λ, beginning at time 0 (for specificity), and we simultaneously observe the evolution of the system. In acting thus on the system, we perform external work, and the amount of work performed from the initial time up to some later time t, is given by [5]

$$w_t = \int_0^t dt' \, \dot{\lambda}_{t'} \, \frac{\partial H}{\partial \lambda} \, (\Gamma_{t'}, \lambda_{t'}) \,. \tag{9}$$

Here λ_t denotes the (externally imposed) time-dependence of λ, and Γ_t is the observed trajectory. The dot denotes, as usual, a total time derivative. (I will not here try to justify the expression for work given in Eq. 9, beyond observing that the rate of change of the internal energy of the system is given by

$$\dot{H} = \dot{\lambda} \frac{\partial H}{\partial \lambda} + \dot{\Gamma} \frac{\partial H}{\partial \Gamma} \,. \tag{10}$$

If we identify the two terms on the right as the rate of work performed on the system, and heat absorbed by the system, respectively, then this equation for $\dot{H}$ amounts to the First Law of Thermodynamics.)

In an experiment involving a system such as the polymer described above, we are typically neither able to prepare the system in a precisely specified initial microstate Γ, nor interested in predicting the subsequent evolution of every degree of freedom. Therefore we adopt a statistical attitude when analysing such a situation: instead of posing the question, "How exactly will each atom evolve?", we ask, "What are the possible microscopic histories consistent with the known preparation and manipulation of the system?" We refer to these potential microscopic histories as *realizations*, and the set of them as a *statistical ensemble*.

We can picture this ensemble of realizations as a swarm of points moving about in $6N$-dimensional phase space. A natural description of such an evolving swarm is the time-dependent density $f(\Gamma, t)$, which can be defined by the usual box-counting prescription, as follows. Imagine $R \gg 1$ trajectories Γ_t^i $(i = 1, \cdots, R)$ evolving independently in phase space, each representing a single realization of our N-body system. At any time t, a "snapshot" of this ensemble of trajectories would appear as a set of R points scattered throughout some region of phase space. Now imagine a tiny box $\mathcal{B}$, of volume $d\Gamma$, centered around a microstate Γ. The density $f(\Gamma, t)$, multiplied by the volume $d\Gamma$, gives us the fraction of the ensemble found inside this box at time t:

$$f(\Gamma, t) \, d\Gamma \approx \frac{1}{R} \sum_{\Gamma_t^i \in \mathcal{B}} 1 \,. \tag{11}$$

The sum on the right is taken over all realizations which happen to be found inside the box $\mathcal{B}$ at time t, and this sum is equal to the total number of such realizations. The approximation in Eq. 11 becomes an equality in the double limit, $R \to \infty$ and $d\Gamma \to 0$, taken in that order. We then obtain

$$f(\Gamma, t) = \lim_{d\Gamma \to 0} \lim_{R \to \infty} \frac{1}{d\Gamma} \frac{1}{R} \sum_{\Gamma_t^i \in \mathcal{B}} 1 = \Big\langle \delta(\Gamma - \Gamma_t) \Big\rangle \,, \tag{12}$$

where Γ_t denotes the microstate (at time t) of one particular realization in our infinite ensemble, and the angular brackets denote an average over the ensemble.

3 Central Result – A Prediction

We now have in place the basic theoretical elements required to address the central issue of this talk, namely, the response of a system driven (arbitrarily far!) away from equilibrium. Consider the following thought experiment. We prepare the system depicted in Fig. 1 in thermal equilibrium with a reservoir at temperature T, with the parameter fixed at some initial value. Then, beginning at $t = 0$, we vary the parameter according to some pre-determined "protocol", λ_t, and we observe the resulting microscopic trajectory Γ_t. For instance, we start with the traps separated by a distance λ_A, then we stretch our polymer by pulling the traps apart at a constant rate $\dot{\lambda}$ while observing the motion of each atom. Here, the protocol for varying the parameter, λ_t, represents "what we do to the system", whereas the trajectory Γ_t specifies "how the system responds" to this external manipulation.

Since we want to analyze this situation statistically, let us imagine that we repeat the above experiment R times, always starting in equilibrium, and always subsequently varying the external parameter according to the same protocol, λ_t. We will then observe R different trajectories

$$\Gamma_t^1, \Gamma_t^2, \cdots, \Gamma_t^R . \tag{13}$$

In the limit of infinitely many realizations ($R \to \infty$), we can construct the time-dependent density $f(\Gamma, t)$ describing the statistical response of the system (Eqs. 11,12). By the assumption of equilibrium, the initial density is

$$f(\Gamma, t = 0) = p_A^{BG}(\Gamma) = \frac{1}{Z_A} e^{-\beta H(\Gamma, A)} . \tag{14}$$

(I will often use the symbol A as compact notation for λ_A, and similarly B for λ_B.) Now, if we happen to vary λ very slowly (quasistatically), then we expect the system to remain in equilibrium. More precisely, the ensemble of realizations continuously responds so as to "keep up" with the slowly changing value of λ:

$$f(\Gamma, t) = \frac{1}{Z_{\lambda_t}} e^{-\beta H(\Gamma, \lambda_t)} \qquad \text{(quasistatic)} . \tag{15}$$

However, if we vary λ at a finite rate – possibly very rapidly – then we will drive the system out of equilibrium, and for $t > 0$ the density $f(\Gamma, t)$ will *not* be the Boltzmann-Gibbs distribution corresponding to the current value of λ.

Thus, as long as our system remains in equilibrium, we have simple, universal expressions for its statistical state (Eqs. 14,15.) Can we write down a comparably simple and general formula for $f(\Gamma, t)$ when the system is driven out of equilibrium by the rapid variation of λ? Probably not: the phase space density can evolve in a horribly complicated way, moreover this evolution will likely depend on details of the environment, and of the coupling between system and environment. In other words, if we insist on using $f(\Gamma, t)$ to represent the evolving state of our system, then we are stuck with a complicated, non-universal description. However, there exists another statistical representation of the state of our system – which I will call $g(\Gamma, t)$ – which does indeed evolve according to a simple, general formula.

Given our ensemble of realizations, we can use the phase space trajectories (Eq. 13) to obtain a set of evolving values,

$$w_t^1, w_t^2, \cdots, w_t^R, \tag{16}$$

by Eq. 9. Let us think of w_t^i as an evolving auxiliary variable, recording of the work performed on the system during the i'th realization. By transforming these auxiliary variables into statistical weights, $\Omega_t^1, \Omega_t^2, \cdots, \Omega_t^R$, defined by

$$\Omega_t^i \equiv e^{-\beta w_t^i}, \tag{17}$$

we can construct the following *weighted* phase space density:

$$g(\Gamma, t) \equiv \lim_{d\Gamma \to 0} \lim_{R \to \infty} \frac{1}{d\Gamma} \frac{1}{R} \sum_{\Gamma_t^i \in \mathcal{B}} \Omega_t^i = \left\langle e^{-\beta w_t} \, \delta(\Gamma - \Gamma_t) \right\rangle. \tag{18}$$

As in Eq. 11, the above sum is taken over all realizations which happen to be found within a box $\mathcal{B}$ of volume $d\Gamma$ around the phase space point Γ, at time t; the difference is that the unity in the sum in Eq. 11 has been replaced by the weight Ω_t^i. The function $g(\Gamma, t)$ is a kind of "undemocratic" phase space density, in which the contribution of each realization at time t depends, by Eq. 17, on the work performed on the system up to that time. [We can also interpret $g(\Gamma, t)$ as a "mass density" in phase space, by assigning to each realization a fictitious and time-dependent mass $m_t^i = \Omega_t^i$.] By contrast, $f(\Gamma, t)$ is a "democratic" phase space density, as each realization contributes equally (Eq. 11). Note that at $t = 0$ the two statistical descriptions are the same, since $\Omega_0^i = 1$:

$$g(\Gamma, 0) = f(\Gamma, 0) = \frac{1}{Z_A} e^{-\beta H(\Gamma, A)}. \tag{19}$$

Recall that $f(\Gamma, t)$ was introduced as a natural description of the evolving state of our ensemble. We can view $g(\Gamma, t)$ as simply an alternative statistical representation (which happens to incorporate information about the work performed during each realization) of the same ensemble. What makes this alternative description interesting is the following prediction. *Even if the system is driven far from equilibrium by rapidly varying λ, the weighted density $g(\Gamma, t)$ evolves in a very simple way:*

$$g(\Gamma, t) = \frac{1}{Z_A} e^{-\beta H(\Gamma, \lambda_t)} \qquad t > 0. \tag{20}$$

This statement is the central result around which this lecture is organized, and represents an answer to the question posed by the title.

4 Derivations

A number of derivations of Eq. 20 (and closely related results) have by now appeared in the literature [6–15]. These differ from one another in the underlying assumptions that are made regarding the dynamics of the system. Here I will go through the details of three such derivations. The first of these, based on Hamiltonian evolution, applies to the special case of isolated systems. In the second derivation, the presence of a heat reservoir is modeled with Langevin dynamics (white noise plus dissipation). The third derivation applies to arbitrary thermal Markov processes.

In each case the analysis will begin with an evolution equation for the ordinary phase space distribution $f(\Gamma,t)$, e.g. the Liouville equation for Hamiltonian dynamics (Eq. 21), the Smoluchowski equation for Langevin dynamics (Eq. 34), etc. Following that, we will consider $h(\Gamma,w,t)$, the joint probability distribution for observing a microstate Γ and work value w at time t, and the evolution equation for h will be obtained from that for f by the simple addition of a continuity term $-\dot{w}\partial h/\partial w$. Finally, the "undemocratic" distribution $g(\Gamma,t)$ will be expressed as a weighted projection of $h(\Gamma,w,t)$ (Eq. 26), and an explicit evolution equation for g will be obtained. In all three cases we will find that Eq. 20 is an exact solution of the evolution equation for g, thus proving our central assertion for the dynamics considered.

4.1 Hamiltonian Evolution

Let us modify the thought experiment described above, by assuming that the polymer in Fig. 1 is isolated from any thermal environments as we pull apart the laser traps. Thus, we first equilibrate the system with a reservoir at temperature T, holding $\lambda = A$ fixed, and then *we remove the reservoir*, start the clock running, and implement the protocol λ_t. The trajectory Γ_t then evolves under Hamilton's equations – as appropriate for an isolated classical system – with a time-dependent Hamiltonian $H(\Gamma,\lambda_t)$. Given a statistical ensemble of realizations, the phase space density $f(\Gamma,t)$ satisfies the Liouville equation,

$$\frac{\partial f}{\partial t} = \{H,f\} = -\dot{\mathbf{q}}\frac{\partial f}{\partial \mathbf{q}} - \dot{\mathbf{p}}\frac{\partial f}{\partial \mathbf{p}}\,, \tag{21}$$

with initial conditions given by the Boltzmann-Gibbs distribution, Eq. 14. In the above equation, $\{\cdot,\cdot\}$ represents the Poisson bracket; $H = H(\Gamma,\lambda_t)$; and $\Gamma = (\mathbf{q},\mathbf{p})$, where

$$\mathbf{q} = (\mathbf{r}_1,\cdots,\mathbf{r}_N) \qquad \text{and} \qquad \mathbf{p} = (\mathbf{p}_1,\cdots,\mathbf{p}_N)\,, \tag{22}$$

specify the configurations and corresponding momenta, satisfying Hamilton's equations:

$$\dot{\mathbf{q}} = \partial H/\partial \mathbf{p}\,, \qquad \dot{\mathbf{p}} = -\partial H/\partial \mathbf{q}\,. \tag{23}$$

If we keep track of the work w_t performed during each realization, then we can represent the evolution of the system by a trajectory (Γ_t, w_t) evolving in an "expanded phase space" which includes the auxiliary variable w. A statistical ensemble of such trajectories is described by joint probability distribution $h(\Gamma, w, t)$, defined as

$$h(\Gamma, w, t) = \Big\langle \delta(\Gamma - \Gamma_t)\, \delta(w - w_t) \Big\rangle , \tag{24}$$

by analogy with Eq. 12. Making use of the fact that $\dot{w}_t = \dot{\lambda}\, \partial H / \partial \lambda$ (from Eq. 9), we can immediately write down the evolution equation

$$\frac{\partial h}{\partial t} = \{H, h\} - \dot{\lambda} \frac{\partial H}{\partial \lambda} \frac{\partial h}{\partial w} = -\dot{\mathbf{q}} \frac{\partial h}{\partial \mathbf{q}} - \dot{\mathbf{p}} \frac{\partial h}{\partial \mathbf{p}} - \dot{w} \frac{\partial h}{\partial w} . \tag{25}$$

Just as Eq. 21 is simply a continuity equation in phase space – describing the deterministic flow generated by Hamilton's equations – so Eq. 25 is a continuity equation in the expanded phase space.

Generically, we will not be able to obtain simple expressions for either $f(\Gamma, t)$ or $h(\Gamma, w, t)$. Indeed, if the underlying classical dynamics shows evidence of chaos, then these distributions will become stretched and folded with time in a very complicated manner. However, let us now consider the following function:

$$g(w, t) = \int dw\, h(\Gamma, w, t)\, e^{-\beta w} . \tag{26}$$

This is exactly the weighted distribution defined earlier (see Eqs. 18 and 24). With this expression, the equation of motion for g follows after a single integration by parts from Eq. 25:

$$\frac{\partial g}{\partial t} = \{H, g\} - \beta \dot{\lambda} \frac{\partial H}{\partial \lambda} . \tag{27}$$

In sharp contrast with the equations of motion for f and h, Eq. 27 affords a simple solution, namely,

$$g(\Gamma, t) = \frac{1}{Z_A}\, e^{-\beta H(\Gamma, \lambda_t)} , \tag{28}$$

as predicted. Here we have assumed initial equilibrium (Eq. 19), and have made use of the identity $\{H, e^{-\beta H}\} = 0$, which follows from the chain rule for Poisson brackets:

$$\{H, e^{-\beta H}\} = -\beta e^{-\beta H} \{H, H\} = 0 . \tag{29}$$

Eq. 20 is thus satisfied for isolated Hamiltonian systems.

4.2 Langevin Evolution

Let us now assume that the system of interest is in thermal contact with a heat reservoir while we vary λ, and let us model the evolution of the system as a Langevin process [16]. In other words, instead of explicitly including the water molecules in the analysis, we mock up their presence by incorporating randomness into the evolution of the polymer. Specifically, we assume that the Hamiltonian for the polymer itself takes has the usual kinetic + potential form,

$$H(\mathbf{q},\mathbf{p},\lambda) = \frac{\mathbf{p}^2}{2m} + V(\mathbf{q},\lambda)\,, \tag{30}$$

(where $\mathbf{p}^2 = \sum_{i=1}^N p_i^2$, and the potential energy V includes the confining harmonic traps for atoms 1 and N) and we model the evolution of the polymer with the following equations of motion:

$$\dot{q}_\mu = \frac{p_\mu}{m}\,, \qquad \dot{p}_\mu = -\frac{\partial V}{\partial q_\mu} - \gamma p_\mu + \xi_\mu\,. \tag{31}$$

Here, q_μ and p_μ are the μ'th components of the $3N$-dimensional vectors $\mathbf{q}$ and $\mathbf{p}$, γ is a friction coefficient, and ξ_μ represents *white noise*, with an autocorrelation function

$$\overline{\xi_\mu(t_1)\,\xi_\nu(t_2)} = D\,\delta_{\mu\nu}\,\delta(t_2 - t_1)\,. \tag{32}$$

These evolution equations (Eq. 31) are just Hamilton's equations, modified by the addition of friction ($-\gamma p_\mu$) and noise (ξ_μ). These terms represent a straightforward method for modeling the "random kicks" received by the polymer from the surrounding water molecules, and are related by a fluctuation-dissipation relation,

$$\gamma = \beta D/2m\,. \tag{33}$$

A statistical ensemble of trajectories evolving under Eq. 31 is described by the Smoluchowski equation for the time-dependent phase space distribution [16]:

$$\frac{\partial f}{\partial t} = \{H,f\} + \gamma\frac{\partial}{\partial \mathbf{p}}\cdot\Big(\mathbf{p}f\Big) + \frac{D}{2}\nabla_p^2 f\,, \tag{34}$$

where $\nabla_p^2 \equiv \sum_\mu \partial^2/\partial p_\mu^2$, and $H = H(\mathbf{q},\mathbf{p},\lambda)$, as always.

We now proceed exactly as we did in Section 4.1. That is, given an ensemble of trajectories Γ_t governed by this stochastic process, we imagine that we keep track of the work w_t for each realization, and we construct the joint probability distribution $h(\Gamma,w,t)$ defined by Eq. 24. This distribution satisfies the evolution equation

$$\frac{\partial h}{\partial t} = \{H,h\} + \gamma\frac{\partial}{\partial \mathbf{p}}\cdot\Big(\mathbf{p}h\Big) + \frac{D}{2}\nabla_p^2 h - \dot{\lambda}\frac{\partial H}{\partial \lambda}\frac{\partial h}{\partial w}\,. \tag{35}$$

As before, we have simply added a continuity term, $-\dot{w}\partial h/\partial w$, to account for the auxiliary variable w_t.

Since neither the evolution of f nor that of h is simple[2], we consider $g = \int dw\, h e^{-\beta w}$ (Eq. 26), and we use Eq. 35 to obtain:

$$\frac{\partial g}{\partial t} = \{H,g\} + \gamma\frac{\partial}{\partial \mathbf{p}}\cdot\Big(\mathbf{p}g\Big) + \frac{D}{2}\nabla_p^2 g - \beta\dot{\lambda}\frac{\partial H}{\partial \lambda}g\,. \tag{36}$$

Simple inspection reveals that a solution of this equation is given by $g = Z_A^{-1}\exp -\beta H(\Gamma,\lambda_t)$.

[2] except in the quasistatic limit, where we get $f = p_{\lambda_t}^{BG}(\Gamma)$ and $h = p_{\lambda_t}^{BG}(\Gamma)\,\delta(w - F_{\lambda_t})$.

4.3 Thermal Markov Dynamics

The Langevin evolution considered above is an example of a *thermal Markov process*. In this section I briefly discuss the defining features of Markov processes in general [17], and then establish the validity of Eq. 20 for arbitrary *thermal* Markov processes. An alternative approach in this context has been taken by Hummer and Szabo [14], who have recently pointed out that Eq. 20 follows directly from the Feynman-Kac theorem [18].

A *stochastic* process is simply one which contains an element of randomness. When modeling a system in contact with a thermal environment, this randomness acts as a substitute for explicitly including the degrees of freedom of the reservoir. Thus, acknowledging that the atoms in Fig. 1 are continuously buffeted by water molecules, we can model these collisions as random "kicks" and attempt to construct a reasonable prescription for generating them. Langevin dynamics (Eq. 31) represents one approach, with the kicks modeled as a combination of white noise and friction. Another approach would be to occasionally – with probability-per-unit-time α – select one of the atoms (i) at random and replace its momentum with one sampled randomly from the Maxwell-Boltzmann distribution ($\propto \exp -\beta p_i^2/2m$); between such "thermalizing collisions", the evolution obeys Hamilton's equations. This scheme is known as the *Andersen thermostat* [19].

The Markov assumption is easily illustrated in the context of a numerical simulation, which inherently involves some prescription for obtaining a new microstate Γ at time $t + dt$, from an old microstate Γ' at time t, where dt is the small but finite numerical time step. Typically this prescription utilizes some deterministic integrator (such as the Runge-Kutta algorithm [20]), but if the process being simulated is stochastic, then there will be additional terms requiring the generation of random numbers. *The process is Markovian if the new microstate Γ is uniquely determined from the old microstate Γ', along with whatever random numbers are generated at time t.* In particular, the random numbers generated at previous time steps must not enter directly into the determination of Γ. For instance, the Andersen thermostat might be implemented as follows. At time t we first integrate Hamilton's equations for a time step dt, and then we generate a random number ζ_t between 0 and 1. If $\zeta_t \leq \alpha\, dt$, then this signifies a thermalizing collision, in which case we generate further random numbers to choose a specific atom and assign its new momentum. This process is Markovian. However, if we were to impose a "dead time" after every collision, e.g. by prohibiting two collisions at successive time steps, then this would violate the Markov assumption, since the probability for generating a collision at time t would depend explicitly on whether or not $\zeta_{t-dt} \leq \alpha\, dt$.

Formally, a Markov process occurring in discrete time steps dt is completely specified by a function $P(\Gamma, t + dt|\Gamma', t)$, which gives the probability distribution of obtaining Γ at time $t + dt$, given Γ' at time t. A continuous-time Markov process is obtained in the limiting case of infinitesimal time steps, and is specified by a *transition function*

$$\mathcal{L}_t(\Gamma' \to \Gamma) = \lim_{dt\to 0} \frac{1}{dt}\Big[P(\Gamma, t+dt|\Gamma', t) - P(\Gamma, t|\Gamma', t)\Big] , \tag{37}$$

which gives the instantaneous rate for making transitions to microstate Γ, starting from Γ' at time t. [Note that $P(\Gamma, t|\Gamma', t) = \delta(\Gamma - \Gamma')$.] The evolution of an ensemble of trajectories governed by such a Markov process is described by the following equation for the time-dependent phase space density:

$$\frac{\partial f}{\partial t}(\Gamma, t) = \int d\Gamma' \, f(\Gamma', t)\, \mathcal{L}_t(\Gamma' \to \Gamma) , \tag{38}$$

or more compactly,

$$\frac{\partial f}{\partial t} = \hat{\mathcal{L}}_t f , \tag{39}$$

where $\hat{\mathcal{L}}_t$ is a time-dependent linear operator acting on the space of distributions f.

In the context of the problem we are considering, Eq. 39 represents – statistically – the evolution of our system as we vary the parameter λ in the presence of a heat reservoir. Assuming the characteristics of the reservoir itself (e.g. its temperature, density, etc.) are constant, the time-dependence of the transition operator $\hat{\mathcal{L}}_t$ enters only through the value of λ. Hence, we will change notation slightly and write

$$\frac{\partial f}{\partial t} = \hat{\mathcal{L}}_{\lambda_t} f , \tag{40}$$

where $\hat{\mathcal{L}}_\lambda$ is a *parameter*-dependent transition operator. Eq. 40 is a *master equation*: given an initial distribution $f(\Gamma, 0)$ and a protocol λ_t for varying the external paramter, Eq. 40 determines the subsequent evolution of the distribution $f(\Gamma, t)$.

The transition operator formalism is quite generally applicable to Markov processes. However, in order to obtain any meaningful results from the application of this formalism to the problem at hand, we must incorporate the specifically *thermal* nature of the random forces acting on the system. We do so by imposing the following constraint on our transition operator:

$$\hat{\mathcal{L}}_\lambda e^{-\beta H(\Gamma,\lambda)} = 0 , \tag{41}$$

or, equivalently, $\hat{\mathcal{L}}_\lambda p_\lambda^{BG} = 0$. Combined with Eq. 40, this simply says that if we begin with our ensemble in the Boltzmann-Gibbs distribution corresponding to a given value of λ, and then we keep the external parameter fixed at that value, then the ensemble will not change with time. In other words, *the Boltzmann-Gibbs distribution is stationary when λ is held fixed.* If a Markov process satisfies Eq. 41, then we will call it a *thermal* Markov process.

With these definitions in place, we now proceed to derive Eq. 20 for arbitrary thermal Markov dynamics. As in Sections 4.1 and 4.2, for an ensemble of trajectories Γ_t evolving under this stochastic process, we write down an evolution equation for the joint probability distribution $h(\Gamma, w, t)$:

$$\frac{\partial h}{\partial t} = \hat{\mathcal{L}}_{\lambda_t} h - \dot{\lambda} \frac{\partial H}{\partial \lambda} \frac{\partial h}{\partial w} . \tag{42}$$

The first term on the right describes the evolution of the ensemble of trajectories Γ_t, the second is a continuity term accounting for the auxiliary variable w_t. [3]

Given Eq. 42, the weighted distribution $g = \int dw\, h e^{-\beta w}$ satisfies

$$\frac{\partial g}{\partial t} = \int dw \frac{\partial h}{\partial t} e^{-\beta w} \tag{43}$$

$$= \int dw \left(\hat{\mathcal{L}}_{\lambda_t} h - \dot{\lambda} \frac{\partial H}{\partial \lambda} \frac{\partial h}{\partial w} \right) e^{-\beta w} \tag{44}$$

$$= \left(\hat{\mathcal{L}}_{\lambda_t} - \beta \dot{\lambda} \frac{\partial H}{\partial \lambda} \right) g . \tag{45}$$

Now making use of Eq. 41, we once again find that a solution of this evolution equation is given by $g(\Gamma, t) = Z_A^{-1} \exp -\beta H(\Gamma, \lambda_t)$.

4.4 Other Derivations

In addition to the derivations presented above, two other approaches to establishing the validity of Eq. 20 deserve mention, although details of the analyses will not be provide here.

In the dynamics considered above, time was taken to be a continuous variable. However, it is common to model thermal processes as occurring in discrete time steps. The discreteness of time is not to be viewed here as a necessary evil related to practical issues of numerical simulation, but rather as intrinsic to the process. Assuming that this evolution is Markovian, a given realization of the process is specified by a *Markov chain* (an ordered sequence of microstates, $\Gamma_0 \to \Gamma_1 \to \cdots$). For such discrete-time Markov processes, the validity of Eq. 20 is easily established [7,12,15], given an appropriate definition of work, along with a thermal assumption akin to Eq. 41.

As mentioned earlier, stochastic processes (Markovian or otherwise) are ultimately convenient models which allow us to avoid explicitly dealing with the degrees of freedom of the heat reservoir. As such, these models are only approximations of physical reality, and it would be desirable to have a derivation of Eq. 20 based on more fundamental equations of motion. An obvious approach is to proceed along the lines taken in Section 4.1, but to drop the assumption that the system is isolated, and instead include *all* microscopic degrees of freedom – e.g. of both the polymer and the surrounding water – in the analysis. This is in fact not very difficult: treating the system and reservoir together as a large, isolated system, one can simply apply the analysis of Section 4.1 to this larger system,

[3] Integrating both sides of Eq. 42, and then performing an integration by parts on the last term, we recover Eq. 39.

and then project out the reservoir degrees of freedom [21]. A certain amount of care is needed if the physical interaction between the system and reservoir is non-negligible in magnitude (in effect, the Hamiltonian of the polymer needs to be "renormalized" to include a free energy of solvation, but at the end of the day one again obtains Eq. 20. Moreover, this approach can be extended to more general (non-isothermal) irreversible processes [11].

5 The Nonequilibrium Work Relation

The central prediction of this lecture, Eq. 20, is admittedly somewhat abstract, not the sort of result which typically inspires experimentalists to rush off to the laboratory to test its validity! Let us now assume (for specificity) that from $t = 0$ to $t = \tau$ we vary our parameter from the initial value λ_A to a final value λ_B, and let $W = w_\tau$ denote the total work we perform during this process. By setting $t = \tau$ in Eq. 20, integrating both sides of the equation over all of phase space, and invoking Eq. 8, we obtain the *nonequilibrium work relation*:

$$\left\langle e^{-\beta W} \right\rangle = e^{-\beta \Delta F} , \tag{46}$$

where $\Delta F = F_B - F_A$. This result relates irreversible work (l.h.s.) to an equilibrium free energy difference (r.h.s.), and – as with Eq. 20 – *remains valid no matter how quickly or slowly we vary* λ.

As formulated above, the nonequilibrium work relation applies regardless of whether or not the system itself ever reaches equilibrium state B: if we simply pull the laser traps apart at a constant, finite rate, then at the moment we reach λ_B the polymer will be in some nonequilibrium state.[4] However, we are perfectly within our rights to consider a process whereby λ is first varied rapidly from λ_A to λ_B, and then held fixed until the system relaxes to equilibrium state B, over a total time interval from $t = 0$ to $t = \tau$. In this scenario we have an irreversible process from one equilibrium state to another ($A \to B$), and Eq. 46 relates the work performed during this process to the free energy difference between the initial and final states of the system.

The nonequilibrium work relation is both easier to state and experimentally more accessible than the more abstract result from which it was derived. Fundamentally, while Eq. 20 makes a prediction regarding a (weighted) distribution in a $6N$-dimensional phase space, Eq. 46 involves a distribution on the one-dimensional work axis, in the following sense. Given infinitely many repetitions of our pulling experiment with the polymer, let $\rho(W)$ denote the distribution of values of work: $\rho(W)\, dW$ is the fraction of realizations for which the work value fell between W and $W + dW$. Then the nonequilibrium work relation predicts that

$$\int dW\, \rho(W)\, e^{-\beta W} = e^{-\beta \Delta F} , \tag{47}$$

[4] We should in this case interpret F_B as a free energy associated with the final value of λ (Eqs. 7 and 8), rather than as "the final free energy of the system".

even if the system is driven far from equilibrium as we vary the parameter from λ_A to λ_B.

To test this prediction experimentally, of course, one does not need to actually construct $\rho(W)$. Rather, given R repetitions of the pulling experiment, we expect

$$\frac{1}{R}\sum_{i=1}^{R} e^{-\beta W^i} \approx e^{-\beta \Delta F} , \tag{48}$$

where W^i is the work performed during the i'th realization. There is a catch, however: because the quantity whose average is being taken is highly nonlinear, very many repetitions might be needed in order for this approximation to be a good one. Specifically, if $\rho(W)$ is much wider than β^{-1}, then the above average will be dominated by values of work which are in the far left tail of the distribution [6]. In that case R must be extremely large in order to sample that region of the W axis with good statistics, making Eq. 46 experimentally inaccessible, though formally correct.

While these considerations seem to rule out laboratory tests using macroscopic systems (roughly speaking, we can expect the dispersion of work values to grow with the size of the system), microscopic systems are a different story. In recent experiments using a laser trap and a micromechanical cantilever to pull on the two ends of an RNA strand, as described briefly in Section 2, Liphardt *et al* [4] have found good agreement with Eq. 46. These results have provided the first experimental evidence of the validity of the nonequilibrium work relation, and by extension of Eq. 20, suggesting that these theoretical predictions might ultimately become offer useful tools in the analysis of experimental data. Indeed, Hummer and Szabo [14] had earlier proposed a scheme, based on Eq. 20, for extracting equilibrium free energy information – specifically a potential of mean force – from micromanipulation experiments.

The nonequilibrium work relation might also provide a useful tool for estimating free energy differences from numerical simulations. A traditional method of computing ΔF involves simulating the system under consideration, at fixed temperature, as an external parameter λ is varied from and initial value (A) to a final value (B). If the simulation proceeds sufficiently slowly that the system remains very close to equilibrium from start to finish, then the work performed on the system gives a good estimate of ΔF (Eq. 2). However, for many systems of practical interest – such as large biomolecules, where the estimation of free energy differences plays a central role in rational drug design [22] – such nearly-reversible simulations are prohibitively time-consuming. The nonequilibrium work relation offers a potential method for getting around this problem, by using a number of irreversible simulations and then either directly taking the "exponential average" of the work, $\langle e^{-\beta W}\rangle$, to obtain ΔF, or else using an approximation derived from Eq. 46. Recent results have suggested that this may indeed be a practical approach to numerical free energy estimation [23–27].

6 Relation to Second Law of Thermodynamics

Both the nonequilibrium work relation and the Second Law of Thermodynamics have something to say about irreversible-isothermal processes connecting two equilibrium states A and B. The former claims that $\langle e^{-\beta W}\rangle = e^{-\beta\Delta F}$, the latter that $W > \Delta F$ (Eq. 3). What is the relationship between these two statements? Indeed, are they even compatible? On the face of it, the answer to the second question seems to be negative: if $W > \Delta F$ for *every* realization, then the $\langle e^{-\beta W}\rangle$ would necessarily be less than $e^{-\beta\Delta F}$. However, we know that the Second Law must ultimately be interpreted *statistically*, thus allowing for the possibility that occasionally (though extremely rarely, for macroscopic systems) we will observe a realization for which $W < \Delta F$. With this in mind we now show that the nonequilibrium work relation implies two inequalities which are closely related to the Second Law.

Jensen's inequality [28] states that $\overline{\exp x} \geq \exp \overline{x}$, where the overbar denotes an average over any set of values of a real variable x. If we combine this mathematical result with Eq. 46, we get $\exp -\beta\Delta F = \langle \exp -\beta W\rangle \geq \exp -\beta\langle W\rangle$, i.e.

$$\langle W\rangle \geq \Delta F\,. \tag{49}$$

In other words, the nonequilibrium work relation immediately and rigorously implies that the average work performed, over an ensemble of realizations of an irreversible-isothermal process, is no less than ΔF. This is certainly consistent with the Second Law, but in fact we can say more, as we now show.

Given a particular irreversible-isothermal process, suppose we observe a realization for which $W = \Delta F - \epsilon$, where $\epsilon > 0$ has units of energy. The value of ϵ is a measure of the degree to which the Second Law was "violated" for this particular realization. Using the nonequilibrium work relation, we now rigorously derive a prediction regarding the probability distribution of such putatively illegal events. With $\rho(W)$ as defined as in Section 5, let

$$\mathrm{Prob}(W < \Delta F - \epsilon) = \int_{-\infty}^{\Delta F-\epsilon} dW\,\rho(W)\,, \tag{50}$$

denote the probability of observing a work value less than $\Delta F - \epsilon$. Using the chain of inequalities,

$$\begin{aligned}\int_{-\infty}^{\Delta F-\epsilon} dW\,\rho(W) &\leq \int_{-\infty}^{\Delta F-\epsilon} dW\,\rho(W)e^{\beta(\Delta F-\epsilon-W)}\\ &\leq e^{\beta(\Delta F-\epsilon)}\int_{-\infty}^{+\infty} dW\,\rho(W)e^{-\beta W}\,,\end{aligned} \tag{51}$$

and then invoking Eq. 47, we get

$$\mathrm{Prob}(W < \Delta F - \epsilon) \leq e^{-\beta\epsilon}\,. \tag{52}$$

This result states that the probability of observing a violation of at least magnitude ϵ, decays exponentially (or faster) with ϵ/k_BT. This in turn implies that the chance of observing a macroscopic violation ($\epsilon \gg k_BT$) is fantastically small, in complete agreement with the empirical evidence.

7 Nonequilibrium Stationary States

Up to this point, this lecture has concerned systems driven away from equilibrium by the external variation of some parameter. The underlying assumption has been that, if the parameter is held fixed, the system relaxes to and then remains in a state of thermal equilibrium. However, recent work by Hatano and Sasa [29,30] has revealed that much of the above formalism can be generalized to arbitrary stationary states, as I now briefly discuss.

Suppose we have a system whose dynamics is a Markov process specified by a parameter-dependent transition function $\mathcal{L}_\alpha(\Gamma' \to \Gamma)$. If the parameter α is held fixed, then a statistical ensemble of realizations of this process obeys the master equation

$$\frac{\partial f}{\partial t} = \hat{\mathcal{L}}_\alpha f \,. \tag{53}$$

Let us assume that this Markov process has a unique stationary distribution, $f_\alpha^S(\Gamma)$, for any value of α. This stationary distribution satisfies

$$\hat{\mathcal{L}}_\alpha f_\alpha^S = 0 \,, \tag{54}$$

but we do *not* assume that this distribution is a Boltzmann-Gibbs distribution. Rather, $f_\alpha^S(\Gamma)$ represents a *nonequilibrium stationary state* of the system.

Let us now define $\phi(\Gamma, \alpha)$ by

$$f_\alpha^S(\Gamma) = e^{-\phi(\Gamma,\alpha)} \,, \tag{55}$$

and introduce an auxiliary variable

$$y_t = \int_0^t dt' \, \dot{\alpha}_{t'} \, \frac{\partial \phi}{\partial \alpha} \left(\Gamma_{t'}, \alpha_{t'}\right) , \tag{56}$$

by analogy with our earlier definition of work (Eq. 9). Here we have assumed some protocol α_t for varying the external parameter.

Now consider the following situation. Beginning with the system in some stationary state $f_{\alpha_A}^S$, we change the parameter from α_A to a new value α_B according to a protocol α_t; we observe the evolution of the system, Γ_t; and from this trajectory we construct the evolution of the auxiliary variable, y_t. Then we repeat this experiment infinitely many times, always employing the same protocol α_t. From this data, assigning a time-dependent statistical weight e^{-y_t} to each trajectory, we finally construct the weighted distribution

$$g(\Gamma, t) = \left\langle e^{-y_t} \, \delta(\Gamma - \Gamma_t) \right\rangle . \tag{57}$$

You can probably guess what I will now tell you: no matter how slowly or quickly we vary α, the evolution of this weighted distribution obeys a very simple equation, namely,

$$g(\Gamma, t) = e^{-\phi(\Gamma, \alpha_t)} \,. \tag{58}$$

(Since the derivation of this result is essentially identical to that presented in Section 4.3, I won't reproduce it here.)

From this result we can obtain predictions entirely analogous to the nonequilibrium work relation [30], as well as the inequalities given by Eqs. 49 and 52. Thus, the entire mathematical structure of Sections 4.3, 5, and 6 carries over to the context considered here. Unfortunately, a serious obstacle currently prevents us from transforming this abstract analysis into predictions which might be tested experimentally: *there exists no general microscopic statistical theory of nonequilibrium stationary states.* In other words, for a specific physical situation, we will not generally have an explicit expression for $f_\alpha^S(\Gamma)$, hence we will not know $\phi(\Gamma, \alpha)$, and will not be able to deduce the evolution of the auxiliary variable y_t from knowledge of the trajectory Γ_t.

8 Conclusions

In this lecture I have tried to summarize recent research related to the statistical physics of systems driven away from an initial state of thermal equilibrium. The central point which I have stressed is that there exists a particular statistical representation of such processes which is simple, largely independent of the details of the process, and valid even far from equilibrium. Moreover, this formalism leads to predictions which are both potentially useful and experimentally testable.

The extension of these results to the regime of quantum mechanics represents an obvious potential avenue for future research. Yukawa [13] has made progress in this direction with respect to the nonequilibrium work relation, but the existence of a quantal version of Eq. 20 remains an open question.

Finally, the generalization discussed in Section 7 is, in my opinion, very intriguing. It tells us that if some day a statistical theory of nonequilibrium stationary states (analogous to the Boltzmann-Gibbs formalism for thermal equilibrium) will be constructed, then we will automatically be able to make concrete predictions about systems driven away from such states. In particular, the generalization of Eqs. 49 and 52 suggests that transitions between nonequilibrium stationary states obey an inequality analogous to the Clausius inequality (Eq. 4) for transitions between equilibrium states. Indeed, Hatano and Sasa [30] have related their work in along these lines to the phenomenological, macroscopic framework introduced by Oono and Paniconi [31] for the study of nonequilibrium stationary states. However, the story will not really be complete until, as suggested above, a genuine microscopic undertanding of such states is developed.

Acknowledgment

This research is supported by the U.S. Department of Energy, under contract W-7405-ENG-36, and by the Polish-American Maria Sklodowska-Curie Joint Fund II, under project PAA / DOE-98-343.

References

1. A. Einstein, Ann. Phys. (Leipzig) **17**, 549 (1905).
2. M. Smoluchowski, Ann. Phys. (Leipzig) **21**, 756 (1906).
3. L. Onsager, Phys. Rev. **37**, 405 (1931); Phys.Rev. **38**, 2265 (1931).
4. J. Liphardt *et al*, submitted for publication.
5. The essence of this interpretation seems to date back to Erwin Schrödinger's lectures: *Statistical Thermodynamics* (Cambridge, 1962); see the paragraphs found between Eqs. 2.13 and 2.14. For a more recent reference, see K. Sekimoto, Prog. Theor. Phys. Suppl. **130**, 17 (1998).
6. C. Jarzynski, Phys. Rev. Lett. **78**, 2690 (1997).
7. C. Jarzynski, Phys. Rev. **E56**, 5018 (1997).
8. G.E. Crooks, J. Stat. Phys. **90**, 1481 (1998).
9. C. Jarzynski, Acta Phys. Polon. B **B29**, 1609 (1998).
10. G.E. Crooks, Phys. Rev. **E60**, 2721 (1999).
11. C. Jarzynski, J. Stat. Phys. **96**, 415 (1999).
12. G.E. Crooks, Phys. Rev. **E61**, 2361 (2000).
13. S. Yukawa, J. Phys. Soc. Japan **69**, 2367 (2000).
14. G. Hummer and A. Szabo, Proc. Natl. Acad. Sci. (USA) **98**, 3658 (2001).
15. R.M. Neal, Statistics and Computing **11**, 125 (2001).
16. H. Risken, *The Fokker-Planck Equation*, Springer Verlag, 1989.
17. C.W. Gardiner, *Handbook of Stochastic Methods*, Springer Verlag, 2001.
18. Z. Schuss, *Theory and Applications of Stochastic Differential Equations*, Wiley, 1980.
19. H.C. Andersen, J. Chem. Phys. **72**, 2384 (1980).
20. W.H. Press *et al*, *Numerical recipes in C: the art of scientific computing*, Cambridge University Press, 1988.
21. Ref. [6] discusses this approach in the context of the nonequilibrium work relation.
22. M.R. Reddy and M.D. Erion, eds, *Free Energy Calculations in Rational Drug Design*, Kluwer Publishing, 2001.
23. D.A. Hendrix and C. Jarzynski, J. Chem. Phys. **114**, 5974 (2001).
24. G. Hummer, J. Chem. Phys. **114**, 7330 (2001).
25. H. Nanda, D.M. Zuckerman and T.B. Woolf, Biophys. J. **82**, 328A (2002).
26. D.M. Zuckerman and T.B. Woolf, Chem. Phys. Lett. **351**, 445 (2002).
27. M.O. Jensen *et al*, Proc. Natl. Acad. Sci. (USA) **99**, 6731 (2002).
28. D. Chandler, *Intoduction to Modern Statistical Mechanics*, Oxford University Press, 1987, Section 5.5.
29. T. Hatano, Phys. Rev. **E60**, R5017 (1999).
30. T. Hatano and S. Sasa, Phys. Rev. Lett. **86**, 3463 (2001).
31. Y. Oono and M. Paniconi, Prog. Theor. Phys. Suppl. **130**, 29 (1998).

Non-equilibrium Statistical Mechanics of Classical and Quantum Systems

D. Kusnezov[1], E. Lutz[1], and K. Aoki[2]

[1] Center for Theoretical Physics, Sloane Physics Lab, Yale University, New Haven, CT 06520-8120, USA
[2] Dept. of Physics, Keio University, 4—1—1 Hiyoshi, Kouhoku–ku, Yokohama 223–8521, Japan

Abstract. We study the statistical mechanics of classical and quantum systems in non-equilibrium steady states. Emphasis is placed on systems in strong thermal gradients. Various measures and functional forms of observables are presented. The quantum problem is set up using random matrix techniques, which allows for the construction of the master equation. Special solutions are discussed.

1 Introduction

The efforts to develop statistical mechanics of non-equilibrium systems are easily traced back to the foundations of the subject. As statistical mechanics was developed, it was a natural step to consider how to extend the theory. One can see this in the handwritten notes of J.W. Gibbs [1], in which he formulated the theory of statistical mechanics. Gibbs certainly expended some effort to understand how he might characterize the non-equilibrium steady-state. Gibbs certainly puzzled over how to define entropy for systems in non-equilibrium steady states, as well as how to define the statistical mechanics of a gas in a pressure or temperature gradient. However, no solutions were found. The field has remained active since that time. In the past ten to fifteen years, the emergence of results in classical chaos has led to renewed interest and many new ideas have come about [2].

Non-equilibrium statistical mechanics is replete with unanswered questions. While many theoretical techniques have been suggested to treat transport, a general understanding of the statistical mechanics is still lacking. In these lectures we survey some recent results on non-equilibrium steady states in classical and quantum systems using dynamical boundary conditions.

2 Some Results from Classical Non-equilibrium Statistical Mechanics

We examine model systems in which a Hamiltonian is coupled to heat reservoirs at two endpoints, as shown in Fig. 1. The dynamics will be that of the given Hamiltonian, except at the boundaries where it is coupled to the reservoirs. The boundary conditions can be implemented in a variety of ways [3]. Before we discuss quantum systems, we present some results for classical systems, which summarizes results in Refs. [3]-[9] and references therein.

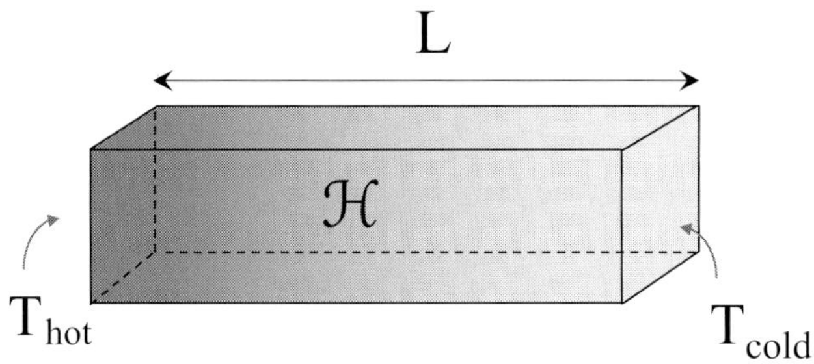

Fig. 1. Model system of a Hamiltonian H coupled to two thermal reservoirs, at temperatures $T_{cold} = T_1^0$ and $T_{hot} = T_2^0$, separated by a distance L in the $x-$direction.

Thermal Conductivity

The transport coefficient in this case is the thermal conductivity κ. There are two approaches that are used to compute κ. The first is the Green-Kubo approach, where the non-equilibrium transport is computed in terms of equilibrium auto-correlation functions. In this case:

$$\kappa(T) = \frac{1}{T^2} \int_0^\infty dt \int dx \; \langle J(x,t) J(x_0, 0) \rangle_{EQ} \; . \tag{1}$$

The heat flow J can be readily computed from the energy-momentum tensor, $J = \mathcal{T}^{01}$, for heat flow in the $x-$direction. Alternately, one can compute the conductivity directly in a non-equilibrium steady state using Fourier's Law. Measuring the local heat flow J inside the system, and the local temperature $T(x)$, leads to

$$\kappa(T) = -\frac{\langle J \rangle_{NE}}{\nabla T(x)} \; . \tag{2}$$

For many systems $\kappa(T)$ behaves as a power law, or can be approximated by a power law in temperature ranges of interest:

$$\kappa(T) = \frac{A}{T^\gamma} \; . \tag{3}$$

We should make a few notes concerning Fourier's law.

- A constant gradient inside a system does not guarantee that Fourier's law holds. When γ is very small, as in the 1-d Fermi-Pasta-Ulam (FPU) $\beta-$model around $T \sim 1$, the system can already be too far from equilibrium for Fourier's law to hold, yet the nearly constant κ still provides a near constant gradient and the illusion that the law holds. In the case of the FPU model one finds that Fourier's law is not valid even locally when the system has a near constant temperature gradient.
- A strongly-curved temperature profile $T(x)$ (as in Fig. 2) does not mean that Fourier's law is not valid. It has been shown that even strongly curved profiles can be derived analytically if Fourier's law is satisfied locally. So to

contrast with the previous point, the curvature of $T(x)$ does in itself provide no information on whether Fourier's law is valid.

- A system does not need to have a bulk limit to satisfy Fourier's law. In the 1-d FPU model, there is no bulk limit, and κ depends on the system size L. Never the less, Fourier's law is valid even for reasonably strong thermal gradients.
- Fourier's law (2) and the Green-Kubo result (1) have been seen to agree and describe the properties of systems even very far from equilibrium, for cases where $T(x)$ is strongly curved. Hence linear response theory predictions are quite robust and can be expected to hold in strong non-equilibrium environments.

Temperature Profile

When we discuss the temperature profile inside, $T(x)$, we must distinguish the temperatures of the heat reservoirs, denoted T_1^0 and T_2^0, and the temperatures just inside the system T_1 and T_2. This is indicated in Fig. 2. The boundary jumps δT can be quantitatively understood, as we mention below. The temperature inside $T(x)$ is given most directly in terms of the extrapolated temperatures $T_{1,2}$.

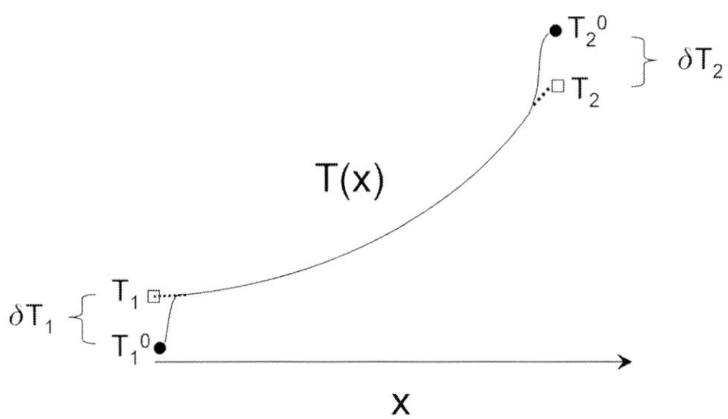

Fig. 2. Temperature profile of a system such as that depicted in Fig. 1. The boundary conditions impose temperatures $T_{cold} = T_1^0$ and $T_{hot} = T_2^0$ on the edges, $x = 0, L$. Just inside the system, there is a boundary jump δT. Extrapolating the smooth temperature profile $T(x)$ to the edges defines the temperatures T_1 and T_2.

We can understand the curvature in the temperature profile using Fourier's law. When the thermal conductivity can be described by a power law in the temperature range of interest (T_1, T_2), such as $\kappa(T) = AT^{-\gamma}$, we can integrate Fourier's law and re-express the result in terms of extrapolated temperatures

(T_1, T_2):

$$T(x) = \begin{cases} T_1 \left[1 - \left(1 - \left(\frac{T_2}{T_1} \right)^{1-\gamma} \right) \frac{x}{L} \right]^{\frac{1}{1-\gamma}} & \gamma \neq 1\,, \\ T_1 \left(\frac{T_2}{T_1} \right)^{x/L} & \gamma = 1\,. \end{cases} \tag{4}$$

Note that $\kappa(T)$ need not have a bulk behavior, nor does it need to behave globally as a power law to describe $T(x)$ in non-equilibrium systems. These formulas are quite robust.

Boundary Jumps

An interesting and often overlooked property of non-equilibrium steady states is the presence of boundary jumps. In fluids which are sheared by moving walls, one observes that the velocity just inside the fluid can be different from the velocity of the wall. For systems in a thermal gradient, there is an analogous jump in the value of the temperature just inside the system. This jump, δT, is illustrated in Fig. 2.

The jump arises due to a mismatch of the mean free path with the edge of the system. It has the general form

$$\delta T_i = T_i^0 - T_i = \eta \left. \frac{\partial T}{\partial n} \right|_{boundary}, \tag{5}$$

where η is a parameter on the order of the mean free path ℓ and n denotes the normal to the boundary [10]. This relation has been verified in two independent manners.

In the first, the boundary jumps δT have been shown to be linearly related to the normal derivative of the temperature profile $T(x)$, extrapolated to the edges of the system. The coefficient η can then be extracted. This relation has been verified in several systems. In the ϕ^4 model in 1-d, for example,

$$\eta(T) = (6.1 \pm 0.5) T^{-1.5 \pm 0.1}. \tag{6}$$

To check if we have the correct understanding, we can make an estimate of η from kinetic theory. Recall that the thermal conductivity is related to the mean free path by $\kappa \simeq C_V c \ell$ by a standard kinetic theory argument. For systems where C_V and c are largely temperature independent, such as in various lattice models, we expect $\kappa \sim \ell$. Dimensionally, the coefficient η should be on order of the mean free path ℓ, so that $\eta \sim \kappa$. Both the magnitude and power law behavior are consistent with the thermal conductivity obtained in the ϕ^4 model in 1-3 D as well as the FPU model.

An independent verification of the behavior of the jumps can be made by studying how the jumps depend on the heat flux. We let $\eta = \alpha\kappa$, where α is a constant to be determined. From Eq. (5), we can then associate the heat flux with the right side of Eq. (5):

$$T_i - T_i^0 \simeq \alpha \langle J \rangle_{NE} \sim \alpha (T_2^0 - T_1^0) \frac{\kappa(T_{av})}{L} + \cdots. \tag{7}$$

α is a coefficient which measures the efficacy of the boundary conditions. The linear relation between δT and $\langle J \rangle_{NE}$ has been verified. α has been extracted for a variety of boundary thermostats, and can show strong dependence on those boundary conditions. The understanding of these jumps together with the temperature profile (4) provides a complete description of $T(x)$ in terms of the boundary temperatures, (T_1^0, T_2^0).

Observables

If we probe an observable somewhere in the system in the non-equilibriums steady state, can we understand generically how we expect these to behave? It is a natural idea that a physical observable $\mathcal{A}$ will deviate from its value in local equilibrium as we move further away from equilibrium. In our case, a temperature gradient ∇T provides a measure of how far we are from equilibrium. The deviation of an observable $\delta\mathcal{A}$ from its local equilibrium value is expected to behave as

$$\delta_{\mathcal{A}} \equiv \frac{\delta\mathcal{A}}{\mathcal{A}} = C_{\mathcal{A}} \left(\frac{\nabla T}{T}\right)^2 + C'_{\mathcal{A}} \left(\frac{\nabla T}{T}\right)^4 + \dots \,. \tag{8}$$

This expansion is in even powers since the result should be insensitive to which side of the box is taken as the hot side and which is cold. (Using Fourier's law, we can equally well expand in terms of the current J.) While this seems natural, there has been some contention on analogous expansions in sheared fluids. There non-analytic dependencies on the shear rate have been observed in some numerical simulations, but the situation has not been entirely clarified [11]. The expansion coefficients $C_{\mathcal{A}}, C'_{\mathcal{A}}, \dots$ are in principle dependent on physical quantities, such as T and L, the size of the lattice in the direction of the gradient. If the departures from local equilibrium are local, then the the coefficients would be expected to be independent of L. It has been shown that even in systems with bulk transport behavior, the coefficients do indeed depend on the system size.

Far from Equilibrium Spatial Dependence

The expansion in (8) has the advantage that we can combine it with (4) and obtain the spatial information on how observables vary from local equilibrium values within a system:

$$\frac{\delta\mathcal{A}}{\mathcal{A}} = C_{\mathcal{A}} \left(\frac{J}{\kappa T}\right)^2 = C_{\mathcal{A}} \left(\frac{1}{a+bx}\right)^2 , \tag{9}$$

where

$$a = \frac{T_1^{1-\gamma} A}{\langle J \rangle_{NE}} , \qquad b = \gamma - 1 \,. \tag{10}$$

So knowing $C_{\mathcal{A}}$, we can also predict the spatial variation of the non-equilibrium observable $\mathcal{A}$. This has been tested and verified in classical lattice models in $d = 1 - 3$ dimensions.

The form of (9) brings to light an interesting relation to coarse graining. Let us assume we have an observable which is locally non-equilibrium, so that $\delta A/A$ at some point x' is non-zero. Using traditional arguments, we would take a larger box containing x' and assume that on the larger scale, the behavior is closer to local equilibrium. In examining (9) we see that it is a positive definite function of position (up to the overall sign of $C_{\mathcal{A}}$). So any averaging over a larger box will not necessarily converge to $\delta A/A = 0$, required for local equilibrium.

Breaking of Local Equilibrium

Local equilibrium is a property which is usually invoked to allow application of equilibrium physics to a problem of interest. It is seldom quantified. Using the expansion (8), it is possible to quantify the 'quality' of the local equilibrium assumption. The local temperature at position $\mathbf{x}$ in the model of Fig. 1 is given by $T_k = \langle p_k^2 \rangle$. A natural measure for the deviations from local equilibrium is the deviation of the momentum distribution from the Maxwellian distribution. Consider

$$\mathcal{A} = \langle p^4 \rangle \,. \tag{11}$$

In local equilibrium, we expect a Gaussian distribution of momenta, thermalized at the local temperature $T(x)$ (or T_k), so that

$$\mathcal{A}_{eq} = \langle p^4 \rangle_{eq} = 3\langle p^2 \rangle^2 = 3T(x)^2 \,. \tag{12}$$

Hence a measure of local equilibrium is given by

$$\delta \langle p_k^4 \rangle = \langle p_k^4 \rangle - 3\langle p_k^2 \rangle^2 = \langle\langle p_k^4 \rangle\rangle \,, \tag{13}$$

where $\langle\langle p_k^4 \rangle\rangle$ is the cumulant at location k. In principle, we can examine the behavior of other cumulants are measures of the breaking of local equilibrium, such as $\langle\langle p_k^6 \rangle\rangle = \langle p_k^6 \rangle - 15\langle p_k^4 \rangle\langle p_k^2 \rangle + 30\langle p_k^2 \rangle^3$. It is convenient to normalize the cumulants by the local temperature, $\langle\langle p^n \rangle\rangle/T^{n/2}$. These provide a quantitative measure on how far we are from local equilibrium. In local equilibrium, $\langle\langle p^n \rangle\rangle = 0$ ($n > 2$).

Here a few properties which have been observed in lattice models:

- Contrary to naive intuition, the steepest gradient in Fig. 2 does *not* lead to the system being furthest from local equilibrium. In fact, the converse is true — the system is furthest from equilibrium in the flattest region.
- It has been found that $\nabla T/T$ provides a good measure of how far we are from equilibrium and that the cumulants behave as

$$\delta_{LE} = \frac{\langle\langle p^4 \rangle\rangle}{3T^2} = C_{LE} \left(\frac{\nabla T}{T} \right)^2 \,. \tag{14}$$

In the ϕ^4 and FPU models, $C_{LE}^{\phi} = 1.1(8)L^{0.9(2)}$ ($T = 1$) and $C_{LE}^{FPU} = 4.3(4)L^{0.99(2)}$ ($T = 8.8$). There seems to be a weak T dependence for C_{LE} which is difficult to establish.

- These results are consistent with $d > 1$ in lattice models at the same temperatures.
- Using C_{LE}^{ϕ}, C_{LE}^{FPU}, one can predict the shape of $\langle\langle p^4\rangle\rangle/T^2$ in the system using (9), which has been verified numerically.
- As the system moves away from equilibrium by increasing the difference in the boundary temperatures, each point in the interior deviates from local equilibrium in a predictable manner, *without* any threshold. Away from equilibrium, local equilibrium is an approximation that is quite good for small gradients since the deviations from it only vary as $(\nabla T)^2$.
- Similar results hold for higher momentum cumulants.

There are many open questions here that related to measurements in lattice models. The naive expectation would be that since the gradients and the cumulants are local, their relationship should not depend on the system size L. There does seem to be size dependence, even in systems with bulk transport, many mean free paths away from the boundaries.

Corrections to Linear Response

There is an interesting relation between linear response predictions for transport coefficients and the notion of local equilibrium. For instance, if there is a departure of a predicted transport coefficient from a measured value for a system far from equilibrium, is this an indication that *(a)* corrections to linear response are needed or *(b)* local equilibrium is no longer a valid assumption. If the latter is true, temperature can no longer be unambiguously defined. Near global equilibrium, Fourier's law is satisfied *globally* so that there is a constant thermal gradient:

$$J_0 = -\kappa(T)(T_2 - T_1)/L \simeq -\kappa(T)(T_2^0 - T_1^0)/L\,. \tag{15}$$

As the temperature difference increases, curvature in $T(x)$ usually develops, and Fourier's law is satisfied locally. For $\kappa \sim T^{-\gamma}$, Fourier's law can be integrated to obtain the next leading order correction to J_0, due to curvature in $T(x)$. Denote this heat flow, J_{LR}, which includes the curvature correction. It behaves as:

$$J_{LR} = J_0\left[1 + \frac{\gamma(\gamma+1)}{24}L^2\left(\frac{\nabla T}{T}\right)^2 + \cdots\right]. \tag{16}$$

Notice that the curvature correction to J_0 behaves as L^2.

Using numerical simulations on lattice models, one can push the system very far from equilibrium and examine how J behaves. It is found that as the gradient increases, the energy that can be pumped through the system becomes less than that predicted by linear response theory *even when it is applied locally*. The 'violation' of linear response is defined as δ_{LR} and is found to behave as

$$\delta_{LR} = \frac{J - J_{LR}}{J_{LR}} = C_{LR}\left(\frac{\nabla T}{T}\right)^2. \tag{17}$$

In the ϕ^4 and FPU models, it is found that $C_{LR}^{\phi} = -4(3)L^{1.0(2)}$ $(T = 1)$ and $C_{LR}^{FPU} = -6.6(8)L^{0.9(1)}$ $(T = 8.8)$. Notice that this will naively give rise to decreasing current J with increasing gradient. However, when the gradient is this large, the higher order terms in $(\nabla T/T)$ becomes as important. It is believed that the current will eventually saturate under extremely large gradients.

A very interesting picture emerges. Within error, the violation of linear response and local equilibrium are closely connected, occurring in the same manner:

$$\delta_{LE} \sim \delta_{LR} . \tag{18}$$

Further, local equilibrium and linear response have no threshold, and break down in a similar manner as the system moves away from global thermal equilibrium.

Non-equilibrium Equation of State

For a simple one-component theory, the equation of state will have a simple form such as $P_{eq}(T)$ or $E_{eq}(T)$, where P and E are the equilibrium pressure and energy density. In the non-equilibrium steady state, one expects new variables to emerge. For thermal gradients, ∇T will become an independent variable. What is surprising, is that even in systems with bulk behavior, there can develop a system size dependence for the non-equilibrium 'equation of state'. For ϕ^4 and FPU lattice models, it is found that

$$P(T, \nabla T, L) = P_{eq}(T)\left[1 + C_P\left(\frac{\nabla T}{T}\right)^2\right] , \tag{19}$$

where $C_P^{\phi} = 1.5(1.2)L^{0.9(2)}$ $(T = 1)$, $C_P^{FPU} = 4.1(6)L^{0.30(4)}$ $(T = 8.8)$. Notice that the L dependence makes the equation of state *non-local* even though it is measured locally, in a system with bulk behavior. If we study the energy density, we have a similar result:

$$E(T, \nabla T, L) = E_{eq}(T)\left[1 + C_E\left(\frac{\nabla T}{T}\right)^2\right] , \tag{20}$$

where $C_E^{\phi} = 0.5(3)L^{0.9(2)}$ $(T = 1)$, $C_E^{FPU} = 1.7(7)L^{0.3(1)}$ $(T = 8.8)$. In contrast, Extended Irreversible Thermodynamics (EIT), predicts a local behavior for $P - P_{eq}$ in contrast to our observations.

Dimensional Loss

A consistent picture that emerges from classical simulations of non-equilibrium steady states is that the accessible volume of phase space contracts onto a fractal set. The dimensionality is obtained from the Kaplan-Yorke estimate, which involves calculating the full spectrum of Lyapunov exponents. There is a corresponding loss of dimension ΔD. An important question that arises in non-equilibrium systems is whether the dimensional loss is extensive, and persists in the bulk, or simply a manifestation of low dimensionality. In systems *with bulk behavior*, we expect ΔD to behave "extensively". That is, ΔD should remain relatively the same under the same local non-equilibrium conditions when we change the size of the system. Extensivity has been shown in the following form:

$$\frac{\Delta D}{V_{in}} = C_D J^2 \,, \tag{21}$$

where V_{in} is the volume of the system inside, excluding the thermostatted degrees of freedom, while D includes all degrees of freedom. The coefficient is given by

$$C_D = \frac{1}{\kappa \lambda_{max}^{eq} T^2} \left(1 + \frac{2\alpha\kappa}{V_{in}} \right) \overset{V_{in} \to \infty}{\longrightarrow} \frac{1}{\kappa \lambda_{max}^{eq} T^2} \,. \tag{22}$$

Here λ_{max}^{eq} is the maximum Lyapunov exponent of the full system plus bath, in thermal equilibrium. Some results are

- Extensivity is explicitly related to the macroscopic transport properties through entropy production, which leads to (22). The first systematic study of the role of dynamical thermostats on dimensional loss was also shown in [8].
- While it was emphasized that the extensivity of ΔD is not compatible with local equilibrium so that it is questionable [12]. As was mentioned above, violations of local equilibrium appear in the same manner as dimensional loss. Hence there is no conflict.
- An estimate of dimensional loss in an ideal gas gives an idea of the magnitude of the effect. Using the standard estimates of λ_{max} [2] yields $C_D \simeq [9\rho v^2 \ln(4l/d)/2]^{-1}$, where ρ is the density, v is the average particle velocity, l is the mean free path and d is the particle diameter. For $\nabla T/T \sim 0.01\,\mathrm{m}^{-1}$, $\Delta D \sim 10^9\,\mathrm{m}^{-3}$ at room temperature — quite large, yet far smaller than the total number of degrees of freedom.
- The results are satisfying from the physics point of view. ΔD pertains to the whole system, it includes the temperature profile which is curved in general, boundary temperature jumps and the various types of thermostats. Yet, ΔD can be related to macroscopic transport with the thermostat dependence cleanly separated into λ_{max}. Furthermore, ΔD has extensive behavior with respect to the internal volume wherein the the system is manifestly in non-equilibrium. We have seen that ΔD and λ_{max}^{eq} both depend strongly on the thermostat used, while their product $\Delta D \lambda_{max}^{eq}$ is thermostat *independent* and related to macroscopic quantities.
- λ_{max}^{eq} is *not* unique: In global thermal equilibrium, different choices of heat-baths can lead to very different values. The result is that dimensional loss is not unique either, only the product behaves macroscopically and can be related to thermodynamic quantities

Non-equilibrium Temperature Renormalization

In expansions such as (8), it is natural to consider whether other powers are important, such as $(\nabla^n T)/T$. In the region where the temperature dependence of the thermal conductivity can be described by a power law, $\kappa = AT^{-\gamma}$, one can show, using Fourier's law, that

$$\frac{\nabla^n T}{T} = \frac{(\nabla T)^n}{T^n}(n-1)!\prod_{k=1}^{n-1}\left(\gamma - \frac{k-1}{k}\right) . \tag{23}$$

While Fourier's law holds only close to local equilibrium, the deviations from it is of order $(\nabla T/T)^2$ so that the difference is a higher order correction in the expansion in $(\nabla T/T)$. Hence the expansion in (8) is sufficiently general.

When local equilibrium is broken, the definition of T is no longer unique. For the observables discussed above, it is possible see that certain redefinitions of T leave the observables invariant to leading order. For the generic redefinition

$$T = T' + \nu\left(\frac{\nabla T}{T}\right)^2 , \qquad C'_{\mathcal{A}} = C_{\mathcal{A}} + \nu\left(\frac{dA}{dT}\right) , \tag{24}$$

the non-equilibrium deviations of J, P, E are affected covariantly. Further, the local equilibrium violations seen in $\langle\langle p^n\rangle\rangle/T^{n/2}$ are *invariant* under such redefinitions, up to the order we consider. The physics, of course, is invariant under any redefinition in temperature.

Summary of Classical Results

We have seen that we can begin to understand the behavior of systems both near and far from equilibrium. Certainly many questions remain, including the behavior of phase transitions in strongly non-equilibrium steady-state environments, which seem to display interesting differences with the equilibrium understanding of phase transitions [7]. But we would now like to turn to the treatment of the quantum analogs of the above results.

3 Dynamical Quantum Heat Baths

Before we explore the quantum analog of Fig. 1 and the results for classical non-equilibrium steady states, let us first consider how to formulate a Hamiltonian description of a heat bath. We would like to use chaos in a similar manner; the (chaotic) interaction with the baths should cause the system of interest to thermalize. We decompose the system+heat bath problem as:

$$\begin{aligned} H(XP, xp) &= H_S(XP) + H_1(X; xp) \\ &= H_S(XP) + \overbrace{H_B(xp) + W(X; xp)} . \end{aligned} \tag{25}$$

The system, $H_S(XP)$, and the bath, $H_B(xp)$, Hamiltonians are coupled through the interaction $W(X; xp)$. In this formulation, for simplicity, the interaction does not include any system momentum dependence. The density of states of the bath+interaction is defined by H_1 through

$$\rho(E) = \overline{\mathrm{Tr}\delta(H_1 - E)} . \tag{26}$$

In order to use chaotic methods, we should make a few remarks concerning Random Matrix Theory (RMT), which has arisen as the quantum counterpart to classical chaos.

What is a random matrix? A matrix with random matrix elements [13,14], often taken as Gaussian random numbers. This is convenience rather than necessity, as the Gaussian model is far more tractable and well developed than other choices. The motivation is simple. The Schrödinger equation, $H|\psi_i\rangle = E_i|\psi_i\rangle$, cannot be solved for complex many–body systems. This lead Wigner (1951) to adopt a stochastic description instead of a dynamical description and to replace the Hamilton operator by a random operator.

How are the matrix elements of a random matrix H_{ij} distributed? Assume that

- H is invariant under "rotation" (no preferred basis in Hilbert space),
- the matrix elements H_{ij} $(1 \leq j \leq N)$ are independent.

It can then be shown that H_{ij} are Gaussian random variables. Depending on the symmetries of the problem, there are only three possible "rotations" in Hilbert space: orthogonal, unitary and symplectic transformations. Accordingly, there are only three Gaussian random matrix ensembles denoted GOE, GUE and GSE (note that generalizations exist, for example chiral ensembles in QCD). For GOE, $P(\{H_{ij}\}) = K \exp[-\sum_{ij} H_{ij}^2/\langle H_{ij}H_{ji}\rangle]$. In general the variance $\langle H_{ij}H_{ji}\rangle$ depends on i, j.

The classical analogy is the velocity distribution in the kinetic theory of gases. Newton's equations describing the (large number of) molecules of an ideal gas cannot be solved. This motivated Maxwell (1859) to use, for the first time in physics, a statistical approach to treat a dynamical problem ($\rightarrow$ random vector). He assumed that

- space is isotropic (no preferred direction in velocity space),
- the components v_i $(i = 1, 2, 3)$ of the velocity vector are independent.

He then found that the velocities are Gaussian distributed (the well–known Maxwell distribution), $P(\{v_i\}) = K \exp[-\sum_i v_i^2/\langle v_i^2\rangle]$, with $\langle v_i^2\rangle = 2kT/M$.

Heat Bath Model

We will use the eigenbasis of $H_B(xp)$ to define the matrix elements of W. If we take the interaction to be chaotic, it is reasonable to choose the statistical ensemble to be defined through W according to the GOE ensemble. As the Gaussian ensemble is entirely characterized by its first two moments, this is taken to be in the form:

$$\overline{[W(X)]_{kl}} = 0 \qquad \overline{[W(X)]_{kl}[W(Y)]_{mn}} = [\delta_{km}\delta_{nl} + \delta_{kn}\delta_{lm}]\mathcal{G}_{kl}(X - Y) \,. \tag{27}$$

The physics is put into the properties of the correlations. We use the form:

$$\mathcal{G}_{ij}(X) = \frac{\Gamma^{\downarrow}}{2\pi\sqrt{\rho(\varepsilon_i)\rho(\varepsilon_j)}} e^{-(\varepsilon_i-\varepsilon_j)^2/2\kappa_0^2} G\left(\frac{X}{X_0}\right) \tag{28}$$

$$\equiv V_{ij}^2 G(X/X_0) \,. \tag{29}$$

This type of parameterization was introduced some time ago in the study of heavy ion collisions [15] and widely used afterwards [16–19]. The physical motivation is:

- The interaction is expected to couple only neighboring energy states $\rightarrow W =$ random band–matrix with bandwidth $\kappa_0 \ll N$. For an ordinary random matrix (not banded), $\overline{\mathcal{G}_{ab}}^2 =$ constant.
- With increasing energy, the states $|a\rangle$ and $|b\rangle$ become more and more complex. Consequently, their overlap is expected to decrease. This is the physical origin of the factor containing the density of states of the bath.
- It is possible that correlations exists between W at different values of position X. This information is included in the correlation function $G(X/X_0)$, where X_0 is the characteristic length scale.

We emphasize that the coupling to the random matrix bath is fully determined by the variance.

The parameters that enter are: $\Gamma^{\downarrow}$ – the spreading width; X_0 – correlation length measuring how rapidly the system decorrelates due to the bath; κ_0 – band width of the random matrix W; $\rho = \rho_0 \exp[\beta\varepsilon]$ – density of states of the bath used to define the thermodynamic temperature through $\beta = 1/T = d\log\rho/d\varepsilon$.

Influence Functional Approach

The interaction of the system H_S with a chaotic bath $H_B + W$ can act to thermalize the system. The evolution equation for the system can be obtained by integrating over the bath. This is most readily achieved using the influence functional [20]. The density matrix of the system+bath evolves according to

$$\rho(t) = \mathcal{J}(t,t')\rho(t') . \tag{30}$$

The evolution operator $\mathcal{J}$ has a simple path integral representation which is the squared amplitude:

$$\mathcal{J}(X,x,X',x',t|Y,y,Y',y',t') = K(X,x,t|Y,y,0)K^*(X',x',t'|Y',y',0) \tag{31}$$

$$= \int_y^x Dx \int_Y^X DX \exp\left[\frac{i}{\hbar}S(X,x)\right] \tag{32}$$

$$\times \int_{y'}^{x'} Dx' \int_{Y'}^{X'} DX' \exp\left[-\frac{i}{\hbar}S(X',x')\right] .$$

S is the action for the system+bath. The density matrix can be expressed in terms of this path integral averaged over the bath degrees of freedom [20]

$$\rho(X,X',t) = \int dX_0 dY_0 \rho_0(X_0,X_0') \int_{X(0)=X_0}^{X(t)=X} \mathcal{D}X(t) \int_{X'(0)=X_0'}^{X'(t)=X'} \mathcal{D}X'(t)$$

$$\times \exp\left[\frac{i}{\hbar}\left[S_0(X(t)) - S_0(X'(t))\right]\right] \mathcal{L}(X(t),X'(t),t) , \tag{33}$$

with the influence functional given by:

$$\mathcal{L}(X(t), X'(t), t) = \overline{\left\langle i \right| \left\{ T_a \exp\left[\frac{i}{\hbar} \int_0^t dt'' H_1(X'(t'')) \right] \right\}} \times \overline{\left\{ T \exp\left[-\frac{i}{\hbar} \int_0^t dt' H_1(X(t')) \right] \right\} \left| i \right\rangle} . \tag{34}$$

Here T (T_a) is the (anti) time ordering operator. At this point we can expand the propagators in the influence function, and use the statistical properties of the matrix elements. This allows a systematic approach to include finite temperature effects into the quantum dynamics. To first order in $\beta = 1/T$ one finds

$$\mathcal{L}(X, X') = \exp\left\{ \frac{\Gamma^{\downarrow}}{\hbar} \int_0^t ds \left[G\left(\frac{X(s) - X'(s)}{X_0} \right) - 1 \right] \right\} \times \exp\left\{ \frac{i\beta\Gamma^{\downarrow}}{4X_0} \int_0^t ds [\dot{X}(s) + \dot{X}'(s)] G'\left(\frac{X(s) - X'(s)}{X_0} \right) \right\} , \tag{35}$$

where $G'(x) = dG(x)/dx$. From this one can derive the effective time evolution of the density matrix for the system:

$$\begin{aligned} i\hbar\partial_t \rho(X, X', t) = \Bigg\{ & \frac{P_X^2}{2M} - \frac{P_{X'}^2}{2M} + U(X) - U(X') \\ & - \frac{\beta\Gamma^{\downarrow}\hbar}{4X_0 M} G'\left(\frac{X - X'}{X_0} \right) (P_X - P_{X'}) \\ & + i\Gamma^{\downarrow} \left[G\left(\frac{X - X'}{X_0} \right) - 1 \right] \Bigg\} \rho(X, X', t) , \end{aligned} \tag{36}$$

with an arbitrary initial condition $\rho(X, X', 0) = \rho_0(X, X')$. This equation provides the quantum dynamics of the system at finite temperature [20]. The initial conditions can be taken as gaussian, $\psi_0(X) = \exp[-X^2/4\sigma^2]/[2\pi\sigma^2]^{1/4}$, so that the initial density matrix is

$$\rho_0(X, X') = \frac{1}{\sqrt{2\pi\sigma^2}} e^{-(X^2 + X'^2)/4\sigma^2} = \frac{1}{\sqrt{2\pi\sigma^2}} e^{-(4r^2 + s^2)/8\sigma^2} , \tag{37}$$

where

$$r = \frac{X + X'}{2} , \quad s = X - X' . \tag{38}$$

In order to learn about the transport behavior, it is convenient to define the characteristic function $d(s, k, t)$, whose moments give the transport coefficients:

$$\rho(r, s, t) = \int \frac{dk}{2\pi\hbar} \exp\left(\frac{ikr}{\hbar} \right) d(s, k, t) . \tag{39}$$

Then the cumulants are extracted as follows:

$$\ln d(s, k, t)|_{s=0} = \sum_{n=1}^{\infty} \frac{1}{n!} \left(\frac{ik}{\hbar} \right)^n \langle\langle X^n \rangle\rangle , \tag{40}$$

$$\ln d(s, k, t)|_{k=0} = \sum_{n=1}^{\infty} \frac{1}{n!} \left(\frac{s}{i\hbar} \right)^n \langle\langle P^n \rangle\rangle . \tag{41}$$

The first cumulant is just the average, $\langle\langle r\rangle\rangle = \langle r\rangle$, the second is $\langle\langle r^2\rangle\rangle = \langle r^2\rangle - \langle r\rangle^2$, the third is usually referred to as kurtosis, and so forth. The cumulants are closely related to transport coefficients.

4 Free Particle Strongly Coupled to the Reservoir

The master equation can be solved in many limits. To see the effects of the bath, we take the potential $U(X) = 0$. Then density matrix is expressed as:

$$\rho(r,s) = e^{ikr/\hbar}\Psi(s)\,. \tag{42}$$

This leads to the simplified master equation:

$$\left\{\frac{i\hbar k}{M}\frac{\partial}{\partial s} - i\gamma\hbar X_0 G'\left(\frac{s}{X_0}\right)\frac{\partial}{\partial s} - i\Gamma^{\downarrow}\left[G\left(\frac{s}{X_0}\right) - 1\right]\right\}\Psi(s) = 0\,, \tag{43}$$

which can be solved through direct integration. For a general correlation function G:

$$\rho(r,s) = \exp\left\{\frac{ikr}{\hbar} - \frac{M\Gamma^{\downarrow}}{\hbar k}\int_0^s ds'\frac{1-G}{1-(\gamma X_0 M/k)G'}\right\}\,. \tag{44}$$

Consider some examples.

Exponential Correlator: $G(x) = \exp(-|x|)$

Defining $a = \gamma X_0 M/k$ and $b = M\Gamma^{\downarrow}/\hbar k$

$$\Psi(s) = \left(\frac{1+a}{a+\exp(s)}\right)^{-(1+a)b/a} e^{sb/a}\,. \tag{45}$$

Cosine Correlator: $G(x) = \cos(x)$

In this case, the solution is of the form

$$\Psi(s) = \left(\frac{1+(a+\sqrt{a^2-1}\tan s/2)}{1+(a-\sqrt{a^2-1}\tan s/2)}\right)^b\,. \tag{46}$$

From these explicit solutions to the strongly coupled master equation, one can compute the coordinate and momentum cumulants and demonstrate how the quantum system evolves in time in the presence of this type of bath. As our interest is in the weak coupling limit where the bath looks closer to an idealized heat bath, we will not pursue this here.

5 Random Matrix Theory Master Equation: Weak Coupling Plus One Bath

A. Hamiltonian

We consider a system S coupled to a heat bath B. The corresponding Hamiltonian is given by

$$H = H_S + H_B + W \tag{47}$$

where $H_S = P^2/2M + U(X)$ describes the system [a particle moving in a potential $U(x)$], H_B describes the bath and $W = X \otimes V$ is the coupling between the system and the bath. The coupling Hamiltonian W is taken linear in the position x of the system and V is an operator acting on the bath. We denote by ε_a the energy eigenstates of the bath, $H_B|a\rangle = \varepsilon_a|a\rangle$, and assume that the heat bath is complex (chaotic). It is then justifiable to model V by a random matrix: For a complex system, the states $|a\rangle$ are expected to be highly complex and to display random features; it is thus reasonable to assume that V_{ab} is a stochastic function of a and b.

B. Master Equation

Our aim in this section is to provide an alternate derivation of a quantum master equation which gives the time evolution of the (ensemble averaged) reduced density operator of the system, $\rho_S(t) = \mathrm{tr}_B\rho(t)$, when the coupling is weak. Here $\rho(t)$ is the total density operator for system plus bath. We shall employ a method well–known in quantum optics (see for instance [21]). Starting point of the derivation is the von Neumann equation for $\rho(t)$ written in the Interaction Picture,

$$\frac{d\widetilde{\rho}(t)}{dt} = -i\,[\widetilde{W}(t), \widetilde{\rho}(t)] \tag{48}$$

with $\widetilde{A}(t) = \exp(iH_0t)A\exp(-iH_0t)$ and $H_0 = H_S + H_B$. Equation (48) can be formally integrated to give

$$\widetilde{\rho}(t) = \widetilde{\rho}(0) - i\int_0^\infty dt'\,[\widetilde{W}(t'), \widetilde{\rho}(t')]\ . \tag{49}$$

We now substitute this expression for $\widetilde{\rho}(t)$ inside the commutator of (48) and obtain

$$\frac{d\widetilde{\rho}(t)}{dt} = -i\,[\widetilde{W}(t), \widetilde{\rho}(0)] - \int_0^\infty dt'\,[\widetilde{W}(t), [\widetilde{W}(t'), \widetilde{\rho}(t')]]\ . \tag{50}$$

This equation is still exact. We next assume that initially the system and the bath are not correlated and that the latter is in thermal equilibrium, i.e., $\rho(0) = \rho_s(0) \otimes \rho_B(0)$ with $\rho_B(0) = Z_B^{-1}\exp(-\beta H_B)$. We also assume (Born approximation) that the bath remains in equilibrium at all times so that we can write $\widetilde{\rho}(t') = \widetilde{\rho}_S(t') \otimes \widetilde{\rho}_B(0)$. Clearly, this approximation is true provided the coupling between the system and the bath is weak. We now take the trace over the bath and note that $\mathrm{tr}_B\widetilde{\rho}(t) = \exp(iH_St)\mathrm{tr}_B\rho(t)\exp(-iH_St) = \widetilde{\rho}_S(t)$. After ensemble averaging we eventually arrive at

$$\begin{aligned}\frac{d\overline{\rho}_S(t)}{dt} = &- \int_0^t d\tau\, \overline{K}(\tau)\Big\{\widetilde{x}(t)\widetilde{x}(t-\tau)\overline{\rho}_S(t-\tau) - \widetilde{x}(t-\tau)\overline{\rho}_S(t-\tau)\widetilde{x}(t)\Big\} \\ &+ \int_0^t d\tau\, \overline{K}(-\tau)\Big\{\overline{\rho}_S(t-\tau)\widetilde{x}(t-\tau)\widetilde{x}(t) - \widetilde{x}(t)\overline{\rho}_S(t-\tau)\widetilde{x}(t-\tau)\Big\}\end{aligned} \tag{51}$$

where we have used $\widetilde{W}(t) = \widetilde{x}(t) \otimes \widetilde{V}(t)$ and introduced the bath correlation function $K(t) = K'(t) + iK''(t) = \mathrm{tr}_B[\widetilde{V}(t)\widetilde{V}(0)\rho_B(0)]$. Its ensemble average is given by the Fourier transform of $\overline{V_{ab}{}^2}$ with respect to ε_b [22],

$$\overline{K}(t) = \int_{-\infty}^{+\infty} d\varepsilon_b\, \rho(\varepsilon_b)\, \overline{V_{ab}{}^2}\, e^{i(\varepsilon_a-\varepsilon_b)t} \ . \tag{52}$$

Transforming Eq. (50) back into the Schrödinger Picture yields,

$$\begin{aligned}\frac{d\overline{\rho}_S(t)}{dt} = &- i\,[H_S, \overline{\rho}_S(t)] - \int_0^\infty d\tau\, [x, [\widetilde{x}(-\tau), \overline{\rho}_S(t)]]\, \overline{K}'(\tau) \\ &- i\int_0^\infty d\tau\, [x, \{\widetilde{x}(-\tau), \overline{\rho}_S(t)\}]\, \overline{K}''(\tau) \ .\end{aligned} \tag{53}$$

Equation (53) is the ensemble averaged master equation we were looking for. Let us consider the limit of large bandwidth and high temperature, $1 \ll \kappa_0 \ll T$. Using Eq. (27) we find

$$\overline{K}(\tau) = \frac{\Gamma}{\sqrt{2\pi}}\kappa_0 \exp\Big[-\frac{\kappa_0^2}{2}\Big(\tau + i\frac{\beta}{2}\Big)^2\Big] \underset{\kappa_0\to\infty}{\longrightarrow} \Gamma\,\delta(\tau + i\frac{\beta}{2}) \underset{\beta\to 0}{\longrightarrow} \Gamma\,\delta(\tau) + i\Gamma\frac{\beta}{2}\,\delta'(\tau) \tag{54}$$

so that $\overline{K}'(\tau) = \Gamma\,\delta(\tau)$ and $\overline{K}''(\tau) = \Gamma\beta\,\delta'(\tau)/2$. We have accordingly,

$$\int_0^\infty d\tau\, \widetilde{x}(-\tau)\, \overline{K}'(\tau) = \frac{\Gamma}{2}x = Dx\,, \tag{55}$$

where we have defined the diffusion coefficient $D = \Gamma/2$, and

$$\int_0^\infty d\tau\, \widetilde{x}(-\tau)\, \overline{K}''(\tau) = -\Gamma\frac{\beta}{4}\frac{d\widetilde{x}(-\tau)}{d\tau}\Big|_{\tau=0} = i\Gamma\frac{\beta}{4}\,[H_S, x] = \gamma p \ . \tag{56}$$

Here we have used $[H_S, x] = [p^2, x]/2M = -ip/M$ and defined the friction coefficient $\gamma = \Gamma\beta/4M$. Collecting everything, we finally obtain

$$\frac{d\overline{\rho}_S(t)}{dt} = -i\,[H_S, \overline{\rho}_S(t)] - D[x, [x, \overline{\rho}_S(t)]] - i\gamma[x, \{p, \overline{\rho}_S(t)\}] \ . \tag{57}$$

The master equation (57) is often referred to as the Caldeira–Leggett master equation [23,24]. It consists of three parts: a von Neumann part which describes the free motion without the coupling to the environment, a diffusive part and a dissipative part. Note that the diffusion and friction coefficients satisfy the Einstein relation, $D = 2MT\gamma$, which expresses the fact that diffusion of the particle and damping of its energy have a common physical origin, namely the coupling with the heat bath. The Einstein relation is an example of the more general fluctuation–dissipation theorem. We also mention that Eq. (57) is often derived from an oscillator bath model where the system is coupled to a set of independant harmonic oscillators (thus an integrable system). It is quite remarkable that the coupling to a complex random matrix environment leads to the same equation. This shows that in the limit considered here (weak coupling, high temperature), the master equation (57) is independent of the specific structure of the bath as well as of the specific form of the coupling and therefore universal [19].

Coordinate representation and semiclassical limit
The coordinate representation of the master equation (57) reads

$$\begin{aligned}\frac{\partial \overline{\rho}_S(x,x',t)}{\partial t} &= \Big[-\frac{i}{\hbar}\Big(H_S(x) - H_S(x')\Big) - \frac{D}{\hbar^2}\,(x-x')^2 \\ &\quad - \frac{i}{\hbar}\gamma\,(x-x')(p_x - p_{x'})\Big]\overline{\rho}_S(x,x',t)\,, \end{aligned} \tag{58}$$

where for convenience we have reintroduced the constant $\hbar$. This corresponds to the master equation derived from the influence functional in the limit

$$G(x) \sim 1 - \frac{1}{2}x^2\,. \tag{59}$$

The corresponding classical transport equation is obtained by taking the Wigner transform of the density matrix

$$f(q,p,t) = \frac{1}{2\pi\hbar}\int_{-\infty}^{\infty} dr \exp\Big[-\frac{ipr}{\hbar}\Big]\rho\Big(q+\frac{r}{2}, q-\frac{r}{2}, t\Big)\,. \tag{60}$$

Keeping only terms in leading orders of $\hbar$, this leads to

$$\frac{\partial f}{\partial t} = -\frac{p}{M}\frac{\partial f}{\partial q} + U'(q)\frac{\partial f}{\partial p} + 2\gamma\frac{\partial}{\partial p}(pf) + D\frac{\partial^2 f}{\partial p^2}\,. \tag{61}$$

This is the usual Klein–Kramers equation.

C. Relation with the Oscillator Bath Model

We now turn to a direct comparison of the oscillator bath model and the random matrix model [22]. We still place ourselves in the weak coupling limit. The oscillator bath model consists of a particle coupled to a large number of independent harmonic oscillators (mass m_i and frequency ω_i). The Hamiltonian of the composite system is then (see [24] and references therein)

$$H = H_S + \sum_{i=1}^{N}\left(\frac{p_i^2}{2m_i} + \frac{m_i\omega_i^2}{2}\left[x_i - \frac{c_i}{m_i\omega_i^2 x}\right]^2\right)\,. \tag{62}$$

In this model, the coupling Hamiltonian $W = x\sum_i c_i x_i$ is bilinear in the position of the system and the the positions of the harmonic oscillators. Here c_i are coupling constants. The coupling to the harmonic bath is fully characterized by the spectral density function,

$$I(\omega) = \pi \sum_i \frac{c_i^2}{2m_i\omega_i}\, \delta(\omega - \omega_i)\ . \tag{63}$$

In the following we shall relate $I(\omega)$ and the variance $\overline{V_{ab}}^2$. To this end it is useful to rewrite Eq. (29) in the form

$$\overline{V_{ab}}^2 = [\rho(\varepsilon_a)\rho(\varepsilon_b)]^{-\frac{1}{2}}\, \mathrm{f}(\omega_{ab} = \varepsilon_b - \varepsilon_a) \tag{64}$$

where we have introduced the band form factor $\mathrm{f}(\omega)$. It can be shown that with the form (64) the bath correlation function satifies the KMS condition $\overline{K}(-t) = \overline{K}(t - i\beta)$ which defines the thermal equilibrium state of the bath. The bath correlation function can further be expressed in terms of the spectral density function as

$$K(t) = \int_0^\infty \frac{d\omega}{\pi} I(\omega) \left(\coth\left(\frac{\beta\omega}{2}\right)\cos(\omega t) - i\sin(\omega t)\right)\ . \tag{65}$$

We see from Eq. (65) that the imaginary part $K''(t)$ of the correlation function is simply the Fourier sine transform of the spectral density $I(\omega)$. Comparing the imaginary parts of Eqs. (53) and (65), we find,

$$I(\omega) = 2\pi \sinh\left(\frac{\beta\omega}{2}\right) \mathrm{f}(\omega)\ . \tag{66}$$

This equation provides the desired link between the two models. It shows that there is a one–to–one correspondence between the density function $I(\omega)$ of the oscillator bath model and the variance $\overline{V_{ab}}^2$ of the random band–matrix model. It is important to note that both the oscillator bath model and the random matrix model, in the limit of weak coupling we consider here, are Gaussian. This means that in these two models the dynamics of the bath is is entirely characterized by the two–point correlation function $K(t)$ and it is therefore not necessary to consider higher–order correlation functions.

A straightforward application of Eq. (66) using the variance Eq. (29) leads to

$$I(\omega) = \Gamma \sinh\left(\frac{\beta\omega}{2}\right) \exp\left[-\frac{\omega^2}{2\kappa_0^2}\right]\ . \tag{67}$$

In the limit $\omega \to 0$, Eq. (67) reduces to $I(\omega) \sim M\gamma\omega$ which defines the so–called Ohmic regime. In this regime, as we already saw in the derivation of the master equation above, the bath correlation function is delta–correlated in time, $\overline{K}'(t) = \Gamma\delta(t)$, which means there are no memory effects. The process in thus Markovian. We can further easily check that the band form factor

$$\mathrm{f}(\omega) = \frac{\Gamma}{2\pi}|w|^{\alpha-1} \exp\left[-\frac{\omega^2}{2\kappa_0^2}\right] \tag{68}$$

yields $I(\omega) \sim \pi\beta\omega \mathrm{f}(\omega) \sim \omega^{\alpha}$ as $\omega \to 0$. This corresponds to fractal (non–Ohmic) environment. Here, contrary to the Ohmic case, the bath correlation is given by a inverse power law,

$$\overline{K'}(t) = \frac{\Gamma}{\pi}\Gamma(\alpha)\cos(\frac{\alpha\pi}{2})\, t^{-\alpha} \quad \overline{K''}(t) = \frac{\beta}{2}\frac{d\overline{K'}}{dt} \tag{69}$$

in the limit $1 \ll \kappa_0 \ll T$. As a result, there are long–time (memory) effects (the process in now non-Markovian) and the dynamics of the particle becomes anomalous. We shall develop that point in more detail in the next section. But before we would like to show that the band form factor $\mathrm{f}(\omega)$ can be directly related to the velocity autocorrelation function (VACF) $\langle v(0)v(t)\rangle$ of the system. This is done by using the first fluctuation–dissipation theorem

$$C_v[z] = \Big(z + \gamma[z]\Big)^{-1}, \tag{70}$$

which relates the Laplace transform $C_v[z]$ of the normalized VACF $C_v(t) = \langle v(0)v(t)\rangle/\langle v(0)^2\rangle$ to the Laplace transform of the damping kernel $\gamma(t) = \overline{K}'(t)/MT$ (see below). From Eq. (52), we have

$$\gamma[z] = \frac{\beta}{M}\int_{-\infty}^{\infty} d\omega\, e^{-\beta\omega}\mathrm{f}(\omega)\frac{z}{z^2+\omega^2}\ . \tag{71}$$

In the limit $z \to 0$, the last factor in Eq. (71) reduces to a delta function and we obtain

$$\lim_{z\to 0}\gamma[z] = \frac{\pi\beta}{M}\lim_{z\to 0}\mathrm{f}(z)\,. \tag{72}$$

The final value theorem of the theory of Laplace Transform tells us that

$$\lim_{t\to\infty} C_v(t) = \lim_{z\to 0} z\, C_v[z] \tag{73}$$

and we can therefore write

$$\lim_{t\to\infty} C_v(t) = \lim_{z\to 0}\Big(1 + \frac{\pi\beta}{M}\, z^{-1}\mathrm{f}(z)\Big)^{-1}\ . \tag{74}$$

We thus see that the long–time behavior of the VACF is determined by the shape of the band form factor of the origin.

D. Langevin Equation

Consider a system operator P (for instance the momentum operator p of the particle). Our aim in the present section is to derive a quantum Langevin equation for this operator in the limit of weak coupling. The time evolution of the operator P is given by $P(t) = \exp(iHt)\,P\,\exp(-iHt)$ (Heisenberg representation). Accordingly,

$$\dot{P}(t) = i[H_S(t), P(t)] + i[x(t), P(t)]V(t)\ . \tag{75}$$

This equation is exact. We shall now look for an approximate expression for $[x(t), P(t)] = \exp(iHt)\,[x, P]\,\exp(-iHt)$ and $V(t) = \exp(iHt)\,V\,\exp(-iHt)$. In the limit of weak coupling, the time evolution operator $\exp(-iHt)$ can be written in a Dyson series as,

$$\exp(-iHt) \simeq \exp(-iH_0 t) - i\int_0^t d\tau\, \exp(-iH_0(t-\tau))\,W\,\exp(-iH_0\tau)\,. \tag{76}$$

In lowest order in V, this leads to

$$[x(t), P(t)] = [\widetilde{x}(t), \widetilde{P}(t)] + i\int_0^t d\tau\,[\widetilde{x}(t-\tau), [\widetilde{x}(t), \widetilde{P}(t)]]\widetilde{V}(t) \tag{77}$$

and

$$V(t) = \widetilde{V}(t) + i\int_0^t d\tau\,\widetilde{x}(t)\Big(\widetilde{V}(t-\tau)\widetilde{V}(t) - \widetilde{V}(t)\widetilde{V}(t-\tau)\Big)\,. \tag{78}$$

Inserting Eqs. (77) and (78) into Eq. (75) we then obtain up to second order in V,

$$\begin{aligned}\dot{P}(t) = i[H_S(t), P(t)] &- \int_0^t d\tau\,[\widetilde{x}(t-\tau), [\widetilde{x}(t), \widetilde{P}(t)]]\mathrm{Re}[\widetilde{V}(t-\tau)\widetilde{V}(t)] \\ &- i\int_0^t d\tau\,\{\widetilde{x}(t-\tau), [\widetilde{x}(t), \widetilde{P}(t)]\}\mathrm{Im}[\widetilde{V}(t-\tau)\widetilde{V}(t)] \\ &+ i[\widetilde{x}(t), \widetilde{P}(t)]\widetilde{V}(t)\,.\end{aligned} \tag{79}$$

This equation is valid for any system operator P. We now consider the case where $P = p$ the momentum operator of the particle. Using the commutation relation, $[x, p] = i$, we can rewrite Eq. (78) in the form

$$\dot{p}(t) = -U'(x(t)) + 2\int_0^t d\tau\,\mathrm{Im}[\widetilde{V}(t-\tau)\widetilde{V}(t)]\widetilde{x}(t-\tau) - \widetilde{V}(t)\,. \tag{80}$$

Equation (80) is almost a Langevin equation. We further note that $\widetilde{V}(t)$ depends on the (thermal) initial conditions of the heat bath. This makes the force operator $\xi(t) = -\widetilde{V}(t)$ a fluctuating quantity. Introducing first the thermal average over the bath and then the average over the random matrix ensemble, we get

$$\overline{\langle\xi(t)\rangle} = 0, \qquad \overline{\langle\xi(t)\xi(0)\rangle} = \overline{\langle\widetilde{V}(t)\widetilde{V}(0)\rangle} = \overline{K}(t)\,. \tag{81}$$

Moreover, after partial integration, we have

$$2\int_0^t d\tau\,\overline{K}''(-\tau)\overline{x}(t-\tau) \simeq -\beta\int_0^t d\tau\,\overline{K}'(t-\tau)\dot{\overline{x}}(\tau) \tag{82}$$

where we have used Eq. (69). In the limit of weak coupling and high temperature, the Langevin equation can therefore be written as

$$M\ddot{\overline{x}}(t) + M\int_0^t d\tau\,\gamma(t-\tau)\dot{\overline{x}}(\tau) + U'(\overline{x}(t)) = \overline{\xi}(t) \tag{83}$$

where we have introduced the damping kernel $\gamma(t)$ which obeys $MT\gamma(t) = \overline{K'}(t)$. This last relation is often referred to as the second fluctuation–dissipation theorem. We note that the Langevin equation is completely determined by the real part $\overline{K'}(t)$ of the bath correlation function. Introducing the Riemann–Liouville fractional derivative $(-1<\lambda<0)$ [26],

$$\frac{\partial^\lambda f(t)}{\partial t^\lambda} = \frac{1}{\Gamma(-\lambda)} \int_0^t \frac{f(\tau)\, d\tau}{(t-\tau)^{\lambda+1}} \tag{84}$$

we can rewrite (83) in the form of a fractional Langevin equation [25]

$$M\ddot{x}(t) + M\gamma_\alpha \frac{\partial^{\alpha-1}}{\partial t^{\alpha-1}} \dot{x}(t) + U'(x(t)) = \xi(t) \tag{85}$$

where we have defined $\gamma_\alpha = \Gamma\beta/(2M\sin(\alpha\pi/2))$ and dropped the overline. The fractional Langevin equation (84) can be easily solved for the case of a free particle $U(x) = 0$ by making use of the Laplace transform. We find

$$x(t) = x_0 + v_0 B_v(t) + \int_0^t d\tau\, B_v(t-\tau)\xi(\tau) \tag{86}$$

where (x_0, v_0) are the initial coordinates of the particle and $B_v(t) = \int_0^t C_v(t')dt'$ is the integral of the (normalized) velocity autocorrelation function $C_v(t)$. Since $\gamma[z] = \gamma_\alpha z^{\alpha-1}$, the Laplace of the VACF is given by

$$C_v[z] = \frac{1}{z + \gamma_\alpha z^{\alpha-1}} \, . \tag{87}$$

By taking the inverse Laplace transform, the velocity autocorrelation function can be written as

$$C_v(t) = E_{2-\alpha}(-\gamma_\alpha t^{2-\alpha}) \, . \tag{88}$$

Here we have introduced the Mittag–Leffler function $E_\alpha(t)$, which is defined by the series expansion [27]

$$E_\alpha(t) = \sum_{n=0}^{\infty} \frac{t^n}{\Gamma(\alpha n + 1)} \, . \tag{89}$$

The function $E_\alpha(t)$ reduces to the exponential when $\alpha = 1$. The asymptotic behavior of the Mittag–Leffler function (89) for short and long times is respectively given by $\sim \exp(t)$ and $\sim -(t\,\Gamma(1-\alpha))^{-1}$, $0<\alpha<1$ and $1<\alpha<2$. For the velocity autocorrelation function (88) this yields a typical stretched exponential behavior at short times

$$C_v(t) \sim \exp \frac{-\gamma_\alpha t^{2-\alpha}}{\Gamma(3-\alpha)} \, , \qquad t \ll \frac{1}{(\gamma_\alpha)^{1/\alpha}} \tag{90}$$

and an inverse power–law tail at long times

$$C_v(t) \sim \frac{t^{\alpha-2}}{\gamma_\alpha \Gamma(\alpha-1)} \, , \qquad t \gg \frac{1}{(\gamma_\alpha)^{1/\alpha}} \, . \tag{91}$$

After time integration, we finally get from Eq. (88)

$$B_v(t) = t\,E_{2-\alpha,2}(-\gamma_\alpha t^{2-\alpha}) \tag{92}$$

where we have used the generalized Mittag–Leffler function $E_{\alpha,\beta}(t)$ defined as [27]

$$E_{\alpha,\beta}(t) = \sum_n^\infty \frac{t^n}{\Gamma(\alpha n + \beta)} \; . \tag{93}$$

In the long–time limit, the generalized Mittag–Leffler function satisfies $E_{\alpha,\beta}(t) \sim -(t\,\Gamma(\beta - \alpha))^{-1}$. Accordingly, $B_v(t)$ exhibits a decay of the form

$$B_v(t) \sim \frac{t^{\alpha-1}}{\gamma_\alpha \Gamma(\alpha)} \qquad \text{when} t \to \infty \; . \tag{94}$$

We emphasize that the solution (86) of the fractional Langevin equation in the force free case is completely specified by the knowledge of the function $B_v(t)$.

The mean displacement and the mean–square displacement are readily deduced from Eq. (86). We find

$$\langle x \rangle = x_0 + v_0\, t\, E_{2-\alpha,2}(-\gamma_\alpha t^{2-\alpha}) \underset{t\to\infty}{\sim} \frac{v_0}{\gamma_\alpha} \frac{t^{\alpha-1}}{\Gamma(\alpha)} \tag{95}$$

and

$$\langle x^2 \rangle = \frac{2T}{M} t^2 E_{2-\alpha,3}(-\gamma_\alpha t^{2-\alpha}) \underset{t\to\infty}{\sim} \frac{2T}{\gamma_\alpha M} \frac{t^\alpha}{\Gamma(1+\alpha)} \; . \tag{96}$$

In the last equation, thermal initial conditions have been assumed ($x_0 = 0$, $v_0^2 = T/M$). Equation (96) shows that the coupling to a fractal heat bath leads in general to anomalous diffusion, $\langle x^2 \rangle \sim t^\alpha$, $\alpha \neq 1$. The band form factor (68) gives rise to subdiffusion when $\alpha < 1$ and to superdiffusion when $1 < \alpha < 2$. In general a wide range of anomalous transport behaviors can be realized through RMT (quantum chaotic) environments, with the character of the diffusion related to the microscopic properties of the quantum environment [20].

6 Random Matrix Theory Master Equation: Weak Coupling Plus Two Baths

We are now in a position to formulate the master equation for the quantum analog of Fig. 1, a system coupled to two heat baths B_1 and B_2. The corresponding Hamiltonian is given by

$$H = H_S + H_{B_1} + H_{B_2} + Q_1(x) \otimes V_1 + Q_2(x) \otimes V_2 \tag{97}$$

where $H_S = \sum_i p_i^2/2m_i + U(x_i, x_j)$ is a N-particle system Hamiltonian, $Q_1(x_1, ..., x_N)$ and $Q_2(x_1, ..., x_N)$ are two (arbitrary) system operators and V_1 and V_2 are two random matrix bath operators. The variance of these two random operators is given by Eq. (29) with the respective parameters of the two baths given by T_k, Γ_k and κ_{0k}, $k = 1, 2$. A quantum master equation for the system can now be derived using the method described in section 5. This leads to the ensemble averaged equation

$$\begin{aligned} \frac{d\overline{\rho}_S(t)}{dt} = & -i\,[H_S, \overline{\rho}_s(t)] \\ & - \int_0^\infty d\tau\,[Q_1, [\widetilde{Q_1}(-\tau), \overline{\rho}_S(t)]]\,\overline{K}_1'(\tau) \\ & -i \int_0^\infty d\tau\,[Q_1, \{\widetilde{Q_1}(-\tau), \overline{\rho}_S(t)\}]\,\overline{K}_1''(\tau) \\ & - \int_0^\infty d\tau\,[Q_2, [\widetilde{Q_2}(-\tau), \overline{\rho}_S(t)]]\,\overline{K}_2'(\tau) \\ & - i \int_0^\infty d\tau\,[Q_2, \{\widetilde{Q_2}(-\tau), \overline{\rho}_S(t)\}]\,\overline{K}_2''(\tau)\,. \end{aligned} \tag{98}$$

In the following we take for the system a 1D harmonic crystal of unit masses and frequency ω. We attach the two heat baths at both ends, x_1 and x_N, of the linear chain. We thus have $H_S = \sum_i p_i^2 + \omega^2/2 \sum_{i,j} G_{ij} x_i x_j$, where the matrix $\mathbf{G}$ is defined as $G_{ij} = 2\delta_{ij} - \delta_{i+1,j} - \delta_{i-1,j}$. The system operators are respectively given by $Q_1(x_1, ..., x_N) = x_1$ and $Q_2(x_1, ..., x_N) = x_N$. In the Ohmic regime, the corresponding master equation can be easily written in coordinate representation in the form

$$\begin{aligned} \frac{d}{dt}\overline{\rho}_S(x_i, x_i', t) = & \Bigg\{ \frac{i}{2} \sum_i \Big(\frac{\partial^2}{\partial x_i^2} - \frac{\partial^2}{\partial x_i'^2} \Big) - \frac{i}{2}\omega^2 \sum_{ij} G_{ij} \Big(x_i x_j - x_i' x_j' \Big) \\ & - D_1\,(x_1 - x_1')^2 - \gamma_1\,(x_1 - x_1') \Big(\frac{\partial}{\partial x_1} - \frac{\partial}{\partial x_1'} \Big) \\ & - D_2\,(x_N - x_N')^2 - \gamma_2\,(x_N - x_N') \Big(\frac{\partial}{\partial x_N} - \frac{\partial}{\partial x_N'} \Big) \Bigg\}\, \overline{\rho}_S(x, x', t) \end{aligned} \tag{99}$$

with $D_k = 2T_k\gamma_k$. Taking the Wigner transform (60) we further obtain

$$\begin{aligned} \frac{\partial f}{\partial t}(q_i, p_i, t) = & - \sum_{i=1}^{N} p_i \frac{\partial f}{\partial q_i} + \sum_{i,j=1}^{N} \frac{\partial U}{\partial q_i} \frac{\partial f}{\partial p_j} + 2\gamma_1 \frac{\partial}{\partial p_1}(p_1 f) + 2\gamma_1 T_1 \frac{\partial^2 f}{\partial p_1^2} \\ & + 2\gamma_2 \frac{\partial}{\partial p_N}(p_N f) + 2\gamma_2 T_2 \frac{\partial^2 f}{\partial p_N^2}\,. \end{aligned} \tag{100}$$

Introducing the notation, $x_i = q_i$, $x_{i+N} = p_i$, $i = 1, ..., N$, Eq. (100) can be rewritten in the compact form

$$\frac{\partial f}{\partial t}(x_i, t) = \sum_{i=1}^{2N} \frac{\partial}{\partial x_i}(\xi_i f) + \frac{1}{2} \sum_{i,j=1}^{2N} \frac{\partial^2}{\partial x_i \partial x_j}(d_{ij} f) \tag{101}$$

with $\xi_i = \sum_{ij} a_{ij} x_j$. The two $2N \times 2N$ matrices $\mathbf{a}$ and $\mathbf{d}$ are given by

$$\mathbf{a} = \begin{pmatrix} 0 & -\mathbf{I} \\ \omega^2 \mathbf{G} & \mathbf{R} \end{pmatrix} \quad \text{and} \quad \mathbf{d} = \begin{pmatrix} 0 & 0 \\ 0 & \boldsymbol{\varepsilon} \end{pmatrix} \tag{102}$$

where 0 and I denote the null and unit $N \times N$ matrices, and the two $N \times N$ matrices $\mathbf{R}$ and $\boldsymbol{\varepsilon}$ obey $R_{ij} = (2\gamma_1\delta_{1i} + 2\gamma_2\delta_{Ni})\delta_{ij}$ and $\varepsilon_{ij} = 2T_i R_{ij}$. The classical generalized Klein–Kramers equation (101) has been studied by Rieder, Lebowitz and Lieb [28] (the quantum problem has been treated using a Langevin approach in [29]). Of interest here is the stationary non–equilibrium solution $\partial f_s(x_i)/\partial t = 0$ for a small temperature difference, $T_1 = T + \Delta T/2$, $T_2 = T - \Delta T/2$, $\Delta T \ll T$. It is given by

$$f_s(\mathbf{x}) = (2\pi)^{-N} \mathrm{Det}[\mathbf{b}^{-\frac{1}{2}}] \exp\left[-\frac{1}{2} \sum_{i,j=1}^{2N} b_{ij}^{-1} x_i x_j\right] \tag{103}$$

where $\mathbf{b}$ is the $2N \times 2N$ covariance matrix, $b_{ij} = \langle x_i x_j \rangle = \int f_s(\mathbf{x}) x_i x_j d\mathbf{x}$. It is useful to write the covariance matrix in the form

$$\mathbf{b} = \begin{pmatrix} \mathbf{x} & \mathbf{z} \\ \mathbf{z}^\dagger & \mathbf{y} \end{pmatrix} \tag{104}$$

where the $N \times N$ matrices $\mathbf{x}$, $\mathbf{y}$ and $\mathbf{z}$ give, respectively, the correlations between the coordinates, the momenta and between the coordinates and momenta, $x_{ij} = \langle q_i q_j \rangle$, $y_{ij} = \langle p_i p_j \rangle$, $z_{ij} = \langle q_i p_j \rangle$. In the limit of large N and small coupling $\gamma = \gamma_1 = \gamma_2$, one finds

$$\mathbf{x} = \frac{T}{\omega^2}\mathbf{G}^{-1}, \mathbf{y} = \begin{pmatrix} T(1) & & & \\ & T(2) & & \\ & & \ddots & \\ & & & T(N) \end{pmatrix}, \mathbf{z} = \begin{pmatrix} 0 & \frac{\gamma \Delta T}{\omega^2} & & & \\ -\frac{\gamma \Delta T}{\omega^2} & 0 & \frac{\gamma \Delta T}{\omega^2} & & \\ & -\frac{\gamma \Delta T}{\omega^2} & \ddots & \frac{\gamma \Delta T}{\omega^2} & \\ & & -\frac{\gamma \Delta T}{\omega^2} & 0 & \end{pmatrix}. \tag{105}$$

Here $T(1) = T + \Delta T/2$, $T(N) = T - \Delta T/2$ and $T(i) = T$ otherwise. We see that the temperature profile in constant in the bulk and presents a discontinuity at the edges. By inverting the covariance matrix $\mathbf{b}$ we eventually arrive at

$$f_s(q_i, p_i) \sim \exp\left[-\sum_i \frac{p_i^2}{2T(i)} - \frac{U}{2T} - \frac{\gamma \Delta T}{2T} \sum_i \frac{p_i}{\omega^2 T(i)} \left(\frac{\partial U}{\partial q_{i-1}} - \frac{\partial U}{\partial q_{i+1}}\right)\right]. \tag{106}$$

This equation is valid in the limit $\Delta T, \gamma \to 0$. We note that Eq. (106) reduces to the equilibrium Boltzmann distribution when $\Delta T = 0$. The last term in (106) is proportional to the heat flux $J = \gamma \Delta T/2$ (note that J is not proportional to the temperature gradient). Taking the inverse Wigner transform we obtain the stationary density operator in the form

$$\overline{\rho}_S(q_i + \frac{r_i}{2}, q_i - \frac{r_i}{2}) = \prod_i \exp\left[\frac{1}{2T(i)}\left(\frac{\gamma \Delta T}{2T}\frac{1}{\omega^2}\left(\frac{\partial U}{\partial q_{i-1}} - \frac{\partial U}{\partial q_{i+1}}\right) - iT(i)r_i\right)^2\right] \times \exp\left[-\frac{U}{2T}\right] . \quad (107)$$

This form of the solution gives us a guide on how to better solve the general problem of Fig. 1 using a general quantum master equation.

7 Conclusions

The transport properties of classical and quantum systems in non-equilibrium steady states were formulated and discussed. While the situation for classical non-equilibrium systems does not seem very well understood yet, it is clearly far better understood than the quantum counterpart. At the moment one can formulate a master equation for the non-equilibrium density matrix, however, aside from the trivial harmonic case discussed above, almost nothing is known. It would be very desirable to see if one can derive Fourier's law or the temperature profile from Random Matrix models, and generally how far one can push the matrix models which seem to provide a tractable approach to quantum thermalization.

Acknowledgments

This work was supported by the Office of Naval Research under contract # N00014-01-1-0594. DK thanks R. Olkiewicz and P. Garbaczewki, the organizers of the 38th Winter School of Theoretical Physics.

References

1. Gibbs-Van Name Papers, Beinecke Rare Book and Manuscript Library, Yale University.
2. W.G. Hoover, *Computational Statistical Mechanics* (Elsevier, Amsterdam, 1991);
   P. Gaspard, *Chaos, Scattering and Statistical Mechanics*, (Cambridge, New York, 1998);
   J.R. Dorfman, *An Introduction to Chaos in Nonequilibrium Statistical Mechanics*, Cambridge Univ. Press (1999);
   W.G. Hoover, *Time Reversibility, Computer Simulation, and Chaos*, (World Scientific, Singapore, 1999).
3. K. Aoki, D. Kusnezov, Ann. Phys. **295**, 50 (2002).
4. K. Aoki, D. Kusnezov, Phys. Lett. **B477**, 348 (2000).
5. K. Aoki, D. Kusnezov, *"On the Violations of Local Equilibrium and Linear Response"* `nlin.CD/0105063`.
6. K. Aoki, D. Kusnezov, Phys. Rev. Lett. **86** (2001) 4029.

7. D.Kusnezov, K. Aoki, *"Statistical Mechanics of Non-Equilibrium Field Theory: From Transport to Phase Transitions"*, in *Dynamics of Gauge Fields*, Frontiers Science Series No. 33, (Universal Academy Press, Tokyo, 2001).
8. K. Aoki, D. Kusnezov, *"Lyapunov Exponents, Transport and the Extensivity of Dimensional Loss"*, nlin.CD/0204015.
9. K. Aoki, D. Kusnezov, Phys. Lett., 250 **A265** (2000).
10. E. M. Lifshits, L.P. Pitaevskii, *Physical Kinetics*, (Pergamon Press, New York, 1981).
11. G. Marcelli, B.D. Todd and R. Sadus, Phys. Rev. **E63** (2001) 021204; J.P.Ryckaert, A.Bellemans, G.Ciccotti, G.V. Paolini, Phys. Rev. Lett. **60**, 128 (1988);
S. Rastogi, N. Wagner and S. Lustig, J. Chem. Phys. **104**, 9234 (1996);
D. Evans and H.J.M. Hanley,Phys. Lett. **80A**, 175 (1980).
12. N.I. Chernov, G.L. Eyink, J.L. Lebowitz and Ya.G. Sinai, Phys. Rev. Lett. **70**, (1993) 2209.
13. M.L. Mehta, *Random Matrices*, (Academic, New York, 1991).
14. T. Guhr, A. Mueller-Groeling and H.A. Weidenmüller, Phys. Rep. **299**, 189 (1998).
15. C.M. Ko, H.J. Pirner and H.A. Weidenmueller, Phys. Lett. **B62**, 248 (1976).
16. H.A. Weidenmüller, in *Progress in Particle and Nuclear Physics* Vol.3 (Pergamon, Oxford, 1980).
17. P.A. Mello, P. Peyrera and N. Kumar, J. Stat. Phys. **51**, 77 (1988); P. Pereyra, J. Stat. Phys. **65**, 773 (1990).
18. A. Bulgac. G. Do Dang and D. Kusnezov, Phys. Rev. **E54**, 3468 (1996).
19. E. Lutz and H.A. Weidenmüller, Physica **A 267**, 354 (1999).
20. D. Kusnezov, A. Bulgac and G. DoDang, Phys. Rev. Lett. **82** (1999) 1136;
D. Kusnezov, Czech. J. Phys. **49** (1999) pp.35-87;
A. Bulgac, G. DoDang, D. Kusnezov, Phys. Rev. **E58** 196 (1998);
D. Kusnezov, A.Bulgac, G. DoDang, Phys. Lett. **234A** (1997) 103.
21. H. Carmichael, *An Open System Approach to Quantum Optics*, Lectures Notes in Physics m18, (Springer, Berlin, 1993).
22. E. Lutz, Europhys. Lett. **54**, 293 (2001).
23. A.O. Caldeira and A.J. Leggett, Physica **A121**, 587 (1983).
24. U. Weiss, *Quantum Dissipative Systems*, (World Scientific, Singapore, 1999).
25. E. Lutz, Phys. Rev. **E64**, 51106 (2001).
26. A.I Saichev, G.M. Zaslavsky, Chaos **7**, 753 (1997);
S.G Samko, A.A. Kilbas, and O.I. Marichev, *Fractional Integrals and Derivatives, Theory and Applications* (Gordon and Breach, Amsterdam, 1993)
27. A. Erdélyi, *Higher Transcendantal Functions. Vol. 3.* (McGraw–Hill, New York, 1955).
28. Z. Rieder, J.L. Lebowitz and E. Lieb, J. Math. Phys. **8**, 1073 (1967).
29. U. Zürcher and P. Talkner, Phys. Rev. **A42**, 3267 (1990), 3278 (1990).

Chapter II

Dynamics of Relaxation and Chaotic Behaviour

Dynamical Theory of Relaxation in Classical and Quantum Systems

P. Gaspard

Center for Nonlinear Phenomena and Complex Systems,
Faculté des Sciences, Université Libre de Bruxelles,
Campus Plaine, Code Postal 231, B-1050 Brussels, Belgium

Abstract. An overview is given of recent work on the dynamical foundations of relaxation toward the thermodynamic equilibrium. The hydrodynamic modes play a central role in the relaxation. Recent advances show that exponentially decaying modes can be explicitly constructed at the microscopic level of description thanks to the concept of Pollicott-Ruelle resonance if the dynamics is hyperbolic. This construction is developed in detail for the irreversible processes of diffusion in the Lorentz gases and the multibaker models. These diffusion models are shown to obey the second law of thermodynamics according to which entropy production corresponds to the dissipation of energy.

1 Introduction

The advances in our knowledge of classical and quantum dynamics shed new lights on our understanding of the time evolution of the systems of interacting particles which compose matter. In particular, the discovery of chaotic dynamical systems has shown that a random time evolution is not incompatible with a deterministic law as given by Newton's equations. This remarkable result has suggested that many random processes in Nature may be governed by an underlying deterministic dynamics if this latter is chaotic.

In this perspective, the chaotic hypothesis appears as a principle of order. Indeed, a chaotic system naturally develops a milder dynamical randomness than the stochastic systems commonly used to model random processes. If the stochastic models are valid on relatively large spatial and temporal scales their validity becomes questionable on smaller scales where the deterministic dynamical rule manifests itself. In this regard, the chaotic hypothesis brings a missing link between the dynamic and the stochastic descriptions. Mathematical tools have been developed in order to characterize the dynamical randomness as well as the dynamical instability such as the Kolmogorov-Sinai (KS) and the ε-entropies per unit time and the Lyapunov exponents [1–3]. These characteristic quantities of chaos allow us to classify the different models of time evolution, in particular, in statistical mechanics [4].

Furthermore, the long-time relaxation to an asymptotic invariant statistical state can nowadays be characterized at the microscopic level of description in many systems of relevance to statistical mechanics. The long-time relaxation can already be characterized by the concept of mixing, which distinguishes between nondecaying and decaying time correlation functions, albeit ergodicity questions the unicity of the asymptotic invariant state [5]. The mixing property guarantees the statistical independency between two events separated by a long-time interval.[1] This statistical decorrelation turns out to play a fundamental role in nonequilibrium statistical mechanics because the transport coefficients are directly given in terms of the time correlation functions of the microscopic currents according to the Green-Kubo formulas [6,7].

Moreover, the new concept of Pollicott-Ruelle resonance allows us to define at the microscopic level the exponential decay of statistical ensembles in systems of hyperbolic type [8–14] as well as in nonhyperbolic systems [15,16]. It is at this place that a close connection appears with the macroscopic theory of transport and other irreversible processes where exponential decay are very common. Indeed, thanks to the Pollicott-Ruelle resonances, the exponential relaxation to the thermodynamic equilibrium can today be explained in great detail at the microscopic level of description in the phase space of spatially extended systems sustaining irreversible transport and reactive processes at their macroscopic level. In these systems, the relaxation to the thermodynamic equilibrium corresponds to the dissipation of work.

Besides, the Pollicott-Ruelle resonances also appear in the escape-rate theory of transport [17–19]. In this theory, the system is set out of equilibrium by absorbing boundary conditions. Accordingly, trajectories escape out of the system and the decay is characterized by an escape rate which is the leading Pollicott-Ruelle resonance of the system. The escape rate turns out to be given in terms of the positive Lyapunov exponents and the KS entropy, which establishes a relationship between transport and chaotic properties.

The Pollicott-Ruelle resonances have also been shown to be of importance to understand the relaxation in wave and quantum systems [20–25]. In this context, a new regime of semiclassical relaxation, referred to in some papers as the "Lyapunov decay regime", has been identified beside the previously known deep quantum "golden rule decay regime" [26–28].

The purpose of these notes is to explain in detail the physical and mathematical bases of the connections between the macroscopic irreversible properties of relaxation and the underlying laws of time evolution of the microscopic dynamics and their characteristic features. These notes are organized in two different parts. A first part composed of Secs. 2 to 4 deals with the classical systems, in the context of which the characteristic quantities of chaos as well as the Pollicott-Ruelle resonances have been developed. Section 2 describes the macroscopic phenomenology of the approach to the thermodynamic equilibrium with the hydrodynamic modes. Section 3 summarizes the tools of dynamical systems theory which are important to analyze and characterize the microscopic dynamics. The microscopic construction of the hydrodynamic modes and nonequilibrium steady states is overviewed in Sec. 4. The connection to thermodynamics is explained in Subsec. 4.6 for the processes of diffusion in the Lorentz gases, in which relaxation

[1] The mixing is the temporal analog of the clustering property which assumes the statistical independency of two spatially separated random events or variables.

by diffusion is shown to correspond to the dissipation of work. A second part composed of Sec. 5 is concerned by the possible extension of these considerations to the understanding of relaxation in quantum systems.

2 Macroscopic Description: The Hydrodynamic Modes

Let us start from the phenomenological description of macroscopic systems in order to explain how the concept of hydrodynamic mode is central to our understanding of the approach to the thermodynamic equilibrium.

The time evolution of macroscopic systems is described by the continuous-media partial differential equations such as the heat and diffusion equations, the Navier-Stokes equations, as well as the reaction-rate equations in chemical kinetics. The macroscopic equations are established starting from the laws of local conservation of mass, energy, linear momentum, and also angular momentum. Closed equations for the mean local variables are obtained by using phenomenological relations: firstly, for the equilibrium equations of states and, secondly, for the nonequilibrium relations between the currents and the gradients (or thermodynamic forces). The former phenomenological relations obey the laws of equilibrium thermodynamics while the latters obey those of irreversible thermodynamics and, especially, the second law of thermodynamics which requires the transport and reaction-rate coefficients of proportionality between the currents and the thermodynamic forces to be positive [29,30].

The so-established macroscopic equations describe the large-scale properties of systems such as fluids, solids, reactive media, plasmas, etc... under equilibrium or nonequilibrium constraints. The constraints are imposed on the boundaries of the system and they determine whether the system will evolve to the state of thermodynamic equilibrium or not. An example of a system under equilibrium constraints is a glass of water which has been initially stirred. In this example, the fluid is transiently out of equilibrium but will eventually reach a state of macroscopic rest since the velocity field vanishes after a long enough time by the effect of viscosity. The description of this approach to the equilibrium is macroscopically described by linearizing the Navier-Stokes equations. The linearized Navier-Stokes equations must be solved by using the boundary conditions that the velocity field vanishes at the interface between the fluid and the glass. These boundary conditions select different modes of exponential decay known as the hydrodynamic modes. It is a very general result that the macroscopic equations predict such an approach to the equilibrium rest state by a linear combination of exponentially decaying modes.

This property also holds for spatially periodic modes in translationally invariant macroscopic systems. These spatially periodic modes are characterized by a wavelength λ or, equivalently, by a vector $\boldsymbol{k}$ of wavenumber $k = 2\pi/\lambda$ giving the wavelength in terms of its magnitude and the direction of the spatial periodic modulation by its orientation $\boldsymbol{k}/k$. These modes are the solutions of the linearized macroscopic equations behaving as

$$\exp\left(s_{\boldsymbol{k}}\, t + \mathrm{i}\, \boldsymbol{k}\cdot\boldsymbol{r}\right) , \tag{1}$$

and they can be considered as the Fourier modes introduced by the spatial Fourier transform of the local variables describing the system. The time evolution of each mode is determined by the so-called *dispersion relation*

$$s_{\boldsymbol{k}} = \mathrm{Re}\, s_{\boldsymbol{k}} + \mathrm{i}\, \mathrm{Im}\, s_{\boldsymbol{k}} \,. \tag{2}$$

The real part of $s_{\boldsymbol{k}}$ gives the decay rate of the modes and $\mathrm{Re}\, s_{\boldsymbol{k}} \leq 0$ because the system is expected to reach equilibrium for $t \to +\infty$. The imaginary part of $s_{\boldsymbol{k}}$ gives the frequency of oscillation of the modes. This frequency is nonvanishing for the propagative modes such as the sound modes but vanishes for the purely dissipative modes such as the thermal and diffusive modes.

It should be emphasized that the existence of spatially periodic and exponentially decaying solutions of the macroscopic equations does not mean that all the solution are exponentially decaying. Indeed, it is well-known that linear combinations of slower and slower exponential decays can lead to power-law decay. An example is given by the diffusion equation for which the dispersion relation is given by $s_{\boldsymbol{k}} = -\mathcal{D}k^2$. In this case, the integration of the Fourier modes (1) over vanishing wavenumbers $\boldsymbol{k}$ leads to the power-law decay $t^{-d/2}$ for $t \to \infty$, where d is the spatial dimension.

The modes can be distinguished by their behavior at large wavelength. The modes for which the rate $s_{\boldsymbol{k}}$ vanishes as the wavelength becomes infinite, $\lim_{\boldsymbol{k}\to 0} s_{\boldsymbol{k}} = 0$, are called the *hydrodynamic modes*. This is the case for the modes associated with the conserved quantities such as mass, energy, and momentum. Each globally conserved quantity is defined as the integral over the whole system of some combination of the local variables describing the system. Accordingly, the conserved quantity is given in terms of the spatial Fourier modes at zero wavenumber. Whereupon the time evolution of the modes associated with the conserved quantities must be frozen in the infinite wavelength limit so that $s_{\boldsymbol{k}=0} = 0$ for these modes.

An example of macroscopic systems illustrating this situation is a normal fluid composed of spherical particles carrying no angular momentum.[2] The five locally conserved quantities which are mass, energy, and the three components of linear momentum give five corresponding hydrodynamic modes [32–35]

- Two longitudinal sound modes which are propagative;
- Two shear modes which are purely dissipative;
- One thermal mode which is purely dissipative.

Their dispersion relation is schematically depicted in Fig. 1.

It should be noticed that kinetic modes may exist beside the hydrodynamic modes. The kinetic modes have the property that their damping rate does not vanish for $\boldsymbol{k} = 0$ so that they cannot be associated with conserved quantities.

[2] Here we ignore the hydrodynamic modes associated with the local conservation of angular momentum. If the particles are assumed to be spherical their instrinsic angular momentum can be neglected and the modes of angular momentum derive from those of linear momentum [31].

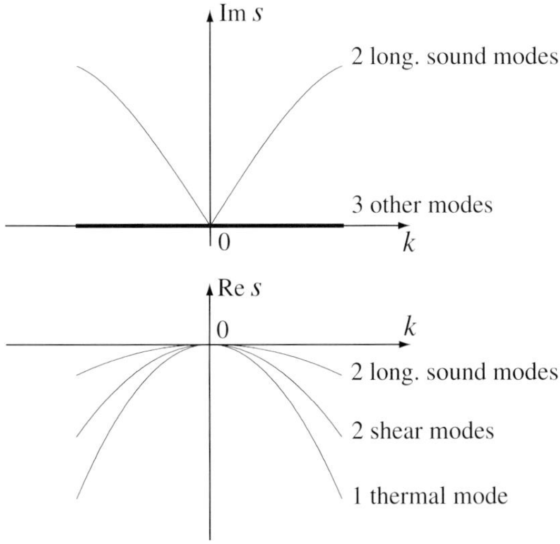

Fig. 1. Schematic dispersion relations of the five hydrodynamic modes of a normal fluid.

Usually, the kinetic modes are not easily identified at the macroscopic level of description but at the kinetic level by using for instance the Boltzmann equation in the case of dilute gases. The spectrum of kinetic modes depends on the interaction between the particles (as well as on the internal dynamics of the colliding particles such as rotation and vibration in the case of molecules). For a gas of spherical particles without internal structure, the kinetic modes have typically a damping rate of the order of the collision frequency. It is important to notice that the dynamics of relaxation to a local Maxwell-Boltzmann velocity distribution is controlled by the kinetic modes. Instead, the hydrodynamic modes rule the long-time evolution of the local macroscopic variables which are the mean local fluid velocity, density and temperature, the velocity distribution having locally reached its equilibrium Maxwell-Boltzmann distribution after a few intercollisional times. In the case of dilute gases, the spectrum of hydrodynamic and kinetic modes is obtained by linearizing the Boltzmann equation and finding the eigenvalues of the collision operator [31,34,36]. At very short wavelength of the order of the distance between the particles in dense fluids, the propagative hydrodynamic modes may present propagation gaps where the oscillation frequency vanishes, $\mathrm{Im}\, s_{\boldsymbol{k}} = 0$, as shown in Fig. 2 [37].

In systems with a mixture of several species of particles further hydrodynamic modes exist which correspond to the local conservation of the different species. If the fluid is composed of n species there are $(n-1)$ such modes of mutual diffusion [38].

If the different species undergo chemical reactions they are no longer conserved so that some of the hydrodynamic modes become kinetic modes with a nonvanishing decay rate at zero wavenumber. To illustrate this situation, we

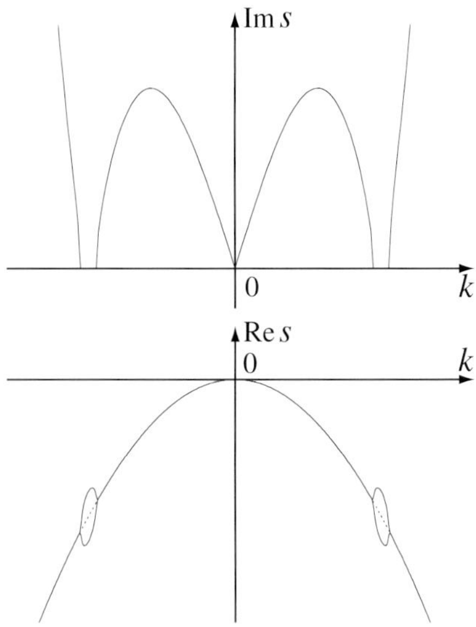

Fig. 2. Example of propagation gap in the dispersion relation of a sound mode.

may consider the reaction of isomerization, $A \leftrightarrow B$, between two species A and B. The particles A and B are supposed to diffuse in a solvent of inert particles, the whole fluid is at rest. This system is a ternary mixture in which we expect the five standard hydrodynamic modes plus two modes of mutual diffusion in absence of reaction. If the fluid is at rest and the reaction isothermal, the system can be modeled by two coupled diffusion-reaction equations [39–42]

$$\partial_t \rho_A = \mathcal{D}_{AA} \nabla^2 \rho_A + \mathcal{D}_{AB} \nabla^2 \rho_B - \kappa_+ \rho_A + \kappa_- \rho_B \,, \tag{3}$$

$$\partial_t \rho_B = \mathcal{D}_{BA} \nabla^2 \rho_A + \mathcal{D}_{BB} \nabla^2 \rho_B + \kappa_+ \rho_A - \kappa_- \rho_B \,. \tag{4}$$

Supposing solutions of the form (1) for Eqs. (3)-(4), we obtain two modes: a diffusive mode and a reactive mode. The diffusive mode has the dispersion relation:

$$s_{\boldsymbol{k}} = -\mathcal{D}\, k^2 + O(k^4) \,, \tag{5}$$

with the diffusion coefficient

$$\mathcal{D} = \frac{\kappa_+ (\mathcal{D}_{BB} + \mathcal{D}_{AB}) + \kappa_- (\mathcal{D}_{AA} + \mathcal{D}_{BA})}{\kappa_+ + \kappa_-} \,. \tag{6}$$

The decay rate of this mode vanishes at $\boldsymbol{k} = 0$ so that it corresponds to the hydrodynamic mode of conservation of total mass of species A and B. The reactive mode has the dispersion relation:

$$s_{\boldsymbol{k}} = -\kappa_+ - \kappa_- - \mathcal{D}^{(\mathrm{r})}\, k^2 + O(k^4) \,, \tag{7}$$

with the reaction rate $-s_0 = \kappa_+ + \kappa_-$ and the reactive diffusion coefficient

$$\mathcal{D}^{(\mathrm{r})} = \frac{\kappa_+ (\mathcal{D}_{AA} - \mathcal{D}_{AB}) + \kappa_- (\mathcal{D}_{BB} - \mathcal{D}_{BA})}{\kappa_+ + \kappa_-} \,. \tag{8}$$

Both dispersion relations are depicted in Fig. 3. We notice that the reactive mode becomes diffusive again if the reaction rates vanish, $\kappa_+ = \kappa_- = 0$, in which case the ternary mixture has two diffusive modes beside its five standard hydrodynamic modes, as expected. Accordingly, we conclude that some hydrodynamic diffusive modes turn into kinetic modes with $s_0 \neq 0$ in reacting systems. Very often these reactive modes are observed at the macroscopic level because their decay rates are of the order of the inverse of reaction rates in the range of seconds, minutes, or more. This is the case for the typical reactions studied in chemical kinetics at the beginning of the XXth century [43]. In the second half of the XXth century, faster and faster chemical reactions have been studied bringing the inverse reaction rates down to the time scale of the picoseconds and below [44]. Such ultrafast reactions evolve on the picosecond time scale of the kinetic modes of inert fluids and even on faster time scales. It is therefore appropriate to consider the reactive modes as some kind of kinetic modes.

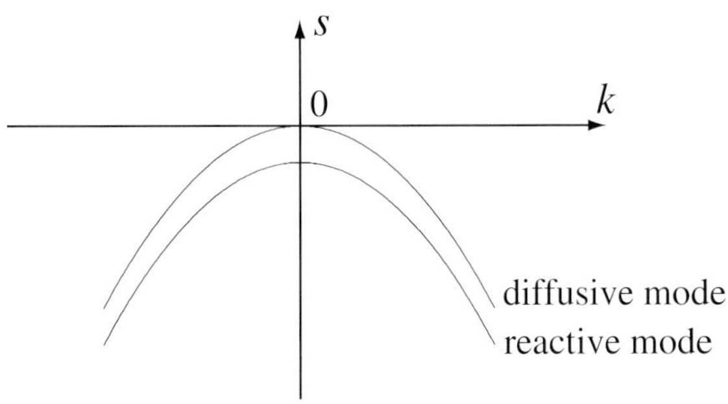

Fig. 3. Schematic dispersion relations of the diffusive and reactive modes of a reaction of isomerization $A \leftrightarrow B$ taking place in a fluid solvent at rest.

Hydrodynamic modes, i.e., modes with a vanishing dispersion relation at infinite wavelength, also arises in systems undergoing the spontaneous breaking of a continuous symmetry. These hydrodynamic modes are often referred to as Goldstone modes [45]. For instance, a breaking of continuous symmetry occurs in the fluid-solid transition at which the system looses its continuous translational symmetry in the three spatial directions. Accordingly, solids have three extra hydrodynamic modes beside those associated with the five fundamental conserved quantities which are the three components of linear momentum, energy, and mass. Solids have therefore eight hydrodynamic modes which are known to be [46]

- Two longitudinal sound modes which are propagative;
- Four transverse sound modes which are propagative;
- One thermal mode which is purely dissipative;
- One mode of vacancy diffusion which is purely dissipative.

Another example of continuous symmetry breaking occurs in the fluid-superfluid transition. In the superfluid phase, the superfluid coexists with the normal fluid and is characterized by an order parameter which breaks a continuous gauge symmetry. Since the order parameter is a scalar one Goldstone modes adds to the five hydrodynamic modes of a normal fluid, leading to a total of the six following hydrodynamic modes [47]

- Two first sound modes which are propagative;
- Two second sound modes which are propagative;
- Two shear modes which are purely dissipative.

Their dispersion relation is schematically depicted in Fig. 4.

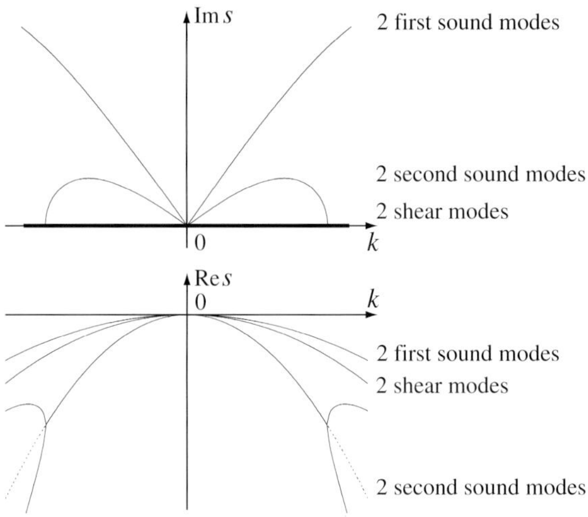

Fig. 4. Schematic dispersion relations of the six hydrodynamic modes of a superfluid.

These different examples show the fundamental importance of the hydrodynamic modes for the understanding of the time evolution of different materials and, especially, the approach to the equilibrium rest state in each system. The existence of these hydrodynamic modes emphasize the importance of the exponential decay to the thermodynamic equilibrium. The exponential decay is one of the most compeling expression of the irreversibility of the approach to the thermodynamic equilibrium. We may wonder how such exponential decay are compatible with the underlying microscopic dynamics which is known to be Hamiltonian and time-reversal symmetric. This question brings us to the study of the microscopic dynamics.

3 Generalities About the Microscopic Dynamics

All the systems of Nature are observed to be of Hamiltonian type at their microscopic level. It turns out that the observation of nonHamiltonian behavior has always been found to have its origin in the existence of previously ignored degrees of freedom and a Hamiltonian description has always been recovered by including the neglected degrees of freedom. We should probably understand the importance of the Hamiltonian framework as due to the fact that time translations are naturally generated by the energy operator in quantum mechanics.

The Hamiltonian of electromagnetic interaction, which rules the low-energy macroscopic world, is known to be time-reversal symmetric whereupon the motion of atoms and molecules is assumed to be ruled by a time-reversal symmetric Hamiltonian. Fluids are very well described by the Hamiltonian of atoms or molecules interacting through the van der Waals interactions[3]

$$H = \sum_{i=1}^{N} \frac{\boldsymbol{p}_i^{\,2}}{2m} + \frac{1}{2} \sum_{i \neq j} V(\|\boldsymbol{r}_i - \boldsymbol{r}_j\|) , \tag{9}$$

where $V(r)$ is the interaction potential. This system has $f = Nd$ degrees of freedom if the physical space has dimension d. The time evolution is time-reversal symmetric because

$$H(\{\boldsymbol{r}_i, -\boldsymbol{p}_i\}) = H(\{\boldsymbol{r}_i, \boldsymbol{p}_i\}) . \tag{10}$$

At room temperature, the de Broglie wavelength of the particles is much shorter than their mean free path so that classical mechanics provides an excellent description according to Hamilton's equations:

$$\begin{cases} \dot{\boldsymbol{r}}_i = +\dfrac{\partial H}{\partial \boldsymbol{p}_i} , \\[2ex] \dot{\boldsymbol{p}}_i = -\dfrac{\partial H}{\partial \boldsymbol{r}_i} , \end{cases} \tag{11}$$

with $i = 1, 2, ..., N$.

We notice that the Hamiltonian assumption is stronger than the fulfilment of the conservation of energy since it imposes a symplectic structure to the microscopic time evolution. One of the consequences of this symplectic structure is Liouville's theorem that volumes $d^f r d^f p$ are preserved in phase space [48].

The time integration of Hamilton's equations uniquely determines the trajectories $\Gamma_t = \{\boldsymbol{r}_i(t), \boldsymbol{p}_i(t)\}$ in terms of their initial conditions $\Gamma_0 = \{\boldsymbol{r}_i(0), \boldsymbol{p}_i(0)\}$. We denote by $\Gamma_t = \Phi^t \Gamma_0$ the *flow* induced in phase space by Hamilton's equations.

3.1 The Poincaré Map

A very powerful tool in the analysis of continuous-time dynamical systems is provided by the *Poincaré map* [3,30]. The idea is to intersect the trajectories of the flow Φ^t by a surface of section introduced in phase space. Each trajectory will thus generate a sequence of intersection points $\{\Gamma_n\}$. Since the system is

[3] The internal rotation and vibration of the molecules can also be included in the Hamiltonian.

deterministic the knowledge of an intersection point determines the next one. Accordingly, the flow induces the Poincaré map:

$$\begin{cases} \Gamma_{n+1} = \Phi^{T(\Gamma_n)}(\Gamma_n) \equiv \phi(\Gamma_n)\,, \\ \\ t_{n+1} = t_n + T(\Gamma_n)\,, \end{cases} \tag{12}$$

where $\{t_n\}$ denote the intersection times which are controled by the first-return time function $T(\Gamma)$ [3]. The Poincaré map provides a description of the system which is strictly equivalent to the flow, given that each segment of trajectory between two successive intersections can be reconstructed.

An example of Poincaré map is the Birkhoff map in billiards where the trajectory moves in free flight between the collisions [3]. The Birkhoff map is the mapping from each collision to the next. In two-dimensional billiards, each collision can be represented by the so-called Birkhoff coordinates: the arc of perimeter $r_\parallel$ of the impact point and the momentum $p_\parallel$ tangent to the border of the billiard at the impact. The free flight from one collision to the next induces a mapping of Birkhoff's coordinates, $(r_{\parallel,n+1}, p_{\parallel,n+1}) = \phi(r_{\parallel,n}, p_{\parallel,n})$, which is known to be area preserving. In this way, the continuous-time dynamics of a billiard such as the two-dimensional Lorentz gas can be described by an area-preserving mapping, which can greatly simplify the analysis while keeping the strict equivalence with the continuous-time dynamics [3].

3.2 Statistical Ensembles and Their Time Evolution

The Newtonian scheme does not fix the initial conditions to be used for Newton's equations. In many experiments, the initial conditions are fixed by the preparation of the system under study. For instance, in artillery, the initial condition of the cannon-ball is determined by the cannon, which is a mechanical device *external* to the system determining the parabolic trajectory of the cannon-ball. It is in general impossible to achieve the perfect reproducibility of the initial condition from one experiment to the next. Therefore, the initial conditions are always submitted to some uncertainty, which can be modeled by repeating many times the experiment and establishing the statistics of the initial conditions. The statistical description defines the probability density of initial conditions as

$$f_0(\Gamma) = \lim_{M\to\infty} \frac{1}{M} \sum_{m=1}^{M} \delta(\Gamma - \Gamma_0^{(m)})\,, \tag{13}$$

where M is the number of performed experiments with the idea that this number should be large enough in order for the initial density $f_0(\Gamma)$ to be represented by a smooth function of Γ.

In laboratory experiments on matter, the initial conditions are also submitted to uncertainty. Often, the experiment starts at a given initial time when the system is in a state of stationarity for as many quantities as can possibly be monitored by measuring devices. Since the resolution of the measuring devices is always finite the initial conditions are not reproducible and will vary from one experiment to the next. This nonreproducibility requires the use of statistical considerations and the description of the initial conditions by a probability density such as (13).

From the initial condition, the system is supposed to be left to itself so that the trajectory evolves according to Hamilton's equations in each repetition of the experiment. Because of the sensitivity to initial conditions which is the feature of chaotic systems, the prediction of the time evolution of the individual trajectories is vain beyond the time of divergence of trajectories from nearby initial conditions [30]. Instead, predictions remain fruitful for statistical quantities such as averages or correlation functions which are described in terms of the probability density of the trajectories defined as

$$f_t(\Gamma) = \lim_{M\to\infty} \frac{1}{M} \sum_{m=1}^{M} \delta(\Gamma - \Gamma_t^{(m)}) \,. \tag{14}$$

The time evolution of this probability density is induced by Hamilton's equations and is known to be ruled by the Liouville equation

$$\partial_t f = \hat{L} f \equiv -\mathrm{div}\left(\dot{\boldsymbol{\Gamma}} f\right) = \{H, f\} \,, \tag{15}$$

where $\{\cdot,\cdot\}$ denotes the Poisson bracket of classical mechanics. The time integral of Liouville's equation defines the Frobenius-Perron operator which is given by

$$f_t(\Gamma) = \hat{P}^t f_0(\Gamma) = \mathrm{e}^{\hat{L}t} f_0(\Gamma) = f_0(\Phi^{-t}\Gamma) \,, \tag{16}$$

for time-independent systems. The last equality is a consequence of Liouville's theorem. The probability density can then be used to calculate the statistical averages of the physical observables:

$$\langle A \rangle_t \ = \ \int A(\Gamma)\, f_t(\Gamma)\, d\Gamma = \langle A|\hat{P}^t|f_0\rangle \,. \tag{17}$$

Very often, the system under study is not isolated but in contact with other systems through walls. This is the case for a fluid in a container. After a time interval, the bulk of the fluid becomes determined by the state of the walls of the container and, in particular, by their temperature. At the wall, the atoms of the fluid interact with the atoms of the wall via short-ranged interactions. The description of this interaction implies to suppose that the whole system composed of the fluid plus its surrounding is Hamiltonian. If the description is reduced to the sole degrees of freedom of the fluid, the equations of motion typically become Langevin stochatic differential equations of the form of Hamilton's equations plus extra forces describing the stochastic forcing of the fluid degrees of freedom by the surrounding and the associated dissipation from the fluid to the surrounding [49].

If the temperature of the surrounding is uniform and if the system does not sustain a net flux of matter, the system is expected to reach a state of thermodynamic equilibrium for typical statistical ensembles of initial conditions. This is the case for the initially stirred glass of water described here above. The description of such an experiment of approach to the thermodynamic equilibrium is most often simplified by ignoring the stochastic forcing at the wall and supposing that the fluid particles undergo elastic collisions on the wall of the container. In this way, the boundary stochastic forcing is replaced by *reflecting boundary conditions* which require that the probability current ingoing the wall exactly compensate the outgoing current so that no particle remains stuck at the wall. For reflecting boundary conditions, the system is typically expected to evolve toward a microcanonical statistical state of thermodynamic equilibrium at the total energy of the initial conditions. The convergence to the microcanonical state should be understood as a weak convergence holding only for statistical averages such as (17).

For the purpose of computing the viscosity or other quantities of interest by molecular dynamics, *periodic boundary conditions* are often used to simplify the dynamics. These boundary conditions are implemented by supposing that the system is composed of N particles moving on a torus obtained by identifying two-by-two the opposite boundaries of a cell tessellating the physical space. For periodic boundary conditions, the system is typically expected to evolve toward a microcanonical statistical state of thermodynamic equilibrium at the total energy and linear momentum of the initial conditions.

In many systems of interest, the surrounding is not in equilibrium, for instance, because different walls are at different temperatures or, more generally, because there is an exchange of matter under the form of a flux of particles from one reservoir to another. Typical experiments in which such fluxes are essential are the laminar or turbulent flows in pipes or around an object as studied in fluid mechanics, or flows of chemical species in a continuously stirred tank reactor as used in nonequilibrium chemical kinetics [30]. Another example is given by the study of diffusion of tracer particles through slabs of fluids or even solids sandwitched between two reservoirs [50]. In these different systems, a fictitious boundary delimits an open system through which matter is exchanged. The number of particles inside the system fluctuates so that the state of the system should be described by a sort of classical analog of the Fock space introduced in the study of many-body quantum systems. Indeed, the system may contain a varying number of particles and, for each number N of particles, the system is described by a probability density such as (14):

$$\boldsymbol{f} = \left\{ f^{(0)}, f^{(1)}(\boldsymbol{r}_1, \boldsymbol{p}_1), f^{(2)}(\boldsymbol{r}_1, \boldsymbol{p}_1, \boldsymbol{r}_2, \boldsymbol{p}_2), ..., f^{(N)}(\boldsymbol{r}_1, \boldsymbol{p}_1, ..., \boldsymbol{r}_N, \boldsymbol{p}_N), ... \right\} . \tag{18}$$

The probability that the system contains N particles is given by

$$P_N = \int f^{(N)}(\boldsymbol{r}_1, \boldsymbol{p}_1, ..., \boldsymbol{r}_N, \boldsymbol{p}_N)\, d\boldsymbol{r}_1\, d\boldsymbol{p}_1 \cdots d\boldsymbol{r}_N\, d\boldsymbol{p}_N , \tag{19}$$

which is normalized by

$$\sum_{N=0}^{\infty} P_N = 1\,. \tag{20}$$

Under similar circumstances as the stochastic forces due to the surrounding can be modeled by reflecting or periodic boundary conditions at equilibrium, we can here also take into account of the nonequilibrium constraints from the surrounding by appropriate boundary conditions which fix the probability currents ingoing the system at its boundaries. The advantage of this description is that the Hamiltonian character of the time evolution is maintained in the whole description. A typical trajectory of the system is a sequence of segments of trajectories of Hamiltonian N-body systems, N being the number of particles in the system during each segment of trajectory. During each segment, all the particles are contained inside the boundaries delimiting the open system and the trajectory belongs to some phase space $\mathcal{M}^{(N)}$ which is limited in position.

Each probability density in Eq. (18) obeys a N-body Liouvillian equation:

$$\partial_t f^{(N)} = \hat{L}^{(N)} f^{(N)} = \{H^{(N)}, f^{(N)}\}\,, \tag{21}$$

with appropriate boundary conditions. The probability that the system contains N particles evolves in time according to

$$\begin{aligned}
\frac{dP_N}{dt} &= \int_{\mathcal{M}^{(N)}} d\Gamma^{(N)} \{H^{(N)}, f^{(N)}\} &(22)\\
&= -\int_{\mathcal{M}^{(N)}} d\Gamma^{(N)} \,\mathrm{div}\left(\dot{\boldsymbol{\Gamma}}^{(N)} f^{(N)}\right) &(23)\\
&= -\int_{\partial\mathcal{M}^{(N)}} d\boldsymbol{\Sigma}^{(N)} \cdot \dot{\boldsymbol{\Gamma}}^{(N)}\, f^{(N)} &(24)\\
&= \underbrace{-\int_{\partial\mathcal{M}^{(N)}_{\mathrm{in}}} d\boldsymbol{\Sigma}^{(N)} \cdot \dot{\boldsymbol{\Gamma}}^{(N)}\, f^{(N)}}_{\geq 0} \underbrace{-\int_{\partial\mathcal{M}^{(N)}_{\mathrm{out}}} d\boldsymbol{\Sigma}^{(N)} \cdot \dot{\boldsymbol{\Gamma}}^{(N)}\, f^{(N)}}_{\leq 0}\,, &(25)
\end{aligned}$$

where the first term of the last equality is fixed by the boundary conditions.

These considerations provide a fairly general framework to describe a large variety of nonequilibrium systems while preserving the Hamiltonian and Liouvillian properties. Open systems exchanging matter with its surrounding can be described by imposing *flux boundary conditions* that the ingoing probability currents at the boundaries be fixed according to the concentrations and velocity distributions of the different species of ingoing particles. We notice that such a framework has already been developed for solving Boltzmann's equation in nonequilibrium systems. Such considerations are here generalized to Liouville's equation. Nonequilibrium systems between heat reservoirs at different temperatures can also be modeled with flux boundary conditions by supposing that the ingoing particles have a Maxwellian velocity distribution at the temperature of the heat reservoir. Since Liouville's theorem is strictly preserved in the present framework we could here speak about *Liouvillian thermostats*. Systems such as a fluid undergoing the Rayleigh-Bénard convection or nonlinear chemical reactions sustaining nonequilibrium oscillations can be described in this framework.

After a long time, the system is supposed to reach a stationary state which is a nonequilibrium steady state if the boundary conditions are of nonequilibrium type. In such a stationary state driven by *flux boundary conditions*, each probability (19) should reach a nonvanishing constant value so that the ingoing and outgoing fluxes of probability given by the two terms of the last line of Eq. (25) balance each other. Such a stationary state should in general be described by a vector of probability densities (18) which is very different from the equilibrium densities. At stationarity, this vector defines the invariant probability measure describing the statistical steady state of the system.

Among the different possible boundary conditions, the *absorbing boundary conditions* play a special role. These boundary conditions may be imposed on a system with a fixed number N of particles by supposing that there is no ingoing flux of probability (or particle). Therefore, the first term in the last line of Eq. (25) vanishes and the probability $P_N(t)$ monotonously decays to zero as time increases. Here, the interesting properties manifest themselves in the way the decay of $P_N(t)$ proceeds. If the decay is asymptotically exponential the decay rate defines the so-called *escape rate* used in the escape-rate theory of transport. Under certain circumstances, an invariant probability measure can still be defined in the presence of absorbing boundary conditions as explained in Subsection 4.3.

3.3 Ergodicity and Mixing

The construction of the invariant measure mentioned in the previous discussion is an important issue in order to understand the long-time approach of a system to a stationary statistical state.

In closed deterministic systems in which the number of particles is conserved, a natural invariant measure is selected by time averaging the quantities of interest over a given trajectory. Birkhoff's ergodic theorem, that

$$\lim_{T\to\infty} \frac{1}{T} \int_0^T A(\Phi^t \Gamma_0)\, dt = \int A(\Gamma)\, \mu(d\Gamma)\,, \tag{26}$$

for μ-almost all initial conditions Γ_0 [5], provides a decomposition of the phase space into ergodic components which are the support of the different invariant measures μ constructed by the time average. A system is commonly said to be ergodic if its smallest ergodic components are the energy shells or the energy-momenta shells. The corresponding invariant measures μ define the statistical ensembles which can be identified with the thermodynamic equilibria.

In open systems with absorbing boundary conditions, a unique ergodic normalizable measure can be defined on the invariant set of trajectories which are forever trapped inside the system under the condition that this set is transitive, i.e., contains a dense trajectory. Otherwise, the invariant set should be decomposed into different ergodic components in a similar way as for closed systems.

In open systems with flux boundary conditions, ergodic invariant measures can be defined by reference to the random process induced by the dynamics and the boundary conditions.

In closed or open systems, the relaxation toward a supposed unique invariant measure μ is guaranteed by the property of *mixing* which has been introduced by Gibbs [51] and which is defined by the condition that [5]

$$\lim_{t\to\infty} \mu(\Phi^{-t}\mathcal{A}\cap\mathcal{B}) = \mu(\mathcal{A})\,\mu(\mathcal{B})\,, \tag{27}$$

where $\mathcal{A}$ and $\mathcal{B}$ are two sets in phase space. The mixing property guarantees the statistical independency between the random events that the trajectory be in the set $\mathcal{B}$ at time $t = 0$ and in the set $\mathcal{A}$ at time t, in the long-time limit $t \to \infty$. The definition (27) implies that the time correlation functions defined by statistical average over the invariant measure μ decays to their stationary values as

$$\lim_{t\to\infty} \langle A(\Phi^t\Gamma)\,B(\Gamma)\rangle_\mu = \langle A\rangle_\mu\,\langle B\rangle_\mu\,, \tag{28}$$

where $\langle X\rangle_\mu = \int X(\Gamma)\,\mu(d\Gamma)$.

3.4 Dynamical Instability

Besides the mere knowledge of the trajectories it is often important to characterize their stability and, in particular, their linear stability which is the behavior of infinitesimal perturbations $\delta\boldsymbol{\Gamma}_t = \{\delta\boldsymbol{r}_i(t), \delta\boldsymbol{p}_i(t)\}$ on each trajectory. These perturbations evolves in time under the linearized Hamiltonian equations:

$$\begin{cases} \delta\dot{\boldsymbol{r}}_i = +\sum_{j=1}^N \left(\dfrac{\partial^2 H}{\partial\boldsymbol{p}_i\partial\boldsymbol{r}_j}\cdot\delta\boldsymbol{r}_j + \dfrac{\partial^2 H}{\partial\boldsymbol{p}_i\partial\boldsymbol{p}_j}\cdot\delta\boldsymbol{p}_j\right)\,, \\[2ex] \delta\dot{\boldsymbol{p}}_i = -\sum_{j=1}^N \left(\dfrac{\partial^2 H}{\partial\boldsymbol{r}_i\partial\boldsymbol{r}_j}\cdot\delta\boldsymbol{r}_j + \dfrac{\partial^2 H}{\partial\boldsymbol{r}_i\partial\boldsymbol{p}_j}\cdot\delta\boldsymbol{p}_j\right)\,, \end{cases} \tag{29}$$

with $i = 1, 2, ..., N$. The infinitesimal perturbations may remain bounded or grow as a power of time, or even exponentially. The exponential growth is the fastest possible growth in a system described by ordinary differential equations. This form of dynamical instability is characterized by the so-called *Lyapunov exponents* [1]

$$\lambda = \lim_{t\to\infty}\frac{1}{t}\ln\frac{\|\delta\boldsymbol{\Gamma}_t\|}{\|\delta\boldsymbol{\Gamma}_0\|}\,, \tag{30}$$

in closed systems with a constant number N of particles. We may have as many different Lyapunov exponents as there are different directions in phase space so that the total number of Lyapunov exponents is equal to the phase-space dimension $2f = 2Nd$. The Lyapunov exponents may vary from one trajectory to another.

The Hamiltonian character implies a pairing rule among the different Lyapunov exponents due to the symplectic property: To each Lyapunov exponent λ_i, there corresponds another one which is $-\lambda_i$ [3].

Several Lyapunov exponents vanish which are associated with the conserved quantities such as energy or linear momentum when they exist. The pairing rule implies that, to each vanishing Lyapunov exponent associated with a conserved quantity, there corresponds another vanishing Lyapunov exponent. In the case of the vanishing Lyapunov exponent associated with energy, the other vanishing Lyapunov exponent is the one associated with the direction of the flow.[4] In Hamiltonian systems, the number of vanishing Lyapunov exponents is thus always even by the symplectic property.

Systems with positive Lyapunov exponents are dynamically unstable. Typical Hamiltonian systems have dynamically unstable trajectories, in particular, by the formation of homoclinic orbits. However, typical Hamiltonian systems also present quasiperiodic trajectories with all their Lyapunov exponents vanishing. In Hamiltonian systems with two degrees of freedom, these quasiperiodic trajectories exist inside elliptic islands coexisting with zones filled with unstable trajectories. The quasiperiodic trajectories occupy the largest phase-space volume in systems close to integrability according to the KAM theorem, which is the case for systems such as Hamiltonian chains of coupled anharmonic oscillators at very low energy or temperature [52,53]. However, at high energy or temperature, the quasiperiodic trajectories are numerically observed to shrink while the unstable trajectories fill the largest part of phase space. This phenomenon is observed, in particular, in the standard map where the elliptic islands become tiny and practically inobservable at large values of the parameter although they exist for a dense set of parameter values [54].

Systems are said to be *hyperbolic* if the positive Lyapunov exponents associated with the different directions and trajectories of the system remain separated from zero by a gap: $\lambda_i \geq \lambda_{\min} > 0$. However, typical Hamiltonian systems such as the standard map are nonhyperbolic because the quasiperiodic trajectories have vanishing Lyapunov exponents and, furthermore, trajectories close to the quasiperiodic ones may have arbitrarily small positive Lyapunov exponents.

The spectrum of Lyapunov exponents is best known for billiards with convex boundaries such as the Sinai billiard, the Lorentz gas, or the finite hard-sphere gases. In these systems, almost all the trajectories have positive Lyapunov exponents which are well separated from zero. However, there exist sets of zero Lebesgue measure of trajectories with all their Lyapunov exponents vanishing because these trajectories never bounce on a convex boundary. Under certain conditions, these special trajectories do not exist as in the periodic Lorentz gas on a triangular lattice with a finite horizon [3].

Beside the dynamically unstable billiards, we also find the polygonal billiards [55] such as the wind-tree model [56], in which the boundaries are flat so that all the Lyapunov exponents vanish. In polygonal billiards, infinitesimal perturbations grow as a power of time but never exponentially.

Beyond the closed systems, the open systems are of great importance in nonequilibrium conditions and, especially, in scattering experiments [3,20,14]. In

[4] The Lyapunov exponent associated with the direction of the flow is always vanishing in flows without fixed point according to a general argument [3].

scattering systems, the majority of trajectories are coming from and returning to infinity. These systems are typically out of equilibrium with a possible steady state fixed by the velocity distribution and the spatial profile of the beams of incoming particles. An example of scattering systems is given by a particle moving in a one-dimensional potential with a maximum above two valleys of asymptotically free motion. Other examples are the disk scatterers. In these open systems, the trajectories which control the long-time escape are those which are trapped inside the system. The forever trapped trajectories are unstable with positive Lyapunov exponents and form an invariant set which is of zero Lebesgue measure. Open systems can be dynamically unstable without being chaotic. This is the case for the point particle on the maximum of a potential as well as for the particle periodically bouncing on the line joining the centers of two disks in the two-disk scatterer.

For the system to be chaotic, we need more than a positive Lyapunov exponent: we need a random time evolution, the characterization of which requires to introduce another quantity than the Lyapunov exponent.

3.5 Dynamical Randomness

The definition of dynamical randomness requires the introduction of the so-called Kolmogorov-Sinai (KS) entropy per unit time [1,5]. The KS entropy is defined as the supremum of the decay rates of the multiple time probabilities for the trajectory to successively visit different cells in phase space. Let us suppose that the phase space is partitioned into cells forming a partition $\mathcal{P}$. If ω_n denotes the label of the cell visited by the trajectory at time $t = n\Delta t$, the multiple time probabilities would decay exponentially if the motion of the system is random:

$$\mu(\omega_0\omega_1 \cdots \omega_{n-1}) \sim \exp\left(-n \; \Delta t \; h_{\mathcal{P}}\right) \qquad \text{for } n \to \infty \, . \tag{31}$$

The exponential decay here goes with the number n of visited cells and not with the time interval Δt. The rate $h_{\mathcal{P}}$ is the entropy per unit time of the partition $\mathcal{P}$ and the KS entropy is defined as [1,5]

$$h_{\mathrm{KS}} = \mathrm{Sup}_{\mathcal{P}} \; h_{\mathcal{P}} \, . \tag{32}$$

The KS entropy can equivalently be introduced in terms of the Chaitin-Kolmogorov-Solomonoff algorithmic complexity K_t, which is the length of the shortest program running on a universal Turing machine and able to reproduce the trajectory over a time interval t. The length of the program can be measured as the binary length or in the base of natural logarithms as assumed from now on. If the trajectory is periodic, it is enough to store the pattern of one period and the number of periods over the time t so that the length of this program grows as $\ln t$. However, if the trajectory is random there is no other mean than to store the whole segment of trajectory in which case the length of the program grows proportionally to the time t, which is the fastest possible growth. The growth rate of the algorithmic complexity is equal to the KS entropy

$$h_{\mathrm{KS}} = \lim_{t\to\infty} \frac{1}{t} \, K_t \, , \tag{33}$$

for μ-almost all the trajectories [57]. This result shows that the KS entropy per unit time characterizes the dynamical randomness developed by the system during its time evolution.

A system is defined to be *chaotic* if it is deterministic (i.e., governed by a mapping or a differential equation) and its KS entropy per unit time is positive [3].

The KS entropy is related to the sum of positive Lyapunov exponents and the rate of escape γ of trajectories out of an open system according to a generalization of Pesin's theorem [1]

$$h_{\rm KS} = \sum_{\lambda_i > 0} \lambda_i - \gamma \,. \tag{34}$$

In closed systems, the escape rate vanishes $\gamma = 0$ and the KS entropy is given by the sum of positive Lyapunov exponents, which is the content of Pesin's theorem. If there exists at least one positive Lyapunov exponent the KS entropy is also positive and the time evolution is therefore random. In this case, we recover the usual definition of a chaotic system as a deterministic system with at least one positive Lyapunov exponent.

However, if the system is open, the dynamics can be nonchaotic if the escape rate is itself equal to the sum of positive Lyapunov exponents, $\gamma = \sum_{\lambda_i>0} \lambda_i$, because the KS entropy vanishes in this case. This situation happens in the system composed of a particle moving in a potential with a maximum, as well as in the two-disk scatterer where there is only one unstable periodic orbit which remains trapped at finite distance.

We notice that there exists the so-called *sporadic* systems which are intermediate between the periodic and the chaotic ones [3,58]. In these systems, the trajectories have an algorithmic complexity which grows faster than the logarithm $(\ln t)$ but slower than the time t, for instance as $t/(\ln t)$ or t^ν with $0 < \nu < 1$. Their KS entropy per unit time vanishes because the algorithmic complexity is not extensive in time. Nevertheless, the sporadic systems still develop a dynamical randomness but slower than in chaotic systems.

For deterministic systems with finitely many particles, the KS entropy is typically finite. However, the KS entropy may become infinite for systems with infinitely many particles. The extensivity of the entropy per unit time with the number of particles is an important issue in statistical mechanics. We can here define an entropy per unit time and per unit volume, which is positive if randomness is generated not only in time but also in space. If the particles do not interact as in an ideal gas or in a harmonic solid, the entropy per unit time and unit volume vanishes. On the other hand, this space-time entropy is positive in systems of interacting particles such as the hard-sphere gas [3]. The space-time entropy is also positive in probabilistic cellular automata [59–61], which therefore belong to the same class of random processes as the chaotic ones.

If the system is described as a random process generating continuously distributed variables, the KS entropy becomes infinite and no longer provides a useful characterization of the randomness. In this case, we should keep the notion of entropy per unit time associated with a partition into cells of diameter

ε, defining in this way an ε-entropy per unit time [4]. Processes in which a finite number of outcomes of continuously distributed variables happens per unit time typically have their ε-entropy per unit time diverging as $\ln(1/\varepsilon)$ for $\varepsilon \to 0$. An example is the random processes described by the linearized Boltzmann equation which has an ε-entropy per unit time and unit volume diverging as $\ln(1/\varepsilon)$ for $\varepsilon \to 0$, although the deterministic Hamiltonian equations generate a random process with a finite entropy per unit time and unit volume [3,4]. On the other hand, Langevin processes generate continuously distributed variables continuously in time, leading to an ε-entropy typically diverging as ε^{-2} for $\varepsilon \to 0$ [3,4]. Langevin processes are thus qualitatively more random than the aforementioned processes.

The ε-entropy per unit time depends on the kind of boundary conditions which are considered. In finite systems with reflecting or periodic boundary conditions, the determinism is preserved throughout the whole time evolution so that the dynamical randomness is characterized by a finite KS entropy per unit time. If the system is in contact with reservoirs, the interaction of the particles with the walls is in general stochastic and described by a nondeterministic Langevin process in a thin boundary layer near the wall of the reservoir [49]. In this case, the process is characterized by an ε-entropy per unit time which can be estimated as $Vh^{(\text{time,volume})} + cA\varepsilon^{-2}$ where V and A are respectively the volume and area of the system, $h^{(\text{time,volume})}$ is the bulk entropy per unit time and unit volume, and c some constant characteristic of the Langevin process. If the stochastic layer has a negligible thickness with respect to the size of the system and if the contact with the reservoir is described by a flux boundary condition for Liouville's equation, the ε-entropy per unit time can be estimated as $Vh^{(\text{time,volume})} + \nu A \ln(1/\varepsilon)$ where ν is the frequency of income of new particles at the boundary (with the same notations as before). The logarithmic divergence of the ε-entropy per unit time results from the fact that each new incoming particle brings five continuously distributed random variables: two positions for the point of crossing with the wall and three velocities. The bulk entropy per unit time and unit volume $h^{(\text{time,volume})}$ can be estimated as the sum of positive Lyapunov exponents divided by the volume V for systems close to equilibrium: $h^{(\text{time,volume})} = 0$ for an ideal gas, while $h^{(\text{time,volume})} > 0$ for a hard-sphere gas. Accordingly, the dynamical randomness is dominated by the bulk dynamical instability in systems of interacting particles [3].

4 Microscopic Approach to the Decay and Relaxation of Classical Systems

4.1 Pollicott-Ruelle Resonances

In a mixing system, the time correlation functions decay to their stationary values. It is an important issue to characterize the decay and, if possible, to decompose the decay in terms of exponentials. In order to investigate this question, we must consider the time asymptotics of the average quantities ruled by the

Frobenius-Perron operator, i.e., the long-time behavior of the statistical averages (17) before they practically reach their stationary values. Recent work has shown the remarkable result that exponential decays are compatible with Hamiltonian dynamical systems [3,14–16]. Hamiltonian systems exist for which:

$$\langle A\rangle_t \simeq_{t\to+\infty} \langle A|\Psi_{\rm st}\rangle\,\langle\tilde{\Psi}_{\rm st}|f_0\rangle + \sum_j \exp(s_j t)\,\langle A|\Psi_j\rangle\,\langle\tilde{\Psi}_j|f_0\rangle + ({\rm Jb})\,, \tag{35}$$

where s_j are the so-called Pollicott-Ruelle resonances [8–13] associated with the eigenstates

$$\hat{P}^t\Psi_j = \exp(s_j t)\,\Psi_j\,, \tag{36}$$

$$\hat{P}^{t\dagger}\tilde{\Psi}_j = \exp(s_j^* t)\,\tilde{\Psi}_j\,, \tag{37}$$

of the Frobenius-Perron operator $\hat{P}^t$ here considered as the evolution operator of the forward semigroup. Since the evolution operator of the semigroup is not unitary the resonances are in general complex and, moreover, there is the possibility of formation of Jordan blocks denoted by (Jb) in Eq. (35) [3]. Jordan-block structures arise from multiple eigenvalues in the spectrum and lead to time behavior as $t^{m_j-1}\exp(s_j t)$ where m_j is the multiplicity of the resonance s_j. Jordan-block structures are not exceptional especially in Hamiltonian systems, but we shall here focus on the eigenstates.

In classical dynamical systems, the eigenstates Ψ_j or $\tilde{\Psi}_j$ are generally given by mathematical distributions so that the associated measures are singular contrary to what is usually assumed in kinetic theory. These singular measures can be constructed by a kind of semigroup renormalization obtained by rewritting Eq. (36) in the following form [3,62]

$$\Psi_j = \lim_{t\to+\infty} \exp(-s_j t)\,\hat{P}^t\Psi_j\,. \tag{38}$$

As a simple example of time asymptotics, let us consider a particle moving in a one-dimensional potential under the Hamiltonian

$$H = \frac{p^2}{2M} + V(r)\,. \tag{39}$$

The potential $V(r)$ is supposed to have a unique maximum at $r=0$ and to monotonously decrease to a constant value at large distances $r\to\pm\infty$ so that $V(0) > V(r) > V(+\infty) = V(-\infty)$ (see Fig. 5a).

The phase space of this one-degree-of-freedom system is the plane of the variables $\Gamma = (r,p)$. The point $r=p=0$ is an unstable fixed point of the flow. Two branches of unstable manifold are issued from the fixed point and are asymptotic to the momenta $p = \pm\sqrt{2M\,[V(0)-V(\infty)]}$ for $r\to\pm\infty$ respectively (see Fig. 5b). By time-reversal symmetry, there exist two branches of stable manifold of trajectories incoming from infinity and reaching asymptotically the fixed point as $t\to+\infty$. The stable and unstable manifolds form a separatrix between the low-energy trajectories which always remain on one side of the maximum and the

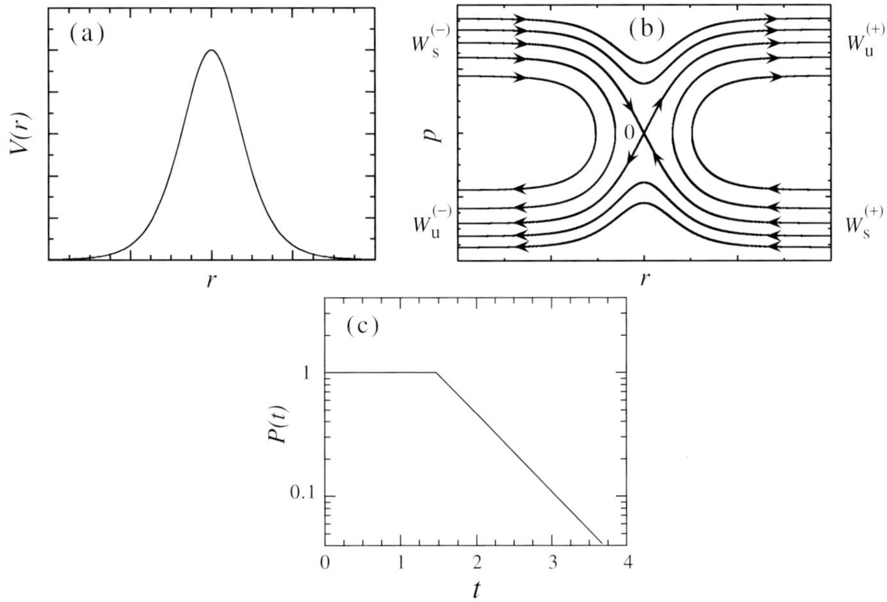

Fig. 5. Particle moving down a hill under the Hamiltonian (39): (a) the potential; (b) the phase portrait; (c) a typical survival probability versus time.

high-energy trajectories which move from one side to the other. If the statistical ensemble of initial conditions is located around the fixed point, the number of trajectories which remain in the vicinity of the fixed point is observed to decay exponentially as $t \to +\infty$ (see Fig. 5c). The rate of decay is the so-called escape rate γ, which is equal to the Lyapunov exponent λ of the fixed point in this simple system. The reason for this equality is that the trajectories escaping after a long time are coming from very close to the maximum and their dynamics is controlled by the linear stability of the fixed point. This simple system thus provides an example of a nonchaotic system with a positive Lyapunov exponent but a vanishing KS entropy because $\gamma = \lambda$. We notice that the positivity of the Lyapunov exponent implies the exponential decay so that there is here a close connection between the hyperbolicity of the dynamics near the fixed point and the existence of an exponential decay. During the decay, the cloud of points tends to stretch along both branches of unstable manifold as $t \to +\infty$. Asymptotically in time, the statistical ensemble will be distributed as a kind of Dirac distribution having the branches $\{\Gamma_{\rm u}^{(\epsilon)}\}_{\epsilon=\pm}$ of the unstable manifold for support. This intuition is supported by the fact that the following expression,

$$\Psi(\Gamma) = \sum_{\epsilon=\pm} \int_{-\infty}^{+\infty} \exp(\gamma\tau)\, \delta\left[\Gamma - \Gamma_{\rm u}^{(\epsilon)}(\tau)\right] \, d\tau \,, \tag{40}$$

defines an exact eigenstate of the Frobenius-Perron operator associated with the escape rate γ because

$$\hat{P}^t\Psi(\Gamma) \;=\; \Psi(\Phi^{-t}\Gamma) \;=\; \exp(-\gamma t)\,\Psi(\Gamma)\,, \tag{41}$$

since the flow is area-preserving and $\Phi^t\Gamma_{\rm u}^{(\epsilon)}(\tau) = \Gamma_{\rm u}^{(\epsilon)}(\tau+t)$. This result shows that the leading Pollicott-Ruelle resonances of this simple system is essentially given by the escape rate: $s = -\gamma$ [3,62].

A complete asymptotic expansion of statistical averages in terms of decaying exponential functions is given in Appendix 6 for the simple case of the inverted harmonic potential. Such complete asymptotic expansions have also been constructed in simple flows undergoing a pitchfork and a Hopf bifurcations [63,64]. In these bifurcating systems, the spectrum of Pollicott-Ruelle resonances collapses onto a continuous spectrum at the bifurcation, leading to power-law decay at criticality.

Exponential relaxation to the invariant measure can also be proved in simple chaotic systems such as the baker map, as shown in Appendix 6. Besides the simple models which are analytically tractable, the Pollicott-Ruelle resonances can be numerically obtained thanks to the periodic-orbit theory.

4.2 Periodic-Orbit Theory

The periodic-orbit theory of the classical systems is based on the Cvitanović-Eckhardt trace formula which expresses the trace of the Frobenius-Perron operator of the flow at energy E as a sum over the periodic orbits $\{p\}$ and their repetitions $r = 1, 2, 3, \ldots$

$$\mathrm{Tr}_E\, \mathrm{e}^{\hat{L}t} \equiv \int_E \delta(X - \Phi_E^t X)\, dX = \sum_p \sum_{r=1}^{\infty} T_p \, \frac{\delta(t - rT_p)}{|\det(\mathbf{I} - \mathbf{m}_p^r)|}\,,$$

where X are the phase-space coordinates inside the energy shell, T_p is the prime period of the periodic orbit p, and $\mathbf{m}_p$ is the linearized Poincaré map evaluated on the periodic orbit [65].

The Laplace transform of the trace of the Frobenius-Perron operator is formally identical to the trace of the resolvent of the Liouvillian operator and it can be expressed in terms of the so-called classical Zeta function $Z_{\rm cl}$ for a hyperbolic dynamical system in which all the periodic orbits are isolated and unstable:

$$\int_0^\infty \mathrm{e}^{-st}\, \mathrm{Tr}_E\, \mathrm{e}^{\hat{L}t}\, dt = \mathrm{Tr}_E\, \frac{1}{s - \hat{L}} = \frac{\partial}{\partial s} \ln Z_{\rm cl}(s;E)\,. \tag{42}$$

For a two-degree-of-freedom hyperbolic system, the classical Zeta function is a product over all the unstable periodic orbits:

$$Z_{\rm cl}(s;E) \;=\; \prod_p \prod_{m=0}^{\infty} \left\{ 1 \;-\; \frac{\exp\left[-sT_p(E)\right]}{|\Lambda_p(E)|\,\Lambda_p(E)^m} \right\}^{m+1}\,, \tag{43}$$

where Λ_p is the (instability) eigenvalue of the linearized Poincaré map of the periodic orbit p. Since each periodic orbit is unstable we have that $|\Lambda_p| > 1$. The Pollicott-Ruelle resonances s_j are given as the zeros of the Zeta function:

$$Z_{\rm cl}(s_j; E) = 0\,. \tag{44}$$

The periodic-orbit theory has been applied to different systems in order to characterize the decay of a statistical ensemble of trajectories [14]. A class of systems which have been studied in detail are the disk scatterers.

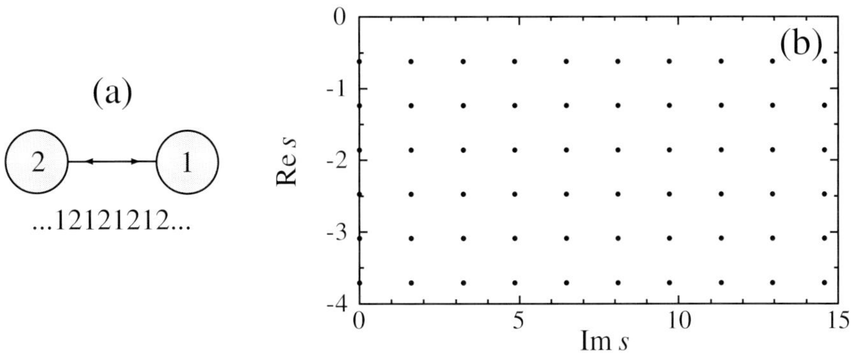

Fig. 6. Two-disk scatterer: (a) configuration of the system; (b) spectrum of Pollicott-Ruelle resonances.

In the two-disk scatterer, there is a single unstable periodic orbit bouncing between the two disks (see Fig. 6a). This unique periodic orbit forms the so-called repeller of this system, which is the set of trajectories which are forever trapped at finite distance. This periodic orbit has a positive Lyapunov exponent λ which is proportional to the velocity of the particle. The instability eigenvalue is given by $\Lambda = \exp(\lambda T)$ where T is the period which is inversely proportional to the velocity. The Pollicott-Ruelle resonances of the two-disk scatterer are the zeros of the Zeta function:

$$Z_{\rm cl}(s) \;=\; \prod_{m=0}^{\infty} \left[1 \;-\; \frac{\exp\left(-sT\right)}{|\Lambda|\,\Lambda^m}\right]^{m+1} = 0\,. \tag{45}$$

This infinite product converges absolutely so that the Pollicott-Ruelle resonances are given by

$$s_{mn} = -\frac{m+1}{T}\ln \Lambda + {\rm i}\,\frac{2\pi n}{T}\,, \tag{46}$$

with $m = 0, 1, 2, 3, \ldots$ and $n = 0, \pm 1, \pm 2, \pm 3, \ldots$ The resonances form a half periodic array extending toward negative values of Re s and separated from Re $s = 0$ by a gap given by the escape rate

$$\gamma = \frac{1}{T}\ln \Lambda = \lambda\,, \tag{47}$$

which is here equal to the Lyapunov exponent (see Fig. 6b). Accordingly, the KS entropy vanishes and the system is nonchaotic as expected.

In order to obtain a chaotic system, it is enough to add a third disk forming an equilateral triangle with the two previous disks, which defines the three-disk scatterer (see Fig. 7a).

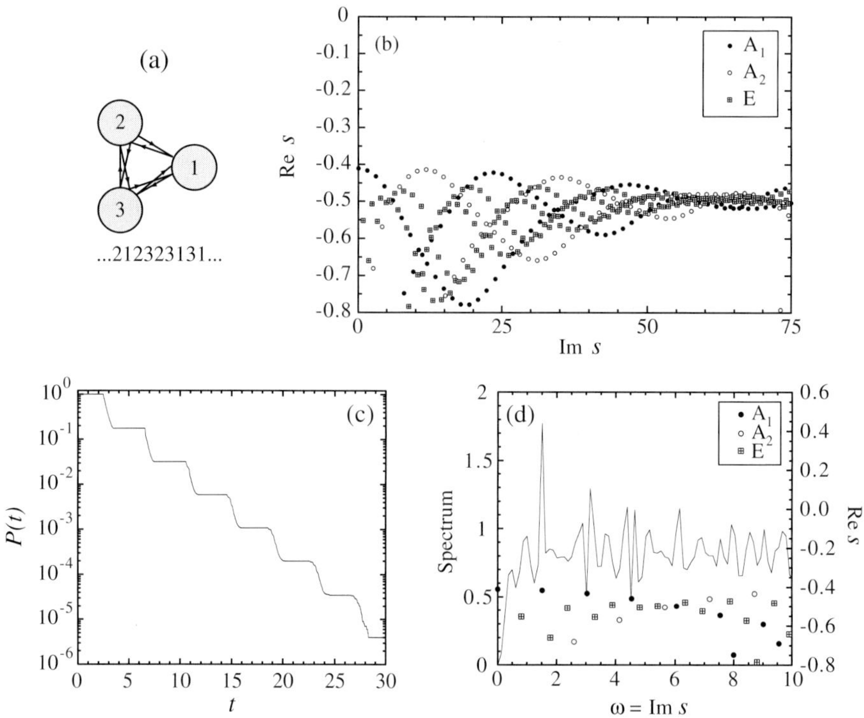

Fig. 7. Three-disk scatterer: (a) configuration of the system; (b) spectrum of Pollicott-Ruelle resonances; (c) typical survival probability versus time; (d) Fourier transform of the modulations of the survival probability versus the frequency, compared with the spectrum of Pollicott-Ruelle resonances.

In this open system, the set of trapped trajectories forms a fractal and chaotic repeller. If the disks are sufficiently far apart, the forever trapped trajectories are in one-to-one correspondence with a symbolic dynamics based on the labels of the disks with the constraint that no two consecutive labels are equal because the particle cannot bounce twice on the same disk without having an intermediate collision on another disk [20]. This constraint means that the symbolic dynamics is effectively dyadic, leading to a twofold proliferation of orbits with the period. All the periodic orbits can be listed in order of increasing period or instability. This listing helps to numerically compute the Zeta function by truncation after a given period, which amounts to neglect the effect of the most unstable trajectories. This method, called the cycle expansion [66], is excellent in a system such as the three-disk scatterer and allows a high-precision calculation of the different characteristic quantities of chaos with four or five digits [14]. In particular, it is possible to calculate the spectrum of Pollicott-Ruelle resonances (see Fig. 7b). Because of the threefold symmetry of the equilateral three-disk scatterer the Pollicott-Ruelle resonances fall into the different irreducible representations (A_1, A_2, and E) of the C_{3v} symmetry group. The spectrum is separated from the imaginary axis by a gap given by the escape rate γ which is the leading Pollicott-Ruelle resonances.

Since the system is chaotic, the escape rate is no longer equal to the maximum Lyapunov exponent but it is lowered by the KS entropy per unit time according to: $\gamma = \lambda - h_{\mathrm{KS}}$. The different Pollicott-Ruelle resonances control the decay of a statistical ensemble of trajectories escaping from the scatterer. The survival probability, i.e., the probability that a trajectory is still in the vicinity of the scatterer, decays in a gross exponential way at a rate given by the escape rate. But on top of the decay curve, we observe modulations at certain frequencies which are precisely given by the imaginary parts of the Pollicott-Ruelle resonances (see Figs. 7c and 7d) [14].

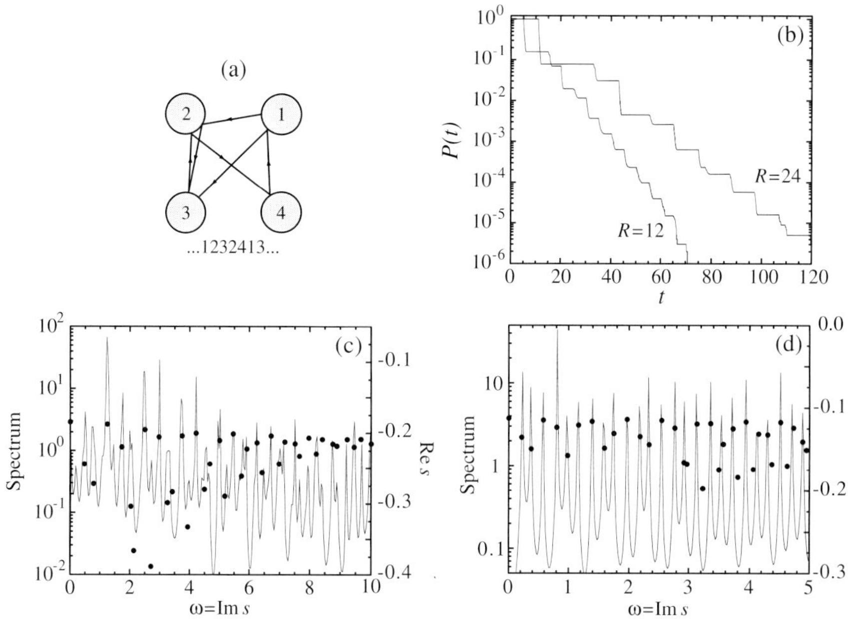

Fig. 8. Four-disk scatterer composed of four disks of unit radius at the vertices of a square of size R: (a) configuration of the system; (b) survival probabilities versus time for the configurations with $R = 12$ and $R = 24$; (c) Fourier transform of the modulations of the survival probability versus the frequency, compared with the spectrum of Pollicott-Ruelle resonances in the case $R = 12$ and (d) in the case $R = 24$.

This phenomenon is also clearly evidenced in Fig. 8 which depicts the survival probability in two geometries of the four-disk scatterer, composed of four disks forming a square of side $R = 12$ and $R = 24$ respectively, the radius of the disks being taken as unity. In these systems, the modulations of the survival probability are very pronounced and their Fourier transform presents peaks at the frequencies of the Pollicott-Ruelle resonances as can be seen in Fig. 8 [14]. These results show that the Pollicott-Ruelle resonances control the decay under classical dynamics in these open systems.

4.3 Escape-Rate Theory of Transport

The purpose of the escape-rate theory of transport is to express the transport coefficients in terms of the escape rate of some fractal repeller [3,17–19]. The fractal repeller is defined by considering some absorbing boundary conditions in the phase space of the system. The absorbing boundary conditions set up a first passage problem for the Liouvillian equation ruling the time evolution of the statistical ensemble of trajectories. As explained here above, the probability density $f(\Gamma)$ evolves in time according to Liouville's equation. Since this latter is a partial differential equation, it requires boundary conditions to be solved. The absorbing boundary conditions impose that there is a vanishing flux of incoming trajectories at the border of the system so that the probability (19) defines the survival probability and decays monotonously according to Eq. (25).

In general, the decay can be a power law (possibly corrected by logarithms) or an exponential. In finite ergodic hyperbolic systems with positive Lyapunov exponents, the decay turns out to be exponential in the long-time limit $t \to \infty$ with a positive escape rate related to the positive Lyapunov exponents and the KS entropy according to Eq. (34). The set of trapped trajectories form a fractal and chaotic repeller, which can be characterized in terms of the partial information dimensions $\{d_i\}$ associated with the different unstable directions in phase space [1]

$$h_{\mathrm{KS}} = \sum_{\lambda_i>0} d_i \, \lambda_i \, . \tag{48}$$

The codimensions are defined by $c_i = 1-d_i$ so that the escape rate can be written in terms of the positive Lyapunov exponents and the partial codimensions as [1]

$$\gamma = \sum_{\lambda_i>0} \lambda_i - h_{\mathrm{KS}} = \sum_{\lambda_i>0} c_i \, \lambda_i \, . \tag{49}$$

The transport properties such as diffusion, viscosity or heat conductivity are characterized by coefficients which enter the macroscopic equations of motion such as the diffusion or heat equations and the Navier-Stokes equations. In the fifties and early sixties, these transport coefficients have been expressed in terms of the Helfand moments $G^{(\alpha)}$ or the microscopic currents $J^{(\alpha)} = dG^{(\alpha)}/dt$ [6,7,67]. In particular, the transport coefficients are given as the integral of the time autocorrelation function of each microscopic current by the Green-Kubo formula [6,7] or, equivalently, as the linear growth rate of the variance of the Helfand moment [67]

$$\alpha = \int_0^\infty \langle J_0^{(\alpha)} J_t^{(\alpha)} \rangle \, dt = \lim_{t\to\infty} \frac{1}{2t} \, \langle (G_t^{(\alpha)} - G_0^{(\alpha)})^2 \rangle \, . \tag{50}$$

The Helfand moments associated with the different transport properties are given in Table I.

Table 1. Helfand's moments of different transport properties.

transport property	*moment*
self-diffusion	$G^{(\mathcal{D})} = x_i$
shear viscosity	$G^{(\eta)} = \frac{1}{\sqrt{V k_\mathrm{B} T}} \sum_{i=1}^N x_i \, p_{iy}$
bulk viscosity ($\psi = \zeta + \frac{4}{3}\eta$)	$G^{(\psi)} = \frac{1}{\sqrt{V k_\mathrm{B} T}} \sum_{i=1}^N x_i \, p_{ix}$
heat conductivity	$G^{(\kappa)} = \frac{1}{\sqrt{V k_\mathrm{B} T^2}} \sum_{i=1}^N x_i \, (E_i - \langle E_i \rangle)$
electric conductivity	$G^{(\mathrm{e})} = \frac{1}{\sqrt{V k_\mathrm{B} T}} \sum_{i=1}^N eZ_i \, x_i$

The Einstein-like formula in Eq. (50) shows that the Helfand moments have a diffusive motion. A first passage problem can be set up by imposing absorbing boundary conditions in the space of variation of the Helfand moment [18]. These absorbing boundary conditions select all the trajectories of the whole system for which the Helfand moment never reaches the absorbing boundaries:

$$-\frac{\chi}{2} \le G_t^{(\alpha)} \le +\frac{\chi}{2} \, . \tag{51}$$

However, most of the trajectories reach the absorbing boundaries and escape out of the domain delimited by these boundaries. Accordingly, the trajectories forever trapped inside the absorbing boundaries form a set of zero probability in the form of a fractal repeller. This repeller is characterized by an escape rate which can be evaluated by using the equation of diffusive motion of the Helfand moment $g = G_t^{(\alpha)}$

$$\frac{\partial p}{\partial t} = \alpha \, \frac{\partial^2 p}{\partial g^2} \, , \tag{52}$$

with the corresponding absorbing boundary conditions $p(g = \pm\chi/2, t) = 0$. The solution of this equation is given by

$$p(g,t) = \sum_{j=1}^{\infty} a_j \, \exp(-\gamma_j t) \, \sin\left(\frac{j\pi g}{\chi} + \frac{j\pi}{2}\right) \, , \tag{53}$$

so that the decay rates of the survival probability are

$$\gamma_j = \alpha \left(\frac{j\pi}{\chi} \right)^2 \qquad \text{and} \quad j = 1, 2, 3, \ldots . \tag{54}$$

The escape rate is the slowest decay rate which is thus related to the transport coefficient α according to

$$\gamma \simeq \gamma_1 = \alpha \left(\frac{\pi}{\chi} \right)^2 \qquad \text{for} \quad \chi \to \infty . \tag{55}$$

Substituting this result in the escape-rate formula (49) we obtain the following relationships between the transport coefficients and the characteristic quantities of chaos which are the Lyapunov exponents λ_i, the KS entropy $h_{\rm KS}$, or the partial codimensions c_i as [18]

$$\alpha = \lim_{\chi, V \to \infty} \left(\frac{\chi}{\pi} \right)^2 \left(\sum_{\lambda_i > 0} \lambda_i - h_{\rm KS} \right)_\chi = \lim_{\chi, V \to \infty} \left(\frac{\chi}{\pi} \right)^2 \left(\sum_{\lambda_i > 0} c_i \, \lambda_i \right)_\chi . \tag{56}$$

In the limit $\chi \to \infty$, the partial information codimensions can be replaced by the Hausdorff codimensions in hyperbolic systems.

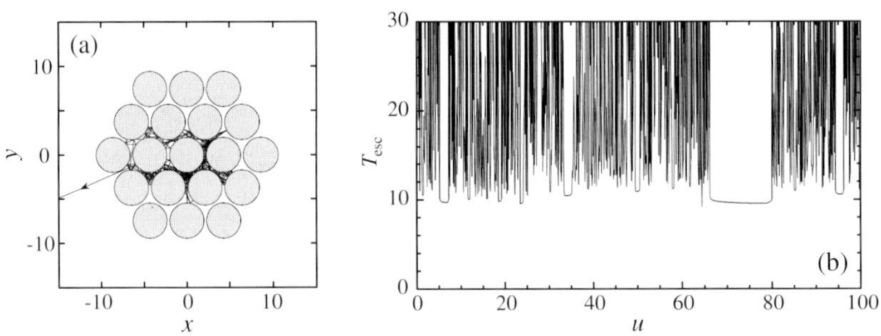

Fig. 9. Open Lorentz gas: (a) configuration of the system with a typical trajectory; (b) time taken by the particle to escape versus the initial angle $\theta = 0.75\pi - u\pi 10^{-10}$ of velocity of the particle starting from the central disk. The escape time becomes infinite on each stable manifold of a trapped trajectory.

Equation (56) has been applied to the transport by diffusion in the open two-dimensional periodic Lorentz gas (see Fig. 9a) [68]. In this system, the particles have a deterministic diffusive motion induced by the elastic collisions on immobile scatterers until the particles escape upon reaching absorbing boundaries in the form of a hexagon, a circle, or two parallel lines separated by a large distance L. The absorbing boundary conditions select a fractal repeller of trapped trajectories which can be numerically evidenced by plotting the time to escape as a function of initial condition (see Fig. 9b). In this two-degree-of-freedom system, there is only one unstable direction and, thus, one positive

Lyapunov exponents and associated codimension. Accordingly, Eq. (56) shows that the diffusion coefficient is given by [17,68]

$$\mathcal{D} = \lim_{L\to\infty} \left(\frac{L}{\pi}\right)^2 (\lambda - h_{\rm KS})_L = \lim_{L\to\infty} \left(\frac{L}{\pi}\right)^2 (c\,\lambda)_L \,, \qquad (57)$$

where c is either the information or the Hausdorff partial codimension associated with the unstable direction. The open random Lorentz gases have also been investigated [69].

The escape-rate theory has also been applied to reaction-diffusion processes in which point particles diffuse in a Lorentz gas until they reach catalytic disks where they are annihilated [70]. In this reactive process, absorbing boundary conditions are set up on the border of the catalytic disks and select a fractal repeller of trajectories which never hit a catalyst. In this reaction-diffusion system, the reaction rate is given by the escape rate, which is in turn related to the characteristic quantities of chaos [70].

Equation (56) can also be applied to transport in many-body systems such as the diffusion of a Brownian or tracer particle in a fluid of other particles or to the shear viscosity (see Fig. 10). In these many-body systems, there is a spectrum of Lyapunov exponents as well as of partial codimensions. Typically, the spectrum of codimensions is peaked on the maximum Lyapunov exponent in multidimensional fractal repellers.

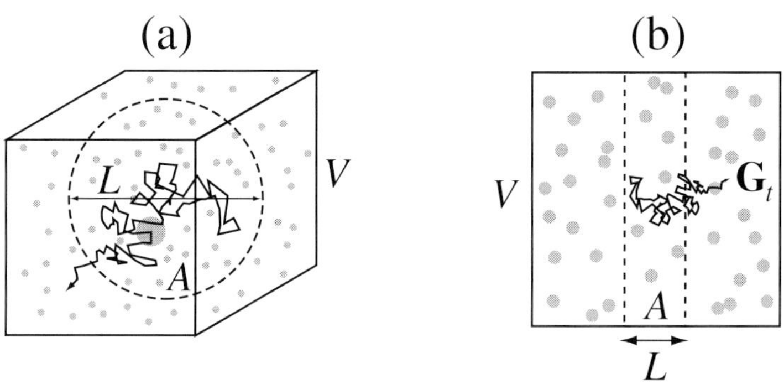

Fig. 10. Schematic illustration of the application of the escape-rate theory to (a) the diffusion of a Brownian particle escaping out of a sphere A of diameter L in a fluid of volume V and (b) the viscosity in a fluid of volume V in which the Helfand moment $\mathbf{G}_t$ escapes out of a domain A of width L.

The escape-rate theory can also be considered for the determination of the mobility coefficient in the biased diffusion of a charged particle in a static external electric field [3,71], or for heating by the diffusivity of total energy in systems driven by a time-periodic external force.

4.4 Hydrodynamic and Reactive Modes in Spatially Extended Systems

In spatially extended systems, the relaxation to the equilibrium state is ruled by the hydrodynamic modes. These modes can be explicitly constructed in periodic dynamical systems which are spatially extended by considering infinitely many images of the system in the form of a checkerboard [72–74].

A simple example is the periodic Lorentz gas in which the unit cell contains a single scatterer. The system is spatially extended by tessellating the plane with the unit cell. The motion of the point particle bouncing on the disk can be considered either inside the unit cell which forms a torus, or in the array formed by the infinitely many images of the unit cell. Diffusion is conceived as the irregular motion on the infinite spatially extended system. When the particle performs a bounded chaotic motion inside the unit cell (or torus) its images diffuses on the lattice made of the infinitely many images of the unit cell. The probability density of finding the particle on the lattice can be Fourier transformed and decomposed into different components associated with the different values of the wavenumber $\boldsymbol{k}$ ranging in the first Brillouin zone of the lattice. Each $\boldsymbol{k}$-component of the probability density has its own time evolution under an operator $\hat{Q}^t_{\boldsymbol{k}}$ which is obtained by Fourier transforming the Frobenius-Perron operator $\hat{P}^t$ over the whole lattice. This new operator may admit Pollicott-Ruelle resonances which also depend on the wavenumber as for the associated eigenstates:

$$\hat{Q}^t_{\boldsymbol{k}}\, \Psi_{\boldsymbol{k}} = \mathrm{e}^{s_{\boldsymbol{k}} t}\, \Psi_{\boldsymbol{k}}\,. \tag{58}$$

If $\hat{T}^{\boldsymbol{l}}$ denotes the operator of translation by the lattive vector $\boldsymbol{l}$, the eigenstates satisfy the *quasiperiodic boundary conditions*:

$$\hat{T}^{\boldsymbol{l}}\, \Psi_{\boldsymbol{k}} = \mathrm{e}^{\mathrm{i}\boldsymbol{k}\cdot\boldsymbol{l}}\, \Psi_{\boldsymbol{k}}\,. \tag{59}$$

When the wavenumber vanishes $\boldsymbol{k} = 0$, we recover the Frobenius-Perron operator only ruling the dynamics on the torus. If this dynamics is ergodic and mixing as it is the case for the Lorentz gas or the Sinai billiard one of the Pollicott-Ruelle resonances must be identical to the vanishing Liouvillian eigenvalue $s_0 = 0$ associated with the unique invariant measure. This suggests that, at least, one of the Pollicott-Ruelle resonances vanishes with the wavenumber: $\lim_{\boldsymbol{k}\to 0} s_{\boldsymbol{k}} = 0$. Since the Pollicott-Ruelle resonances rule the relaxation to the invariant measure we can identify that particular Pollicott-Ruelle resonance with the dispersion relation of the hydrodynamic mode of diffusion [3,72]. This dispersion relation is given by the van Hove formula as [73]

$$s_{\boldsymbol{k}} = \lim_{t\to\infty} \frac{1}{t} \ln\langle \exp[\mathrm{i}\boldsymbol{k}\cdot(\boldsymbol{r}_t - \boldsymbol{r}_0)]\rangle = -\mathcal{D}\, k^2 + O(k^4)\,, \tag{60}$$

which is a consequence of Eqs. (58) and (59).

At the microscopic level of description, the hydrodynamic mode is thus represented by the corresponding eigenstate (58) of the evolution operator. This eigenstate is typically a mathematical distribution represented by a cumulative function which is continuous but nondifferentiable.

In two-degree-of-freedom systems such as the two-dimensional Lorentz gases, the cumulative functions of the hydrodynamic modes of diffusion can be defined in terms of the density $\Psi_{\boldsymbol{k}}$ of the eigenstate as the one-dimensional integral over a line in phase space

$$F_{\boldsymbol{k}}(\xi) = \int_0^{\xi} d\xi' \, \Psi_{\boldsymbol{k}}(\Gamma_{\xi'}) \, . \tag{61}$$

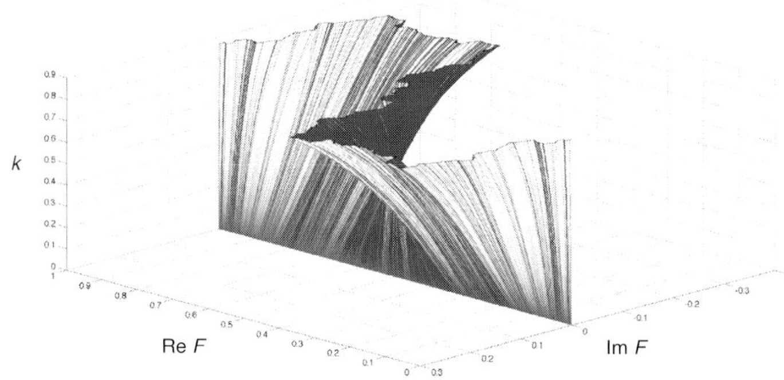

Fig. 11. Cumulative functions of the hydrodynamic modes of diffusion in the hard-disk Lorentz gas. The disks form a triangular lattice, their centers are separated by the distance $d = 2.3$, and their radius is unity. The cumulative functions (61) are depicted in the complex plane $(\mathrm{Re}\, F_{\boldsymbol{k}}, \mathrm{Im}\, F_{\boldsymbol{k}})$ versus their wavenumber $\boldsymbol{k} = k\, \boldsymbol{e}_x$ of magnitude varying in the interval $0 \leq k < 0.9$. ξ is here the angle of position on the border of a disk. The thermodynamic equilibrium corresponds to the vanishing wavenumber $k = 0$.

For $\boldsymbol{k} \neq 0$, this function is complex and forms a fractal curve in the complex plane (see Fig. 11) [73]. Its Hausdorff dimension depends on the wavenumber $\boldsymbol{k}$, the diffusion coefficient $\mathcal{D}$, and the positive Lyapunov exponent λ as [73,75]

$$D_{\mathrm{H}}(\boldsymbol{k}) = 1 + \frac{\mathcal{D}}{\lambda} \, k^2 + O(k^4) \, . \tag{62}$$

This formula can be rewritten to express the diffusion coefficient in terms of the Hausdorff dimension and the other characteristic quantities of chaos as [73,75]

$$\mathcal{D} = \lim_{\boldsymbol{k} \to 0} \left(\frac{1}{k} \right)^2 \, \Big[\lambda \, (D_{\mathrm{H}} - 1) \Big]_{\boldsymbol{k}} \, , \tag{63}$$

which has a structure very similar to the escape-rate formula (57). Indeed, in the problem with absorbing boundary conditions, the role of the wavenumber is played by $k = \pi/L$, the Lyapunov exponent is essentially the same for both problems in the limits $L \to \infty$ or $\boldsymbol{k} \to 0$ in which the Lyapunov exponent reaches the value obtained by averaging over the equilibrium invariant measure. On the other hand, the partial Hausdorff codimension of the fractal repeller is given by $c_{\rm H} = 1 - d_{\rm H}$ which is positive since the Hausdorff dimension of a fractal object on a line is known to be bounded as $0 \leq d_{\rm H} \leq 1$. The role of the partial Hausdorff dimension of the repeller is here played by the Hausdorff dimension of the fractal cumulative curve in the two-dimensional complex plane, which is bounded as $1 \leq D_{\rm H} \leq 2$. We therefore find the positive number $D_{\rm H} - 1$ in Eq. (63) in the place of the codimension c appearing in Eq. (57).

The hydrodynamic modes have been explicitly constructed and shown to have fractal cumulative functions in the hard-disk and Yukawa potential Lorentz gases [73], as well as in the multibaker model of diffusion (see Appendix 6). Similar considerations apply to the reactive modes of reaction-diffusion processes in which colored particles diffuse in a Lorentz gas and change their color upon colliding on catalytic disks [40–42]. The reaction is here an isomerization $A \leftrightarrow B$. The microscopic analysis of this reaction allows us to explicitly construct the diffusive and reactive modes of relaxation toward the thermodynamic equilibrium. These modes have fractal cumulative functions and their dispersion relations obey the macroscopic equations (4)-(8).

4.5 Explicit Construction of a Nonequilibrium Steady State

The nonequilibrium steady state can be explicitly constructed in systems of diffusion between two reservoirs at different concentrations [3,76]. To fix the idea, let us consider the open periodic Lorentz gas with flux boundary conditions at both sides separated by a distance L. From the left-hand side (resp. right-hand side), particles are incoming from a reservoir of phase-space density f_- (resp. f_+). The particle moves under the deterministic dynamics of elastic collisions inside the Lorentz gas. We suppose that the density has reached its stationary value between both reservoirs. At each point inside the Lorentz gas, the stationary density can then be determined by integrating the trajectory issued from this point backward in time till the point of entrance in the domain between both reservoirs. The density is f_- (resp. f_+) if the point of entrance is at the boundary with the left-hand side (resp. right-hand side) reservoir [76]. We notice that the backward integration can run for $t \to -\infty$ between both reservoirs in the case where the point belongs to the unstable manifold of one trapped trajectory of the fractal repeller of the escape-rate theory. Accordingly, the stationary density is a piecewise constant function taking the values f_+ or f_-. The discontinuities of the stationary density occurs on the unstable manifolds of the fractal repeller. This invariant density can be expressed by the formula

$$f_{\rm st}(\Gamma) = \frac{f_+ + f_-}{2} + \boldsymbol{g} \cdot \left[\boldsymbol{r}(\Gamma) + \int_0^{T_{\rm in}(\Gamma)} \boldsymbol{v}(\Phi^t \Gamma)\, dt \right] , \tag{64}$$

where $T_{\rm in}(\Gamma)$ is the negative time of entrance of the trajectory going through the point Γ and $\boldsymbol{g} = \boldsymbol{e}_x (f_+ - f_-)/L$ is the gradient of phase-space concentration here taken in the x-direction. Indeed, if we integrate the velocity from the initial time $t = 0$ backward to the time of entrance, we get

$$f_{\rm st}(\Gamma) \;=\; \frac{f_+ + f_-}{2} + \boldsymbol{g} \cdot \Big[\boldsymbol{r}(\Gamma) + \boldsymbol{r}_{\rm in}(\Gamma) - \boldsymbol{r}(\Gamma)\Big] \,. \tag{65}$$

Since the position of entrance is $\boldsymbol{r}_{\rm in} = \boldsymbol{e}_x(\pm L/2)$, we finally obtain that

$$f_{\rm st}(\Gamma) \;=\; \frac{f_+ + f_-}{2} + \frac{f_+ - f_-}{L}\,\frac{\pm L}{2} = f_\pm \,, \tag{66}$$

as announced. The three terms in Eq. (64) can be interpreted as follows: The first term is simply the concentration in the middle of the system which is a constant. The second term represents the average linear profile of concentration expected for the nonequilibrium steady state of a diffusive system. It is remarkable that this linear profile arises simply from this formula. The third term represents the velocity fluctuations of deterministic origin around the mean linear profile.

In the limit where the reservoirs are at large distances while keeping the gradient of concentration constant, the stationary density in the middle of the system becomes wildely fluctuating between the concentrations of both reservoirs. The density relative to the average density is no longer defined as a function and becomes a distribution defined by a singular measure of invariant density

$$\Psi_{\boldsymbol{g}}(\Gamma) \;=\; \boldsymbol{g} \cdot \left[\boldsymbol{r}(\Gamma) + \int_0^{-\infty} \boldsymbol{v}(\Phi^t \Gamma)\; dt\right] = \Psi_{\boldsymbol{g}}(\Phi^t \Gamma) \,. \tag{67}$$

It is remarkable that the invariant density (67) is equivalently obtained in terms of the hydrodynamic modes by considering the following zero wavenumber limit [3,76]

$$\Psi_{\boldsymbol{g}}(\Gamma) \;=\; -{\rm i}\, \boldsymbol{g} \cdot \frac{\partial \Psi_{\boldsymbol{k}}(\Gamma)}{\partial \boldsymbol{k}}\bigg|_{\boldsymbol{k}=0} \,, \tag{68}$$

as shown elsewhere [77].

As for the hydrodynamic modes, the invariant density (67) is singular and the nonequilibrium steady state should be represented by its cumulative function as

$$F_{\boldsymbol{g}}(\xi) = \int_0^{\xi} d\xi' \; \Psi_{\boldsymbol{g}}(\Gamma_{\xi'}) \,. \tag{69}$$

In the Lorentz gases and the multibaker models of diffusion, this cumulative function has been shown to be continuous but nondifferentiable [72,77]. In the multibaker models of diffusion, this function is known as the Takagi function in the name of the Japanese mathematician who invented this function in 1903 as an example of continuous but nondifferentiable function [77]. In the Lorentz gases, the Takagi function generalizes into similar functions representing, at the microscopic phase-space level, the nonequilibrium steady states of diffusion corresponding to a linear average profile of concentration between both reservoirs (see Fig. 12) [62]. The singular character of the nonequilibrium steady states has very simply its origin in the mixing property of the deterministic dynamics which has the effect of mixing the different concentrations coming from both reservoirs.

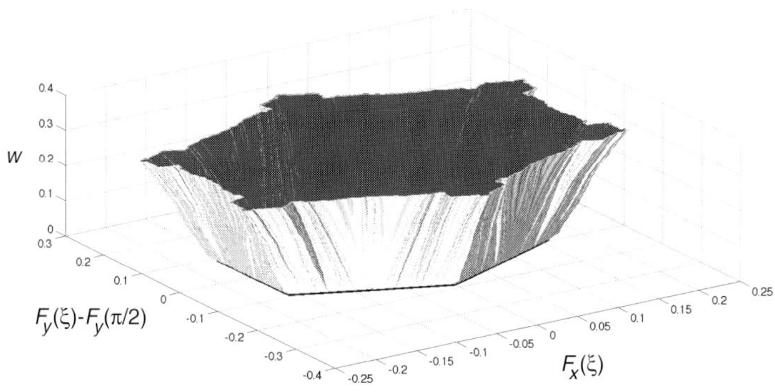

Fig. 12. Cumulative functions of the nonequilibrium steady states of diffusion in the hard-disk Lorentz gas. The scatterers are disks of unit radius forming a triangular lattice and separated by the gap w. The gap takes the values $0 \leq w < (4/\sqrt{3}) - 2$ in the finite-horizon case. In Eq. (69), ξ is the angle of position on the border of a disk and the gradient $\boldsymbol{g}$ is either in the x-direction or in the y-direction. The cumulative functions are depicted in the plane of the quantities $F_x(\xi)$ and $F_y(\xi) - F_y(\pi/2)$ versus the gap w between the disks.

4.6 Connection to Thermodynamics

The connection between dynamics and thermodynamics has always been a difficult problem for the reason that many different irreversible processes take place in a fluid as explained in Sec. 2. Progress in the theoretical understanding has always proceeded by the analysis of simplified models such as the Ising model to understand critical phenomena in phase transitions. In the study of irreversible processes, the Lorentz gases and the multibaker models play a similar role because they only sustain one transport process, namely, diffusion. The Lorentz gas model has historically been introduced by Lorentz in 1905 as a model for transport of electrons in a solid [78]. Today, the Lorentz gas is considered as a model of diffusion of light particles among heavy ones [79,80]. The heavy particles are immobile on the time scale of motion of the light particles. At each elastic collision between light and heavy particles, energy is nearly conserved although momentum is not. Therefore, the randomization of the velocity direction is very fast albeit the randomization of energy occurs on a much longer time scale. This decoupling of time scales does not preclude the existence of a well-defined diffusion coefficient on each energy shell. Indeed, Bunimovich and Sinai proved in 1980 that the two-dimensional hard-disk periodic Lorentz gas has a positive and finite diffusion coefficient if the horizon is finite, i.e., if no trajectory exists running through the lattice of hard disks without elastic collision [81]. Besides, the hard-disk Lorentz gas is ergodic and mixing. Similar results have been proved by Knauf for the two-dimensional periodic Lorentz gas composed of scatterers with Yukawa potentials under the condition that the energy of the moving particle is large enough [82]. The mixing property holds on each energy shell and leads to the establishment of a *local equilibrium in velocity direction*. The dynamics of elastic collision on the scatterers has the effect of randomizing the velocity *direction* so that the velocity distribution asymptotically becomes uniform in the *velocity angle*. This local equilibrium is sufficient for a transport by diffusion to occur which is compatible with the laws of irreversible thermodynamics.

In the perspective of Sec. 2, the Lorentz gases are binary fluids composed of light particles diffusing among heavy particles which remain immobile. It is enough that the light particles interact with the heavy ones in order to induce diffusion. The interaction among the light particles is therefore neglected in the Lorentz gases. The light particles thus constitute a noninteracting gas of particles moving in a nonuniform potential due to the immobile heavy particles. This potential is such that the motion of each light particle is chaotic. This chaotic dynamics is mixing and induces the relaxation to the local equilibrium in the velocity direction as well as the diffusion across the lattice. On large spatial scales, the diffusion occurs on each energy shell so that the density $p(\boldsymbol{r},\epsilon)$ of light particles with energy ϵ at position $\boldsymbol{r}$ obeys the diffusion equation

$$\partial_t\, p(\boldsymbol{r},\epsilon) \simeq \mathcal{D}(\epsilon)\ \nabla^2 p(\boldsymbol{r},\epsilon)\,, \tag{70}$$

where $\mathcal{D}(\epsilon)$ is the diffusion coefficient on the shell of energy ϵ [83]. At each elastic collision energy is conserved so that the initial distribution of energy is preserved by the dynamics. According to the zeroth law of thermodynamics, the temperature T of a system is fixed by putting it in contact with a heat reservoir before further operation. The implementation of the zeroth law of thermodynamics thus fixes the energy distribution of the ideal gas of light particles to a Boltzmann distribution at the given temperature which is thereafter invariant under the dynamics of the Lorentz gas. The large-scale density is thus given by

$$p(\boldsymbol{r},\epsilon) = n(\boldsymbol{r})\ \frac{\mathrm{e}^{-\epsilon/k_{\mathrm{B}}T}}{\int_0^\infty \mathrm{e}^{-\epsilon/k_{\mathrm{B}}T}\ d\epsilon}\,, \tag{71}$$

where $n(\boldsymbol{r})$ is the number of particles per unit volume around the spatial point $\boldsymbol{r}$. The particle density $n(\boldsymbol{r})$ obeys the diffusion equation

$$\partial_t n(\boldsymbol{r}) \simeq \bar{\mathcal{D}}\ \nabla^2 n(\boldsymbol{r})\,, \tag{72}$$

with the mean diffusion coefficient

$$\bar{\mathcal{D}} = \frac{\int_0^\infty \mathrm{e}^{-\epsilon/k_{\mathrm{B}}T}\mathcal{D}(\epsilon)\ d\epsilon}{\int_0^\infty \mathrm{e}^{-\epsilon/k_{\mathrm{B}}T}\ d\epsilon}\,. \tag{73}$$

The equilibrium thermodynamics of the hard-disk Lorentz gas is established as follows. Let us consider a gas of N noninteracting light particles in a spatial domain $\mathcal{Q}$ delimited by the hard disks and by a wall enclosing the Lorentz gas in a square. The volume of the domain $\mathcal{Q}$ is $V = \mathrm{Vol}(\mathcal{Q})$. The phase space of this gas is

$$\mathcal{M} = \Big\{(\boldsymbol{r}_1,\boldsymbol{p}_1,...,\boldsymbol{r}_N,\boldsymbol{p}_N)|\boldsymbol{r}_i\in\mathcal{Q},\boldsymbol{p}_i\in\mathbb{R}^d, x_1<x_2<\cdots<x_N\Big\}\,, \tag{74}$$

where the last conditions provide the ordering of indices required in the case of identical particles. The phase-space volume element is

$$d\Gamma = d\boldsymbol{r}_1 \ \cdots \ d\boldsymbol{r}_N \, d\boldsymbol{p}_1 \ \cdots \ d\boldsymbol{p}_N \,, \tag{75}$$

and the invariant distribution is given in terms of the Maxwellian velocity distribution at the temperature T fixed by the zeroth law of thermodynamics:

$$f(\Gamma) = \frac{N!}{V^N} \frac{\exp\left(-\sum_{j=1}^N \frac{\boldsymbol{p}_j^2}{2mk_\mathrm{B}T}\right)}{(2\pi m k_\mathrm{B} T)^{Nd/2}} \,, \tag{76}$$

which is normalized according to

$$\int_{\mathcal{M}} f(\Gamma) \, d\Gamma = 1 \,. \tag{77}$$

The entropy can be calculated as Gibbs' coarse-grained entropy

$$S = -k_\mathrm{B} \sum_i p_i \ln p_i \,, \tag{78}$$

by coarse graining the phase space into cells of size $\Delta\Gamma = \Delta^{Nd}r \ \Delta^{Nd}p = (2\pi\hbar)^{Nd}$. The probability for the system to belong to the ith cell is

$$p_i = \int_{i^{\mathrm{th}} \ \mathrm{cell}} f \, d\Gamma \simeq f_i \, \Delta\Gamma \,, \tag{79}$$

so that

$$S \simeq k_\mathrm{B} \left\langle \ln \frac{1}{f(\Gamma) \, \Delta\Gamma} \right\rangle \,, \tag{80}$$

and we obtain the Sackur-Tetrode formula

$$S \simeq k_\mathrm{B} \, N \, \ln \frac{V \mathrm{e}^{\frac{d+2}{2}} (2\pi m k_\mathrm{B} T)^{\frac{d}{2}}}{N \, (2\pi\hbar)^d} \,, \tag{81}$$

in the limit $N, V \to \infty$ keeping the particle density $n = N/V$ constant [84,85]. This result shows that the consistency of our previous assumptions. The energy of the ideal gas is known to be

$$E = N \, \frac{d k_\mathrm{B} T}{2} = N \, \bar{\epsilon} \,, \tag{82}$$

with the mean kinetic energy of each particle $\bar{\epsilon} = d k_\mathrm{B} T/2$. As a consequence of both the Sackur-Tetrode formula (81) for entropy and the energy equation of state (82) we obtain Gibbs' relation [29]

$$dE = T \, dS - P \, dV + \mu \, dN \,, \tag{83}$$

with the pressure equation of state

$$P \, V = N \, k_\mathrm{B} \, T \,, \tag{84}$$

and the chemical potential

$$\mu = k_{\rm B}\, T\, \ln \frac{N(2\pi\hbar)^d}{V(2\pi m k_{\rm B} T)^{\frac{d}{2}}} \,. \tag{85}$$

If we define the local densities of energy and entropy respectively as

$$e = \frac{E}{V} \qquad \text{and} \qquad s = \frac{S}{V} \,, \tag{86}$$

we obtain the local Gibbs' relation

$$de = T\, ds + \mu\, dn \,, \tag{87}$$

of local thermodynamic equilibrium at the basis of irreversible thermodynamics [29]. It is then a consequence of the diffusion equation (72) that the local density of entropy obeys the equation

$$\partial_t\, s = -\boldsymbol{\nabla} \cdot \boldsymbol{J}_s + \sigma_s \,, \tag{88}$$

with an entropy current $\boldsymbol{J}_s$ and the local entropy production

$$\sigma_s = \bar{\mathcal{D}}\, \frac{(\boldsymbol{\nabla} n)^2}{n} \geq 0 \,, \tag{89}$$

which is always nonnegative by the second law of thermodynamics. The calculation can be carried out *ab initio* at the microscopic level of description starting from the hydrodynamic modes of diffusion which has been discussed in the previous sections, as shown elsewhere [74,85–88]. The detailed calculation shows that the positivity of the entropy production comes from the singular character of the hydrodynamic modes of diffusion and of the nonequilibrium steady state. This result is very natural because the singular character of the nonequilibrium states has its origin in the mixing property of the dynamics at the microscopic level as aforementioned. Our result therefore appears as the development of the intuitive idea that the positivity of the entropy production has its origin in the mixing induced in phase space, in particular, by the stretching and folding mechanism of a chaotic dynamics.

The entropy production (89) is physical as the following reasoning shows. Let us consider a slab of the Lorentz gas of width l between two large cubic reservoirs of the same volume $V = L^d$ containing respectively N_1 and N_2 light particles so that the particle densities in the reservoirs are respectively $n_1 = N_1/V$ and $n_2 = N_2/V$ (see Fig. 13). The velocity distribution can be taken to be a Maxwell-Boltzmann distribution at the same temperature T across the whole system. Since the diffusion process is isothermal the temperature T remains constant through the whole process and there is no heat produced. However, there is a current of particles equal to $j = \bar{\mathcal{D}}(n_1 - n_2)/l$ from the high-density reservoir to the low-density one because of the difference of concentrations across the Lorentz slab. This is an example of a process of exchange of matter as they are

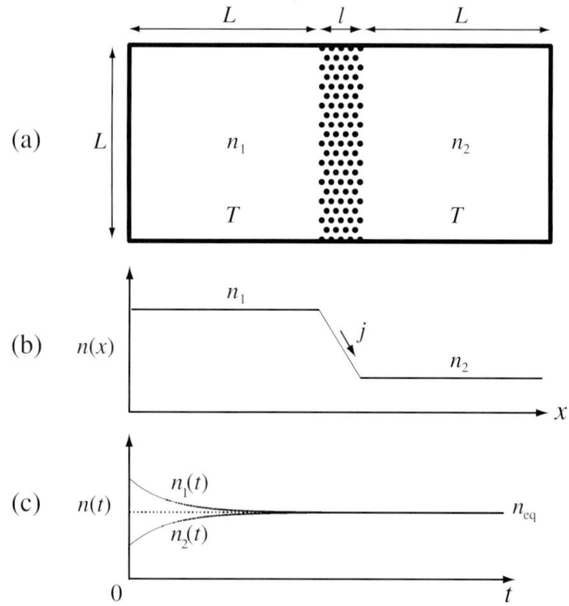

Fig. 13. Irreversible process of diffusion across a Lorentz slab leading to the equilibration of densities between both reservoirs: (a) configuration of the system; (b) initial density profile across the system; (c) time evolution of the densities which decay to their equilibrium value in each reservoir.

encountered in chemical thermodynamics. In such processes, a current is possible without simultaneous heat production. The current of particles decreases the difference of densities between both reservoirs. As long as there is a difference of densities the system is out of equilibrium until the thermodynamic equilibrium is reached when $n_1(t = \infty) = n_2(t = \infty) = n_{\rm eq} = (N_1 + N_2)/(2V)$ so that the current vanishes.

The irreversible process of equilibration by the diffusion of particles across the Lorentz slab leads to a production of entropy. Indeed, in the initial situation, the entropy of the whole system is equal to the entropies of both reservoirs if the Lorentz slab is supposed to be smaller than the reservoirs ($l \ll L$) so that its entropy is negligible:

$$S_{\rm initial} = S_1 + S_2 \qquad \text{with} \quad S_1 = k_{\rm B} N_1 \ln \frac{cV}{N_1} \quad \text{and} \quad S_2 = k_{\rm B} N_2 \ln \frac{cV}{N_2} \,, \tag{90}$$

with a constant c depending on temperature. The entropy after equilibration is given by

$$S_{\rm final} = k_{\rm B} (N_1 + N_2) \ln \frac{c\, 2V}{N_1 + N_2} \,, \tag{91}$$

so that the total entropy production is

$$\varDelta S = S_{\text{final}} - S_{\text{initial}} = k_{\text{B}} N_1 \ln \frac{2N_1}{N_1 + N_2} + k_{\text{B}} N_2 \ln \frac{2N_2}{N_1 + N_2} \tag{92}$$

$$\geq k_{\text{B}} N_1 \left(1 - \frac{N_1 + N_2}{2N_1}\right) + k_{\text{B}} N_2 \left(1 - \frac{N_1 + N_2}{2N_2}\right) = 0\,, \tag{93}$$

because of the inequality $\ln(1/x) \geq 1 - x$. Accordingly, the entropy production (93) which can be rewritten in the form

$$\varDelta S = k_{\text{B}} N_1 \ln \frac{n_1}{n_{\text{eq}}} + k_{\text{B}} N_2 \ln \frac{n_2}{n_{\text{eq}}} \tag{94}$$

is positive in the above isothermal process. The process is thus irreversible in the sense of the second law of thermodynamics.

Moreover, the entropy production (94) corresponds to some work which could otherwise be used if we replaced the Lorentz slab by a movable piston of equal volume and if the system was put in contact with a heat reservoir (see Fig. 14).

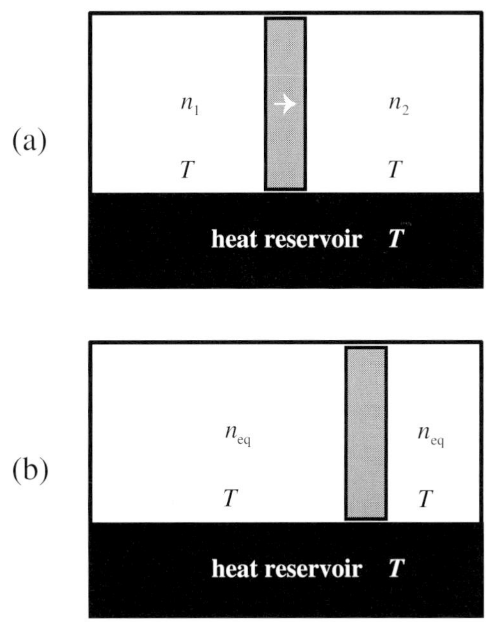

Fig. 14. Replacement of the Lorentz slab by a movable piston so that work can be extracted from the difference of densities if the whole system is in contact with a heat reservoir: (a) initial configuration of the system; (b) final configuration after the work W has been extracted.

In the initial situation where there is a difference of particle densities between both sides, the piston is submitted to a force due to the difference of pressures

$$P_{1,\text{initial}} = \frac{N_1 k_B T}{V} \qquad \text{and} \qquad P_{2,\text{initial}} = \frac{N_2 k_B T}{V} \,. \tag{95}$$

This force can produce a work if the piston is slowly moved until both pressures equilibrate. We suppose that the piston is moved so slowly that the process be isothermal. The equilibrium of pressures is reached when the high-density, high-pressure fluid has expanded and compressed the fluid on the other side so that both have the same equilibrium density n_{eq}, the temperature T being kept constant. The volumes of both fluids have changed so that

$$n_{\text{eq}} = \frac{N_1}{V_1} = \frac{N_2}{V_2} = \frac{N_1 + N_2}{2V} \quad \text{and} \quad P_{1,\text{final}} = \frac{N_1 k_B T}{V_1} = P_{2,\text{final}} = \frac{N_2 k_B T}{V_2} \,, \tag{96}$$

with $V_1 + V_2 = 2V$, the numbers of particles being here constant in each fluids. The work extracted in this process is

$$W = \int_{\text{initial}}^{\text{final}} P_1 \, dV_1 + P_2 \, dV_2 = k_B T \left(N_1 \ln \frac{V_1}{V} + N_2 \ln \frac{V_2}{V} \right) . \tag{97}$$

Since the ratios of volumes are equal to the ratios of densities

$$\frac{V_1}{V} = \frac{n_1}{n_{\text{eq}}} \qquad \text{and} \qquad \frac{V_2}{V} = \frac{n_2}{n_{\text{eq}}} \,, \tag{98}$$

we find that the extracted work is equal to the temperature multiplied by the entropy production of the previous process:

$$W = k_B T \left(N_1 \ln \frac{n_1}{n_{\text{eq}}} + N_2 \ln \frac{n_2}{n_{\text{eq}}} \right) = T \, \Delta S \,. \tag{99}$$

Since the initial and final energies of the particles in both fluids are equal by the equation of state (82), the work W has been pumped from the heat reservoir. Accordingly, we conclude that the entropy (94) produced by diffusion corresponds to some work which is lost in the process of diffusion. In this sense, the diffusion process dissipates an energy which could otherwise be used, which confirms that diffusion in the Lorentz gas is an irreversible process obeying the second law of thermodynamics.

5 Decay and Relaxation in Quantum Systems

Recent work has shown that different regimes exist in the relaxation of a quantum system [21–28]. One regime is the famous golden rule regime in the case of weak coupling. This relaxation proceeds slowly between the energy levels of a noninteracting system because of the weak coupling between these levels due to a perturbing interaction.

Another regime is the semiclassical regime in classically chaotic systems. In the semiclassical limit, the time correlation functions of the quantum dynamics have an early decay which follows the classical dynamics and its Pollicott-Ruelle resonances. Therefore, the decay rate of the quantum time correlation function becomes equal to the decay rate of the leading Pollicott-Ruelle resonance. In simple systems such as the dyadic baker map or the standard map at large values of the parameter K the leading nontrivial Pollicott-Ruelle resonance is essentially given by the positive Lyapunov exponent. For this reason, this regime is sometimes referred to as the "Lyapunov regime" [27,28].

In the following, we shall focus on the semiclassical regime and explain how the Pollicott-Ruelle resonances emerge from the quantum dynamics.

Let us consider a quantum time correlation function of an observable $\hat{A}$ defined with respect to an equilibrium density matrix $\hat{\rho} = p(\hat{H})$ where $\hat{H}$ is the Hamiltonian operator of the system:

$$\langle \hat{A}(t)\hat{B}(0)\rangle \equiv \mathrm{tr}\, p(\hat{H})\ \hat{A}(t)\ \hat{B}(0)\,. \tag{100}$$

The time evolution is determined by the Hamiltonian according to the Schrödinger equation as

$$\hat{X}(t) = \exp(+\mathrm{i}\hat{H}t/\hbar)\ \hat{X}\ \exp(-\mathrm{i}\hat{H}t/\hbar)\,. \tag{101}$$

The correlation function (100) is related to the one defined on the microcanonical ensemble

$$C_E(t) \equiv \mathrm{tr}\ \delta(E-\hat{H})\ \hat{A}(t)\ \hat{B}(0)\,, \tag{102}$$

by

$$\langle \hat{A}(t)\hat{B}(0)\rangle = \int dE\ p(E)\ C_E(t)\,. \tag{103}$$

If the system is closed the energy spectrum is discrete and the time evolution can be decomposed on the energy eigenstates

$$\hat{H}|n\rangle = E_n|n\rangle\,, \tag{104}$$

so that the time correlation function becomes

$$\langle \hat{A}(t)\hat{B}(0)\rangle = \sum_{m,n} p(E_m)\ A_{mn}\ B_{nm}\ \mathrm{e}^{+\frac{\mathrm{i}}{\hbar}(E_m-E_n)\,t}\,. \tag{105}$$

This is an almost-periodic function of time presenting an early decay followed by quantum fluctuations due to the discreteness of the energy spectrum. The quantum fluctuations manifest themselves on time scales longer than the Heisenberg time which is the time of resolution of the energy levels by using a time Fourier transform. For shorter times, the individual energy levels cannot be resolved and the time correlation function decays as if the spectrum was continuous. Since the level density is huge in many-body systems the Heisenberg time is more than cosmological and the energy spectrum can be assumed to be quasi-continuous.

At room temperature, the de Broglie wavelength is smaller than the mean free path and the semiclassical approximation is most appropriate in order to evaluate the time correlation function [89–92]. This is performed by first using the Weyl-Wigner expansion in which the operators $\hat{X}$ are transformed into functions $X_{\rm cl}(\{\boldsymbol{r}_i, \boldsymbol{p}_i\}_{i=1}^N)$ defined on the phase space of positions and momenta. Quantum oscillations are then taken into account by Gutzwiller's periodic-orbit corrections [93]. The semiclassical expansion can be carried out on the correlation function (102) defined for the microcanonical ensemble to get [91,92]

$$
\begin{aligned}
C_E(t) = & \int \frac{d^f r d^f p}{(2\pi\hbar)^f}\, \delta(E - H_{\rm cl})\, A_{\rm cl}(t)\, B_{\rm cl}(0) + O(\hbar^{-f+1}) \\
& + \frac{1}{\pi\hbar} \sum_p \sum_{r=1}^{\infty} \frac{\cos\left[\frac{r}{\hbar} S_p(E) - r\frac{\pi}{2}\mu_p\right]}{|\det(\mathbf{I} - \mathbf{m}_p^r)|^{\frac{1}{2}}} \\
& \times \oint_p A_{\rm cl}(\tau + t) B_{\rm cl}(\tau) d\tau + O(\hbar^0)\,,
\end{aligned}
\tag{106}
$$

where the classical observables evolve according to

$$
X_{\rm cl}(t) = \exp(-\hat{\mathcal{L}}_{\rm cl} t)\, X_{\rm cl}(0)\,, \tag{107}
$$

with the classical Liouvillian operator $\hat{\mathcal{L}}_{\rm cl} \equiv \{H_{\rm cl}, \cdot\}$. The leading term in Eq. (106) is precisely the classical correlation function which decays under the effect of the Frobenius-Perron operator and of its Pollicott-Ruelle resonances [22,23,91,92]. Beside the leading term, there is a Weyl-Wigner series of quantum corrections in powers of the Planck constant which can be computed by a similar method as the leading term. Beyond the Weyl-Wigner series, there are the periodic-orbit corrections given by a sum over all the unstable periodic orbits and their repetitions $r = 1, 2, 3, ...$, the system being assumed to be hyperbolic. The correction of the periodic orbit p depends on its reduced action $S_p(E) = \sum_{i=1}^N \oint_p \boldsymbol{p}_i \cdot d\boldsymbol{r}_i$, its Maslov index μ_p, its linearized Poincaré map $\mathbf{m}_p$, as well as the time correlation function calculated over the periodic orbit p. When the number of degrees of freedom $f = Nd$ increases the periodic-orbit corrections become negligible with respect to the leading terms of the Weyl-Wigner series.

A similar expansion holds for the quantum survival probability in open quantum systems such as the disk scatterers [20] and the open quantum graphs [24]. The survival probability of a particle in a domain $\mathcal{S}$ of indicator function $\chi(\boldsymbol{r})$ is defined by

$$
P(t) \equiv \int_{\mathcal{S}} |\psi_t(\boldsymbol{r})|^2\, d\boldsymbol{r} = {\rm tr}\, \chi(\hat{\boldsymbol{r}})\, {\rm e}^{-\frac{i}{\hbar}\hat{H}t}\, \hat{\rho}_0\, {\rm e}^{+\frac{i}{\hbar}\hat{H}t}\,, \tag{108}
$$

with the initial density matrix $\hat{\rho}_0 = |\psi_0\rangle\langle\psi_0|$. The survival probability can be written in the form of a time correlation function as

$$
P(t) = \int dE\, {\rm tr}\, \delta(E - \hat{H})\, \hat{\rho}_0\, {\rm e}^{+\frac{i}{\hbar}\hat{H}t}\, \chi(\hat{\boldsymbol{r}})\, {\rm e}^{-\frac{i}{\hbar}\hat{H}t}\,. \tag{109}
$$

In the semiclassical limit, the survival probability becomes

$$\begin{aligned} P(t) &= \int \frac{d^f r\, d^f p}{(2\pi\hbar)^f}\, \rho_{0\mathrm{cl}}\, \mathrm{e}^{-\hat{\mathcal{L}}_{\mathrm{cl}} t} \chi_{\mathrm{cl}} + O(\hbar^{-f+1}) \\ &+ \frac{1}{\pi\hbar} \int dE \sum_p \sum_{r=1}^{\infty} \frac{\cos\left[\frac{r}{\hbar} S_p(E) - r\frac{\pi}{2}\mu_p\right]}{|\det(\mathbf{I} - \mathbf{m}_p^r)|^{1/2}} \\ &\times \oint_p \rho_{0\mathrm{cl}}\, \mathrm{e}^{-\hat{\mathcal{L}}_{\mathrm{cl}} t} \chi_{\mathrm{cl}}\, d\tau + O(\hbar^0)\,. \end{aligned} \tag{110}$$

Here also, the time evolution of the leading term is ruled by the classical Frobenius-Perron operator so that the early decay of the survival probability is determined by the Pollicott-Ruelle resonances [20–24].

This result applies in particular to the scattering of waves on disk scatterers [14,20]. These systems have been studied in remarkable microwave experiments by Sridhar and coworkers, who have been able to obtain not only the escape rate but also the next Pollicott-Ruelle resonances of the two-, three- and four-disk scatterers [21].

We can thus say that the classical decay and its Pollicott-Ruelle resonances emerge out of the wave-mechanical dynamics. The Pollicott-Ruelle resonances are therefore important to understand the relaxation of a system not only in the classical context but also in the quantum one.

In simple systems such the K-adic baker maps defined as

$$(x_{n+1}, y_{n+1}) = \left(K x_n - l,\, \tfrac{y_n + l}{K}\right) \qquad \text{for} \qquad \frac{l}{K} < x_n < \frac{l+1}{K}\,, \qquad (l = 0, 1, 2, ..., K-1)\,, \tag{111}$$

the Pollicott-Ruelle resonances are integer multiples of the positive Lyapunov exponent $\lambda = \ln K$: $s_j = -j\lambda$ with $j = 0, 1, 2, 3, ...$ The reason is that the dynamics acts by a uniform stretching in the K-adic baker maps. The leading nontrivial Pollicott-Ruelle resonance of the standard map is also well approximated by the Lyapunov exponent at large values of the parameter K: $s_1 \approx -\lambda$. However, for general baker maps without a uniform stretching, the Pollicott-Ruelle resonances differ from multiples of the Lyapunov exponent [25]. Therefore, the connection between the decay rates and the Lyapunov exponent is not general. Instead, the decay rates are directly given by the Pollicott-Ruelle resonances which may become very small in spatially extended systems where they can connect to the hydrodynamics, although the Lyapunov exponents keep a value corresponding to the time scale of kinetics.

6 Conclusions

Thanks to the recent advances in dynamical systems theory, we have a better understanding of the dynamical basis of the macroscopic behavior of matter and, especially, of the mechanisms of relaxation toward the thermodynamic equilibrium. This understanding also concerns the nonequilibrium steady states.

An important role is played by Gibbs' property of mixing in the phase space of the underlying microscopic system. Dynamical chaos characterized by positive Lyapunov exponents and KS entropy induces the stretching and folding of phase-space volumes and provides an effective mechanism of mixing. This chaotic mixing generates fractal structures at the phase-space level of description when an explicit construction is carried out of the nonequilibrium objects or states such as the fractal repeller selected by absorbing boundary conditions, the hydrodynamic modes selected by quasiperiodic boundary conditions, or the nonequilibrium steady states selected by flux boundary conditions.

Different nonequilibrium states are thus selected by different boundary conditions which mathematically express the way nonequilibrium constraints are imposed on the system. As explained in Subsection 3.2, the nonequilibrium constraint can be imposed within the Liouvillian formulation of statistical mechanics in agreement with a Hamiltonian volume-preserving microscopic dynamics obeying Liouville's theorem of classical mechanics. In this way, Liouvillian thermostats are defined which can describe at the microscopic statistical level a broad variety of nonequilibrium systems from linearly conducting regimes close to the thermodynamic equilibrium to far-from-equilibrium regimes sustaining the Rayleigh-Bénard convection or other macroscopic fluid instabilities, as well as nonlinear chemical instabilities [94].

The Liouvillian statistical formulation leads, in particular, to the construction of the hydrodynamic and reactive modes which rule the long-time relaxation toward the thermodynamic equilibrium. The dispersion relations of the hydrodynamic modes can be given in terms of the Pollicott-Ruelle resoances. The modes themselves are the eigenstates associated with the Pollicott-Ruelle resonances and they can be constructed by a kind of renormalization semigroup generated by the time evolution of statistical ensembles of trajectories from the initial time to infinity [42]. The point-like nature of classical mechanics and the mixing property only allow a weak convergence to a statistical state of thermodynamic equilibrium. Thanks to the renormalization semigroup, modes of pure exponential decay can be constructed in hyperbolic systems with positive Lyapunov exponents.[5] These modes are represented by a singular density which is not a regular function as is the case in stochastic processes but a mathematical distribution or generalized function in the sense of Schwartz and Gel'fand [95,96]. If the underlying dynamics is chaotic (i.e., has a positive KS entropy), the cumulative function of these modes are fractal. In this context, relationships can be established between the irreversible properties and the characteristic quantities of chaos. These relationships have been obtained, in particular, for diffusion and reaction-diffusion in the Lorentz gases as well as in the multibaker models. These processes of exchange of matter obey the second law of thermodynamics

[5] Exponential decay does not proceed until arbitrarily along time in nonchaotic models where the decay is numerically observed to deviate from exponential beyond a finite time [56].

consistently at the macroscopic level as explained in Subsection 4.6 and at the microscopic level because of the singular character of the nonequilibrium states [74,85–88].

At the quantum-mechanical level of description, the emergence of classical relaxation can be understood thanks to the semiclassical method and Gutwiller's trace formula. The quantum time correlation functions can be expressed by a Weyl-Wigner series corrected by periodic-orbit contributions. The leading term of the Weyl-Wigner series is the classical time correlation function which decays according to the Pollicott-Ruelle resonances as explained in Section 5. In simple models, the Pollicott-Ruelle resonances correspond to the "Lyapunov decay regime" which have been recently investigated beside the "golden rule decay regime" [26–28]. All these results show that various processes of decay and relaxation can effectively be understood thanks to the recent advances in dynamical systems theory.

Appendix A: Exponential Decay in the Inverted Harmonic Potential

The inverted harmonic potential appears for instance in the vicinity of the maximum of the potential in the Hamiltonian system (39) [3]. If this Hamiltonian is restricted to quadratic terms and if a canonical transformation is carried out to the new canonical coordinates which are the x-coordinate along the unstable manifold and the y-coordinate along the stable manifold, we obtain the Hamiltonian

$$H = \lambda\, x\, y\,. \tag{A1}$$

The solution of Hamilton's equations

$$\begin{cases} \dot{x} = +\frac{\partial H}{\partial y} = +\lambda\, x\,, \\ \\ \dot{y} = -\frac{\partial H}{\partial x} = -\lambda\, y\,, \end{cases} \tag{A2}$$

is the flow

$$\Phi^t(x,y) = \left(\mathrm{e}^{+\lambda t}x, \mathrm{e}^{-\lambda t}y\right)\,. \tag{A3}$$

In the long-time limit $t \to +\infty$, $\exp(-\lambda t)$ is a small parameter in terms of which we can carried out a Taylor expansion of the statistical averages according to [3]

$$\langle A\rangle_t = \int dx\; dy\; f_0(x,y)\; A\Big(\underbrace{\mathrm{e}^{+\lambda t}x}_{=x'}, \mathrm{e}^{-\lambda t}y\Big) \tag{A4}$$

$$= \mathrm{e}^{-\lambda t} \int dx'\; dy\; f_0\left(\mathrm{e}^{-\lambda t}x', y\right)\; A\left(x', \mathrm{e}^{-\lambda t}y\right) \tag{A5}$$

$$= \mathrm{e}^{-\lambda t} \int dx'\; dy\; \sum_{l=0}^{\infty} \frac{1}{l!}\, \mathrm{e}^{-\lambda l t}\, x'^{l}\, \partial_x^l f_0(0,y)$$

$$\times \sum_{m=0}^{\infty} \frac{1}{m!} \, \mathrm{e}^{-\lambda m t} \, y^m \, \partial_y^m A(x',0) \tag{A6}$$

$$= \sum_{l,m=0}^{\infty} \mathrm{e}^{-\lambda(l+m+1)t} \, \frac{1}{m!} \int dx' \, x'^l \, \partial_y^m A(x',0)$$

$$\times \frac{1}{l!} \int dy \, y^m \, \partial_x^l f_0(0,y) \tag{A7}$$

$$= \sum_{l,m=0}^{\infty} \mathrm{e}^{-\lambda(l+m+1)t} \, \langle A|\Psi_{lm}\rangle \, \langle\tilde{\Psi}_{lm}|f_0\rangle \, , \tag{A8}$$

where the eigenstates can be identified as

$$\Psi_{lm}(x,y) = \frac{1}{m!} \, x^l \, (-\partial_y)^m \, \delta(y) \, , \tag{A9}$$

$$\tilde{\Psi}_{lm}(x,y) = \frac{1}{l!} \, y^m \, (-\partial_x)^l \, \delta(x) \, . \tag{A10}$$

The eigenstates are given by the derivatives of the Dirac distribution. The right-eigenstates Ψ_{lm} have the unstable manifold $y = 0$ for support, while the left-eigenstates $\tilde{\Psi}_{lm}$ have the stable manifold $x = 0$ for support. We can check that these distributions are respectively the eigensolutions of the Liouvillian operator and of its adjoint:

$$\hat{L} \, \Psi_{lm} = -\lambda \, (l+m+1) \, \Psi_{lm} \, , \tag{A11}$$

$$\hat{L} \, \tilde{\Psi}_{lm} = -\lambda \, (l+m+1) \, \tilde{\Psi}_{lm} \, . \tag{A12}$$

Accordingly, the Pollicott-Ruelle resonances of the inverted harmonic potential are simply given by the integer multiples of the Lyapunov exponent λ:

$$s_{lm} = -\lambda \, (l+m+1) \, , \tag{A13}$$

with $l, m = 0, 1, 2, 3, \ldots$

Appendix B: Exponential Relaxation in the Baker Map

The baker map is defined as the following area-preserving map of the unit square onto itself [52]

$$\phi(x,y) = \begin{cases} \left(2x, \frac{y}{2}\right) & 0 \le x \le \frac{1}{2} \, , \\ \left(2x-1, \frac{y+1}{2}\right) & \frac{1}{2} < x \le 1 \, , \end{cases} \tag{B1}$$

The unstable manifold are parallel to the x-axis, and the stable manifolds to the y-axis. The baker map is dynamically unstable with the positive Lyapunov exponent $\lambda = \ln 2$. Since the system is closed its KS entropy is equal to the positive Lyapunov exponent $h_{\mathrm{KS}} = \ln 2$ so that the baker map is chaotic. This map is known to be mixing with respect to the Lebesgue invariant measure on the unit square. Periodic-orbit theory shows that the Pollicott-Ruelle resonances of the baker map are

$$s_j = -\, j \,\ln 2 \qquad \text{with} \quad j = 0, 1, 2, 3, \ldots , \tag{B2}$$

of multiplicity $m_j = j + 1$. The spectral decomposition involves Jordan blocks for the multiple resonances.

The Jordan blocks do not arise in the case where the initial probability density f_0, as well as the observable A, do not depend on the x-coordinate. We therefore assume for simplicity that

$$f_0(x, y) \;=\; \begin{cases} 1 & 0 \le y < \frac{1}{2} \,, \\ 0 & \frac{1}{2} \le y \le 1 \,, \end{cases} \tag{B3}$$

and $A(x, y) = A(y)$. In this case, as a consequence of the Euler–Maclaurin expansion [97], we find that [3]

$$\begin{aligned} \langle A \rangle_t \;&=\; \langle A | \hat{P}^t | f_0 \rangle \\ &=\; \frac{1}{2} \int_0^1 A(y)\; dy \;+\; \sum_{j=1}^{\infty} \frac{A^{(j-1)}(1) - A^{(j-1)}(0)}{2^{jt}\; j!} \\ &\quad\times \int_0^{1/2} B_j(y)\; dy \end{aligned} \tag{B4}$$

$$=\; \frac{1}{2} \int_0^1 A(y)\; dy \;-\; \frac{A(1) - A(0)}{2^{t+3}} \;+\; O(2^{-3t}) \,, \tag{B5}$$

where $A^{(j)}(y) = d^j A / dy^j$ and $B_j(y)$ is the jth Bernoulli polynomial [97]. The first term of the asymptotic expansion of the statistical average is nothing but the long-time limit expected by the mixing property. The next term gives the slowest exponential decay. We notice that the coefficients of the exponential decays involve the values and the derivatives of the observable $A(y)$ at the ends of the unit interval where the baker map is defined, which is an unconventional feature of these expansions related to the fact that the eigenstates are mathematical distributions instead of regular functions.

Appendix C: Hydrodynamic Modes of Diffusion in the Multibaker Map

The multibaker map is a model of diffusion constructed by simplifying the Birkhoff map of the Lorentz gas. Different multibaker models have been constructed corresponding to different Lorentz models (see Fig. 15) [3,71,77,86,87,98–101].

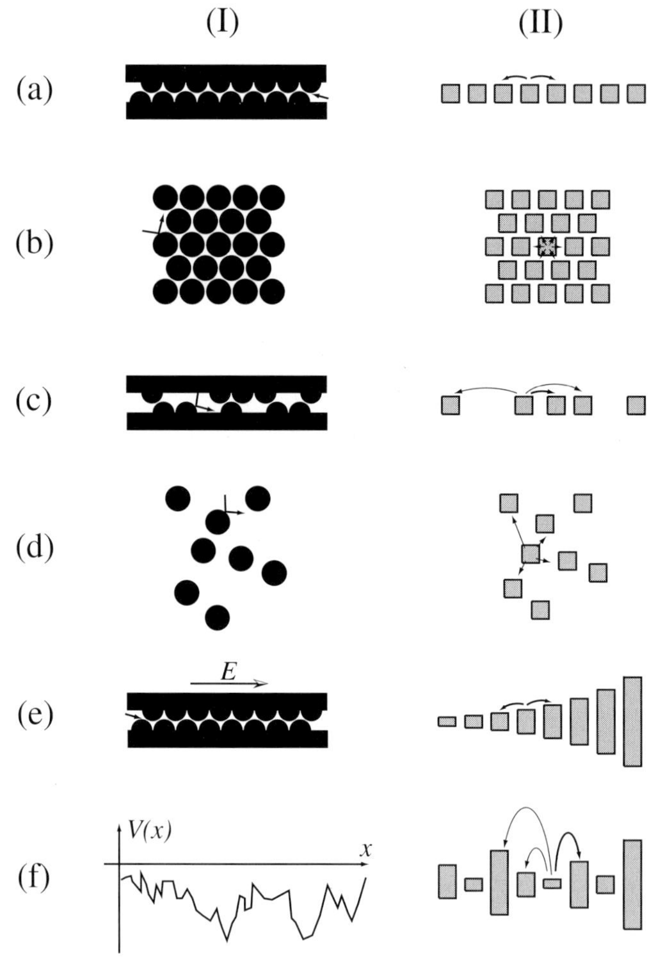

Fig. 15. Schematic representation of different Lorentz gases (I) and their corresponding multibaker model (II) for configurations of scatterers which are: (a) one-dimensional and periodic; (b) two-dimensional and periodic; (c) one-dimensional and disordered; (d) two-dimensional and disordered; (e) with a constant electric field E; (f) with a disordered potential $V(x)$.

A dyadic area-preserving version of the multibaker map is defined by the following mapping which acts on a chain of unit squares [77]

$$\phi(\ell, x, y) = \begin{cases} \left(\ell - 1, 2x, \frac{y}{2}\right) & 0 \leq x \leq \frac{1}{2}\,, \\ \\ \left(\ell + 1, 2x - 1, \frac{y+1}{2}\right) & \frac{1}{2} < x \leq 1\,, \end{cases} \tag{C1}$$

where $\ell \in \mathbb{Z}$ denotes the label of a square which is stretched, cut in two pieces, and mapped on the next-neighboring squares $\ell \pm 1$. This phase-space dynamics induces jumps of the particle along the chain of squares. These jumps generate a symmetric random walk with diffusion coefficient $\mathcal{D} = 1/2$. The dynamics of the multibaker is time-reversal symmetric under the involution $I(\ell, x, y) = (\ell, 1-y, 1-x)$ so that $\phi^{-1} = I \circ \phi \circ I$. The dyadic multibaker (C1) is hyperbolic with the positive Lyapunov exponent $\lambda = \ln 2$ and chaotic with the KS entropy $h_{\rm KS} = \lambda = \ln 2$. Moreover, the multibaker is mixing as the hard-disk Lorentz gas.

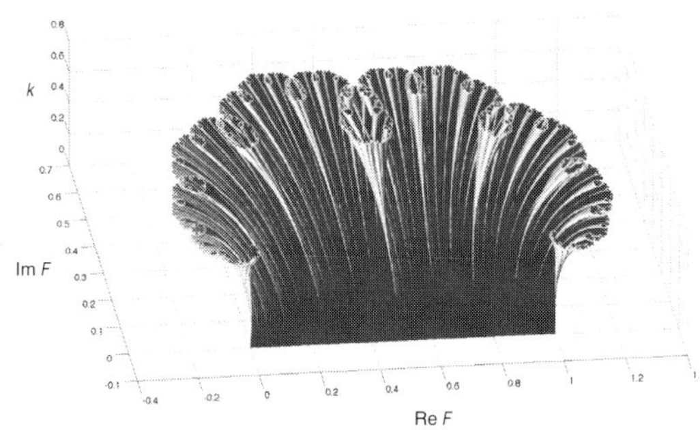

Fig. 16. Cumulative function of the hydrodynamic modes of diffusion of the dyadic multibaker map (C1). The cumulative functions $F_k(y)$ are depicted in the complex plane $(\text{Re } F_k, \text{Im } F_k)$ versus their wavenumber k varying in the interval $0 \leq k < 0.7$. The thermodynamic equilibrium corresponds to the vanishing wavenumber $k = 0$ for which the cumulative function is $F_0(y) = y$ with $0 \leq y \leq 1$.

The time evolution of statistical ensembles of noninteracting particles moving in the multibaker chain can be studied by spatial Fourier transform of the Frobenius-Perron operator. The resulting Pollicott-Ruelle resonances are given by

$$s_{j,k} = -j \ \ln 2 + \ln \cos k \,, \tag{C2}$$

where $k \in (-\pi, +\pi)$ is the wavenumber and $j = 0, 1, 2, 3, \ldots$ The hydrodynamic modes correspond to the leading Pollicott-Ruelle resonance with $j = 0$ so that the dispersion relation of these modes is given by

$$s_{0,k} = \ln \cos k = -\frac{1}{2}\, k^2 - \frac{1}{12}\, k^4 + \cdots \,, \tag{C3}$$

which allows us to identify the diffusion coefficient $\mathcal{D} = 1/2$ as well as the super-Burnett coefficient $\mathcal{B} = -1/12$. For small nonvanishing wavenumber k, the associated eigenstates have a singular density but a continuous cumulative function given by a de Rham iterative equation [3,102–104]. The cumulative functions form fractal curves in the complex plane (see Fig. 16) for the same reason as in the hard-disk and the Yukawa-potential Lorentz gases.

Acknowledgments

The author thanks Professor G. Nicolis for support and encouragement in this research. The author and this research are financially supported by the National Fund for Scientific Research (F. N. R. S. Belgium).

References

1. J.-P. Eckmann and D. Ruelle, Rev. Mod. Phys. **57**, 617 (1985).
2. J. R. Dorfman, *An Introduction to Chaos in Nonequilibrium Statistical Mechanics* (Cambridge University Press, Cambridge UK, 1999).
3. P. Gaspard, *Chaos, Scattering and Statistical Mechanics* (Cambridge University Press, Cambridge UK, 1998).
4. P. Gaspard and X.-J. Wang, Phys. Rep. **235**, 291 (1993).
5. I. P. Cornfeld, S. V. Fomin, and Ya. G. Sinai, Ergodic Theory (Springer, Berlin, 1982).
6. M. S. Green, J. Chem. Phys. **20**, 1281 (1952); **22**, 398 (1954).
7. R. Kubo, J. Phys. Soc. Jpn. **12**, 570 (1957).
8. M. Pollicott, Invent. Math. **81**, 413 (1985).
9. M. Pollicott, Invent. Math. **85**, 147 (1986).
10. D. Ruelle, Phys. Rev. Lett. **56**, 405 (1986).
11. D. Ruelle, J. Stat. Phys. **44**, 281 (1986).
12. D. Ruelle, J. Diff. Geom. **25**, 99, 117 (1987).
13. D. Ruelle, Commun. Math. Phys. **125**, 239 (1989).
14. P. Gaspard and D. Alonso Ramirez, Phys. Rev. A **45**, 8383 (1992).
15. M. Khodas and S. Fishman, Phys. Rev. Lett. **84**, 2837 (2000).
16. J. Weber, F. Haake, and P. Seba, Phys. Rev. Lett. **85**, 3620 (2000).
17. P. Gaspard and G. Nicolis, Phys. Rev. Lett. **65**, 1693 (1990).
18. J. R. Dorfman and P. Gaspard, Phys. Rev. E **51**, 28 (1995).
19. P. Gaspard and J. R. Dorfman, Phys. Rev. E **52**, 3525 (1995).
20. P. Gaspard and S. A. Rice, J. Chem. Phys. **90**, 2225, 2242, 2255 (1989).
21. W. Lu, M. Rose, K. Pance, and S. Sridhar, Phys. Rev. Lett. **82**, 5233 (1999); W. Lu, L. Viola, K. Pance, M. Rose, and S. Sridhar, Phys. Rev. E **61**, 3652 (2000); K. Pance, W. Lu, and S. Sridhar, Phys. Rev. Lett. **85** (2000).
22. F. Haake, *Quantum Signatures of Chaos* (Springer, Berlin, 2000).
23. O. Agam, Phys. Rev. E **61**, 1285 (2000).
24. F. Barra and P. Gaspard, Phys. Rev. E **63**, 066215 (2001); Phys. Rev. E **65**, 016205 (2001).
25. A. Jordan and M. Srednicki, *The Approach to Ergodicity in the Quantum Baker's Map*, preprint `chao-dyn/0108024` (2001).
26. R. A. Jalabert and H. M. Pastawski, Phys. Rev. Lett. **86**, 2490 (2001).
27. Ph. Jacquod, P. G. Silvestrov, and C. W. J. Beenakker, Phys. Rev. E **64** 055203(R) (2001).
28. Ph. Jacquod, I. Adagideli, and C. W. J. Beenakker, *Decay of the Loschmidt echo for quantum states with sub-Planck scale structures*, preprint `nlin.CD/0203052` (2002).
29. S. de Groot and P. Mazur, *Non-equilibrium Thermodynamics* (North-Holland, Amsterdam, 1962); reprinted by (Dover Publ. Co., New York, 1984).

30. G. Nicolis, *Introduction to Nonlinear Science* (Cambridge University Press, Cambridge UK, 1995).
31. S. Chapman and T. G. Cowling, *The Mathematical Theory of Non-Uniform Gases*, 3rd Edition (Cambridge University Press, Cambridge UK, 1970).
32. L. Kadanoff and P. C. Martin, Ann. Phys. (N.Y.) **24**, 419 (1963).
33. M. H. Ernst and J. R. Dorfman, Physica **61**, 157 (1972);
J. Stat. Phys. **12**, 311 (1975).
34. R. Balescu, *Equilibrium and Nonequilibrium Statistical Mechanics* (Wiley, New York, 1975).
35. J.-P. Boon and S. Yip, *Molecular Hydrodynamics* (Dover, New York, 1980).
36. P. Résibois and M. De Leener, *Classical Kinetic Theory of Fluids* (Wiley, New York, 1977).
37. E. G. D. Cohen, I. M. de Schepper, and M. J. Zuilhof, Physica B **127**, 282 (1984).
38. J. O. Hirschfelder, C. F. Curtiss, and R. B. Bird, *Molecular Theory of Gases and Liquids* (Wiley, New York, 1954).
39. P. Gaspard and R. Klages, Chaos **8**, 409 (1998).
40. I. Claus and P. Gaspard, J. Stat. Phys. **101**, 161 (2000).
41. I. Claus and P. Gaspard, *The fractality of the relaxation modes in deterministic reaction-diffusion systems*, preprint `cond-mat/0204264` to appear in Physica D (2002).
42. P. Gaspard and I. Claus, Phil. Trans. Roy. Soc. London A **360** (2002).
43. K. J. Laidler, *Theories of Chemical Reaction Rates* (McGraw-Hill, New York, 1969).
44. A. H. Zewail, *Femtochemistry: Ultrafast Dynamics of the Chemical Bond*, Vols. I-II (World Scientific, Singapore, 1994).
45. D. Forster, *Hydrodynamic Fluctuations, Broken Symmetry, and Correlation Functions* (Benjamin/Cummings, Reading, 1975).
46. P. D. Fleming and C. Cohen, Phys. Rev. B **13**, 500 (1975).
47. P. C. Hohenberg and P. C. Martin, Ann. Phys. (N.Y.) **34**, 291 (1965).
48. H. Goldstein, *Classical Mechanics* (Addison-Wesley, Reading, 1950).
49. J.-P. Eckmann, C.-A. Pillet, and L. Rey-Bellet, J. Stat. Phys. **95**, 305 (1999).
50. S. Berry, S. A. Rice, and J. Ross, Physical Chemistry (Wiley, New York, 1980).
51. J. W. Gibbs, *Elementary Principles in Statistical Mechanics* (Yale University Press, New Haven, 1902); reprinted by (Dover Publ. Co, New York, 1960).
52. V. I. Arnold and A. Avez, *Ergodic Problems of Classical Mechanics* (Benjamin, New York, 1968).
53. R. Abraham and J. E. Marsden, *Foundations of Mechanics* (Benjamin/Cummings Publ., Reading, 1978).
54. B. V. Chirikov, Phys. Reports **52**, 263 (1979).
55. A. Katok, Commun. Math. Phys. **111**, 151 (1987);
E. Gutkin, Physica D **19**, 311 (1986);
E. Gutkin, J. Stat. Phys. **83**, 7 (1996);
S. Troubetzkoy, Chaos **8**, 242 (1998).
56. C. P. Dettmann and E. G. D. Cohen, J. Stat. Phys. **101**, 775 (2000).
57. A. K. Zvonkin and L. A. Levin, Russian Math. Surveys **25**, 83 (1970).
58. P. Gaspard and X.-J. Wang, Proc. Natl. Acad. Sci. (USA) **85**, 4591 (1988).
59. J. L. Lebowitz, C. Maes, and E. R. Speer, J. Stat. Phys. **59**, 117 (1990).
60. J. P. Boon, D. Dab, R. Kapral, and A. Lawniczak, Phys. Reports **273**, 55 (1996).
61. J.-P. Rivet and J. P. Boon, *Lattice Gas Hydrodynamics* (Cambridge University Press, Cambridge UK, 2001).

62. P. Gaspard, Int. J. Mod. Phys. B **15**, 209 (2001); reprinted in: L. H. Y. Chen, J. Packer Jesudason, C. H. Lai, C. H. Oh, K. K. Phua & Eng-Chye Tan, Eds., *Challenges for the 21st Century, Proceedings of the International Conference on Fundamental Sciences: Mathematics and Theoretical Physics, Singapore, 13-17 March 2000* (World Scientific, Singapore, 2001) pp. 398-429.
63. P. Gaspard, G. Nicolis, A. Provata, and S. Tasaki, Phys. Rev. E **51**, 74 (1995).
64. P. Gaspard and S. Tasaki, Phys. Rev. E **64**, 056232 (2001).
65. P. Cvitanović and B. Eckhardt, J. Phys. A: Math. Gen. **24**, L237 (1991).
66. R. Artuso, E. Aurell, and P. Cvitanović, Nonlinearity **3**, 325, 361 (1990).
67. E. Helfand, Phys. Rev. **119**, 1 (1960).
68. P. Gaspard and F. Baras, Phys. Rev. E **51**, 5332 (1995).
69. H. van Beijeren, A. Latz, and J. R. Dorfman, Phys. Rev. E **63**, 016312 (2000).
70. I. Claus and P. Gaspard, Phys. Rev. E **63**, 036227 (2001).
71. T. Tél, J. Vollmer, and W. Breymann, Europhys. Lett. **35**, 659 (1996).
72. P. Gaspard, Phys. Rev. E **53**, 4379 (1996).
73. P. Gaspard, I. Claus, T. Gilbert, and J. R. Dorfman, Phys. Rev. Lett. **86**, 1506 (2001).
74. J. R. Dorfman, P. Gaspard, and T. Gilbert, *Entropy production of diffusion in spatially periodic deterministic systems*, preprint `nlin.CD/0203046` (2002).
75. T. Gilbert, J. R. Dorfman, and P. Gaspard, Nonlinearity **14**, 339 (2001).
76. P. Gaspard, Physica A **240**, 54 (1997).
77. S. Tasaki, and P. Gaspard, J. Stat. Phys. **81**, 935 (1995).
78. H. A. Lorentz, Proc. Roy. Acad. Amst. **7**, 438, 585, 684 (1905).
79. H. van Beijeren, Rev. Mod. Phys. **54**, 195 (1982).
80. C. P. Dettmann, *The Lorentz gas as a paradigm for nonequilibrium stationary states*, in: D. Szasz, Editor, *Hard Ball Systems and Lorentz Gas*, Encycl. Math. Sci. (Springer, Berlin, 2000).
81. L. A. Bunimovich, and Ya. G. Sinai, Commun. Math. Phys. **78**, 247, 479 (1980).
82. A. Knauf, Commun. Math. Phys. **110**, 89 (1987);
Ann. Phys. (N. Y.) **191**, 205 (1989).
83. D. Alonso, R. Artuso, G. Casati, and I. Guarneri, Phys. Rev. Lett. **82**, 1859 (1999).
84. R. K. Pathria, *Statistical Mechanics* (Pergamon, Oxford, 1972).
85. P. Gaspard, J. Stat. Phys. **88**, 1215 (1997).
86. S. Tasaki and P. Gaspard, Theoretical Chemistry Accounts **102**, 385 (1999).
87. S. Tasaki and P. Gaspard, J. Stat. Phys. **101**, 125 (2000).
88. T. Gilbert, J. R. Dorfman, and P. Gaspard, Phys. Rev. Lett. **85**, 1606 (2000).
89. M. Wilkinson, J. Phys. A: Math. Gen. **20**, 2415 (1987).
90. K. Richter, Europhys. Lett. **29**, 7 (1995);
B. Mehlig and K. Richter, Phys. Rev. Lett. **80**, 1936 (1998).
91. P. Gaspard and S. R. Jain, Pramana-J. of Phys. **48**, 503 (1997).
92. P. Gaspard, in: J. Karkheck, Editor, *Dynamics: Models and Kinetic Methods for Non-Equilibrium Many Body Systems*, NATO ASI Series (Kluwer, Dordrecht, 2000) pp. 425-456.
93. M. C. Gutzwiller, *Chaos in Classical and Quantum Mechanics* (Springer, New York, 1990).
94. I. Claus, *Microscopic Chaos, Fractals, and Reaction-Diffusion Processes* (Ph. D. Thesis, Université Libre de Bruxelles, Faculty of Sciences, February 2002).
95. N. Dunford and J. T. Schwartz, *Linear operators*, Vols. I-III (Interscience-Wiley, New York, 1958, 1963, 1971).

96. I. Gel'fand and G. Shilov, *Generalized Functions*, Vol. 2 (Academic Press, New York, 1968).
97. M. Abramowitz and I. A. Stegun, *Handbook of Mathematical Functions* (Dover, New York, 1972).
98. P. Gaspard, J. Stat. Phys. **68**, 673 (1992).
99. G. Radons, *Festkürperprobleme/Advances in Solid State Physics* **38**, 339 (1999).
100. T. Tél and J. Vollmer, *Entropy Balance, Multibaker Maps, and the Dynamics of the Lorentz gas*, in: D. Szász, Editor, *Hard Ball Systems and the Lorentz Gas* (Springer-Verlag, Berlin, 2000) pp. 367-418.
101. J. Vollmer, T. Tél, and L. Mátyás, J. Stat. Phys. **101**, 79 (2000).
102. H. H. Hasegawa and D. J. Driebe, Phys. Rev. **E50**, 1781 (1994).
103. D. J. Driebe, *Fully Chaotic Maps and Broken Time Symmetry* (Kluwer, Dordrecht, 1999).
104. S. Tasaki and P. Gaspard, Condensed-Matter Theory **66**, 23 (1996).

Relaxation and Noise in Chaotic Systems

S. Fishman and S. Rahav

Department of Physics, Technion, Haifa 32000, Israel

Abstract. For a class of idealized chaotic systems (hyperbolic systems) correlations decay exponentially in time. This result is asymptotic and rigorous. The decay rate is related to the Ruelle-Pollicott resonances. Nearly all chaotic model systems, that are studied by physicists, are not hyperbolic. For many such systems it is known that exponential decay takes place for a long time. It may not be asymptotic, but it may persist for a very long time, longer than any time of experimental relevance. In this review a heuristic method for calculation of this exponential decay of correlations in time is presented. It can be applied to model systems, where there are no rigorous results concerning this exponential decay. It was tested for several realistic systems (kicked rotor and kicked top) in addition to idealized systems (baker map and perturbed cat map). The method consists of truncation of the evolution operator (Frobenius-Perron operator), and performing all calculations with the resulting finite dimensional matrix. This finite dimensional approximation can be considered as coarse graining, and is equivalent to the effect of noise. The exponential decay rate of the chaotic system is obtained when the dimensionality of the approximate evolution operator is taken to infinity, resulting in infinitely fine resolution, that is equivalent to vanishing noise. The corresponding Ruelle-Pollicott resonances can be calculated for many systems that are beyond the validity of the Ruelle-Pollicott theorem.

1 Introduction

The purpose of the lectures that are summarized in this review is to describe the behavior of ensembles of chaotic systems. As in the case of statistical mechanics, the dynamics of ensembles turns out to be simpler then the one of individual systems. For chaotic systems [1–7] the long time asymptotic behavior is rigorously known to exhibit exponential decay of correlations only for a class of idealized systems (hyperbolic systems). Physical model systems do not belong to this class. In what follows *nonrigorous* methods, that enable the exploration of the long time behavior of chaotic systems, are presented and their application is demonstrated for several systems. The review is pedagogical and descriptive in nature, and is intended for an overview of the subject. The reader should consult the references for the details and the precise statements.

The dynamics of the systems considered in this review is of Hamiltonian nature. The dynamics of continuous systems is determined by the Hamilton equations:

$$\begin{aligned} \dot{p} &= -\frac{\partial \mathcal{H}}{\partial q} , \\ \dot{q} &= \frac{\partial \mathcal{H}}{\partial p} , \end{aligned} \tag{1}$$

where $\mathcal{H}$ is the Hamiltonian, while q and p are the position and momentum respectively. The phase space density $\rho(\mathbf{x})$, where $\mathbf{x} = (q, p)$ are the phase space points, satisfies the Liouville theorem,

$$\frac{d\rho}{dt} = 0\,, \tag{2}$$

or

$$\frac{\partial \rho}{\partial t} = \{\mathcal{H}, \rho\}\,, \tag{3}$$

where $\{...\}$ are the Poisson brackets. The main issue that will be discussed is the way ρ spreads in phase space for chaotic systems. For classical chaotic systems finer and finer structures develop for longer and longer times. These structures are reflected in ρ. Coarse graining, with some fixed scale of resolution, results in the truncation of the evolution of these very fine structures. On the finite (but arbitrarily small) resolution scale the equilibrium uniform phase space density is approached. In this review the asymptotic (in time) relaxation to this density will be discussed and in the end the coarse graining scale will be taken to 0. The results differ from the ones obtained without coarse graining, since the limits of infinite time and vanishing coarse graining do not commute. The approach, where a finite resolution scale is used and then the limit where this scale tends to zero is taken, is relevant for experimental realizations where the idealized classical description on the finest scales is destroyed resulting of the coupling to the environment.

Maps are transformations of phase space in discrete time, denoted by n. A map $\mathbf{F}$ is a transformation

$$\mathbf{x}_{n+1} = \mathbf{F}(\mathbf{x}_n)\,. \tag{4}$$

We study maps since they are easier to handle analytically and numerically and they reproduce the most interesting results found for systems evolving continuously in time.

Maps can be derived from Hamiltonians of the form:

$$\mathcal{H} = \frac{p^2}{2} + V(q) \sum_n \delta(t-n)\,. \tag{5}$$

For such maps the phase space area is preserved. Maps can also be defined with no reference to a Hamiltonian. The dynamics of phase space densities will be explored here for several area preserving maps: a. Kicked Rotor (Standard Map), b. Kicked Top, c. Arnold Cat Map and d. Baker Map.

a. Kicked Rotor (Standard Map)

A planar rotor that is periodically kicked is modeled by the Hamiltonian [1,3,7]

$$\mathcal{H} = \frac{J^2}{2} + K \cos\theta \sum_n \delta(t-n)\,, \tag{6}$$

where θ is the coordinate and J is the conjugate momentum. The Hamilton equations (1) are:

$$\begin{aligned} \dot{\theta} &= J \\ \dot{J} &= K \sin\theta \sum_n \delta(t-n) . \end{aligned} \tag{7}$$

Integration with respect to time results in the Standard Map

$$\begin{aligned} \theta_{n+1} &= \theta_n + J_n \\ J_{n+1} &= J_n + K \sin\theta_{n+1} , \end{aligned} \tag{8}$$

where θ_n and J_n are the angle and angular momentum just after the n-th kick. The equations (8) define a map of the form (4)

$$(\theta_{n+1}, J_{n+1}) = \mathbf{F}(\theta_n, J_n) . \tag{9}$$

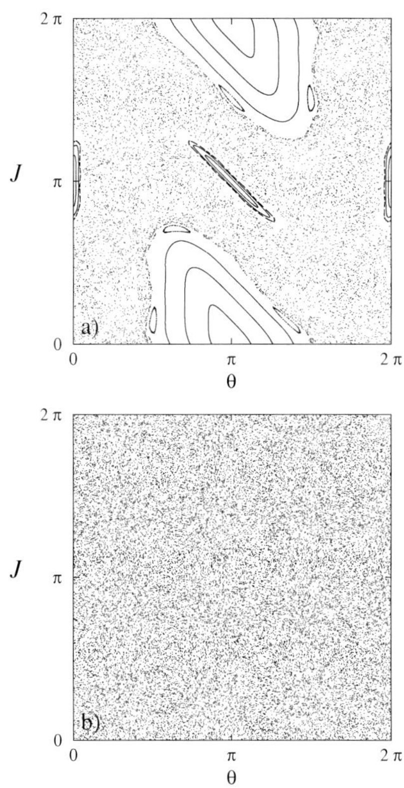

Fig. 1. Phase space portraits of the weakly chaotic kicked rotor with stochasticity parameter $K = 2$ (a), and of the strongly chaotic kicked rotor with $K = 10$ (b).

It is easily checked that the map is area preserving. It becomes more chaotic as the stochasticity parameter K increases. The phase space is plotted in Fig. 1.

A variant of this map provides a good description of driven laser cooled atoms [8] and of beams deflected by dielectrics with a modulated index of refraction [9].

b. Kicked Top

A large spin can be described by a classical vector

$$\mathbf{J} = j\;(\sin\theta\cos\varphi, \sin\theta\sin\varphi, \cos\theta)\,, \tag{10}$$

where θ and φ are the polar angles. The kicked top map is defined by the transformation [3]

$$\mathbf{F} = R_z(\tau\cos\theta)R_z(\beta_z)R_y(\beta_y)\,, \tag{11}$$

applied to $\mathbf{J}$ where $R_i(\beta)$ is the rotation around the axis i by the angle β. The nonlinearity results of the dependence of a rotation on the angle θ. The chaoticity of the map increases with τ that is the stochasticity parameter. The canonical phase space variables for this map are $q = \varphi$ and $p = \cos\theta$. The phase space is plotted in Fig. 2.

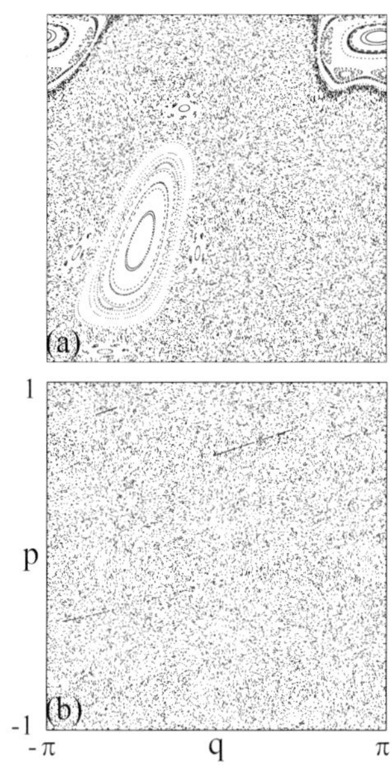

Fig. 2. Phase space portraits of the weakly chaotic kicked top with stochasticity parameter $\tau = 2.1$ (a), and of the strongly chaotic kicked top with $\tau = 10.2$ (b). (Fig. 1 of [10]).

c. Arnold Cat Map

It is defined in the $[0,1] \times [0,1]$ square of the (x, y) phase plane by [2,4]

$$\begin{aligned} x_{n+1} &= x_n + y_n \qquad \bmod 1\,, \\ y_{n+1} &= x_n + 2y_n \qquad \bmod 1\,. \end{aligned} \tag{12}$$

The evolution is demonstrated in [2,4].

d. Baker map

It is defined in the $[0,1] \times [0,1]$ square of the (x,y) phase plane by [4,6,1]

$$(x_{n+1}, y_{n+1}) = \mathbf{F}(x_n, y_n) = \begin{cases} (2x_n, y_n/2)\,, & \text{for } 0 \le x < \frac{1}{2}\,, \\ (2x_n - 1, (y_n+1)/2)\,, & \text{for } \frac{1}{2} \le x < 1\,. \end{cases} \tag{13}$$

This transformation is demonstrated in Fig. 3.

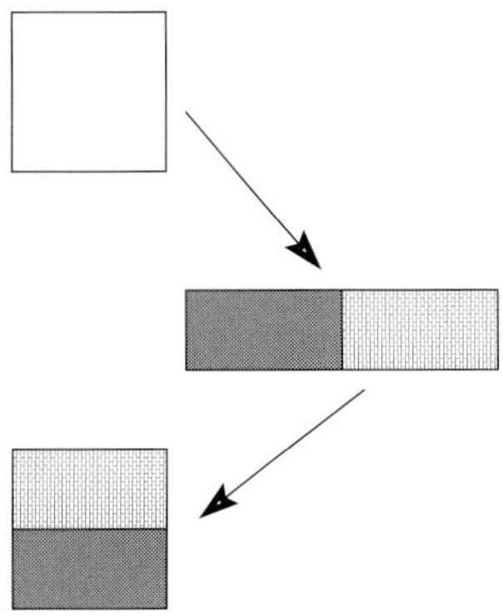

Fig. 3. The baker map.

The systems (a) and (b) model physical problems. Their phase space is mixed, namely the dynamics in some parts is chaotic and in some parts it is regular. Systems (c) and (d) are very idealized. For these systems the motion is chaotic in the entire phase space.

A system is chaotic if points, that are initially close, spread exponentially in phase space. For chaotic systems the separation between points grows with the number of steps as

$$\epsilon(n) = \epsilon(0)\Lambda^n\,, \tag{14}$$

in the limit $n \to \infty$ and $\epsilon(0) \to 0$. The Lyapunov exponent is

$$\lambda = \ln \Lambda \tag{15}$$

and Λ is sometimes called the Lyapunov number. A system that evolves in a bounded region in phase space is called *chaotic* if its Lyapunov exponent λ according to the definition (14,15) is positive. (More generally the Lyapunov numbers are the eigenvalues of the tangent map. The Lyapunov number defined by (14) is equal to the magnitude of the largest eigenvalue.) One can study the local expansion and contraction in the various directions. If for a map, for any point of phase space, in each direction, either expansion or contraction takes place, or in other words if none of the real parts of the local Lyapunov exponents vanishes, the system is called *hyperbolic*.

Exponential spread of trajectories in phase space does *not* imply exponential decay of correlations in time. Such correlations may result of sticking to small structures in phase space, as is the case for the stadium billiard and for the Sinai billiard as well as for chaotic trajectories of mixed systems, such as the kicked rotor and the kicked top. For hyperbolic systems, such as the baker map, typically correlations in time decay exponentially.

Chaotic motion is reminiscent of a random walk. Therefore it is instructive to review the dynamics of the probability density of random walkers. Assume a random walk on a one dimensional lattice, with equal probability of 1/2 to make a step to the left or to the right. The path of each walker is complicated but the evolution of a probability density of random walkers is simple. It is simple in particular in the continuum approximation (the limit of vanishing lattice spacing) where it is described by the diffusion equation

$$\frac{\partial}{\partial t}\rho(x,t) = D\frac{\partial^2}{\partial x^2}\rho(x,t), \tag{16}$$

D is the diffusion coefficient and $\rho(x,t)$ is the density of random walkers at the point x at time t. Let us assume that the random walk is confined to an interval of length s, for example $-s/2 < x < s/2$. The current at the ends of the interval vanishes, resulting in the boundary conditions

$$\frac{\partial \rho}{\partial x}\Big|_{x=\pm s/2} = 0 \tag{17}$$

on (16). This problem is similar to the standard quantum mechanical problem of a particle in an infinite square well, but with the boundary conditions (17) and with no $i\hbar$ on the left hand side of (16). Therefore the probability density can be expanded as:

$$\rho(x,t) = \sum_{k=0}^{\infty} a_k e^{-\gamma_k t} u_k(x)\,, \tag{18}$$

where

$$u_k = \begin{cases} \sqrt{\frac{2}{s}}\left[\cos\frac{\pi x}{s}k + \left(\sqrt{\frac{1}{2}}-1\right)\delta_{k0}\right], & k \text{ even}\,, \\ \sqrt{\frac{2}{s}}\sin\frac{\pi x}{s}k\,, & k \text{ odd}\,, \end{cases} \tag{19}$$

form an orthonormal basis and

$$\gamma_k = \left(\frac{\pi k}{s}\right)^2 D\,. \tag{20}$$

The units of a_k are of $1/\sqrt{[\text{length}]}$. The expansion coefficients a_k are determined by the initial density $\rho(x,t=0)$. The equilibrium density is

$$\lim_{t\to\infty}\rho(x,t) = a_0\sqrt{\frac{1}{s}}\,, \tag{21}$$

and the asymptotic approach to this density is

$$\rho(x,t) - \frac{a_0}{\sqrt{s}} \sim a_1 e^{-\gamma_1 t} u_1(x)\,, \tag{22}$$

with the relaxation rate $\gamma_1 = \left(\frac{\pi}{s}\right)^2 D$. The density-density correlation function is

$$C(t) = \int_{-s/2}^{s/2} dx \rho(x,0)\rho(x,t) - a_0^2 = \sum_{k=1}^{\infty} |a_k|^2 e^{-\gamma_k t} \sim |a_1|^2 e^{-\gamma_1 t} . \tag{23}$$

Hence also the correlations decay with the rate γ_1.

The example of the probability density of random walkers demonstrates that although the trajectories of specific random walkers are complicated their probability density ρ follows simple dynamics. We turn now to study the phase space probability density $\rho(\mathbf{x})$, that for Hamiltonian dynamics satisfies (3). The introduction of such probability densities, that are sufficiently smooth, is actually a coarse graining over some fine scale. It is an averaging over phase space of the same type that was performed for random walkers. For any given area preserving map $\mathbf{F}$ it is instructive to introduce the one step evolution operator $\hat{U}$,

$$\rho_{n+1}(\mathbf{x}) = \hat{U}\rho_n(\mathbf{x}) , \tag{24}$$

where $\rho_n(\mathbf{x})$ is the phase space density after n steps of the map. If the operator is defined on a space of sufficiently smooth functions it is called the *Frobenius-Perron operator*. If $\mathbf{F}$ is area preserving and invertible, namely if $\mathbf{F}^{-1}$ exists, then $\hat{U}$ is unitary. All eigenvalues of $\hat{U}$ are on the unit circle in the complex plane, they take the form $e^{-i2\pi\alpha}$, with real α. The corresponding "eigenfunctions" are not square integrable. An example of such a function will be presented now [11]. Assume the map $\mathbf{F}$ has a periodic orbit of period n, that is:

$$\begin{aligned} \mathbf{x}_{j+1} &= \mathbf{F}(\mathbf{x}_j) , \qquad\qquad j = 1, 2, ...n , \\ \mathbf{x}_1 &= \mathbf{F}(\mathbf{x}_n) . \end{aligned} \tag{25}$$

The function

$$\psi_\alpha(\mathbf{x}) = \sum_{j=1}^{n} \delta(\mathbf{x} - \mathbf{x}_j) e^{i2\pi\alpha j} , \tag{26}$$

with $\alpha = l/n$, where l is an integer, is an "eigenfunction" of $\hat{U}$ with the eigenvalue $e^{-i2\pi\alpha}$. Here ψ_α is called "eigenfunction" if

$$\hat{U}\psi_\alpha = e^{-i2\pi\alpha}\psi_\alpha , \tag{27}$$

and no requirement is made that it belongs to the space where $\hat{U}$ is defined. For example if $\hat{U}$ is defined on the space of square integrable functions, ψ_α of (26) does not belong to this space and is not a function in the usual sense (it is a distribution).

The operator $\hat{U}$ can be represented by an infinite dimensional matrix. In physics applications it is natural to use finite dimensional approximations. Let us assume that a basis that is ordered by increased resolution is used (the basis states may be for example sines and cosines or orthogonal polynomials). For the truncation of $\hat{U}$, namely a finite dimensional approximation, the matrix is not unitary, and its eigenvalues are inside the unit circle in the complex plane. They may vary with the dimension of the matrix. The natural question is: Do these eigenvalues approach the unit circle in the limit of an infinite dimensional matrix? It will be demonstrated in what follows that this is *not* the case for chaotic systems and values *inside* the unit circle are approached in this limit, in spite of the fact that the infinite dimensional matrix representing $\hat{U}$ is unitary. A heuristic justification for this behavior was proposed by F. Haake [10]. Chaotic systems, in contrast with regular ones, exhibit phase space structures on all scales. These structures are revealed during the evolution. The operator $\hat{U}$ couples the fine scales via its matrix elements that couple to states with high resolution. As a result of the truncation, probability that was originally transferred to the fine scales is lost, resulting in nonunitarity of $\hat{U}$. Convergence of the eigenvalues to values inside the unit circle in the complex plane, in the limit of infinite dimension of the matrix, results of the asymptotic self similarity of the chaotic dynamics. For regular motion, on the other hand, as the dimension of the matrix is increased the eigenvalues approach the unit circle [10] (see Sec. 4).

It will be shown that for chaotic systems the finite dimensional approximations of $\hat{U}$, in the limit of large dimension, describe the decay of correlations. The phase space density-density correlation function, in analogy with (23) is

$$\begin{aligned} C(n) &= \int d\mathbf{x} \rho(\mathbf{x},0)\rho(\mathbf{x},n) - \rho_\infty^2 \Omega \\ &= \int d\mathbf{x} \rho(\mathbf{x},0)\hat{U}^n \rho(\mathbf{x},0) - \rho_\infty^2 \Omega \,, \end{aligned} \tag{28}$$

where Ω is the volume of the chaotic component in phase space. The equilibrium density is $\rho_\infty = \lim_{n\to\infty} \rho(\mathbf{x},n)$ that is independent of position in phase space. A more general correlation function is:

$$C^{(A,B)}(n) = \int d\mathbf{x} A(\mathbf{x})\hat{U}^n B(\mathbf{x}) \,, \tag{29}$$

where we assumed

$$\lim_{n\to\infty} \hat{U}^n A(\mathbf{x}) = \lim_{n\to\infty} \hat{U}^n B(\mathbf{x}) = 0 \,. \tag{30}$$

This can always be obtained by the subtraction of the asymptotic value. For simplicity it will be assumed that both A and B are real.

It is useful to study the Laplace transforms of the correlation functions. For this purpose we introduce the resolvent

$$\hat{R}(z) = \frac{1}{z}\sum_{j=0}^{\infty} \hat{U}^j z^{-j} = \frac{1}{z-\hat{U}} \tag{31}$$

where z is a complex number. Since $\hat{U}$ is unitary the sum is convergent for $|z| > 1$. This is analogous to the usual definition used in quantum mechanics:

$$\hat{R}(E) = \frac{1}{i\hbar}\int_0^\infty dt \hat{U} \exp\left(\frac{i}{\hbar}Et - \frac{\epsilon}{\hbar}t\right) , \tag{32}$$

where

$$\hat{U} = \exp\left(-\frac{i}{\hbar}\hat{\mathcal{H}}t\right) \tag{33}$$

is the evolution operator, leading to

$$\hat{R}(E) = \frac{1}{E - \hat{\mathcal{H}} + i\epsilon} \tag{34}$$

for $\epsilon > 0$. The convergence of the integral (32) requires $\epsilon > 0$. In analogy the convergence of the sum (31) requires $|z| > 1$ in (31). The discrete Laplace (one sided Fourier) transform of the correlation function (29) is

$$\tilde{C}^{(A,B)}(z) = \sum_{n=0}^{\infty} C^{(A,B)}(n) z^{-n} = \int d\mathbf{x} A(\mathbf{x}) \left[z\hat{R}(z)\right] B(\mathbf{x}) \,, \tag{35}$$

as one finds from (31).

Let ψ_i be an "eigenfunction" of $\hat{U}$ with the eigenvalue z_i,

$$\hat{U}\psi_i = z_i \psi_i \,. \tag{36}$$

By "eigenfunction" we mean here that it satisfies (36), but it may not be a function in the usual sense (for example it may be a distribution). Then z_i is a pole of the matrix elements of the resolvent as one can see from (31). The correlation function, involving an "eigenfunction", takes the form

$$\tilde{C}^{(A,\psi_i)}(z) = \langle A|\psi_i\rangle \frac{z}{z - z_i} \,, \tag{37}$$

where the Dirac notation

$$\langle A|B\rangle = \int d\mathbf{x} A^*(\mathbf{x}) B(\mathbf{x}) \tag{38}$$

is used (A^* is the complex conjugate of A). Here A and B were assumed to be real. Let us expand $B(\mathbf{x})$ in terms of the "eigenfunctions" of $\hat{U}$ as

$$B(\mathbf{x}) = \sum_i b_i \psi_i(\mathbf{x}) \,. \tag{39}$$

Then

$$\tilde{C}^{(A,B)}(z) = \frac{r_i}{z - z_i} \,, \tag{40}$$

where the poles z_i depend only on $\hat{U}$ while the residues

$$r_i = b_i \langle A|\psi_i\rangle z \,, \tag{41}$$

depend also on A and B.

Because of the unusual nature of the "eigenfunctions" ψ_i the manipulations leading to (40) are only heuristic. The Ruelle-Pollicott theorem justifies the expression (40) for *hyperbolic* systems if A and B are typical smooth functions (for a precise statement of the theorem that is transparent for physicists see [12]). Moreover the theorem assures that $|z_i| < 1$, except for the eigenvalue $z_0 = 1$ corresponding to the equilibrium density ψ_0, that is independent of $\mathbf{x}$. The z_i are called the Ruelle-Pollicott resonances.

The existence of poles with $|z_i| < 1$ implies the decay of correlations with the rate $\ln |z_i|$. This results of

$$C^{(A,B)}(n) = \frac{1}{2\pi i} \oint_{|z|=1} \frac{dz}{z} \tilde{C}^{(A,B)}(z) z^n \tag{42}$$

and the application of the residue theorem to (40).

For any approximation of dimension N of the Frobenius-Perron operator $\hat{U}$, that will be denoted by $\hat{U}^{(N)}$ in what follows, the eigenfunctions $\psi_i^{(N)}$, corresponding to the eigenvalues $z_i^{(N)}$, are well defined. For hyperbolic systems, where the Ruelle-Pollicott theorem applies, it is reasonable to assume that in the limit $N \to \infty$, the eigenvalues $z_i^{(N)}$ approach the Ruelle-Pollicott resonances, for a typical choice of the basis. Inspired by the Ruelle-Pollicott theorem the following heuristic scheme for the calculation of the decay rates of correlations is proposed:

1. Introduce an orthogonal basis where the basis states are ordered by resolution. These states, for example, may be orthogonal polynomials or trigonometric functions $(\cos 2\pi kx, \sin 2\pi kx)$ or exponentials $\exp i2\pi kx$.
2. Calculate the matrix elements of $\hat{U}$ in this basis.
3. Introduce a truncation of dimension N of this matrix, that will be denoted as $\hat{U}^{(N)}$.
4. Calculate the eigenvalues $z_i^{(N)}$ and the eigenfunctions $\psi_i^{(N)}$ of $\hat{U}^{(N)}$. The eigenfunctions $\psi_i^{(N)}$ are finite linear combinations of the basis states, and therefore are smooth.
5. Take the limit $N \to \infty$. In this limit $z_i^{(N)} \to z_i$ and $\psi_i^{(N)} \to \psi_i$.

From the Ruelle-Pollicott theorem it is expected, for hyperbolic systems, that the Ruelle-Pollicott resonances are obtained in this way, if a typical basis is used. The limiting functions ψ_i are singular [13–15]. This scheme was applied to various systems, including also to mixed systems [10,16–20]. Limiting eigenvalues z_i satisfying $|z_i| < 1$ were obtained in this way, implying exponential relaxation of correlations with the rate $\ln |z_i|$ that takes place for a very long time (for some nonhyperbolic systems it is known that eventually the correlations decay as a power-law, but the time required to obtain this power-law behavior may be too long for any physical relevance). Although the Hamiltonian dynamics of trajectories is reversible the phase space densities exhibit relaxation that is irreversible, and is similar to the behavior of probability densities of random walkers that follow irreversible dynamics.

In Sec. 2 the heuristic scheme for the calculation of the z_i will be demonstrated for the baker map that is hyperbolic and consequently the Ruelle-Pollicott theorem applies. Exploration of a hyperbolic system that is a modification of the cat map is briefly mentioned in Sec. 3. In Sec. 4 the scheme will be applied to the kicked top and in Sec. 5 it will be applied to the kicked rotor, that are mixed systems, where the Ruelle-Pollicott theorem is not valid. Also the limitations on the validity of this scheme for such systems will be demonstrated. The main conclusions are summarized in Sec. 6. The review will follow closely references [10,13,19], that are marked by ** in the list of references. A detailed discussion on the relaxation of correlations in chaotic systems and on related topics can be found in [5,6]. The scheme for the use of truncation in the calculation of the Ruelle-Pollicott resonances is discussed in a wider context in [14].

2 The Frobenius-Perron Operator for the Baker Map, a Demonstration

The heuristic prescription for the analysis of the Frobenius-Perron operator that was outlined in the Introduction will be demonstrated for the baker map (13), where all the results are exactly known. The analysis will make use of exact results obtained in [13].

The basis states that will be used are

$$\langle xy|kl\rangle = \tilde{P}_k(x)\tilde{P}_l(y) , \tag{43}$$

where

$$\tilde{P}_k(x) = \sqrt{2k+1}P_k(1-2x) = \frac{\sqrt{2k+1}}{k!}\frac{d^k}{dx^k}x^k(1-x)^k , \tag{44}$$

are the modified Legendre polynomials while P_k are the Legendre polynomials. A basis of orthogonal polynomials is natural, since the order of the polynomial is not affected by the map (13). The orthonormality of the basis

$$\langle kl|k'l'\rangle = \delta_{kk'}\delta_{ll'} , \tag{45}$$

follows from the orthonormality of the modified Legendre polynomials

$$\int_0^1 dx \tilde{P}_k(x)\tilde{P}_{k'}(x) = \delta_{kk'} . \tag{46}$$

The basis is naturally ordered by increased resolution, since this is a property of orthogonal polynomials. The action of $\hat{U}$ of (24) on any phase space density is

$$\hat{U}\rho(\mathbf{x}) = \rho(\mathbf{F}^{-1}(\mathbf{x})) , \tag{47}$$

since the points that are at $\mathbf{x}$ at time step $n+1$ were at $\mathbf{F}^{-1}(\mathbf{x})$ at the time step n. For the baker map (13)

$$\rho(\mathbf{F}^{-1}(x,y)) = \begin{cases} (\rho(x/2, 2y) , & \text{for } 0 \le y < \frac{1}{2} , \\ (\rho((x+1)/2, 2y-1) , & \text{for } \frac{1}{2} \le y < 1 . \end{cases} \tag{48}$$

From (47) and (48) one finds that the matrix elements of $\hat{U}$ are

$$\langle kl|\hat{U}|k'l'\rangle = \frac{1}{2}\left[1 + (-1)^{k+k'+l+l'}\right] I_{kk'} I_{l'l} , \tag{49}$$

where

$$I_{kk'} = \int_0^1 dx \tilde{P}_k(x) \tilde{P}_{k'}(x/2) . \tag{50}$$

Using (44) and integrating (50) by parts k times one finds that if $k > k'$,

$$I_{kk'} = 0 . \tag{51}$$

This results in the *nonrecurrence* property of $\hat{U}$

$$\langle kl|\hat{U}|k'l'\rangle = 0 , \qquad \text{for } k > k' \text{ or } l' > l . \tag{52}$$

During the evolution, probability is transformed from states with k' to states with k only if $k \leq k'$, therefore after application of $\hat{U}$ the density becomes more uniform because the weight of lower order Legendre polynomials is increased. This is expected since x is the unstable direction, where stretching takes place, making the density more uniform. In the y direction, on the other hand, the density is transformed from l' to l only if $l \geq l'$, therefore during the evolution, the weight of the high order Legendre polynomials increases. This results of the fact that y is the stable direction, where contraction takes place, resulting in complexity that increases with time.

For $k \leq k'$ one finds [13]

$$I_{kk'} = \frac{[(2k+1)(2k'+1)]^{1/2}}{2^k} \sum_{l=0}^{k'-k} \left(-\frac{1}{2}\right)^l \frac{(k'+k+l)!}{(k'-k-l)!(2k+l+1)!l!} . \tag{53}$$

In particular $I_{kk} = 2^{-k}$ and the diagonal matrix elements are

$$\langle kl|\hat{U}|kl\rangle = \frac{1}{2^{k+l}} . \tag{54}$$

The matrix elements of $\hat{U}^n$ can be calculated with the help of the nonrecurrence property (52). Introducing the resolution of the identity one finds

$$\langle kl|\hat{U}^n|k'l'\rangle = \sum_{[k_i;l_i]} \langle kl|\hat{U}|k_1 l_1\rangle \langle k_1 l_1|\hat{U}|k_2 l_2\rangle ..... \quad ...\langle k_{n-1} l_{n-1}|\hat{U}|k'l'\rangle , \tag{55}$$

where $[k_i; l_i] \equiv [k_1, \ k_2 .....k_{n-1}; \ l_1, \ l_2, ...l_{n-1}]$ and the nonrecurrence property implies $k \leq k_1 \leq k_2 ..... \quad .. \leq k'$ and $l \geq l_1 \geq l_2 .... \quad ... \geq l'$. In particular the diagonal matrix elements satisfy

$$\langle kl|\hat{U}^n|kl\rangle = \langle kl|\hat{U}|kl\rangle^n = \left(\frac{1}{2^{k+l}}\right)^n . \tag{56}$$

In the limit $n \to \infty$ the off-diagonal matrix elements are dominated by powers of the diagonal matrix elements, since the nonrecurrence property limits the number of non-diagonal matrix elements in (55).

The diagonal matrix elements of the resolvent can be easily calculated with the help of (31). For $|z| > 1$ the sum in (31) is convergent and using (56) one finds

$$\langle kl|\hat{R}(z)|kl\rangle = \sum_{n=0}^{\infty} z^{-(n+1)}\langle kl|\hat{U}|kl\rangle^n = \frac{1}{z - 2^{-(k+l)}} . \tag{57}$$

These matrix elements are singular for $z = 2^{-(k+l)}$, and except for $k = l = 0$ all these singular points are inside the unit circle in the complex plane. This is also the case for the off-diagonal matrix elements. Therefore the Ruelle-Pollicott resonances are

$$z_m = 2^{-m} \tag{58}$$

and their degeneracy is $m + 1$. They are related to the decay of correlations by (40). They were obtained by analytic continuation from $|z| > 1$ where the resolvent is defined. One can show [13] that the matrix elements exhibit a cut at $|z| = 1$ and the physically relevant poles are on the Riemann sheet continued from $|z| > 1$.

What happens if the Frobenius-Perron operator is restricted to $k, k' \leq k_{max}$ and $l, l' \leq l_{max}$ and $\hat{U}$ is approximated by $\hat{U}^{(N)}$, an $N = k_{max}l_{max}$ dimensional matrix, resulting of the truncation of $\hat{U}$? The nonrecurrence property holds for $\hat{U}^{(N)}$ and consequently (56) holds. Therefore for the N dimensional matrix $\hat{U}^{(N)}$ the diagonal matrix elements are zeros of the characteristic polynomial, that is of order N. To see this, note that by the Hamilton-Cayley theorem, $\hat{U}^{(N)}$ satisfies its characteristic polynomial, namely $\sum_{j=0}^{N} p_j \hat{U}^{(N)j} = 0$, and calculate the diagonal matrix elements of this expression with the help of (56). Consequently the diagonal matrix elements are eigenvalues of $\hat{U}^{(N)}$, taking the values 2^{-m}. The multiplicity of the eigenvalue 2^{-m} is $m + 1$ if $m \leq k_{max}, l_{max}$. By a similarity transformation the matrix can be transformed to the canonical Jordan form. Another way to see that the eigenvalues of $\hat{U}^{(N)}$ coincide with the diagonal matrix elements is by using the indices $l_{max} - l$ and $l_{max} - l'$ instead of l and l'. With these indices, $\hat{U}^{(N)}$ is upper triangular, therefore the eigenvalues coincide with the diagonal matrix elements. The eigenvalues are independent of k_{max} and l_{max}, therefore these are of the form 2^{-m} also in the limit $k_{max} \to \infty$ and $l_{max} \to \infty$, in spite of the fact that $\hat{U}$ is unitary. For $\hat{U}^{(N)}$ the eigenfunctions are finite combinations of Legendre polynomials and are therefore smooth. In the limit $k_{max} \to \infty$ and $l_{max} \to \infty$ the right eigenfunctions are independent of x and are polynomials of y, while the left eigenfunctions tend to distributions [13].

The main results that were obtained for the baker map are:

1. A basis that is ordered by increased resolution was introduced. A truncation of dimension N in this basis was implemented.
2. The Ruelle-Pollicott resonances were calculated from $\hat{U}^{(N)}$, the N dimensional truncation of $\hat{U}$. As N increases more eigenvalues are revealed. The values of the eigenvalues do not depend on the truncation dimension N, and therefore also in the limit of infinite N they remain at values that were obtained for finite N.

3. The poles of the matrix elements of the resolvent (31) and (57) were obtained by analytic continuation from $|z| > 1$.
4. The eigenfunctions of $\hat{U}$ are functions of y only (y is the stable direction).

In this section the procedure for the calculation of Ruelle-Pollicott resonances from $\hat{U}^{(N)}$, a truncated matrix approximating $\hat{U}$, was demonstrated for a system where the results are known exactly. In the following sections this method will be applied to systems where there is no exact theory.

The basis used here was of Legendre polynomials. If a basis of sines and cosines is used it is impossible to find the Ruelle-Pollicott resonances in this way, since each of the basis states collapses to 0 after a finite number of applications of $\hat{U}$ [13,21]. On the basis of the Ruelle-Pollicott theorem, we believe that for a typical basis one does not encounter problems of this nature.

3 The Frobenius-Perron Operator of a Modified Cat Map

For the Arnold cat map, defined by (12), correlations in time decay faster than exponentially. A modification, where the function $f(x_{n+1})$, that is defined by

$$f(x) = \frac{K_0}{2\pi} \left[\cos(2\pi x) - \cos(4\pi x)\right] , \tag{59}$$

is added to the equation for y_{n+1} in (12), is a hyperbolic system if K_0 is sufficiently small [22]. The correlations for this system decay exponentially in time. It was studied with the help of the truncated Frobenius-Perron operator, according to the scheme that was outlined in the introduction. The leading Ruelle-Pollicott resonances were found with the help of a variational approach [16].

4 The Frobenius-Perron Operator for the Kicked Top

We turn now to apply the method where $\hat{U}$ is approximated by a finite dimensional matrix to mixed systems (where in some parts of phase space the motion is chaotic and in other parts it is regular). The Ruelle-Pollicott theorem does not apply to such systems. It will be demonstrated, however, that the Ruelle-Pollicott resonances are meaningful and describe the decay of correlations in the chaotic component of mixed systems for a time that may be very long. Asymptotically power-law decay takes place, due to sticking to regular islands in phase space.

In this section the kicked top [3], that was defined by (10) and (11), will be analyzed. It is a summary of results that were presented in [10] (see also [17] and [18]). The phase space is shown in Fig. 2.

A natural basis for this problem is of spherical harmonics

$$Y_{lm}(\theta, \varphi) = (-1)^m \sqrt{\frac{2l+1}{4\pi} \frac{(l-m)!}{(l+m)!}} P_l^m(\cos\theta) e^{im\varphi} . \tag{60}$$

Alternatively one can choose a basis of real functions with $e^{im\varphi}$ replaced by $\sin m\varphi$ and $\cos m\varphi$, resulting in a real matrix for $\hat{U}$. It is naturally ordered by increased resolution. A truncation at $l = l_{max}$ is introduced. The dimension of the truncated matrix $\hat{U}^{(N)}$, approximating $\hat{U}$ is $N = (l_{max} + 1)^2$. As l_{max} is increased the eigenvalues of $\hat{U}^{(N)}$ converge to values inside the unit circle in the complex plane, as can be seen from Table 1. The corresponding eigenfunctions

Table 1. Eigenvalues of $\hat{U}^{(N)}$ for $\tau = 10.2$ truncated at $l_{\max} = 30, 40, 50$ and 60 (Table I of [10]).

$l_{\max} = 30$	$l_{\max} = 40$	$l_{\max} = 50$	$l_{\max} = 60$
0.7700	0.7688	0.7523	0.7696
0.3075 ±i 0.5740	0.3429 ±i 0.6140	0.3523 ±i 0.6211	0.3550 ±i 0.6199
−0.3170 ±i 0.6003	−0.3348 ±i 0.6272	−0.3444 ±i 0.6283	−0.3388 ±i 0.6243
−0.0042 ±i 0.7161	−0.0002 ±i 0.7133	−0.0100 ±i 0.6930	−0.0058 ±i 0.7080
−0.7025	−0.7228	−0.7155	−0.7165
0.6544	0.6230	0.6495	0.6480
...	−0.5619	−0.5753	−0.5667

are presented in Fig. 4. Dark-shaded regions in phase space indicate large amplitudes of the eigenfunctions. In Figs. 5 (a) and (b) it is demonstrated that as l_{max} increases finer details of the eigenfunctions are revealed. This is a result of increased resolution. Comparison with Fig. 6 (b) demonstrates that the eigenfunctions are large and tend to be uniform on the unstable manifold. In Fig. 7 the density-density correlation function (proportional to (28)) is plotted. It is demonstrated that also here the Ruelle-Pollicott resonances describe the decay of correlations for a very long time.

The behavior of the eigenfunctions localized in regular regions and the corresponding eigenvalues is very different. Eigenfunctions that are localized in islands of regular motion presented in Fig. 2 (a) are depicted in Fig. 8. The eigenfunctions follow elliptic orbits inside the regular island and therefore do not involve the resolution of very fine structures. The corresponding eigenvalues approach the unit circle in the complex plane as l_{max} increases.

It is instructive to examine the evolution operator $\hat{U}^{-1}$ corresponding to the inverse map $\mathbf{F}^{-1}$. Unitarity implies

$$\hat{U}^{-1} = \hat{U}^{\dagger} . \tag{61}$$

A basis can be chosen so that $\hat{U}$ is real and $\hat{U}^{\dagger} = \hat{U}^{T}$, where T denotes transpose. The eigenvalues of $\hat{U}^{T}$ are equal to the ones of $\hat{U}$. Introducing the truncation one finds

$$\hat{U}^{T(N)} = \hat{U}^{(N)T} , \tag{62}$$

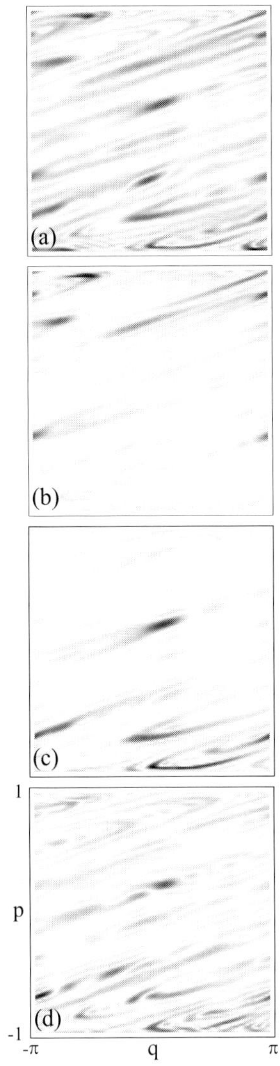

Fig. 4. The eigenfunctions corresponding to the eigenvalues 0.7696 (a), $-0.3388 \pm$ i0.6243 (b), $-0.0058 \pm$ i0.7080 (c), 0.6480 (d) of $\hat{U}^{(N)}$ for $\tau = 10.2$, and $l_{\max} = 60$ (Fig. 4 of [10]).

where $\hat{U}^{T(N)}$ is the truncated $\hat{U}^T$ and $\hat{U}^{(N)T}$ is the transpose of $\hat{U}^{(N)}$, the truncated $\hat{U}$. The eigenfunctions of $\hat{U}^{T(N)}$ are localized on the unstable manifold of $\mathbf{F}^{-1}$, that is the stable manifold of $\mathbf{F}$, as is clearly demonstrated comparing Fig. 5 (c) with Fig. 6 (c). By eigenfunctions we meant so far the right eigenfunctions. The left eigenfunctions of $\hat{U}$ (and $\hat{U}^{(N)}$) are the right eigenfunctions of $\hat{U}^T$ (and $\hat{U}^{T(N)}$).

The main results that were obtained for the kicked top are:

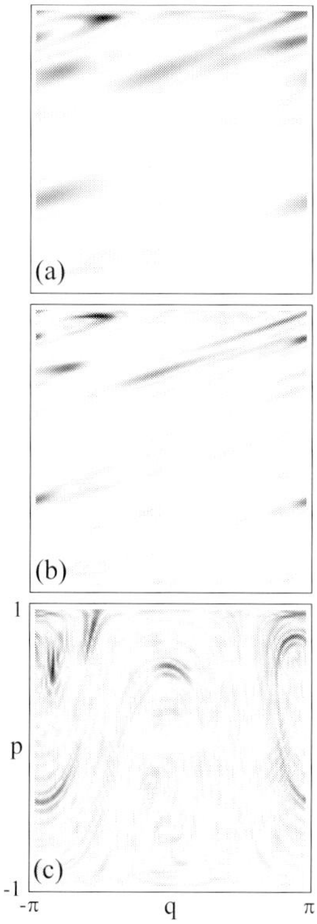

Fig. 5. As phase-space resolution increases from $l_{\max} = 30$ (a) to $l_{\max} = 60$ (b), the eigenfunction of $\hat{U}^{(N)}$ for the eigenvalue $-0.3388 \pm \mathrm{i}0.6243$ (for $\tau = 10.2$) gains new structures on finer scales. The corresponding eigenfunction (c) of $\hat{U}^{T(N)}$ (resolution $l_{\max} = 60$) is localized at the same periodic orbit as the eigenfunction (b) but with stable and unstable manifolds interchanged. See also Fig. 6 (a) for the periodic orbits, Fig. 6(b) for the unstable and Fig. 6(c) for the stable manifolds (Fig. 5 of [10]).

1. As N or l_{max} increases the eigenvalues of $\hat{U}^{(N)}$, corresponding to eigenfunctions localized in the chaotic component of phase space, converge to values inside the unit circle. Unlike the case of the baker map, the eigenvalues $z_i^{(N)}$ depend on N and $\lim_{N\to\infty} z_i^{(N)} = z_i$ with $|z_i| < 1$, except the unit eigenvalue corresponding to the equilibrium density.
2. These eigenvalues determine the decay of correlations in the chaotic component of phase space.
3. In the chaotic component of phase space the right eigenfunctions of the truncated Frobenius-Perron operator $\hat{U}^{(N)}$ are localized on the unstable manifold of $\mathbf{F}$, and tend to be uniform along this manifold. The left eigenfunctions are localized on the stable manifold. These manifolds are not as simple as for the baker map.
4. The eigenvalues of $\hat{U}^{(N)}$ corresponding to eigenfunctions localized in regular regions approach the unit circle as N or l_{max} increases.

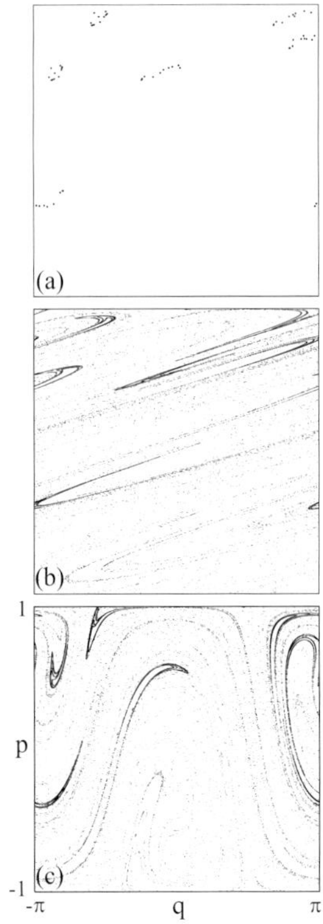

Fig. 6. (a): The 12 orbits of primitive length 6 that can be identified for the resonances $0.3550 \pm i0.6199$ and $-0.3388 \pm i0.6243$ related to the eigenfunction shown in Figs. 4(b) and 5(b). The unstable manifolds of these orbits are shown in (b), and the stable manifolds in (c) (Fig. 6 of [10]).

5 The Frobenius-Perron Operator for the Kicked Rotor

In this section the decay of correlations in time will be studied for the kicked rotor, defined by the Hamiltonian (6), and its dynamics is given by the standard map (8) [1,3,7]. The phase space is presented in Fig. 1. It is a mixed system (where in some regions of phase space the motion is chaotic and in other regions it is regular). In this section we summarize results that were presented in [19] (see also [20], [23] and [16]).

For sufficiently large values of the stochasticity parameter K (that takes a typical value), the spread of angular momentum is diffusive, to a good approximation, namely

$$\left\langle (J_n - J_0)^2 \right\rangle \approx 2Dn\,, \tag{63}$$

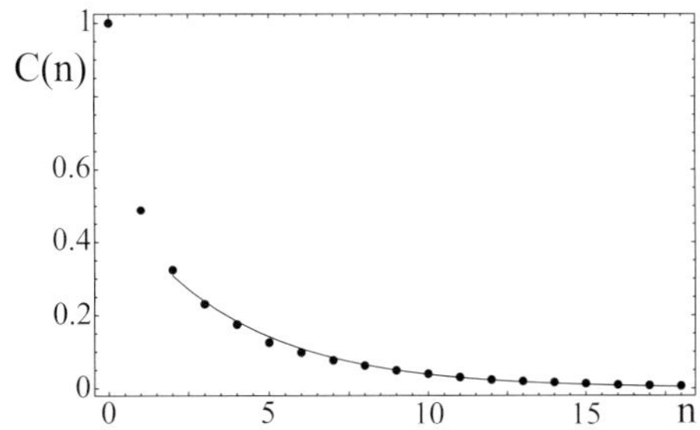

Fig. 7. The decay of $C(n)$ (dots) for $\rho(0)$ corresponding to the eigenfunction shown in Fig. 4(b) (also 5(b)). The numerical fit (line) yields a decay factor 0.7706 (compared to $|z_i^{(N)}| \approx 0.7103$). (Fig. 7 of [10]).

for large n. The average is over the initial points and is denoted by $\langle .....\rangle$. A very crude way to get this result is to iterate (8) n times to obtain

$$J_n - J_0 = K \sum_{i=1}^{n} \sin\theta_i\,, \tag{64}$$

then squaring and averaging with the assumption of absence of angular correlations

$$\langle \sin\theta_i \sin\theta_j \rangle = \frac{1}{2}\delta_{ij}\,, \tag{65}$$

one obtains (63) with $D = \frac{1}{4}K^2$. A more careful calculation, that takes some of the angular correlations into account results in [24]

$$D(K) = \frac{K^2}{4}\left(1 - 2J_2(K) + ...\right)\,. \tag{66}$$

It is actually an expansion in powers of $1/\sqrt{K}$. Because of the mixed nature of phase space it is expected that diffusion may not take place asymptotically in time [25]. Moreover, there are values of K (near integer multiples of 2π) where acceleration in momentum is found for some initial conditions, resulting in $\left\langle (J_n - J_0)^2 \right\rangle \sim n^2$. In what follows the Frobenius-Perron operator will be used to study whether correlations in angular momentum decay as expected for a true diffusion process, described in the Introduction (see (16)-(23)). We will analyze also the decay of angular correlations.

The Frobenius-Perron operator, corresponding to the map (8) on the torus

$$(0 \le J < 2\pi s)\,,$$

$$(0 \le \theta < 2\pi)\,, \tag{67}$$

where s is integer, will be studied. For the usual kicked rotor $s \to \infty$. Noise is added to the standard map by the addition of a random variable ξ_n with variance σ^2 on the right hand side of the equation for θ_{n+1} in (8) following [24] and [26,27]. The noise leads to truncation of the Frobenius-Perron operator. The natural basis for the analysis is

$$\langle J\theta|km\rangle = \frac{1}{\sqrt{2\pi}}\frac{1}{\sqrt{2\pi s}}\exp(im\theta)\exp\left(i\frac{kJ}{s}\right), \tag{68}$$

where k and m are integers. Note that the functions $\langle J\theta|k, m = 0\rangle$ form the basis of eigenstates of the diffusion operator in the angular momentum J. In this basis, in presence of noise, the Frobenius-Perron operator is

$$\langle km|\hat{U}^{(\sigma)}|k'm'\rangle = J_{m-m'}\left(\frac{k'K}{s}\right)\exp\left(-\frac{\sigma^2}{2}m^2\right)\delta_{k-k',ms}\,. \tag{69}$$

For $\sigma \neq 0$ the operator is not unitary. The effective truncation is $m < 1/\sigma$, $|k - k'| < s/\sigma$, $|m - m'| < k'K/s$. The $\sigma \to 0$ limit, corresponding to the $N \to \infty$ limit of the previous sections, will be taken in the end of calculation. The analysis will be performed for

$$\begin{aligned} s &\gg 1\,, \\ K &\gg 1\,, \\ \frac{kK}{s} &\ll 1\,, \end{aligned} \tag{70}$$

and the limit $\sigma \to 0$ will be taken in the end of the calculation.

The Ruelle resonances are identified from the matrix elements of the resolvent (31). It is useful to define a variant of the resolvent

$$\hat{R}'(z) = \sum_{j=0}^{\infty}\hat{U}^j z^j = \frac{1}{1 - z\hat{U}}, \tag{71}$$

that is related to $\hat{R}(z)$ via

$$\frac{1}{z}\hat{R}'\left(\frac{1}{z}\right) = \hat{R}(z)\,. \tag{72}$$

Also the matrix elements of $\hat{R}$ and $\hat{R}'$ satisfy (72). Let

$$R_{12} = \langle k_1 m_1|\hat{R}(z)|k_2 m_2\rangle \tag{73}$$

and

$$R_{12}' = \langle k_1 m_1|\hat{R}'(z)|k_2 m_2\rangle \tag{74}$$

be the matrix elements of $\hat{R}$ and $\hat{R}'$ respectively. Therefore a singularity of R_{12} at z_c implies a singularity of R_{12}' at $1/z_c$ and vice versa. The series (31) for $\hat{R}(z)$ converge for $|z| > 1$ while the series (71) for $\hat{R}'(z)$ converge for $|z| < 1$. Therefore the Ruelle-Pollicott resonance (located inside) that is closest to the unit circle corresponds to the singularity of R'_{12} that is closest to the unit circle (located outside). This singularity is the radius of convergence of the series

$$R'_{12} = \sum_{j=0}^{\infty} a_j z^j \,, \tag{75}$$

where

$$a_j = \langle k_1 m_1 | \hat{U}^j | k_2 m_2 \rangle \,. \tag{76}$$

According to the Cauchy-Hadamard theorem (see [28]) the inverse of the radius of convergence is given by

$$r^{-1} = \lim_{j\to\infty} \sup \sqrt[j]{|a_j|} \tag{77}$$

and asymptotically

$$|a_j| \sim \frac{\text{const.}}{r^j} \,. \tag{78}$$

Since the radius of convergence is the singularity of $R'_{12}(z)$ that is the closest to the unit circle it satisfies $r = 1/z_c$ and $\sqrt[j]{a_j} \to z_c$ in the limit $j \to \infty$ (at least for some subsequence of $\{j\}$).

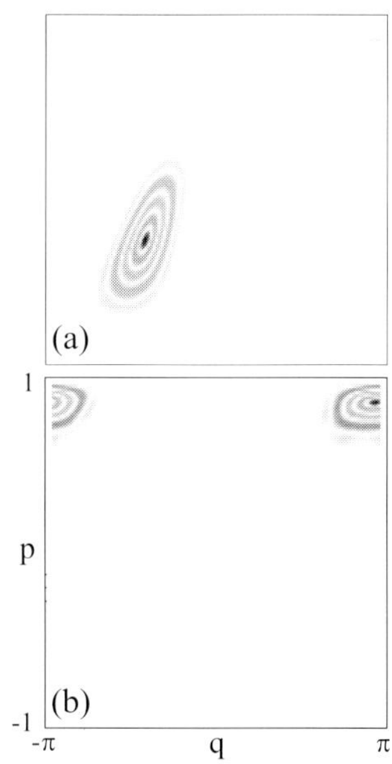

Fig. 8. Eigenfunctions of $\hat{U}^{(N)}$ with $l_{\max} = 60$ that are localized on elliptic islands shown in Fig. 2 (a) for the weakly chaotic case $\tau = 2.1$. The corresponding eigenvalues 0.999976 (a) and 0.999974 (b) are almost at unity (Fig. 3 of [10]).

The limit of the series (76) was found in the leading order in perturbation theory in the small parameters implied by (70). The calculation can be performed separately for the various values of k, resulting in

$$z_k = \exp\left(-\frac{k^2K^2}{4s^2}\left(1 - 2J_2(K)e^{-\sigma^2}\right)\right), \tag{79}$$

corresponding to diffusion modes in presence of noise. Taking the limit $\sigma \to 0$ one finds

$$z_k = e^{-k^2 D(K)/s^2}, \tag{80}$$

where $D(K)$ is given by (66). This relaxation is similar to the one that is found for usual diffusion (see (16-23)). The relaxation rates are

$$\gamma_k = \frac{k^2}{s^2} D(K), \tag{81}$$

that correspond to (20). These describe the decay of correlations in angular momentum J. The equilibrium density corresponds to z_0. Exploring the subspace involving $|k = 0, m\rangle$, one finds that the Ruelle-Pollicott resonances, corresponding to the slowest mode of the relaxation of correlations in angle, take four values

$$\pm\tilde{z}, \qquad \pm i\tilde{z}, \tag{82}$$

with

$$\tilde{z} = \sqrt{|J_{2m^*}(m^*K)| \exp\left(-\sigma^2 m^{*2}/2\right)}. \tag{83}$$

Taking the limit $\sigma \to 0$ one finds

$$\tilde{z} = \sqrt{|J_{2m^*}(m^*K)|}, \tag{84}$$

where m^* is the integer m that maximizes $|J_{2m}(mK)|$ for a given value of K. A similar result was found in [16] by a variational approach. The corresponding relaxation rate is

$$\tilde{\gamma} = \ln \tilde{z}. \tag{85}$$

In order to test the analytical results, the correlation function of the form (29)

$$C_{fg}(n) = \langle f|\hat{U}^n|g\rangle, \tag{86}$$

where f and g are taken to be basis states of (68), was computed numerically. The results are plotted in Fig. 9. Clear exponential decay is found. From the slopes, the relaxation rates γ_k are obtained for the various modes, k. The values of D implied by γ_k are calculated from (81) and compared to the ones obtained from (66) in Fig. 10. Good agreement was found. In Figs. 9 and 10 large values of the stochasticity parameter, $K \geq 10$, were used. In this regime the regular regions are extremely small (they are invisible in Fig. 1), therefore the theory presented in this review is expected to work very well, as is indeed found in Fig. 10. In Fig. 11 the correlation function (86) is presented for $K \leq 20$. From these results the relaxation rates γ_k were calculated. The resulting values of the diffusion coefficient obtained from (81) are compared to the ones found from (66) in Fig. 12, for $K \leq 20$.

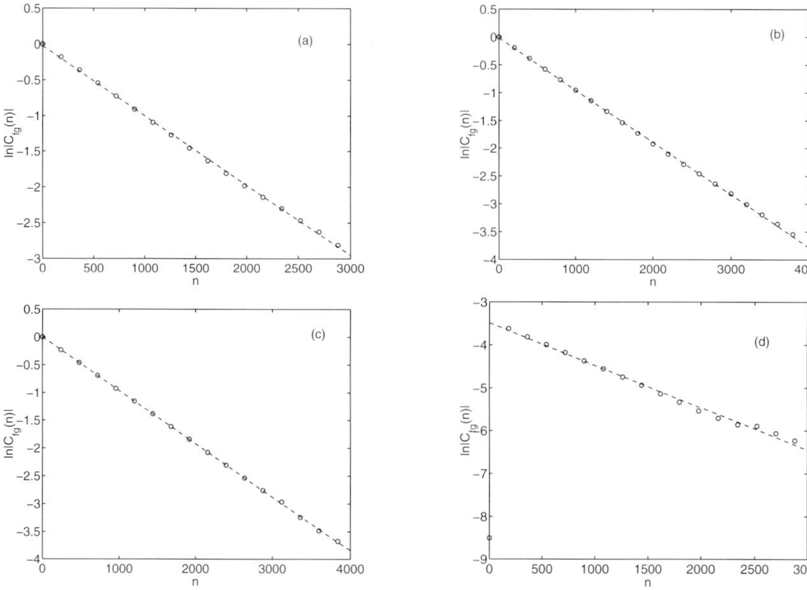

Fig. 9. The function $C_{fg}(n)$ for: (a) $K = 20$; (b) $K = 30$; (c) $K = 40$; (d) $K = 27$, for various functions f and g and for various values of s (Fig. 1 of [19]).

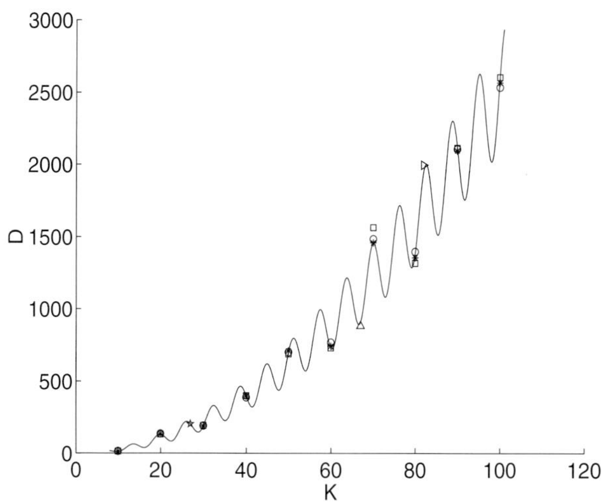

Fig. 10. The diffusion coefficient D for $K \geq 10$ as found from plots like the ones presented in Fig. 9. Various symbols represent results found for different modes k while the solid line represents the analytical value (66) (Fig. 2 of [19]).

Here, in contrast to the regime of large K presented in Fig. 10, appreciable deviations are found. These are related to sticking to regular structures, that is not taken into account in the theory presented in this review.

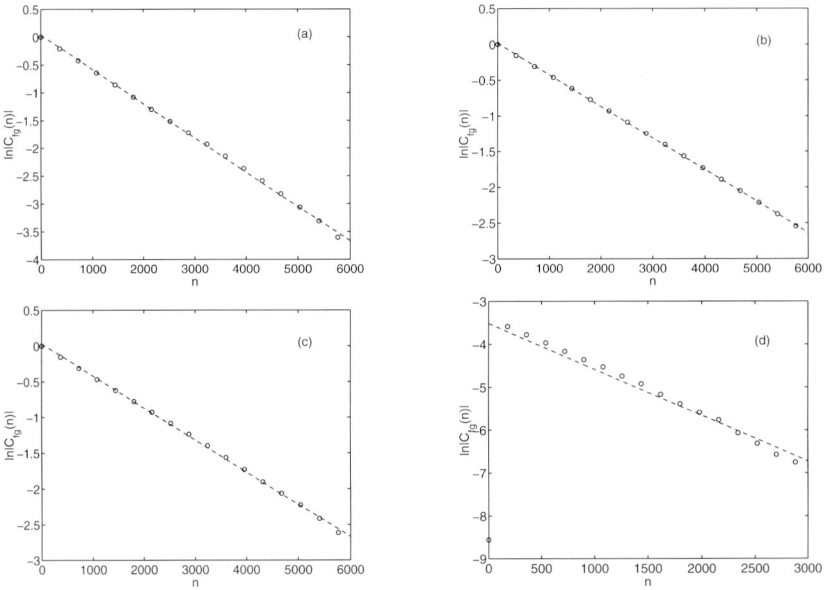

Fig. 11. The function $C_{fg}(n)$ for: (a) $K = 7$; (b) $K = 8$; (c) $K = 3$; (d) $K = 17$, for various functions f and g and for various values of s (Fig. 3 of [19]).

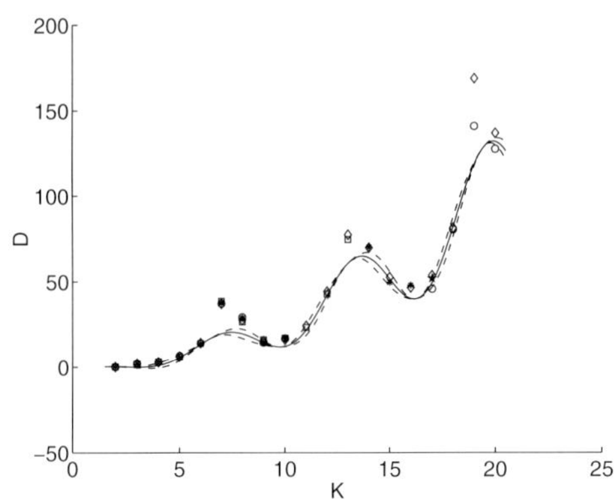

Fig. 12. The diffusion coefficient D for $K \leq 20$ as found from plots like the ones presented in Fig. 11. Various symbols represent results found for different modes k while the solid line represents the analytical value (66). The dashed line represents the approximate error, resulting of the truncation of the perturbation theory expansion. The values of D obtained by direct simulation of propagation of trajectories are marked by diamonds (Fig. 4 of [19]).

To test the decay of correlations in the angle variable, the correlation function (86) was computed for $|f\rangle$ and $|g\rangle$ that are basis states of the form $|k = 0, m\rangle$, and some of the results are presented in Fig. 13.

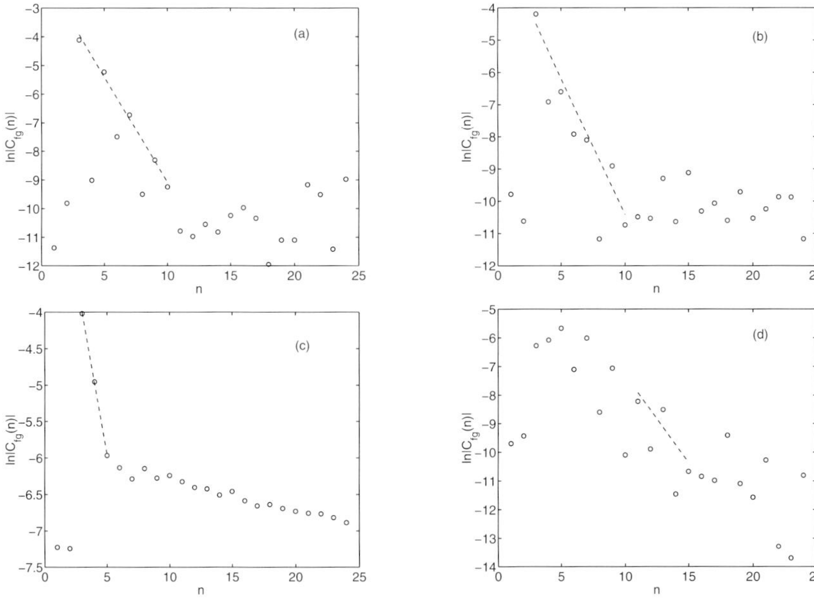

Fig. 13. The function $C_{fg}(n)$ for $s = 1$ with $|f\rangle = |01\rangle, |g\rangle = |02\rangle$ and (a) $K = 16.3$; (b) $K = 19.5$; (c) $K = 12$; (d) $K = 16$. The dashed line represents the best fit to the data (Fig. 5 of [19]).

The relaxation is very fast, therefore it is very difficult to estimate numerically the relaxation rate.

The numerical results are compared to the analytical prediction (85) in Fig. 14.

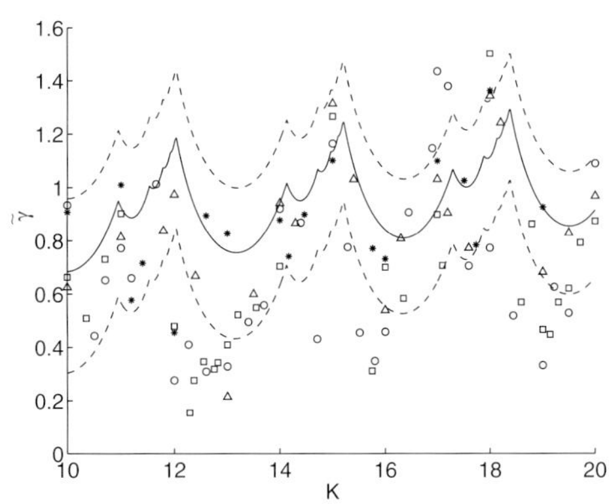

Fig. 14. The fast relaxation rates $\tilde{\gamma}$ as found from plots like Fig. 13 for various functions f and g with $k = 0$, marked by various symbols. The analytical value (85) is represented by a solid line while the dashed lines denote the analytically estimated error, resulting of the truncation of the perturbation theory expansion (Fig. 7 of [19]).

The analytical formula provides a reasonable estimate of the relaxation rate and of its dependence on the stochasticity parameter K. There are however deviations that are significant. Also here some of the most significant deviations were found to be related to regular structures.

The main results that were obtained for the kicked rotor are:

1. Addition of noise results in the effective truncation of the Frobenius-Perron operator. In the limit of vanishing noise the Ruelle-Pollicott resonances are found. The resonances depend on the noise, and therefore on the truncation, as is the case for the kicked top, and approach their limiting values inside the unit circle in the limit of vanishing noise, except the unit eigenvalue corresponding to the equilibrium density.
2. The Ruelle-Pollicott resonances determine the decay of correlations.
3. Deviations from exponential decay of correlations may be found after some time as a result of the existence of regular structures.
4. The following physical picture emerges. First the correlations in the angle variable θ decay. Then the correlations in angular momentum J decay as for a true diffusion process. For finite s a uniform equilibrium density is reached. For $s = \infty$, after the decay of the angular correlations, diffusive spreading takes place.
5. At some time, that is increasing with the stochasticity parameter K, sticking to regular structures becomes important, and the picture based on Ruelle-Pollicott resonances and exponential decay of correlations, breaks down.

6 Summary

It was demonstrated that Ruelle-Pollicott resonances are relevant even when the conditions for the Ruelle-Pollicott theorem do not hold. A method for the calculation of these resonances by a finite truncation of the Frobenius-Perron operator was presented. The relation between the resonances and the decay of correlations was established under these conditions. For mixed systems (as well as for other nonhyperbolic systems) the picture breaks down after some time, that may be very long, as a result of sticking to regular structures.

In order to evaluate the relevance of the Ruelle-Pollicott resonance picture for realistic systems it is instructive to introduce several time scales:

1. If γ_1 is the slowest decay rate (corresponding to the resonance that is closest to the unit circle), $t_{chaos} = 1/\gamma_1$ is the decay time of correlations and the relaxation time to the invariant density.
2. The time scale when sticking to regular structures becomes important will be denoted by t^*.
3. In presence of external noise correlations are destroyed on the time scale $t_c \sim 1/\sigma^2$, where σ^2 is the variance of the noise.

In order to observe the exponential decay of correlations it is required that

$$t_{chaos} \ll t_c, \; t^* . \tag{87}$$

Destruction of sticking to regular structures requires $t_c < t^*$. In the regime

$$t_{chaos} \ll t_c < t^* , \tag{88}$$

the decay of correlations is expected to be similar to the one of hyperbolic systems and one should be able to explore it in the framework of the method that was outlined in the present review. Application of the method to specific examples is of great interest.

Acknowledgments

It is our great pleasure to thank Fritz Haake for many illuminating discussions during the preparation of the lectures and of the review, for the files of the figures from [10] and for the permission to use them. Part of the review is based on work in collaboration with Oded Agam and Maxim Khodas, that is acknowledged with great pleasure. We would like to thank Christopher Manderfeld and Joachim Weber for useful discussions and communications. SF thanks Andreas Buchleitner for the hospitality at the Max Planck Institute for the Physics of Complex Systems in Dresden, where the lectures were prepared. This research was supported in part by the US-Israel Binational Science Foundation (BSF), by the Minerva Center of Nonlinear Physics of Complex Systems, by the Max Planck Institute for the Physics of Complex Systems in Dresden, and by the fund for Promotion of Research at the Technion.

References

Most of the results presented in the review can be found in references marked by **

1. E. Ott, *Chaos in Dynamical Systems*, (Cambridge University Press, Cambridge, 1997).
2. V.I. Arnold and A. Avez, *Ergodic Problems of Classical Mechanics*, (Addison-Wesley NY, 1989).
3. F. Haake, *Quantum Signatures of Chaos*, (Springer-Verlag, Berlin, 1991).
4. B.G Schuster, *Deterministic Chaos, An Introduction*, (Physik-Verlag, Weinheim, 1984).
5. P. Gaspard, *Chaos, Scattering and Statistical Mechanics*, (Cambridge Press, Cambridge 1998).
6. J.R. Dorfman, *An Introduction to Chaos in Non-Equilibrium Statistical Mechanics*, (Cambridge Press, Cambridge 1999).
7. A.J. Lichtenberg and M.A. Lieberman, *Regular and Stochastic Motion* (Springer, NY 1983).
8. F. L. Moore, J. C. Robinson, C. F. Bharucha, Bala Sundaram and M. G. Raizen, Phys. Rev. Lett. **75**, 4598 (1995);
   C. F. Bharucha, J. C. Robinson, F. L. Moore, Qian Niu, Bala Sundaram and M. G. Raizen, Phys. Rev. **E 60**, 3881 (1999).
9. B. Fischer, A. Rosen, A. Bekker and S. Fishman, Phys. Rev. **E**, R4694 (2000).
10. **J. Weber, F. Haake, P.A. Braun, C. Manderfeld and P. Šeba, J. Phys. **A34**, 7195 (2001).

11. M. V. Berry, in *New trends in Nuclear Collective Dynamics*, eds: Y. Abe, H. Horiuchi, K. Matsuyanagi, Springer proceedings in Physics. vol **58**, 183-186 (1992).
12. D. Ruelle, Phys. Rev. Lett. **56**, 405 (1986).
13. **H.H. Hasegawa and W.C. Saphir, Phys. Rev. **A46**, 7401 (1992).
14. S. Fishman, in *Supersymmetry and Trace Formulae, Chaos and Disorder*, edited by I.V. Lerner, J.P. Keating, D.E. Khmelnitskii (Kluwer Academic / Plenum Publishers, New York, 1999).
15. A. Jordan and M. Srednicki, The approach to Ergodicity in the Quantum Baker's Map, nlin.CD/0108024 (2001), This paper contains also some results on the classical baker's map, that are relevant for the present review.
16. G. Blum and O. Agam, Phys. Rev. **E62**, 1977 (2000).
17. J. Weber, F. Haake and P. Šeba, Phys. Rev. Lett. **85**, 3620 (2000).
18. C. Manderfeld, J. Weber and F. Haake, J. Phys. **A34**, 9893 (2001).
19. **M. Khodas, S. Fishman and O. Agam, Phys. Rev. **E62**, 4769 (2000).
20. M. Khodas and S. Fishman, Phys. Rev. Lett. **84**, 2837 (2000); Erratum **84**, 5918 (2000).
21. We thank E. Bogomolny, M. Saraceno, M. Srednicki and A. Jordan for illuminating comments.
22. M. Basilio de Matos and A.M. Ozorio de Almeida, Ann. Phys. (N.Y.) **237**, 46 (1995).
23. R. Balescu, *Statistical Dynamics, Matter out of Equilibrium*, (Imperial College Press,Singapore, 1983).
24. A.B. Rechester and R.B. White, Phys. Rev. Lett. **44**, 1586 (1980);
A.B. Rechester M.N. Rosenbluth and R.B. White, Phys. Rev. **A23**, 2664 (1981);
E. Doron and S. Fishman, Phys. Rev. **A37**, 2144 (1988).
25. B. Sundaram and G.M. Zaslavsky, Phys. Rev. **E59**, 7231 (1999);
G.M. Zaslavsky, M. Edelman and B.A. Niyazov, Chaos **7**, 159 (1997).
26. P. Gaspard, G. Nicolis, A. Provata and S. Tasaki, Phys. Rev. **E 51**, 74 (1995).
27. A. Ostruszka and K. Życzkowski, Phys. Lett. **A 289**, 306 (2001).
28. K. Knopp, *Infinite Sequences and Series*, (Dover publ. NY, 1956).

Fractal Structures in the Phase Space of Simple Chaotic Systems with Transport

J.R. Dorfman

Institute for Physical Science and Technology, and
Department of Physics,
University of Maryland
College Park, Maryland, 20910, USA

Abstract. Recent work by several authors has shown that non-equilibrium processes in simple, classical, chaotic systems can be described in terms of fractal structures that develop in the system's phase space. These structures form exponentially rapidly in phase space as an initial non-equilibrium distribution evolves in time. Since the motion of a region in phase space, for a Hamiltonian system, is measure preserving, the phase space distribution is advected as a passive scalar in the motion of the phase points. Due to the chaotic nature of the motion, the stretching and folding motion in phase space produces very complicated fractal distributions which may vary greatly over regions of small measure. This mechanism is responsible for the formation of the fractals under discussion. Here we illustrate this phenomenon for a few simple models with deterministic diffusion. The origin of the fractals is explained and connected to the microscopic properties of the hydrodynamic modes of the system. These hydrodynamic modes are, in turn, closely related, on averaging, to the van Hove intermediate scattering function. Further we describe the connections of the properties of the fractals with important quantities for transport - transport coefficients and irreversible entropy production. One interesting result is a connection between the coefficient of diffusion for the moving particle in a chaotic Lorentz gas and the Hausdorff dimension of the hydrodynamic modes of diffusion at small wave numbers.

1 Introduction

One of the most interesting results of recent studies of diffusive transport in chaotic systems is the demonstration that microscopic hydrodynamic modes of diffusion have a fractal structure [1–3]. This result, described in some detail by Gaspard and Tasaki in their study of multi-baker maps, has led to a deeper understanding of the connections between macroscopic hydrodynamic processes in simple low-dimensional diffusive systems, and the underlying microscopic motions that lead to macroscopic diffusion [4,5]. Moreover, one can begin to understand, in a quantitative way, the Gibbs picture of the approach to equilibrium as a weak limit of a very fine grained distribution in phase space. Here too the fractal nature of hydrodynamic modes has been shown to give both qualitative and quantitative descriptions of the approach of a diffusive system to a uniform equilibrium state [6,7].

The origin and role of these fractals, for transport in classical, chaotic systems, is rather easy to understand. Consider, for example, a non-equilibrium distribution in phase space corresponding to some non-uniform distribution of particles in a system with a uniform equilibrium state. In accordance with Liouville's equation, the value of the initial distribution function at each phase point will be carried along with the phase point as it travels in the phase space of the system. If the system is chaotic, two initially close points, with nearly identical phase space densities, will separate exponentially rapidly with time. The stretching and folding motions of small regions of phase space will lead to a complicated distribution of values of the phase space density in any small region of phase space, as the initial distribution function is *advected* by the dynamically chaotic motion of the phase points. For sufficiently long times, the variations of the phase space density over any small region of phase space will be well described by a fractal function. Eventually, if the system approaches an equilibrium state, the variations of the phase space distribution over any small region will include all possible values of the initial distribution. For a system that is mixing in Gibbs' sense, the weak limit of this fractal-like distribution function will be a uniform, equilibrium distribution.

In this paper we will describe this process in some detail for the low dimensional systems that have been analyzed so far. We begin by treating a deterministic random walk in one-dimension, and showing how a microscopic analysis of the time displacement of the moving particle naturally leads to a fractal function, characteristic of the long wavelength modes of diffusion in the system. The diffusion coefficient can be expressed in terms of this fractal structure, in at least two ways: (1) It can be expressed as an integral [8] of a fractal function, called a Takagi function after the mathematician who first described this class of functions; and (2) It appears as a coefficient in the wave-number expansion of the Hausdorff dimension of the fractal hydrodynamic mode of diffusion [4]. The latter result is especially interesting, as it provides a connection between macroscopic and microscopic properties of the diffusive motion of the particle, the microscopic properties being the Hausdorff dimension of the fractal and the positive Lyapunov exponent describing the chaotic dynamics of the system. After treating the one-dimensional case, we move on to similar results for two-dimensional, periodic Lorentz gases, in which a moving particle travels in an array of fixed scatterers placed at vertices of a two dimensional Bravais lattice. Finally we show that an explanation of the phenomenon of positive macroscopic entropy production can be based upon the properties of fractal hydrodynamic modes [6,7,9]. We conclude with a discussion of a number of open questions and of directions for future research.

2 Diffusive Mode Fractals for a One-Dimensional, Deterministic, Random Walk

One of the simplest systems that can be used to illustrate the deterministic diffusion of a particle, is a model of an unbiased random walk on a line. Here we will use this model to illustrate a number of points that hold also for more general cases. In particular, we will use this model to develop the theory for microscopic, fractal diffusive modes and to illustrate their relation to the van Hove intermediate scattering function.

We begin be considering the line $-\infty \leq x \leq \infty$, and expressing it as the union of unit intervals $n \leq x < n+1$, where $n \in \mathbf{Z}$. We construct a map, $M(x)$, that acts at discrete times t, with $t \in \mathbf{N}$, which replaces the position of the point at time t, x_t by its position at time $t+1, x_{t+1}$, so that $x_{t+1} = M(x_t)$, given by

$$\begin{aligned} x_{t+1} = M(x_t) &= n(x_t) + 1 + 2[x_t] \text{ for } 0 \leq [x_t] < 1/2 \\ &= n(x_t) - 1 + 2[x_t] - 1 \text{ for } 1/2 \leq [x_t] < 1 \,. \end{aligned} \tag{1}$$

Here $x = n(x) + [x]$, where $n(x)$ is the largest integer less than, or equal to, x, and $[x]$ is the fractional part of x. This is a measure preserving, deterministic map, although it is not reversible. Many of its properties are discussed elsewhere [3,8]. The macroscopic diffusion of this particle is identical to that of a particle making an unbiased random walk on the line, from one integer lattice site to either of its nearest neighbors at each step. On the microscopic level, one can describe the diffusion of the particle in terms of its mean square displacement as evaluated in terms of its exact positions at each time, that is $< (x_t - x_0)^2 > \to 2Dt$ for large times, or in terms of the mean square displacement as measured in terms of lattice sites, $< (n(x_t) - n(x_0))^2 > \to 2Dt$, for large times. Here the angular brackets denote an average over a uniform distribution of starting points, x_0, taken in some unit interval. The difference between these two expressions is insignificant for large t. We use the description in terms of the lattice sites, recognizing that we will need to follow the exact locations of the particle, in order to determine the proper lattice site at each time.

It is very convenient to use the van Hove intermediate scattering function to describe the diffusive process [3,8,10]. This function, denoted by $V_k(t)$, is defined as

$$V_k(t) = \left\langle e^{[ik(n(x_t) - n(x_0))]} \right\rangle , \tag{2}$$

where the average is taken over a uniform distribution of initial points in some unit interval, as above. For this simple process one can show that $V_k(t)$ is given by

$$V_k(t) = e^{s(k)t} , \tag{3}$$

where $s(k) = \ln\cos(k)$ is the wave-number dependent decay rate for diffusive motion. Thus, for wave numbers, k much smaller than the inverse of a lattice spacing, in this case, unity, this function takes the form

$$s(k) = -Dk^2 + \tilde{D}k^4 + O(k^6) , \tag{4}$$

where $D = 1/2$ is the diffusion coefficient, $\tilde{D}$ is the super-Burnett diffusion coefficient, and so on. Notice that $V_k(t)$ satisfies the Fourier transform of a generalized diffusion equation,

$$\frac{dV_k(t)}{dt} = -Dk^2 V_k(t) + \tilde{D}k^4 V_k(t) + \cdots , \tag{5}$$

where all higher order diffusion coefficients are retained.

The point to be emphasized is that the averaging process used in Eq. (2), and in similar expressions for more realistic diffusive systems, hides a great deal of interesting and useful physics! This is most evident in the case that the underlying microscopic motion is chaotic, which is the case we shall discuss in this paper. In order to see the physics obscured by the averaging, for a chaotic system, consider the microscopic quantity that is averaged, namely, $\exp\left[ik(n(x_t) - n(x_0))\right]$. Since the map, $M(x)$, that determines the motion of the point is chaotic, with positive Lyapunov exponent, $\lambda_+ = \ln 2$, the displacement over a time t, $n(x_t) - n(x_0)$, is a very rapidly varying function of the initial point, x_0. As a result, the exponential containing this displacement will be a rapidly oscillating function of the initial point. To capture these oscillations and to explore the fractal structure of the exponential function, we consider a partially averaged quantity, which we will call a normalized, incomplete van Hove function, $I_k(x,t)$, defined by

$$I_k(x,t) = e^{-s(k)t} \int_0^x dx' e^{[ik(n(x'_t) - n(x'))]} \,. \tag{6}$$

Here we can take initial points to be uniformly distributed in one of the unit intervals, *e.g.* the interval $0 \le x < 1$, since all unit intervals are identical and the displacement is independent of the particular unit interval in which the particle begins it motion. Further, the fully integrated value of $I_k(1,t) = 1$, due to the normalization factor in front of the integral in Eq. (6). Using Eq. (1), we can see that the incomplete van Hove function satisfies the recursion relation

$$\begin{aligned} I_k(x,t+1) &= \frac{1}{2} e^{[-s(k)+ik]} I_k(2x,t) \text{ for } x < 1/2 \\ &= \frac{1}{2} e^{-s(k)} \left[e^{ik} + e^{-ik} I_k(2x-1,t)\right] \text{ for } 1/2 \le x < 1 \,. \end{aligned} \tag{7}$$

It is clear from its definition that $I_k(x,0) = x$, so that this expression may be used to start the iterations which determine all $I_k(x,t)$ for $t = 1, 2, \ldots$. The iterations converge rapidly so that the variations of the function from one time to the next take place on scales that become exponentially small with time t, as 2^{-t}. Therefore, for large enough times, we may replace $I_k(x,t)$ by its asymptotic value given by the solution of the recursion relation

$$\begin{aligned} I_k(x) &= \frac{1}{2} e^{-s(k)+ik} \, I_k(2x) \text{ for } x < 1/2 \\ &= \frac{1}{2} e^{-s(k)} \left[e^{ik} + e^{-ik} I_k(2x-1)\right] \text{ for } 1/2 \le x < 1 \,. \end{aligned} \tag{8}$$

The exponential function in the integrand of the incomplete van Hove function is referred to as the fractal hydrodynamic mode of diffusion, but this exponential is difficult to visualize because of its wild variations. The incomplete van Hove function is, in fact, the cumulative function of this hydrodynamic mode, and it is helpful to think of the incomplete normalized van Hove function as a representation of the hydrodynamic mode for this diffusive process. In Fig. 1 we plot the real and imaginary parts of $I_k(x)$ for a few values of k. For very small

k, we can expand $I_k(x)$ as

$$I_k(x) = x + ik\,T(x) + O(k^2)\,, \tag{9}$$

where the first imaginary term is, apart from a factor of k, a self-similar fractal function, $T(x)$, called a Takagi function [1,3]. It satisfies a de Rham type recursion relation

$$\begin{aligned} T(x) &= x + \frac{1}{2}T(2x) \;\text{ for } 0 \le x < 1/2 \\ &= 1 - x + \frac{1}{2}T(2x-1) \;\text{ for } 1/2 \le x < 1\,. \end{aligned} \tag{10}$$

This Takagi function is illustrated in Fig. 1, if one examines the curves for small, positive values of k.

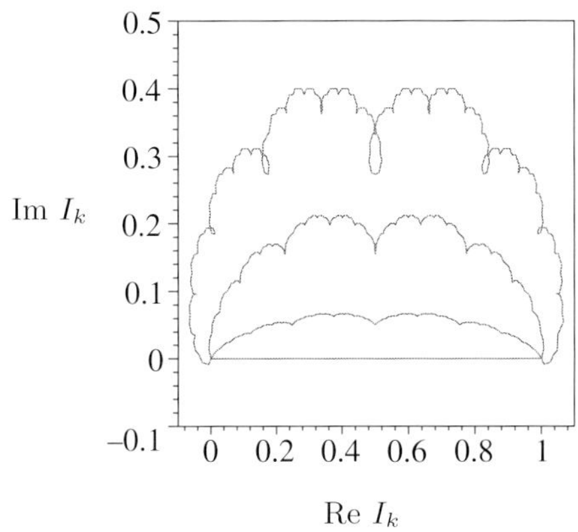

Fig. 1. The normalized, incomplete van Hove function for the one-dimensional deterministic walk, Eq. (1). Curves with wave numbers $k = 0, 0.1, 0.3$ and 0.5 (from bottom to top) are shown.

An important property of these curves is a striking connection between the fractal Hausdorff dimension, $\mathcal{D}_H$, of the curve $(\mathrm{Re}\, I_k(x), \mathrm{Im}\, I_k(x))$, for the incomplete van Hove function, illustrated in Fig. 1, for small k, the diffusion coefficient, D, and the positive Lyapunov exponent, λ, characterizing the chaotic process underlying the diffusive motion of the particle [4,5]. This connection is provided by the equation

$$\mathcal{D}_H(k) = 1 + \frac{D}{\lambda}k^2 + O(k^4)\,. \tag{11}$$

A brief derivation of this equation in a more general context is given in the Appendix. This equation illustrates the fact that for chaotic models such as the one discussed here, the incomplete van Hove function encodes in its structure both a macroscopic property of the system, the diffusion coefficient, and a microscopic property, the Lyapunov exponent. Such interesting connections are important for a deeper understanding of the microscopic foundations of transport processes in fluids.

3 Diffusive Mode Fractals in Periodic Lorentz Gases

The considerations in the previous section have been extended to periodic Lorentz gases. In this class of models, a moving point particle moves in a two-dimensional array of scatterers, fixed at the vertices of a Bravais lattice. One then can study, either analytically or by computer simulations, the diffusive motion of the particle as well as the corresponding incomplete, normalized van Hove function. Two chaotic examples have been studied recently. These are the periodic hard disk Lorentz gas, where the disks are placed, without overlapping, at the vertices of a triangular lattice at a sufficiently high density that the moving particle has a finite horizon. That is, there is a finite maximum length for the free path of the moving particle between any two successive collisions. The interactions of the particle with the scatterers are elastic, specular, and instantaneous so that the kinetic energy remains constant throughout the particle's motion. It is well known that the motion of the moving particle is chaotic [11,12]. The other model studied has the scatterers fixed at the vertices of a square lattice, and the interactions between the particle and scatterers is due to an attractive Yukawa potential [13]. Knauf has proved that the motion of the particle is chaotic provided that its energy is sufficiently high. Both of these models are time reversible, measure preserving, have a four dimensional phase space, and a three dimensional constant energy surface. Their chaotic properties are characterized by two non-zero Lyapunov exponents, of equal magnitude but opposite sign. The positive Lyapunov exponent is associated with the unstable manifold on the tangent space of the constant energy surface, where two nearby points separate exponentially rapidly, the negative exponent with the stable manifold, where two nearby points separate exponentially upon time reversal of the motion. There are zero Lyapunov exponents associated with the direction of a trajectory and with the direction perpendicular to the energy surface.

In these cases, the incomplete van Hove function is defined by partial phase space averages of the quantity $\exp\left[i\boldsymbol{k}\cdot(\boldsymbol{r}(t)-\boldsymbol{r}(0))\right]$, determined by the projection of the particle's displacement in time, $(\boldsymbol{r}(t)-\boldsymbol{r}(0))$, along a wave vector $\boldsymbol{k}$. The displacement is, of course, a function of both the initial position and initial velocity of the moving particle. Due to the periodicity of the lattice of scatterers, one can consider just the trajectories starting from points within one unit cell, and having the same total energy. For numerical studies, a convenient way to define a normalized, incomplete van Hove function is to consider all trajectories that start from a point on a circle around a scatterer, with velocities along the radial direction from the center of the scatterer to the initial point, and directed away from the circle. In the case of hard disk scatterers, one can choose the circle to be just outside the scatterer in the unit cell, and for the Yukawa case, one can take the circle to be close to a scatterer, with an appropriate choice of total energy so as to be in the chaotic region. An incomplete, normalized van Hove function, $F_{\boldsymbol{k}}(\theta,t)$ is then defined by

$$F_{\boldsymbol{k}}(\theta,t)=\frac{\int_0^{\theta}d\theta' e^{i\boldsymbol{k}\cdot(\boldsymbol{r}(\theta',t)-\boldsymbol{r}(\theta',0))}}{\int_0^{2\pi}d\theta' e^{i\boldsymbol{k}\cdot(\boldsymbol{r}(\theta',t)-\boldsymbol{r}(\theta',0))}}\,. \tag{12}$$

Here $0 \leq \theta < 2\pi$ is an angle around the circle of initial points, with the zero chosen in some convenient direction. For each model, all trajectories are taken to be on the same constant energy surface. As in the one dimensional case, the incomplete van Hove function is a cumulative function of the hydrodynamic mode described by the exponential displacement function.

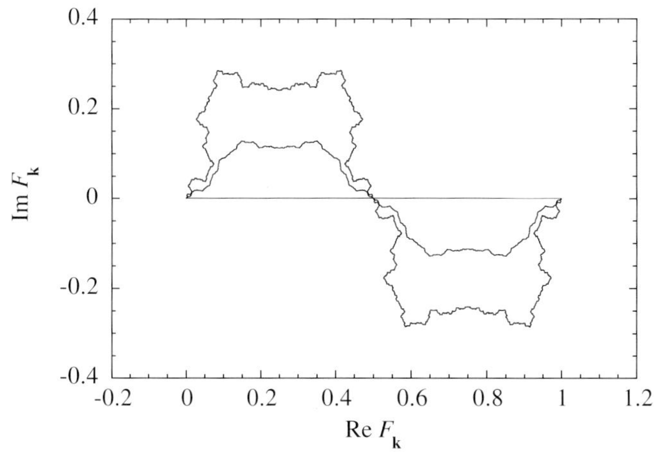

Fig. 2. Real *vs.* imaginary parts of F_k for a periodic Lorentz gas where a point particle of unit mass and velocity undergoes elastic collisions on hard disks of unit radius forming a triangular lattice with interdisk distance $d = 2.3$: Curves are shown of the cumulative functions of the hydrodynamic modes of wavenumber $k_x = 0.0$, 0.5, and 0.9 with $k_y = 0$. Note that the fractality increases with the wavenumber. The curves are constructed by averaging over 10^6 points.

In Figs. 2 and 3, we show curves representing $(\mathrm{Re}\, F_{\boldsymbol{k}}(\theta, t), \mathrm{Im}\, F_{\boldsymbol{k}}(\theta, t))$, obtained by Claus and Gaspard, for the hard disk and the attractive Yukawa Lorentz gases, respectively. The curves shown in each figure correspond to a range of values of $\boldsymbol{k}$. The time is chosen to be sufficiently long that the curves no longer change as t increases. For hard disks, 30 collisions or so is a sufficient time for the curves to stabilize. One should notice that the fractal structures of the curves become more pronounced as the wave number increases. In the Appendix, we show that the fractal dimension of these curves satisfy exactly the same equation as that given in the previous section, Eq. (11), namely

$$\mathcal{D}_H(\boldsymbol{k}) = 1 + \frac{D}{\lambda}\boldsymbol{k}^2 + O(\boldsymbol{k}^4) \,. \tag{13}$$

Figure 4 shows the comparison between the theoretical result for this fractal dimension, and the results obtained from an analysis of the computer studies of the incomplete van Hove function defined in Eq. (12). The agreement is excellent. The Ph. D. thesis of I. Claus has further, beautiful illustrations of the incomplete

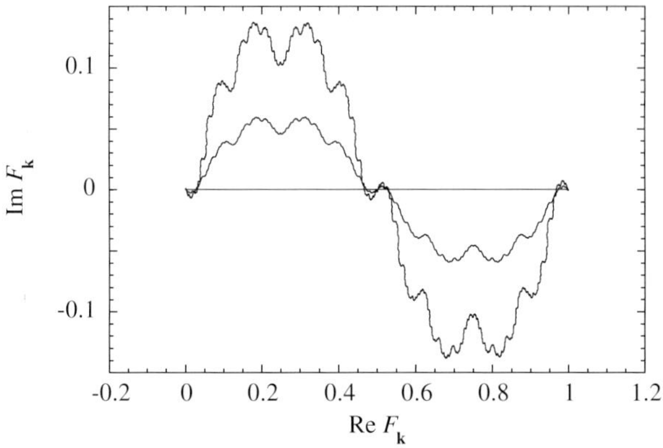

Fig. 3. Real *vs.* imaginary parts of F_k for the periodic Lorentz gas where a charged particle of unit mass and energy $E = 3$ moves in a square lattice of screened Coulomb potentials with $\alpha = 2$: Curves are shown of the cumulative functions of the hydrodynamic modes of wavenumber $k_x = 0.0$, 0.2, and 0.4 with $k_y = 0$. Note that here also the fractality increases with the wavenumber. The curves are constructed by averaging over 10^6 points.

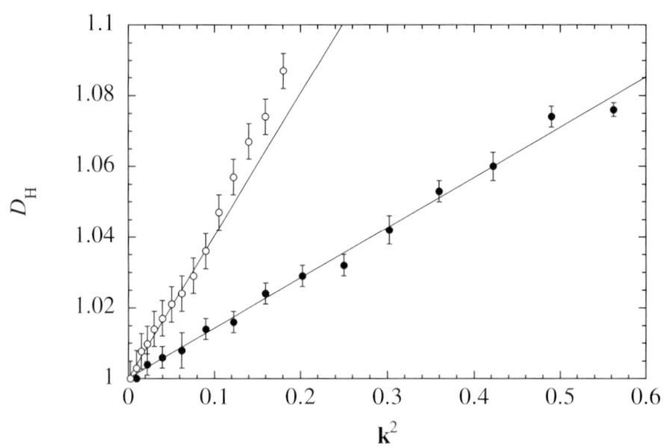

Fig. 4. Hausdorff dimension D_{H} of the hydrodynamic modes versus $\mathbf{k}^2 = k_x^2$ ($k_y = 0$) for both periodic Lorentz gases with hard-disk scatterers (filled circles) and Coulomb scatterers (open circles). Both solid lines have slopes equal to $\mathcal{D}/\lambda$ for the respective diffusion coefficient $\mathcal{D}$ and Lyapunov exponent λ of the Lorentz gases.

van Hove functions for these systems, and includes studies of reactive Lorentz gases, as well [14]. We mention also that Viscardy and Gaspard have obtained closely related curves in their study of a simple model of a two dimensional viscous flow [15].

4 The Rate of Entropy Production in the Approach to Equilibrium: SRB Measures and Fractal Structures

Having demonstrated that the diffusive modes for a few chaotic models have a fractal structure, we now turn to a discussion of the approach to equilibrium as an example of the chaotic advection of a passive scalar. To repeat, the values of any initially non-uniform phase space density will be distributed through phase space as a result of the chaotic motion of the phase points, and the fact that the phase space density remains constant on the trajectory of any phase point. If the system is mixing, so that the phase space consists of only one ergodic component, not a union of a number of disjoint components, eventually any small region of phase space, with finite Liouville measure, will contain points with almost every initial value of the phase space density. Then any coarse graining of the density will lead to a uniform value over the entire phase space, which is the equilibrium density. For the simple, low dimensional systems discussed in the previous section, periodic Lorentz gases, and for two-dimensional multi-baker maps, this picture can be verified in some detail, and the approach to equilibrium can be characterized by a positive rate of entropy production that agrees with the predictions of irreversible thermodynamics. The extension of this work to higher dimensional systems still remains to be done, as we will mention in the Conclusion.

4.1 SRB Measures

In order to give a proper description of the theory for irreversible entropy production, as it applies to the systems under discussion, it is helpful to review some of the physical and mathematical properties of what are known as SRB (Sinai-Ruelle-Bowen) measures. We recall that in the tangent space for almost every point on the constant energy surface for two-dimensional Lorentz gases, there are three directions defined by the dynamics. These are an unstable direction where two infinitesimally close points separate exponentially with a rate determined by the positive Lyapunov exponent, a stable direction where two infinitesimally close points exponentially approach each other - more properly, they exponentially separate in their time reversed motion, and a neutral direction tangent to the direction of the phase space trajectory, where two infinitesimally close points neither separate nor approach each other, but rather follow along together like the cars in a train. This picture has strong consequences for the behavior of any initially non-uniform phase space density. The stretching motion in unstable directions will cause uniformities in the phase space density to become less and less pronounced with time, along the unstable directions, much like the smoothing of wrinkles by stretching a fabric. On the other hand, the "squeezing" motion in stable directions will cause variations in the phase space density to take place over finer and finer scales as time goes on, actually producing more and more wrinkles in the density on arbitrarily fine scales. As a result, the phase space density becomes very smooth in unstable directions and very irregular in stable directions. After a long enough time, the irregularities in the stable direction become strong enough that it is useful to consider partially integrated values of the phase space density, in much that same way as we have considered partially integrated values of the hydrodynamic modes to obtain the incomplete van Hove functions of the previous sections. These partially integrated densities can be used to define a time dependent measure on the phase space which, for long enough time, is well approximated by a dynamically invariant SRB measure. These SRB measures are smooth in unstable directions and may be fractal in stable directions due to variations on arbitrarily fine scales. The equilibrium micro-canonical ensemble is described by a particular invariant SRB measure which is smooth in both stable and unstable directions. However this smooth measure is often the weak limit of an SRB measure with fractal properties in stable directions.

Fractal SRB measures have been explicitly calculated for some simple two dimensional dynamical models which are based upon variations of the bakers transformation [16]. In these cases, the SRB measures can be expressed as a direct product of a smooth measure in the unstable direction with a singular measure in the stable direction, proportional to either a Lebesgue singular function or a Takagi-like function. When the SRB measure is singular in a stable direction, it is said to be "not absolutely continuous" with respect to Lebesgue measure. An absolutely continuous measure, $\mu(A)$ for a set A has the property that

$$\mu(dA) = \rho(P_A)dA\,, \tag{14}$$

where dA represents both an infinitesimally small set in phase space about a point P_A, as well as the Lebesgue measure of the set. Also $\rho(P_A)$ is the density associated with the measure at the point P_A. For reasons given above, the SRB measures associated with the relaxation of a non-uniform phase space distribution function to equilibrium typically are not absolutely continuous with respect to Lebesgue measure.

4.2 Entropy Production in the Periodic Lorentz Gas

We now outline the calculation of the rate of entropy production for a non-equilibrium distribution function for the moving particles on a periodic Lorentz gas, as the system approaches total equilibrium [6,9]. We refer to the recent paper of Dorfman, Gilbert, and Gaspard for the details of this calculation, which we only sketch here [7]. We begin by considering the constant energy surface for such models and suppose that moving particles are distributed on the periodic lattice in such a way that the distribution can be described by an initial ensemble density $\rho(\boldsymbol{l}, \Gamma, 0)$, where ρ denotes the number of systems per unit phase space volume. We take this density to be close to that of total equilibrium and write it in the form

$$\rho(\boldsymbol{l}, \Gamma, 0) = \rho_{\mathrm{eq}}\,\left[1 + R(\boldsymbol{l}, \Gamma, 0)\right]\,, \tag{15}$$

where the equilibrium distribution ρ_{eq} is, for Lorentz gases, uniform, having the same value in each of the unit cells of the lattice, labeled by the vector $\boldsymbol{l}$, and within each cell, uniform with respect to the phase variables $\Gamma = (\boldsymbol{r}, \boldsymbol{v})$. The

Lebesgue measure within any cell, $d\Gamma$, is preserved by the flow, which we denote by Φ^t, which is the time displacement operator that replaces any phase point Γ by its image obtained by following the dynamics over a time t. One can use a technique from condensed matter physics for the analysis of periodic systems, whereby distribution functions and related quantities are conveniently expressed in terms of lattice Fourier transforms [17]. Using the lattice Fourier transform, we can express the deviation of the distribution from equilibrium in the form

$$R(\boldsymbol{l}, \Gamma, 0) = \frac{1}{|\mathcal{B}|} \int_{\mathcal{B}} d\boldsymbol{k}\; e^{i\boldsymbol{k}\cdot\boldsymbol{l}}\; \tilde{R}(\boldsymbol{k}, \Gamma)\,, \tag{16}$$

where $\mathcal{B}$ represents the first Brillouin zone of the reciprocal lattice, and $|\mathcal{B}|$ is the volume of this zone. We will simplify matters a bit by assuming that the initial deviation from equilibrium $R(\boldsymbol{l}, \Gamma, 0)$ depends only upon the cell $\boldsymbol{l}$ but not on the initial phase Γ of the system within the cell. In this case the Fourier transform $\tilde{R}(\boldsymbol{k}, \Gamma)$ does not depend upon Γ, either, and will just be denoted by $R_{\boldsymbol{k}}$.

The time dependent distribution function $\rho(\boldsymbol{l}, \Gamma, t)$ is the solution of Liouville's equation and is given by

$$\begin{aligned}\rho(\boldsymbol{l}, \Gamma, t) &= \rho_{\mathrm{eq}}\,[1 + R(\boldsymbol{l}, \Gamma, t)] \\ &= \rho_{\mathrm{eq}} \left\{1 + \frac{1}{|\mathcal{B}|} \int_{\mathcal{B}} d\boldsymbol{k}\; R_{\boldsymbol{k}}\; e^{i\boldsymbol{k}\cdot\boldsymbol{L}\left[\Phi^{-t}(\boldsymbol{l},\Gamma,0)\right]}\right\}\,.\end{aligned} \tag{17}$$

Here $\boldsymbol{L}$ denotes the projection on the lattice coordinate. Thus $\boldsymbol{L}\,[\Phi^{-t}(\boldsymbol{l}, \Gamma, 0)]$ is the lattice vector of the cell in which a moving particle would be located at time $-t$ if it were in cell $\boldsymbol{l}$ at time $t = 0$, with phase Γ. We can express the time dependent deviation from total equilibrium in the form

$$R(\boldsymbol{l}, \Gamma, t) = \frac{1}{|\mathcal{B}|} \int_{\mathcal{B}} d\boldsymbol{k}\; R_{\boldsymbol{k}}\; e^{i\boldsymbol{k}\cdot[\boldsymbol{l}+\boldsymbol{d}(\Gamma,t)]}\,, \tag{18}$$

where the *backward* displacement of the lattice vector $\boldsymbol{l}$ over a time interval t is defined by

$$\boldsymbol{d}(\Gamma, t) = \boldsymbol{L}\left[\Phi^{-t}(\boldsymbol{l}, \Gamma, 0)\right] - \boldsymbol{l}\,. \tag{19}$$

We note that $\boldsymbol{d}(\Gamma, t)$ depends upon the time interval $(-t, 0)$ and upon the phase point Γ, at the initial time, but not upon the initial cell $\boldsymbol{l}$. In other words, as we saw in the one-dimensional case, the periodicity of the lattice and the dynamics produce a "winding number" $\boldsymbol{d}(\Gamma, t)$ that does not depend upon the cell in which the trajectory is located at the initial time. We remark here that the long time limit of the displacement vector $\boldsymbol{d}(\Gamma, t)$ may be a wildly varying function of the phase coordinate Γ. Thus we expect that the decaying modes of the time dependent distribution Eq. (18) are fractal functions of the phase coordinates.

To make the connection with the macroscopic description of this diffusive process, we define the time dependent density of the moving particle by integrating the phase space density over the coordinates Γ of an elementary phase space cell of the lattice, $\mathcal{M}_{\boldsymbol{l}}$:

$$n(\boldsymbol{l},t) \equiv \int_{\mathcal{M}_{\boldsymbol{l}}} d\Gamma\ \rho(\boldsymbol{l},\Gamma,t) = \mu_t(\mathcal{M}_{\boldsymbol{l}})\,. \tag{20}$$

Using Eq. (17), we obtain the density as

$$n(\boldsymbol{l},t) = \nu(\mathcal{M}) \times \left[1 + \frac{1}{|\mathcal{B}|}\int_{\mathcal{B}} d\boldsymbol{k}\ R_{\boldsymbol{k}}\ e^{i\boldsymbol{k}\cdot\boldsymbol{l}}\ \frac{\int_{\mathcal{M}} d\Gamma \rho_{\mathrm{eq}}\ e^{i\boldsymbol{k}\cdot\boldsymbol{d}(\Gamma,t)}}{\int_{\mathcal{M}} d\Gamma\ \rho_{\mathrm{eq}}}\right]\,. \tag{21}$$

Here $\nu(A)$ is the equilibrium measure of a set A, and $\nu(\mathcal{M})$ is the equilibrium measure of a unit cell. We consider times t that are long compared to the mean time between collisions of the moving particles, but short compared to the time needed for the system to relax to total equilibrium. For such times, we expect the time dependent deviation from total equilibrium to decay exponentially with a rate given by the van Hove relation

$$\begin{aligned} s_k &\equiv \lim_{t\to\infty} \frac{1}{t} \ln\langle e^{i\boldsymbol{k}\cdot\boldsymbol{d}(\Gamma,t)}\rangle_{\mathcal{M}} \\ &= \lim_{t\to\infty} \frac{1}{t} \ln \frac{\int_{\mathcal{M}} d\Gamma \rho_{\mathrm{eq}}\ e^{i\boldsymbol{k}\cdot\boldsymbol{d}(\Gamma,t)}}{\int_{\mathcal{M}} d\Gamma \rho_{\mathrm{eq}}}\,, \end{aligned} \tag{22}$$

which gives the decay rate of a hydrodynamic mode of diffusion of wave number $\boldsymbol{k}$. An expansion in powers of the wave number gives

$$s_k = -D\boldsymbol{k}^2 + O(\boldsymbol{k}^4)\,, \tag{23}$$

with diffusion coefficient D. For finite times t, it is convenient to introduce a sub-exponential function of time, $C(\boldsymbol{k},t)$, that approaches unity for long times, by

$$\frac{\int_{\mathcal{M}} d\Gamma \rho_{\mathrm{eq}}\ e^{i\boldsymbol{k}\cdot\boldsymbol{d}(\Gamma,t)}}{\int_{\mathcal{M}} d\Gamma \rho_{\mathrm{eq}}} = C(\boldsymbol{k},t)\ e^{s_k t}\,. \tag{24}$$

In accordance with our previous discussions, we try to calculate the rate of entropy production in terms of the non-equilibrium SRB measures that characterize the strong variations of the phase space density over very small sets in phase space. The point to be emphasized here is that the non-equilibrium measures of sets in phase space are, in fact, partially integrated exponentials that we have associated with incomplete van Hove functions. For chaotic systems these incomplete van Hove functions become well approximated by their asymptotic forms, as time increases, since the chaoticity leads to changes in the density over finer and finer, exponentially small, scales, as the time increases.

For our calculation of the rate of entropy production in a unit cell of the periodic lattice, we use a partition of the total phase space into the small disjoint sets A_j chosen in such a way that *all* of the points in a given set A_j travel through the *same* sequence of unit cells over some large time interval $-\mathcal{T} < t < \mathcal{T}$. In such a case the set of points $\Phi^{-\mathcal{T}} A_j$ are all in the same cell with location denoted by $\boldsymbol{l} + \boldsymbol{d}(\Gamma_{A_j}, \tau)$ which is determined by the backward evolution of an arbitrary phase point Γ_{A_j} in the set A_j. We also suppose that the same partition is used for each unit cell. The phase space cell located at the lattice vector $\boldsymbol{l}$ is decomposed by this partition as:

$$\mathcal{M}_{\boldsymbol{l}} = \cup_{A_j \subset \mathcal{M}_{\boldsymbol{l}}} A_j \,. \tag{25}$$

This partition can be seen as a translationally invariant grid extending over the whole phase space. The time dependent measure, $\mu_t(A_j)$, of a set of the partition, A_j is defined by an integration of the phase space density over the set. That is

$$\mu_t(A_j) = \int_{A_j} d\Gamma \rho(\boldsymbol{l}, \Gamma, t) = \nu(A_j) + \delta\mu_t(A_j)\,, \tag{26}$$

where $\nu(A_j)$ is the equilibrium measure of the set, and

$$\delta\mu_t(A_j) = \int_{A_j} d\Gamma \rho_{eq} R(\boldsymbol{l}, \Gamma, t)\,, \tag{27}$$

where $R(\boldsymbol{l}, \Gamma, t)$ is given by Eq. (18).

We define the entropy of the lattice cell $\mathcal{M}_{\boldsymbol{l}}$ at time t as the *coarse-grained entropy of this cell with respect to the partition* $\{A_j\}$:

$$\begin{aligned} S_t(\mathcal{M}_{\boldsymbol{l}}|\{A_j\}) \equiv -\sum_{A_j \subset \mathcal{M}_{\boldsymbol{l}}} \mu_t(A_j) \, \ln \frac{\mu_t(A_j)}{\nu(A_j)} \\ + S_{\mathrm{eq}}(\mathcal{M}_{\boldsymbol{l}}|\{A_j\})\,, \end{aligned} \tag{28}$$

where we have set Boltzmann's constant equal to unity, $k_{\mathrm{B}} = 1$. It is important to note that we always define the entropy with respect to a given partition of the phase space. The first term on the right hand side of Eq. (28) is the non-equilibrium relative entropy with respect to the equilibrium entropy for this partition. The equilibrium entropy is given by

$$S_{\mathrm{eq}}(\mathcal{M}_{\boldsymbol{l}}|\{A_j\}) = -\sum_{A_j \subset \mathcal{M}_{\boldsymbol{l}}} \nu(A_j) \, \ln \frac{\nu(A_j)}{c}\,, \tag{29}$$

where c is a constant which fixes the absolute value of the equilibrium entropy. The time variation of the entropy over a time interval τ is of course only due to the change in the relative entropy, and is defined as the difference

$$\begin{aligned} \Delta^\tau S(\mathcal{M}_{\boldsymbol{l}}) &\equiv S_t(\mathcal{M}_{\boldsymbol{l}}|\{A_j\}) - S_{t-\tau}(\mathcal{M}_{\boldsymbol{l}}|\{A_j\}) \\ &= S_t(\mathcal{M}_{\boldsymbol{l}}|\{A_j\}) - S_t(\Phi^\tau \mathcal{M}_{\boldsymbol{l}}|\{\Phi^\tau A_j\})\,. \end{aligned} \tag{30}$$

To obtain the second line of this equation we used Liouville's theorem for the conservation of the measure of a set, which is expressed as

$$\mu_{t-\tau}(A) = \mu_t(\Phi^\tau A)\,. \tag{31}$$

On the other hand, the *entropy flow* is defined as the difference between the entropy which enters the cell $\mathcal{M}_l$ and the entropy which exits that cell:

$$\begin{aligned}\Delta_{\mathrm{e}}^{\tau} S(\mathcal{M}_l) &\equiv S_{t-\tau}(\Phi^{-\tau}\mathcal{M}_l|\{A_j\}) - S_{t-\tau}(\mathcal{M}_l|\{A_j\}) \\ &= S_t(\mathcal{M}_l|\{\Phi^{\tau} A_j\}) - S_t(\Phi^{\tau}\mathcal{M}_l|\{\Phi^{\tau} A_j\}) .\end{aligned} \tag{32}$$

Accordingly, the *entropy production* over a time τ is defined as

$$\begin{aligned}\Delta_{\mathrm{i}}^{\tau} S(\mathcal{M}_l) &\equiv \Delta^{\tau} S(\mathcal{M}_l) - \Delta_{\mathrm{e}}^{\tau} S(\mathcal{M}_l) \\ &= S_t(\mathcal{M}_l|\{A_j\}) - S_t(\mathcal{M}_l|\{\Phi^{\tau} A_j\}) .\end{aligned} \tag{33}$$

Equation (33) gives the expression of the entropy production as the difference between the entropy with respect to the original partition into sets A_j and the entropy with respect to a partition into sets which are the images, $\Phi^{\tau} A_j$, of the sets A_j after time τ. Each set $\Phi^{\tau} A_j$ belongs to a single unit cell because of the way we constructed the sets of the partition. We see here that in this approach, the production of entropy is determined by the difference between two time dependent entropies of the same cell, but taken with respect to two different partitions. Because of the dynamics, the partitions differ in their resolutions of the phase space density along the stable and unstable directions. With respect to the partition$\{A_j\}$, the partition $\{\Phi^{\tau} A_j\}$ is stretched in the unstable direction and compressed in the stable direction.

Written out in full, this entropy production is

$$\begin{aligned}\Delta_{\mathrm{i}}^{\tau} S(\mathcal{M}_l) = &- \sum_{A_j \subset \mathcal{M}_l} \mu_t(A_j) \ln \frac{\mu_t(A_j)}{\nu(A_j)} \\ &+ \sum_{\phi^{\tau} A_j \subset \mathcal{M}_l} \mu_t(\phi^{\tau} A_j) \ln \frac{\mu_t(\phi^{\tau} A_j)}{\nu(A_j)} ,\end{aligned} \tag{34}$$

where we have used $\nu(\phi^{\tau} A_j) = \nu(A_j)$. Next we expand in powers of the deviations of the measures from their equilibrium values and find

$$\begin{aligned}\Delta_{\mathrm{i}}^{\tau} S(\mathcal{M}_l) = &\frac{1}{2} \sum_{\phi^{\tau} A_j \subset \mathcal{M}_l} \frac{[\delta\mu_t(\phi^{\tau} A_j)]^2}{\nu(A_j)} \\ &- \frac{1}{2} \sum_{A_j \subset \mathcal{M}_l} \frac{[\delta\mu_t(A_j)]^2}{\nu(A_j)} + O\left(\delta\mu_t^3\right) .\end{aligned} \tag{35}$$

So far this has all been formal. To proceed further, we need to take advantage of the fact that the deviations of the measure from equilibrium values, $\delta\mu_t(A_j)$, are, apart from some obvious factors, normalized, incomplete van Hove functions, since they are obtained by integration of the exponential expression, $\exp[i\boldsymbol{k} \cdot \boldsymbol{d}(\Gamma, t)]$ over the set A_j. We can see this immediately by writing the deviation in measure for a set A, in the unit cell denoted by $\boldsymbol{l}$, as

$$\begin{aligned}\delta\mu_t(A_{\boldsymbol{l}}) &= \int_{A_{\boldsymbol{l}}} d\Gamma\ \rho_{\rm eq} R(\boldsymbol{l},\Gamma,t)\\ &= \int_{A_{\boldsymbol{l}}} d\Gamma \rho_{\rm eq} \frac{1}{|\mathcal{B}|}\int_{\mathcal{B}} d\boldsymbol{k}\ R_{\boldsymbol{k}}\ e^{i\boldsymbol{k}\cdot[\boldsymbol{l}+\boldsymbol{d}(\Gamma,t)]}\\ &= \frac{\nu(\mathcal{M})}{|\mathcal{B}|}\int_{\mathcal{B}} d\boldsymbol{k}\ R_{\boldsymbol{k}}\ e^{i\boldsymbol{k}\cdot\boldsymbol{l}}\ \frac{\int_{\mathcal{M}} d\Gamma\ \rho_{\rm eq}\ e^{i\boldsymbol{k}\cdot\boldsymbol{d}(\Gamma,t)}}{\int_{\mathcal{M}} d\Gamma \rho_{\rm eq}}\\ &\quad\times \frac{\int_A d\Gamma \rho_{\rm eq}\ e^{i\boldsymbol{k}\cdot\boldsymbol{d}(\Gamma,t)}}{\int_{\mathcal{M}} d\Gamma \rho_{\rm eq}\ e^{i\boldsymbol{k}\cdot\boldsymbol{d}(\Gamma,t)}}\,,\end{aligned} \tag{36}$$

where we have used the property that the backward displacement $\boldsymbol{d}(\Gamma,t)$ is independent of the initial lattice vector $\boldsymbol{l}$ so that the integral over the set $A_{\boldsymbol{l}}$ is equal to the integral over the the corresponding sets for all cells of the lattice. In the last line, we have factorized the exponential decay according to Eq. (22), which is independent of the set A, from a further factor, which depends on the set A but which is expected to have a well-defined limit for $t \to \infty$ because both its numerator and denominator are expected to decay exponentially as $\exp(s_k t)$. This structure of the measure in Eq. (36) motivates the introduction of a quantity, $\chi_{\boldsymbol{k}}(A,t)$, which we call the hydrodynamic measure of a set, and is an incomplete van Hove function of the type discussed in previous sections. Here it is defined for a particular set, A, of the partition, by

$$\chi_{\boldsymbol{k}}(A,t) \equiv \nu(\mathcal{M})\ \frac{\int_A d\Gamma \rho_{\rm eq}\ e^{i\boldsymbol{k}\cdot\boldsymbol{d}(\Gamma,t)}}{\int_{\mathcal{M}} d\Gamma \rho_{\rm eq}\ e^{i\boldsymbol{k}\cdot\boldsymbol{d}(\Gamma,t)}}\,. \tag{37}$$

We emphasize that the hydrodynamic measures are independent of the cell location $\boldsymbol{l}$. We notice that the hydrodynamic measures are complex measures because of the lattice Fourier transform. These hydrodynamic measures can be expanded in powers of the wave vector as

$$\chi_{\boldsymbol{k}}(A,t) = \nu(A) + i\boldsymbol{k}\cdot\boldsymbol{T}(A,t) + \boldsymbol{k}\boldsymbol{k} : \overset{\leftrightarrow}{g}(A,t) + \cdots\,, \tag{38}$$

where the functions $\boldsymbol{T}(A,t)$ are the analogs of Takagi functions adapted here to the periodic Lorentz gas. Finally we can write an expression for the measure of a set A in cell $\mathcal{M}_{\boldsymbol{l}}$:

$$\begin{aligned}\mu_t(A_{\boldsymbol{l}}) &= \nu(A) + \delta\mu_t(A_{\boldsymbol{l}})\\ &= \nu(A)\\ &\quad+\frac{1}{|\mathcal{B}|}\int_{\mathcal{B}} d\boldsymbol{k}\ R_{\boldsymbol{k}}\ e^{i\boldsymbol{k}\cdot\boldsymbol{l}}\ C(\boldsymbol{k},t)\ e^{s_k t}\ \chi_{\boldsymbol{k}}(A,t)\,.\end{aligned} \tag{39}$$

The conservation of phase space measure, Eq. (31), implies that the hydrodynamic measures satisfy [7]

$$e^{s_k\tau}\ \chi_{\boldsymbol{k}}(\phi^\tau A) = e^{i\boldsymbol{k}\cdot\boldsymbol{d}(\Gamma_{\phi^\tau A},\tau)}\ \chi_{\boldsymbol{k}}(A)\,, \tag{40}$$

and that the Takagi-like functions, in turn, satisfy

$$\boldsymbol{k} \cdot \boldsymbol{T}(\phi^{\tau} A) = \boldsymbol{k} \cdot \boldsymbol{T}(A) + \nu(A)\, \boldsymbol{k} \cdot \boldsymbol{d}(\Gamma_{\phi^{\tau} A}, \tau)\,, \tag{41}$$

as well as

$$\sum_j [\boldsymbol{d}_j\, \boldsymbol{T}(A_j) + \boldsymbol{T}(A_j)\, \boldsymbol{d}_j + \boldsymbol{d}_j\, \boldsymbol{d}_j\, \nu(A_j)] = 2D\tau\, \nu(\mathcal{M})\, \overset{\leftrightarrow}{1}\,. \tag{42}$$

Here $\overset{\leftrightarrow}{1}$ is the unit tensor. Equation (42) is a relation between the diffusion coefficient, the Takagi-like functions, and the mean square displacement of the moving particle

We can now substitute the explicit forms for the measures $\delta\mu_t(A)$ given by Eq. (39) into the right hand side of Eq. (35). After some algebra, and the use of the conservation of measures, we find that the right hand side of Eq. (35) becomes

$$\begin{aligned}\Delta_{\mathrm{i}}^{\tau} S(\mathcal{M}_{\boldsymbol{l}}) = \frac{1}{2}\, \frac{1}{|\mathcal{B}|^2} \int_{\mathcal{B}} d\boldsymbol{k}_1\, R_{\boldsymbol{k}_1} \int_{\mathcal{B}} d\boldsymbol{k}_2\, R_{\boldsymbol{k}_2} e^{i\boldsymbol{l}\cdot(\boldsymbol{k}_1+\boldsymbol{k}_2)}\, C(\boldsymbol{k}_1, t)\, C(\boldsymbol{k}_2, t) \\ \times e^{(s_{k_1}+s_{k_2})t} \sum_j \frac{1}{\nu(A_j)} \boldsymbol{k}_1 \boldsymbol{k}_2 : [\boldsymbol{T}(A_j)\boldsymbol{T}(A_j) \\ - \boldsymbol{T}(\phi^{\tau} A_j)\boldsymbol{T}(\phi^{\tau} A_j)]\,.\end{aligned} \tag{43}$$

Here the summation is over the sets A_j that form a partition of the unit cell $\mathcal{M}_{\boldsymbol{l}}$. Now we use the identity Eq. (41) and the summation formula Eq. (42) to obtain our central result

$$\begin{aligned}\Delta_{\mathrm{i}}^{\tau} S(\mathcal{M}_{\boldsymbol{l}}) &= -D\,\tau\, \frac{\nu(\mathcal{M})}{|\mathcal{B}|^2} \int_{\mathcal{B}} d\boldsymbol{k}_1\, R_{\boldsymbol{k}_1} \int_{\mathcal{B}} d\boldsymbol{k}_2\, R_{\boldsymbol{k}_2}\, \boldsymbol{k}_1 \cdot \boldsymbol{k}_2 \\ &\quad \times e^{i\boldsymbol{l}\cdot(\boldsymbol{k}_1+\boldsymbol{k}_2)}\, C(\boldsymbol{k}_1, t)\, C(\boldsymbol{k}_2, t)\, e^{(s_{k_1}+s_{k_2})t} \\ &\simeq D\,\tau\, \frac{1}{n_{\mathrm{eq}}} \left[\frac{\partial n(\boldsymbol{l}, t)}{\partial \boldsymbol{l}}\right]^2,\end{aligned} \tag{44}$$

since $n_{\mathrm{eq}} = \nu(\mathcal{M})$.

Here we have used the isotropy of the motion, in order to eliminate correlations between the displacements in orthogonal directions. This results in the factor $\boldsymbol{k}_1 \cdot \boldsymbol{k}_2$ appearing in the integrand in Eq. (44). We have also implied a scaling limit in order to write the last line of this equation. Equation (44) is exactly the result one obtains by applying the methods of irreversible thermodynamics [18] to this diffusive process. It completes our brief, if somewhat incomplete survey of the role of fractal functions in the description of non-equilibrium transport in simple, low-dimensional, chaotic systems. The reader should be aware that this approach to entropy production is the subject of some controversy [19], and there are issues still to be resolved, the two most important are mentioned below.

5 Conclusion

This work has been devoted to an explanation of the appearance of fractal structures in the phase spaces of low-dimensional, classical, chaotic systems where diffusive processes take place. The fractal structures appear as the result of the mixing of phase space regions, and the concomitant mixing of values of a non-equilibrium distribution function due to the chaotic dynamics taking place in the system. These structures have some remarkable properties: The wave number dependence of the fractal dimension of the incomplete van Hove functions are expressed, for small wave numbers, in terms of macroscopic diffusion coefficients and microscopic positive Lyapunov exponents. Further, as we have seen in the previous section, the rate of entropy production in the diffusive processes described here can be viewed as a consequence of the formation and properties of these fractals. We hasten to point out that in an absolute sense there are no fractal structures in phase space for any finite time, no matter how long. However the difference between the actual structures and their asymptotic forms are only visible on such exponentially small scales that their is no essential difference between the actual structures and their fractal limiting forms, for sufficiently long times.

This work describes just the early stages of what promises to be an interesting investigation of non-equilibrium fractal forms in transport theory. One can immediately identify a number of problems whose resolution would be of great value. For example:

1. Is there an analogous treatment of more complicated transport processes, such as viscous or heat flow in many particle systems, with correspondingly high dimensional phase spaces? It is well known that there are quantities which play a role analogous to the van Hove functions for diffusion [20,21]. Do the cumulative, or partially integrated, versions of these functions have a fractal structure as well? If so, is there an analogous way to connect their fractal dimensions to transport coefficients and lyapunov exponents, or Kolmogorov-Sinai entropies, via a wave number expansion? Can one relate these structures to the production of entropy in the transport processes, as one can for diffusion in low dimensional systems? Some progress in this direction is described in Ref. ([7]).
2. It is well known that diffusion can take place in systems which are not chaotic, such as wind-tree models [22,23]. Clearly the picture developed here which relies to a great extent on the chaoticity of the underlying dynamics, cannot be directly applied to these systems. Are there analogous structures, forming perhaps on algebraic rather than exponential time scales, that play similar roles as the fractal structures do for chaotic systems?

The exploration of these and related questions will be an interesting adventure.

Appendix: Derivation of the Dimension Formula, Eq. (11)

The derivation of Eq. (11) is most simply done by using one of the main new ideas provided by the application of dynamical systems theory to statistical mechanics. That is, one can replace a calculation of equilibrium averages using a microcanonical ensemble by averages over trajectories of points in phase space with an appropriate weighting of the trajectories [24,25]. Each trajectory must be weighted by the inverse of its stretching factor $\Lambda_t^{(j)}$ which characterizes its dynamical instability. For the systems described in this paper, with only one positive Lyapunov exponent, the stretching factors are related to the positive Lyapunov exponent of the trajectory by $\lambda^{(j)} = \lim_{t\to\infty}(1/t)\ln|\Lambda_t^{(j)}|$. Under these circumstances, the average of a quantity A can be expressed in the limit $t\to\infty$ as $\langle A\rangle \sim \sum_j |\Lambda_t^{(j)}|^{-1}A^{(j)}$, where $A^{(j)}$ is the quantity evaluated for the jth trajectory. Using this idea we can transform Eqs. (2) and (22) into the condition

$$\sum_j |\Lambda_t^{(j)}|^{-1} \exp(-s_{\boldsymbol{k}}t) \exp\left\{i\boldsymbol{k}\cdot\left[\boldsymbol{l}^{(j)}(t)-\boldsymbol{l}^{(j)}(0)\right]\right\} \sim_{t\to\infty} 1\,. \tag{45}$$

We obtain Eq. (22) by taking the logarithm of Eq. (45), dividing by t, and taking the limit $t\to\infty$.

To simplify matters we treat here the case of the one-dimensional deterministic walk, since the case of the Lorentz gas is very similar and given elsewhere [5]. Suppose that the sum in Eq. (45) is restricted to the trajectories issued from initial conditions in the interval $[0,x]$. In this case, we obtain at time t a polygonal approximation of the cumulative function Eq. (6) because, the integral $\int_0^x$ is an average over trajectories with initial conditions in $[0,x]$ and the denominator is proportional to the factor $\exp(s_{\boldsymbol{k}}t)$, so that the incomplete van Hove function is normalized, $I_k(1)=1$. At time t, the curve $(\text{Re } I_k, \text{Im } I_k)\in\mathbf{C}$ can be approximated by a polygon of sides given by the small complex vectors

$$\Delta I_k^{(j)} = |\Lambda_t^{(j)}|^{-1} \exp(-s_k t) \exp\left\{ik\left[n^{(j)}(x_t)-n^{(j)}(x_0)\right]\right\}\,. \tag{46}$$

Each side has the length

$$\varepsilon_j = |\Delta I_k^{(j)}| = |\Lambda_t^{(j)}|^{-1} \exp(-\text{Re } s_k\; t)\,, \tag{47}$$

so that the polygon can be covered by balls of diameter ε_j. In the limit $t\to\infty$, this polygon converges to a fractal curve characterized by a Hausdorff dimension given by $\sum_j \varepsilon_j^{D_{\text{H}}} \sim 1$. Accordingly, the Hausdorff dimension of the hydrodynamical mode should satisfy the condition

$$\sum_j |\Lambda_t^{(j)}|^{-D_{\text{H}}} \exp(-D_{\text{H}}\,\text{Re } s_k\; t) \sim_{t\to\infty} 1\,. \tag{48}$$

The next step is to relate this equation to another important quantity from dynamical systems theory, the topological pressure of Ruelle. The topological pressure is defined in dynamical systems theory by [3,12,26]

$$P(\beta) \equiv \lim_{t\to\infty}\frac{1}{t}\,\ln\langle|\Lambda_t|^{1-\beta}\rangle \tag{49}$$

and $P(1) = 0$. The mean positive Lyapunov exponent of the system is given by $\lambda = -dP/d\beta|_{\beta=1}$. Equation (49) can be rewritten

$$\sum_j |\Lambda_t^{(j)}|^{-\beta} \exp[-P(\beta)\,t] \sim_{t\to\infty} 1\,. \tag{50}$$

Comparing with Eq. (48), we infer that for $\beta = D_{\mathrm{H}}$

$$P(D_{\mathrm{H}}) = D_{\mathrm{H}}\;\mathrm{Re}\,s_k\;, \tag{51}$$

which is already a striking result, relating the decay rate of the hydrodynamic mode to the Hausdorff dimension of the fractal curve and to the topological pressure evaluated at the Hausdorff dimension. This formula generalizes Bowen's formula $P(D_{\mathrm{H}}) = 0$ for the Hausdorff dimension of a fractal invariant set of trajectories [27]. If the wavenumber **k** vanishes, $s_{k=0} = 0$, Eq. (9) shows that the cumulative function becomes $I_{k=0}(x) = x$, which forms a straight line in the complex plane. In this equilibrium limit, Eq. (51) reduces to $P(D_{\mathrm{H}}) = 0$ so that we get the dimension $D_{\mathrm{H}} = 1$, as it should be for a straight line.

For a non-vanishing but small wavenumber, the Hausdorff dimension is expected to deviate from unity. Inserting $D_{\mathrm{H}} = 1 + \delta$ and the dispersion relation Eq. (4) in Eq. (51), we can expand both sides in powers of the wavenumber by using the properties of the topological pressure. This straightforward calculation shows that the Hausdorff dimension of the hydrodynamic mode is given by

$$D_{\mathrm{H}}(k) = 1 + \frac{\mathcal{D}}{\lambda}\,k^2 + \mathcal{O}(k^4)\;, \tag{52}$$

as given by Eq. (11).

Acknowledgments

The author would like to thank his colleagues P. Gaspard, T. Gilbert, I. Claus, and S. Viscardy, for allowing him to present some of their work at this Conference, and for all of their help and contributions to their work with him. He would also like to thank D. Wojcik for his help in the preparation of this paper. The author would also like to acknowledge support from the National Science Foundation (USA) under Grant PHY 98-20824.

References

1. S. Tasaki and P. Gaspard, J. Stat. Phys. **81**, 935 (1995).
2. P. Gaspard, Phys. Rev. E **53**, 4379 (1996).
3. P. Gaspard, *Chaos, Scattering and Statistical Mechanics* (Cambridge University Press, Cambridge, 1998).
4. T. Gilbert, J. R. Dorfman and P. Gaspard, Nonlinearity, **14**, 339, (2001).
5. P. Gaspard, I. Claus, T. Gilbert and J. R. Dorfman, Phys. Rev. Lett., **86**, 1506, (2001).

6. T. Gilbert, J. R. Dorfman and P. Gaspard, Phys. Rev. Lett. **85**, 1606 (2000).
7. J. R. Dorfman, P. Gaspard and T. Gilbert, "Entropy production of diffusion in spatially periodic deterministic systems", arXiv:nlin.CD/0203046; (to appear in Phys. Rev. E).
8. J. R. Dorfman, *An Introduction to Chaos in Nonequilibrium Statistical Mechanics*, (Cambridge University Press, Cambridge UK, 1999).
9. P. Gaspard, J. Stat. Phys., **88**, 1215, (1997).
10. L. Van Hove, Phys. Rev. **95**, 249 (1954).
11. Ya. G. Sinai, Russian Math. Surveys **25**, 137 (1970).
12. P. Gaspard and F. Baras, Phys. Rev. **E51**, 5332 (1995).
13. A. Knauf, Commun. Math. Phys. **110**, 89 (1987); Ann. Phys. (N. Y.) **191**, 205 (1989).
14. I. Claus, Ph. D. Thesis: *Microscopic Chaos, Fractals and Reaction-Diffusion Processes*, Universite Libre de Bruxelles, (2002).
15. S. Viscary, Mémoire: *Viscosité et chaos dans un modèle à deux disques durs*, Universite Libre de Bruxelles, (2000).
16. S. Tasaki, T. Gilbert and J. R. Dorfman, CHAOS, **8**, 424, (1998).
17. P. M Chaikin and T. C. Lubensky, *Principles of Condensed Matter Physics*, (Cambridge University Press, Cambridge, 1995).
18. S. R. de Groot and P. Mazur, *Nonequilibrium Thermodynamics*, (Dover Publishing Co., New York, 1984).
19. References can be found in the paper of L. Rondoni in this volume.
20. D. McQuarrie, *Statistical Mechanics*, (Harper and Row, New York, 1976).
21. R. Kubo, M. Toda and N. Hashitume, *Statistical Physics II*, 2nd. Edition, (Springer-Verlag, Berlin, 1991).
22. E. H. Hauge in *Transport Phenomena*, G. Kirczenow and J. Marro, eds., (Springer-Verlag, Berlin, 1974)
23. C. P. Dettmann and E. G. D. Cohen, J. Stat. Phys., **101**, 775, (2000); **103**, 589, (2001).
24. Ya. G. Sinai, Russian Math. Surveys **27**, 21 (1972).
25. R. Bowen and D. Ruelle, Invent. Math. **29**, 181 (1975).
26. D. Ruelle, *Thermodynamic Formalism* (Addison-Wesley, Reading MA, 1978).
27. R. Bowen, Publ. Math. IHES **50**, 11 (1976).

Chapter III

Dynamical Semigroups

Markov Semigroups and Their Applications ⋆

R. Rudnicki[1,2], K. Pichór[2], and M. Tyran-Kamińska[1]

[1] Institute of Mathematics, Polish Academy of Sciences, Bankowa 14, PL-40-007 Katowice, Poland
[2] Institute of Mathematics, Silesian University, Bankowa 14, PL-40-007 Katowice, Poland

Abstract. Some recent results concerning asymptotic properties of Markov operators and semigroups are presented. Applications to diffusion processes and to randomly perturbed dynamical systems are given.

1 Introduction

Markov operators were introduced to study dynamical systems and dynamical systems with stochastic perturbations. These systems describe a movement of points. If we look at such a system statistically, then we observe the evolution of a probability measure describing the distribution of points on the phase space X. In this way we obtain a transformation P defined on the space of probability measures. Assume that P is defined by a transition probability function, i.e. the transformation of Dirac measures δ_x determines P. Then P is linear. If there is some standard measure m on the space X, then we can only consider measures which are absolutely continuous with respect to m. In that case instead of the transformation of measures we consider the transformation of densities of these measures. In this way we obtain a linear transformation of the space of integrable functions which preserves the set of densities. Such a transformation is called a Markov operator.

It should be noted that also nonlinear Markov operators and semigroups appear in applications. For example Boltzmann equation [2,65] and its simplified version Tjon-Wu equation [30,61] generate a nonlinear Markov semigroups. Also coagulation-fragmentation processes are described by nonlinear Markov semigroups [4,13,25]. Though it is a little easier to study Markov operators on densities, sometimes it is more convenient to consider Markov operators on measures. Such a situation appears in constructions of fractal measures [5,29,31].

The main subject of our paper are Markov operators and Markov semigroups acting on the set of densities. Such operators and semigroups have been intensively studied because they play a special role in applications. The book of Lasota and Mackey [27] is an excellent survey of many results on this subject. Semigroups of Markov operators are generated by partial differential equations (transport equations). Equations of this type appear in the theory of stochastic

⋆ This research was partially supported by the State Committee for Scientific Research (Poland) Grant No. 2 P03A 010 16 and by the Foundation for Polish Science.

processes (diffusion processes and jump processes), in the theory of dynamical systems and in population dynamics.

In this paper we present recent results in the theory of Markov operators and semigroups and illustrate them by some physical and biological applications. Presented results are based on papers [44–47,54].

The organization of the paper is as follows. Section 2 contains the definitions of a Markov operator and a Markov semigroup. Then we give examples of Markov operators connected with dynamical systems and dynamical systems with stochastic perturbations and of Markov semigroups generated by generalized Fokker-Planck equations and transport equations. In Section 3 we study asymptotic properties of Markov operators and semigroups: asymptotic stability and sweeping. Theorems concerning asymptotic stability and sweeping allow us to formulate the Foguel alternative. This alternative says that under suitable conditions a Markov operator (semigroup) is asymptotically stable or sweeping. Then we define a notion called a Hasminskiĭ function. This notion is very useful in proofs of asymptotic stability of Markov semigroups. In Section 4 we give some applications of general results to differential equations connected with diffusion and jump processes. In Section 5 we present some results concerning other asymptotic properties of Markov operators: completely mixing and limit distribution [8,52,53].

2 Markov Operators and Semigroups

2.1 Definitions

Let the triple (X, Σ, m) be a σ-finite measure space. Denote by D the subset of the space $L^1 = L^1(X, \Sigma, m)$ which contains all densities

$$D = \{f \in L^1 : \ f \geq 0, \ \|f\| = 1\}.$$

A linear mapping $P : L^1 \to L^1$ is called a *Markov operator* if $P(D) \subset D$.

One can define a Markov operator by means of a *transition probability function*. We recall that $\mathcal{P}(x, A)$ is a transition probability function on (X, Σ) if $\mathcal{P}(x, \cdot)$ is a probabilistic measure on (X, Σ) and $\mathcal{P}(\cdot, A)$ is a measurable function. Assume that $\mathcal{P}$ has the following property

$$m(A) = 0 \Longrightarrow \mathcal{P}(x, A) = 0 \text{ for } m\text{-a.e. } x\,. \tag{1}$$

Then for every $f \in D$ the measure

$$\mu(A) = \int f(x)\mathcal{P}(x, A)\, m(dx)$$

is absolutely continuous with respect to the measure m and the formula $Pf = d\mu/dm$ defines a Markov operator $P : L^1 \to L^1$. Moreover, if $P^* : L^\infty \to L^\infty$ is the adjoint operator of P then $P^*g(x) = \int g(y)\, \mathcal{P}(x, dy)$. There are Markov operators which are not given by transition probability functions [17]. But if

X is a Polish space (i.e. a complete separable metric space), $\Sigma = \mathcal{B}(X)$ is the σ-algebra of Borel subsets of X, and m is a probability Borel measure on X then every Markov operator on $L^1(X, \Sigma, m)$ is given by a transition probability function [23].

A family $\{P(t)\}_{t\geq 0}$ of Markov operators which satisfies conditions:

(a) $P(0) = \mathrm{Id}$,
(b) $P(t+s) = P(t)P(s) \quad \text{for} \quad s, t \geq 0$,
(c) for each $f \in L^1$ the function $t \mapsto P(t)f$ is continuous

is called a *Markov semigroup.*

Now we give some examples of Markov operators and Markov semigroups.

2.2 Frobenius–Perron Operator

This operator describes statistical properties of simple point to point transformations [27]. Let (X, Σ, m) be a σ-finite measure space and let S be a measurable transformation of X. If a measure μ describes the distribution of points in the phase space X, then the measure ν given by the formula $\nu(A) = \mu(S^{-1}(A))$ describes the distribution of points after the action of the transformation S. Assume that the transformation S is non-singular, that is if $m(A) = 0$ then $m(S^{-1}(A)) = 0$. If the measure μ is absolutely continuous with respect to the measure m, then the measure ν is also absolutely continuous. If f is the density of μ and if g is the density of ν then we define the operator P_S by $P_S f = g$. This operator can be extended to a linear operator $P_S : L^1 \to L^1$. In this way we obtain a Markov operator which is called the *Frobenius–Perron operator* for the transformation S.

Remark 1. Frobenius–Perron operators can be successfully used to study ergodic properties of transformations [27]. The general rule is: the better ergodic properties a transformation has the stronger convergence of the iterates of Frobenius–Perron operator is. Namely, if the measure m is probabilistic and invariant with respect to S then S is ergodic, mixing or exact if for each density f the sequence $P^n f$ is, respectively, Cesàro, weakly or strongly convergent to $\mathbf{1}_X$.

2.3 Iterated Function System

Let $S_1, \ldots, S_n$ be non-singular transformations of the space X. Let $P_1, \ldots, P_n$ be the Frobenius–Perron operators corresponding to the transformations $S_1, \ldots, S_n$. Let $p_1(x), \ldots, p_n(x)$ be non-negative measurable functions defined on X such that $p_1(x) + \cdots + p_n(x) = 1$ for all $x \in X$. We consider the following process. Take a point x. We choose a transformation S_i with probability $p_i(x)$ and $S_i(x)$ describes the position of x after the action of the system. The evolution of densities of the distribution is described by the Markov operator

$$Pf = \sum_{i=1}^{n} P_i(p_i f).$$

2.4 Integral Operator

If $k : X \times X \to [0, \infty)$ is a measurable function such that

$$\int_X k(x, y)\, m(dx) = 1\,,$$

for almost all $y \in X$, then

$$Pf(x) = \int_X k(x, y) f(y)\, m(dy) \tag{2}$$

is a Markov operator. The function k is called a *kernel* of the operator P.

Many biological and physical processes can be modelled by means of randomly perturbed dynamical systems whose stochastic behaviour is described by integral Markov operators. Such systems are generally of the form

$$X_{n+1} = S(X_n, \xi_{n+1})\,, \tag{3}$$

where $(\xi_n)_{n=1}^{\infty}$ is a sequence of independent random variables (or elements) with the same distribution and the initial value of the system X_0 is independent of the sequence $(\xi_n)_{n=1}^{\infty}$. Studying systems of the form (3) we are often interested in the behaviour of the sequence of the measures (μ_n) defined by

$$\mu_n(A) = \operatorname{Prob}(X_n \in A)\,.$$

The evolution of these measures can be described by a Markov operator P given by $\mu_{n+1} = P\mu_n$. The operator P is defined on the space of probability measures. Assume that for almost all y the distribution μ_y of the random variable $S(y, \xi_n)$ is absolutely continuous with respect to m. Let $k(x, y)$ be the density of μ_y and the operator P be given by (2). Then P describes the evolution of the system (3).

Integral Markov operators appear in a two phase model of cell cycle proposed by J. Tyrcha [63] which generalizes the model of Lasota–Mackey [28] and the tandem model of Tyson–Hannsgen [64].

2.5 Fokker-Planck Equation

Consider the Stratonovitch stochastic differential equation

$$dX_t = \sigma(X_t) \circ dW_t + \sigma_0(X_t)\, dt\,, \tag{4}$$

where W_t is a m-dimensional Brownian motion, $\sigma(x) = [\sigma_j^i(x)]$ is a $d \times m$ matrix and $\sigma_0(x)$ is a vector in $\mathbb{R}^d$ with components $\sigma_0^i(x)$ for every $x \in \mathbb{R}^d$. We assume that for all $i = 1, ..., d$, $j = 0, ..., m$ the functions σ_j^i are sufficiently smooth and have bounded derivatives of all orders, and the coefficients of the matrix σ are also bounded. Recall that the Itô equivalent equation is of the form

$$dX_t = \sigma(X_t)\, dW_t + b(X_t)\, dt\,, \tag{5}$$

where $b_i = \sigma_0^i + \frac{1}{2}\sum_{k=1}^m \sum_{j=1}^d \sigma_k^j \frac{\partial \sigma_k^i}{\partial x_j}$. Assume that X_t is a solution of (4) or (5) such that the distribution of X_0 is absolutely continuous and has the density $v(x)$. Then X_t has also the density $u(x,t)$ and u satisfies the Fokker-Planck equation:

$$\frac{\partial u}{\partial t} = \sum_{i=1}^d \frac{\partial}{\partial x_i}\left(\sum_{j=1}^d a_{ij}(x)\frac{\partial u}{\partial x_j}\right) - \sum_{i=1}^d \frac{\partial(\sigma_0^i(x)u)}{\partial x_i}, \tag{6}$$

where $a_{ij}(x) = \frac{1}{2}\sum_{k=1}^m \sigma_k^i(x)\sigma_k^j(x)$. Equation (6) can be written in another equivalent form

$$\frac{\partial u}{\partial t} = \sum_{i,j=1}^d \frac{\partial^2(a_{ij}(x)u)}{\partial x_i \partial x_j} - \sum_{i=1}^d \frac{\partial(b_i(x)u)}{\partial x_i}. \tag{7}$$

Note that the $d \times d$-matrix $a = [a_{ij}]$ is symmetric and nonnegative definite, i.e. $a_{ij} = a_{ji}$ and

$$\sum_{i,j=1}^d a_{ij}(x)\lambda_i\lambda_j \geq 0, \tag{8}$$

for every $\lambda \in \mathbb{R}^d$ and $x \in \mathbb{R}^d$, so we only assume weak ellipticity of the operator on the right hand side of equation (6). Let consider the operator

$$Af = \sum_{i=1}^d \frac{\partial}{\partial x_i}\left(\sum_{j=1}^d a_{ij}(x)\frac{\partial f}{\partial x_j}\right) - \sum_{i=1}^d \frac{\partial(\sigma_0^i(x)f)}{\partial x_i},$$

on the set $E = \{f \in L^1(\mathbb{R}^d)\cap C_b^2(\mathbb{R}^d) : Af \in L^1(\mathbb{R}^d)\}$, where $C_b^2(\mathbb{R}^d)$ denotes the set of all twice differentiable bounded functions whose derivatives of order ≤ 2 are continuous and bounded. If $v \in C_b^2(\mathbb{R}^d)$ then equation (6) has in any time interval $[0,T]$ a unique classical solution u which satisfies the initial condition $u(x,0) = v(x)$ and this solution and its spatial derivatives up to order 2 are uniformly bounded on $[0,T]\times\mathbb{R}^d$ (see [59], [21]). But if the initial function has a compact support, i.e. $v \in C_c^2(\mathbb{R}^d)$, then the solution $u(x,t)$ of (6) and its spatial derivatives converge exponentially to 0 as $\|x\| \to \infty$. From the Gauss-Ostrogradski theorem it follows that the integral $\int u(x,t)\,dx$ is constant. Let $P(t)v(x) = u(x,t)$ for $v \in C_c^2(\mathbb{R}^d)$ and $t \geq 0$. Since the operator $P(t)$ is a contraction on $C_c^2(\mathbb{R}^d)$ it can be extended to a contraction on $L^1(\mathbb{R}^d)$. Thus the operators $\{P(t)\}_{t\geq 0}$ form a Markov semigroup. We have $P(t)(C_c^2(\mathbb{R}^d)) \subset C_b^2(\mathbb{R}^d)$ for $t \geq 0$. According to Proposition 1.3.3 of [18] the closure of the operator A generates the semigroup $\{P(t)\}_{t\geq 0}$. The adjoint operators $\{P^*(t)\}_{t\geq 0}$ form a semigroup on $L^\infty(\mathbb{R}^d)$ given by the formula

$$P^*(t)g(x) = \int_{\mathbb{R}^d} g(y)\mathcal{P}(t,x,dy) \quad \text{for } g \in L^\infty(\mathbb{R}^d),$$

where $\mathcal{P}(t,x,A)$ is the transition probability function for the diffusion process X_t, i.e. $\mathcal{P}(t,x,A) = \mathrm{Prob}(X_t \in A)$ and X_t is a solution of equation (4) with the initial condition $X_0 = x$.

2.6 Liouville Equation

If we assume that $a_{ij} \equiv 0$ in equation (7), then we obtain the *Liouville equation*

$$\frac{\partial u}{\partial t} = -\sum_{i=1}^{d} \frac{\partial}{\partial x_i} (b_i(x)u) \,. \tag{9}$$

As in the previous example, equation (9) generates a Markov semigroup given by $P(t)v(x) = u(x,t)$, where $v(x) = u(x,0)$. The semigroup $\{P(t)\}_{t\ge 0}$ can be given explicitly. Namely, for each $\bar{x} \in X$ denote by $\pi_t \bar{x}$ the solution $x(t)$ of the equation

$$x'(t) = b(x(t)) \,, \tag{10}$$

with the initial condition $x(0) = \bar{x}$. Then

$$P(t)f(x) = f(\pi_{-t}x) \det \Big[\frac{d}{dx}\pi_{-t}x\Big] \quad \text{for } f \in L^1 \,,$$

is the Frobenius-Perron operator corresponding to the map $x \mapsto \pi_t x$. Equation (9) has the following interpretation. In the space $\mathbb{R}^d$ we consider the movement of points given by equation (10). We look at this movement statistically, that is, we consider the evolution of densities of the distribution of points. Then this evolution is described by (9).

2.7 Transport Equations

If the equation $\frac{\partial u}{\partial t} = Au$ generates a Markov semigroup $\{S(t)\}_{t\ge 0}$, K is a Markov operator, and $\lambda > 0$, then the equation

$$\frac{\partial u}{\partial t} = Au - \lambda u + \lambda K u \,, \tag{11}$$

also generates a Markov semigroup. From the Phillips perturbation theorem [15], equation (11) generates a Markov semigroup $\{P(t)\}_{t\ge 0}$ on L^1 given by

$$P(t)f = e^{-\lambda t} \sum_{n=0}^{\infty} \lambda^n S_n(t) f \,, \tag{12}$$

where $S_0(t) = S(t)$ and

$$S_{n+1}(t)f = \int_0^t S(t-s) K S_n(s) f \, ds, \quad n \ge 0 \,.$$

Equations of this type appear in such diverse areas as population dynamics [36,40], in the theory of jump processes [49,62], and in astrophysics – where describes the fluctuations in the brightness of the Milky-Way [12].

In many applications A is the operator from equation (9) and the Markov operator K corresponds to some transition probability function $\mathcal{P}(x,E)$. In this case equation (11) has an interesting probabilistic interpretation. Consider a collection of particles moving under the action of the equation $x' = b(x)$. This motion is modified in the following way. In every time interval $[t, t+\Delta t]$ a particle with the probability $\mathcal{P}(x, E)\Delta t + o(\Delta t)$ changes its position from x to a point from the set E. Then any solution of (11) is the probability density function of the position of the particle at time t.

Time and size dependent models of populations can be described by a transport equation of the form (11), namely

$$\frac{\partial u}{\partial t} + \frac{\partial (V(x)u)}{\partial x} = -u(x,t) + Ku(x,t)\,. \tag{13}$$

Here the function $V(x)$ is the velocity of the growth of the size of a cell and K is a Markov operator describing the process of replication. If we assume that the size of a daughter cell is exactly a half of the size of the mother cell, then $Kf(x) = 2f(2x)$. If we consider unequal division then K is some integral operator.

It is interesting that more advanced models of population dynamics lead to equations similar to (13), but instead of the operator $K-I$ on the right-hand side of (13) appears a non-bounded linear operator (e.g. [16]). Also these equations often generate Markov semigroups [57].

Equation (11) also describes the distribution of the solutions of a Poisson driven stochastic differential equation ([62]):

$$dX_t = b(X_t)\,dt + f(X_t)\,dN_t\,,$$

where N_t is the Poisson process. Here $Au = -\sum_{i=1}^{d} \frac{\partial}{\partial x_i}\big(b_i(x)u\big)$ and K is the Frobenius-Perron operator corresponding to the transformation $T(x) = x + f(x)$.

2.8 Randomly Flashing Diffusion

Consider the stochastic equation

$$dX_t = (Y_t\sigma(X_t))\,dW_t + b(X_t)\,dt\,, \tag{14}$$

where Y_t is a homogeneous Markov process with values 0 and 1 independent of W_t and X_0. Equation (14) describes the process which randomly jumps between stochastic and deterministic states. Such processes appear in transport phenomena in sponge–type structures [3,10,35]. This process also generates a Markov semigroup but on the space $L^1(\mathbb{R} \times \{0,1\})$. The densities of the distribution of this process satisfies the following system of equations

$$\begin{cases} \dfrac{\partial u_1}{\partial t} = -pu_1 + qu_0 + \dfrac{\partial^2}{\partial x^2}\big(a(x)u_1\big) - \dfrac{\partial}{\partial x}\big(b(x)u_1\big)\,, \\ \dfrac{\partial u_0}{\partial t} = pu_1 - qu_0 - \dfrac{\partial}{\partial x}\big(b(x)u_0\big)\,. \end{cases} \tag{15}$$

In a similar way we can introduce a notion of a multistate diffusion process on $\mathbb{R}^d$ and check that it generates a Markov semigroup [54]. Let Y_t be a continuous time Markov chain on the phase space $\Gamma = \{1, \dots, k\}$, $k \geq 2$, such that the transition probability from the state j to the state $i \neq j$ in time interval Δt equals $p_{ij}\Delta t + o(\Delta t)$. We assume that $p_{ij} > 0$ for all $i \neq j$. Let b be a d–dimensional vector function defined on $\mathbb{R}^d \times \Gamma$. Let X_0 be a d–dimensional random variable independent of Y_t. Consider the stochastic differential equation

$$dX_t = \sigma(X_t, Y_t)\, dW_t + b(X_t, Y_t)\, dt\,.$$

The pair (X_t, Y_t) constitutes a Markov process on $\mathbb{R}^d \times \Gamma$. We assume that the random variable X_0 has an absolutely continuous distribution. Then the random variable X_t has an absolutely continuous distribution for each $t > 0$. Define the function u by the formula

$$\mathrm{Prob}((X_t, Y_t) \in E \times \{i\}) = \int_E u(x, i, t)\, dx\,.$$

Denote by A_l the differential operators

$$A_l f = \sum_{i,j=1}^{d} \frac{\partial^2 (a_{ij}(x, l) f)}{\partial x_i \partial x_j} - \sum_{i=1}^{d} \frac{\partial (b_i(x, l) f)}{\partial x_i}\,.$$

Let $p_{ii} = -\sum_{j \neq i} p_{ji}$ and denote by M the matrix $[p_{ij}]$. We use the notation $u_i(x, t) = u(x, i, t)$ and $u = (u_1, \dots, u_k)$ is a vertical vector. Then the vector u satisfies the following equation

$$\frac{\partial u}{\partial t} = Mu + Au\,, \tag{16}$$

where $Au = (A_1 u_1, \dots, A_k u_k)$ is also a vertical vector. The operator A_l generates a semigroup $\{S(t)(l)\}_{t \geq 0}$ of Markov operators on the space $L^1(\mathbb{R}^d, \mathcal{B}(\mathbb{R}^d), \mu)$, where μ is the Lebesgue measure.

Let $\mathcal{B}(\mathbb{R}^d \times \Gamma)$ be the σ–algebra of Borel subsets of $\mathbb{R}^d \times \Gamma$ and let m be the product measure on $\mathcal{B}(\mathbb{R}^d \times \Gamma)$ given by $m(B \times \{i\}) = \mu(B)$ for each $B \in \mathcal{B}(\mathbb{R}^d)$ and $1 \leq i \leq k$. The operator A generates a Markov semigroup $\{S(t)\}_{t \geq 0}$ on the space $L^1(\mathbb{R}^d \times \Gamma, \mathcal{B}(\mathbb{R}^d \times \Gamma), m)$ given by the formula

$$S(t)f = (S(t)(1)f_1, \dots, S(t)(k)f_k)\,,$$

where $f_i(x) = f(x, i)$ for $x \in \mathbb{R}^d$, $1 \leq i \leq k$. Now, let λ be a constant such that $\lambda = \max\{-p_{11}, \dots, -p_{kk}\}$ and $K = \lambda^{-1}M + I$. Then (16) can be written in the form

$$\frac{\partial u}{\partial t} = Au - \lambda u + \lambda K u\,, \tag{17}$$

and the matrix K is a Markov operator on $L^1(\mathbb{R}^d \times \Gamma, \mathcal{B}(\mathbb{R}^d \times \Gamma), m)$. Equation (17) has the form (11) and generates a Markov semigroup $\{P(t)\}_{t \geq 0}$ given by (12).

If $\sigma \equiv 0$ then the process X_t describes the movement of points under the action of k dynamical systems $\pi_t^i(x)$ corresponding to the equations $x' = b(x, i)$,

$i = 1, \dots, k$. The Markov chain Y_t decides which dynamical system acts at time t. We will call such a stochastic process a *randomly controlled dynamical system* and we will study it in subsection 4.6. Let E be a Borel subset of $\mathbb{R}^n$. If for each $1 \le i \le k$ and for all $t \ge 0$ we have $\pi_t^i(E) \subset E$, then the operator A generates a semigroup $\{S(t)\}_{t\ge 0}$ of Markov operators on the space $L^1(E \times \Gamma, \mathcal{B}(E \times \Gamma), m)$.

3 Asymptotic Properties of Markov Operators and Semigroups

Now we introduce some notions which characterize the behaviour of Markov semigroups $\{P(t)\}_{t\ge 0}$ when $t \to \infty$ and powers of Markov operators P^n when $n \to \infty$. Since the powers of Markov operators also form a (discrete time) semigroup we will consider only Markov semigroups.

3.1 Asymptotic Stability

Consider a Markov semigroup $\{P(t)\}_{t\ge 0}$. A density f_* is called *invariant* if $P(t)f_* = f_*$ for each $t > 0$. The Markov semigroup $\{P(t)\}_{t\ge 0}$ is called *asymptotically stable* if there is an invariant density f_* such that

$$\lim_{t\to\infty} \|P(t)f - f_*\| = 0 \quad \text{for} \quad f \in D\,.$$

If the semigroup $\{P(t)\}_{t\ge 0}$ is generated by some differential equation then the asymptotic stability means that all solutions of the equation starting from a density converge to the invariant density.

In order to formulate the main result of this section we need an auxiliary definition. A Markov semigroup $\{P(t)\}_{t\ge 0}$ is called *partially integral* if there exist $t_0 > 0$ and a measurable non-negative function $q(x,y)$ such that

$$\int_X \int_X q(x,y)\, m(dx)\, m(dy) > 0 \tag{18}$$

and

$$P(t_0)f(x) \ge \int_X q(x,y) f(y)\, m(dy) \quad \text{for every} \;\; f \in D\,. \tag{19}$$

The main result of this part is the following

Theorem 1 ([54]). *Let $\{P(t)\}_{t\ge 0}$ be a partially integral Markov semigroup. Assume that the semigroup $\{P(t)\}_{t\ge 0}$ has an invariant density f_* and has no other periodic points in the set of densities. If $f_* > 0$ a.e. then the semigroup $\{P(t)\}_{t\ge 0}$ is asymptotically stable.*

The proof of Theorem 1 is based on the theory of Harris operators given in [19,24]. Now we formulate corollaries which are often used in applications. Let f be a measurable function. The *support* of f is defined up to a set of measure zero by the formula

$$\operatorname{supp} f = \{x \in X : \; f(x) \neq 0\}\,.$$

We say that a Markov semigroup $\{P(t)\}_{t\geq 0}$ *spreads supports* if for every set $A \in \Sigma$ and for every $f \in D$ we have

$$\lim_{t\to\infty} m(\operatorname{supp} P(t)f \cap A) = m(A)$$

and *overlaps supports* if for every $f, g \in D$ there exists $t > 0$ such that

$$m(\operatorname{supp} P(t)f \cap \operatorname{supp} P(t)g) > 0\,.$$

Corollary 1 ([54]). *A partially integral Markov semigroup which spreads supports and has an invariant density is asymptotically stable.*

Corollary 2 ([54]). *A partially integral Markov semigroup which overlaps supports and has an invariant density $f_* > 0$ a.e. is asymptotically stable.*

These corollaries generalize some earlier results [6,37,50,53] for integral Markov semigroups. Another proof of Corollary 2 is given in [7].

Corollary 1 remains true also for the Frobenius-Perron operators. Precisely, let S be a double-measurable transformation of a probabilistic measure space (X, Σ, m). If S preserves the measure m and the Frobenius-Perron operator P_S spreads supports, then the powers of P_S are asymptotically stable [54]. It is interesting that if we assume only that a Markov operator (or semigroup) P has an invariant density f_* and spreads supports, then P is weakly asymptotically stable (*mixing*). It means that for every $f \in D$ the sequence $P^n f$ converges weakly to f_*. One can expect that we can omit in Corollary 1 the assumption that the semigroup is partially integral. But it is not longer true. Indeed, in [56] we construct a Markov operator $P : L^1[0,1] \to L^1[0,1]$ which spreads supports and $P\mathbf{1} = \mathbf{1}$ but it is not asymptotically stable.

If $\{P(t)\}_{t\geq 0}$ is a continuous time Markov semigroup then we can strengthen considerably Theorem 1.

Theorem 2 ([47]). *Let $\{P(t)\}_{t\geq 0}$ be a continuous time partially integral Markov semigroup. Assume that the semigroup $\{P(t)\}_{t\geq 0}$ has the only one invariant density f_*. If $f_* > 0$ a.e. then the semigroup $\{P(t)\}_{t\geq 0}$ is asymptotically stable.*

Remark 2. In applications we often replace the assumption that the invariant density is unique by the following one. We assume that there does not exist a set $E \in \Sigma$ such that $m(E) > 0$, $m(X \setminus E) > 0$ and $P(t)E = E$ for all $t > 0$. Here $P(t)$ is the operator acting on the σ-algebra Σ given by: if $f \geq 0$, $\operatorname{supp} f = A$ and $\operatorname{supp} Pf = B$ then $PA = B$.

3.2 Sweeping

A Markov semigroup $\{P(t)\}_{t\geq 0}$ is called *sweeping* with respect to a set $A \in \Sigma$ if for every $f \in D$

$$\lim_{t\to\infty} \int_A P(t)f(x)\, m(dx) = 0\,. \tag{20}$$

The notion of sweeping was introduced by Komorowski and Tyrcha [26]. The crucial role in theorems concerning sweeping plays the following condition.

(KT) There exists a measurable function f_* such that: $0 < f_* < \infty$ a.e., $P(t)f_* \le f_*$ for $t \ge 0$, $f_* \notin L^1$ and $\int_A f_*\, dm < \infty$.

Theorem 3 ([26]). *Let $\{P(t)\}_{t\ge 0}$ be an integral Markov semigroup which has no invariant density. Assume that the semigroup $\{P(t)\}_{t\ge 0}$ and a set $A \in \Sigma$ satisfy condition (KT). Then the semigroup $\{P(t)\}_{t\ge 0}$ is sweeping with respect to A.*

In paper [54] it was shown that Theorem 3 holds for a wider class of operators than integral ones. In particular, the following result was proved (see [54] Corollary 4 and Remark 6).

Theorem 4. *Let $\{P(t)\}_{t\ge 0}$ be a Markov semigroup which overlaps supports. Assume that the semigroup $\{P(t)\}_{t\ge 0}$ and a set $A \in \Sigma$ satisfy condition (KT). Then the semigroup $\{P(t)\}_{t\ge 0}$ is sweeping with respect to A.*

The main difficulty in applying Theorems 3 and 4 is to prove that a Markov semigroup satisfies condition (KT). Now we formulate a criterion for sweeping which will be useful in applications.

Theorem 5 ([54]). *Let X be a metric space and Σ be the σ–algebra of Borel sets. We assume that a Markov semigroup $\{P(t)\}_{t\ge 0}$ has the following properties:*
(a) for every $f \in D$ we have $\int_0^\infty P(t)f\, dt > 0$ a.e. or $\sum_{n=0}^\infty P^n f > 0$ a.e. if $\{P(t)\}_{t\ge 0}$ is a discrete time semigroup,
(b) for every $y_0 \in X$ there exist $\varepsilon > 0$ and a measurable function $\eta \ge 0$ such that $\int \eta\, dm > 0$ and

$$q(x,y) \ge \eta(x)\mathbf{1}_{B(y_0,\varepsilon)}(y),$$

where q is a function satisfying (18) and (19). If the semigroup $\{P(t)\}_{t\ge 0}$ has no invariant density then it is sweeping with respect to compact sets.

3.3 Foguel Alternative

We say that a Markov semigroup $\{P(t)\}_{t\ge 0}$ satisfies the *Foguel alternative* if it is asymptotically stable or sweeping from a sufficiently large family of sets. For example this family can be all compact sets.

From Corollary 1 and Theorem 5 it follows immediately

Theorem 6. *Let X be a metric space and Σ be the σ–algebra of Borel sets. Let $\{P(t)\}_{t\ge 0}$ be a Markov semigroup. We assume that there exist $t > 0$ and a continuous function $q : X \times X \to (0,\infty)$ such that*

$$P(t)f(x) \ge \int_X q(x,y)f(y)\, m(dy) \quad \textit{for } f \in D. \tag{21}$$

Then this semigroup is asymptotically stable or is sweeping with respect to compact sets.

Using Theorem 6 one can check that the Foguel alternative holds for multistate diffusion processes [35,45,54], diffusion with jumps [46] and transport equations (11) [44].

More general results concerning Foguel alternative can be found in [54]. These results were applied to the Markov operator P considered in the cell cycle model [63].

3.4 Hasminskiĭ Function

Now we consider only continuous time Markov semigroups. Sometimes we know that a given semigroup satisfies the Foguel alternative. We want to prove that this semigroup is asymptotically stable. In order to exclude sweeping we introduce a notion called a Hasminskiĭ function.

Consider a Markov semigroup $\{P(t)\}_{t\geq 0}$ and let A be the infinitesimal generator of $\{P(t)\}_{t\geq 0}$. Let $\mathcal{R} = (I - A)^{-1}$ be the resolvent operator at point 1. A measurable function $V : X \to [0, \infty)$ is called a *Hasminskiĭ function* for the Markov semigroup $\{P(t)\}_{t\geq 0}$ and a set $Z \in \Sigma$ if there exist $M > 0$ and $\varepsilon > 0$ such that

$$\int_X V(x)\mathcal{R}f(x)\,dm(x) \leq \int_X (V(x) - \varepsilon)f(x)\,dm(x) + \int_Z M\mathcal{R}f(x)\,dm(x)\,. \quad (22)$$

Theorem 7. *Let $\{P(t)\}$ be a Markov semigroup generated by the equation*

$$\frac{\partial u}{\partial t} = Au\,.$$

Assume that there exists a Hasminskiĭ function for the semigroup $\{P(t)\}_{t\geq 0}$ and a set Z. Then the semigroup $\{P(t)\}$ is not sweeping with respect to the set Z.

In application we take V such that the function A^*V is "well defined" and it satisfies the following condition $A^*V(x) \leq -c < 0$ for $x \notin Z$. Then we check that V satisfies inequality (22). This method was applied to multistate diffusion processes [45] and diffusion with jumps [46], where inequality (22) was proved by using some generalization of the maximum principle. This method was also applied to transport equations (11) in [44] but the proof of inequality (22) is different and based on an approximation of V by a sequence of elements from the domain of the operator A^*.

The function V was called a Hasminskiĭ function because he showed [22] that the semigroup generated by the Fokker-Planck equation (7) has an invariant density if there exists a positive function V such that $A^*V(x) \leq -c < 0$ if $\|x\| \geq r$.

4 Applications

4.1 The Fokker-Planck Equation

If we assume that the functions a_{ij} satisfy the uniform elliptic condition

$$\sum_{i,j=1}^{d} a_{ij}(x)\lambda_i\lambda_j \ge \alpha|\lambda|^2 , \tag{23}$$

for some $\alpha > 0$ and every $\lambda \in \mathbb{R}^d$ and $x \in \mathbb{R}^d$ then the Markov semigroup generated by the Fokker-Planck equation (6) is an integral semigroup. That is

$$P(t)f(x) = \int_{\mathbb{R}^d} q(t,x,y)f(y)\,dy, \quad t > 0 ,$$

and the kernel q is continuous and positive. From the Foguel alternative follows

Corollary 3. *Let $\{P(t)\}_{t\ge 0}$ be a Markov semigroup generated by the Fokker-Planck equation. Then this semigroup is asymptotically stable or is sweeping with respect to compact sets.*

It is easy to check that if this semigroup is not asymptotically stable, then it is sweeping with respect to the family of sets with finite Lebesgue measures.

The operator A^* is given by the formula

$$A^*V = \sum_{i,j=1}^{d} a_{ij}\frac{\partial^2 V}{\partial x_i \partial x_j} + \sum_{i=1}^{d} b_i \frac{\partial V}{\partial x_i} .$$

If there exist a non-negative C^2-function V, $\varepsilon > 0$ and $r \ge 0$ such that

$$A^*V(x) \le -\varepsilon \qquad \text{for} \qquad \|x\| \ge r ,$$

then the Markov semigroup generated by the Fokker-Planck equation is asymptotically stable. This theorem generalizes earlier results [14,58].

Now we give an example of application of Theorem 4 to study sweeping property. Consider the Fokker-Planck equation

$$\frac{\partial u}{\partial t} = \frac{\partial^2 u}{\partial x^2} - \frac{\partial(b(x)u)}{\partial x} . \tag{24}$$

Let $\{P(t)\}_{t\ge 0}$ be a Markov semigroup generated by equation (24) and let $f_*(x) = \exp\{\int_0^x b(s)\,ds\}$. Observe that if $\int_{-\infty}^{\infty} f_*(x)\,dx < \infty$ then the semigroup $\{P(t)\}_{t\ge 0}$ has an invariant density $f_*/\|f_*\|$ and consequently it is asymptotically stable. If $\int_{-\infty}^{\infty} f_*(x)\,dx = \infty$ then the semigroup $\{P(t)\}_{t\ge 0}$ is sweeping from bounded sets. But if additionally $\int_0^{\infty} f_*(x)\,dx < \infty$ then the semigroup $\{P(t)\}_{t\ge 0}$ is also sweeping from intervals $[c,\infty)$, $c \in \mathbb{R}$. Indeed, since $f_* > 0$, $Af_* \le 0$ and $\int_c^{\infty} f_*(x)\,dx < \infty$ the semigroup $\{P(t)\}_{t\ge 0}$ and the set $[c,\infty)$ satisfy condition (KT). Thus Theorem 4 implies that the semigroup $\{P(t)\}_{t\ge 0}$ is sweeping from $[c, \infty)$. Theorems 3 and 4 can be applied to study the sweeping property in the cell cycle model ([34],[54]).

Now we consider degenerate diffusion processes. Here instead of (23) we assume (8). The fundamental theorem on the existence of smooth densities of the transition probability function for degenerate diffusion processes is due to Hörmander. In a series of papers [38,39] Malliavin has developed techniques, called Malliavin calculus, to give probabilistic proof of this fact. Now we recall some results from this theory. Let $a(x)$ and $b(x)$ be two vector fields on $\mathbb{R}^d$. The Lie bracket $[a, b]$ is a vector field given by

$$[a,b]_j(x) = \sum_{k=1}^{d} \left(a_k \frac{\partial b_j}{\partial x_k}(x) - b_k \frac{\partial a_j}{\partial x_k}(x) \right) .$$

Consider the Stratonovitch stochastic differential equation (4), i.e. equation

$$dX_t = \sigma(X_t) \circ dW_t + \sigma_0(X_t)\, dt .$$

Let $\sigma_j(x)$ be a vector in $\mathbb{R}^d$ with components $\sigma_j^i(x)$ for every $x \in \mathbb{R}^d$. We assume Hörmander's condition as in [42]

(H) For every $x \in \mathbb{R}^d$ vectors

$$\sigma_1(x), \ldots, \sigma_m(x),\ [\sigma_i, \sigma_j](x)_{0\le i,j\le m},\ [\sigma_i, [\sigma_j, \sigma_k]](x)_{0\le i,j,k\le m}, \cdots ,$$

span the space $\mathbb{R}^d$.

Note that the vector σ_0 appears only through brackets. The reason why σ_0 does not appear in condition (H) can be seen by considering $(X^1(t), X^2(t)) = (W(t), t)$, which certainly does not have a density in $\mathbb{R}^2$.

Theorem 8 (Hörmander). *Under hypothesis (H) the transition probability function $\mathcal{P}(t, x, A)$ has a density $k(t, y, x)$ and $k \in C^\infty((0,\infty) \times \mathbb{R}^d \times \mathbb{R}^d)$.*

Remark 3. Note that in the uniformly elliptic case the vectors $\sigma_1(x)$, $\sigma_2(x)$,..., $\sigma_m(x)$ span $\mathbb{R}^d$, so that the hypothesis (H) is satisfied and a smooth transition density exists.

To formulate the Foguel alternative for the semigroup $\{P(t)\}_{t\ge0}$ generated by equation (4) we need the following condition

(I) For every open set $U \subset \mathbb{R}^d$ and every measurable set A with a positive Lebesgue measure there exists $t > 0$ such that

$$\int_U \int_A k(t, x, y)\, dx\, dy > 0 . \tag{25}$$

Theorem 9. *Assume that conditions (H) and (I) hold. Then the semigroup $\{P(t)\}_{t\ge0}$ is asymptotically stable or is sweeping with respect to compact sets. Moreover, if there exist a nonnegative C^2-function V and $r > 0$ such that*

$$\sup_{\|x\|>r} A^* V(x) < 0 , \tag{26}$$

then the semigroup $\{P(t)\}_{t\ge0}$ is asymptotically stable.

Proof. From (H) it follows that the semigroup $\{P(t)\}_{t\ge 0}$ is integral and given by

$$P(t)f(x) = \int_{\mathbb{R}^d} k(t,x,y)f(y)\,dy\,,$$

for $f \in L^1(\mathbb{R}^d)$. Let f be a density. Then for $t > 0$ the function $P(t)f$ is continuous and condition (I) implies that

$$\int_0^\infty P(t)f\,dt > 0 \text{ a.e.} \tag{27}$$

If the semigroup $\{P(t)\}_{t\ge 0}$ has an invariant density f_* then from (I) it follows that f_* is a unique invariant density and $f_* > 0$ a.e. According to Theorem 2 the semigroup $\{P(t)\}_{t\ge 0}$ is asymptotically stable. If the semigroup $\{P(t)\}_{t\ge 0}$ has no invariant density then according to Theorem 5 this semigroup is sweeping with respect to compact sets. Using similar arguments to that of [45] one can check that V is a Hasminskiĭ function for the semigroup and the closed ball $\{x : \|x\| \le r\}$, which completes the proof.

In order to verify condition (I) we describe a method based on support theorems ([1,9,60]) for checking positivity of k. Let $U(x_0,T)$ be the set of all points y for which we can find a $\phi \in L^2([0,T];\mathbb{R}^m)$ such that there exists a solution of the equation

$$x_\phi(t) = x_0 + \int_0^t \big(\sigma(x_\phi(s))\phi(s) + \sigma_0(x_\phi(s))\big)\,ds\,, \tag{28}$$

satisfying the condition $x_\phi(T) = y$. From the support theorem for diffusion processes it follows that the topological support of the measure $\mathcal{P}(T,x_0,\cdot)$ coincides with closure in $\mathbb{R}^d$ of the set $U(x_0,T)$. Let $D_{x_0,\phi}$ be the Frechét derivative of the function $h \mapsto x_{\phi+h}(T)$ from $L^2([0,T];\mathbb{R}^m)$ to $\mathbb{R}^d$. By $\widetilde{U}(x_0,T)$ we denote all points y such that $x_\phi(T) = y$ and the derivative $D_{x_0,\phi}$ has rank d. Then

$$\widetilde{U}(x_0,T) = \{y : k(T,y,x_0) > 0\} \quad \text{and} \quad \operatorname{cl}\widetilde{U}(x_0,T) = \operatorname{cl}U(x_0,T)\,,$$

where cl = closure. The derivative $D_{x_0,\phi}$ can be found by means of the perturbation method for ordinary differential equations. Let

$$\Lambda(t) = \frac{d\sigma_0}{dx}(x_\phi(t)) + \sum_{i=1}^m \frac{d\sigma_i}{dx}(x_\phi(t))\phi_i(t)\,, \tag{29}$$

and let $Q(t,t_0)$, for $T \ge t \ge t_0 \ge 0$, be a matrix function such that $Q(t_0,t_0) = I$ and $\dfrac{\partial Q(t,t_0)}{\partial t} = \Lambda(t)Q(t,t_0)$. Then

$$D_{x_0,\phi}h = \int_0^T Q(T,s)\sigma(x_\phi(s))h(s)\,ds\,. \tag{30}$$

Example. Consider the Newton equation with stochastic perturbation

$$\frac{d^2x}{dt^2} + \beta\frac{dx}{dt} + \psi(x) = \frac{dW_t}{dt}, \tag{31}$$

where $\beta > 0$. Equation (31) describes the dynamics of mechanical systems perturbed by white noise [27]. Let $X_t = x(t)$ and $Y_t = x'(t)$. Then equation (31) is equivalent to the system

$$dX_t = Y_t\,dt, \quad dY_t = dW_t - (\beta Y_t + \psi(X_t))\,dt\,. \tag{32}$$

Then $\sigma_1 \equiv [0,1]$, $\sigma_0(x,y) = [y, -\beta y - \psi(x)]$, $[\sigma_0, \sigma_1] \equiv [1, -\beta]$ and condition (H) holds. System (28) corresponding to (32) can be written in the following way

$$x'_\phi = y_\phi, \quad y'_\phi = \phi - \beta y_\phi - \psi(x_\phi)\,. \tag{33}$$

For given $x_0, x_1, y_0, y_1 \in \mathbb{R}$ and $T > 0$ there exist functions ϕ and x_ϕ of the form $x_\phi(t) = a_0 + a_1 t + a_2 t^2 + a_3 t^3$ such that x_ϕ and $y_\phi = x'_\phi$ satisfy system (33) and boundary conditions $x_\phi(0) = x_0$, $x_\phi(T) = x_1$, $y_\phi(0) = y_0$, $y_\phi(T) = y_1$. In our case

$$A(t) = \begin{bmatrix} 0 & 1 \\ -\psi'(x_\phi(t)) & -\beta \end{bmatrix}, \quad \sigma = \sigma_1 = \begin{bmatrix} 0 \\ 1 \end{bmatrix}.$$

Let $\varepsilon \in (0,T)$ and $h = \mathbf{1}_{[T-\varepsilon,T]}$. Since $Q(T,s) = I - A(T)(T-s) + o(T-s)$, from (29) we obtain

$$D_{x_0,y_0;\phi}h = \varepsilon\sigma_1 - \tfrac{1}{2}\varepsilon^2 A(T)\sigma_1 + o(\varepsilon^2)\,. \tag{34}$$

Since $\sigma_1 \equiv [0,1]$ and $A(T)\sigma_1 \equiv [1, -\beta]$, these vectors are linearly independent and the derivative $D_{x_0,y_0;\phi}$ has rank 2. Thus the system (32) generates an integral Markov semigroup $\{P(t)\}_{t\ge 0}$ with a continuous and strictly positive kernel k. Consider the Fokker-Planck equation corresponding to (32):

$$\frac{\partial u}{\partial t} = \frac{1}{2}\frac{\partial^2 u}{\partial y^2} - \frac{\partial}{\partial x}(yu) + \frac{\partial}{\partial y}[(\beta y + \psi(x))u]\,. \tag{35}$$

Let $\Psi(x) = \int_0^x \psi(s)\,ds$. Then the function $u_*(x,y) = \exp\{-\beta y^2 - 2\beta\Psi(x)\}$ is a stationary solution of (35). If $\int_{-\infty}^{\infty} e^{-2\beta\Psi(x)}\,dx < \infty$ then, according to Corollary 1, the semigroup $\{P(t)\}_{t\ge 0}$ is asymptotically stable. If $\int_{-\infty}^{\infty} e^{-2\beta\Psi(x)}\,dx = \infty$ then the function u_* satisfies condition (KT) for every set of the form $A(L) = \{(x,y) : x \in [-L,L],\ y \in \mathbb{R}\}$ because $P(t)u_* = u_*$ and $\iint_{A(L)} u_*(x,y)\,dx\,dy < \infty$. According to Theorem 4 this semigroup is sweeping from the sets $A(L)$.

Remark 4. Since a lot of transport equations generates a partially integral semigroup which spreads supports we can obtain similar results for these equations. Consider, for example, a stochastic equation $dX_t = -\lambda X_t\,dt + dC_t$, where C_t is a Cauchy process [20]. The semigroup generated by this equation is an integral semigroup with a continuous and positive kernel. From the Foguel alternative this semigroup is asymptotically stable or sweeping from compact sets. If $\lambda > 0$ then $f_*(x) = \lambda/\pi(\lambda^2 x^2 + 1)$ is an invariant density for semigroup $\{P(t)\}_{t\ge 0}$ and consequently it is asymptotically stable.

4.2 Diffusion with Jumps

Consider the following equation

$$\frac{\partial u}{\partial t} = Au - \lambda u + \lambda P u\,, \tag{36}$$

where $\lambda > 0$,

$$Au = \sum_{i,j=1}^{d} \frac{\partial^2 (a_{ij} u)}{\partial x_i \partial x_j} - \sum_{i=1}^{d} \frac{\partial (b_i u)}{\partial x_i}\,, \tag{37}$$

and P is a Markov operator corresponding to the iterated function system

$$(S_1(x), \ldots, S_N(x), p_1(x), \ldots, p_N(x))\,.$$

The probabilistic interpretation of equation (36) is similar to that of equation (11). We assume that for each j we have

$$\lim_{\|x\| \to \infty} \|S_j(x)\| = \infty\,.$$

Assume that

$$\lim_{\|x\| \to \infty} 2\langle x, b(x) \rangle + \lambda \sum_{j=1}^{n} p_j(x) \left(\|S_j(x)\|^2 - \|x\|^2 \right) = -\infty\,,$$

where $\langle \cdot, \cdot \rangle$ is the scalar product in $\mathbb{R}^d$. Then a Markov semigroup $\{P(t)\}_{t \geq 0}$ generated by equation (36) is asymptotically stable [46].

4.3 Randomly Interrupted Diffusion

This process was described by the following system of equations

$$\begin{cases} \dfrac{\partial u_1}{\partial t} = -p u_1 + q u_0 + A_1 u_1\,, \\ \dfrac{\partial u_0}{\partial t} = p u_1 - q u_0 + A_0 u_0\,. \end{cases}$$

A semigroup generated by this system satisfies the Foguel alternative. In order to prove asymptotic stability it is sufficient to construct a proper Hasminskiĭ function. One can check that if there exist non-negative C^2-functions V_1 and V_2 such that

$$-p(x)V_1(x) + p(x)V_2(x) + A_1^* V_1(x) \leq -\varepsilon\,,$$
$$q(x)V_1(x) - q(x)V_2(x) + A_2^* V_2(x) \leq -\varepsilon\,,$$

for $\|x\| \geq r$, then the corresponding Markov semigroup is asymptotically stable [45].

4.4 Transport Equation

Consider a partial differential equation with an integral perturbation

$$\frac{\partial u}{\partial t} + \lambda u = -\sum_{i=1}^{d} \frac{\partial (b_i u)}{\partial x_i} + \lambda \int k(x,y) u(y,t)\, dy\,. \tag{38}$$

If $k(x,y)$ is a continuous and strictly positive function and there exists a C^1-function $V : X \to [0,\infty)$ such that

$$\sum_{i=1}^{d} b_i \frac{\partial V}{\partial x_i} - \lambda V(x) + \lambda \int k(y,x) V(y)\, dy \le -c < 0\,,$$

for $\|x\| \ge r$, $r > 0$, then a Markov semigroup $\{P(t)\}_{t\ge 0}$ generated by equation (38) is asymptotically stable [44].

Remark 5. Consider the transport equation

$$\frac{\partial u}{\partial t} + \lambda u = Au + \lambda K u\,, \tag{39}$$

where A is a generator of the Markov semigroup $\{S(t)\}_{t\ge 0}$ and K is a Markov operator. If the semigroup $\{S(t)\}_{t\ge 0}$ is partially integral or the operator K is partially integral then from (12) it follows that the semigroup $\{P(t)\}_{t\ge 0}$ is partially integral. From (12) and continuity of the semigroups $\{S(t)\}_{t\ge 0}$ and $\{P(t)\}_{t\ge 0}$ it follows that for a measurable set E we have $P(t)E \subset E$ for all $t \ge 0$ if and only if $KE \subset E$ and $S(t)E \subset E$ for all $t \ge 0$. Let $\mathcal{P}(x,E)$ be the transition probability function corresponding to K. Then $KE \subset E$ if and only if $\mathcal{P}(x,E) = 1$ for a.e. $x \in E$. In the next subsections we consider two examples of random movement of this type. In these examples both the semigroup $\{S(t)\}_{t\ge 0}$ and the operator K are singular (have no integral parts) but the semigroup $\{P(t)\}_{t\ge 0}$ is partially integral. Moreover we give sufficient conditions for asymptotic stability of these semigroups which are based on Theorem 2.

4.5 Jump Process

We consider equation (36) but instead of the operator (37) we consider the Liouville operator

$$Au = -\sum_{i=1}^{d} \frac{\partial (b_i u)}{\partial x_i}\,.$$

The probabilistic interpretation of this equation was given in Subsection 2.7.

Theorem 10 ([47]). *Assume that the semigroup $\{P(t)\}_{t\ge 0}$ has a non-zero invariant function and has no non-trivial invariant sets. Let $(i_1,\dots,i_d)$ be a given sequence of integers from the set $\{1,\dots,k\}$. Let $x_0 \in X$ be a given point and let $x_j = S_{i_j}(x_{j-1})$ for $j = 1,\dots,d$. Set*

$$v_j = S'_{i_d}(x_{d-1}) \dots S'_{i_j}(x_{j-1}) b(x_{j-1}) - b(x_d)\,,$$

for $j = 1, \dots, d$. *Assume that* $p_{i_j}(x_{j-1}) > 0$ *for all* $j = 1, \dots, d$ *and suppose that the vectors* $v_1, \dots, v_d$ *are linearly independent. Then the semigroup* $\{P(t)\}_{t\ge 0}$ *is asymptotically stable.*

4.6 Randomly Controlled Dynamical System

Now we consider a stochastic process introduced in the end of subsection 2.8. We recall that we have k dynamical systems $\pi_t^i(x)$ corresponding to the equations $x' = b(x, i)$, $i = 1, \dots, k$ and we exchange their randomly. Denote by $\{P(t)\}_{t\ge 0}$ the semigroup corresponding to this system. Let $(i_1, \dots, i_{d+1})$ be a sequence of integers from the set $\Gamma = \{1, \dots, k\}$. For $x \in X$ and $t > 0$ we define the function $\psi_{x,t}$ on the set $\Delta_t = \{\tau = (\tau_1, \dots, \tau_d) : \tau_i > 0, \tau_1 + \dots + \tau_d \le t\}$ by

$$\psi_{x,t}(\tau_1, \dots, \tau_d) = \pi^{i_{d+1}}_{t-\tau_1-\tau_2-\dots-\tau_d} \circ \pi^{i_d}_{\tau_d} \circ \dots \circ \pi^{i_2}_{\tau_2} \circ \pi^{i_1}_{\tau_1}(x)\,.$$

Theorem 11 ([47]). *Assume that the semigroup* $\{P(t)\}_{t\ge 0}$ *has a non-zero invariant function and has no non-trivial invariant sets. Suppose that for some* $x_0 \in X$, $t_0 > 0$ *and* $\tau^0 \in \Delta_{t_0}$ *we have*

$$\det \left[\frac{d\psi_{x_0,t_0}(\tau^0)}{d\tau} \right] \ne 0\,. \tag{40}$$

Then the semigroup $\{P(t)\}_{t\ge 0}$ *is asymptotically stable.*

Remark 6. A measurable set $E \subset X \times \Gamma$ is invariant with respect to the semigroup $\{P(t)\}_{t\ge 0}$ if and only if E is of the form $E = E_0 \times \Gamma$ and

$$\pi_t^i(E_0) = E_0 \qquad \text{for } t \ge 0 \text{ and } i = 1, \dots, k\,.$$

Remark 7. Condition (40) can be formulated using Lie brackets. Let $b_i(x) = b(x, i)$. If vectors

$$b_2(x_0) - b_1(x_0), \dots, b_k(x_0) - b_1(x_0), [b_i, b_j](x_0)_{1\le i,j\le k}, [b_i, [b_j, b_l]](x_0)_{1\le i,j,l\le k}, \dots,$$

span the space $\mathbb{R}^d$ then (40) holds.

4.7 Population Dynamics Equation

Some models of size-structured cell populations lead to transport equations similar to (11), but these equations do not generate Markov semigroups. Also in these cases we can often apply results presented in Section 3. We consider here a model derived in [57], which generalized some earlier models of cell populations (e.g. [16]).

We assume that a cell is fully characterized by its size x which ranges from $x = a$ to $x = 1$. The cell size grows according to equation $x' = g(x)$. Cells can die or divide with rates $\mu(x)$ and $b(x)$. We assume that the cells cannot divide before they have reached a minimal maturation $a_0 \in (a, 1)$. Since the cells have to divide before they reach the maximal size $x = 1$, we assume that $\lim_{x\to 1} \int_a^x b(\xi)\, d\xi = \infty$. If $x \geq a_0$ is the size of a mother cell at the point of cytokinesis, then a new born daughter cell has the size which is randomly distributed in the interval $(a, x - h]$, where h is a positive constant. We denote by $\mathcal{P}(x, [x_1, x_2])$ the probability for a daughter cell born from a mother cell of size x to have a size between x_1 and x_2.

The function $N(x,t)$ describing the distribution of the size satisfies the following equation

$$\frac{\partial N}{\partial t} = -\frac{\partial (gN)}{\partial x} - (\mu + b)N + 2P(bN)\,, \tag{41}$$

where $P : L^1(a,1) \to L^1(a,1)$ is a Markov operator such that $P^*\mathbf{1}_B(x) = \mathcal{P}(x, B)$. The main result concerning equation (41) is the following

Theorem 12. *There exist $\lambda \in \mathbb{R}$ and continuous and positive functions f_* and w defined on the interval $(a,1)$ such that $e^{-\lambda t}N(\cdot,t)$ converges to $f_*\Phi(N)$ in $L^1(a,1)$, where $\Phi(N) = \int_a^1 N(x,0)w(x)\, dx$.*

The proof of Theorem 12 goes as follows. Equation (41) can be written as an evolution equation $N'(t) = AN$. First we show that A is an infinitesimal generator of a continuous semigroup $\{T(t)\}_{t\geq 0}$ of linear operators on $L^1(a,1)$. Then we prove that there exist $\lambda \in \mathbb{R}$ and continuous and positive functions v and w such that $Av = \lambda v$ and $A^*w = \lambda w$. From this it follows that the semigroup $\{P(t)\}_{t\geq 0}$ given by $P(t) = e^{-\lambda t}T(t)$ is a Markov semigroup on the space $L^1(X, \Sigma, m)$, where m is a Borel measure on the interval $[a,1]$ given by $m(B) = \int_B w(x)\, dx$. Moreover, for some $c > 0$ the function $f_* = cv$ is an invariant density with respect to $\{P(t)\}_{t\geq 0}$. Finally, from Theorem 1 we conclude that this semigroup is asymptotically stable. Since the Lebesgue measure and the measure m are equivalent we obtain that $e^{-\lambda t}N(\cdot,t)$ converges to $f_*\Phi(N)$ in $L^1(a,1)$.

5 Other Asymptotic Properties

In this section we give some results concerning other asymptotic properties of Markov operators: completely mixing and limit distribution.

5.1 Completely Mixing

Semigroup $\{P(t)\}_{t\geq 0}$ is called completely mixing if for any two densities f and g we have

$$\lim_{t\to\infty} \|P(t)f - P(t)g\| = 0\,. \tag{42}$$

This notion has the following probabilistic interpretation. Let $\{P(t)\}_{t\geq 0}$ be the Markov semigroup corresponding to a diffusion process. Assume that this process describes a movement of particles. Then condition (42) means that particles are mixed in such a way that after a long time their distribution does not depend on the initial distribution. If there exists an invariant density f_* then completely mixing is equivalent to asymptotic stability. However, the semigroup $\{P(t)\}_{t\geq 0}$ can be completely mixing, but it can have no invariant density. For example, the heat equation $\dfrac{\partial u}{\partial t} = \Delta u$ generates the semigroup which is completely mixing and has no invariant density.

Completely mixing property of the semigroup $\{P(t)\}_{t\geq 0}$ is strictly connected with the notion of the relative entropy. The relative entropy can be written down in the following way

$$H(t) = \int P(t)f(x) \ln \frac{P(t)f(x)}{P(t)g(x)}\, dx, \qquad f, g \in D\,.$$

It is not difficult to check that if $\lim_{t\to\infty} H(t) = 0$ then the semigroup $\{P(t)\}_{t\geq 0}$ is completely mixing (see [33] for a more general result). It is also easy to check that completely mixing implies that all fixed points of the semigroup $\{P^*(t)\}_{t\geq 0}$ are constant functions.

Completely mixing property for the Fokker-Planck equation (7) was studied in the papers [11,51]. The most general result in this direction was received in [8]. They proved that if the coefficients in the Fokker-Planck equation are bounded with their first and second partial derivatives, the diffusion term satisfies uniform elliptic condition (23) and all fixed points of the semigroup $\{P^*(t)\}_{t\geq 0}$ are constant functions then the semigroup $\{P(t)\}_{t\geq 0}$ is completely mixing. In other words the semigroup $\{P(t)\}_{t\geq 0}$ is completely mixing if and only if all bounded solutions of the elliptic equation

$$\sum_{i,j=1}^{n} a_{ij}(x) \frac{\partial^2 u}{\partial x_i \partial x_j} + \sum_{i=1}^{n} b_i(x) \frac{\partial u}{\partial x_i} = 0\,,$$

are constant. It is worth pointing out that even in one-dimensional case with constant diffusion the assumption that the drift coefficient is bounded cannot be replaced with the assumption that it grows linearly [51].

Remark 8. Let P_S be the Frobenius–Perron operator for a measurable transformation S of a σ-finite measure space (X, Σ, m). Then P_S is completely mixing if and only if $\bigcap_{n=1}^{\infty} S^{-n}\Sigma = \{\emptyset, X\}$ ([32]). If additionally the measure m is invariant then the transformation S is exact. In the paper [48] we give an example of a piecewise linear and expanding transformation of the interval $[0,1]$ which is completely mixing but for every density f the iterations $P_S^n f$ converge weakly to the standard Cantor measure. This transformation has similar properties to the Smale horseshoe.

5.2 Limit Distribution

Let $S = \{x \in \mathbb{R}^d : \|x\| = 1\}$ and A be a measurable subset of S. Denote by $K(A)$ the cone spanned by A:

$$K(A) = \{x \in \mathbb{R}^d : x = \lambda y,\ y \in A,\ \lambda > 0\}\,.$$

Consider a Markov semigroup $\{P(t)\}_{t\geq 0}$ corresponding to a diffusion process. Then the function

$$p_A(t) = \int_{K(A)} P(t)f(x)\,dx, \qquad f \in D,$$

describes the mass of particles which are in the cone $K(A)$. If the semigroup $\{P(t)\}_{t\geq 0}$ is completely mixing then the asymptotic behaviour of $p_A(t)$ does not depend on f. It is interesting when there exists the limit $p_A = \lim_{t\to\infty} p_A(t)$. If $\{P(t)\}_{t\geq 0}$ is sweeping then nearly all particles are in a neighborhood of ∞ for large t and p_A measures the *sectorial limit distribution* of particles.

The problem of finding the limit distribution for arbitrary diffusion process in d-dimensional space is difficult. Some partial results can be obtained under additional assumption that all functions a_{ij} and b_i are periodic with the same periods (we recall that a function $f : \mathbb{R}^d \to \mathbb{R}$ is periodic if there exist independent vectors $v_1, \dots, v_d$ such that $f(x + v_i) = f(x)$ for each $x \in \mathbb{R}^d$ and $i = 1, \dots, d$).

In one-dimensional space we can consider the function $p_+(t) = \int_c^\infty u(x,t)\,dx$ which describes the mass of particles in the interval (c, ∞). The paper [52] provides a criterion for the existence of the limit $\lim_{t\to\infty} p_+(t)$. In the same paper we construct an equation such that the following condition holds

$$\limsup_{t\to\infty} \frac{1}{t}\int_0^t p_+(s)\,ds = 1 \qquad \text{and} \qquad \liminf_{t\to\infty} \frac{1}{t}\int_0^t p_+(s)\,ds = 0. \tag{43}$$

In this example $a(x) = 1$ and $b(x) \to 0$ as $|x| \to \infty$. Condition (43) means that particles synchronously oscillate between $+\infty$ and $-\infty$.

Remark 9. Many abstract results concerning completely mixing property can be found in books [41,43]. Completely mixing property of an integral Markov operator appearing in a model of cell cycle was studied in [55]. If a Markov semigroup has no invariant density one can investigate a property of convergence after rescaling. We say that a Markov semigroup $\{P(t)\}_{t\geq 0}$ is *convergent after rescaling* if there exist a density g and functions $\alpha(t)$, $\beta(t)$ such that

$$\lim_{t\to\infty} \int_X |\alpha(t)P(t)f\big(\alpha(t)x + \beta(t)\big) - g(x)|\,dx = 0 \qquad \text{for every } f \in D. \tag{44}$$

Condition (44) implies completely mixing property. One of the weak versions of this condition is the central limit theorem. In papers [49,50] it is shown that semigroups connected with processes with jumps satisfy condition (44), precisely, these processes are asymptotically log-normal.

References

1. S. Aida, S. Kusuoka and D. Strook, *On the support of Wiener functionals* in Asymptotic problems in probability theory: Wiener functionals and asymptotic, K. D. Elworthy and N. Ikeda (eds.), pp. 3–34, Pitman Research Notes in Math. Series **284**, Longman Scient. Tech., 1993.

2. L. Arkeryd, R. Espositio and M. Pulvirenti, Comm. Math. Phys. **111**, 393–407 (1987).
3. V. Balakrishnan, C. Van den Broeck and P. Hanggi, Phys. Rev. **A38**, 4213–4222 (1988).
4. J.M. Ball and J. Carr, J. Statist. Phys. **61**, 203–234 (1990).
5. M.F. Barnsley, *Fractals Everywhere*, Academic Press, New York, 1993.
6. K. Baron and A. Lasota, Ann. Polon. Math. **58**, 161–175 (1993).
7. W. Bartoszek and T. Brown, Bull. Pol. Ac.: Math. **45**, 17–24 (1997).
8. C.J.K. Batty, Z. Brzeźniak and D.A. Greenfield, Studia Math. **121**, 167–183 (1996).
9. G. Ben Arous and R. Léandre, Probab. Theory Relat. Fields **90**, 377–402 (1991).
10. V. Bezak, J. Phys. A: Math. Gen. **25**, 6027–6041 (1992).
11. Z. Brzeźniak and B. Szafirski, Bull. Pol. Ac.: Math. **39**, 1–10 (1991).
12. S. Chandrasekhar and G. Münch, Astrophys. J. **125**, 94–123 (1952).
13. F.P. da Costa, J. Math. Anal. Appl. **192**, 892–914 (1995).
14. T. Dłotko and A. Lasota, *Statistical stability and the lower bound function technique*, in Semigroups theory and applications, Vol I, H. Brezis, M. Crandall and F. Kappel (eds.), pp. 75–95, Longman Scientific & Technical, 1987.
15. N. Dunford and J.T. Schwartz, *Linear Operators, Part I*, Interscience Publ., New York, 1968.
16. O. Diekmann, H.J.A. Heijmans and H.R. Thieme, J. Math. Biol. **19**, 227–248 (1984).
17. J. Dieudonne, Ann. Univ. Grenoble **23**, 25–53 (1948).
18. S. Ethier and T. Kurtz, *Markov Processes: Characterization and Convergence*, John Wiley, New York, 1986.
19. S.R. Foguel, *The Ergodic Theory of Markov Processes*, Van Nostrand Reinhold Comp., New York, 1969.
20. P. Garbaczewski and R. Olkiewicz, J. Math. Phys. **41**, 6843–6860 (2000).
21. I.I. Gihman and A.V. Skorohod, *Stochastic Differential Equations*, Springer-Verlag, New York, 1972.
22. R.Z. Hasminskiĭ, Teor. Verojatn. Primenen. **5**, 196–214 (1960) (in Russian).
23. P. Hennequin and A. Tortrat, *Theorie des probabilities et quelques applications*, Masson et Cie, Paris, 1965.
24. B. Jamison and S. Orey, Z. Wahrsch. Verw. Gebiete **8**, 41–48 (1967).
25. I. Jeon, Comm. Math. Phys. **194**, 541–567 (1998).
26. T. Komorowski and J. Tyrcha, Bull. Pol. Ac.: Math. **37**, 221–228 (1989).
27. A. Lasota and M.C. Mackey, *Chaos, Fractals and Noise. Stochastic Aspects of Dynamics*, Springer Applied Mathematical Sciences **97**, New York, 1994.
28. A. Lasota and M.C. Mackey, J. Math. Biol. **19**, 43–62 (1984).
29. A. Lasota and J. Myjak, Bull. Pol. Acad. Sci.: Math. **44**, 5–21 (1996).
30. A. Lasota and J. Traple, J. Diff. Equations **159**, 578–596 (1999).
31. A. Lasota and J.A. Yorke, Random and Computational Dynamics **2**, 41–77 (1994).
32. M. Lin, Z. Wahrsch. Verw. Gebiete **19**, 231–242 (1971).
33. K. Łoskot and R. Rudnicki, Ann. Pol. Math. **53**, 139–145 (1991).
34. K. Łoskot and R. Rudnicki, Bull. Pol. Ac.: Math. **37**, 229–235 (1989).
35. J. Łuczka and R. Rudnicki, J. Statist. Phys. **83**, 1149–1164 (1996).
36. M.C. Mackey and R. Rudnicki, J. Math. Biol. **33**, 89–109 (1994).
37. J. Malczak, Rend. Sem. Mat. Univ. Padova **87**, 281–297 (1992).
38. P. Malliavin, *Stochastic calculus of variations and hypoelliptic operators*, in Proc. Intern. Symp. Stoch. Diff. Equations of Kyoto 1976, K. Itô (eds.), pp. 195–263, John Wiley, New York, 1978.

39. P. Malliavin, *C^k-hypoellipticity with degeneracy,* in Stochastic Analysis, A. Friedman and M. Pinsky (eds.), pp. 199–214, Acad. Press, New York 1978.
40. J.A.J. Metz and O. Diekmann, *The Dynamics of Physiologically Structured Populations*, Springer Lecture Notes in Biomathematics **68**, New York, 1986.
41. J. van Neerven, *The Asymptotic Behaviour of a Semigroup of Linear Operators*, Birkhäuser, Basel, 1996.
42. J. Norris, *Simplified Malliavin calculus*, in Séminaire de probabilitiés XX, Lecture Notes in Mathematics, **1204**, pp.101-130, Springer, 1986.
43. E. Nummelin, *General Irreducible Markov Chains and Non-negative Operators*, Cambridge Tracts in Mathematics **83**, Cambridge University Press, Cambridge, 1984.
44. K. Pichór, Ann. Pol. Math. **68**, 83–96 (1998).
45. K. Pichór and R. Rudnicki, J. Math. Anal. Appl. **215**, 56–74 (1997).
46. K. Pichór and R. Rudnicki, Bull. Pol. Ac.: Math. **45**, 379–397 (1997).
47. K. Pichór and R. Rudnicki, J. Math. Anal. Appl. **249**, (2000)668–685.
48. R. Rudnicki, Ann. Pol. Math. **54**, 47–153 (1991).
49. R. Rudnicki, Ann. Pol. Math. **57**, 45–55 (1992).
50. R. Rudnicki, Bull. Pol. Ac.: Math. **40**, 111–128 (1992).
51. R. Rudnicki, J. Diff. Equations **102**, 391–401 (1993).
52. R. Rudnicki, Ann. Pol. Math. **58**, 37–45 (1993).
53. R. Rudnicki, *Asymptotic properties of the Fokker-Planck equation*, in Chaos – The interplay between stochastics and deterministic behaviour, Karpacz'95 Proc., P. Garbaczewski, M. Wolf, A. Weron (eds.), pp. 517–521, Lecture Notes in Physics **457**, Springer, Berlin, 1995.
54. R. Rudnicki, Bull. Pol. Ac.: Math. **43**, 245–262 (1995).
55. R. Rudnicki, Integ. Equat. Oper. Th. **24**, 320–327 (1996).
56. R. Rudnicki, Bull. Pol. Ac.: Math. **45**, 1–5 (1997).
57. R. Rudnicki and K. Pichór, J. Biol. Systems **8**, 69–94 (2000).
58. R. Sanders, J. Math. Anal. Appl. **112**, 335–346 (1985).
59. D.W. Stroock and S.R.S. Varadhan, Comm. Pure Appl. Math. **24**, 651–713 (1972).
60. D.W. Stroock and S.R.S. Varadhan, *On the support of diffusion processes with applications to the strong maximum principle,* Proc. Sixth Berkeley Symposium on Mathematical Statistics and Probability, vol. III, pp. 333-360, Univ. Cal. Press, Berkeley, 1972.
61. J.A. Tjon and T.T. Wu, Phys. Rev. **A19**, 883–888 (1987).
62. J. Traple, Bull. Pol. Ac.: Math. **44**, 230–252 (1996).
63. J. Tyrcha, J. Math. Biology **26**, 465–475 (1988).
64. J.J. Tyson and K.B. Hannsgen, J. Math. Biol. **23**, 231–246 (1986).
65. B. Wennberg, Arch. Rational Mech. Anal. **130**, 103–144 (1995).

Invitation to Quantum Dynamical Semigroups

R. Alicki

Institute of Theoretical Physics and Astrophysics, University of Gdańsk, Wita Stwosza 57, PL-80-952 Gdańsk, Poland

Abstract. The theory of quantum dynamical semigroups within the mathematically rigorous framework of completely positive dynamical maps is reviewed. First, the axiomatic approach which deals with phenomenological constructions and general mathematical structures is discussed. Then basic derivation schemes of the constructive approach including singular coupling, weak coupling and low density limits are presented in their higly simplified versions. Two-level system coupled to a heat bath, damped harmonic oscillator, models of decoherence, quantum Brownian particle and Bloch-Boltzmann equations are used as illustrations of the general theory. Physical and mathematical limitations of the quantum open system theory, the validity of Markovian approximation and alternative approaches are discussed also.

1 Introduction

Classical theory of systems interacting with environment employs evolution equations which can be written in the following abstract mathematical form

$$\frac{d}{dt}p_t = \mathcal{L}p_t \quad t \geq 0 \ . \tag{1}$$

Here p_t is a time dependent probability distribution over the relevant system's configuration space Ω. In particular Ω can be either a phase-space of a system, its position or momentum (velocity) space or a discrete set in the case of coarse-grained description or a discretized model. The most general "continuous" form of $\mathcal{L}$ is a sum of a differential operator (Fokker-Planck or diffusion type) and a collisional integral one [1]. For a discrete version we can write (1) as a Pauli master equation

$$\frac{d}{dt}p_t(k) = \sum_l \big(a_{kl}p_t(l) - a_{lk}p_t(k)\big) \ . \tag{2}$$

The "gain – loss" structure of the RHS of (2) guarantees preservation of $\sum_k p_t(k)$ while positivity of the transition probabilities per unit time ($a_{kl} \geq 0$) implies $p_t(k) \geq 0$. Similar mathematical consistency conditions are satisfied for the continuous cases due to the positivity of the diffusion constant (or positively defined diffusion matrix) and/or positivity of the scattering cross-sections. The solution of Eq. (1) is given in terms of a Markov semigroup $T_t = \exp\{t\mathcal{L}\}$ satisfying the composition law

$$T_tT_s = T_{t+s} \quad t, s \geq 0 \tag{3}$$

and preserving positivity and normalization of the probability distribution p_t.

In the end of 60-ties and begining of 70-ties one needed an analogical formalism to describe a variety of irreversible phenomena in quantum optics, solid state physics, magnetic resonance, nuclear and particle physics, etc. [2,3]. The classical probability distribution p_t should be replaced by the density matrix ρ_t satisfying the analog of Eq. (1)

$$\frac{d}{dt}\rho_t = L\rho_t \quad t \geq 0 \ . \tag{4}$$

Besides many heuristic attempts the mathematically sound theory of quantum dynamical semigoups has been developed very soon [4-7]. It is convenient for the further discussion to distinguish two approaches in this theory - the *axiomatic* and the *constructive* one.

The former approach, presented in the next Section, is concentrated on general mathematical properties of Eq. (4) and its solutions which are by no means trivial and lead to the theory of comptelely positive maps and semigroups on operator algebras. The fundamental structural theorems will be given and illustrated by several examples.

In constructive approach one tries to derive Eq. (4) from first principles. Starting with a model of a quantum system interacting with a quantum environment one obtains Markovian master equation (4) as an approximation to the exact reduced dynamics of the open system. The canonical models of reservoirs and basic approximation procedures are discussed in Section III.

In the recent years quantum theory of open systems and in particular quantum dynamical semigroups became again a subject of intensive investigations. This is motivated by the new experimental techniques which allow precise monitoring of decoherence and dissipation in microscopic and mesoscopic quantum systems [8]. Possible future applications of controlled quantum systems in quantum information processing and quantum computing are another sources of this revival [9]. Both mentioned above approaches are necessary to create physically correct, mathematically consistent and convenient for practical applications theory.

The number of publications on this topic is enormous and therefore the presented list of references is far from being complete, often review papers, books and other contributions to this volume are quoted instead of the original papers.

2 Axiomatic Approach

Consider an open quantum system with the associated Hilbert space $\mathcal{H}$ with the time evolution governed by the master equation (4) for its density matrix ρ_t. This is the *Schrödinger picture* of a time evolution which will be used in this paper. We leave as a simple exercise for the reader to translate all the presented results to the *Heisenberg picture*.

In order to give a precise mathematical meaning to Eq. (4) we first extent it to the whole Banach space $\mathcal{T}(\mathcal{H})$ of trace-class operators acting on $\mathcal{H}$ equipped with the norm $\|\sigma\|_1 = \mathrm{tr}(\sigma\sigma^*)^{1/2}$. Then L is a linear, perhaps unbounded, operator defined on its domain $D(L) \subset \mathcal{T}(\mathcal{H})$. The formal solution of Eq. (4) is given by the the one-parameter semigroup $\Lambda_t, t \geq 0$

$$\Lambda_t \sigma = \sum_{n=0}^{\infty} \frac{t^n}{n!} L^n \sigma \,, \tag{5}$$

for a bounded operator L while for an unbounded one

$$\Lambda_t \sigma = \lim_{n \to \infty} (\mathbf{1} - tL)^{-n} \sigma \,, \tag{6}$$

makes sense. The semigroup Λ_t should satisfy the following conditions

$$\Lambda_t \Lambda_s = \Lambda_{t+s} \qquad - \text{ semigroup property} \tag{7}$$

$$\lim_{t \to +0} \Lambda_t \sigma = \sigma \qquad - \text{ continuity} \tag{8}$$

$$\text{for } \sigma \geq 0 \;\; \Lambda_t \sigma \geq 0 \qquad - \text{ positivity} \tag{9}$$

$$\text{for any } \sigma \in \mathcal{T}(\mathcal{H}) \;\; \text{Tr}\,(\Lambda_t \sigma) = \text{Tr}\,\sigma \qquad - \text{ trace preservation .} \tag{10}$$

We shall see that the particular feature of composed quantum systems, namely the existence of *entangled states*, makes us to replace the positivity condition (9) by a stronger one called *complete positivity*.

2.1 Complete Positivity

Consider two well-separated open quantum systems with dynamics given by two dynamical maps $\Lambda^{(1)}$ and $\Lambda^{(2)}$ respectively. Assume also that the initial state of a joined system is determined by the entangled density matrix

$$\rho^{(12)} \neq \sum_j p_j \rho_j^{(1)} \otimes \rho_j^{(2)} \;. \tag{11}$$

One can easily find examples (e.g. for 2 qubits) of positive trace preserving $\Lambda^{(1)}$ and $\Lambda^{(2)}$ and $\rho^{(12)}$ satisfying (11) such that $\Lambda^{(1)} \otimes \Lambda^{(2)} \rho^{(12)}$ is not positively defined [10]. The minimal condition which must be imposed on dynamical maps to allow meaningful tensor products is complete positivity which means that for any $n = 1, 2, 3, ..., \Lambda \otimes I_n$ is positive,where I_n is an identity map acting on $n \times n$ matrices (i.e. trivial dynamical map on n-level quantum system). Completely positive maps on operator algebras were studied already in the 50-ties and the celebrated Stinespring representation [11] leads to a general form of completely positive dynamical map called often Kraus decomposition [12]

$$\Lambda \rho = \sum_\alpha W_\alpha \rho W_\alpha^* \,, \tag{12}$$

where W_α are bounded operators on $\mathcal{H}$ satisfying $\sum_\alpha W_\alpha^* W_\alpha = \mathbf{1}$. The decomposition (12) is highly nonunique, in particular the sum over α can be replaced by an integral. If $\mathcal{H}$ is n-dimensional then one can always find Kraus decomposition in terms of at most n^2 terms.

2.2 Completely Positive Dynamical Semigroups

From now on by a *quantum dynamical semigroup* (QDS) we mean a family of maps $\Lambda_t, t \geq 0$ satisfying conditions (7)-(10) with (9) strengthened to complete positivity. The celebrated result of Lindblad, Gorini, Kossakowski and Sudarshan [4,5] provides us with the most general form of a quantum Markovian master equation (MME) with a *bounded generator*

$$\frac{d}{dt}\rho_t = -i[H, \rho_t] + \sum_j V_j \rho_t V_j^* - \frac{1}{2}\{\sum_j V_j^* V_j, \rho_t\} \tag{13}$$

or in a slightly different but often used form

$$\frac{d}{dt}\rho_t = -i[H, \rho_t] + \frac{1}{2}\sum_j \left([V_j, \rho_t V_j^*] + [V_j \rho_t, V_j^*]\right) . \tag{14}$$

The choice of bounded operators $H = H^*$ and V_j is again not unique and the sum over $\{j\}$ can be replaced by an integral. To simplify the notation we put always $\hbar \equiv 1$ and $k_B \equiv 1$ to have the same units for energy, frequency and temperature.

The non-Hamiltonian part of the generator (13) displays the "gain - loss" structure similar to that of Pauli master equation (2) and determined by a completely positive *quantum transition map*

$$\rho \mapsto \Phi\rho = \sum_j V_j \rho V_j^* , \tag{15}$$

which is a sum of "pure" transition maps $\rho \mapsto V_j \rho V_j^*$ corresponding to elementary irreversible processes [7].

The following expansion for $\Lambda_t = \exp(tL)$ involving only sums (or integrals) and compositions of manifestly completely positive maps Φ (15) and W_t

$$W_t\rho = S_t \rho S_t^* \quad S_t = \exp\{-itH - (t/2)\sum_j V_j^* V_j\} , \tag{16}$$

is often useful

$$\Lambda_t = W_t + \sum_{n=1}^{\infty} \int_0^t dt_n \int_0^{t_n} dt_{n-1} \ldots \int_0^{t_2} dt_1 W_{t-t_n} \Phi W_{t_n - t_{n-1}} \Phi \ldots W_{t_2 - t_1} \Phi W_{t_1} . \tag{17}$$

For open systems with infinite dimensional Hilbert spaces we expect that the generators of QDS are typically unbounded. Although there exists no theorem giving the most general structure of L in this case, the standard form (13)(14) makes sense very often [13,14]. Then the operators H and V_j can be unbounded, in fact V_j can be even operator-valued distributions (e.g. quantum fields) and the sum over $\{j\}$ can be replaced by an integral. The typical mathematical condition which could be imposed is the existence of the contracting semigroup S_t (16)

on the Hilbert space such that the operators $S_t V_j$ become bounded and the expansion (17) makes sense. One of the unexpected features of the solutions to MME with unbounded generators is that despite the formal trace preservation of (13)(14) we can have $\mathrm{Tr}\rho_t < \mathrm{Tr}\rho_0$. This phenomenon is known for classical Pauli master equations with infinite number of states where for certain transition probabilities "the system can escape to infinity in finite time" [15].

2.3 Examples

We present few examples of QDS generators obtained using axiomatic construction based on phenomenological arguments. For other interesting models of dissipative quantum systems studied using numerical computations also, see [16].

2-Level System. The simplest and the oldest example of QDS for a 2-level system (qubit) can be constructed using three elementary transition maps. Assume that $|1>$ and $|2>$ form a basis of Hamiltonian eigenvectors and define the following operators

$$P_k = |k><k| \quad k = 1,2, \quad \sigma^+ = (\sigma^-)^* = |2><1|, \quad \sigma_3 = P_2 - P_1 \,. \tag{18}$$

The transition maps correspond to three different irreversible processes

$$\text{damping} \quad \rho \mapsto \gamma_\downarrow \sigma^- \rho \sigma^+ \tag{19}$$

$$\text{pumping} \quad \rho \mapsto \gamma_\uparrow \sigma^+ \rho \sigma^- \tag{20}$$

$$\text{pure decoherence} \quad \rho \mapsto \delta_1 P_1 \rho P_1 + P_2 \rho P_2 \,. \tag{21}$$

The master equation obtained from (18)-(21) reads

$$\begin{aligned}\frac{d}{dt}\rho_t = &-\frac{i\omega}{2}[\sigma_3, \rho_t] + \frac{\gamma_\downarrow}{2}\big([\sigma^-, \rho_t\sigma^+] + [\sigma^-\rho_t, \sigma^+]\big) \\ &+ \frac{\gamma_\uparrow}{2}\big([\sigma^+, \rho_t\sigma^-] + [\sigma^+\rho_t, \sigma^-]\big) - \frac{\delta}{2}[\sigma_3, [\sigma_3, \rho_t]] \,,\end{aligned} \tag{22}$$

where $\delta = (\delta_1 + \delta_2)/2$. Using the decomposition

$$\rho_t = p_1(t)P_1 + (1 - p_1(t))P_2 + \alpha(t)\sigma^+ + \bar{\alpha}(t)\sigma^- \,, \tag{23}$$

we obtain

$$\begin{aligned} p_1(t) &= p_1(0)e^{-(\gamma_\uparrow + \gamma_\downarrow)t} + \frac{\gamma_\downarrow}{\gamma_\downarrow + \gamma_\uparrow}\big(1 - e^{-(\gamma_\uparrow + \gamma_\downarrow)t}\big) \\ \alpha(t) &= \alpha(0)\exp\{-i\omega t - (\gamma_\downarrow + \gamma_\uparrow + \delta)t/2\} \,. \end{aligned} \tag{24}$$

The diagonal elements of ρ_t evolve independently of the off-diagonal ones and the former satisfy Pauli master equation with a stationary state which can be written in a form of the Gibbs state

$$\rho_{eq} = \Big(2\cosh\frac{\omega}{2T}\Big)^{-1} e^{-\omega\sigma_3/2T} \quad \frac{\gamma_\uparrow}{\gamma_\downarrow} = e^{-\omega/T} \,. \tag{25}$$

Moreover, for any initial state $\rho_t \to \rho_{eq}$ for $t \to \infty$ and the generator satisfies *quantum detailed balance* condition.

The equation (22) written explicitly in terms of matrix elements is called *Bloch equation* and is frequently used in quantum optics and magnetic resonance theory.

Harmonic Oscillator. Another well-known example is a linearly damped and pumped harmonic oscillator with a Hamiltonian $H = \omega a^+ a$ ($[a, a^+] = 1$) and transition maps describing coherent damping and pumping

$$\text{damping} \quad \rho \mapsto \gamma_\downarrow a \rho a^+ \quad \text{pumping} \quad \rho \mapsto \gamma_\uparrow a^+ \rho a \ . \tag{26}$$

The master equation with an unbounded generator

$$\begin{aligned} \frac{d}{dt}\rho_t &= -i\omega[a^+ a, \rho_t] + \frac{\gamma_\downarrow}{2}\big([a, \rho_t a^+] + [a\rho_t, a^+]\big) \\ &+ \frac{\gamma_\uparrow}{2}\big([a^+, \rho_t a] + [a^+ \rho_t, a]\big) \end{aligned} \tag{27}$$

is exactly solvable and the solution can be presented for instance in the form of the quantum generating function of the complex variables $z, \bar{z}$ [7]

$$F_t(z, \bar{z}) = \mathrm{Tr}\big(\rho_t \exp\{za - \bar{z}a^+\}\big) = e^{-A(t)} F_0(z_t, \bar{z}_t) \,, \tag{28}$$

where

$$\begin{aligned} z_t &= z \exp\{-i\omega t - (\gamma_\downarrow - \gamma_\uparrow)t/2\} \\ A(t) &= \frac{|z|^2}{4} \frac{\gamma_\downarrow}{\gamma_\downarrow - \gamma_\uparrow}\big(1 - e^{-(\gamma_\downarrow - \gamma_\uparrow)t}\big) \ . \end{aligned} \tag{29}$$

By differentiation of $F_t(z, \bar{z})$ with respect to z or $\bar{z}$ one obtains time dependence of all moments $\mathrm{Tr}(\rho_t (a^+)^n a^m)$. One can see from (28)(29) that if $\gamma_\downarrow > \gamma_\uparrow$ then $\lim_{t\to\infty} \rho_t = \rho_{eq}$ where

$$\rho_{eq} = \big(1 - e^{-\omega/T}\big) \exp\{-\frac{1}{T}\omega a^+ a\} \quad \frac{\omega}{T} = \log \frac{\gamma_\downarrow}{\gamma_\uparrow} \ . \tag{30}$$

One can also show that the diagonal and off-diagonal matrix elements of ρ_t evolve independently with the former satisfying detailed balance Pauli master equation.

Equation (27) found numerous applications, for instance in quantum optics to describe single mode of radiation in cavity. This is also the simplest example of the important class of the *quasi-free* QDS (see [7,10] and references therein).

Decoherence of Mass Center. Following [17] we briefly present a model based on QDS which can be used to analyse environmental decoherence which cause the transition from quantum to classical world. Consider a mass center of a body described by the Hamiltonian

$$H = \frac{1}{2M}\mathbf{P}^2 + V(\mathbf{X}), \tag{31}$$

where $\mathbf{X}$ and $\mathbf{P}$ are the operators of mass center and total momentum satisfying

$$[X_k, P_l] = i\delta_{kl} \;\; k, l = 1, 2, 3 . \tag{32}$$

We assume that the interaction with an environment can be reduced to processes of scattering, absorption or emission. A collision with a gas particle and emission, absorption or scattering of a photon (or other quasiparticle) causes a momentum transfer $\mathbf{k}$ which changes the total momentum. The following transformation (in the Heisenberg picture)

$$e^{i\mathbf{kX}}\mathbf{P}e^{-i\mathbf{kX}} = \mathbf{P} + \mathbf{k}, \tag{33}$$

describes this momentum transfer independently of the detailed microscopic mechanism of energy redistribution. Therefore the elementary irreversible process is given by the transition map

$$\rho \mapsto e^{-i\mathbf{kX}}\rho e^{i\mathbf{kX}} . \tag{34}$$

Assuming statistical independence of different momentum transfer events (called simply *collisions*) we obtain the following form of the MME

$$\frac{d}{dt}\rho_t = -i[H, \rho_t] + \int_{\mathbf{R}^3} d^3\mathbf{k}\, n(\mathbf{k})\Big(e^{-i\mathbf{kX}}\rho_t e^{i\mathbf{kX}} - \rho_t\Big), \tag{35}$$

where $n(\mathbf{k})$ is a density of collisions per unit time leading to the momentum transfer $\mathbf{k}$.

The generator given by (35) takes into account momentum conservation but the average kinetic energy grows to infinity for $t \to \infty$ and hence the process of ultimate relaxation to equilibrium is not properly described. However, it is expected that in the limit of large mass M and for slowly varying potential $V(\mathbf{x})$ the decoherence time is much shorter than the energy dissipation time scale. Therefore the master equation (35) can be considered as a good approximation for the study of pure decoherence in the relevant regime of large body at slowly varying potential and rare collisions (see Section III. D.3).

Bloch-Boltzmann Equation. In quantum optics the investigation of an active atom interacting with radiation and immersed in the thermal bath of perturbers (typically, noble gas atoms) is a very important topic [18]. To simplify the description one assumes that the spatially uniform distribution of active atoms and strong decoherence due to collisions with perturbers justify the use of the density operators diagonal in momentum (or velocity) representation. Therefore, we consider a family $\rho(\mathbf{v}) \geq 0$ of velocity-dependent positively defined operators (matrices) acting on the Hilbert space describing the internal structure of an active n-level atom. Using a standard form of the master equation (13) one can easily derive the most general evolution equation for QDS which preserves such partially diagonal structure of the density matrix [19]. It reads

$$\frac{d}{dt}\rho_t(\mathbf{v}) = -i\sum_a h_a(\mathbf{v})[S_a, \rho_t(\mathbf{v})] + \sum_{a,b}\int_{\mathbf{R}^3} d^3\mathbf{v}'\mathcal{K}_{ab}(\mathbf{v};\mathbf{v}')S_a\rho_t(\mathbf{v}')S_b^* \\ -\frac{1}{2}\sum_{a,b}\gamma_{ab}(\mathbf{v})\{S_a^*S_b, \rho_t(\mathbf{v})\}, \tag{36}$$

where $\{S_a\}$ form a linear basis in the space of $n \times n$ matrices and

$$\text{for any } \mathbf{v},\mathbf{v}', \quad \mathcal{K}_{ab}(\mathbf{v};\mathbf{v}') \text{ is positively defined}$$

$$\gamma_{ab}(\mathbf{v}) = \int_{\mathbf{R}^3} d^3\mathbf{v}'\mathcal{K}_{ba}(\mathbf{v};\mathbf{v}') . \tag{37}$$

Equation (36) provides an example of a classical-quantum description which combines a generalisation to n-level case of the MME (22), called Bloch equation with a classical linear Boltzmann equation for spatially uniform case. Similar type of equations found applications in the quantum measurement theory [20].

2.4 Ito-Schrödinger Equations

In the classical theory the Fokker-Planck equation describing Brownian motion in terms of probability distribution over the phase-space of Brownian particle is completely equivalent to the Langevin equation (in Ornstein-Uhlenbeck formulation) for the particle's random trajectory. The later equation is a prototype of *stochastic differential equation* which can be written either in Ito or Stratonovitch form [1]. The similar equivalent description exists for quantum dynamical semigroups with the generator (13) in terms of the following *Ito-Schrödinger equation*

$$d\psi(t) = -iH\psi(t)dt - \frac{1}{2}\sum_j V_j^*V_j\psi(t)dt - i\sum_j V_j\psi(t)dB_j(t) . \tag{38}$$

In (38) $B_j(t)$ denote independent and normalized Brownian motion processes and heuristically "$\frac{dB}{dt}$" corresponds to "white noise". The short-hand notation $df(t) = a(t)dt + b(t)dB(t)$ has a meaning of the corresponding integral equation

$$f(t_1) - f(t_2) = \int_{t_1}^{t_2} a(t)dt + \int_{t_1}^{t_2} b(t)dB(t), \tag{39}$$

where the second integral is an "nonanticipating" Ito integral given by the probabilistic limit of the sums $\sum b(s_k)[B(s_{k+1}) - B(s_k)]$ with $t_1 = s_1 < s_2 < \cdots < s_n = t_2$. To perform computations using stochastic integrals one applies Ito rule

$$dB_i(t)dB_j(t) = \delta_{ij}dt \tag{40}$$

and $E[dB_j(t)] = 0$, where $E[\cdot]$ denotes the average with respect to Brownian motions. One should remember that $dB_j(t) = B_j(t+dt) - B_j(t)$ is always statistically independent on the solution of stochastic differential equation up to time t. Applying those rules to the density matrix of the open system defined as an average $\rho(t) = E[|\psi(t) >< \psi(t)|]$ we obtain

$$\begin{aligned} d\rho_t &= dE[|\psi(t) >< \psi(t)|] \\ &= E[|d\psi(t) >< \psi(t)| + |\psi(t) >< d\psi(t)| + |d\psi(t) >< d\psi(t)|] \\ &= -i[H, \rho_t]dt - \frac{1}{2}\{\sum_j V_j^* V_j, \rho_t\}dt + \sum_j V_j \rho_t V_j^* dt\,, \end{aligned} \tag{41}$$

that coincides with MME (13). The Ito-Schrödinger equation (38) for a general case was firstly proposed in [21,22]and then derived by different authors as a useful tool for numerical simulations of quantum open systems [23]. For such applications it is convenient to use an equivalent nonlinear modification of (38) with the normalized solutions $\phi(t) = \psi(t)/\|\psi(t)\|$. One should notice that to obtain (41) we can replace the Brownian motions $B_j(t)$ by arbitrary stochastic processes $X_j(t)$ (continuous or jump processes) with independent increments such that $E[dX_i(t)dX_j(t)] = \delta_{ij}dt$. The different processes represent unitarily equivalent representations of the singular reservoirs constructed in terms of "time ordered Fock spaces" [21]. Possible physical interpretations of these different representations are discussed in [24].

An alternative approach in terms of "quantum noises" has been proposed in [25] and subsequently developed into a mathematically rigorous theory of noncommutative stochastic differential equations.

2.5 Problems and Pitfalls

Although, the axiomatic approach allows to write down the MME (13,14) in a mathematically consistent form, we generally need additional phenomenological insights which could lead to concrete form of the operators $\{V_j\}$. The first problem is the relation between the dissipative part in (13,14) and the Hamiltonian of the open system. Here the demanded structure of the stationary state and detailed balance condition are helpful [7].For example adding a nonlinear term to the Hamiltonian of the harmonic oscillator we have to modify the dissipative part too, in contrast to often used simplified models with linear dissipation. The symmetries of the system and reservoir give another restrictions on the form of generators [26].

For many interesting physical problems we have to include time dependent-external fields acting on the system. Very often it is done by simply adding the proper time-dependent term to the Hamiltonian without changing the dissipative part. This can be justified, however, for weak and quickly oscillating fields while in general case the whole Markovian approximation breaks down [27]. Another extreme case is a slowly varying external potential which leads to a proper time-dependent generator which for any time reproduces, for example, return to an equilibrium state given by a total instant Hamiltonian [28].

A particular care is needed when we put two open systems S_1, S_2 with their "privat" reservoirs R_1, R_2 into contact by adding their interaction Hamiltonian V_{12}. In this case we cannot simply add two dissipative generators obtained previously for decoupled systems S_1, S_2. This is obvious for the case of two heath baths at the same temperature which should drive the interacting system S_1+S_2 into its Gibbs state which of course depends on the new element V_{12}.

3 Constructive Approach

The goal of the constructive approach is to derive, using mathematically controlled approximations, Markovian dynamics of the quantum system interacting with the quantum environment. We restrict ourselves to the derivations which respect complete positivity of the reduced dynamics. In author's opinion the very scheme of quantum open systems approach and particularly Markovian approximations are consistent only under well defined conditions discussed below.

3.1 Basic Assumptions

In the theory of quantum open systems we decompose the Universe into three parts (S, O, R). The first one is an open system S controlled by an "active" observer O and interacting with the "passive" rest R which is called reservoir. If the state of the system $S+R$ is given by the density matrix ρ_{SR} that we asume then the reduced density matrix $\rho_S = \mathrm{Tr}_R \rho_{SR}$ possesses an operational meaning i.e. that the observer O can prepare an arbitrary initial state of S at a certain moment $t_0 = 0$ and than measure the evolved state at the moment $t > 0$. We often assume also that the observer can control to some extend the evolution of S switching-on external time-dependent "fields". The physical structure of O and the description of measurements performed na S is a subject of quantum measurement theory and will be not discussed here.

One should notice that the above scheme contains a tacit assumption that the interaction between S and R is weak. Any preparation of the initial state of S or its state measurement takes time $t_p = \hbar/\delta E$ where δE is an energy resolution of state preparation or measurement. For a quantum system S with a discrete spectrum δE should be smaller that the typical separation between the energy levels while for systems with continuous spectrum (e.g. quantum Brownian particle) t_p is directly related to a chosen time-scale of observation (see Section III.D.3). Obviously, during preparation or measurement processes the system S should not be strongly perturbed by the environment R what implies the inequality

$$t_p << \tau_S \,, \tag{42}$$

where τ_S is a characteristic time scale of relaxation of S due to the interaction with R.

The first step towards mathematical description of the open system consists in writing down the formal "bare" Hamiltonian

$$H_{SR} = H_S^0 + \lambda \sum_k S_k \otimes R_k^0 + H_R \,, \tag{43}$$

where H_S^0 is a bare Hamiltonian of S, the interaction term is proportional to a dimensionless coupling constant λ and contains bare reservoir's operators R_k^0. For some applications we shall add a time-dependent contribution $V(t)$ to the Hamiltonian H_S^0 in order to take into account external fields used by the observer O to control our system S.

The next important assumption is the product structure of the initial state for $S + R$

$$\rho_{SR}(0) = \rho \otimes \omega_R \,, \tag{44}$$

with an arbitrarily prepared initial state of S and a fixed reference state of R. Again this assumption can be justified if the interaction is weak in a defined below sence [29]. Namely, the system $S + R$ for times preceding the preparation at $t_0 = 0$ is expected to be at equlibrium or ground state ω_{SR}. In order to produce the initial state (44) by applying external perturbations to S we must assume that

$$\omega_{SR} \simeq \omega_S \otimes \omega_R \,. \tag{45}$$

The state of S for times $t >> t_p$ is then given by the reduced density matrix (we omit subscript S)

$$\rho_t = \mathrm{Tr}_R\big(U_t \rho \otimes \omega_R U_t^*\big) = \Lambda_t(\rho) \,, \tag{46}$$

where $U_t = \exp\{-itH_{SR}\}$ and Tr_R is a partial trace over the Hilbert space of R. The dynamical maps $\{\Lambda_t; t \geq 0\}$ are by construction completely positive and trace preserving.

However, the practical implementation of the above prescription leads to difficulties. Take as an example, a model of atom interacting with electromagnetic field at zero temperature. The corresponding Hamiltonian (43) is well-defined only if an ultraviolet cut-off at the frequency ω_c is introduced replacing "bare" operators R_k^0 by cut-off dependent regularized R_k. Then a simple lowest order evaluation of (46) shows that the interaction produces a cut-off dependend Hamiltonian correction to the bare Hamiltonian H_S^0. When $\omega_c \to \infty$ corrections diverge and must be cancelled by proper counterterms in the bare Hamiltonian (renormalization procedure). In the case of our example it leads to mass and charge renormalization and slight modification of the Coulomb potential which influences energy levels of the atom (Lamb-shift)[30]. A different example is an electron immersed into a polar solid. The interaction with lattice ions modelled by the electron-phonon coupling (with natural well-defined cut-off) completely changes the propeties of an electron leading to the so-called polaron model. In both cases the initial bare system is strongly modified by a "cloud" which consists of certain degrees of freedom of the reservoir [31]. Hence the decomposition into $S+R$ is not unique and the practical criterion should be the weakness of the residual interaction between the "dressed" system S' and the rest of the reservoir R'. Practically, we describe the sytems S', R' by the same Hilbert spaces as S, R and the Hamiltonian of the bath H_R remains the same too. We introduce, however, a physical (renormalized) Hamiltonian H_S and modify the structure of interaction Hamiltonian introducing suitable formfactors with cut-offs and often leaving only resonant terms ("rotating wave approximation" in quantum optics, Lee models in particle physics) to reduce creation of the "cloud". We shall illustrate these problems later on in the case of 2-level open system (see Sections III.C.2, III.D.2).

3.2 Models of Reservoirs

Similarly to the classical case represented by the standard example of Brownian motion the details of the reservoir should not be essential for the dynamical behaviour of an open system S at least on the proper "coarse-grained" time scale. Instead we expect that the mechanisms illustrated by limit theorems in classical probability should work in the quantum domain also. Indeed, take a sequence of quantum observables $A_1, A_2, ..., A_N$ which are statistically independent (i.e. $< A_k A_l >=< A_k >< A_l >$) or even weakly dependent. Then, one can prove a quantum version of the Central Limit Theorem for $N \to \infty$

$$\frac{1}{\sqrt{N}} \sum_{j=1}^{N} (A_j - < A_j >) \to a(f_A) + a^+(f_A) , \tag{47}$$

where $a(f_A), a^+(f_A)$ are "smeared" bosonic anihilation and creation operators [32,33]. The limit is understand in the sense of convergence of all correlation functions, where for Bose fields we choose a "vacuum" as a reference state. The detailed structure of the Fock space for bosonic field is discussed in [10,32,33] and is not relevant here. One should think about bosonic fields as quantum counterparts of classical Gaussian random fields and the "vacuum" can represent any quasi-free state of bosons as for example arbitrary Gibbs state for noninteracting Bose gas. The index j for A_j can correspond both to different parts of the reservoir and to different time instants. The limit theorem (47) reflects the physical picture of influence from R on S which can be decomposed into a sum of many almost independent small contributions. Such case is usually called *diffusive regime* and due to quantum central limit theorem (47) is well described by the harmonic oscillators bath with a weak linear coupling in the interaction Hamiltonian

$$H_{SR}^{int} = \lambda \sum_k S_k \otimes [a(f_k) + a^+(f_k)] . \tag{48}$$

Diffusive regime does not describe all physically interesting situations. Another, *low density regime* corresponds to rare events of scattering by essentially independent "particles" from R each of the scattering processes takes short time but need not to be weak. This is similar to classical limit theorems leading to

Poisson distributions and processses. The physical model of such reservoir is a free fermionic or bosonic gas and the bilinear interaction given by

$$H_{SR}^{int} = \sum_k S_k \otimes \left[a^+(f_k)a(g_k) + a^+(g_k)a(f_k) \right], \tag{49}$$

with a small parameter being now the density of gas $\sim < a^+(f_k)a(f_k) >$.

For some applications the quantum nature of the reservoir is not very important, like for instance in the *high temperature regime*. In this case one can replace a quantum reservoir R by a time dependent random Hamiltonian or in other words operator-valued (self-adjoint) stochastic process. A particular choice of white-noise process leads to a stochastic Ito-Schrödinger equation (38) with $V_j = V_j^+$.

3.3 Markovian Limits

The general reduced dynamics (46) does not satisfy the semigroup composition law (7). However, one expects that at least for a certain coarse-grained time scale, roughly determined by τ_S (see (42)), this law is often a very good approximation. Physically, it is true if the exact state of the system $S + R$ given by $U_t \rho \otimes \omega_R U_t^*$ does not differ *locally* from the state $\Lambda_t(\rho) \otimes \omega_R$. Here "locality" is determined by the radius of interaction between S and R. Mathematically this condition can be expressed as a sufficiently fast decay of the reservoirs correlation functions

$$R_{kl}(t) = \mathrm{Tr}(\omega_R R_k(s+t) R_l(s)), \tag{50}$$

where $R_k(t) = \exp(itH_R) R_k \exp(-itH_R)$ and we assume that the reservoir's reference state ω_R is stationary with respect to its evolution. We shall argue that those correlation functions contain the total information about reservoirs which is relevant for different Markovian regimes. Introducing the reservoir's relaxation time scale τ_R we may write the standard condition for the validity of the Markovian approximation as

$$\tau_R << \tau_S . \tag{51}$$

However, as we shall see, important examples of reservoirs may not have a natural decay time scale τ_R and one has to include the averaging effect of the Hamiltonian dynamics of S in order to satisfy (51). Moreover, rigorous analysis shows that the Markovian approximation needs also sufficently rapid decay of higher order multitime correlation functions which can be proved for quasi-free models of reservoirs discussed in the previous subsection.

In the present review we are not going to reproduce involved rigorous derivations which can be found in the literature [34-36]. Our aim is to discuss physical assumptions behind different Markovian regimes and proper approximation schemes based on the relevant order of perturbation which yield mathematically consistent results. For notational simplicity we use as an illustration the model of an effective 2-level open system with a physical (renormalized) Hamiltonian H_S and the interaction one H_{int} of the form

$$H_S = \frac{1}{2}\epsilon\sigma_3 \quad H_{int} = \lambda\sigma_1 \otimes R\,, \tag{52}$$

where $\sigma_k, k = 1, 2, 3$ are Pauli matrices. We always assume that

$$\mathrm{Tr}(\omega_R R) = 0\ . \tag{53}$$

In the heuristic derivations of MME the lowest order expansion for quantum dynamical semigroup of the form

$$\rho_t = e^{-itH_S}\big(\rho + \sum_j \int_0^t V_j(s)\rho V_j^*(s) - \frac{1}{2}\{V_j^*(s)V_j(s), \rho\}ds\big)e^{itH_S} + \mathcal{O}(t^2)\,, \tag{54}$$

is compared with the relevant term in the expansion for the reduced dynamics (46). More precisely, it is enough to compare only the transition map term ("gain")

$$\rho \mapsto \sum_j \int_0^t V_j(s)\rho V_j^*(s)ds\,, \tag{55}$$

with the corresponding expression obtained from (46) using different assumption concerning mainly important time scales in the joint system $S + R$. Any Hamiltonian-type corrections are put equal to zero according to the renormalization procedure.

Singular Coupling or White Noise Anzatz. Using second order Dyson expansion for the total dynamics of $S + R$ with the interaction Hamiltonian treated as a perturbation we obtain the following formula for the manifestly completely positive "transition map" which has to be compared with (55)

$$\rho \mapsto \int_0^t ds_1 \int_0^t ds_2 \mathrm{Tr}_R\big(H_{int}(s_1)\rho \otimes \omega_R H_{int}(s_2)\big)\ . \tag{56}$$

For our 2-level system we obtain

$$\begin{aligned}\rho \mapsto \lambda^2 \int_0^t ds_1 \int_0^t ds_2 R(s_2 - s_1)\Big(&e^{i\epsilon(s_2-s_1)}\sigma^-\rho\sigma^+ + e^{-i\epsilon(s_2-s_1)}\sigma^+\rho\sigma^- \\ &+ e^{-i\epsilon(s_2+s_1)}\sigma^-\rho\sigma^- + e^{i\epsilon(s_2+s_1)}\sigma^+\rho\sigma^+\Big)\ .\end{aligned} \tag{57}$$

The simplest approximation which produces the Markovian transition map term (55) is based on the assumption that the correlation time of the reservoir is much shorter than any other relevant time scale i.e.

$$\tau_R << \min\{\tau_S, \tau_H\}\,, \tag{58}$$

where τ_H denotes Heisenberg time scale for S related to the average energy level spacing ΔE by $\tau_H = (\Delta E)^{-1}$. Under the condition (58) we can use the *white-noise Anzatz*

$$R(t) \simeq \hat{R}(0)\delta(t) \;\; \hat{R}(\omega) = \int_{-\infty}^{\infty} R(t)e^{it\omega}dt\,, \tag{59}$$

which leads to the QDS-type of the transition map corresponding to the following MME

$$\frac{d}{dt}\rho = -\frac{i}{2}\epsilon[\sigma_3,\rho] - \frac{\lambda^2}{2}\hat{R}(0)[\sigma_1,[\sigma_1,\rho]]\;. \tag{60}$$

For a general case the substitution of the type (59) leads to MME (13,14) with $V_j = V_j^*$ which can be rewritten in a *double commutator form*. Moreover, the dissipative part does not depend on the Hamiltonian one and we can even add a time-dependent term $V(t)$ to obtain

$$\frac{d}{dt}\rho_t = -i[(H_S + V(t)),\rho_t] - \frac{1}{2}\sum_j [V_j,[V_j,\rho_t]]\;. \tag{61}$$

The dynamics generated by (61) is very special because the dynamical maps Λ_t are *bistochastic*, i.e. $\Lambda_t\mathbf{1} = \mathbf{1}$. It means, depending on the physical context, that the *infinite temperature* or *microcanonical* state is preserved and the H-theorem holds i.e. for any solution of (61)[37]

$$S(\rho_t) \geq S(\rho_s)\;\text{for}\; t \geq s \geq 0 \;\; S(\rho) = -\mathrm{Tr}\rho\ln\rho\;. \tag{62}$$

There are several instances where the equations of type (61) are useful and provide reasonable approximations to the exact dynamics of the open system. These are the cases where the Hamiltonian self-evolution due to $H_S + V(t)$ is very slow or $H_S + V(t)$ commutes with S_k in (43)and therefore the pure decoherence processes dominate over the energy redistribution. Another case is the high temperature limit where $T >> E_{max}$ and E_{max} is a maximal difference of energy levels of the system. The white noise Anzatz can be obtained from the Hamiltonian models by limit procedure called *singular coupling limit* which leads to unphysical limiting reservoirs with unbounded from below Hamiltonians [35].

One should remember, however, that the lack of memory expressed by the δ-like correlations in (59) contradicts the quantum nature of the reservoir (see Section III.D.1) and the noise governing the dissipative part of (61) is essentially classical. Hence, in author's opinion, equations of the type (61) are too rough to describe properly the control of decoherence and dissipation in the context of quantum computations and error correction schemes. The same is true for their discrete-time versions which often appear in the literature on this topic [9].

Weak Coupling Limit. Another approximation scheme takes into account the interplay between the Hamiltonian dynamics of S governed by H_S and the coupling to R. It allows to describe, for instance, the equilibration process leading in the case of a heat bath to a final Gibbs state of S. We assume the following relations between the relevant time scales

$$\tau_H << \tau_S \ \tau_R << \tau_S \ . \tag{63}$$

As a consequence of (63) in the example (57) the last two non-resonant terms can be omitted and the obtained transition map corresponds to the following MME

$$\frac{d}{dt}\rho_t = -\frac{i}{2}\epsilon[\sigma_3,\rho_t] - \frac{\lambda^2}{2}\Big(\hat{R}(\epsilon)\big([\sigma^-,\rho_t\sigma^+] + [\sigma^-\rho_t,\sigma^+]\big) \\ + \hat{R}(-\epsilon)\big([\sigma^+,\rho_t\sigma^-] + [\sigma^+\rho_t,\sigma^-]\big)\Big) \tag{64}$$

which is a special case of (22) with $\delta = 0$. The similar procedure applied to a harmonic oscilator linearly coupled to an environment yields MME (27)

For a general case (43) the MME obtained under the conditions (63) reads

$$\frac{d}{dt}\rho_t = -i[H_S,\rho_t] + \frac{\lambda^2}{2}\sum_{\omega;k,l}\hat{R}_{kl}(\omega)\big([S_k(\omega),\rho_t S_l^*(\omega)] + [S_k(\omega)\rho_t, S_l^*(\omega)]\big)\ , \tag{65}$$

where

$$S_k(t) = e^{itH_S}S_k e^{-itH_S} = \sum_\omega S_k(\omega)e^{-i\omega t} \tag{66}$$

and

$$\hat{R}_{kl}(\omega) = \int_{-\infty}^{\infty} R_{kl}(t)e^{-it\omega}dt\ . \tag{67}$$

Obviously, as for any eigenfrequency ω the matrix $[\hat{R}_{kl}(\omega)]$ is positively defined we can always rewrite (65) in a diagonal form (13,14).

The approximation procedure leading to MME (65) (in the interaction picture) can be made mathematically rigorous introducing the concept of weak coupling or van Hove limit applied to the reduced dynamics in the interaction picture which consists of using the rescaled time $\tau = \lambda^2 t$ for $\lambda \to 0$. The details can be found in [34] see also the related idea of stochastic limit in [33].

MME of the type (65) possess several interesting properties. The dissipative part of the generator commutes with the Hamiltonian one and the diagonal elements of ρ_t in the energy representation evolve independently of the off-diagonal ones. The corresponding transition probabilities which appear in the Pauli master equation for diagonal elements are exactly equal to those calculated using the Fermi Golden Rule [38]. In the frequently used case of the reservoir being at thermal equilibrium (heat bath) the KMS relation of the form

$$\hat{R}_{kl}(-\omega) = e^{-\beta\omega}\hat{R}_{lk}(\omega) \ \ \beta = T^{-1} \tag{68}$$

is valid and implies that that the Gibbs state $\rho_{eq} = Z^{-1}\exp\{-\beta H_S\}$ is an invariant state for MME (65). If for all k,ω, $[S_k(\omega), X] = 0$ implies $X = c\mathbf{1}$ the

Gibbs state is ergodic i.e. for $t \to \infty$ and any initial state the solution of (65) $\rho_t \to \rho_{eq}$. Moreover, the Pauli master equation for the diagonal elements of ρ_t satisfies detailed balance condition. The part of the generator in (65) which is given by terms with $\omega = 0$ does not influence the diagonal elements of ρ_t and describes pure decoherence which is not accompanied by the energy change.

One should notice the important differences between the regimes characterized by inequalities (58) and (63) respectively. The former allows white-noise Anzatz and hence the strictly Markovian approximation entirely due to the memoryless reservoir. In the later case Markovian approximation is valid for the time evolution averaged over many periods of Hamiltonian evolution of the open system. It follows that the dissipative part of the semigroup generator strongly depends of the Hamiltonian H_S and therefore we cannot freely add a time dependent part $V(t)$ to H_S.

Weak Coupling and Thermodynamics. We have already seen that the MME (65) obtained using weak coupling limit satisfies the *0-law of thermodynamics*, namely the heat bath drives an open system S to its equlibrium state at the same temperature. Introducing a time dependent Hamiltonian $H_S(t)$ which varies on the time-scale much longer than all other time-scales in (63) we can apply the same method used to derive MME (65). As a result of such combined weak-coupling and adiabatic approximation we obtain an inhomogeneous in time MME of the form

$$\frac{d}{dt}\rho_t = -i[H_S(t), \rho_t] + L_D(t)\rho_t \equiv L(t)\rho_t \quad t \geq 0 \tag{69}$$

where the dissipative part $L_D(t)$ is computed according to the formulas (65,67) with the Hamiltonian H_S replaced by its temporal value $H_S(t)$. We can now formulate the *I-st law of thermodynamics* as

$$\frac{d}{dt}E(t) = \frac{d}{dt}W(t) + \frac{d}{dt}Q(t)\,, \tag{70}$$

where $E(t) = \mathrm{Tr}(\rho_t H_S(t))$ is an internal energy of S, $W(t) = \int_0^t \mathrm{Tr}(\rho_s \frac{d}{ds}H_S(s))ds$ is the work performed on S by external forces and $Q(t) = \int_0^t \mathrm{Tr}\big[(\frac{d}{ds}\rho_s)H_S(s)\big]ds$ is the heat supplied to S by R [28,39].

In order to illustrate the second law of thermodynamics we use the *relative entropy* for a pair of density matrices ρ, σ

$$S(\rho|\sigma) = \mathrm{Tr}\big(\rho\ln\rho - \rho\ln\sigma\big) \tag{71}$$

and the following inequality valid for any trace preserving completely positive map Λ [37]

$$S(\rho|\sigma) \geq S(\Lambda\rho|\Lambda\sigma)\ . \tag{72}$$

For the time evolution governed by the generator $L(t)$ (Eq. (69)) which possesses a temporal stationary state

$$L(t)\rho_{eq}(t) = 0 \;\; \rho_{eq}(t) = Z^{-1}\exp\{-\beta H_S(t)\} \tag{73}$$

the inequality (72) implies the following form of the second law of thermodynamics for open systems

$$\frac{d}{dt}S(\rho_t) = \sigma[\rho_t] + \frac{1}{T}\frac{dQ}{dt} \tag{74}$$

where $\sigma[\rho_t] \geq 0$ is an entropy production and the second term describes the entropy exchange with the heat baths. The formula (74) can be easily generalized to the case of open system coupled to several heath baths at different temperatures [28,40].

The presented derivation of the three fundamental laws of thermodynamics has been possible within the assumptions of weak coupling between S to R and the adiabatic change of external forces. Beyond this regime, even the unique definitions of the fundamental thermodynamic notions are not obvious. One should mention that the deviations from the thermodynamical behaviour for strongly coupled quantum systems attracted, recently, attention of the scientific community [41].

Low Density Limit. Another situation which justifies the use of MME is an open system S with the discrete spectrum Hamiltonian $H_S = \sum_k \epsilon_k |k><k|$ interacting with a dilute gas. The rigorous results can be found in [33,36] while here we present a heuristic approach. Consider first an abstract formulation of the *impact approximation* for the scattering problem given by the Hamiltonian $H = H_0 + V$. Starting with the identity

$$\begin{aligned} U_t = e^{-itH} &= e^{-itH_0}\Big(\mathbf{1} + \int_0^t e^{isH_0} V e^{-isH} ds\Big) \\ &= e^{-itH_0}\Big(\mathbf{1} + \int_0^t e^{isH_0} V \big[e^{-isH} e^{isH_0}\big] e^{-isH_0} ds\Big), \end{aligned} \tag{75}$$

we replace $e^{-isH}e^{isH_0}$ by $\Omega_+ = \lim_{s\to\infty} e^{-isH}e^{isH_0}$ what makes sense for $t >> \tau_{coll}$ where τ_{coll} is a typical collision time for the discussed model. Hence we have

$$U_t \approx e^{-itH_0}\Big(\mathbf{1} + \int_0^t e^{isH_0} V \Omega_+ e^{-isH_0} ds\Big) \tag{76}$$

and instead of the transition map (56) we obtain its low density counterpart

$$\rho \mapsto \int_0^t ds_1 \int_0^t ds_2 \mathrm{Tr}_R\big(T(s_1)\rho\otimes\omega_R T(s_2)\big) \tag{77}$$

where $T(s) = e^{isH_0} T e^{-isH_0}$ and $T = V\Omega_+$. In the next step we obtain a transition map similar to that derived in the weak coupling limit by averaging over the Hamiltonian evolution of S and eliminating the non-resonant oscillating terms. Take a single particle of the bath being in a state described by the density matrix

$$\omega_R = \ell^{-3} \int_{\mathbf{R}^3} d^3\mathbf{p}\, G(\mathbf{p})|\mathbf{p}><\mathbf{p}| \quad <\mathbf{p}|\mathbf{p}'>= \delta^3(\mathbf{p}-\mathbf{p}') \tag{78}$$

normalized in a cube ℓ^3 and diagonal in the momentum representation. Putting $H = H_S + \int_{\mathbf{R}^3} d^3\mathbf{p}\, E_{\mathbf{p}} |\mathbf{p} >< \mathbf{p}|$ we obtain from (77) and for $t >> \tau_{coll}, \tau_H$ the final form of the transition map which commutes with Hamiltonian evolution

$$\rho \mapsto t\ell^{-3} \sum_{\omega} \int_{\mathbf{R}^3} d^3\mathbf{p} \int_{\mathbf{R}^3} d^3\mathbf{p}\, G(\mathbf{p}) \pi \delta(E_{\mathbf{p}'} - E_{\mathbf{p}} + \omega) T_\omega(\mathbf{p}, \mathbf{p}') \rho T^*_\omega(\mathbf{p}, \mathbf{p}')\,, \quad (79)$$

where

$$T_\omega(\mathbf{p}, \mathbf{p}') = \sum_{\epsilon_k - \epsilon_l = \omega} < k, \mathbf{p}' | T | l, \mathbf{p} > |k >< l|\,. \quad (80)$$

The transition map (79) corresponds to a single-particle scattering during the time interval t. Because in the volume ℓ^3 we have N particles the low density approximation implies additive effect producing the factor $\nu = N/\ell^3$ (instead of ℓ^{-3}) which remains finite in the thermodynamical limit. Therefore, the final form of the MME valid under the conditions

$$\tau_H << \tau_S \;\; \tau_{coll} << \tau_S \quad (81)$$

is the following

$$\begin{aligned} \frac{d}{dt}\rho_t = & -i[H_S, \rho_t] + \nu\pi \sum_{\omega} \int_{\mathbf{R}^6} d^3\mathbf{p} d^3\mathbf{p}'\, G(\mathbf{p}) \delta(E_{\mathbf{p}'} - E_{\mathbf{p}} + \omega) \\ & \times \Big([T_\omega(\mathbf{p}, \mathbf{p}'), \rho_t T^*_\omega(\mathbf{p}, \mathbf{p}')] + [T_\omega(\mathbf{p}, \mathbf{p}')\rho_t, T^*_\omega(\mathbf{p}, \mathbf{p}')] \Big)\,. \end{aligned} \quad (82)$$

Similarly to the weak coupling limit the rigorous derivation of MME (82) involves the limit procedure $\nu \to 0$ for the interaction picture version of the reduced dynamics with the rescaled time $\tau = \nu t$ [36].

It is not difficult to show that the MME (82) has similar properties to MME (65). Namely, the diagonal elements of ρ_t evolve independently of the off-diagonal ones, for the equilibrium momentum distribution of scatterers $G(\mathbf{p}) \sim \exp(-\beta E_{\mathbf{p}})$ the Gibbs state $\rho_{eq} \sim \exp(-\beta H_S)$ is stationary and under natural conditions ergodic. The discussion of the Section III.C.3 can be applied also for inhomogeneous in time versions of MME (82).

Application to Bloch-Boltzmann Equation. The MME (82) can be a starting point for the heuristic derivation of the Bloch-Boltzmann equation (36)[19]. The active atom with finite number of internal levels can be put into a finite box to take an advantage of the dicrete spectrum of its momentum and kinetic energy operators. To such a discretized system of an active atom we can apply the derivation of the previous section. As an initial state we choose a quasi-diagonal density matrix

$$\rho = \sum_{\mathbf{v}} \rho(\mathbf{v}) |\mathbf{v} >< \mathbf{v}|\,, \quad (83)$$

with respect to the orthonormal velocity basis $\{|\mathbf{v} >\}$. Under some reasonable conditions the quasi-diagonal density matrices remains quasi-diagonal during the time evolution and finally we can go with the size of the box to infinite to obtain continuous velocity spectrum. The final result of this procedure which involves also the transition to the center of motion reference frame is the Bloch-Boltzmann equation (36)(37) with (we put here $\hbar$ for readers convenience)

$$H_S = \sum_{j=1}^{n} \epsilon_j |j><j| \ \ [H_S, S_a] = \omega_a S_a \ \ a = 1, 2, ..., n^2 , \tag{84}$$

$$\begin{aligned}\mathcal{K}_{ab}(\mathbf{v};\mathbf{v}') = &(2\pi)^4 \hbar^2 \mu^{-3} N_p \delta_{\omega_a,\omega_b} \int_{\mathbf{R}^3} d^3\mathbf{v}'_r \\ &\times \int_{\mathbf{R}^3} d^3\mathbf{v}_r \delta^3\big(\mathbf{v} - \mathbf{v}' - \frac{\mu}{m}(\mathbf{v}_r - \mathbf{v}'_r)\big)\delta\Big(\frac{\mu}{2}(\mathbf{v}_r^2 - \mathbf{v}'^2_r) + \hbar\omega_a\Big) \\ &\times W(\mathbf{v}' - \mathbf{v}'_r) T_a(\mathbf{v}_r, \mathbf{v}'_r)\overline{T_b(\mathbf{v}_r, \mathbf{v}'_r)} ,\end{aligned} \tag{85}$$

where $\mathbf{v}_r, \mathbf{v}'_r$ denote relative velocities, $W_p(\mathbf{v})$ is the equilibrium velocity distributions of perturbers, m is a mass of an active atom, μ is a reduced mass of an "active atom - perturber" system, N_p is a density of perturbers and the functions $T_a(.,.)$ are related to the T-matrix calculated in the center of motion reference frame by the following expression

$$\sum_a T_a(\mathbf{v}_r, \mathbf{v}'_r) S_a = \sum_{j,j'=1}^{n} <\mathbf{v}_r, j|T|\mathbf{v}'_r, j'> |j><j'| . \tag{86}$$

Due to the presence of Kronecker's delta $\delta_{\omega_a,\omega_b}$ in (85) the matrix $\mathcal{K}_{ab}(\mathbf{v};\mathbf{v}')$ is positively defined as demanded by the conditions (37). One should mention that the obtained Bloch-Boltzmann differs from the existing ones which generally do not preserve positivity of $\rho(\mathbf{v})$ [18].

3.4 Problems and Pitfalls

The derivations of the MME presented in the previous sections are based on certain assumption concerning the decay of correlations in the environment, separation of energy levels of the open systems and the magnitude of the coupling constant or the density of perturbers. These conditions are formulated in terms of relations between the different time scales (42), (51),(63), (81). Now, we discuss briefly some of them and refer to the important physical situations beyond the presented scheme stressing the related difficulties and misconceptions.

Memory Effects. The most demanding condition for the lack of memory of the reservoir is an inequality (58) which allows the white noise Anzatz $R_{kl}(t) \simeq \hat{R}_{kl}(0)\delta(t)$. It means that the spectral density matrix $\hat{R}_{kl}(\omega)$ is weakly dependent on ω in the relevant energy region. Such an assumption essentially contradicts the KMS condition (68) what is the source of the so-called *thermal quantum memory* characterized by the time scale $\tau_T = T^{-1}$.

Another difficulty with the assumption of memoryless reservoir can be illustrated by a model of 2-level atom coupled to the quantum electromagnetic field at the vacuum state ($T = 0$). Puting in Eq. (52) the standard "dipol× electric field" interaction one obtains

$$R(t) \sim \frac{1}{(t + i\omega_c^{-1})^4}\ , \quad \hat{R}(\omega) \sim \omega^3 e^{-\omega/\omega_c}\ , \tag{87}$$

where ω_c is a cut-off frequency assumed to be larger than any energy scale relevant for this model. The decay of reservoir's correlations is powerlike and does not possess any natural time scale. Moreover, the correlation function is singular at the origin and $\hat{R}(0) = 0$ for the removed cut-off ($\omega_c \to \infty$). Fortunately, in the weak coupling regime, the Markovian behaviour can be restored on the coarse-grained time scale determined by τ_S - spontaneous emission time. This is due to the averaging effect of fast Hamiltonian oscillations which allow to replace $R(t)$ by $R(t)e^{i\epsilon t}$ - the function which effectively acts like $\hat{R}(\epsilon)\delta(t)$ under the integral in (57). The situation is different when H_S is replaced by a time-dependent Hamiltonian and /or collective effect for multi-atomic systems are relevant. Namely, that fast variations of the Hamiltonian introduce high frequency contributions increasing the decay rates (see Eq. (87)) and even a system of two atoms possesses degenerated energy levels for which averaging effect described above does not apply. All that implies serious limitations on the use of Markovian approximation in the context of controled quantum open systems, the problem which is crucial for quantum information processing [42].

Decoherence vs. Dissipation. The dynamics of an isolated quantum system is governed by its Hamiltonian H_S and is characterized by two fundamental features: initial pure states remain pure, the average energy is a constant of motion. On the contrary, for an open quantum system S interacting with a quantum environment R which starts its joint evolution from the product state, the entangled states of $S+R$ are developed in the course of time what lead to the appearance of reduced mixed states of S and the energy exchange between S and R. The first phenomenon is called *decoherence* and the second one *dissipation*. Although decoherence and dissipation are usually present at the same time, model calculations show that for large quantum systems approaching the border between quantum and classical worlds decoherence acts on a much faster time scale than dissipation [43,44,17]. As decoherence seems to be a more important agent in the context of quantum measurement theory and quantum information processing [9] it is convenient to discuss models describing *pure decoherence* often called *dephasing* which is not accompanied by the energy exchange.

The pure decoherence is described by the models with Hamiltonians (43) satisfying the condition $[H^0_S, S_k] = 0$ what implies the same condition for the renormalized Hamiltonian $[H_S, S_k] = 0$ for all k. In Markovian approximations the decoherence rates are proportional to $\hat{R}_{kl}(0)$ (see (59,60) and (65-67)) which is typically zero for the systems linearly coupled to bosonic fields. This is the case for electromagnetic interaction (87), the same result holds for linear coupling to phonons. In a general case one can formulate the following "no-go theorem" for pure decoherence:

For quantum open systems linearly coupled to bosonic reservoirs decoherence is always accompanied by dissipation.

The "physical proof" is rather simple. Any irreversible decoherence must be related to an irreversible change in the environment. In the case of linear coupling to bosonic field (48) this change can be realized only by emission or absorption of a boson - the process which changes energy of an open system as well. Obviously, in the case of scattering process governed by the bilinear interaction Hamiltonian (49) we can alter other quantum numbers of the environment's state (e.g. its momentum) keeping the energy of S conserved (elastic scattering).

One can easily find in the literature the models of dephasing based on the linear coupling to bosonic field which are essentially variations of the so-called Caldeira-Leggett model [45-47]. To explain this apparent contradiction with our "no-go theorem" consider the simplest version of the *spin - boson model* defined by the Hamiltonian (here $H_S \equiv 0$)

$$H_{SR} = \lambda\sigma_3 \otimes \int_0^\infty d\omega[f(\omega)a(\omega) + \bar{f}(\omega)a^+(\omega)] + \int_0^\infty d\omega\,\omega\, a^+(\omega)a(\omega) \tag{88}$$

acting on the Hilbert space

$$\mathcal{H}_{SR} = \mathbf{C}^2 \otimes \mathcal{F}_B\big(L^2[0,\infty))\big) \equiv \mathcal{F}_B\big(L^2[0,\infty))\big) \oplus \mathcal{F}_B\big(L^2[0,\infty)\big)\,, \tag{89}$$

where $\mathcal{F}_B\big(L^2[0,\infty))\big)$ is a bosonic Fock space over a single-particle Hilbert space $L^2[0,\infty)$ and $[a(\omega), a^+(\omega')] = \delta(\omega - \omega')$. The unitary Weyl operator U_g acting on fields operators as

$$U_g a(\omega) U_g^* = a(\omega) + g(\omega) \tag{90}$$

exists if and only if $g \in L^2[0,\infty)$. Putting

$$g(\omega) = \lambda\omega^{-1} f(\omega) \tag{91}$$

we can diagonalize H_{SR} (88)

$$\mathbf{U}_g H_{SR} \mathbf{U}_g^* = \int_0^\infty d\omega\,\omega\, a^+(\omega)a(\omega) + \text{const.} \tag{92}$$

where

$$\mathbf{U}_g = \begin{pmatrix} U_g & 0 \\ 0 & U_{-g} \end{pmatrix}\,. \tag{93}$$

The ground states subspace of the diagonalized Hamiltonian (92) is spanned by the vectors $|1> \otimes|\Omega>, |2> \otimes|\Omega>$ and therefore the corresponding degenerated ground states of H_{SR} are given by

$$|1> \otimes|\phi[-g]> \quad |2> \otimes|\phi[g]> \ \text{where}\ |\phi[g]> = U_g|\Omega> \quad g(\omega) = \lambda\omega^{-1} f(\omega)\,. \tag{94}$$

The vectors $|\phi[\pm g]>$ are coherent states in $\mathcal{F}_B\big(L^2[0,\infty)\big)$ and

$$< \phi[-g], \phi[g] >= \exp\{-2\|g\|^2\} \ . \tag{95}$$

Assume now that we would like to describe dephasing of our spin using the model given by(88). Then we should have $\hat{R}(0) = \lambda^2|f(0)|^2 > 0$ what implies due to (91) that the function $g(\omega)$ is not square integrable i.e. $\|g\| = \infty$. The same divergence appears for any "ohmic" or "subohmic" coupling $|f(\omega)|^2 \sim \omega^s$ around 0 with $0 \le s \le 1$. It follows that the diagonalizing transformation of a bosonic field (90)(91) cannot be implemented by the unitary operator on the Fock space and therefore the formal expression (88) does not define a meaningful bounded from above Hamiltonian. It means that for an arbitrary coupling constant λ the model given by (88) is either nonphysical or cannot describe exponential (Markovian) dephasing. In a less mathematical language the large value of $\|g\|$ due either to a large coupling constant or to a large integral $\int_0^\infty d\omega\, \omega^{-2}|f(\omega)|^2$ makes the coherent states $|\phi[\pm g]>$ corresponding to a "cloud" and the vacuum $|\Omega>$ almost orthogonal. As a consequence the standard choice of the initial state as a product state $\rho\otimes|\Omega><\Omega|$ is inappropriate and impossible to prepare. The proper initial state should have a support spanned by the "dressed" ground states (94). The computed lost of coherence for the former choice of the initial state is therefore unphysical and describes the spurious process of a "cloud formation". Again the problem of a proper decomposition of the total system into open system and environment such that the effective interaction between them is weak, is crucial for the physical interpretation of the obtained results.

One should mention that from the mathematical point of view the above example illustrates the subtle problem of nonequivalent representations of canonical commutation relations for systems with infinite number of degrees of freedom [48].

Open Systems with Continuous Spectrum. In the derivations of MME based on the weak coupling or low density limits the discretness of H_S spectrum plays a crucial role. The averaging over Hamiltonian oscillations justifies the Markovian approximation on the coarse-grained time scale and the canceling of non-resonant terms is a necessary condition to preserve complete positivity of the QDS. This interplay between the self-evolution of S and the interaction with R produces the desired properties of QDS like for instance the relaxation to a proper equilibrium state for R being a heat bath.

However, there are important examples of open systems with continuous spectrum of H_S, the most studied is a quantum Brownian particle. For simplicity, we discuss first the case of Brownian motion in one-dimensional space. The most frequently used MME for this case is the so-called Caldeira-Leggett equation [47] of the form

$$\frac{d}{dt}\rho_t = -i[H_S, \rho_t] - i\gamma[X, \{P, \rho_t\}] - 2M\gamma T[X, [X, \rho_t]] \ , \tag{96}$$

where X, P are position and momentum operators, $H_S = P^2/2M + V(X)$, γ is a *friction constant* and M- mass of the Brownian particle. V(X) is a generic potential which can produce both continuous and discrete parts of the energy spectrum.

The MME (96) possesses the following well-known drawbacks:

1) the solution of (96) does not preserve positivity of the density matrix,

2) the Gibbs state (perhaps unnormalized) $\sim \exp\{-H_S/T\}$ is not its stationary state.

The first drawback can be cured by adding the term $-\gamma(8MT)^{-1}[P,[P,\rho_t]]$ which allows to write the corrected MME in a standard form (13,14). However the dissipative part of this new generator corresponds to a damped harmonic oscillator one (27) (with $\gamma_\uparrow = 0$) and the Hamiltonian part gains the correction $\sim (XP + PX)$ with a rather unclear physical interpretation.

In author's opinion the above difficulties have its source in the underlying Hamiltonian model for $S{+}R$ system, the so-called Caldeira-Leggett model, which is essentially a non-zero temperature version of the model (88) with σ_3 replaced by the operator X and the "forbidden" ohmic choice of $|f(\omega)|^2 \sim \omega$ around zero. In other words according to the "no-go theorem" the equation (96) which in the extreme heavy particle limit ($M \to \infty$)describes pure decoherence cannot be derived from a physically admissible model with linear coupling to the bosonic reservoir.

It seems, that for a continuous spectrum of H_S the lack of a natural time scale provided by the Hamiltonian evolution makes impossible to find a single MME which accurately describes all relevant stages of the evolution of S [49]. We do not mean here the well-known and expected deviations from the Markovian (exponential) behaviour for very short times typically $\sim \omega_c^{-1}$ (ω_c - cut-off frequency) and very long ones (due to the boundness from below of the Hamiltonian of R). In the simplest case of a Brownian particle in the free space described by the center of mass position $\mathbf{X}$ and the total momentum $\mathbf{P}$ the situation can be summarized in a following way.

The proper model of an environment should involve interactions of the type (49) describing scattering process with particles of the medium. Beside the direct collisions with atoms, molecules, photons etc. the other "bilinear" processes are possible within this model like for instance absorption of a foton followed by the excitation of the internal degrees of freedom of the Brownian particle and the time-reversed process [17].

The following different but approximatively Markovian stages of the evolution can be singled out:

1) *Pure decoherence stage*, when the decay of macroscopically distinguishable quantum superpositions into mixed states dominates over energy thermalization. For the low density medium and/or small Brownian particle this stage is well-described by the Eq. (35) on the time scale determined by $\tau_{dec} = \left(\int d^3\mathbf{k}\, n(\mathbf{k})\right)^{-1}$. For dense media or/and large particle τ_{dec} can be comparable or shorter than the already mentioned time scale ω_c^{-1} what demands a different theoretical treatement [50].

2) *Thermalization stage*, when the density matrix of the Brownian particle written in momentum representation is close to diagonal i.e. $\rho_t(\mathbf{p},\mathbf{p}')$ differs

essentially from zero for $|\mathbf{p} - \mathbf{p}'| \leq \sqrt{2MT}$. The detailed analysis of this regime and the manifest standard form of MME which depends on the so-called dynamic structure factor characterizing the environment are presented in [51].

Another completely different situation where the continuous spectrum of H_S appears is the theory of many-body open systems. An ensemble of interacting quantum spins coupled to an infinite heat bath and described in the thermodynamic limit is a perfect example. The mathematical formalism used in this context to construct proper MME, difficulties and partial solutions of the problems are discussed in [52].

Acknowledgments

The work is supported by the Gdańsk University Grant BW-5400-5-0234-2.

References

1. C.W. Gardiner, *Handbook of stochastic methods*, (Springer, Berlin, 1983).
2. G.S Agarwal, *Quantum Optics. Quantum Statistical Theories of Spontaneous Emission and their Relation to Other Approaches*, (Springer, Berlin 1971).
3. F. Haake, *Statistical treatment of open systems by generalized master equations*, (Springer, Berlin, 1973).
4. V. Gorini, A. Kossakowski and E.C.G. Sudarshan, J. Math. Phys. **17**, 821 (1976).
5. G. Lindblad, Commun. Math. Phys., **48**, 119 (1976).
6. E.B. Davies, *Quantum Theory of Open Systems*, (Academic Press, London 1976).
7. R. Alicki and K. Lendi, *Quantum Dynamical Semigroups and Applications*, (Springer, Berlin, 1987).
8. Ph. Blanchard et.al. (eds), *Decoherence: Theoretical, Experimental and Conceptual Problems*, (Springer, Berlin, 2000).
9. M.A. Nielsen and I.L. Chuang, *Quantum Computation and Quantum Information*, (Cambridge University Press, Cambridge, 2000).
10. R. Alicki and M. Fannes,*Quantum Dynamical Systems*, (Oxford University Press, Oxford, 2000).
11. W.F. Stinespring, Proc. Am. Math. Soc., **6**, 211 (1955).
12. K. Kraus, Annals of Physics, **64**, 311 (1971).
13. E.B. Davies, Rep. Math. Phys., **11**, 169 (1977).
14. A.M. Chebotarev and F. Fagnola, J. Funct. Anal., **153**, 382 (1998).
15. W. Feller, *An introduction to probability theory and its applications* Vol.2, (Wiley, New York, 1966).
16. M. Fannes, *Contribution to this volume.*
17. R. Alicki, Phys. Rev. **A65**, 034104-1 (2002).
18. S.G. Rautian and A.M. Shalagin, *Kinetic problems of non-linear spectroscopy* (North-Holland, Amsterdam, 1991).
19. R. Alicki and S. Kryszewski, *Completely positive Bloch-Boltzmann equations*, (physics/0202001)
20. Ph. Blanchard and A. Jadczyk, Ann. der Physik **4**, 583 (1995)
21. R. Alicki and M. Fannes, Lett. Math. Phys. **11**, 259 (1986).
22. R. Alicki and M. Fannes, Commun. Math. Phys. **108**, 353 (1987).

23. H.J. Carmichael, *An Open Systems Approach to Quantum Optics*, (Springer, Berlin, 1993).
24. W. Strunz, *Contribution to this volume.*
25. R.L. Hudson and K.R. Parthasarathy, Commun. Math. Phys. **93**, 301 (1984).
26. A.S. Holevo, J. Math. Phys. **37**, 1812 (1996).
27. E.B. Davies and H. Spohn, J. Stat. Phys. **19**, 511 (1978).
28. R. Alicki, J. Phys. **A 12**, L103 (1979).
29. P. Pechukas, Phys. Rev. Lett. **73**, 1060 (1994); **75**, 3021 (1995); R. Alicki, *ibid.* **75**, 3020 (1995).
30. S. Weinberg, *The Quantum Theory of Fields*, (Cambridge University Press, Cambridge, 2000).
31. A.J. Leggett et al., Rev. Mod. Phys. **59**, 1 (1987).
32. D. Goderis, A. Verbeure and P. Vets, Probability Theory and Related Fields **82**, 527 (1989).
33. L. Accardi et. al., in *Quantum Probability and Related Topics*,Vol.6, 237, (World Scientific, Singapore, 1992).
34. E.B. Davies, Commun. Math. Phys. **39**, 91 (1974).
35. V. Gorini and A. Kossakowski, J. Math. Phys. **17**, 1298 (1976).
36. R. Dümcke, Commun. Math. Phys. **97**, 331 (1985).
37. G. Lindblad, Commun. Math. Phys. **40**, 147 (1975).
38. R. Alicki, Int. J. Theor. Phys. **16**, 351 (1977).
39. W. Pusz and S.L. Woronowicz, Commun. Math. Phys. **58**, 273 (1978).
40. H. Spohn and J. Lebowitz, Adv. Chem. Phys. **38**, 109 (1978).
41. A.E. Allahverdyan and Th.M. Niewenhuizen, Phys. Rev. **E 64**, 056117 (2001).
42. R. Alicki, M. Horodecki, P. Horodecki and R. Horodecki, Phys.Rev. **A** (in print 2002).
43. E. Joos and H.D. Zeh, Z.Phys. **B 59**, 223 (1985).
44. W.H. Zurek, Physics Today **40**, 36 (1991)
45. W.G. Unruh, Phys. Rev. **A 51**, 992 (1995).
46. L. Viola and S. Lloyd, Phys. Rev. **A 58**, 2733 (1998).
47. A.O. Caldeira and A.J. Leggett, Phys. Rev. **A 31**, 1057 (1985)
48. G.G Emch, *Algebraic Methods in Statistical Mechanics and Quantum Field Theory.* (Wiley, New York, 1972).
49. R. Alicki, Phys. Rev. **A 40**, 4077 (1989).
50. F. Haake, *Contribution to this volume.*
51. B. Vacchini, J. Math. Phys. **42**, 4291 (2001).
52. W.A. Majewski, *Contribution to this volume.*

Finite Dissipative Quantum Systems

M. Fannes

Instituut voor Theoretische Fysica, K. U. Leuven, Celestijnenlaan 200D, B-3001 Heverlee, Belgium

Abstract. The subject of these notes is friction in classical and quantum mechanics. Its mathematical description in terms of Markovian semigroups is reviewed, with special emphasis on the notions of complete positivity and normality for the quantum evolutions. Damping is then used to stabilise the motion of a kicked oscillator. Some statistical features of the orbits of the kicked quantum oscillator with damping are analysed in the semi-classical regime.

1 Introduction

In a physical context, the term dissipation may have several connotations such as loss of energy, loss of information on initial data, randomness, irreversibility, ... It is a deep and ambitious goal of mathematical physics to establish the connection between the reversible mechanical theories which describe the evolution of a system at the microscopic level, both classical and quantum mechanical, and the irreversible behaviour that we observe on a macroscopic level. Slightly reformulated: how can one deduce the macroscopic transport equations from the microscopic dynamics?

A very different approach consists in using and studying dissipation on a phenomenological level. This will be the case for these notes where I shall use a simple example, the kicked and oscillator with damping, to comment on friction-related phenomena and open problems for small systems. The main concern are quantum systems in the semi-classical limit, i.e. quantum systems considered at time scales properly chosen with respect to $\hbar$ and $\hbar$ very small. Many references on matters related with these notes can be found in [2].

2 Friction in a Classical Context

We shall here briefly consider a closed dissipative classical system. This may be a reasonable approximation in some situations. Even if the origin of the dissipation is due to an interaction with an environment, it is sometimes possible to separate the system from its environment. Often, this introduces stochasticity in the evolution and this leads to irreversible behaviour. We shall still assume, however, that the future is determined by an initial condition of our system at a single time, i.e. that we don't have to specify the full history in order to make predictions about future behaviour. In this sense, we may consider our system well-separated from the outside world.

As we are considering a classical system, we have a geometrical substrate at our disposal: the phase space $\mathfrak{X}$. Typically, we can think about a finite-dimensional, compact, smooth manifold. We shall also select a reference measure $\mathrm{d}x$ on $\mathfrak{X}$ that is absolutely continuous with respect to the Lebesgue measure.

We can now distinguish between two types of evolutions: either in terms of orbits in phase space (type I) or of orbits in the space of probability measures on $\mathfrak{X}$ (type II). The first type is a special case of the second as we can identify a point $x \in \mathfrak{X}$ with the Dirac measure δ_x supported in x.

- Evolutions of type I are given by maps

$$t \in t_0 + \mathbb{R}^+ \mapsto x(t; (x_0, t_0)).$$

  $x(t; (x_0, t_0))$ is the point in phase space reached at time t if we start out at x_0 at time t_0.
- Assuming enough regularity, evolutions of type II are given by maps

$$t \in t_0 + \mathbb{R}^+ \mapsto \mu_t(dx) = \rho_t(x; (x_0, t_0))\, \mathrm{d}x\,.$$

  Here, μ_t gives the probability distribution of the system in phase space corresponding to the initial condition (x_0, t_0). We now rather think of a bunch of possible orbits emerging from each point and randomly chosen by the system. Moreover, we assume the probability distribution absolutely continuous with respect to $\mathrm{d}x$ and therefore given by a probability kernel ρ: $\rho_t(x; (x_0, t_0)) \geq 0$, $\int_{\mathfrak{X}} \mathrm{d}x\, \rho_t(x; (x_0, t_0)) = 1$ and ρ satisfies suitable continuity properties with respect to (x_0, t_0).

Being able to separate the system from the outside world translates into

$$x(t_2; (x_0, t_0)) = x\Bigl(t_2 - t_1; \Bigl(x(t_1 - t_0; (x_0, t_0)), t_1\Bigr)\Bigr)\,,$$

for $t_0 \leq t_1 \leq t_2$ in case I and an analogous relation for case II. Such an evolution still allows for time-dependent external fields or parameters. Often, one deals with completely autonomous systems and we can, using a time-shift, start at time 0:

$$x(t; (x_0, t_0)) = x(t - t_0; x_0) \qquad \text{and} \qquad \rho_t(x; (x_0, t_0)) = \rho_{t-t_0}(x; x_0)\,.$$

We now have the semigroup property of the evolution

$$x(t_1 + t_2; x_0) = x(t_2; x(t_1; x_0))\,,$$
$$\rho_{t_1+t_2}(x; x_0) = \int_{\mathfrak{X}} \mathrm{d}x_1\, \rho_{t_2}(x; x_1)\, \rho_{t_1}(x_1; x_0)\,.$$

As we wish to look at the classical limit of quantum systems for which the notion of phase space is not available, we must describe both classical and quantum systems on a common level namely that of observables. We shall therefore deal with the commutative algebra $\mathfrak{A} := \mathcal{C}(\mathfrak{X})$ of continuous complex functions on $\mathfrak{X}$. It comes with the natural supremum norm

$$\|f\| := \sup_{x\in\mathfrak{X}} |f(x)| ,$$

which satisfies the C*-property $\|f^*f\| = \|f\|^2$ (f^* denotes the complex conjugate of f). Moreover, Gelfand's Theorem tells us that any commutative C*-algebra with unit is of the type $\mathcal{C}(\mathfrak{X})$ with $\mathfrak{X}$ a uniquely determined compact Hausdorff space. We then lift the dynamics to the level of observables.

- In case I, we put

  $$\Gamma_t(f)(x_0) := f(x(t;x_0)) .$$

  It is immediately checked that

  $$\begin{aligned} &\Gamma_t(\alpha f + g) = \alpha\Gamma_t(f) + \Gamma_t(g) , &&\Gamma_t(f^*) = \Gamma_t(f)^* , \\ &\Gamma_t(\mathbb{1}) = \mathbb{1} , &&\Gamma_t(fg) = \Gamma_t(f)\Gamma_t(g)\text{and} \\ &\Gamma_{t_1+t_2} = \Gamma_{t_1}\Gamma_{t_2},\ t_1,t_2 \geq 0 , \end{aligned}$$

  i.e. $\{\Gamma_t \mid t \in \mathbb{R}^+\}$ is a semigroup of endomorphisms of $\mathfrak{A}$.
- In case II

  $$\Gamma_t(f)(x_0) := \int_{\mathfrak{X}} \mathrm{d}x\, f(x)\, \rho_t(x;x_0)) .$$

  A simple application of Schwarz's inequality shows that

  $$|\Gamma_t(f)|^2 \leq \Gamma_t(|f|^2), \tag{1}$$

  which is precisely the necessary and sufficient condition for the unity- preserving Γ_t to be positive. We now obtain a semigroup $\{\Gamma_t \mid t \in \mathbb{R}^+\}$ of unity-preserving, positive, linear transformations of $\mathfrak{A}$.

Often, one has suitable continuity properties of $t \mapsto \Gamma_t$ and considers the generator L of the dynamics: $\Gamma_t = \exp(tL)$. Without entering in domain questions of the mostly unbounded generator L, one obtains derivations in case I

$$L(fg) = fL(g) + gL(f)$$

and dissipations in case II

$$2fL(f) \leq L(f^2), f \text{ real-valued.}$$

Example: Let $\mathfrak{X}$ be the unit circle $\mathcal{S}^1$ in $\mathbb{R}^2$.

- If L is a smooth derivation, i.e. $L(\mathcal{C}^\infty(\mathcal{S}^1)) \subseteq \mathcal{C}^\infty(\mathcal{S}^1)$, then there exists a real function $a \in \mathcal{C}^\infty(\mathcal{S}^1)$ such that $L(f) = af'$.
- If L is a dissipation of the form $L(f) = \sum_{k=1}^N a_k f^{\langle k\rangle}$ with $a_k \in \mathcal{C}^\infty(\mathcal{S}^1)$, then $N = 2$, $a_2 \geq 0$ and a_1 is real. We then obtain the generalised heat equation $L(f) = a_2 f'' + a_1 f'$.

3 Kicked Classical Oscillator with Damping

3.1 Kicked Oscillator

The evolution from kick to kick is given by

$$\begin{aligned} q' &= q\cos\alpha + p\sin\alpha \\ \text{and} & \\ p' &= -q\sin\alpha + p\cos\alpha + \lambda\sin q'. \end{aligned} \tag{2}$$

The frequency of the oscillator and the time between two successive kicks determine α and λ describes the strength of the kick. Clearly, the energy $e = \frac{1}{2}(p^2+q^2)$ is not conserved. Depending on the values of α and λ, the nature of the motion can be quite different. For small λ, the motion is rather regular while for large λ the $\sin q'$ term in (3) behaves more or less like a random term and we obtain some kind of random walk for p. In such a case the energy grows more or less linearly in time. For special choices of initial conditions, there is even a ballistic regime where the energy grows quadratically in time. Fig. 1 shows the growth of the energy in the large λ regime. Obviously, the motion is not confined to some bounded region of phase space.

3.2 The Case with Damping

We now introduce a damping coefficient $0 < \gamma < 1$ and modify the equations of motion to

$$\begin{aligned} q' &= q\gamma\cos\alpha + p\gamma\sin\alpha \\ \text{and} & \\ p' &= -q\gamma\sin\alpha + p\gamma\cos\alpha + \lambda\sin q'. \end{aligned}$$

Because of the damping term, the surface in phase space is no longer conserved but shrinks at each time-step with a factor

$$\frac{\partial(q',p')}{\partial(q,p)} = \gamma^2.$$

We shall then see the appearance of attractors of zero surface measure.

Moreover, the motion becomes effectively compact:

$$2e' = (p')^2 + (q')^2 \le 2e\gamma^2 + \lambda^2 + 2\lambda\gamma\sqrt{2e} = (\gamma\sqrt{2e} + \lambda)^2.$$

Because of this inequality, all initial conditions eventually end up in the circle centred at the origin and with radius determined by $2e = \lambda/(1-\gamma)$. For small λ, all initial conditions are attracted to a single fixed point, the origin. For larger λ, there is a period doubling scheme and for λ sufficiently large, the motion becomes chaotic. Fig. 2 shows the behaviour of the energy in the large λ regime and should be compared with Fig. 1.

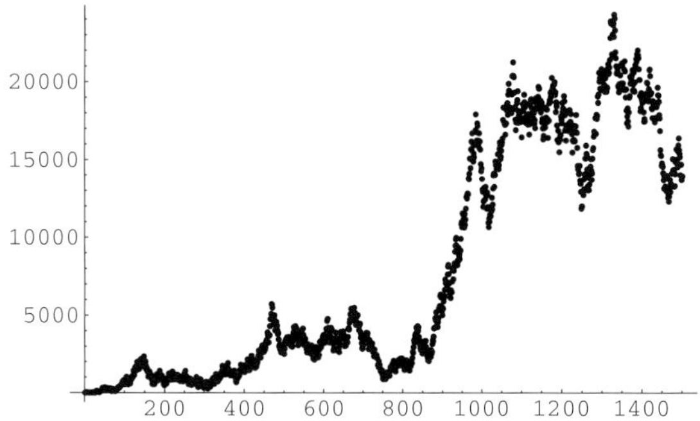

Fig. 1. Energy of kicked oscillator

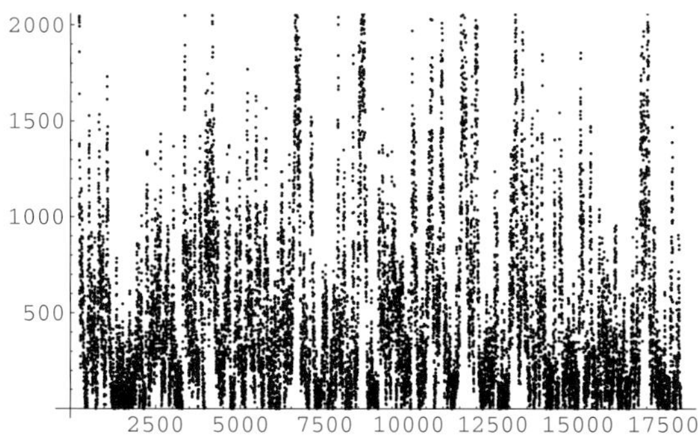

Fig. 2. Energy of damped kicked oscillator in chaotic regime

4 Dissipation in a Quantum Context

The crucial difference between this section and Section 2 is that observables of a quantum system generally don't commute. Indeed, one of the essential features of quantisation is to specify non-trivial commutation relations such as $[Q, P] = i\hbar$ for the usual position and momentum. There are two main formal approaches:

- The observables are given in terms of an abstract, unital (non-commutative) C*-algebra $\mathfrak{A}$, the essential property of the norm being $\|x^*x\| = \|x\|^2$. Normalised positive linear functionals ω on $\mathfrak{A}$ are called states and they associate expectations values to the observables: $x \in \mathfrak{A} \mapsto \omega(x) \in \mathbb{C}$.
- The observables are given in terms of a concrete C^*-algebra acting on a Hilbert $\mathfrak{H}$. In this case, we shall assume that the algebra is also closed in the strong topology and we obtain a von Neumann algebra $\mathfrak{M}$. There are many equivalent characterisations of von Neumann algebras such as von Neumann's Bicommutant Theorem $\mathfrak{M} = \mathfrak{M}''$. (If S is a self-adjoint subset of the bounded linear operators $\mathfrak{B}(\mathfrak{H})$ on $\mathfrak{H}$, then S' is the commutant of S, it consists of all $y \in \mathfrak{B}(\mathfrak{H})$ such that $[y, s] = 0$ for all $s \in S$.)

Focusing again on closed systems with autonomous dynamics, we need to consider semigroups of positive unity-preserving maps. What is their general form? How can one characterise their generators? Both questions are very difficult and have, even for the case of $\mathfrak{B}(\mathfrak{H})$, only partially been answered. Moreover, we need to narrow somewhat further our scope in order to obtain a workable notion of positive map:

- Positivity should be strengthened to complete positivity.
- in the case of von Neumann algebras, the maps should be normal, which is the counterpart of σ-additivity in the classical case.

4.1 Complete Positivity

A linear transformation Γ of a unital C*-algebra $\mathfrak{A}$ is positive and unity-preserving (p.u.p.) if

$$x \geq 0 \Rightarrow \Gamma(x) \geq 0 \qquad \text{and} \qquad \Gamma(\mathbb{1}) = \mathbb{1}\,.$$

A unity-preserving Γ is positive iff it satisfies Schwarz's inequality

$$\Gamma(x)^2 \leq \Gamma(x^2) \text{for} x = x^*.$$

This is, in fact, precisely the same inequality (1) that holds for classical systems. The main problem with mere positivity is that it is not robust with respect to tensor constructions and therefore hardly useful when considering composed systems.

A unity-preserving linear transformation Γ of a C*-algebra $\mathfrak{A}$ is completely positive (c.p.u.p.) if $\Gamma \otimes \mathrm{id}_n$ is positive on $\mathfrak{A} \otimes \mathcal{M}_n$ for all $n \in \mathbb{N}$ (we denote the algebra of complex $n \times n$ matrices by $\mathcal{M}_n$), i.e.

$$[x_{ij}]_{i,j=1,\ldots,n} \geq 0, \quad x_{ij} \in \mathfrak{A} \Rightarrow [\Gamma(x_{ij})]_{i,j=1,\ldots,n} \geq 0.$$

A useful inequality is obtained by choosing $n = 2$, $x \in \mathfrak{A}$ and observing that

$$\begin{pmatrix} \mathbb{1} & x \\ x^* & x^*x \end{pmatrix} = \begin{pmatrix} \mathbb{1} & x \\ 0 & 0 \end{pmatrix}^* \begin{pmatrix} \mathbb{1} & x \\ 0 & 0 \end{pmatrix}$$

is positive. Hence, if Γ is unity-preserving and completely positive (two-positive would suffice)

$$\begin{pmatrix} \mathbb{1} & \Gamma(x) \\ \Gamma(x^*) & \Gamma(x^*x) \end{pmatrix} \geq 0,$$

which is equivalent to

$$\Gamma(x^*)\Gamma(x) \leq \Gamma(x^*x). \tag{3}$$

This two-positivity inequality is strictly stronger than the Schwarz inequality for a truly non-commutative algebra. But, in order to extract additional information from it, we must apply it to non-normal elements x.

Stinespring's Theorem characterises all c.p.u.p. maps $\Gamma : \mathfrak{A} \to \mathfrak{B}(\mathfrak{H})$. To each such Γ there corresponds a unique triple $(\mathfrak{K}, \pi, V)$ where $\mathfrak{K}$ is a Hilbert space, π a representation of $\mathfrak{A}$ on $\mathfrak{K}$, and $V : \mathfrak{H} \to \mathfrak{K}$ an isometry such that

$$\Gamma(x) = V^* \pi(x)\, V \qquad \text{and} \qquad \pi(\mathfrak{A}) V \mathfrak{H} \text{is dense in} \mathfrak{K}.$$

Finding all c.p.u.p. maps therefore essentially amounts to finding all representations of $\mathfrak{A}$.

4.2 Normality

When dealing with a von Neumann algebra $\mathfrak{M} \subset \mathfrak{B}(\mathfrak{H})$, we may use other notions of convergence than just that stemming from the operator norm. E.g.:

- A sequence $\{x_n\}$ in $\mathfrak{M}$ converges strongly to x ($x = \text{s}-\lim_n x_n$) if $\lim_n \|(x - x_n)\varphi\| = 0$ for any choice of $\varphi \in \mathfrak{H}$.
- A sequence $\{x_n\}$ in $\mathfrak{M}$ converges weakly to x ($x = \text{w}-\lim_n x_n$) if $\lim_n \langle\psi, (x - x_n)\varphi\rangle = 0$ for any choice of $\psi, \varphi \in \mathfrak{H}$.

A state ω on $\mathfrak{M}$ is called normal if

$$\left.\begin{array}{l}\text{s}-\lim_n x_n = x \\ \|x_n\| \le 1\end{array}\right\} \Rightarrow \lim_n \omega(x_n) = \omega(x).$$

As with von Neumann algebras, there are many equivalent defining properties for normality such as ω is normal iff there exists a density matrix ρ on $\mathfrak{H}$ such that $\omega(x) = \operatorname{Tr} \rho\, x$. The space of normal linear functionals on $\mathfrak{M}$, i.e. the linear combinations of at most four normal states, is called the predual of $\mathfrak{M}$ and denoted by $\mathfrak{M}_*$. The predual of $\mathfrak{B}(\mathfrak{H})$ is the space of trace-class operators on $\mathfrak{H}$.

A c.p.u.p. transformation Γ of $\mathfrak{M}$ is called normal if

$$\left.\begin{array}{l}\text{s}-\lim_n x_n = x \\ \|x_n\| \le 1\end{array}\right\} \Rightarrow \text{s}-\lim_n \Gamma(x_n) = \Gamma(x).$$

We have now two equivalent options for specifying a dynamical map: the Heisenberg picture with $\Gamma : \mathfrak{M} \to \mathfrak{M}$ normal c.p.u.p. or the Schrödinger picture with $\Gamma_* : \mathfrak{M}_* \to \mathfrak{M}_*$ c.p. and state-preserving. They are connected by

$$\omega(\Gamma(x)) = \Gamma_*(\omega)(x), \quad \omega \in \mathfrak{M}_*,\ x \in \mathfrak{M}.$$

Kraus [16] has obtained the general form of a normal c.p.u.p. transformation Γ of $\mathfrak{B}(\mathfrak{H})$

$$\Gamma(x) = \text{s}-\sum_i v_i^* x\, v_i\,, \tag{4}$$

where $\{v_1, v_2, \ldots\}$ is a sequence in $\mathfrak{B}(\mathfrak{H})$ such that s–$\sum_i v_i^* v_i = \mathbb{1}$. Unfortunately, the sequence $\{v_1, v_2, \ldots\}$ is generally not unique. Some particular cases are given by:

- There is only a single term in (4) determined by a unitary u. In this case $\Gamma(x) = u^* x\, u$ is a general automorphism of $\mathfrak{B}(\mathfrak{H})$.
- If $\{u_1, u_2, \ldots, u_n\}$ is a sequence of elements satisfying

$$u_k u_\ell^* = \delta_{k\ell} \mathbb{1} \qquad \text{and} \qquad \sum_k u_k^* u_k = \mathbb{1},$$

then $x \mapsto \sum_k u_k^* x\, u_k$ is a normal endomorphism of $\mathfrak{B}(\mathfrak{H})$ of Powers index n [18]. The C*-algebra generated by such isometries u_j is a Cuntz algebra [4].

4.3 Semigroups

There is now a continuity issue with respect to the time variable. One may essentially consider three continuity notions:

- Uniform continuity, which is a very strong requirement,

$$\lim_{t \downarrow 0} \sup_{\{x \mid \|x\| \leq 1\}} \bigl\| \Gamma_t(x) - x \bigr\| = 0.$$

  The generator is bounded and, excepted for finite dimensional situations, it will not hold in most applications.
- Strong continuity

$$\lim_{t \downarrow 0} \bigl\| \Gamma_t(x) - x \bigr\| = 0.$$

  This is e.g. a realistic condition in the framework of quantum spins on a lattice.
- Weak continuity, which is useful when considering semigroups of normal c.p.u.p. maps on a von Neumann algebra $\mathfrak{M}$

$$\lim_{t \downarrow 0} \omega(\Gamma_t(x) - x) = 0, \quad \omega \in \mathfrak{M}_*, \; x \in \mathfrak{M}.$$

  This is technically the most delicate notion as e.g. the domain of the generator of such a semigroup will generally not be norm-dense in $\mathfrak{M}$.

Gorini, Kossakowski and Sudarshan [11] and Lindblad [17] have obtained the general form of the generator L of a uniformly continuous semigroup of normal c.p.u.p. maps on $\mathfrak{B}(\mathfrak{H})$. They showed that there exists a bounded $h = h^*$ and a sequence $\{v_1, v_2, \ldots\}$ of bounded operators on $\mathfrak{H}$, with s$-\sum_i v_i^* v_i$ bounded, such that

$$L(x) = i[h, x] + \sum_i \Bigl(v_i^* x \, v_i - \tfrac{1}{2} v_i^* v_i \, x - \tfrac{1}{2} x \, v_i^* v_i \Bigr).$$

The unbounded case has been much studied [5,19,15,3], but no general answer is available yet. A well-known particular case is that of weakly continuous groups $\{\Gamma_t \mid t \in \mathbb{R}\}$ of automorphisms of $\mathfrak{B}(\mathfrak{H})$, where Stone's Theorem yields a, generally unbounded, self-adjoint operator H such that $\Gamma_t(x) = \mathrm{e}^{itH} x \, \mathrm{e}^{-itH}$.

5 Kicked Quantum Oscillator with Damping

We begin this section by introducing some notation and reminding some basic properties of the harmonic oscillator. In terms of the standard position and momentum operators on $\mathfrak{H} := \mathcal{L}^2(\mathbb{R}, dx)$

$$\begin{aligned} Q\varphi &:= \mathrm{id}\, \varphi, & &\varphi \text{and} \, \mathrm{id}\, \varphi \in \mathfrak{H} \\ P\varphi &:= -i\hbar\varphi', & &\varphi \in \mathfrak{H} \text{absolutely continuous with} \varphi' \in \mathfrak{H}, \end{aligned}$$

the Hamiltonian reads

$$H_0 = \frac{1}{2m} P^2 + \frac{m\omega^2}{2} Q^2 - \frac{1}{2}\hbar\omega.$$

H_0 has non-degenerate, pure point spectrum $\mathbb{N}$ and we shall need the orthonormal basis $\{|n\rangle \mid n \in \mathbb{N}\}$ of eigenstates of H_0. The creation and annihilation operators a^* and a are related to Q and P by

$$a^* := \frac{P}{\sqrt{2m\omega\hbar}} + i\sqrt{\frac{m\omega}{2\hbar}} Q \qquad \text{and} \qquad a := \frac{P}{\sqrt{2m\omega\hbar}} - i\sqrt{\frac{m\omega}{2\hbar}} Q. \tag{5}$$

They act on the eigenbasis of H_0 as

$$a^*|n\rangle = \sqrt{n+1}\,|n+1\rangle \qquad \text{and} \qquad a|n\rangle = \sqrt{n}\,|n-1\rangle,$$

satisfy the canonical commutation relation $[a, a^*] = \mathbb{1}$, and the Hamiltonian can be written as $H_0 = \hbar\omega a^* a$.

5.1 Kicked Oscillator

The dynamics of the kicked oscillator is determined by a unitary operator u which specifies the evolution from kick to kick, just after or just before a kick. If two consecutive kicks are separated by a time interval Δt

$$u = \mathrm{e}^{-i\Delta t H_0}\, \mathrm{e}^{-i\lambda \cos Q}.$$

The parameter λ measures the strength of the kick. It is believed that u has a pure point spectrum, meaning that there is an orthonormal basis $\{f_1, f_2, \ldots\}$ of $\mathfrak{H}$ of eigenvectors of u: $uf_j = \lambda_j f_j$ with $|\lambda_j| = 1$. Let $\varphi \in \mathfrak{H}$ be arbitrary, then we find for any $\epsilon > 0$ a finite index set I_ϵ such that

$$\sup_n \Big\| u^n \varphi - \sum_{j \in I_\epsilon} \lambda_j^n \langle f_j, \varphi\rangle f_j \Big\| \le \epsilon.$$

Therefore $n \mapsto u^n\varphi$ is almost periodic and so are all time-dependent correlation functions. Fig. 3 shows the behaviour of $n \mapsto \langle u^n\varphi, H_0 u^n \varphi\rangle$.

5.2 Damped Oscillator

Explicit formulas. In the classical case, the motion is the solution of

$$\dot{q} = p \qquad \text{and} \qquad \dot{p} = -\gamma p - \omega^2 q. \tag{6}$$

The evolution yields a flow on phase space, the energy decreases monotonically in time and all orbits eventually end up in $(0, 0)$. For quantum systems, the energy levels of the oscillator are discrete and the only way to evolve continuously in

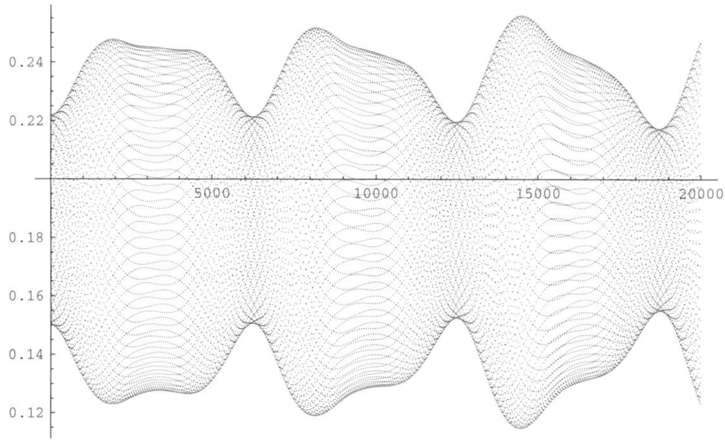

Fig. 3. Energy of kicked oscillator

time from an initial pure state to the vacuum $|0\rangle\langle 0|$ is to pass through mixed states. We shall use the Schrödinger picture and consider a generator of the form

$$L_*(\rho) := -i[H_0, \rho] + \lambda(a\,\rho\,a^* - \tfrac{1}{2}a^*a\,\rho - \tfrac{1}{2}\rho\,a^*a).$$

It is easily seen that, formally, $-i[H_0, \cdot]$ and $L_*^{\mathrm{D}}(\cdot) := a \cdot a^* - \frac{1}{2}a^*a\,\cdot - \frac{1}{2} \cdot a^*a$ commute. So we shall concentrate on the dissipative part L_*^{D} of the generator.

An explicit computation shows that for arbitrary $m, n \in \mathbb{N}$

$$\sigma_{m,n} := \sum_{j=0}^{m\wedge n} (-1)^j \binom{m}{j}^{\frac{1}{2}} \binom{n}{j}^{\frac{1}{2}} \binom{m\vee n}{m\wedge n}^{-\frac{1}{2}} |m-j\rangle\langle n-j| \tag{7}$$

is an eigenvector of L_*^{D} with eigenvalue $-(m+n)/2$. It is also possible to express any matrix unit $|k\rangle\langle\ell|$ in terms of the $\sigma_{m,n}$

$$|m\rangle\langle n| = \sum_{j=0}^{m\wedge n} \binom{m\wedge n}{j} \binom{m\vee n}{m\wedge n}^{\frac{1}{2}} \sigma_{m-j,n-j}\,. \tag{8}$$

The expressions (7) and (8) allow us to explicitly check that $\exp(tL_*^{\mathrm{D}})$ is state preserving and completely positive on $\mathrm{Span}\big(\{|m\rangle\langle n| \mid m,n \in \mathbb{N}\}\big)$. It then extends to the trace-class operators on $\mathfrak{H}$ and defines, by duality, a weakly continuous semigroup of normal c.p.u.p. transformations of $\mathfrak{B}(\mathfrak{H})$. It is surprising that, despite the large set of eigenvectors of L_*^{D} with eigenvalue set $-\frac{1}{2}\mathbb{N}$, the spectrum of $\exp(tL_*^{\mathrm{D}})$ is still the full unit circle in $\mathbb{C}$.

We may quickly see, in a very formal way, that the generator L_*^{D} indeed returns in the classical limit the flow generated by (6). Denoting by F and G quantisations of classical functions f and g, we just apply the standard rules

$$\lim_{\hbar\to 0} Q = q, \qquad \lim_{\hbar\to 0} P = p, \qquad \text{and} \qquad \lim_{\hbar\to 0} -\frac{i}{\hbar}[F,G] = \{f,g\}.$$

Using the expression (5) for a and a^*, we obtain

$$\begin{aligned}\lim_{\hbar\to 0} a^* F\, a - \tfrac{1}{2}a^* a\, F - \tfrac{1}{2}F\, a^* a &= \lim_{\hbar\to 0} \tfrac{1}{2}a^*[F,a] - \tfrac{1}{2}[F,a^*]\, a \\ &= \tfrac{1}{2}p\frac{\partial f}{\partial p} - \tfrac{1}{2}q\frac{\partial f}{\partial q}\,.\end{aligned}$$

Weak-coupling limit [1]. The dynamics of the damped oscillator can be obtained as an approximation of a deterministic evolution where the oscillator (the small system) is coupled to a large heat bath (a boson field in its vacuum state in this case). The Hilbert space of the composed system is $\mathfrak{H}^{\mathrm{sys}} \otimes \mathfrak{H}^{\mathrm{bath}}$, the Hamiltonian is of the form

$$H^{\mathrm{tot}} = H^{\mathrm{sys}} \otimes \mathbb{1} + \mathbb{1} \otimes H^{\mathrm{bath}} + \lambda H^{\mathrm{int}},$$

and we start out with an initial state $\sigma \otimes \rho^{\mathrm{bath}}$. After some time t, the initial state σ of the system has evolved into a state σ_t

$$\sigma_t(x) := \sigma \otimes \rho^{\mathrm{bath}}\Big(\mathrm{e}^{itH^{\mathrm{tot}}}(x \otimes \mathbb{1})\mathrm{e}^{-itH^{\mathrm{tot}}}\Big).$$

The map $\sigma \mapsto \sigma_t$ is in general not a semigroup because of memory effects. However, in the weak-coupling regime $\lambda \to 0$, $t \to \infty$, $\lambda^2 t = \tau$, it becomes a semigroup in the rescaled time τ. In order for this approximation to work, we need some additional requirements such as:

- The interaction term has to be suitably chosen.
- The spectrum of H^{sys} should be pure point and well-separated.
- One has to get rid of infinitely rapidly oscillating phases by turning to the interaction picture.
- ρ^{bath} should have sufficiently rapidly decreasing time autocorrelation functions.

For the oscillator, a very explicit construction is possible because of its linearity. For the heat bath we introduce a boson field in its vacuum state. Let $\Gamma(\mathfrak{K})$ denote the symmetric Fock space on $\mathfrak{K} = \mathcal{L}^2(\mathbb{R}^+, \mathrm{d}\varepsilon)$, $b^*(\varphi)$ and $b(\varphi)$ the canonical smeared field operators acting on $\Gamma(\mathfrak{K})$:

$$\varphi \mapsto b^*(\varphi)\text{is linear}, \qquad [b(\varphi), b(\psi)] = 0, \qquad \text{and} \qquad [b(\varphi), b^*(\psi)] = \langle\varphi, \psi\rangle,$$

and Ω the vacuum vector ($b(\varphi)\Omega = 0,\ \forall\varphi$). We then choose

$$\begin{aligned}H^{\mathrm{sys}} &= \hbar\omega a^* a\,,\\ H^{\mathrm{bath}} &= \int_{\mathbb{R}^+} \mathrm{d}\varepsilon\, \varepsilon b^*(\varepsilon) b(\varepsilon)\,,\\ H^{\mathrm{int}} &= a^* \otimes b(\xi) + a \otimes b^*(\xi)\,.\end{aligned}$$

A direct computation yields

$$[H^{\mathrm{tot}}, \mu a^* + b^*(\varphi)] = \hbar\omega\mu a^* + b^*(h^{\mathrm{bath}}\varphi) + \lambda\mu b^*(\xi) + \lambda\langle\xi, \varphi\rangle a^*,$$

with

$$(h^{\mathrm{bath}}\varphi)(\varepsilon) := \varepsilon\varphi(\varepsilon).$$

Such evolutions are called (quasi-)free and are fully determined by

$$\begin{pmatrix} \mu \\ \varphi \end{pmatrix} \mapsto \begin{pmatrix} \hbar\omega\mu + \lambda\langle\xi, \varphi\rangle \\ h^{\mathrm{bath}}\varphi + \lambda\mu\xi \end{pmatrix} =: h^{\mathrm{tot}} \begin{pmatrix} \mu \\ \varphi \end{pmatrix}.$$

In order to compute the weak-coupling limit, we need to control

$$\lim_{\substack{t\to\infty \\ \lambda\to 0 \\ \lambda^2 t=\tau}} \left[\mathrm{e}^{ith^{\mathrm{tot}}} \mathrm{e}^{-ith^0} \right]_{11},$$

with

$$h^{\mathrm{tot}} := \begin{pmatrix} \hbar\omega & \lambda\langle\xi, \cdot\rangle \\ \lambda\xi & h^{\mathrm{bath}} \end{pmatrix} \qquad \text{and} \qquad h^0 := \begin{pmatrix} \hbar\omega & 0 \\ 0 & h^{\mathrm{bath}} \end{pmatrix}.$$

This is a scattering type problem and it can be handled by using the Dyson expansion. We essentially need that

$$t \mapsto \int_{\mathbb{R}+} \mathrm{d}\varepsilon \, \langle\xi, \mathrm{e}^{it\varepsilon}\xi\rangle$$

converges sufficiently rapidly to 0, e.g. that $t^\delta\langle\xi, \mathrm{e}^{it\varepsilon}\xi\rangle$ be integrable for a $\delta > 0$.

5.3 Kicked Oscillator with Damping

Because the classical motion of the kicked oscillator with damping is effectively compact and each quantum state occupies a volume $\hbar$ in phase space, one may expect that only a finite number of states will eventually be relevant for the time evolution. One can therefore expect that arbitrary initial states relax to a unique invariant state. Let Γ denote the evolution of the damped harmonic oscillator during a time Δt followed by a kick. Fig. 4 shows the (absolute) values of the expansion coefficients of the invariant state in terms of the matrix units $|m\rangle\langle n|$.

Existence of normal invariant state. The existence of a normal invariant state for a semigroup of normal c.p.u.p. maps is in no way guaranteed. The simplest approach, starting out with an initial density matrix ρ_0, putting $\rho_n := \Gamma^n(\rho_0)$ and trying to show the existence of $\lim_{n\to\infty}\rho_n$ might fail, e.g. when the ρ_n converge to some periodic cycle. It is therefore better to look at limits of averages of ρ_n. Putting $\overline{\rho}_N := \frac{1}{N}\sum_{j=0}^{N-1}\rho_j$, we have by weak-$*$ compactness that $\{\overline{\rho}_N \mid N \in \mathbb{N}_0\}$ contains a convergent subnet A: $\overline{\rho}_A(x) := \lim_{\alpha\in A}\rho_\alpha(x)$ exists for each $x \in \mathfrak{B}(\mathfrak{H})$ and is a Γ-invariant state on $\mathfrak{B}(\mathfrak{H})$. A possible problem is that $\overline{\rho}_A$ might fail to be normal, i.e. not given by a density matrix on $\mathfrak{H}$. If the sequence $\{\overline{\rho}_N\}$ is tight, however, then normality of limit points of the sequence is ensured. Tightness means that the $\overline{\rho}_N$ have to stay almost local: for any $\epsilon > 0$ there must exist a fixed finite dimensional projector P_ϵ such that $\overline{\rho}_N(P_\epsilon) \geq 1 - \epsilon$. Tightness criteria for the case of weakly continuous semigroups recently appeared in [7].

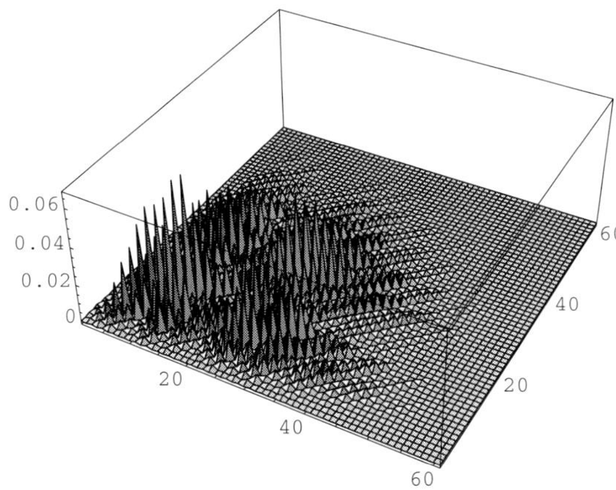

Fig. 4. Expansion coefficients of invariant state

In order to obtain tightness in our case we consider the energy distribution of a density matrix

$$\rho \mapsto (\langle n|, \rho|n\rangle)_{n\in\mathbb{N}}.$$

There are now two opposite effects: the damping brings the probability distribution $\{\langle n|, \rho|n\rangle\}$ down to zero, while the kicks might raise the energy, but altogether not too much. Tightness will hold when we are able to find an upper bound for the energy distribution that is invariant under the joint action of damping and kicking and that has an integrable tail.

Uniqueness of invariant state. We just present here a general argument that ensures uniqueness of an invariant state. The origins of this criterion date back to [20] and it is more or less still the only general technique available, see e.g. [8].

Suppose that Γ is a c.p.u.p. transformation of a C*-algebra $\mathfrak{A}$ and that ω_0 is Γ-invariant and faithful in the sense that $a \geq 0$ and $\omega_0(a) = 0$ imply $a = 0$. Consider now an element x such that $\Gamma(x) = x$. Using the two-positivity inequality (3) and the given invariances, we find

$$\omega_0(x^*x) = \omega_0(\Gamma(x^*)\Gamma(x)) \leq \omega_0(\Gamma(x^*x)) = \omega_0(x^*x).$$

Hence, $\omega_0(\Gamma(x^*x) - \Gamma(x^*)\Gamma(x)) = 0$ and, because of the assumed faithfulness of ω_0, $\Gamma(x^*x) = \Gamma(x^*)\Gamma(x)$. For any $y \in \mathfrak{A}$ and $\epsilon \in \mathbb{C}$ we have

$$\Gamma(x + \epsilon y)^*\Gamma(x + \epsilon y) \leq \Gamma((x + \epsilon y)^*(x + \epsilon y)).$$

This can only hold if

$$\Gamma(y^*)\Gamma(x) = \Gamma(y^*x) \qquad \text{and} \qquad \Gamma(x)\Gamma(y^*) = \Gamma(x\,y^*),$$

i.e. the Γ-invariant elements form a subalgebra of $\mathfrak{A}$.

Consider now the case of $\mathfrak{B}(\mathfrak{H})$, with a normal Γ and a faithful normal invariant ρ_0. By Kraus's result (4), we have

$$\Gamma(x) = \sum_i v_i^* x \, v_i \qquad \text{and} \qquad \sum_i v_i^* v_i = \mathbb{1}.$$

Supposing $\Gamma(x) = x$

$$\sum_i [v_i, x]^* [v_i, x] = x^* x - x^* \Gamma(x) - \Gamma(x)^* x + \Gamma(x^* x) = 0,$$

because the invariant elements form an algebra. Hence $x \in \{v_i, v_i^* \mid i\}'$. In particular, if the commutant of the v_i turns out to be trivial, which is often easily verified, then x has to be a multiple of the identity. This implies that ρ_0 is the unique invariant state under Γ_*.

6 Statistical Properties of Orbits in the Semi-classical Limit

When considering the (semi-)classical limit of the damped kicked oscillator one could study spectral properties of the evolution. In particular, due to the damping, the discrete spectrum of the kicked oscillator will now be unfolded (spiral towards 0) and one could certainly wonder about level statistics [12,21].

We are following another path, however, concentrating on the orbits traced out by the system. The procedure to follow for unitary evolutions is by now rather clear. Analytical results have been obtained for a random model and more realistic systems such as baker maps or kicked tops have been studied numerically [9]. For dissipative evolutions, there still remain many open questions. This section will therefore mainly deal with unitary evolutions.

The paths traced out by a chaotic classical dynamical system in discrete time look like points generated by a Poisson process. This is a striking difference with the behaviour in regular regions. Recent results have established a typical distribution for the return times to a small region in phase space [14]. One may wonder about the statistical properties of return times for quantum systems and their relation with chaotic or regular behaviour of their classical limit. Because of the almost periodicity that governs the evolution of finite quantum systems, one should consider them at a suitable time scale with respect to the quantisation parameter.

Let $\boldsymbol{\varphi} := \{\varphi_0, \varphi_1, \ldots, \varphi_{K-1}\}$ be a sequence of K vectors traced out in the Hilbert space of a system by a discrete unitary dynamics or by some random process. Instead of considering the overlap $n \mapsto |\langle \varphi_n, \varphi_0 \rangle|^2$ with the initial state as a function of time, we might look at all joint overlaps in the sequence by considering the spectrum of the Gram matrix $G(\boldsymbol{\varphi})$ of the sequence

$$G(\boldsymbol{\varphi}) := [\langle \varphi_i, \varphi_j \rangle]_{ij}.$$

$G(\boldsymbol{\varphi})$ is a positive definite matrix and we associate with $\boldsymbol{\varphi}$ the empirical eigenvalue distribution

$$\rho_{\boldsymbol{\varphi}}(\lambda)\mathrm{d}\lambda := \frac{1}{K}\sum_{i=1}^{K}\delta(\lambda-\lambda_i)\,\mathrm{d}\lambda,$$

where the λ_i are the eigenvalues of $G(\boldsymbol{\varphi})$. It is not hard to see that few large overlaps in the sequence will lead to a $\rho_{\boldsymbol{\varphi}}$ concentrated around 1, while many large overlaps will contribute to $\rho_{\boldsymbol{\varphi}}$ with a few large eigenvalues and many others close to zero. For a classical sequence, identified with vectors picked from an orthonormal basis possibly with repetitions, $\rho_{\boldsymbol{\varphi}}$ encodes all the information about the relative frequencies of the elements in the sequence.

Consider next a random vector model by choosing $K = \tau N$ normalised vectors in $\mathbb{C}^N$, independently and uniformly distributed. The empirical measure of such a random sequence is a random probability measure on $\mathbb{R}^+$, but when $N \to \infty$ it converges to a definite probability measure: the Marchenko-Pastur distribution with density τ [6]

$$\rho_\tau(\mathrm{d}t) = \begin{cases} \frac{\sqrt{4\tau t-(t+\tau-1)^2}}{2\pi\tau t}\,\mathrm{d}t\,, & 0<\tau\le 1\,, \\ \frac{\tau-1}{\tau}\delta(t)\,\mathrm{d}t + \frac{\sqrt{4\tau t-(t+\tau-1)^2}}{2\pi\tau t}\,\mathrm{d}t\,, & 1<\tau\,. \end{cases}$$

The dimension N of the Hilbert space measures the total number of available states and plays the role of Heisenberg time. If a completely random choice of vectors mimics extreme chaoticity, this result shows the typical eigenvalue distribution of the Gram matrix for that case. A very regular choice, where only a small fraction of the available states is visited would, at the Heisenberg time scale, lead to a degenerate distribution.

A more realistic model, the kicked top [13], exhibits this kind of behaviour. The classical system is a spin of length 1 in $\mathbb{R}^3$ and has two adjustable parameters which allow to vary the motion from regular to almost global chaos. For each $N \in \mathbb{N}$ one has a quantisation of the top in terms of the irreducible spin-$(N-1)/2$ representation of SU(2). In Fig. 5, the eigenvalue distribution of the Gram matrix is plotted for a spin 100 representation, starting with a coherent state centred around a randomly chosen direction of the classical spin. We let the system evolve for rescaled times $\tau = 0.5$ and $\tau = 1$, first for a parameter choice corresponding to a very regular classical motion, next to almost global chaos. For the chaotic case, the Marchenko-Pastur distribution with the same τ is superimposed on the plots.

If we start with a sequence $\boldsymbol{\rho} = \{\rho_0, \rho_1, \dots, \rho_{K-1}\}$ of mixed states instead of pure ones, then it is not so obvious how to determine mutual overlaps. A possible approach is to jointly purify the ρ_j, i.e. to write for all j

$$\operatorname{Tr}\rho_j x = \langle \varphi_j, x\otimes \mathbb{1}\varphi_j\rangle \qquad \text{with} \qquad \varphi_j \in \mathfrak{H}\otimes\mathfrak{K}$$

and to use the Gram matrix for the purifications. Unfortunately, the resulting empirical eigenvalue distribution depends on the chosen purification. We need

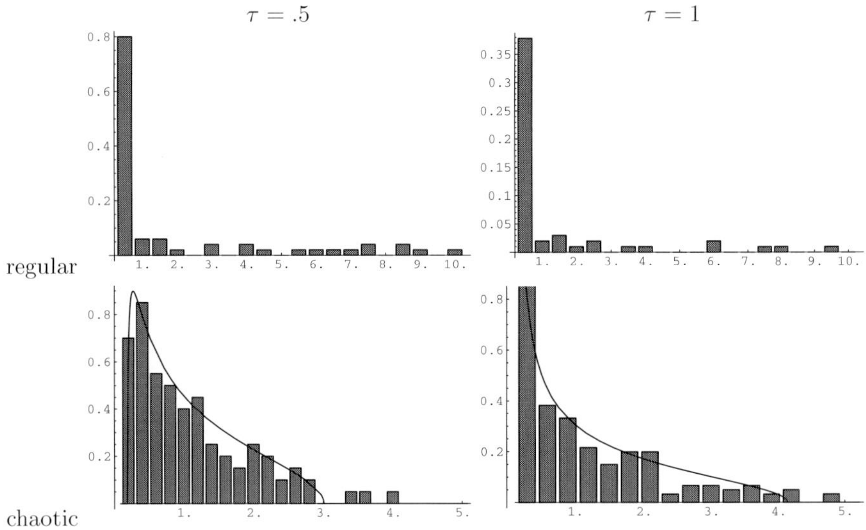

Fig. 5. Eigenvalue densities for the quantum kicked top in regular and chaotic regime

therefore a selection criterion, such as minimising the volume of the parallelotope spanned by the φ_j. For faithful ρ_j, we have to minimise

$$(u_0, u_1, \ldots, u_{K-1}) \mapsto \det[\mathrm{Tr}\, \rho_i^{\frac{1}{2}} \rho_j^{\frac{1}{2}} u_j^* u_i]_{ij} \tag{9}$$

over unitaries u_j. For two density matrices ρ and σ, one recovers the Uhlmann fidelity [22]

$$\Big(\mathrm{Tr}\sqrt{\rho^{\frac{1}{2}}\sigma\rho^{\frac{1}{2}}}\Big)^2$$

and for pure states the usual Gram matrix of above emerges. The variational principle (9) looks forbidding but there is some hope to resolve at least the case of mutually commuting density matrices [10].

References

1. R. Alicki, *Contribution to this volume.*
2. R. Alicki and M. Fannes, *Quantum Dynamical Systems*, Oxford University Press, Oxford (2001).
3. A. M. Chebotarev and F. Fagnola, J. Funct. Anal., **153**, 382–404 (1998).
4. J. Cuntz, Commun. Math. Phys., **57**, 173–185 (1977).
5. E.B. Davies, Rep. Math. Phys., **11**, 169–188 (1977).
6. M. De Cock, M. Fannes and P. Spincemaille, J. Phys. A: Math. Gen., **32**, 6547–6571 (1999).
7. F. Fagnola and R. Rebolledo, J. Math. Phys., **42**, 1296–1308 (2001).
8. M. Fannes, B. Nachtergaele and R.F. Werner, J. Funct. Anal., **120**, 511–534 (1994).
9. M. Fannes and P. Spincemaille, Phys. Lett., **A294**, 74–78 (2002).

10. M. Fannes and P. Spincemaille, Work in progress.
11. V. Gorini, A. Kossakowski and E.C.G. Sudarshan, J. Math. Phys., **17**, 821–825 (1976).
12. F. Haake, *Quantum Signatures of Chaos*, Springer, Berlin (1991).
13. F. Haake, M. Kuś and R. Scharf, Zeit. Physik, **B65**, 381–395 (1987).
14. M. Hirata, B. Sausol and S. Vaienti, Commun. Math. Phys., **206**, 33–55 (1999).
15. A.S. Holevo, Prob. Theory and Related Fields, **104**, 483–500 (1996).
16. K. Kraus, *States, Effects, and Operations*, Lect. Notes in Phys., **190** Springer, Berlin (1983).
17. G. Lindblad, Commun. Math. Phys., **48**, 119–130 (1976).
18. R.T. Powers, Canadian J. Math., **40**, 86–114 (1988).
19. K. Sinha, Operator Theory: Advances and Applications, **70**, 161–9, Basel. Birkhäuser. (1994).
20. H. Spohn, Lett. Math. Phys., **2**, 33 (1977).
21. J.-J. Stöckmann, *Quantum Chaos, An Introduction*, Cambridge University Press, Cambridge (1999).
22. A. Uhlmann, Rep. Math. Phys., **9**, 273–279 (1976).

Complete Positivity in Dissipative Quantum Dynamics

F. Benatti, R. Floreanini, and R. Romano

[1] Dipartimento di Fisica Teorica, Università di Trieste Strada Costiera 11, 34014 Trieste, Italy
[2] Istituto Nazionale di Fisica Nucleare, Sezione di Trieste, Dipartimento di Fisica Teorica, Università di Trieste Strada Costiera 11, 34014 Trieste, Italy
[3] Dipartimento di Fisica Teorica, Università di Trieste Strada Costiera 11, 34014 Trieste, Italy Istituto Nazionale di Fisica Nucleare, Sezione di Trieste

Abstract. The property of complete positivity is reviewed in connection with dissipation in elementary particle physics and neutron interferometry.

1 Introduction

The time-evolution of closed quantum systems is described by groups of linear maps that transform pure states, that is orthogonal projectors, into pure states. However, quantum systems can be treated as closed only if the influences of the enviroment which surrounds them can be neglected, otherwise they must be treated as open quantum systems. Typically, the environment contributes with background effects embodied by semigroups of linear maps that transform pure states into density matrices, that is into mixtures of projectors. These semigroups describe the so-called reduced dynamics of the open system by taking into account the double role of the environment as a source of dissipation and of noise. Such an approach has been applied to a large variety of different contexts ranging from quantum relaxation [1], to quantum optics [2-4], to foundational aspects of quantum mechanics [5-6] and lately to elementary particle physics [7-12].

We shall denote by S the subsystem in interaction with an environment E; physics requires the reduced dynamics of S to consist of linear maps Γ_t on its states ρ that preserve the positivity of the eigenvalues; such linear maps are called positive. In this way the spectra of $\Gamma_t[\rho]$ are at all times $t \geq 0$ probability distributions and the operators $\Gamma_t[\rho]$ can consistently be used as quantum states, in agreement with the statistical interpretation of quantum mechanics. However, the derivation of a reduced dynamics is based on Markov approximations that, in general, do not lead to positive Γ_t [13-17] and, to ensure positivity, one ends up with Γ_t that are not only positive, but *completely positive* [13,18-19].

Complete positivity of Γ_t is stronger than positivity; it ensures that, given an arbitrary N-level system S_N, an initial state ρ_{S+S_N} of the compound sysytem $S+S_N$ and the joint time-evolution maps $\Gamma_t \otimes \mathbf{1}_N$, then the spectrum of $(\Gamma_t \otimes \mathbf{1}_N)[\rho_{S+S_N}]$ remains positive for whatsoever initial state ρ_{S+S_N} of $S+S_N$. In the above expression, $\mathbf{1}_N$ means that the N-level system is left unaffected by the joint dynamics, whereas its factorized form $\Gamma_t \otimes \mathbf{1}_N$ indicates that the coupling between S and S_N is not dynamical, but only statistical, namely, there can be correlations between S and S_N described by entangled states of $S + S_N$. Entanglement is crucial [20]: positivity of Γ_t is sufficient to guarantee positivity of $\Gamma_t \otimes \mathbf{1}_N$ on separable states, that is on (mixtures of) tensor products of a state of S with a state of S_N, but not on entangled ones. In other words, if Γ_t is only positive, but not completely positive, then there are initial entangled states of $S + S_N$ whose spectra do not remain positive under $\Gamma_t \otimes \mathbf{1}_N$ and cannot then be interpretable as probability distributions.

Noticeably, the standard time-evolution of closed quantum systems is automatically completely positive, while this is not so for open system dynamics; the formal justification of why the latter should be completely positive is based on the possibility that the open system S has got entangled with an arbitrary, dynamically silent, N-level system S_N so that it effectively evolves in time not according to Γ_t, rather to $\Gamma_t \otimes \mathbf{1}_N$. Such an argument looks somewhat abstract and complete positivity is often criticised as a mathematical artifact [21], technically useful because of its formal consequences on the form of the generator of the time-evolution which must be of the Kossakowski-Lindblad form [22,23], but phyisically not strictly necessary or even not corresponding to the actual experimental expectations [14-16].

Much more appealing would be a physical context where entanglement is not only an abstract, uncontrollable possibility, but an actual experimental fact concerning two systems evolving in time within a same environment according to a factorized dynamics $\Gamma_t \otimes \Gamma_t$. This is exactly what occurs in experiments conducted on neutral K or B mesons produced in singlet-like entangled pairs in the so-called ϕ- or B-factories [9-12].

On the other hand, when the statistical coupling is not between two distinct systems, but between two degrees of freedom of a same physical system, only one of which is affected by some noisy background, then also arguing about the positivity of $\Gamma_t \otimes \mathbf{1}_N$ assumes a concrete physical meaning [24].

In the following we will discuss in detail the issue of complete positivity of linear maps by comparing it with positivity and presenting some concrete physical examples drawn from elementary particle phenomenology and quantum interferometry.

2 Complete Positivity

Although we shall later deal with physical systems that can be described by 2-dimensional Hilbert spaces, for sake of generality, we will introduce the notion of complete positivity for d-level systems S with observables from the algebra $M_d(\mathbf{C})$ of $d \times d$ matrices. Technically speaking, complete positivity is a property of linear maps Λ on $M_d(\mathbf{C})$.[1]

[1] The case of infinite dimensional systems is discussed in [23].

Definitions 2.1
a) $X \in M_d(\mathbf{C})$ is positive, $X \geq 0$, if $\langle\psi|X|\psi\rangle \geq 0$ for all $\psi \in \mathbf{C}^d$.
b) A linear map $\Lambda : M_d(\mathbf{C}) \longmapsto M_d(\mathbf{C})$ is *positive* if $X \geq 0 \Longrightarrow \Lambda[X] \geq 0$.
c) Given an arbitrary N-level system with observable algebra $M_N(\mathbf{C})$, Λ is *completely positive* if $\Lambda \otimes \mathbf{1}_N$ is positive on $M_d(\mathbf{C}) \otimes M_N(\mathbf{C})$ for all N, where $\mathbf{1}_N$ is the identity map.

Remarks 2.1
i) In the scheme of above, positivity corresponds to coupling S with 1-level systems, which is trivial; instead complete positivity concerns the coupling of S with N-level systems S_N it is not trivial and stronger than positivity. A theorem of Choi [25] states that it is enough to check positivity for $N = d$; namely, if S is a d-level system and $\Lambda \otimes \mathbf{1}_d$ is positive on $M_d(\mathbf{C}) \otimes M_d(\mathbf{C})$, then $\Lambda \otimes \mathbf{1}_N$ is positive for all N and thus Λ is completely positive.
ii) The structure of completely positive maps is fully characterized by a theorem of Stinespring [26]: a linear map $\Lambda : M_d(\mathbf{C}) \longmapsto M_d(\mathbf{C})$ is completely positive if and only if

$$\Lambda[X] = \sum_j V_j\, X\, V_j^* \tag{2.1}$$

where the V_j's and their adjoints V_j^* are $d \times d$ matrices with the possibly infinite sum converging in the norm topology.

The Heisenberg standard time-evolution of observables of the system S is given by a group of maps Λ_t

$$X \longmapsto \Lambda_t[X] := \mathrm{e}^{i\,H\,t}\, X\, \mathrm{e}^{-i\,H\,t} \tag{2.2}$$

generated by a Hamiltonian operator through

$$\partial_t X(t) = i\Big[H\,,\, X(t)\Big]\ . \tag{2.3}$$

Then, according to (2.1), Λ_t is completely positive.

The states of S are described by density matrices, that is by linear convex combinations of orthogonal projectors, and are represented by hermitian positive $d \times d$ matrices,

$$\rho = \sum_k \lambda_k |\psi_k\rangle\langle\psi_k| \quad \lambda_k \geq 0\ \sum_k \lambda_k = 1\ . \tag{2.4}$$

They can be diagonalized,

$$\rho = \sum_j r_j |r_j\rangle\langle r_j| \ \ \langle r_j|r_k\rangle = \delta_{jk} \tag{2.5}$$

and their spectrum $\{r_j\}$ is a discrete probability distribution, $r_j \geq 0$, $\mathrm{Tr}\rho = \sum_j r_j = 1$, with corresponding uncertainty measured by the von Neumann entropy

$$S(\rho) := -\mathrm{Tr}\rho\log\rho = \sum_j r_j \log r_j\ . \tag{2.6}$$

Given a state ρ, the mean values of observables X are calculated as

$$\mathrm{Tr}\Big(\rho X\Big) = \sum_k \lambda_k \langle\psi_k|X|\psi_k\rangle = \sum_j r_j \langle r_j|X|r_j\rangle \ . \tag{2.7}$$

From the time-evolution (2.2), the Schrödinger time-evolution of states, $\rho \longmapsto \Gamma_t[\rho]$, follows by duality,

$$\mathrm{Tr}(\Lambda_t[X]\,\rho) = \mathrm{Tr}(X\Gamma_t[\rho]) \ . \tag{2.8}$$

Remark 2.2. It is a consequence of (2.7) that positive observables must have positive mean values with respect to all states. Further, from (2.8) it follows that physically consistent time-evolutions must preserve the positivity of the spectrum of states at all times. Varying freely X among the orthogonal projections $|\psi\rangle\langle\psi|$, the positivity of Λ_t on the observables is equivalent to the positivity of its dual Γ_t on the states. Indeed, Λ_t positive implies $\langle\psi|\Gamma_t[\rho]|\psi\rangle = \mathrm{Tr}(\rho\Lambda_t[|\psi\rangle\langle\psi|]) \geq 0$ for all vector states ψ.

If the system S is coupled to an N-level system S_N, the two are statistically independent if their joint state is a tensor product state $\rho_S \otimes \rho_{S_N}$, or a mixture of them,

$$\rho_{S+S_N} = \sum_{ij} \lambda_{ij}\, \rho^i_S \otimes \rho^j_{S_N} \ , \quad \lambda_{ij} \geq 0 \ \sum_{ij} \lambda_{ij} = 1 \ . \tag{2.9}$$

Instead, correlations between S and S_N are described by *entangled* states of $S+S_N$, when they cannot be separated as in (2.9). Further, the dual of $\Lambda_t \otimes \mathbf{1}_N$ is $\Gamma_t \otimes \mathbf{1}_N$.

Proposition 2.1. If Λ_t is not completely positive, there exists an entangled state ρ_{S+S_d} of $S+S_d$ whose positivity is not preserved by $\Gamma_t \otimes \mathbf{1}_d$.

In fact, if Λ_t is not completely positive, because of Choi's theorem (see Remark 2.1), there must exist a positive observable $0 \leq X \in M_d(\mathbf{C}) \otimes M_d(\mathbf{C})$ such that $\Lambda_t \otimes \mathbf{1}_d[X]$ has a negative eigenvalue. Let ρ be the corresponding eigenprojection, then

$$\mathrm{Tr}\Big((\Lambda_t \otimes \mathbf{1}_d)[X]\,\rho\Big) = \mathrm{Tr}\Big(X(\Gamma_t \otimes \mathbf{1}_d)[\rho]\Big) \leq 0 \ . \tag{2.10}$$

Thus, $(\Gamma_t \otimes \mathbf{1}_d)[\rho]$ must have negative eigenvalues.

Remark 2.3. Positivity of Γ_t is sufficient to ensure that separable (tensor product) states remain positive: $\sum_{ij} \lambda_{ij}\Gamma_t(\rho^i_S) \otimes \rho^j_{S_d} \geq 0$. It is the presence of quantum entaglement that asks for complete positivity, that is only entangled states of $S+S_d$ can lose their positivity under $\Gamma_t \otimes \mathbf{1}_d$ when Γ_t is not completely positive (for a more general perspective, see [20]).

The dynamical maps Γ_t must be positive in order to guarantee that the eigenvalues of $\Gamma_t[\rho]$ be interpretable as probabilities at all times. Instead, Proposition

2.1 warns that if Γ_t is only positive, but not completely positive, then entangled states of $S+S_d$ may develop negative eigenvalues and thus lose physical meaning while evolving under $\Gamma \otimes \mathbf{1}_d$. Since the standard unitary time-evolution (2.2) is completely positive, problems can only arise if the presence of an enviroment makes Γ_t dissipative and not of the standard closed form dual to (2.2).

Being the other d-level system S_d, to which the S of interest may be coupled, totally arbitrary and dynamically silent, the necessity of positivity of maps of the form $\Gamma_t \otimes \mathbf{1}_d$, and thus of complete positivity of Γ_t, looks rather abstract. Consequently, complete positivity is often refused as a technical artifact rather than a physical constraint [14-16,21]. In the following section, a more physical argument will be given of why complete positivity is physically necessary.

3 Bidimensional Systems: Positivity

We now fix $d=2$, so that density matrices are represented by

$$\rho = \begin{pmatrix} \rho_1 & \rho_3 \\ \rho_4 & \rho_2 \end{pmatrix} \qquad \rho_4 = \rho_3^* \tag{3.1}$$

with $\mathrm{Tr}(\rho) = \rho_1 + \rho_2 = 1$ and $\mathrm{Det}(\rho) = \rho_1\rho_2 - |\rho_3|^2 \geq 0$ to guarantee normalization and positivity. From (2.8) and (2.2), the standard (Schrödinger) time-evolution of states of S is

$$\rho \longmapsto \rho(t) := \Gamma_t[\rho] = \mathrm{e}^{-i\,H\,t}\,\rho\,\mathrm{e}^{i\,H^*\,t} \tag{3.2}$$

where we allow for possibly non-hermitian Hamiltonians, $H \neq H^*$, in order to describe later on unstable particles; the generating equation is

$$\partial_t \rho(t) = -i\,H\,\rho(t) \;+\; i\,\rho(t)\,H^* \;. \tag{3.3}$$

The commutator that appears in (3.3) when $H = H^*$ is typical of a closed system time-evolution with conservation of energy and reversible behaviour. In case the system S interacts with an environment E, it is the system $S+E$ which is closed, with total Hamiltonian $H = H_S + H_E + H_I$, where $H_{S,E}$ are the Hamiltonian operators of system and environment alone, whereas H_I describes the interaction among the two of them. The (infinitely) many degrees of freedom of the environment can hardly be controlled; when the system S is weakly coupled to E, it is possible to extract a reduced, open system dynamics for S alone consisting of a semigroup of linear maps Γ_t, satisfying $\Gamma_t \circ \Gamma_s = \Gamma_{t+s}$ for $t, s \geq 0$ only [13-19].

We will not be concerned in this presentation with the difficult problem of a microscopic derivation of the generator of the reduced dynamics, rather we assume that it be of the form

$$\partial_t \rho(t) = -i\Big[H_S + H_D \;\; \rho(t)\Big] + L_D[\rho(t)] \tag{3.4}$$

and refer to the literature for a justification of this assumption [17-19].

Remark 3.1. The presence of the environment contributes with a redefinition (Lamb shift) H_D of the system Hamiltonian and with a linear map $\rho \longmapsto L_D[\rho]$ which cannot be written as in (3.3). $L_D[\cdot]$, that will be referred to as the "dissipative term", breaks time-reversal and allows projections, for which $\rho^2 = \rho$ at $t = 0$, to evolve into mixtures, such that $(\Gamma_t[\rho])^2 \neq \Gamma_t[\rho]$. Furthermore, the elimination of the degrees of freedom of the enviroment does not contribute to loss of probability in the subsystem S, so that $\mathrm{Tr}(\Gamma_t[\rho]) = \mathrm{Tr}\rho$ or $\mathrm{Tr}(L_D[\rho]) = 0$.

As previously remarked, physical consistency demands that Γ_t keep positive the states of S; in order to study the resulting constraints on L_D, it is convenient to express the states of S by means of the Pauli matrices σ_j, $j = 1, 2, 3$, plus the identity matrix $\sigma_0 := \begin{pmatrix} 1 & 0 \\ 0 & 1 \end{pmatrix}$ and represent ρ as a 4-dimensional ket-vector $|\rho\rangle$,

$$\rho = \begin{pmatrix} \rho_1 & \rho_3 \\ \rho_4 & \rho_2 \end{pmatrix} = \sum_{\mu=0}^{3} \rho^\mu \sigma_\mu \longmapsto |\rho\rangle = \begin{pmatrix} \rho^0 \\ \rho^1 \\ \rho^2 \\ \rho^3 \end{pmatrix} \tag{3.5a}$$

$$\rho^0 = \frac{\rho_1 + \rho_2}{2} \quad \rho^1 = \frac{\rho_3 + \rho_4}{2} \quad \rho^2 = \frac{\rho_4 - \rho_3}{2i} \quad \rho^3 = \frac{\rho_1 - \rho_2}{2} \; . \tag{3.5b}$$

The component ρ^0 is half the trace of ρ and the projections relative to the positive eigenvalues of the Pauli matrices σ_j correspond to the vectors $|j\rangle$, where $|1\rangle = 1/2(1,1,0,0)$, $|2\rangle = 1/2(1,0,1,0)$, $|3\rangle = 1/2(1,0,0,1)$, while the most mixed state $\sigma_0/2$ corresponds to $|0\rangle = 1/2(1,0,0,0)$.

Being linear operators on the state space, both the commutator in (3.4), that we will denote by L_H, and the dissipative term L_D act on $|\rho\rangle$ as 4×4 matrices, $\mathcal{H}$ and $\mathcal{D}$ respectively, so that the equation of motion reads

$$\partial_t |\rho(t)\rangle = \Big(\mathcal{H} + \mathcal{D}\Big) |\rho(t)\rangle \tag{3.6}$$

and it is solved by the semigroup of 4×4 matrices $\mathcal{G}_t = \exp(t(\mathcal{H} + \mathcal{D}))$.

Proposition 3.1. Necessary condition for the dynamical maps Γ_t to be positivity preserving is that

$$\mathcal{H} = -2 \begin{pmatrix} 0 & 0 & 0 & 0 \\ 0 & 0 & h^3 & -h^2 \\ 0 & -h^3 & 0 & h^1 \\ 0 & h^2 & -h^1 & 0 \end{pmatrix} \quad \text{and} \quad \mathcal{D} = -2 \begin{pmatrix} 0 & 0 & 0 & 0 \\ u & a & b & c \\ v & b & \alpha & \beta \\ w & c & \beta & \gamma \end{pmatrix} \tag{3.7a}$$

be real matrices, with

$$\mathcal{D}^{(3)} := \begin{pmatrix} a & b & c \\ b & \alpha & \beta \\ c & \beta & \gamma \end{pmatrix} \tag{3.7b}$$

positive-definite.

The proof of the previous assertion goes as follows; since the environment E is supposed to affect the dynamics of S by the addition of a linear contribution L to the commutator with H_S, in the vector representation (3.5), L becomes a 4×4 matrix $\mathcal{L}=[\mathcal{L}_{\mu\nu}]$, while the commutator $-i[H_S\ \cdot]$, acts as an antisymmetric real matrix $\mathcal{H}_S$. Probability conservation asks that $\rho^0(t)$ (the trace of $\rho(t)$) remains costant, whence $\sum_{\nu=0}^{3}\mathcal{L}_{0\nu}\rho^\nu=0$ at time $t=0$ and $\mathcal{L}_{0\nu}=0$ for $\nu=0,1,2,3$, follows from varying arbitrarily the state components ρ^ν.

In order to preserve the hermiticity of $\rho(t)$ the components $\rho^j(t)$ must be real, so that differentiating $\rho^j(t)-(\rho^j)^*(t)=0$ at $t=0$ and varying the initial state it follows that the entries $\mathcal{L}_{\mu j}$ must be real.

The real 3×3 submatrix $\mathcal{L}^{(3)}:=[\mathcal{L}_{ij}]$, $i,j=1,2,3$, can be decomposed as the sum of an antisymmetric component $\mathcal{H}_D:=[(\mathcal{L}_{ij}-\mathcal{L}_{ji})/2]$, corresponding to the commutator with a Hamiltonian, and a symmetric one $\mathcal{L}_D^{(3)}:=[(\mathcal{L}_{ij}+\mathcal{L}_{ji})/2]$. The dynamics is thus generated by the antisymmetric 4×4 matrix $\mathcal{H}:=\mathcal{H}_S+\mathcal{H}_D$ plus a 4×4 real matrix $\mathcal{L}_D$.

Since the trace is conserved it follows that $\rho^0(t)=1/2$, therefore positivity of the spectrum is preserved at all times if and only if $\mathrm{Det}(\rho(t))=1/4-\sum_{j=1}^{3}(\rho^j(t))^2\geq 0$. We now set $\rho:=\rho(0)$ and use

$$\frac{\mathrm{dDet}(\rho(t))}{\mathrm{d}t}\bigg|_{t=0}=-2\sum_{ij=1}^{3}(\mathcal{L}_D)_{ij}\rho^i\rho^j-\frac{1}{2}\sum_{j=1}^{3}\rho^j\mathcal{L}_{j0} \tag{3.8}$$

to which $\mathcal{H}$ does not contribute because of antisymmetry. If $|\rho\rangle=(1/2,\rho^1,\rho^2,\rho^3)$ is a pure state, $\mathrm{Det}(\rho)=0$ and the right hand side of (3.8) cannot be negative, otherwise a negative eigenvalue would appear for $t>0$. The same inequality must hold for the state $|\tilde\rho\rangle=(1/2,-\rho^1,-\rho^2,-\rho^3)$, whence, adding the two inequalities, the inhomogeneous term in (3.8) drops out. By varying ρ^j while keeping $\sum_j(\rho^j)^2=1/4$, from $\sum_{ij=1}^{3}(\mathcal{L}_D)_{ij}\rho^i\rho^j\geq 0$ one gets that the real symmetric matrix $\mathcal{D}^{(3)}$ must necessarily be positive, therefore that the following inequalities must be fulfilled,

$$\begin{cases}a\geq 0\\ \alpha\geq 0\\ \gamma\geq 0\end{cases}\quad\begin{cases}a\alpha\geq b^2\\ a\gamma\geq c^2\\ \alpha\gamma\geq\beta^2\end{cases}\quad \mathrm{Det}\mathcal{D}^{(3)}\geq 0\ . \tag{3.9}$$

These conditions are not sufficient for the positivity of the evolving state $\rho(t)$ to be preserved because of the presence of the terms $\mathcal{L}_{j0}$ that are related to the existence of time-invariant states; indeed, for those states

$$(\mathcal{H}+\mathcal{D})|\rho(t)\rangle=0\ . \tag{3.10}$$

If $u=v=w=0$ the only time-invariant state is the most mixed state $\tau=1/2(1,0,0,0)$ with von Neumann entropy (2.6) $S(\tau)=\log 2$. Vice versa, if the first column of $\mathcal{D}$ is not zero, then the most mixed state is not time-invariant and its entropy must decrease. We thus conclude

Proposition 3.2. Necessary and sufficient condition for the dynamics $\mathcal{G}_t$ generated by (3.6) to preserve the positivity of states of two-level systems S and to increase their entropy is that

$$\mathcal{H} = -2 \begin{pmatrix} 0 & 0 & 0 & 0 \\ 0 & 0 & h^3 & -h^2 \\ 0 & -h^3 & 0 & h^1 \\ 0 & h^2 & -h^1 & 0 \end{pmatrix} \quad \text{and} \quad \mathcal{D} = -2 \begin{pmatrix} 0 & 0 & 0 & 0 \\ 0 & a & b & c \\ 0 & b & \alpha & \beta \\ 0 & c & \beta & \gamma \end{pmatrix} \tag{3.11}$$

with $\mathcal{D}^{(3)} = \begin{pmatrix} a & b & c \\ b & \alpha & \beta \\ c & \beta & \gamma \end{pmatrix}$ positive definite.

4 Bidimensional Systems: Complete Positivity

In this section we consider positive and entropy increasing dynamical maps Γ_t generated by $\mathcal{H} + \mathcal{D}$ as in Proposition 3.2 and investigate which further constraints complete positivity imposes upon the $3 + 6$ real parameters that characterize the dissipative time-evolution.

Proposition 4.1. Necessary and sufficient conditions to be fulfilled by the parameters a, b, c, α, β and γ in order that the time-evolution $\mathcal{G}_t$ generated by (3.6) be completely positive and entropy increasing are

$$\begin{aligned} &2R \equiv \alpha + \gamma - a \geq 0 \qquad RS \geq b^2 , \\ &2S \equiv a + \gamma - \alpha \geq 0 \qquad RT \geq c^2 , \\ &2T \equiv a + \alpha - \gamma \geq 0 \qquad ST \geq \beta^2 , \\ &RST \geq 2\, bc\beta + R\beta^2 + Sc^2 + Tb^2 . \end{aligned} \tag{4.1}$$

The proof is based on Lindblad's theorem [22,23] which states that Γ_t is completely positive **if and only if** L_D in (3.4) is of the Kossakowski-Lindblad form

$$L_D[\rho] = \sum_{i,j=0}^{3} C_{ij} \Big[\sigma_i \rho \sigma_j - \frac{1}{2} \Big\{ \sigma_j \sigma_i \ \rho \Big\} \Big] \tag{4.2}$$

with $[C_{ij}]$ a positive 3×3 matrix; further, increasing von Neumann entropy (2.6) forces the matrix $[C_{ij}]$ to be real symmetric [27]. In terms of the entries C_{ij}, the matrix $\mathcal{D}$ in (3.11) reads

$$\mathcal{D} = \begin{pmatrix} 0 & 0 & 0 & 0 \\ 0 & C_{22} + C_{33} & -C_{12} & -C_{13} \\ 0 & -C_{12} & C_{11} + C_{33} & -C_{23} \\ 0 & -C_{13} & -C_{23} & C_{11} + C_{22} \end{pmatrix} \ ; \tag{4.3}$$

thus, the positivity of $[C_{ij}]$, which is necessary and sufficient for the complete positivity of Γ_t generated by (3.4), results in the necessary and sufficient inequalities (4.1).

Vice versa, given a 4×4 matrix $\mathcal{D}$ as in (3.11), one can write down the dissipative operator

$$L_D[\rho] = \sum_{i,j=1}^{3} C_{ij} \left[\sigma_j \rho \sigma_i - \frac{1}{2} \left\{ \sigma_j \sigma_i \ \rho \right\} \right] \tag{4.4}$$

with

$$[C_{ij}] = \frac{1}{2} \begin{pmatrix} \alpha + \gamma - a & -2b & -2c \\ -2b & a + \gamma - \alpha & -2\beta \\ -2c & -2\beta & a + \alpha - \gamma \end{pmatrix} . \tag{4.5}$$

Thus, if the inequalities (4.1) are satisfied, $[C_{ij}]$ is positive definite and Γ_t generated by (3.8) completely positive because of Lindblad's theorem.

Remark 4.1. Complete positivity does not constrain the parameters $h_{1,2,3}$ that specify the Hamiltonian term in (3.4). Indeed, as already observed, the time-evolution generated by the commutator with a Hamiltoninan operator is automatically completely positive.

From the previous discussion, it appears that complete positivity has the advantage of fixing, through Lindblad's theorem, the Kossakowski-Lindblad form (4.2) of the dissipative term in the generator of the reduced dynamics. The question remains, however, why should the request of complete positivity be imposed upon Γ_t? The motivation provided by Proposition 2.1 refers to the necessity of avoiding that the dynamical maps $\Gamma_t \otimes \mathbf{1}_2$ become inconsistent when S is coupled to another arbitrary 2-level system. More physically, one might consider two two-level systems S that are prepared in an initial entangled state and then let evolve independently within a same environment. In such a case the states of the system $S+S$ transform in time according to the tensor product $\Omega_t := \Gamma_t \otimes \Gamma_t$ of the single-particle dynamical semigroups Γ_t and physical consistency demands that the semigroups of dynamical maps Ω_t preserve the positivity of all states of $S+S$.

Remark 4.2. If Γ_t is completely positive, then not only $\Gamma_t \otimes \Gamma_t$, but all dissipative time-evolutions on compound systems $S + S + \cdots + S$ of the form $\Gamma_t \otimes \Gamma_t \otimes \cdots \Gamma_t$ are automatically positive. Instead, if we ask a generic tensor product transformation $\Lambda \otimes \Lambda$ on $S+S$ to be positive, this does not imply Λ to be completely positive as we show below.

We consider the transposition on 2×2 matrices; in the representation where σ_3 is diagonal, it acts as follows

$$\mathcal{T} : \ \rho = \begin{pmatrix} \rho_1 & \rho_3 \\ \rho_4 & \rho_2 \end{pmatrix} \mapsto \mathcal{T}[\rho] = \begin{pmatrix} \rho_1 & \rho_4 \\ \rho_3 & \rho_2 \end{pmatrix} . \tag{4.6}$$

Denoting by $|\uparrow\rangle = \begin{pmatrix} 1 \\ 0 \end{pmatrix}$ and $|\downarrow\rangle = \begin{pmatrix} 0 \\ 1 \end{pmatrix}$ the eigenstates of σ_3, let us consider the compound system consisting of two two-level systems S and the projector onto the antisymmetric state $(|\uparrow\downarrow\rangle - |\downarrow\uparrow\rangle)/\sqrt{2}$,

$$P_A := \frac{1}{2}\left[\begin{pmatrix}1&0\\0&0\end{pmatrix}\otimes\begin{pmatrix}0&0\\0&1\end{pmatrix}+\begin{pmatrix}0&0\\0&1\end{pmatrix}\otimes\begin{pmatrix}1&0\\0&0\end{pmatrix}\right.$$
$$\left.-\begin{pmatrix}0&1\\0&0\end{pmatrix}\otimes\begin{pmatrix}0&0\\1&0\end{pmatrix}-\begin{pmatrix}0&0\\1&0\end{pmatrix}\otimes\begin{pmatrix}0&1\\0&0\end{pmatrix}\right]=\frac{1}{2}\begin{pmatrix}1&0&0&-1\\0&0&0&0\\0&0&0&0\\-1&0&0&1\end{pmatrix}. \quad (4.7)$$

One finds that

$$\mathcal{T}\otimes\mathbf{1}_2[P_A]=\frac{1}{2}\begin{pmatrix}1&0&0&0\\0&0&-1&0\\0&-1&0&0\\0&0&0&1\end{pmatrix} \quad (4.8)$$

has eigenvalues $\pm 1/2$ and positivity is not preserved. However, $\mathcal{T}\otimes\mathcal{T}[P_A]=P_A$. In fact, $\mathcal{T}\otimes\mathcal{T}$ is a positive linear map as can be proved in d-dimensions by fixing an orthonormal basis $|j\rangle$, $j=1,2,\ldots,d$, and acting on the projection onto a generic vector state $|\psi\rangle=\sum_{i,j=1}^{d}\psi(i,j)|i\otimes j\rangle$:

$$\begin{aligned}\mathcal{T}\otimes\mathcal{T}[|\psi\rangle\langle\psi|] &= \sum_{i,j,\ell,k=1}^{d}\psi(i,j)\psi(\ell,k)^*\mathcal{T}\otimes\mathcal{T}[|i\otimes j\rangle\langle\ell\otimes k|]\\ &= \sum_{i,j,\ell,k=1}^{d}\psi(i,j)\psi(\ell,k)^*|\ell\otimes k\rangle\langle i\otimes j| = |\psi^*\rangle\langle\psi^*| \end{aligned}\quad (4.9)$$

where $|\psi^*\rangle=\sum_{i,j=1}^{d}\psi(i,j)^*|i\otimes j\rangle$.

It thus follows that, while the positivity of $\Gamma_t\otimes\mathbf{1}_d$ on the states of $S+S_d$ is equivalent to complete positivity of Γ_t, it need not be true, in general, that the positivity of $\Lambda\otimes\Lambda$ on $S+S_d$ implies the complete positivity of Λ. However, considering a semigroup of Γ_t, we observe that each of its element is continuously connected with the identity, while the transposition cannot. Indeed, by using the semigroup property and the continuity (in t) of Γ_t, we have, in the two-dimensional case,

Proposition 4.2. Given the time-evolution maps Γ_t generated by (3.4) with $\mathcal{D}$ as in (3.11), the tensor product $\Omega_t=\Gamma_t\otimes\Gamma_t$ preserves the positivity of states of $S+S$, with S a two-level sysytem, if and only if Γ_t is completely positive.

As observed in Remark 4.2, one implication is immediate; in order to check the other, we consider again the one-dimensional projector P_A in (4.7). Given another vector state ψ such that $P_A|\psi\rangle=0$, we must have

$$\frac{\mathrm{d}}{\mathrm{d}t}\langle\psi|\Omega_t[P_A]|\psi\rangle\geq 0 \quad (4.10)$$

at $t=0$. In fact, were this not true, the expectation value $\langle\psi|\Omega_t[P_A]|\psi\rangle$ which is 0 at $t=0$ would become negative despite the positivity of P. Varying freely ψ

in the three dimensional linear space orthogonal to P_A, one gets the inequalities (4.1), whence the complete positivity of Γ_t.

Remark 4.3. The above result can be extended to more general non-symmetric dissipative matrix $\mathcal{D}$ of the form (3.7), without asking for entropy increase [28].

5 Complete Positivity in Particle Physics

Physical realizations of what has been discussed in the previous section are provided by K or B neutral mesons which can be phenomenologically described by two-level systems. Such particles can be generated in entangled pairs with their anti-particles flying away in opposite directions without interacting with each other. In the hypothesis that at Planck's scale, that is at energies of the order of the Planck's mass $M_P = 10^{19}$ GeV, the fluctuating space-time geometry affects the time-development of elementary particles as a noisy background, states of entangled neutral mesons may be thought to transform according to the tensor product $\Omega_t = \Gamma_t \otimes \Gamma_t$ of two dynamical maps describing the same dissipative reduced dynamics.

Such a possibility may be checked in actual experiments, *e.g.* experiments on violations of assumed discrete symmetries of nature like *CP* and even *CPT* [7-12]. Indeed, the parameters a, b, c, α, β and γ that specify the dissipative contribution $\mathcal{D}$ to the matrix form of L_D, can be roughly estimated to be of the order M^2/M_P, where M is the mass of the neutral meson considered, $M_K = 0.5$ GeV for the neutral Kaons, $M_B = 5$ GeV for the B-mesons. The accuracies of actual and planned experiments based on K and B neutral mesons are such that noisy effects of the order M^2/M_P are testable and with them whether Γ_t should or might be completely positive.

In the following, we consider neutral K-mesons K^0-$\overline{K^0}$ whose physical states are effectively associated with (mixtures of) projectors onto the vectors of the Hilbert space $\mathbf{C}^2$ with basis

$$|K^0\rangle = \begin{pmatrix}1\\0\end{pmatrix} \qquad |\overline{K^0}\rangle = \begin{pmatrix}0\\1\end{pmatrix} . \tag{5.1}$$

Their time-evolution as decaying systems in vacuum is phenomenologically described by the non-hermitian Weisskopf-Wigner Hamiltonian, $H_{\text{eff}} = M - \frac{i}{2}\Gamma$, through the equation of motion (3.3).

Remark 5.1. Since they are in the form (2.1), the maps Γ_t in (3.2) are completely positive also when $H \neq H^*$. In this latter case, there is loss of probability in time as it should be since particles decay; indeed, $\partial_t \text{Tr}\rho(t) = -\text{Tr}(\rho(t)\Gamma) \leq 0$. However, vector states $|\psi\rangle$ are mapped into $|\psi_t\rangle := \exp(-i\,H\,t)|\psi\rangle$, so that, despite their decreasing norms, they remain vector states. Indeed, to allow for pure states being turned into mixed states, one has to go beyond the standard equations of motion.

In presence of a noisy background, possibily of gravitational origin, which causes dissipation and noise through an entropy increasing mechanism, the equation of motion can be taken to be [9-12]

$$\partial_t \rho(t) = -i\, H_{\rm eff}\, \rho(t) \;+\; i\, \rho(t)\, H_{\rm eff}^\dagger \;+\; L_D[\rho(t)] \tag{5.2}$$

where L_D, as a 4×4 matrix $\mathcal{D}$, is as in (3.11) and fulfils inequalities (4.1). We remark that in [7,8] only positivity, and not complete positivity, was enforced.

Remark 5.2. The request that the von Neumann entropy should always increase comes from Hawking's picture of the foamy structure of space-time geometry at Planck's scale, according to which virtual black holes should absorb information and increase uncertainty.

The physics of K-mesons is studied by measuring observables corresponding *e.g.* to Kaons transforming into pions, $K \to 2\pi$, $K \to 3\pi$, or into semileptonic states $K \to \pi\ell\nu$. Such observables are proportional to projectors

$$X_f = A_f \begin{bmatrix} 1 & r_f \\ r_f^* & |r_f|^2 \end{bmatrix} \qquad r_f = \frac{\mathcal{A}(\overline{K^0} \to f)}{\mathcal{A}(K^0 \to f)} \tag{5.3}$$

where $\mathcal{A}(K^0 \to f)$ is the complex transition-amplitude for K^0 decaying into the final state f. Their mean values can be experimentally studied by measuring so-called asymmetries, of which an example is

$$A(t) = \frac{\mathrm{Tr}\Big[\rho_{\overline{K}^0}(t) X_{\bar f}\Big] - \mathrm{Tr}\Big[\rho_{K^0}(t) X_f\Big]}{\mathrm{Tr}\Big[\rho_{\overline{K}^0}(t) X_{\bar f}\Big] + \mathrm{Tr}\Big[\rho_{K^0}(t) X_f\Big]}\ ; \tag{5.4}$$

it determines the differences in the rates of decays of K^0 into f and anti-K^0 into anti-f.

In the so-called ϕ-factories, many events consist of spin 1 ϕ-mesons decaying into two neutral Kaons; because of angular momentum conservation, the latter fly apart in the antisymmetric state

$$|\psi_A\rangle = \frac{1}{\sqrt{2}}\Big(|K^0, -p\rangle \otimes \overline{K^0}, p\rangle - |\overline{K^0}, -p\rangle \otimes |K^0, p\rangle\Big)\ . \tag{5.5}$$

Setting $P_1 := |K^0\rangle\langle K^0|$, $P_2 := |\overline{K^0}\rangle\langle\overline{K^0}|$, $P_3 := |K^0\rangle\langle\overline{K^0}|$ and $P_4 := |\overline{K^0}\rangle\langle K^0|$, the corresponding density matrix is as in (4.7),

$$P_A = \frac{1}{2}\Big[P_1 \otimes P_2 \;+\; P_2 \otimes P_1 \;-\; P_3 \otimes P_4 \;-\; P_4 \otimes P_3\Big]\ . \tag{5.6}$$

The two K-mesons evolve independently and may decay at different times giving rise to a double-decay statistics which is fully described by mean values with respect to the operator

$$\begin{aligned} P_A(t_1, t_2) := \big(\Gamma_{t_1} \otimes \Gamma_{t_2}\big)\big[P_A\big] = \frac{1}{2}\Big[&P_1(t_1) \otimes P_2(t_2) + P_2(t_1) \otimes P_1(t_2) \\ &- P_3(t_1) \otimes P_4(t_2) - P_4(t_1) \otimes P_3(t_2)\Big]\ . \end{aligned} \tag{5.7}$$

Remark 5.3. According to Lindblad's result [23], once the dissipative term fulfils inequalities (4.1), despite the non-hermitian character of the Wisskopf-Wigner Hamiltonian, the time-evolution generated by (5.2) is completely positive so that $P_A(t_1, t_2)$ is a density matrix at all times t_1, t_2. This is not generally true for Γ_t not completely positive.

From the previous considerations, it turns out that a ϕ-factory provides a quantum interferometer that allows to measure correlation functions of the form

$$\mathcal{P}(f_1, t_1; f_2, t_2) := \mathrm{Tr}\Big[\Big(X_{f_1} \otimes X_{f_2}\Big) P_A(t_1, t_2)\Big] \tag{5.8}$$

where one can vary times and decay product states. Without dissipation and noise, the antisymmetric character of P_A is preserved by the Weisskopf-Wigner equation of motion (3.3) and, choosing equal times and equal decay products, it turns out that $\mathcal{P}(f, t; f, t) = 0$. This is no longer true in presence of a coupling to a background; as a consequence, it is possible to determine, or at least to put an upper bound to the phenomenological parameters a, b, c, α, β and γ that describes the coupling.

Moreover, correlated neutral Kaons may evidentiate the kind of physical inconsistencies that would plague the time-evolution in case the single-Kaon dissipative dynamics is not completely positive. For sake of simplicity, in (5.2) we let $H_{\mathrm{eff}} = 0$ and $a = b = c = 0$ and choose $\alpha \neq \gamma$; then, the inequalities (4.1) cannot be satisfied and the single-Kaon time-evolution Γ_t cannot be completely positive while, with $\beta^2 \leq \alpha\gamma$, Γ_t preserves positivity (see Proposition 3.2). The action of Γ_t is explicitly given by

$$\Gamma_t[\rho] = \begin{pmatrix} \rho_0(t) \\ \rho_1(t) \\ \rho_2(t) \\ \rho_3(t) \end{pmatrix} = \begin{bmatrix} 1 & 0 & 0 & 0 \\ 0 & 1 & 0 & 0 \\ 0 & 0 & A(t) & B(t) \\ 0 & 0 & B(t) & C(t) \end{bmatrix} \begin{pmatrix} \rho_0(0) \\ \rho_1(0) \\ \rho_2(0) \\ \rho_3(0) \end{pmatrix} \tag{5.9}$$

where

$$\begin{aligned} A(t) &= \frac{1}{\lambda_+ - \lambda_-}\left[(\lambda_+ + 2\alpha)\, e^{\lambda_-\, t} - (\lambda_- + 2\alpha)\, e^{\lambda_+\, t}\right] , \\ B(t) &= \frac{2\beta}{\lambda_+ - \lambda_-}\left(e^{\lambda_-\, t} - e^{\lambda_+\, t}\right) , \\ C(t) &= \frac{1}{\lambda_+ - \lambda_-}\left[(\lambda_+ + 2\alpha)\, e^{\lambda_+\, t} - (\lambda_- + 2\alpha)\, e^{\lambda_-\, t}\right] \end{aligned} \tag{5.10}$$

and $\lambda_\pm = -(\alpha + \gamma) \pm \sqrt{(\alpha - \gamma)^2 + 4\beta^2}$.

Physical consistency requires the mean values with respect to the evolving entangled state $P_A(t, t)$ of any observable to remain positive; in particular, this must hold for the projectors $|\Phi_+\rangle\langle\Phi_+|$ and $|\Psi_+\rangle\langle\Psi_+|$ onto the entangled states

$$|\Phi_+\rangle = \frac{1}{\sqrt{2}}\left[\begin{pmatrix} 1 \\ 0 \end{pmatrix} \otimes \begin{pmatrix} 1 \\ 0 \end{pmatrix} + \begin{pmatrix} 0 \\ 1 \end{pmatrix} \otimes \begin{pmatrix} 0 \\ 1 \end{pmatrix}\right]$$

$$|\Psi_+\rangle = \frac{1}{\sqrt{2}}\left[\begin{pmatrix}1\\0\end{pmatrix}\otimes\begin{pmatrix}0\\1\end{pmatrix}+\begin{pmatrix}0\\1\end{pmatrix}\otimes\begin{pmatrix}1\\0\end{pmatrix}\right]. \tag{5.11}$$

Instead, one computes

$$\langle\Phi_+|P_A(t,t)|\Phi_+\rangle = -\langle\Psi_+|\rho_A(t,t)|\Psi_+\rangle = \frac{1}{2}\big(A^2(t) - C^2(t)\big) \tag{5.12}$$

which is a contradiction unless $\alpha = \gamma$.

Remark 5.4. The above argument is slightly oversimplified with respect to the actual setting at ϕ-factories; it can nevertheless be adjusted to account for non-zero Weisskopf-Wigner Hamiltonians and a larger variety of parameters a, b, c, α, β and γ [11]. It turns out that the consequences of lack of complete positivity are the same: from the given time-evolution one predicts negative mean values for positive observables which is physically unacceptable. Moreover, such predictions are not mathematical artifacts, rather they appear to be in reach of actual and future experiments at ϕ or B-factories.

6 Complete Positivity in Neutron Interferometry

Recently, experiments based on neutron interferometry have been proposed [29] to test the hypothesis of non-contextuality in quantum mechanics; the idea is to check whether a Bell-like inequality of the Clauser-Horne-Shimony-Holt form is violated or not. Such an inequality is derived from the assumption that the measured values of physical observables are completely specified by the state of the system prior to measurement and that the actual measurement outcomes do not depend on the context, namely on whether other commuting observables are simultaneously measured.

Differently from Bell-locality tests based on entangled physical systems, the above experiments involve two degrees of freedom of a same physical system; one degree of freedom is translational, related to the two possible paths followed by neutrons inside the interferometer, the other is the spin. If the neutron spin is coupled to a classical, fluctuating external field, its time-evolution is that of an open system described by a dynamical semigroup Γ_t [14,27]; since the other degree of freedom is the spatial one which is not affected by any external field, the physical meaning of complete positivity comes to the fore when we study a time-evolution of the form $\mathbf{1}_2 \otimes \Gamma_t$. With respect to previous sections, in the above time-evolution the non-trivial factor is now the second one, but this does not alter any of the considerations concerning positivity and complete positivity.

Concretely, one is concerned with what happens if the interferometric apparatus, proposed in [29] for non-contextuality tests, is placed in weak gaussian stochastic magnetic fields, that can be modulated to provide different experimental realizations of open system dynamics. In particular, one can give rise to reduced dynamics $\mathbf{1}_2 \otimes \Gamma_t$ with Γ_t completely positive, positive, but not completely positive, and also not even positive [24].

The experimental setting is as follows [29]: an incoming beam of neutrons with spin along the positive z-direction passes through a beam splitter with transmission and reflection coefficients p and q with $|p|^2 + |q|^2 = 1$. The beam gets divided into two components that follow two spatially separated paths u and d; both the spin and the translational degree of freedom are described by 2-dimensional Hilbert spaces, the former with basis vectors $|\uparrow_z\rangle$ and $|\downarrow_z\rangle$, the latter with basis vectors $|\psi_u\rangle$ and $|\psi_d\rangle$ corresponding to the two possible macroscopic paths.

One can choose $q = -p = 1/\sqrt{2}$ and, by making the beam u-component undergo a spin-flip $|\uparrow_z\rangle \mapsto |\downarrow_z\rangle$, prepare an initial beam state

$$|\Psi\rangle = \frac{|\psi_u\rangle \otimes |\downarrow_z\rangle \;-\; |\psi_d\rangle \otimes |\uparrow_z\rangle}{\sqrt{2}} \tag{6.1}$$

where now two-degrees of freedom of a same system, and not two different systems, are entangled. Again, the vector state (6.1) corresponds to the one-dimensional projector in (4.7)

$$P_A = \frac{1}{2}\Big[P_1 \otimes Q_2 + P_2 \otimes Q_1 + P_3 \otimes Q_4 + P_4 \otimes Q_3\Big] \tag{6.2}$$

with

$$\begin{aligned} P_1 &:= |\psi_u\rangle\langle\psi_u| & P_2 &:= |\psi_d\rangle\langle\psi_d| \\ P_3 &:= |\psi_u\rangle\langle\psi_d| & P_4 &:= |\psi_d\rangle\langle\psi_u| \end{aligned} \tag{6.3a}$$

$$\begin{aligned} Q_1 &:= |\uparrow_z\rangle\langle\uparrow_z| & Q_2 &:= |\downarrow_z\rangle\langle\downarrow_z| \\ Q_3 &:= |\uparrow_z\rangle\langle\downarrow_z| & Q_4 &:= |\downarrow_z\rangle\langle\uparrow_z| \,. \end{aligned} \tag{6.3b}$$

The initial beam moves through the interferometer where its spin degree of freedom feels the effects of the stochastic magnetic field, while the translational degree of freedom is untouched; therefore, the intial pure state gets transformed into a statistical mixture

$$P_A(t) := (\mathbf{1}_2 \otimes \Gamma_t)[P_A] = \sum_{i,j=1}^{4} \rho_{ij}(t)\, P_i \otimes Q_j \,. \tag{6.4}$$

At the exit of the interferometer, a second beam splitter recombines the translational components and shifts the u component by an angle φ,

$$\begin{pmatrix} |\psi_u\rangle \\ |\psi_d\rangle \end{pmatrix} \longmapsto \begin{pmatrix} |\psi_u(\vartheta,\varphi)\rangle \\ |\psi_d(\vartheta,\varphi)\rangle \end{pmatrix} = \begin{pmatrix} \mathrm{e}^{-i\varphi}\sin\vartheta & \cos\vartheta \\ \mathrm{e}^{-i\varphi}\cos\vartheta & -\sin\vartheta \end{pmatrix} \begin{pmatrix} |\psi_u\rangle \\ |\psi_d\rangle \end{pmatrix} \tag{6.5}$$

with reflection and transmission probabilities $\cos^2\vartheta$ and $\sin^2\vartheta$. Consequently, at the end of the whole process the neutron beam state is

$$P_A \longmapsto \rho(t) := (U(\vartheta,\varphi) \otimes \mathbf{1}_2)\, P_A(t)\, (U^*(\vartheta,\varphi) \otimes \mathbf{1}_2) \tag{6.6}$$

where $U(\vartheta,\varphi)$ is the unitary matrix in (6.5) and $U^*(\vartheta,\varphi)$ is its adjoint. The two components of the exiting beam are then intercepted by two counters $C_{u,d}$ plus spin-analyzers $S_{\mathbf{n},-\mathbf{n}}$ that record how many neutrons reach them with spins polarized along suitable space-directions $\pm\mathbf{n} = \pm(n_1, n_2, n_3)$. The frequencies of counts give the expectations

$$\mathcal{O}_t^{j,\mathbf{n}}(\vartheta,\varphi) := \mathrm{Tr}\Big(P_A(t)\, P_j(\vartheta,\varphi) \otimes Q_{\mathbf{n}}\Big) \tag{6.7}$$

where $P_j(\vartheta,\varphi) := U^*(\vartheta,\varphi)\, P_j\, U(\vartheta,\varphi)$, P_j, $j=1,2$, are as in (6.3a) and represent observables that are chosen by setting the angles ϑ, φ characteristic of the beam splitter. Further, $Q_{\mathbf{n}} := |\uparrow_{\mathbf{n}}\rangle\langle\uparrow_{\mathbf{n}}|$ projects onto a state with spin polarized along $\mathbf{n}$.

By setting appropriately the angles ϑ and φ of the beam-splitter and the polarization direction $\mathbf{n}$ of the spin-analyzers, the entries of $P_A(t)$ in (6.4) can be measured. From these entries, one has access to the eigenvalues of $P_A(t)$ and thus to the physical effects of the stochastic field inside it. Indeed, using (6.6), one readily computes [24]

$$\begin{aligned} \rho_{11}(t) &= \mathcal{O}_t^{1,z}(0,0)\,, \quad \rho_{12}(t) = \mathcal{O}_t^{1,-z}(0,0)\,, \\ \rho_{21}(t) &= \mathcal{O}_t^{2,z}(0,0)\,, \quad \rho_{22}(t) = \mathcal{O}_t^{2,-z}(0,0)\,. \end{aligned} \tag{6.8}$$

Instead, being the $P_i \otimes Q_j$ with $i,j=3,4$, not self-adjoint, their expectations cannot be measured directly, but one has to use the projectors

$$P_\pm := P_{1,2}(\frac{\pi}{4},0) \;\; P_{\pm i} := P_{1,2}(\frac{\pi}{4},-\frac{\pi}{2}) \tag{6.9a}$$

$$\begin{aligned} Q_{\pm x} &:= \frac{|\uparrow_z\rangle \pm |\downarrow_z\rangle}{\sqrt{2}}\,\frac{\langle\uparrow_z| \pm \langle\downarrow_z|}{\sqrt{2}} \\ Q_{\pm y} &:= \frac{|\uparrow_z\rangle \pm i|\downarrow_z\rangle}{\sqrt{2}}\,\frac{\langle\uparrow_z| \mp i\langle\downarrow_z|}{\sqrt{2}}\,. \end{aligned} \tag{6.9b}$$

Then, from

$$\begin{aligned} P_3 &= \frac{P_+ - P_- + i\,P_{+i} - i\,P_{-i}}{2} = P_4^* \\ Q_3 &= \frac{Q_x - Q_{-x} + i\,Q_y - i\,Q_{-y}}{2} = Q_4^* \end{aligned} \tag{6.10}$$

one obtains expressions for all other entries $\rho_{i\,{j \atop k}}(t) := \mathrm{Tr}\Big(\rho(t) P_i^* \otimes Q_{j,k}^*\Big)$; one of them looks as follows

$$\begin{aligned} \rho_{43}(t) = &\frac{\mathcal{O}_t^{1,x}(\frac{\pi}{4},0) - \mathcal{O}_t^{1,-x}(\frac{\pi}{4},0)}{4} + i\,\frac{\mathcal{O}_t^{1,y}(\frac{\pi}{4},0) - \mathcal{O}_t^{1,-y}(\frac{\pi}{4},0)}{4} \\ &- \frac{\mathcal{O}_t^{2,x}(\frac{\pi}{4},0) - \mathcal{O}_t^{2,-x}(\frac{\pi}{4},0)}{4} - i\,\frac{\mathcal{O}_t^{2,y}(\frac{\pi}{4},0) - \mathcal{O}_t^{2,-y}(\frac{\pi}{4},0)}{4} \end{aligned}$$

$$+i\,\frac{\mathcal{O}_t^{1,x}(\frac{\pi}{4},-\frac{\pi}{2})-\mathcal{O}_t^{1,-x}(\frac{\pi}{4},-\frac{\pi}{2})}{4}-\frac{\mathcal{O}_t^{1,y}(\frac{\pi}{4},-\frac{\pi}{2})-\mathcal{O}_t^{1,-y}(\frac{\pi}{4},-\frac{\pi}{2})}{4}$$
$$-i\,\frac{\mathcal{O}_t^{2,x}(\frac{\pi}{4},-\frac{\pi}{2})-\mathcal{O}_t^{2,-x}(\frac{\pi}{4},-\frac{\pi}{2})}{4}$$
$$+\,\frac{\mathcal{O}_t^{2,y}(\frac{\pi}{4},-\frac{\pi}{2})-\mathcal{O}_t^{2,-y}(\frac{\pi}{4},-\frac{\pi}{2})}{4}\;; \tag{6.11}$$

the remaining ones are collected in [24].

We assume the time-dependent Liouville-Von Neumann evolution equation for the 2×2 density matrix Σ describing the spin degree of freedom travelling through the interferometer to be of the form

$$\begin{aligned}\partial_t \Sigma(t) &= (L_0+L_t)[\Sigma(t)]\\ L_0[\Sigma(t)] :&= -i\Big[\frac{\omega_0}{2}\sigma_3\,,\,\Sigma(t)\Big]\\ L_t[\Sigma(t)] :&= -i\Big[\mathbf{V}(t)\cdot\boldsymbol{\sigma}\,,\,\Sigma(t)\Big]\end{aligned} \tag{6.12}$$

where $\boldsymbol{\sigma}=(\sigma_1,\sigma_2,\sigma_3)$ is the vector of Pauli matrices, $\mathbf{V}(t)=(V_1(t),V_2(t),V_3(t))$ is proportional to gaussian stochastic magnetic field and $H_0 := \dfrac{\omega_0}{2}\sigma_3$ is due to the coupling to a static magnetic field along the z-direction. Furthermore, we assume $\mathbf{V}(t)$ to have zero mean, $\langle\mathbf{V}(t)\rangle=0$, and stationary, real, positive-definite covariance matrix $\mathcal{W}(t)=[W_{ij}(t)]$ with entries

$$W_{ij}(t-s)=\langle V_i(t)V_j(s)\rangle = W^*_{ij}(t-s)=W_{ji}(s-t)\ . \tag{6.13}$$

Because of the stochastic field $\boldsymbol{V}(t)$, the solution $\Sigma(t)$ is stochastic and an effective spin density matrix $\rho(t):=\langle\Sigma(t)\rangle$ is obtained by averaging over the noise. At time $t=0$ we may suppose spin and noise to decouple so that the initial state is $\rho:=\langle\Sigma(0)\rangle=\Sigma(0)$. In order to derive an effective time-evolution for $\rho(t)$, the so-called convolutionless approach developed and used in [14] leads to the equation of motion

$$\partial_t\rho(t) = -i\Big[H_0\ \rho(t)\Big] - \sum_{i,j=1}^{3} C_{ij}(t)\Big[\sigma_i,\Big[\sigma_j\ \rho(t)\Big]\Big] \tag{6.14}$$

where $C_{ij}(t):=\sum_{\ell=1}^{3}\int_0^t \mathrm{d}s\, W_{i\ell}(s)\,U_{\ell j}(-s)$ and $\mathcal{U}(t):=\begin{pmatrix}\cos\omega_0 t & -\sin\omega_0 t & 0\\ \sin\omega_0 t & \cos\omega_0 t & 0\\ 0 & 0 & 1\end{pmatrix}$ is the unitary matrix $\mathcal{U}(t)=[U_{ij}(t)]$ such that $\mathrm{e}^{-tL_0}[\sigma_i]=\sum_{j=1}^{3}U_{ij}(t)\sigma_j$.

The matrix $\mathcal{C}(t):=[C_{ij}(t)]$ is real and can be decomposed into symmetric and antisymmetric components. Correspondingly, the second term on the right hand side of (6.14) splits into a commutator with a Hamiltonian and a purely dissipative contribution

$$\sum_{i,j=1}^{3} C_{ij}(t)\Big[\sigma_i,\Big[\sigma_j\ \rho(t)\Big]\Big] = i\Big[\sum_{i,j,k=1}^{3} C^A_{ij}(t)\,\epsilon_{ijk}\,\sigma_k\ \rho(t)\Big] \tag{6.15a}$$

$$+\sum_{i,j=1}^{3} 2C^S_{ij}(t)\Big(\frac{1}{2}\Big\{\sigma_i\sigma_j\ \rho(t)\Big\}\ -\ \sigma_j\,\rho(t)\,\sigma_i\Big) \tag{6.15b}$$

where $C^{A,S}_{ij}(t) = (C_{ij}(t) \mp C_{ji}(t))/2$. When the coupling between system and stochastic field is weak, the memory effects are expected to be negligible so that the use of a Markov approximation is in general justified; practically, one extends to $+\infty$ the upper limit of the integral in the entries $C_{ij}(t)$ so that time t no longer appears explicitly in the equation of motion, which reads

$$\partial_t\rho(t) = -i\Big[H\ \rho(t)\Big]\ +\ L_D[\rho(t)] \tag{6.16}$$

with $H = H_0 + H_D$, $H_D := \sum_{i,j,k=1}^{3} C^A_{ij}\epsilon_{ijk}\,\sigma_k$,

$$L_D[\rho(t)] := \sum_{i,j=1}^{3} C^D_{ij}\Big(-\frac{1}{2}\Big\{\sigma_i\sigma_j\ \rho(t)\Big\}\ +\ \sigma_j\,\rho(t)\,\sigma_i\Big) \tag{6.17}$$

where C^A_{ij} and L^D_{ij} are the entries of the real matrices

$$\mathcal{C}_A := \int_0^{+\infty} \mathrm{d}s\, \frac{\mathcal{W}(s)\,\mathcal{U}(-s)\ -\ \mathcal{U}(s)\,\mathcal{W}(-s)}{2} \tag{6.18a}$$

$$\mathcal{C}_D := \int_0^{+\infty} \mathrm{d}s\,\Big(\mathcal{W}(s)\,\mathcal{U}(-s) + \mathcal{U}(s)\,\mathcal{W}(-s)\Big)\ . \tag{6.18b}$$

The properties of the open dynamics of the neutron beam within the interferometer clearly depend on those of the stochastic magnetic field; in the sequel we will focus on the case in which the covariance matrix (6.13) has

$$\langle V_1(t)V_1(s)\rangle = \langle V_2(t)V_2(s)\rangle = g^2B_1^2\mathrm{e}^{-\lambda|t-s|}\quad \langle V_3(t)V_3(s)\rangle = g^2B_3^2\mathrm{e}^{-\mu|t-s|} \tag{6.19}$$

and the other entries zero, with B_i constant magnetic field intensities and g proportional to the neutron magnetic moment. Then,

$$\mathcal{C}_A = \frac{g^2\omega_0B_1^2}{\lambda^2+\omega_0^2}\begin{pmatrix}0 & 1 & 0\\ -1 & 0 & 0\\ 0 & 0 & 0\end{pmatrix}\quad \mathcal{C}_D = 2g^2\begin{pmatrix}\frac{\lambda B_1^2}{\lambda^2+\omega_0^2} & 0 & 0\\ 0 & \frac{\lambda B_1^2}{\lambda^2+\omega_0^2} & 0\\ 0 & 0 & \frac{B_3^2}{\mu}\end{pmatrix}\ . \tag{6.20}$$

Setting

$$\varDelta\omega := \frac{4g^2B_1^2\omega_0}{\lambda^2+\omega_0^2}\ ,\quad \gamma := \frac{4g^2B_1^2\lambda}{\lambda^2+\omega_0^2}\quad a := \frac{B_3^2}{\mu} + \frac{2g^2B_1^2\omega_0}{\lambda^2+\omega_0^2} \tag{6.21}$$

the matrix $\mathcal{C}_D$ becomes

$$\mathcal{C}_D = \frac{1}{2}\begin{pmatrix} \gamma & 0 & 0 \\ 0 & \gamma & 0 \\ 0 & 0 & 2a-\gamma \end{pmatrix} . \tag{6.22}$$

Consequently, the entries of the spin density matrix $\rho(t) = \begin{pmatrix} \rho_1(t) & \rho_3(t) \\ \rho_4(t) & \rho_2(t) \end{pmatrix}$ satisfy the Bloch-Redfield equations [1]

$$\dot{\rho}_1 = -\gamma\rho_1 + \gamma\rho_2 \qquad \dot{\rho}_3 = -i(\omega_0 + \Delta\omega)\rho_3 - 2a\rho_3 \tag{6.23}$$

and $\dot{\rho}_2 = -\dot{\rho}_1$, $\dot{\rho}_4 = (\dot{\rho}_3)^*$. The coefficients 2γ and $2a$ are the inverse of the relaxation times T_1 and T_2 of the diagonal, respectively off-diagonal elements of $\rho(t)$; the inequalities (4.1) would now require $1/T_2 \geq 1/2T_1$.

Remark 6.1. The above order relation between relaxation times is a definite physical constraint on the behaviour of the open system and comes about because of the request of complete positivity. In turn, the latter is usually justified by referring to the possible, never excludible, coupling of the same 2-level system with a totally arbitrary system of the same kind, the coupling being not dynamical, but only statistical. This is why the constraints imposed by complete positivity are criticized as mathematical rather than physical. In this section, however, we are considering the entanglement of two degrees of freedom of a same concrete phsyical system.

In [14,15] it is showed that this typical order relation can be reversed. In this case, $1/T_2 < 1/2T_1$ implies $a-\gamma/2 < 0$ and $\mathcal{C}_D$ is no longer positive-definite; as a consequence, the dynamical maps Γ_t generated by (6.16) cannot be completely positive. However, Proposition 3.2 ensures that they are positive; indeed, one finds

$$\mathcal{D}^{(3)} = -2\begin{pmatrix} a & 0 & 0 \\ 0 & a & 0 \\ 0 & 0 & \gamma \end{pmatrix} . \tag{6.24}$$

Analytic solutions of the equation of motion are, in vectorial representation,

$$\begin{aligned} \rho^0(t) &= \rho^0 \\ \rho^1(t) &= \mathrm{e}^{-2at}\left\{\rho^1 \cos\omega t \; - \; \rho^2 \sin\omega t\right\} \\ \rho^2(t) &= \mathrm{e}^{-2at}\left\{\rho^1 \sin\omega t \; + \; \rho^2 \cos\omega t\right\} \\ \rho^3(t) &= \mathrm{e}^{-2\gamma t}\rho^3 \end{aligned} \tag{6.25}$$

with $\omega := \omega_0 + \Delta\omega$. On this level, the difference between complete positivity and positivity shows up in different order relations between the decay diagonal and off-diagonal relaxation, that is either $a \geq \gamma/2$ or $a < \gamma/2$. However, the true physical meaning of complete positivity of Γ_t becomes evident when one considers the time-evolution $\mathbf{1}_2 \otimes \Gamma_t$ of the state P_A in (6.2) within the interferometer. In vectorial representation $|Q_1\rangle = 1/2(1,0,0,1)$, $|Q_2\rangle = 1/2(1,0,0,-1)$, $|Q_3\rangle = 1/2(0,1,i,0)$ and $|Q_4\rangle = 1/2(0,1,-i,0)$, thus, from (6.25) it follows

$$P_A(t) = (\Gamma_t \otimes \mathbf{1}_2)\,[P_A] = \frac{1}{2}\Big(P_1 \otimes Q_2(t) + P_2 \otimes Q_1(t) - P_3 \otimes Q_4(t) - P_4 \otimes Q_3(t)\Big) \tag{6.26}$$

with

$$\begin{aligned} Q_1(t) &= \frac{1}{2}\begin{pmatrix} 1+\mathrm{e}^{-2\gamma t} & 0 \\ 0 & 1-\mathrm{e}^{-2\gamma t} \end{pmatrix}, \quad Q_3(t) = \mathrm{e}^{-t(2a+i\omega)}\begin{pmatrix} 0 & 1 \\ 0 & 0 \end{pmatrix}, \\ Q_2(t) &= \frac{1}{2}\begin{pmatrix} 1-\mathrm{e}^{-2\gamma t} & 0 \\ 0 & 1+\mathrm{e}^{-2\gamma t} \end{pmatrix} \quad Q_4(t) = \mathrm{e}^{-t(2a-i\omega)}\begin{pmatrix} 0 & 0 \\ 1 & 0 \end{pmatrix}. \end{aligned} \tag{6.27}$$

Therefore,

$$\begin{aligned} P_A(t) &= \begin{pmatrix} E_-(t) & 0 & 0 & 0 \\ 0 & E_+(t) & F(t) & 0 \\ 0 & F^*(t) & E_+(t) & 0 \\ 0 & 0 & 0 & E_-(t) \end{pmatrix}, \\ E_\pm &:= \frac{1 \pm \mathrm{e}^{-2\gamma t}}{4}, \quad F := -\frac{\mathrm{e}^{-t(2a-i\omega)}}{2}. \end{aligned} \tag{6.28}$$

The eigenvalues of the state $P_A(t)$ at the exit of the interferometer are $\lambda_{1,2}(t) = E_-(t)$ and

$$\lambda_\pm(t) = \frac{1+\mathrm{e}^{-2\gamma t} \pm 2\mathrm{e}^{-2at}}{4}. \tag{6.29}$$

Let $a < \gamma/2$, that is let Γ_t to be positive, but not completely positive; then, since $\lambda_-(0) = 0$ and $\mathrm{d}\lambda_+(0)/\mathrm{d}t = (2a-\gamma)/2$, there is a whole range of t where $\lambda_+(t) < 0$ and $P_A(t)$ loses physical meaning. On the other hand, if $a \geq \gamma/2$, $\lambda_+(t) \geq 0$, for all t.

Although apparently formal, the previous results are far from being academic: indeed, as already stressed at the beginning of this section, the entries of $P_A(t)$ are directly accessible to the experiment. For instance, it follows from (6.8) that $E_-(t) = \mathcal{O}_t^{1,z}(0,0)$ and $E_+(t) = \mathcal{O}_t^{2,z}(0,0)$, while $F(t)$ coincides with the expression in (6.11).

7 Conclusions

There are two quite opposite attitudes towards quantum open system dynamics in relation to complete positivity: either one starts from Lindblad-type equations of motion that guarantee it from the start, or one igores it thus releasing the strong physical constraints on the relaxation properties of the time-evolution. However, both these attitudes concern the time-evolution of single quantum open systems and thus miss the true meaning of complete positivity which is related to quantum entanglement and thus needs entangled pairs to be fully exposed.

Truely, the formal justification of complete positivity introduces entanglement with another (N-level) system, but this latter system is a somewhat abstract partner, not experimentally controllable. On the contrary, in this contribution we have considered entangled physical systems, neutral mesons, respectively entangled degrees of freedom of single neutrons, in actual experiments, at ϕ factories, respectively in quantum interferometry. In such contexts, lack of complete positivity brings about inconsistencies, essentially the appearance of negative probabilities in the theory, which have physically exposable, experimentally verifiable effects.

Since the presence or absence of complete positivity depends on the Markovian approximations used to derive the reduced dynamics of quantum open systems, it seems appropriate to conclude that, in the cases discussed above, whenever a semigroup composition law is expected, the physically appropriate Markovian approximations are those which lead to reduced dynamics consisting of completely positive maps Γ_t.

References

1. C.P. Slichter, *Principles of Magnetic Resonance* (Springer Verlag, Berlin, 1990).
2. W.H. Louisell, *Quantum Statistical Properties of Radiation*, (Wiley, New York, 1973).
3. C.W. Gardiner and P. Zoller, *Quantum Noise*, 2nd ed., (Springer, Berlin, 2000).
4. M.O. Scully and M.S. Zubairy, *Quantum Optics* (Cambridge University Press, Cambridge, 1997).
5. L. Fonda, G.C. Ghirardi and A. Rimini, Rep. Prog. Phys. **41**, 587 (1978),
6. H. Nakazato, M. Namiki and S. Pascazio, Int. J. Mod. Phys. **B10**, 247 (1996).
7. J. Ellis, J.S. Hagelin, D.V. Nanopoulos and M. Srednicki, Nucl. Phys. **B241**, 381 (1984).
8. P. Huet and M.E. Peskin, Nucl. Phys. **B434**, 3 (1995).
9. F. Benatti and R. Floreanini, Nucl. Phys. **B488**, 335 (1997).
10. F. Benatti and R. Floreanini, Nucl. Phys. **B401**, 550 (1998).
11. F. Benatti and R. Floreanini, Phys. Lett. **B468**, 287 (1999).
12. F. Benatti, R. Floreanini and R. Romano, Nucl. Phys. **B602**, 541 (2001).
13. R. Dümcke and H. Spohn, Z. Physik **B34**, 419 (1979).
14. J. Budimir and J.L. Skinner, J. Stat. Phys. **49**, 1029 (1987).
15. B.B. Laird and J.L. Skinner, J. Chem. phys. **94**, 4405 (1991).
16. A. Suarez, R. Silbey and I. Oppenheim, J. Chem. Phys. **97**, 5101 (1992).
17. E.B. Davies, *Quantum theory of Open systems* (Academic Press, London, 1976).
18. H. Spohn, Rev. Mod. Phys. **52**, 569 (1980).
19. R. Alicki and K. Lendi, *Quantum Dynamical Semigroups and Applications*, Lect. Notes Phys. **286**, (Springer-Verlag, Berlin, 1987).
20. M. Horodecki, P. Horodecki and R. Horodecki, Phys. Lett. **A223**, 1 (1996).
21. P. Pechukas, Phys. Rev. Lett. **73**, 1060 (1994).
22. A. Gorini, A, Kossakowski and E.C.G. Sudarshan, J. Math. Phys. **17**, 821 (1976).
23. G. Lindblad, Comm. Math. Phys. **48**, 119 (1976).

24. F. Benatti, R. Floreanini and R. Romano, *Complete Positivity and Entangled Degrees of Freedom*, Trieste preprint 2002.
25. M. Choi, Linear Alg. Appl. **10**, 285 (1975).
26. K. Kraus, Ann. Phys. **64**, 311 (1971).
27. V. Gorini and A. Kossakowski, J. Math. Phys. **17**, 1298 (1976).
28. F. Benatti and R. Floreanini, in preparation.
29. S. Basu, S. Bandyopadhyay, G. Kar and D. Home, Phys. Lett. **A279**, 281 (2001).

Quantum Stochastic Dynamical Semigroup

W.A. Majewski

Institute of Theoretical Physics, Gdańsk University ul. Wita Stwosza 57, PL-80-592 Gdańsk, Poland

Abstract. We review the quantization of dynamics of stochastic models whose equilibrium states are the classical Gibbs ones. To proceed with the study of correctness of the quantization we indicate how the obtained dynamicals maps are related to quantum correlations and quantum entanglement.

Keywords: quantum dynamical semigroup, quantum entanglement, quantum correlations, non-commutative L_p-spaces.

1 Introduction

In the classical theory of particle systems one of the objectives is to produce, describe and analyze dynamical systems with evolution originated from stochastic processes in such a way that their equilibrium states are Gibbs states (cf. [1]). A well known illustration is a number of papers describing the so called Glauber dynamics [2]. To perform a detailed analysis of dynamical system of that type, it is convenient to use the theory of Markov processes in the context of L_p-spaces. In particular, for the Markov-Feller processes, using the unique correspondence between a process and the corresponding dynamical semigroup one can give a receipe for a construction of Markov generators for this class of processes (see [1]). Recently, this program was carried out in the setting of quantum mechanics [3]-[6]. In particular, guided by the classical theory and applying generalized conditional expectations (in the sense of Accardi-Cechnini), it was possible to define the corresponding Markov generators of the underlying quantum Markov-Feller dynamics. In that way the quantum counterpart of the classical recipe for the construction of quantum Markov generators was obtained. In the sequel, quantum semigroups obtained in that framework will be called *quantum stochastic dynamical semigroups*.

Let us add that such the analysis led to a general scheme for constructing quantum jump processes on a lattice (cf. [3]). We emphasize that interpretation of such quantum processes is the same as in the classical case. Namely, having a transition rate (dependent on the state ρ), one can describe a Markov semigroup corresponding to quantum Markov process (defined by transition rates). Clearly, while describing a physical process in such a scheme, we do not know (explicitly) the interactions which are responsible for the underlying transition rates. Therefore, it is natural to pose the following question: Are the interactions proper (i.e. genuine quantum ones) or not? To answer this question we proceed with a detailed analysis of properly chosen correlation functions. This approach is based on the recently established relation between quantum correlations and entaglement (see [7]). Namely, an analysis of evolution of entanglement can supply information about increment of quantum correlations. In that way we can get a clear indication about existence of non-trivial quantum interactions between two subsystems of the studied quantum system.

The lectures will be organized as follows. Firstly, non-commutative Hilbert spaces associated with a physical system will be introduced. Then, the general strategy of construction of quantum stochastic dynamical semigroups will be given. A special emphasize will be put on the construction of quantum jump-type semigroups. Then, we describe the relation between quantum correlations and entanglement. Finally, some illustrative concrete models will be provided.

2 Non-commutative L_p-Spaces

Let $< X, \mu >$ be a measure space and $p \geq 1$. We denote by $L^p(X, d\mu)$ the set (of equivalence classes) of measurable functions satisfying

$$\|f\|_p \equiv \Big(\int_X |f(x)|^p d\mu(x) \Big)^{\frac{1}{p}} < \infty .$$

The analogue of the concept of L_p-spaces, $p \in [1, \infty]$, of the commutative theory for the pair $(\mathcal{M}, \tau)$ consisting of von Neumann algebra $\mathcal{M}$ and a trace τ can be introduced as follows: define

$$\mathcal{I}_p = \{ x \in \mathcal{M} \quad | \quad \tau(|x|^p) < \infty \}.$$

$\mathcal{I}_p$ is a two sided ideal of $\mathcal{M}$. Further, $\|x\|_p = \tau(|x|^p)^{\frac{1}{p}}$ defines a norm on $\mathcal{I}_p$. The completion of $\mathcal{I}_p$ with respect to the norm $\| \cdot \|_p$ gives $L^p(\mathcal{M}, \tau)$ spaces which can be considered as a generalization of those defined in the commutative case. It is an easy observation that putting $\mathcal{M} = \mathcal{B}(\mathcal{H})$ and $\tau = \mathbf{Tr}$ one obtains the well known Schatten classes [8]. Using this and the Haagerup theory (see [9]) we can introduce quantum L_p spaces for quantum lattice systems.

To this end let $(\mathcal{A}, *, ||\cdot||)$ be C^*-algebra and $\{\mathcal{A}_\Lambda\}_{\Lambda \in \mathcal{F}}$ a family of subalgebras of $\mathcal{A}$; here $\mathcal{F}$ is a direct set, e.g. a family of bounded subsets of a metric separable space; for details of operator algebras in quantum physics we refer the reader to [10]. Furthermore, we assume:

1. $\mathcal{A}_0 = \cup_{\Lambda \in \mathcal{F}} \mathcal{A}_\Lambda$ is dense in $\mathcal{A}$.
2. There exists a family of density operators $\{\varrho_\Lambda \in \mathcal{A}_\Lambda : \varrho_\Lambda > 0, \mathbf{Tr}\varrho_\Lambda = 1\}_{\Lambda \in \mathcal{F}}$ with the compatibility condition $\mathbf{Tr}_{\Lambda_2 \setminus \Lambda_1}\{\varrho_{\Lambda_2}\} = \varrho_{\Lambda_1}$ provided that $\Lambda_1 \subset \Lambda_2$.

Under the above assumptions we can introduce:

- $\omega(f) = \lim_\Lambda \omega_\Lambda(f) \equiv \lim_\Lambda \mathbf{Tr}\{\varrho_\Lambda f\}$,
- $||f||_{L_{p,s}(\omega)} = \lim_\Lambda ||f||_{L_{p,s}(\omega_\Lambda)}$ for $p \in [1, \infty)$, $s \in [0,1]$, where $f \in \mathcal{A}$,
- $||f||_{L_{p,s}(\omega_\Lambda)} = \mathbf{Tr}|\varrho_\Lambda^{1-s/q} f \varrho_\Lambda^{s/q}|^q$.

One can show that $||f||_{L_{p,s}(\omega_\Lambda)}$ is the well defined two-parameter family of norms on $\mathcal{A}$. The same should be done for $||f||_{L_{p,s}(\omega)}$ (see Theorem below).

Having such general scheme let us turn to the basic example. Let $\mathbb{Z}^d$ stand for a d-dimensional integer lattice, and $\mathcal{F}$ denote the family of all finite subsets of $\mathbb{Z}^d$. Finally, by $\mathcal{F}_0$ we denote an increasing sequence of finite volumes covering all the lattice $\mathbb{Z}^d$.

Local algebras are introduced in the following way: $\mathbb{Z}^d \ni z \mapsto \mathbf{M}_z \equiv \mathbf{M}$ where $\mathbf{M}$ is the full finite dimensional matrix algebra. Consequently, $\mathcal{F} \ni \Lambda \mapsto \mathcal{A}_\Lambda = \otimes_{z\in\Lambda}\mathbf{M}$. Then, $\mathcal{A}$ is a C^*-algebra with norm $||\cdot||$ defined as the inductive limit over a finite dimensional complex matrix algebras $\mathbf{M}$.

To define the reference (local) Gibbs state we start with potentials $\Phi \equiv \{\Phi_X \in \mathcal{A}_X\}_{X\in\mathcal{F}}$, i.e. a family of selfadjoint operators such that

$$||\Phi||_\lambda \equiv \sum_{n\geq 0} e^{\lambda n}\Big(\sup_{i\in\mathbb{Z}^d} \sum_{\substack{X\ni i\\ |X|=n+1}} ||\Phi_X||\Big) < \infty.$$

We define a corresponding (local) Hamiltonian H_Λ for $\Lambda \in \mathcal{F}$, by setting

$$H_\Lambda \equiv H_\Lambda(\Phi) \equiv \sum_{X\subset\Lambda} \Phi_X.$$

The just mentioned (local) reference state is given by ρ_Λ - a (local) density matrix:

$$\rho_\Lambda \equiv \frac{e^{-\beta H_\Lambda}}{\mathbf{Tr}e^{-\beta H_\Lambda}}$$

with $\beta \in (0,\infty)$, where β stands for the inverse temperature. Denote by $\omega_\Lambda(\cdot) = \mathbf{Tr}\{\varrho_\Lambda\cdot\}$. Then ω is given by thermodynamic limit of $\{\omega_\Lambda\}_\Lambda$. In [3], [5] it was proved:

Theorem. *For any $p \in (2,\infty)$, $s \in [0,1]$, any local operator $f \in \mathcal{A}_{\Lambda_0}$, $\Lambda_0 \in \mathcal{F}$ and all sets $\Lambda_1, \Lambda_2 \in \mathcal{F}$ such that $\Lambda_0 \subset \Lambda_1 \subset \Lambda_2$, we have*

$$||f||_{\mathbf{L}_p(\omega_{(\Lambda_2)},s)} \leq ||f||_{\mathbf{L}_p(\omega_{(\Lambda_1)},s)}.$$

Thus for any $f \in \mathcal{A}_0$ the limit

$$||f||_{\mathbf{L}_p(\omega,s)} \equiv \lim_{\mathcal{F}_0} ||f||_{\mathbf{L}_p(\omega_{(\Lambda)},s)}$$

exists and is independent of the countable exhaustion $\mathcal{F}_0$ of the lattice.

Corollary. *To every Gibbs state ω on a C^*-algebra $\mathcal{A}$ we can associate an interpolating family of Banach spaces*

$$\{\mathbf{L}_p(\omega,s)\}_{p\in[1,\infty),s\in[0,1]}.$$

3 Quantum Dynamics

Let us consider a quantum lattice system and its C^*-algebra $\mathcal{A}$. We denote by $\mathcal{M}$ the von Neumann algebra given by

$$\mathcal{M} = \pi_\omega(\mathcal{A})'' ,$$

where $\pi_\omega(\cdot)$ stands for the GNS representation while ω is a reference (Gibbs) state on $\mathcal{A}$. The (Markov) quantum stochastic dynamics $P_t : \mathcal{M} \to \mathcal{M}$ will be defined as one parameter family of linear maps such that

1. The map P_t is a strongly continuous map for each $t \in \mathbb{R}^+$;
2. $P_t(\mathbf{1}) = \mathbf{1}$;
3. $P_t(f^* f) \geq 0$ for any $f \in \mathcal{M}$;
4. P_t preserves a given reference state φ on $\mathcal{M}$, i.e. $\varphi(P_t f) = \varphi(f)$ for any t and f; (φ is the weak extension of ω).
5. *Detailed balance condition* of the form $(f, P_t g)_{\mathcal{H}_\varphi} = (P_t f, g)_{\mathcal{H}_\varphi}$ is satisfied, where $\mathcal{H}_\varphi$ is a Hilbert space containing $\mathcal{M}$, i.e. $\mathcal{H}_\varphi = L_p(\mathcal{M}, \varphi)$ (again, here, φ is the weak extension of ω).

We wish to construct P_t of the form $P_t \equiv e^{t\mathcal{L}}$ where

$$\mathcal{L} \equiv \mathcal{E} - \mathbf{1},$$

and $\mathcal{E}$ is a certain map defined on $\mathcal{M}$. To get the explicit construction of $\mathcal{E}$ (we restrict ourselves to jump-type semigroups, for a general case see [3], [5]) we consider a von Neumann algebra $\mathcal{M}$ with a faithful state $\varphi_1 \equiv \varphi$. Let $\mathcal{E}_0$ be a conditional expectation on $\mathcal{M}$, i.e. $\mathcal{E}_0$ satisfies $\mathcal{E}_0(f^* f) \geq 0$, $\mathcal{E}_0(\mathbf{1}) = \mathbf{1}$, $\mathcal{E}_0^2 = \mathcal{E}_0$. Define

$$\varphi_2(\cdot) \equiv \varphi_1 \circ \mathcal{E}_0(\cdot).$$

Suppose, φ_1, φ_2 are faithful states on $\mathcal{M}$.

Then, one can consider the Radon-Nikodym cocycles $V_t \equiv (D\varphi_1 : D\varphi_2)_t$. Suppose, now, there exists a positive constant $c \in (0, \infty)$ such that for any $0 \leq f \in \mathcal{M}$ the following inequalities hold:

$$\frac{1}{c}\varphi_1 \leq \varphi_2(f) \leq c\varphi_1(f).$$

It follows, (cf. [11]) that V_t extends analytically to $-\frac{1}{2} \leq \mathrm{Im} z \leq \frac{1}{2}$ and $\xi \equiv V_{t|t=-\frac{i}{2}}$ is a bounded operator in $\mathcal{M}$.

Theorem. *Assume that $\xi \equiv V_{t|t=-\frac{i}{2}}$ is a bounded operator in $\mathcal{M}$ and define*

$$\mathcal{E}(f) \equiv \mathcal{E}_0(\xi^* f \xi).$$

Then, the generalized conditional expectation $\mathcal{E}(\cdot)$ is well defined and it has the following properties:

1. $\mathcal{E}(\mathbf{1}) = \mathbf{1}$,
2. $\mathcal{E}(f^* f) \geq 0$,
3. $(\mathcal{E}(f), g)_{\varphi_1} = (f, \mathcal{E}(g))_{\varphi_1}$, $f, g \in \mathcal{H}_{\varphi_1}$. *$\mathcal{H}_{\varphi_1}$ is the non-commutative Hilbert space $L_2(\mathcal{M}, \varphi_1)$.*

Corollary: The operator given by:

$$\mathcal{L} \equiv \mathcal{E} - \mathbf{1}$$

is *a well defined Markov generator.*

Here we touch only a few aspects of the presented approach. However to review very briefly some of the recent results in that direction we note: others jump-type dynamics can be defined, e.g. the quantum counterpart of Kawasaki dynamics (see [6]). Secondly, diffusive type can also be defined and studied (see for example [12], [13], [14],[15],[3]). Consequently, a large variety of dynamical maps can be given within the described constructive approach to the description of time evolution.

4 Quantum Correlations

Among the most emblematic concepts in Quantum Mechanics there is the idea of entanglement (cf. [16], [17], [18], [19]). Let us recall that this concept enters in description of quantum correlations between subsystems (cf. [7], [22]). Thus, in particular, it plays a crucial role in quantum mechanical considerations. This explains why the better understanding of this concept is so important.

From a formal point of view, a state of a composite quantum system is called *inseparable (or entangled)* if it cannot be represented as tensor products of states of its subsystems. On the contrary, a density matrix describes a *separable* state if it can be expressed as a convex combination of tensor products of its subsystem states. These definitions stem from the mathematical fact that, in general, the convex hull of product states is not a (weakly) dense subset in the state space of the tensor product of two (von Neumann) C^* algebras provided that both algebras are non-commutative (cf. [20]).

Turning to quantum correlations, we recall that the entanglement is often considered as a signature of quantum correlations. Although, the concept of quantum correlations is essential one for quantum statistical mechanics, its precise characterization was given quite recently ([7]). Namely, to determine in a rigorous way the concept of quantum correlations, guided by the (classical) probability theory with its definition of coefficient of independence, one can define the coefficient of quantum correlations (cf. [7]). The basic idea is not affected if we restrict ourselves to the simplest composite system, i.e. a system consisting of only two subsystems. So, we will consider the simplest composite system. It is worth pointing out that this is exactly the situation considered in the general scheme for quantization stochastic dynamics.

Thus, let $\mathcal{A} = \mathcal{A}_1 \otimes \mathcal{A}_2$ and let us consider a separable state ω on $\mathcal{A} \equiv \mathcal{A}_1 \otimes \mathcal{A}_2$, $\omega(\cdot) \equiv \mathbf{Tr}\{(\sum_i \lambda_i \varrho_i^1 \otimes \varrho_i^2)\cdot\}$. It is an easy observation that, in general, $\omega(a \otimes \mathbf{1} \cdot \mathbf{1} \otimes b) \neq \omega(a \otimes \mathbf{1})\omega(\mathbf{1} \otimes b)$ for $a \in \mathcal{A}_1$ and $b \in \mathcal{A}_2$. Thus, the state ω reflects some correlations. However, as the state ω is separable, these correlations are considered to be of classical nature only. Namely, each (classical) probability measure μ can be (*-weakly) approximated by a net of probability measures with finite support. Moreover, each finitely supported probability measure on the product measure space is the convex hull of products of Dirac's measures. Hence, each (classical) probability measure (so any state) on a composite (classical) system exhibits the basic properties of a separable state.

Therefore, to define "pure" quantum correlations we should "substract" classical correlations. This was done in the following measure of quantum correlations (cf. [7]).

Definition 1.
Assume that $\mathcal{A} = \mathcal{A}_1 \otimes \mathcal{A}_2$. Then, the degree of quantum correlations for the state ϕ evaluated on $a_1 \otimes a_2$ is defined as

$$d(\phi; a_1, a_2) = \inf_{\mu \in M_\phi(\mathcal{S}(\mathcal{A}))} \left|\phi(a_1 \otimes a_2) - \left(\int \xi d(\nu_\mu)(\xi)\right)(a_1 \otimes a_2)\right|, \qquad (1)$$

where $a_i \in \mathcal{A}_i$, $i = 1, 2$; $M_\phi(\mathcal{S}(\mathcal{A}))$ stands for the set of all probability measures on the set of all states of $\mathcal{A}$ with the barycenter ϕ. The definition of the measure ν_μ, which encodes the classical correlations, is based on the theory of restriction maps, see [7] for details.

Clearly, we followed the classical definition of coefficient of independence. The main difference between the classical and quantum approaches (apart from the existence of extra correlations) relies on the fact that in the quantum setting, the set of states does not form a simplex. Therefore, there is no uniqueness in decomposition of a (quantum) state. Hence, to carry out our construction we are forced to use the Choquet theory and to take the infimum operation over the set of all "good" decompositions. Secondly, in the non-commutative case the Fubini theorem, in general, does not hold. This fact was used in our definition of the measure ν_μ.

Turning to separable states one can prove (cf. [7]):

Proposition 1. *A state ϕ is separable one if and only if $d(\phi; a_1, a_2) = 0$ for any a_1, a_2.*

The Proposition may be summarized by saying that any separable state contains classical correlations only. Therefore, an entangled state contains "non-classical" (or quantum) correlations. Consequently, if dynamical maps are able to increse entanglement of a state then quantum correlations are also enhanced. So, such the class of maps should contain non-trivial quantum interactions between $\mathcal{A}_1$ and $\mathcal{A}_2$.

5 Evolution of Entanglement

In this final section we present recent results showing an increse of entanglement for three particular *finite* models of quantum jump-type stochastic dynamics. Here *finite* means that the considered systems are associated with a finite region Λ of the lattice $\mathbb{Z}^d$. Their generators are defined according to the prescription given in the Section 3. In the first example, to define block-spin flip dynamics for *finite region*, $\mathcal{E}_0$ is taken to be of the form $\mathcal{E}_0(\cdot) = \mathbf{Tr}_1(\cdot)$ where $\mathbf{Tr}_1$ stands for the partial trace with respect to the first subsystem. Its infinitesimal generator is defined as:

$$\mathcal{L}(f) = \mathcal{E}(f) - f \,,$$

where

$$\mathcal{E}(f) = \mathbf{Tr}_1\big(\gamma^* f \gamma\big) \quad \text{with} \quad \gamma = \rho^{\frac{1}{2}}\big(\mathbf{Tr}_1\rho\big)^{-\frac{1}{2}}$$

and $f \in \mathcal{A} \equiv \mathcal{A}_\Lambda$. The identification $\mathcal{A} \equiv \mathcal{A}_\Lambda$ follows from our assumption about finitness of dynamical system. ρ is density matrix of the reference state and, again, in the notation of sections 2 and 3 one has $\omega_\Lambda \equiv \omega$.

The time evolution of entanglement for that dynamics was studied in [21]. To carry out that task two different approaches were used. Firstly, taking an explicit form of infinitesimal generator $\mathcal{L}$ it was shown that the block spin flip dynamics leads to the entanglement of the initial separable quantum state. Secondly, a detailed analysis of some properly chosen correlation functions has shown that an enhancement of quantum correlation is *typical* for the considered dynamics.

As the second model we want to mention the exchange type dynamics which is originated from one dimensional Ising model with nearest neighbor interactions only while the third model is also exchange type dynamics but originated from XXZ type hamiltonian. For both cases we took the initial state to be a separable one (see [23]). In other words, both models have only classical correlations for time $t = 0$. Let us present the explicit form of the corresponding infinitesimal generators (for details see [23]). To this end let us consider a finite composite system associated with a region $\Lambda = \Lambda_I \cup \Lambda_{II}$, where $\Lambda_I, \Lambda_{II} \subset \mathbb{Z}^d$. In particular, $\mathcal{A} \equiv \mathcal{A}_\Lambda = \mathcal{A}_I \otimes \mathcal{A}_{II}$ where $\mathcal{A}_I$ is the algebra associated with the region I, and the same for $\mathcal{A}_{II}$. To define exchange type dynamics we write $\mathcal{E}_0$ in terms of local symmetry maps. Namely, let us consider a symmetry transformation (local automorphism) ψ on $\mathcal{A}$ such that

$$\psi(A) = A \quad \text{for} \quad A \in \mathcal{A}_{II}, \quad \psi^2 = \mathbf{1} \,.$$

That means that ψ acts non-trivially only on the subalgebra $\mathcal{A}_I$. Let us add that such the form of ψ gives the possibility to consider the type of symmetries, which are implemented by exchanges of observables between sites of the spin chain. Using transformation ψ one can define a projection $\tau(= \mathcal{E}_0)$ on $\mathcal{A}$ as follows

$$\tau(\cdot) \equiv \frac{1}{2}(\mathbf{1} + \psi)(\cdot).$$

We observe that τ is not a morphism. However, the infinitesimal generator $\mathcal{L}$ can be given as in the section 3, i.e. $\mathcal{L} = \mathcal{E} - \mathbf{1}$ where now

$$\mathcal{E}(A) = \tau(\gamma^* A \gamma),$$

with $\gamma = \rho^{\frac{1}{2}} \left(\tau\rho\right)^{-\frac{1}{2}}$. The reference state (Gibbs) state ρ is of the form $Z^{-1}e^{-\beta H}$, Z is the normalizing factor and H is the concrete hamiltonian. For the second model, H was taken to be nearest neighbor Ising (one dimensional) hamiltonian while for the third model H was taken to be hamiltonian of XXZ model.

Turning to properties of the studied models, let us start with a remark that, in the second model, the hamiltonian of the reference state gives rise to hamiltonian development having the multiperiodic nature for non fixed points (cf. [10]). In that sense, the corresponding hamiltonian model exhibits a behaviour typical for classical interactions. Our analysis of stochastic quantum Ising dynamics clearly has shown, [23], that the transition from that hamiltonian model to quantum stochastic Ising model preserves the above mentioned property. On the other hand, it is well known that quantum hamiltonian XXZ model has much more interesting propagation than the Ising model (cf. [10]). This feature is also reflected in our analysis of its quantum stochastic generalization, [23]). This can be taken as a clear indication that generalized conditional expectations, generating time evolution for that model, contain non-trivial interactions. In other words, the transition rates defining the Markov evolution and determined by the corresponding Gibbs states are correctly designed for XXZ model in the sense that they contain the proper interaction between the subsystem and its environment.

One can summarize the obtained results by saying that *the presented quantum maps can enhanced quantum correlations.* Therefore, the presented strategy of quantization of (classical) stochastic evolution seems to be the correct one.

References

1. T.M. Ligget, Interacting Particle Systems, Springer Verlag, (1985)
2. R.J. Glauber, J. Mat. Phys., **4**, 294, (1963)
3. A.W. Majewski, B. Zegarlinski, Math. Phys. Electronic J. **1**, Paper 2 (1995)
4. A.W. Majewski, B. Zegarlinski, Lett. Math. Phys. **36**, 337 (1996)
5. A.W. Majewski, B. Zegarlinski, Rev. Math. Phys. **8**, 689 (1996)
6. A.W. Majewski, B. Zegarlinski, Markov Proc. and Rel. Fields **2**, 87 (1996)
7. A.W. Majewski, *On entanglement of states and quantum correlations,* LANL preprint, math-phys/0202030
8. R. Schatten, *Norms ideals of completely continuous operators,* 2nd Printing, Springer Verlag, 1970.
9. V. Haagerup, L_p-spaces associated with an arbitrary von Neumann algebra in *Algèbres d'opérateurs et leurs applications en physique mathématique,* Colloques internationaux du CNRS, No. 274, Marseille 20-24 juin 1977, 175-184. Éditions du CNRS, Paris 1979
10. O. Bratteli, D.W. Robinson, *Operator Algebras and Quantum Statistical Mechanics*, Springer Verlag, Vol.I (1979), Vol.II (1981)

11. A. Connes, Bull. Sc. math., 2e série **97**, 253 (1973)
12. F. Cipriani, *Dirichlet Forms and Markovian Semigroups on Standard Forms of von Neumann Algebras*, Thesis Trieste 1992
13. E.B. Davies and M. Lindsay, Math. Zeit. **210** (1992), 379-411
14. S. Goldstein and N. Lindsay, Math. Zeit. **219**, 591-608 (1995)
15. Y.N. Park, Quantum Probability and Related Topics, **3**, 1 (2000)
16. E. Schroedinger, Naturwissenschaften **23**, 807-812; 823-828; 844-849 (1935)
17. A. Einstein, B. Podolsky and N. Rosen, Phys. Rev. **47**, 777 (1935)
18. A. Peres, *Quantum Theory: Concepts and Methods*, Kluwer, Dordrecht, 1993
19. A. Peres, Phys. Rev. Lett **77**, 1413 (1996)
20. R.V. Kadison and J.R. Ringrose, *Fundamentals of the Theory of Operator Algebras I and II*, Pure and Applied Mathematics, vol. 100, Academic Press, New York 1983 and 1986.
21. S. Kozieł, A.W. Majewski, *Evolution of entanglement for spin-flip dynamics* LANL preprint quant-ph/0101033, (2001)
22. A.W. Majewski, Open Systems & Information Dynamics, **6**, 79 (1999)
23. S. Kozieł, A.W. Majewski, Acta Phys. Polonica B, **33**, 1103 (2002)

Chapter IV

Driving, Dissipation and Control in Quantum Systems

Driven Chaotic Mesoscopic Systems, Dissipation and Decoherence

D. Cohen

Department of Physics, Ben-Gurion University, Beer-Sheva, Israel

Abstract. Driven quantum systems, described by Hamiltonian $\mathcal{H}(Q,P,x(t))$ where $x(t)$ is a time dependent parameter, are of interest in mesoscopic physics (quantum dots), as well as in nuclear, atomic and molecular physics. Such systems tend to absorb energy. This irreversible effect is known as dissipation. More generally, x may be a dynamical variable, where the total Hamiltonian is $\mathcal{H}_0(x,p)+\mathcal{H}(Q,P;x)$. In such case the interaction of (x,p) with the environmental degrees of freedom (Q,P) leads to dephasing as well as to dissipation. It should be emphasized that even few (Q,P) degrees of freedom can serve as a miniature heat bath, provided they have chaotic dynamics. We shall introduce a general framework for the analysis of dissipation and dephasing, and we shall clarify the tight connection to recent studies of quantum irreversibility (also referred to as "Loschmidt echo" or as the "fidelity" of quantum computation). Specific model systems that will be presented are: particle in a box driven by moving a wall, and particle in a box/ring driven by electro-motive-force. These two examples are related to studies of nuclear friction and mesoscopic conductance. Specific issues to be discussed are the limitations of kinetic theory, the capabilities of linear response theory, and the manifestation of non-perturbative quantum-mechanical effects. In particular we shall explain that random matrix theory and the semiclassical theory lead to different non-perturbative limits.

1 Introduction

In the following lectures we are interested in systems that are described by a Hamiltonian $\mathcal{H}(Q,P;x(t))$, where (Q,P) is a set of (few) canonical coordinates, and $x(t)$ is a time dependent parameter. We further assume that in the time independent case ($x(t)=\text{const}$) the classical motion is chaotic.

The Quantum mechanical (QM) study of classically chaotic systems is known in the literature as "quantum chaos". The theory of time independent Hamiltonians, in particular their spectral properties, is well documented [1,3,2]. But the QM theory of driven chaotic systems is still a missing chapter.

To avoid misunderstanding we refer here to systems that are chaotic also in the absence of driving. By "driving" we mean that we assume Hamiltonian $\mathcal{H}(Q,P;x(t))$ where $x(t)$ is time dependent, rather than constant. Unlike the case of time independent systems, in case of driven systems the energy distribution evolves with time. What one needs is a theory for this evolution. We shall see that various notions, such as "dissipation", "irreversibility" and "fidelity" just emphasize particular aspects of this evolution.

The available theory for driven systems in the quantum mechanical literature is based mainly upon first order perturbation theory, supplemented by higher order (sometimes non-perturbative) corrections. Depending on the "rate" of the driving, one commonly distinguishes between the "adiabatic" and the "Fermi golden rule" regimes [4,5]. The latter is known also as the "linear response theory" (LRT) or as the "Kubo-Greenwood" regime.

There are circumstances where first order perturbation theory cannot serve as a starting point for the analysis of a driven system. However the well known examples refer to systems that are *not* chaotic in the absence of driving. This includes in particular one-degree-of-freedom systems such as the quantum kicked rotator [6]. Our interest is in generic chaotic systems, therefore we have to consider systems that have at least two degrees of freedom.

The possibility to present a general QM theory for driven chaotic systems [9–12] follows from the simple fact that "chaos" leads to universality. This universality is captured, to some extent, by random matrix theory (RMT). On the other hand we have semiclassical methods. We shall see that there is a clash between RMT and semiclassics, and that they lead to different non-perturbative limits [13,14].

2 Model Systems

The classic example for a driven system is the piston model (Fig. 1), where a gas in confined inside a cylinder, and x is the position of the piston. Our interest is in the case where we have "one particle gas". [Note however that if we know how to solve the problem for one particle, then automatically we can get the solution for many non-interacting particles].

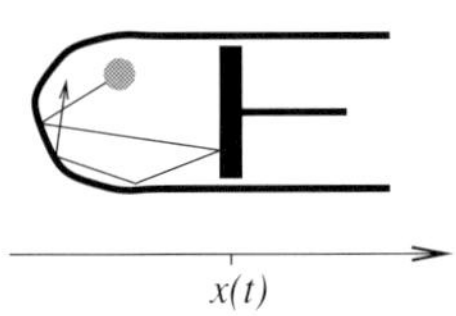

Fig. 1. The prototype piston model. A gas particle is moving chaotically inside a cylinder. The driving is achieved by moving a wall element ("piston").

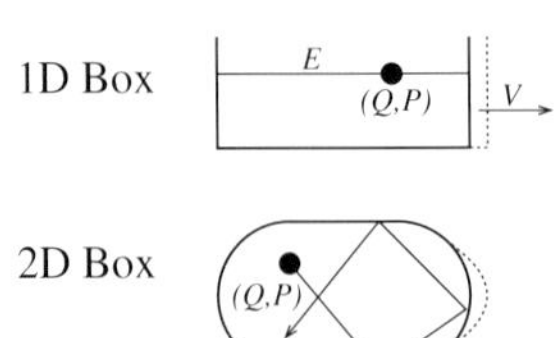

Fig. 2. The 1D version of the piston model (upper panel). The gas particle is moving inside an "infinite well". Its motion is not chaotic. In order to have chaotic motion we should consider at least a 2D box, for example a stadium shaped billiard system (lower panel).

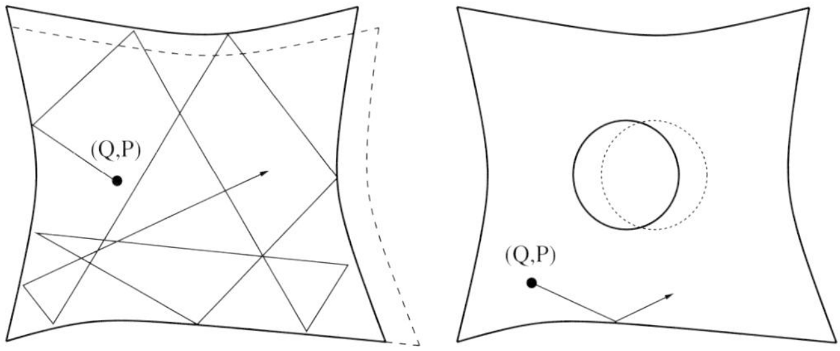

Fig. 3. Other versions of the "piston model". Two examples of "Sinai billiards" are illustrated. Note that in case of the right panel the displacement of the wall element manifestly does not involve a change of volume. This feature simplifies the analysis.

The 1D-box version of this model (Fig. 2a) is known in the literature as the "infinite-well" problem with moving wall [15,16]. Some limited aspects of this problem have been discussed in the literature in connection with the Fermi acceleration problem [17].

A 2D-box variation of the "piston model" is presented in Fig. 2b. Here we have stadium shaped billiard, and the the parameter x controls the deformation of the boundary. Two other variations of the same model are presented in Fig. 3, where the box has the shape of a generalized Sinai billiard.

In the examples so far the parameter x controls the shape of the "box", and $V = \dot{x}$ has the interpretation of wall velocity. The interest in such systems has emerged long time ago in studies of nuclear friction (one-body dissipation) [18,19]. A renewed interest is anticipated in mesoscopic physics where the shape of a quantum dots can be controlled by gate voltages. [Note that in the nuclear physics context the shape is close to spherical, while a quantum dot is typically strongly chaotic].

We can create driving by changing any parameter (or field). In Fig. 4 the driving is achieved by changing the perpendicular magnetic field. Fig. 4a assumes "quantum dot geometry" with homogeneous magnetic field, while Fig. 4b

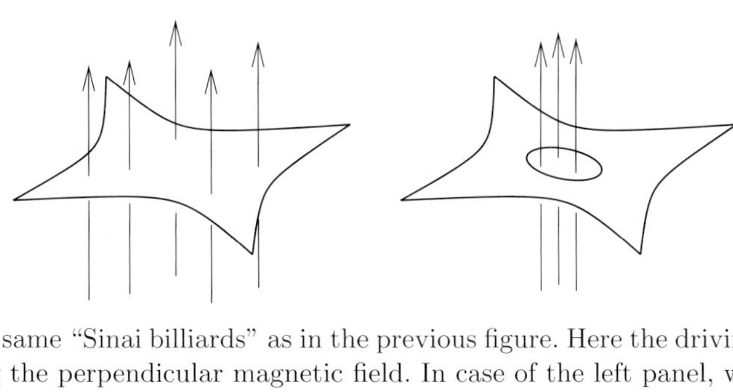

Fig. 4. The same "Sinai billiards" as in the previous figure. Here the driving is achieved by changing the perpendicular magnetic field. In case of the left panel, where the box has a simple "quantum dot" geometry, the magnetic field is assumed to be homogeneous. In case of the right panel, where the box has a Aharonov-Bohm ring topology, the magnetic flux is assumed to be concentrated in the hole.

assumes Aharonov-Bohm ring geometry with magnetic flux that goes via the hole. Let us define x as the total magnetic flux. In such case $V = \dot{x}$ is the electro-motive force (measured in Volts) which is induced in the ring according to Faraday law.

If the variations of the parameter x are classically small, then we can linearize the Hamiltonian as follows

$$\mathcal{H}(Q, P; x) = \mathcal{H}_0(Q, P) + x\mathcal{W}(Q, P), \tag{1}$$

where without loss of generality we have assumed that $x = 0$ is the typical value of x. For generic systems (which means having smooth Hamiltonian that generates a classically chaotic motion), the representation of $\mathcal{W}$, in the ordered $\mathcal{H}_0$ determined basis, is known to be a banded matrix (for details see the next section). A simple example can be found in [20] where

$$\mathcal{H}(Q, P; x) = \tfrac{1}{2}(P_1^2+P_2^2 + Q_1^2+Q_2^2) + (1 + x) \cdot Q_1^2 Q_2^2 . \tag{2}$$

This Hamiltonian describes a particle moving inside a two dimensional anharmonic well (2DW). The shape of the 2DW in controlled by the parameter x. The perturbation is $\mathcal{W}(Q, P) = Q_1^2 Q_2^2$, and its matrix representation $\mathbf{B}_{nm}$ is visualized in the inset of Fig. 5.

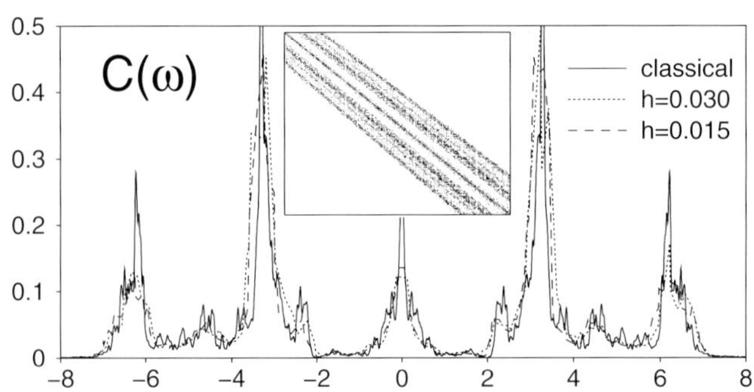

Fig. 5. The band profile $(2\pi\hbar/\Delta) \cdot |\mathbf{B}_{nm}|^2$ versus $\omega = (E_n - E_m)/\hbar$ is compared with $\tilde{C}(\omega)$. See text for further explanations. The calculation is done for the 2DW model of Eq. (2). The inset is an image of a piece of the **B** matrix. Taken from Ref. [20].

The above discussion of generic Hamiltonian models, such as the 2DW model, motivates the definition of a simple artificial model Hamiltonian, that has the same characteristics: This is Wigner model [21,22]. In the following definition of Wigner model we follow closely the notations of [13]. In the standard representation $\mathcal{H}_0 = \mathbf{E}$ is a diagonal matrix whose elements are the ordered energies $\{E_n\}$, with mean level spacing Δ, and $\mathcal{W} = \mathbf{B}$ is a random banded matrix with non-vanishing couplings within the band $0 < |n - m| \leq b$. These coupling elements are zero on the average, and they are characterized by the variance $\sigma = (\langle |\mathbf{B}_{nm}|^2 \rangle)^{1/2}$. Hence the Hamiltonian is

$$\mathcal{H} = \mathbf{E} + x\mathbf{B} . \tag{3}$$

This artificial model can serve as a reference case for testing various theoretical ideas. Moreover, it has been conjectured that such model captures some generic features of chaotic systems. [Note that most of the RMT literature deals with simplified versions of Wigner model, where the bandwidth equals to the matrix size].

3 Quantum Chaos

The notion of chaos in classical mechanics implies that few degree of freedom system, such as the Sinai billiard system, exhibit stochastic-like behavior. This is in contrast to the out-of-date idea that stochasticity and irreversibility are the outcomes of having (infinitely) many degrees of freedom. Chaos means that the motion (e.g. Fig. 6) has exponential sensitivity to any perturbation or change in initial conditions. Another way to characterize a chaotic motion is by its continuous power spectrum (see Fig. 7). This should be contrasted with integrable motion which is characterized by a discrete (rather than continuous) set of frequencies.

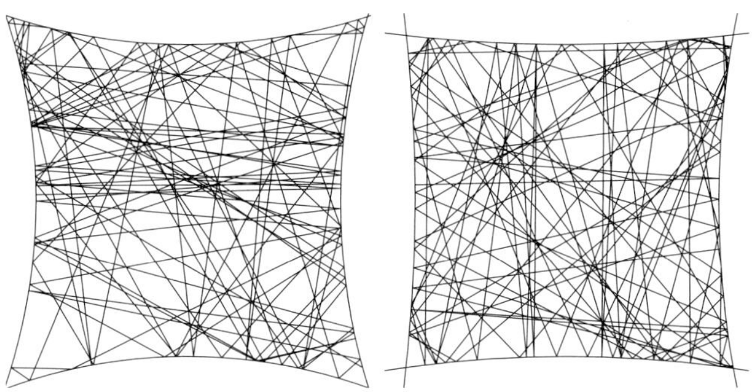

Fig. 6. Chaotic trajectories inside a Sinai billiard. In both examples the motion is completely chaotic. This mean exponential sensitivity to any small change in the initial conditions. This sensitivity can be characterized by the "Lyapunov exponent". In the right illustration the motion is chaotic, but the chaos is weaker, which means smaller Lyapunov exponent.

For sake of later analysis it is useful to define the "power spectrum" of the motion specifically as follows. Let $(Q(t), P(t))$ be an ergodic trajectory that is generated by the time independent Hamiltonian $\mathcal{H}(Q,P;x)$. We can define a fluctuating quantity $\mathcal{F}(t) = -d\mathcal{H}/dx$. In case that x is the displacements of a wall element (e.g. Fig. 3b), the fluctuating $\mathcal{F}(t)$ has the meaning of "Newtonian force". In case that x is the magnetic flux (e.g. Fig. 4b), the fluctuating $\mathcal{F}(t)$ has the meaning of "electric current". In case of the 2DW model we get $\mathcal{F}(t) = -\mathcal{W}(Q(t),P(t)) = -Q_1(t)^2 Q_2(t)^2$. The correlation function of the fluctuating $\mathcal{F}(t)$ will be denoted by $C(\tau)$ and the power spectrum of the fluctuations will be denoted by $\tilde{C}(\omega)$. The latter is the Fourier transform of the former. The variance

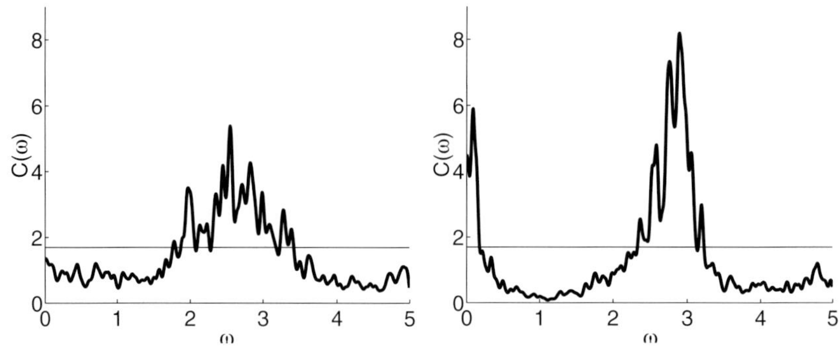

Fig. 7. The power spectrum of the motion for the two examples of the previous figure. It is the power spectrum $\tilde{C}(\omega)$ of the fluctuating quantity $\mathcal{F}(t) = -d\mathcal{H}/dx$. The latter can be described as train of impulses (spikes) due to collisions with the walls. For strongly chaotic motion (left panel) the power spectrum of $\mathcal{F}(t)$ is like that of white noise. In the right panel the bouncing frequency is quite pronounced, and there is also a "diabatic" peak around $\omega = 0$. In both cases, the motion is characterized by a continuous power spectrum, which constitutes an indication for the chaotic nature of the motion.

of the fluctuation is $C(0)$, the intensity of the fluctuations is defined as $\tilde{C}(0)$, and the correlation time is denoted by τ_{cl}.

It is clear that upon quantization we no longer have chaos. Still, the question arise what are the fingerprints of the classical chaos on both the spectral properties of the system, and also on the structure of the eigenstates. This problem was the focus of intensive studies during the last decade [1,3,2], and it has important applications in mesoscopic physics [23–25].

An important observation of "quantum chaos" studies is that Quantum Mechanics introduce *two* additional energy scales into the problem (rather than only one). We can take the 2DW model as a generic example. After rescaling of the classical parameters of the model, we are left with one dimensionless parameter (the dimensionless energy). This parameter controls the nature of the classical dynamics. Upon quantization we have *two* additional (dimensionless) parameters. One energy scale is obviously the mean level spacing Δ, which is proportional to $\hbar^d$. The other energy scale is $\Delta_b = \hbar/\tau_{cl}$, where τ_{cl} is the classical correlation time that characterizes the (chaotic) dynamics. If $\hbar$ is small then the two energy scales are very different ($\Delta \ll \Delta_b$).

The significance of the energy scale Δ_b is a central issue in "quantum chaos". It turns out that the statistical properties of the energy spectrum are universal on the sub-$\hbar$ scale, and obey the predictions of RMT. On the other hand, on large energy scale (compared with Δ_b), non-universal (system specific) features manifest themselves [26]. These features are the fingerprints of the underlying classical dynamics. In the context of ballistic quantum dots, which are in fact billiard systems, Δ_b is also known as the "Thouless energy".

There is another way in which the energy scale $\varDelta_b$ manifests itself. Let $\mathcal{W}(Q,P)$ be some observable, and consider its matrix representation $\mathbf{B}_{nm}$ in the basis which is determined by the (chaotic) Hamiltonian. An example is presented in Fig. 5. It can be argued [27] that $\mathbf{B}_{nm}$ is a banded matrix, and that the bandwidth is $b = \varDelta_b/\varDelta$. This is based on a remarkably robust semiclassical expression that relates the bandprofile to the classical power spectrum:

$$\left\langle |\mathbf{B}_{nm}|^2 \right\rangle \approx \frac{\varDelta}{2\pi\hbar}\, \tilde{C}\left(\frac{E_n - E_m}{\hbar}\right) . \tag{4}$$

We can apply this semiclassical relation to the case where $\mathcal{W}$ is the "perturbation" as defined in Eq. (1). This leads to the interpretation of $\varDelta_b$ as the largest "distance" in energy space that can be realized in a first-order transition. We can also use the semiclassical relation in reverse, in order to find/define the classical correlation function that corresponds to a quantum-mechanical matrix Hamiltonian. In case of the standard Wigner model we get $C(\tau) = b\sigma^2 \sin(\tau/\tau_{cl})/(\tau/\tau_{cl})$, with the correlation time $\tau_{cl} = \hbar/(b\varDelta)$.

4 Parametric Evolution

A more recent development was to consider a parametric set of Hamiltonians, namely $\mathcal{H}(Q,P;x)$ where x is a parameter as in the examples of Section 2. For each value of x we can diagonalize the Hamiltonian, leading to set of (ordered) eigen-energies $E_n(x)$, as in the schematic illustration of Fig. 8. The corresponding eigenstates will be denoted by $|n(x)\rangle$. Their parametric evolution can be characterized by the parametric kernel

$$P(n|m) = |\langle n(x)|m(x_0)\rangle|^2 . \tag{5}$$

We shall use the notation $P(r) = P(n-m) = P(n|m)$, with implicit average over the reference state m. We shall refer to $P(r)$ as the "average spreading profile". This is in fact, up to scaling, the LDOS (local density of states, also known as strength function).

Let us characterize the perturbation by the quantity $\delta x = x - x_0$. The interesting question is how $P(r)$ evolves as we increase the perturbation δx. For the Wigner model the answer is known long ago [21,22,28,29]. $P(r)$ has a standard perturbative structure for very small δx. For larger δx it becomes a chopped Lorentzian, and for even larger δx it becomes a semicircle. We shall denote the border between the standard perturbative regime and the Wigner regime by δx_c, and the border between the Wigner regime and the non perturbative (semicircle) regime will be denoted by $\delta x_{\rm prt}$.

The explicit expressions are:

$$\delta x_c = \frac{\varDelta}{\sigma} \propto \hbar^{(1+d)/2} , \tag{6}$$

$$\delta x_{\rm prt} = \sqrt{b}\frac{\varDelta}{\sigma} = \frac{2\pi\hbar}{\tau_{\rm cl}\sqrt{C(0)}} , \tag{7}$$

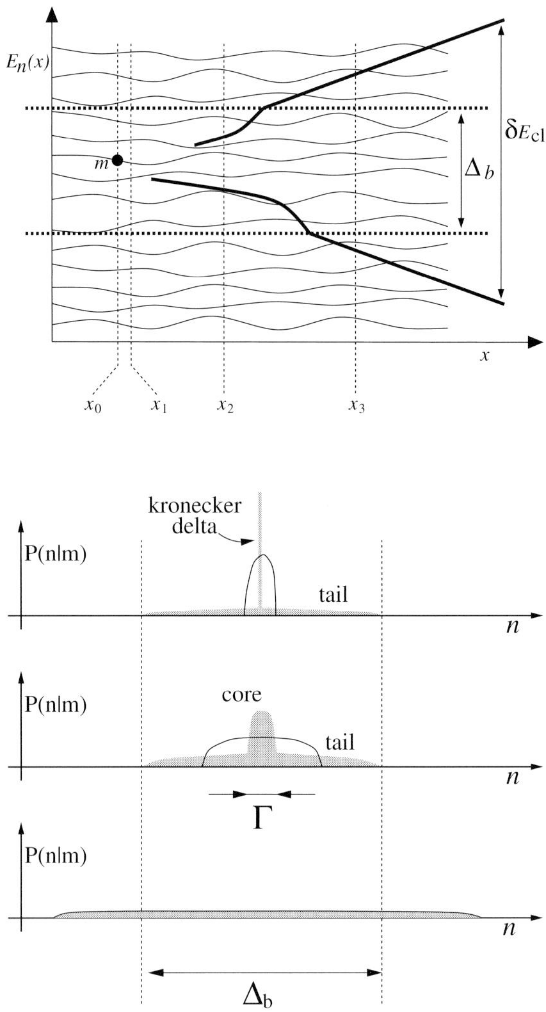

Fig. 8. Upper panel: Schematic illustrations of the set of energies $E_n(x)$ which are obtained via diagonalization of a parameter dependent Hamiltonian. The thick solid line indicates the n-range where 50% of the $P(n|m)$ probability is concentrated (m is fixed). The representative values x_1, x_2 and x_3 correspond to the standard perturbative regime, the core-tail (extended perturbative) regime, and the non-perturbative regime respectively. The corresponding LDOS structures are illustrated (grey shading) in the three plots of the lower panel. The semiclassical approximation (lines) is presented for sake of comparison.

where d is the number of freedoms ($d = 2$ for billiards). In order to determine the $\hbar$ dependence we have used the semiclassical relation Eq. (4), and the proportionality $\Delta \propto \hbar^d$. Note that the latter relation, known as Weyl law, is significant for the determination of δx_c. In contrast to that $\delta x_{\rm prt} \propto \hbar$ is in fact independent of Δ.

The generalization of Wigner scenario has been the subject of our recent research [30,20,31]. In the general case the standard perturbative structure evolves into a "core-tail structure", while for large δx it becomes purely non-perturbative. In the standard perturbative regime ($\delta x \ll \delta x_c$) most of the probability $P(r)$ is concentrated in one level ($n = m$). In the extended perturbative regime most of the probability is concentrated within a "core" whose width is typically $\Gamma(\delta x) \sim (\delta x/\delta x_c)^2$. The "core" is the non-perturbative component which arise due to non-perturbative mixing of nearby levels. The "tail" is the outer perturbative component which is created by first order transitions.

The extended perturbative regime is defined by the requirement of having separation of energy scales $\Gamma(\delta x) \ll \Delta_b$. This condition is trivially satisfied in the "standard perturbative regime" where $\Gamma \sim \Delta$. The condition $\Gamma(\delta x) \ll \Delta_b$ is violated in the non-perturbative regime ($\delta x \gg \delta x_{\text{prt}}$), which in fact leads to the determination of δx_{prt} as in Eq. (7). The theory for $P(r)$ in the non-perturbative regime is not complete yet. However, it can be argued [30] that if δx is large enough, then $P(r)$ becomes of semiclassical nature [32]. The case of billiards with shape deformation requires special considerations and is of particular interest [30,31].

It is important to realize that the border of the standard perturbative regime (δx_c) is related to the energy scale Δ, while the border of the extended perturbative regime (δx_{prt}), which leads to the identification of the non-perturbative regime, is related to the bandwidth Δ_b.

5 Temporal Evolution

After considering the parametric evolution, the next logical stage is to consider the actual (temporal) evolution which is generated by the time dependent Hamiltonian $\mathcal{H}(Q, P, x(t))$. Then, in complete analogy, we can ask how the energy scales Δ and Δ_b are reflected in the actual evolution. We postpone the discussion of the latter issue to Section 11.

The purpose of the present and next sections is to define what does it mean "driving", and how do we quantify the temporal evolution. Without loss of generality we assume $x(0) = 0$. We would like to consider the following driving schemes:

- Linear driving
- One pulse driving cycle
- Periodic driving
- Driving reversal scenario
- Time reversal scenario

In the next section we define the various schemes, some of which are also illustrated in Fig. 9. The evolution is characterized by the obvious generalization of Eq. (5), namely

$$P_t(n|m) = |\langle n(x(t))|U|m(x(0))\rangle|^2 \,. \tag{8}$$

Here U is the evolution operator, with implicit dependence on the time t. The parametric kernel Eq. (5) can be regarded as corresponding to the "sudden" limit where $U \approx 1$. As in the parametric case we can define an average spreading profile $P_t(r)$, where $r = n - m$.

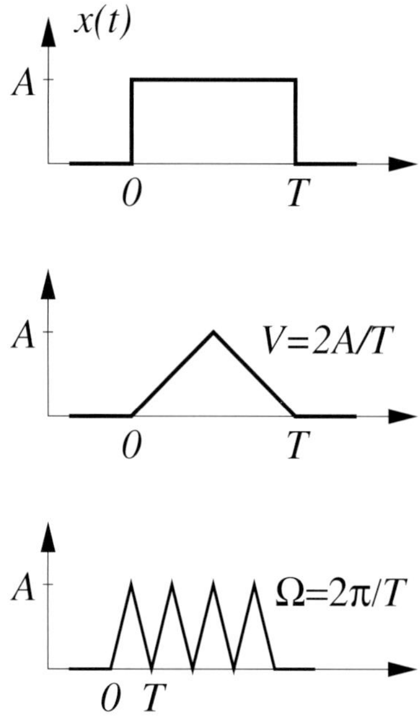

Fig. 9. Various driving schemes: Rectangular pulse of duration T and amplitude A; Triangular pulse which is further characterized by finite driving "velocity" $V = |\dot{x}| = 2A/T$; Periodic driving with frequency $\Omega = 2\pi/T$.

There are various practical possibilities available for the characterization of the distribution $P_t(r)$. It turns out that the major features of this distribution are captured by the following three measures:

$$\mathcal{P}(t) = P_t(r=0)\,, \tag{9}$$

$$\Gamma(t) = 50\% \text{ probability width}\,, \tag{10}$$

$$\delta E(t) = \Delta \times \left(\sum_r r^2 P_t(r) \right)^{1/2} . \tag{11}$$

The first measure is the survival probability $\mathcal{P}(t)$. The second measure $\Gamma(t)$ is the energy width of the central r region that contains 50% of the probability. [For simplicity of presentation we use here a loose definition of Γ as far as prefactors of order unity are concerned.] Finally the energy spreading $\delta E(t)$ is defined as the square-root of the variance.

It is important to realize that the above three measures give different type of information about the nature of the energy spreading profile. In particular the indication for having a core-tail structure is:

$$\Gamma(t) \quad \ll \quad \delta E(t) \quad \ll \quad \Delta_b \,. \tag{12}$$

The core-tail structure (e.g. chopped Lorentzian) is characterized by a "tail" component that contains a vanishingly small probability but still dominates the variance. [Note that $\Delta_b = \infty$, as in the case of un-chopped Lorentzian, would imply $\delta E(t) = \infty$ irrespective of Γ.] In contrast to that a typical semiclassical spreading profile (as well as Wigner's Semicircle) is characterized by

$$\delta E(t) \quad \sim \quad \Gamma(t) \quad \gg \quad \Delta_b \,. \tag{13}$$

In the latter case, in order to avoid confusion, it is better not to use the notation $\Gamma(t)$. [The notation Γ has been adopted in the common diagrammatic formulation of perturbation theory. This formulation is useful in the extended perturbative regime in order to derive Wigner's chopped Lorentzian. In the non-perturbative regime this formulation becomes useless, and therefore the significance of Γ is lost.]

6 Driving Schemes

Linear driving means $x(t) = Vt$, where V is a constant. In such a scenario obviously $x(t) \neq x(0)$. Still it is convenient to assume that the chaotic nature of the dynamics is independent of x, and that changes in x are not associated with changes in phase space volume (no conservative work is being done). This is manifestly the case for the systems which are illustrated in Fig. 3b and Fig. 4b. [Note that the standard Wigner model does not have x invariance property, and therefore the analysis of linear driving for the Wigner model is an ill defined problem. Attempts to overcome this difficulty lead to certain subtleties [11]].

For all the other driving schemes we assume, without loss of generality, that $x(0) = x(T) = 0$, where T is the period of the driving cycle. The simplest driving scheme is a rectangular pulse $x(0 < t < T) = A$, which is characterized by its amplitude A. Does the study of rectangular pulses constitute a good bridge for developing a general theory for driven systems? The answer is definitely not. An essential ingredient in the theory of driven system is the rate V in which the parameter x is being changed in time. Therefore, it is important to consider, for example, a triangular pulse (Fig. 9b). Such pulse is characterized by both amplitude A and driving rate $V = 2A/T$. More generally one may consider (Fig. 9c) a train of such pulses (= periodic driving). In particular one may consider the usual sinusoidal driving $x(t) = A\sin(\Omega t)$ where $\Omega = 2\pi/T$. In the later case we can define the root-mean-square driving rate as $V = \Omega A/\sqrt{2}$.

In all these cases we ask, in complete analogy with the parametric case, what is the evolution of the energy distribution $P_t(r)$. Now the evolution is with respect to time, rather than with respect to δx. The different scenarios are distinguished by the choice of U. We shall use the notation $U[x_A]$ in order to denote the evolution operator that corresponds to driving scheme $x = x_A(t)$.

The case of rectangular pulse is known in the literature as "wavepacket dynamics" [33]. The particle is prepared in an eigenstate of the $\mathcal{H}_0$ Hamiltonian, while the evolution is generated by the perturbed Hamiltonian $\mathcal{H} = \mathcal{H}(Q, P; x = A)$. We may consider more complicated schemes of pulses. For example rectangular pulse $+A$ followed by another rectangular pulse $-A$. The question here is whether the second pulse can compensate the effect of the first pulse. We call such scheme "driving reversal". The evolution operator can be written as $U = U[x(\text{rev})]U[x]$ where $x = A$ represents the rectangular pulse, while $x(\text{rev}) = -A$ is the reversed pulse. The case of triangular pulse can be regarded as another particular variation of driving reversal. In the latter case x represents linear driving with velocity V, and $x(\text{rev})$ is the reversed driving process with velocity $-V$.

The case of "driving reversal" should be distinguished from "time reversal" scheme. The latter notion is explained in the next section.

7 Two Notions of Irreversibility

There are two distinct notions of irreversibility in statistical and quantum mechanics. One is based on the "piston model" paradigm (PMP), while the other [34] is based on the "ice cube in cup of hot water" paradigm (ICP). The recent interest [35–40] in "quantum irreversibility" is motivated by its relevance to quantum computing.

In the PMP case we say that a process is reversible if it is possible to "undo" it by driving reversal. Consider a gas inside a cylinder with a piston (Fig. 1). Let us shift the piston inside. Due to the compression the gas is heated up. Can we undo the "heating" simply by shifting the piston outside, back to its original position? If the answer is yes, as in the case of strictly adiabatic process, then we say that the process is reversible.

In the ICP case we consider the melting process of an ice cube. Let us assume that after some time we reverse the velocities of all the molecules. If the external conditions are kept strictly the same, we expect the ice-cube to re-emerge out of the water. In practice the external conditions (fields) are not exactly the same, and as a result we have what looks like irreversibility.

The mathematical object that should be considered in order to study PMP is just $P_t(r)$ for a driving scheme that involves "driving reversal". Namely, as discussed in the previous section, the evolution operator is

$$U = U[x(\text{rev})]\, U[x] \,. \tag{14}$$

The mathematical object that should be considered in order to study ICP is again $P_t(r)$, but with driving scheme that involves "time reversal". Namely, the evolution operator is defined as

$$U = U[x_B]^{-1}\, U[x_A] \,. \tag{15}$$

In the latter case, if the external conditions are in full control ($x_B = x_A$), then we have complete reversibility ($U = 1$).

It is also important to define precisely what is the *measure* for quantum irreversibility. This is related to the various possibilities which are available for the characterization of the distribution $P_t(r)$. The prevailing possibility is to take the survival probability $\mathcal{P}(t)$ as a measure [34]. Another possibility is to take the energy spreading $\delta E(t)$ as a measure. The latter definition goes well with the PMP, and it has a well defined classical limit. Irreversibility in this latter sense implies diffusion in energy space, which is the reason for having energy absorption (dissipation) in driven mesoscopic systems (see Section 9).

8 Wavepacket Dynamics, Survival Probability and Fidelity

Driving schemes with rectangular pulses are the simplest for both analytical and numerical studies. It is easiest to consider the survival probability in case of a single rectangular pulse. The survival amplitude is defined as

$$\begin{aligned} F(t) &= \langle \Psi_0 | U[A] | \Psi_0 \rangle \\ &= \left| \left\langle \Psi_0 \right| \exp\left(-\frac{i}{\hbar} \mathcal{H} t \right) \left| \Psi_0 \right\rangle \right| . \end{aligned} \tag{16}$$

The survival probability is $\mathcal{P}(t) = |F(t)|^2$, in consistency with the definition of Eq. (9). $F(t)$ is manifestly a Fourier transform of the LDOS, and therefore we can immediately draw a conclusion [30] that the nature of the dynamics is different depending on the parametric regime to which the amplitude A belongs. If $P(r)$ have a core-tail Lorentzian structure, then we get an exponential decay $\mathcal{P}(t) = \exp(-\Gamma t)$. On the other hand if $P(r)$ has a semiclassical structure, then the decay of $\mathcal{P}(t)$ is non-universal (system specific).

A similar picture arise in recent studies of the survival probability for "time reversal" driving scheme. Here one defines the fidelity amplitude as

$$F(t) = \langle \Psi_0 | U[A]^{-1} U[0] | \Psi_0 \rangle . \tag{17}$$

The fidelity, also known as Loschmidt echo, is defined as $\mathcal{P}(t) = |F(t)|^2$. The situation here is more complicated compared with Eq. (16) because we have two LDOS functions [38]: one is the $\mathcal{H}_0$ weighted LDOS, and the other is the Ψ_0 weighted LDOS. The two LDOS functions coincide only if Ψ_0 is an eigenstate of $\mathcal{H}_0$. In the latter case the $F(t)$ of Eq. (17) reduces (up to phase factor) to Eq. (16). It turns out that in case of Eq. (17) there is no simple Fourier Transform relation between $F(t)$ and the LDOS functions. However, the picture "in large" is the same as in the case of Eq. (16) [38]. Namely, one has to distinguish between three regimes of behavior: In the standard perturbative regime ($A < \delta x_c$) one typically encounters a Gaussian decay [34]; In the Wigner regime (also called FGR regime) one typically finds Exponential decay [36]; And in the non-perturbative regime one observes a semiclassical perturbation-independent "Lyapunov decay" [35].

The study of the survival probability, as described above, is only one limited aspect of the temporal evolution. The more general object that should be considered is $P_t(r)$ as defined in Section 5. The major features of this time evolution are captured by the three measures that we have defined in Eq. (9)-(11). In Fig. 10 we display numerical simulations of wavepacket dynamics for the 2DW model. The energy spreading $\delta E(t)$ is plotted as a function of time. The first panel is the classical simulation, which in fact coincides with the "linear response" calculation.

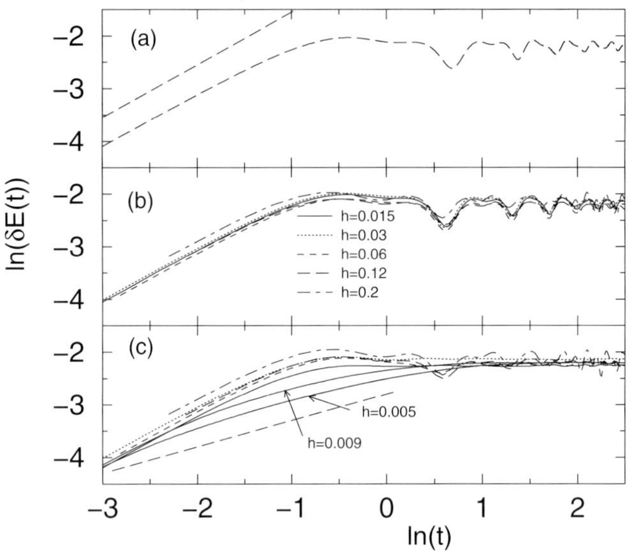

Fig. 10. Energy spreading as a function of time for the 2DW model: (a) classical; (b) quantum mechanical; (c) An effective Wigner model. The energy in these simulations is $E \sim 3$, and $\delta x = 0.2123$. The light dashed lines in (a) and (c), that have slopes 1 and 1/2 respectively, are drawn to guide the eye. In (c) different lines correspond to different values of $\hbar$ as in (b), and additional curves ($\hbar = 0.009, 0.005$) have been added. Taken from Ref. [14].

The input for the LRT calculation is $C(\tau)$, and the result is proportional to the amplitude A. Namely,

$$\delta E(t) = A \times \sqrt{2(C(0) - C(t))} \ . \tag{18}$$

In the second panel we display the results of the quantum mechanical simulations. For smaller $\hbar$ values the agreement with the classical LRT calculation is better. Finally, in the third panel we repeat the quantum mechanical simulations with a sign randomized Hamiltonian. This means that we take Eq. (3), and we randomize the sign of the off-diagonal terms. The bandprofile, and hence $\tilde{C}(\omega)$ are not affected by this procedure, which implies that the LRT calculation gives exactly the same result. But now we see that as $\hbar$ becomes smaller the correspondence with the classical result becomes worse. Specifically: In (a) and (b) we see a crossover from ballistic spreading ($\delta E \propto t$) to saturation ($\delta E \sim \text{const}$) as implied by Eq. (18). Only one time scale ($\tau_{\text{cl}} \sim 1$) is involved. In (c), in contrast to that, we see that as $\hbar \to 0$ an intermediate stage of diffusion ($\delta E \propto \sqrt{t}$) develops.

How can we explain the above results. Obviously we see that for small $\hbar$ we cannot trust LRT. What in fact happens is that we have a crossover from the perturbative regime ($A < \delta x_{\text{prt}}(\hbar)$) to the non-perturbative regime ($A > \delta x_{\text{prt}}(\hbar)$). In the latter case we get either semiclassical behavior, or RMT behavior. In other words, random matrix theory and the semiclassical theory lead to *different* non-perturbative limits. In the semiclassical case the crossover from LRT behavior to non-perturbative behavior cannot be detected by looking on $\delta E(t)$. Still the crossover can be detected by looking on $\Gamma(t)$. See [14] for details.

9 Diffusion in Energy Space and Dissipation

In the following sections we discuss the case of either linear or periodic driving. In such case the long time behavior of the system is characterized by diffusion in energy space. Associated with this diffusion is a systematic increase of the average energy. This irreversible process of energy absorption is known as "dissipation".

There is a satisfactory classical theory for dissipation [41]. Some of the mathematical details are subtle, but the overall physical picture is quite simple. Without loss of generality the main idea can be explained by referring to the billiard example of Fig. 1a. The particle executes chaotic motion, and we may say that each collision has roughly equal probability to be with either the inward-going or with the outward-going wall. As a result the particle either gain or loose kinetic energy. Thus, the dynamics in energy space is like random walk, and it can be described by a diffusion equation. Thus we see that due to the chaos we have stochastic-like energy spreading.

This classical diffusion process is irreversible in the PMP sense. Let us assume that we start with a microcanonical distribution that has definite energy. If, after some time, we reverse the velocity of the walls, then obviously we do not get back the initial microcanonical distribution.

The effect of dissipation is related to the irreversible stochastic-like diffusion in energy space. If the diffusion rate were the same irrespective of the energy, then obviously the average energy would be constant. But this is not the case. The diffusion is stronger as we go up in energy, and as a result the diffusion process is biased. Thus the average energy systematically grows with time, and one can derive a general *diffusion dissipation relation* [44]:

$$\frac{d}{dt}\langle \mathcal{H} \rangle \ = \ -\int_0^\infty dE \ g(E) \ D_{\text{E}} \ \frac{\partial}{\partial E}\left(\frac{\rho(E)}{g(E)}\right) , \tag{19}$$

where $g(E)$ is the density of states, and $\rho(E)$ is the probability distribution (e.g. microcanonical, canonical or Fermi occupation). The diffusion picture is generally valid in the classical case, and it is typically valid also in the quantum mechanical case. [The issue of dynamical localization due to strictly periodic driving [6] is important for driven 1D system, but not so important in the case of driven chaotic systems [11].]

There is a simple linear response (Kubo) expression, that relates the diffusion coefficient to the power spectrum $\tilde{C}(\omega)$ of the fluctuations:

$$D_{\mathrm{E}} \;=\; \frac{1}{2}\tilde{C}(\Omega) \times V^2 \,. \tag{20}$$

The diffusion law for short times is $\delta E(t) = \sqrt{2D_{\mathrm{E}}t}$. This expression is completely analogous to Eq. (18). In both cases the spreading is proportional to the amplitude A. [Recall that for periodic driving we define $V = \Omega A/\sqrt{2}$. In the special case of linear driving the spreading is proportional to V.] Moreover, as in the case of wavepacket dynamics, the LRT result is the *same* classically and quantum-mechanically. But again, as in the case of the wavepacket dynamics, the validity regime of LRT in the quantum mechanical case is much smaller (see Section.11).

If we combine the above Kubo expression with the diffusion-dissipation relation we get

$$\frac{d}{dt}\langle\mathcal{H}\rangle = \mu(\Omega) \times V^2 \,, \tag{21}$$

where μ is related to the power spectrum of the fluctuations. Thus we get a fluctuations-dissipation relation [44]. The standard "thermal" fluctuation-dissipation relation $\mu(0) = \tilde{C}(0)/(2k_BT)$ is obtained from Eq. (19) in case of canonical $\rho(E)$.

Standard textbook formulations [44] takes linear response theory together with thermal statistical assumptions as a package deal. Our presentation provides a more powerful picture. On the one hand we can discuss non-equilibrium situation using LRT combined with an appropriate version of the diffusion-dissipation relation. On the other hand, we may consider situation where LRT does not apply. In such case we may get some (non-perturbative) result for the diffusion, and later use the diffusion-dissipation relation in order to calculate the dissipation rate.

10 Beyond Kinetic Theory

The coefficient μ in Eq. (21) is called the "dissipation coefficient". In the case where x is the displacement of a wall element, it is also known as "friction coefficient", and in the case where x is a magnetic flux it can be called "conductance".

Having dissipation rate proportional to V^2 is known as "ohmic" behavior. In case of "friction" it is just equivalent to saying that there is a friction force proportional to the velocity V, against which the wall is doing mechanical work. This mechanical work is "dissipated" and the gas is "heated up" in a rate proportional to V^2.

In case of "conductance" we may say that there is a drift current proportional to the voltage V. This is in fact "Ohm law". The dissipated energy can either be accumulated by the electrons (as kinetic energy), or it may be eventually transferred to the lattice vibrations (phonons). In the latter case we say that the ring is "heated up". The rate of the heating goes like V^2 which is in fact "Joule law".

The traditional approach to calculate μ is to use a kinetic picture (Boltzmann) which is based on statistical assumptions. This leads in case of friction to the "wall formula" [18,19]:

$$\mu(\Omega) = \frac{N}{\mathsf{V}_{\text{box}}} m v_{\text{E}} \mathsf{A}_{\text{walls}} \tag{22}$$

where N is the number of gas particles (let us say $N = 1$), and $v_{\text{E}} = \sqrt{2E/m}$. We also use the notations V_{box} for the volume of the box, and $\mathsf{A}_{\text{walls}}$ for the effective area of the moving walls. In the latter we absorb some geometric factors [12]. Application of the traditional kinetic (Boltzmann) approach in case of conductance leads to "Drude formula":

$$\mu(\Omega) \sim \frac{N}{\mathsf{A}_{\text{dot}}} \left(\frac{e^2}{m}\tau_{\text{col}}\right) \frac{1}{1 + (\tau_{\text{col}}\Omega)^2} , \tag{23}$$

where A_{dot} is the area of the "quantum dot", and τ_{col} is the average time between collisions with the walls.

Using Linear response theory (Kubo formula), as described in the previous section, we can go beyond the over-simplified picture of kinetic theory. That means to go beyond Boltzmann picture. Below we explain under what assumptions we get the "traditional" kinetic expressions, and what in fact can go wrong with these assumptions.

The interest in friction calculation has started in studies of "one body dissipation" in nuclear physics [18,19]. The "wall formula" assumes that the collisions are totally uncorrelated. In such case the power spectrum $\tilde{C}(\omega)$ of $\mathcal{F}(t)$ is like that of white noise (namely "flat"). By inspection of Fig. 7 we can see that this assumption is apparently reasonable in the limit of very strong chaos. But it is definitely a bad approximation in case of weak chaos. The dynamics of chaotic system is typically characterized by some dominant frequencies. Therefore we have relatively strong response whenever the driving frequency matches a "natural" frequency of the system. This can be regarded as a classical (broad) resonance. By inspection of Fig. 7 we see that a particular feature is having such resonance around $\omega = 0$. This type of resonance, due to bouncing behavior, can be regarded as a "classical diabatic effect" [46].

Even if the chaos is very strong, the "white noise" assumption is not necessarily correct: In [47,45] we explain that for special class of deformations (including translations, rotations and dilations) the low frequency response is suppressed, irrespective of the chaoticity. This is illustrated numerically in Fig. 11.

In case of Drude formula the fluctuating $\mathcal{F}(t)$ has the meaning of "electric current", and therefore the power spectrum $\tilde{C}(\omega)$ is the Fourier transform of

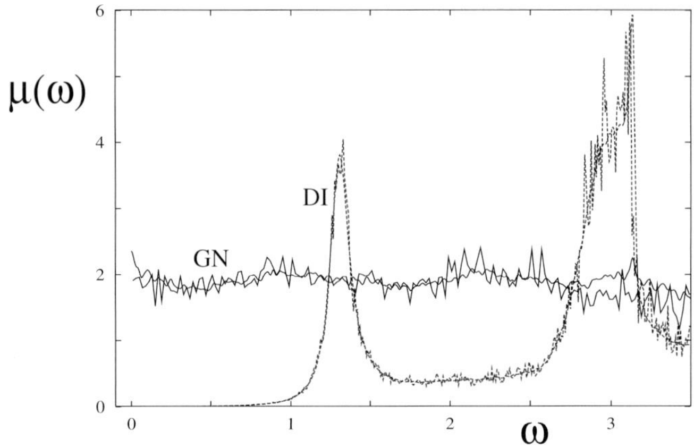

Fig. 11. The dependence of the friction coefficient on the driving frequency, using LRT. "GN" stands for generic deformation of a stadium shaped billiard, while "DI" stands for special deformation (dilation). [In the latter case the friction coefficient vanishes in the low frequency limit.] In both cases the agreement between the classical (solid line) and the quantum-mechanical (dashed line) calculation is remarkable. Taken from Ref. [47].

the current-current (or one may say velocity-velocity) correlation function. Assuming that the velocity-velocity correlation function decays exponentially in time, one obtains the Drude result. A careful analysis of this assumption, and its relation to the "white noise approximation" of the "wall formula", can be found in Section 6 of [45]. Fig. 12 displays a numerical example. We clearly see non-universal deviations from the Drude expression, which reflect the specific geometry of the "quantum dot".

11 Non-perturbative Response

In the classical case, assuming idealized system, the crossover to stochastic energy spreading involves only one time scale, which is τ_{cl}. Gaussian-like spreading profile is obtained only for time t much larger than τ_{cl}. For short times we can use linear analysis in order to calculate the spreading profile. However, this analysis has a breaktime [12] that we call $t_{\text{frc}}(V)$, where V is the rate in which x is being changed. For long times ($t \gg \tau_{\text{cl}}$) we can use stochastic picture. Classical LRT calculation of the diffusion is valid only if the crossover to stochastic behavior is captured by the short time analysis. This leads to the classical slowness condition $\tau_{\text{cl}} \ll t_{\text{frc}}(V)$ which we assume from now on. See specific examples in Sections 13 and 14.

In the quantum mechanical case we follow a similar reasoning. The linear analysis is carried out using perturbation theory. We have presented [12] a careful analysis to determine the breaktime $t_{\text{prt}}(V;\hbar)$ of this analysis. It turns out that

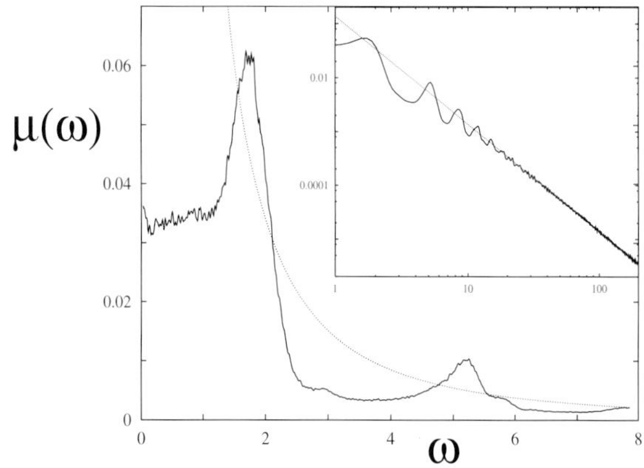

Fig. 12. The dependence of the mesoscopic conductance on the driving frequency. The calculation is done for a Sinai billiard shaped quantum dot, using LRT. The result can be regarded as a mesoscopic version of Drude formula. The specific geometry of the system is reflected in the structure of the response curve. The inset is log-log plot. Taken from Ref. [45].

this breaktime is not related to the mean level spacing Δ, but rather to the much larger energy scale Δ_b.

In complete analogy with the classical analysis, it turns out that the validity of LRT in the quantum domain is restricted by the condition $\tau_{\rm cl} \ll t_{\rm prt}(V;\hbar)$. If this inequality is not satisfied, then we say that we are in a non-perturbative regime. It is important to realize that the $\hbar \to 0$ limit is a non-perturbative limit. This means that the semiclassical regime is contained within the non-perturbative regime.

In the simple examples that are discussed in Sections 13 and 14, the non-perturbative regime is in fact a semiclassical regime. This coincidence does not hold in general [9,10,12]. In case of RMT models, obviously we do not have a semiclassical limit. In such models the non-perturbative response deviates significantly from Kubo formula (Fig. 13). The interest in such models can be physically motivated by considering transport in quantized *disordered* systems. Whether similar deviations from Kubo formula can be found in case of quantized *chaotic* systems is still an open question [11]. In any case, it is important to remember that the rate of dissipation is just one aspect of the energy spreading process. Even if Kubo formula does not fail (thanks to quantum-classical correspondence), still there are other features of the dynamics that are affected by the crossover from the perturbative to the non-perturbative regime. For example: in Sec.19 we are going to show that different results are obtained for the dephasing time, depending whether the process is perturbative or non-perturbative.

We can express the condition for being in the non-perturbative regime as [12]

$$V \gg \frac{\delta x_{\rm prt}}{\tau_{cl}}. \tag{24}$$

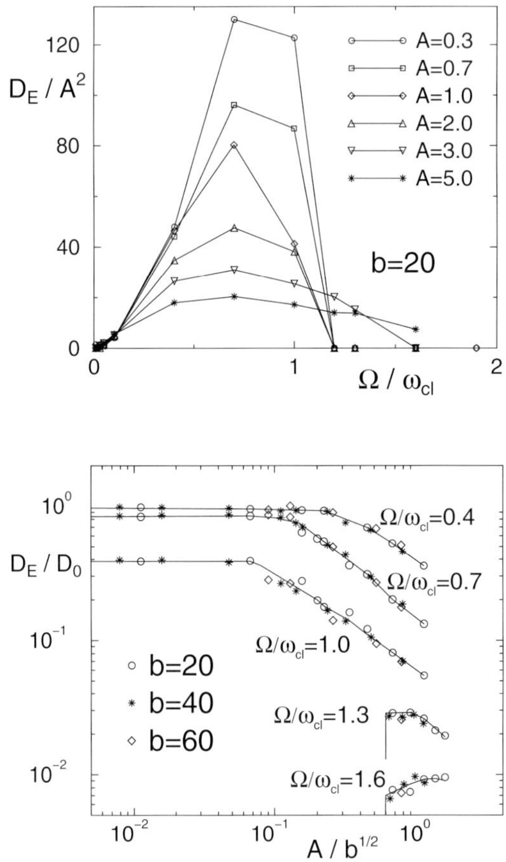

Fig. 13. The response of a quantum mechanical system is displayed as a function of A and Ω. The simulations are done for Wigner model with periodic driving. The units are chosen such that $\Delta = 0.5$ and $\hbar = 1$ and $\sigma = 1$. **Upper panel:** Plots of $D_{\rm E}/A^2$ versus $\Omega/\omega_{\rm cl}$ for few values of A. For small ω the plots coincide as expected from LRT. As A becomes larger the deviations from LRT scaling become more pronounced, and we get response also for $\Omega > \omega_{\rm cl}$. **Lower panel:** Plots of $D_{\rm E}/D_0$ versus $A/\sqrt{b}$ for few values of $\Omega/\omega_{\rm cl}$. LRT implies $D_{\rm E}/D_0 = 1$ for $\Omega/\omega_{\rm cl} < 1$ and $D_{\rm E}/D_0 = 0$ for $\Omega/\omega_{\rm cl} > 1$. The purpose of the horizontal scaling is to demonstrate that $A_{\rm prt}$ rather than A_c is responsible for the deviation from this LRT expectation. Taken from Ref. [11].

The expression in the right hand side scales like $\hbar$, which reflects that this condition is related to Δ_b and not to Δ. In the next section we discuss the definition of the adiabatic regime (very small V) whose existence is related to having finite Δ. A schematic illustration of the three regimes (adiabatic, LRT, non-perturbative) is presented in Fig. 14. Some further reasoning [11] allows to define the non perturbative regime in case of periodic driving. Its location in (A, Ω) space is also illustrated in Fig. 14. Note that for periodic driving we define $V = \Omega A/\sqrt{2}$. The two $V = \text{const}$ curves in the (Ω, A) diagram represent

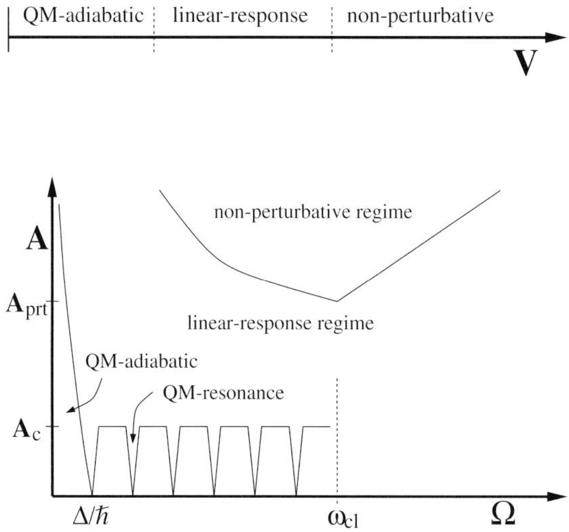

Fig. 14. Upper diagram: The various V regimes in the theory of driven systems for linear driving $x(t) = Vt$. Lower diagram: The various (Ω, A) regimes for periodic driving $x(t) = A\sin(\Omega t)$. We use the notations $\omega_{cl} = 2\pi/\tau_{cl}$ and $A_c = \delta x_c$ and $A_{\rm prt} = \delta x_{\rm prt}$.

the same conditions as in the case of linear driving. Other details of this diagram are discussed in the next section, and in [11].

12 Adiabatic Response and QM Resonances

Let us assume that we are in the perturbative regime (which means that the non-perturbative regime of the previous section is excluded). We ask the following question: can we use the classical Kubo result as an approximation for the quantum mechanical result? The answer is "yes" with the following restrictions: (i) The amplitude of the driving should be large enough; (ii) The frequency of the driving should be large enough. The two conditions are further discussed below. If they are satisfied we can trust the classical result. This follows from the remarkable quantal-classical correspondence which is expressed by Eq. (4). We have an illustration of this remarkable correspondence in Figs. 7 and 11.

Large enough amplitude means $A \gg \delta x_c$. One may say that large-amplitude driving leads to effective "broadening" of the discrete levels, and hence one can treat them as if they form a continuum. This is essential in order to justify the use of Fermi golden rule (FGR) for a small isolated system [11]. Kubo formula can be regarded as a consequence of FGR. If the driving amplitude is not large enough to "mix" levels, we cannot use FGR, but we can still use first order perturbation theory as a starting point. Then we find out, as in atomic physics applications, that the response of the system is vanishingly small unless the driving frequency ω matches energy level spacing. This is called "QM resonance". The strips of QM resonances are illustrated in the schematic diagram of Fig. 14. It is important to realize that higher order of perturbation theory, and possibly non-perturbative corrections, are essential in order to calculated the non-linear response in this regime [4]. Still, first order perturbation theory is a valid starting point, and therefore we do not regard this (non-linear) regime as "non-perturbative".

Large enough frequency means $\omega \gg \Delta/\hbar$. The remarkable quantal-classical correspondence which is expressed by Eq. (4) is valid only on energy scales that are large compared with Δ. If this condition is not satisfied, we have to take into account the level spacing statistics [48,49]. This means that we can have significant difference between the quantal LRT calculation, and the classical LRT calculation.

However, this is not the whole story. If V is small enough, first order perturbation theory implies "QM adiabaticity". The condition for QM adiabaticity is $V \ll \delta x_c/t_{\rm H}$ where $t_{\rm H} = 2\pi\hbar/\Delta$ is the Heisenberg time. A useful way of writing the QM adiabaticity conditions is:

$$V \quad \ll \quad \frac{1}{b^{3/2}} \left(\frac{\delta x_{\rm prt}}{\tau_{cl}} \right) . \tag{25}$$

In the adiabatic regime, first order perturbation theory implies zero probability to make a transition to other levels. Therefore, to the extend that we can trust the adiabatic approximation, all the probability remains concentrated in the initial level. Thus, in this regime, as in the case of small amplitudes, it is essential to use higher orders of perturbation theory, and non-perturbative corrections (Landau-Zener [4]). Still we emphasize that first order perturbation theory is in fact a valid starting point, and therefore we do not regard this (non-linear) regime as "non-perturbative".

13 Driving by Electro-motive Force

Consider a charged particle moving inside a chaotic ring. Let x represent a magnetic flux via the ring. If we change x in time, then by Faraday law $V = \dot{x}$ is the electro-motive force (measured in Volts). The fluctuating quantity $\mathcal{F}(t)$ has the meaning of electric current. The variance of the fluctuations is $C(0) = (ev_{\rm E}/L)^2$, where $v_{\rm E} = \sqrt{2E/m}$, and L is the length of the ring. The correlation time of these fluctuations is the ballistic time $\tau_{cl} = \tau_{\rm col} = L_{\rm col}/v_{\rm E}$.

Having characterized the fluctuations, we can determine the bandwidth $\Delta_b = \hbar/\tau_{\rm col}$. A straightforward calculation leads to the result:

$$b \quad = \quad \left[\frac{L}{L_{\rm col}} \right] \times \left(\frac{L_\perp}{\lambda_{\rm E}} \right)^{d-1} , \tag{26}$$

where $\lambda_{\rm E} = 2\pi\hbar/(mv_{\rm E})$ is the De-Broglie wavelength, and $L_\perp$ is the width of the ring. Using Eq. (7) we can determine the non-perturbative parametric scale:

$$\delta x_{\rm prt} \quad = \quad \left[\frac{L}{L_{\rm col}} \right] \times \frac{\hbar}{e} , \tag{27}$$

which up to a geometric factor equals the quantal flux unit. Note that in order to mix levels a relatively small change in the flux is needed, as implied by comparing Eq. (6) with Eq. (7).

We turn now to the analysis of the spreading in the time dependent case, say for linear driving. The classical "slowness" condition which has been mentioned in section 11 is simply $eV \ll E$ where E is the kinetic energy of the charged particle. Upon quantization we should distinguish the non-perturbative regime using Eq. (24), leading to

$$eV \quad \gg \quad \left[\frac{L}{L_{\mathrm{col}}}\right] \frac{\hbar}{\tau_{\mathrm{col}}} . \tag{28}$$

Disregarding the geometric prefactor, the quantity in the right hand side is the so called Thouless energy. We also should distinguish the QM adiabatic regime using Eq. (25), leading to

$$eV \quad \ll \quad \left(\frac{\lambda_{\mathrm{E}}}{L}\right)^{3/2} \frac{\hbar}{\tau_{\mathrm{col}}} , \tag{29}$$

where we have assumed for simplicity a simple 2D quantum dot geometry as in Fig. 4a.

14 Driving by Moving Walls

There is an ongoing interest [15,16] in the problem of 1D box with moving wall (also known as the infinite well problem with moving wall). If the wall is moving with constant velocity, then it is possible to transform the Schrödinger equation into a time-independent equation, and to look for the stationary states.

We are interested in the dynamics, and therefore we would like to go beyond this limited scope of study. Before we discuss the general case, it is useful to point out the $d > 1$ generalization of the above picture. We can easily show that for any "special deformation" which is executed in either constant "velocity" or "acceleration", we can transform the Schrödinger equation into a time-dependent equation. By "special deformation" we mean either translation or rotation or dilation, or any linear combination of these. The statement is manifestly trivial for translations and rotations (it is like going to a different reference frame), but it is also true for dilations. The 1D box with moving wall is just a special case of dilation.

It is important to realize that in case of generic deformation of chaotic box, we cannot "eliminate" the time dependence. Thus it is not possible to reduce the study of "dynamics" to a search for "stationary solutions".

The determination of $\varDelta_b$ for this system is quite obvious but subtle [31]. As one can expect naively the result is $\varDelta_b = 2\pi\hbar/\tau_{\mathrm{col}}$, where τ_{col} is the mean time between collisions with the moving walls. The subtlety here is that we cannot interpret $\varDelta_b$ as "bandwidth". Formally the correlation time of $\mathcal{F}(t)$ is $\tau_{cl} = \infty$ which implies infinite bandwidth. Still, some non-trivial reasoning [30,31] leads to the conclusion that the naive result (rather than the "formal" one) is in fact effectively correct. A straightforward calculation leads to the result:

$$b = \frac{\mathsf{V}_{\text{box}}}{L_{\text{col}}\lambda_{\text{E}}^{d-1}} = \frac{\mathsf{A}_{\text{walls}}}{\lambda_{\text{E}}^{d-1}}, \tag{30}$$

where $\lambda_{\text{E}} = 2\pi\hbar/(mv_{\text{E}})$ is the De-Broglie wavelength, and V_{box} is the volume of the box, and L_{col} is the mean free path between collisions. As for the effective value of δx_{prt}, again the details are subtle, but the naive guess turns out to be correct. With the proper (natural) choice of units for the displacement parameter x, the result is simply $\delta x_{\text{prt}} = \lambda_{\text{E}}$.

The way to analyze the dynamics for box with moving walls is outlined in [62]. The classical LRT domain is $V \ll v_{\text{E}}$, where $v_{\text{E}} = \sqrt{2E/m}$. Upon quantization we should distinguish the non-perturbative regime using Eq. (24), leading to

$$V \gg \frac{\hbar}{mL_{\text{col}}}. \tag{31}$$

In the non-perturbative regime the dynamics has a semiclassical nature, and the energy spreading process has a resonant random-walk nature. This should be contrasted with the behavior in the perturbative non-adiabatic regime, where Fermi-golden-rule (FGR) picture applies.

We also should distinguish the QM adiabatic regime using Eq. (25), leading to

$$V \ll \left(\frac{\lambda_{\text{E}}^{d-1}}{\mathsf{A}_{\text{walls}}}\right)^{3/2} \frac{\hbar}{mL_{\text{col}}}. \tag{32}$$

In the QM adiabatic regime the spreading is dominated by transitions between near-neighbor levels: This is the so called Landau-Zener spreading mechanism [4]. See also Section 20 of [12], and the numerically related work in [50].

15 Brownian Motion

Brownian motion is a paradigm for the general problem of system that interacts with its environment. (See Fig. 15 and general discussion in the next section). One can imagine, in principle, a "zoo" of models that describe the interaction of a Brownian particle with its environment. However, following Caldeira and Leggett [55], the guiding philosophy is to consider "ohmic models" that give Brownian motion that is described by the standard Langevin equation in the classical limit. Four families of models are of particular interest:

- Interaction with chaos.
- Interaction with many-body bath.
- Interaction with harmonic bath.
- Interaction with random-matrix bath.

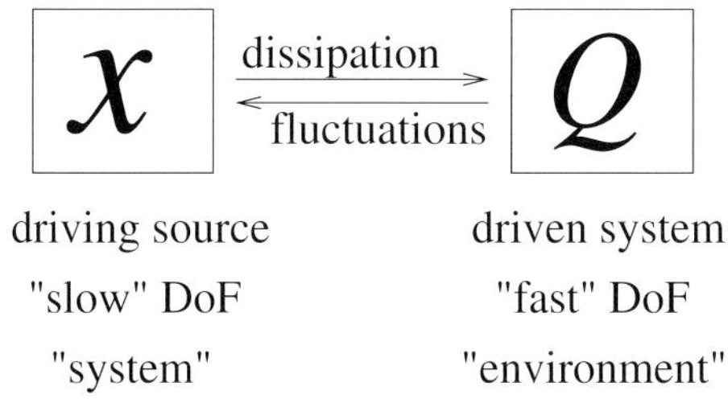

Fig. 15. Block diagram illustrating the interaction between system (x) and environmental (Q) degrees of freedom (DoF). See discussion in the text.

Below we assume that the total Hamiltonian has the following general form

$$\mathcal{H}_{\text{total}} = \mathcal{H}_0(x,p) + \mathcal{H}(Q,P;x)\,, \tag{33}$$

where (x,p) are the system coordinates, and (Q,P) are the environmental degrees of freedom.

Interaction with chaos provides the simplest model for Brownian motion [51]. See Fig. 16a for illustration of the model. The large Brownian particle is described by the canonical coordinates (x,p), while the gas particles are described by the canonical coordinates (Q,P). It is important to realize that in order to have Brownian motion, it is not essential to consider "many particle gas". "One particle gas" in enough, but the motion of the gas particle should be chaotic.

The fluctuations of the environment are in fact (according to our definition in Section 3) the random-like collisions of the gas particle with the Brownian particle. These fluctuations are like "noise". If the motion of the gas particle is strongly chaotic, then the power spectrum of these fluctuations (Fig. 6) is just like that of white noise. [This characterization is meaningful up to a cutoff frequency which is determined by the rate of the collisions.]

On the other hand we have the effect of dissipation. If the particle is launched with a velocity $\dot{x} = V$, then the rate of dissipation is proportional to V^2 as explained in section 9. Having dissipation implies that the Brownian particle experiences friction force which is proportional to the velocity V. This is the reason why the dissipation coefficient is known also as friction coefficient.

Interaction with chaos can be regarded as the "mesoscopic" version of Brownian motion. Our interest in this set of lectures is in this type of interaction. We want to know whether *few* degree of freedoms can serve as a "bath". Before we further get into this discussion we would like to describe the "conventional" point of view regarding Brownian motion. The rest of this section is dedicated for this clarification.

The conventional point of view regarding Brownian motion assumes an interaction with many body bath. We can consider a bath that consists of either Bosons or Fermions [56,57]. The emerging models are quite complicated for analysis, and therefore, as already mentioned above, it is more common to adopt a phenomenological approach.

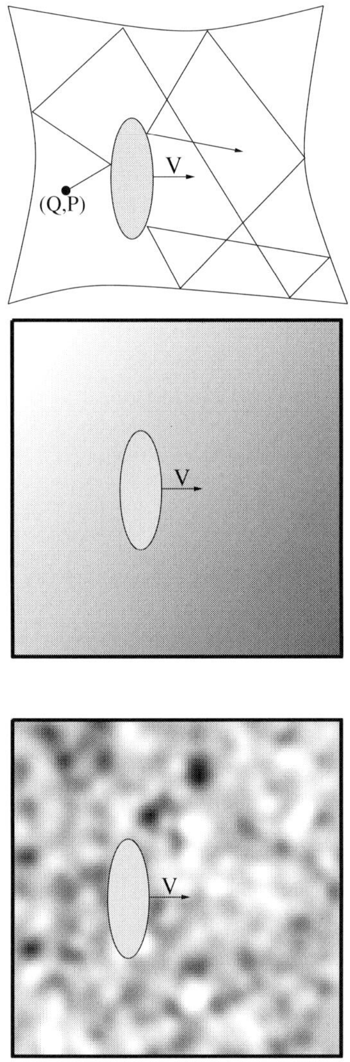

Fig. 16. (a) The Brownian motion is induced due to the interaction with chaotic degrees of freedom. (b) The Brownian particle in the ZCL model experiences a fluctuating homogeneous field of force. (c) In case of the DLD model the fluctuating field is farther characterized by a finite correlation distance. In (b) and (c) the background image is a "snapshot" of the fluctuating environment. Namely, the gray levels correspond to the "height" of an instantaneous potential which is experienced by the Brownian particle.

Interaction with (many body) harmonic bath is not very natural, but yet very popular model for Brownian motion. In order to have "white noise" (at high temperatures or in the classical limit) we should make a special assumption regarding the frequency distribution of the bath oscillators. This is known in the literature as the "ohmic choice". [The characterization of the noise as "white" is valid up to some cutoff frequency. The latter is determined by the specific choice of the frequency distribution.] Also here, as in the case of interaction with chaos, we have fluctuation-dissipation theorem that implies "ohmic" dissipation rate (proportional to V^2).

There is still some freedom left in the definition of the interaction with the harmonic bath. This leads to the introduction of the Diffusion-Localization-Dissipation (DLD) model [52–54]. This model gives in the classical limit Brownian motion which is described by the standard Langevin equation (white noise + ohmic dissipation). The familiar Zwanzig-Caldeira-Leggett (ZCL) model [55] can be regarded as a special limit of the DLD model. The physics of the ZCL and of the DLD model is illustrated in Fig. 16b and Fig. 16c respectively, and the model Hamiltonians can be visualized by the drawings of Fig. 17. The ZCL model describes a motion under the influence of a fluctuating homogeneous field of force which is induced by the environmental degrees of freedom. In case of the DLD model the induced fluctuating field is further characterized by a finite correlation distance.

For completeness we note that random-matrix modeling of the environment, in the regime where it has been solved [58], leads to the same results as those obtained for the DLD model.

16 System Interacting with Environment

The general problem of system that interacts with its environment is of great importance in many fields of physics. The basic ingredients of this interaction are illustrated in Fig. 15. On the one hand we have the effect of dissipation, meaning that energy is lost by the "system" (Brownian particle) and is absorbed by the "environment" (gas particles). On the other hand the environment induces fluctuations that acts like "noise" on the system. The "noise" has two significant effects: One is to pump "thermal" energy into the system, and the other is to spoil quantum coherence. The latter effect is called decoherence.

In case of bounded system, in the absence of external time dependent fields, the interplay between "noise" and "dissipation" leads eventually to "thermalization". One may say that in the thermal state the effect of dissipation is balanced by the energy which is pumped by the noise. Thus, both classically and quantum mechanically we have to distinguish between a "damping" scenario and an "equilibrium" situation. The thermalization process is traditionally described as "irreversible". On the other hand we have the issue of "recurrences". We discuss the latter issue in Section 20.

A systematic approach for the study of the dynamics of a "system", taking into account the influence of its environment, has been formulated by Feynman and Vernon [59]. The state of the system is represented by the probability matrix $\rho(x', x'')$. It is assumed that initially the "system" is prepared is some arbitrary state. Its state at a later time is obtained by a propagator $K(x', x''|x'_0, x''_0)$ which acts on the initial preparation. The calculation of this propagator involves a double path integral over all the possible trajectories $x_A(t)$ and $x_B(t)$ that connect

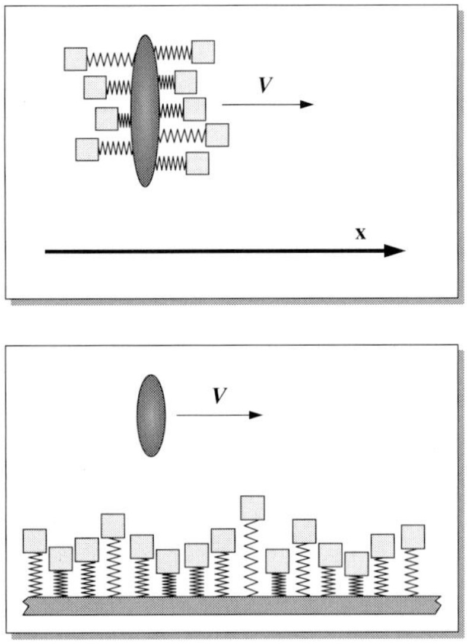

Fig. 17. Schematic illustration of the ZCL model (upper panel) and the DLD model (lower panel). The Hamiltonian of these "spring systems" is literally the ZCL model and the DLD model respectively. In the latter case the height of the masses should be interpreted as the "height" of the potential which is experienced by the particle.

(x'_0, x''_0) with (x', x''). This double path integral incorporates the effect of the environment via an "influence functional" which is defined as follows:

$$F[x_A, x_B] \quad = \quad \langle\, U[x_B]\Psi_0 \mid U[x_A]\Psi_0\, \rangle\,. \tag{34}$$

Here Ψ_0 is the initial state of the environment. If the environment is in "mixed" state, typically a canonical (thermal) state, then the influence functional should be averaged accordingly.

The absolute value of the influence functional can be *re-interpreted* as arising from the interaction with a c-number noisy field (with no back reaction). The "phase" of the influence functional can be regarded as representing the effect of "friction" (back reaction). Thus there is one to one correspondence between the Feynman-Vernon formalism, and the corresponding classical Langevin approach. Note however that the distinction between "noise" and "friction" is a matter of "taste". Some people regard this distinction meaningless.

It should be realized that the calculation of the influence functional for a given environment takes us back to the more restricted problem of considering a "driven system". The influence functional $F[x_A, x_B]$ is nothing but the survival amplitude for a driving scheme that involves "time reversal" (Eq. (15)).

17 Entanglement, Decoherence and Irreversibility

The definition of decoherence is not a trivial matter conceptually. There are several equivalent ways to think about decoherence. The most "robust" approach is to define decoherence as the irreversible entanglement of the system with the environment: Let us describe the the state of the system using the probability matrix $\rho(x', x'')$. If the system is prepared in pure state then trace$(\rho^2) = 1$. Due to the interaction with the environment the system gets entangled with the environment, and as a result we will have trace$(\rho^2) \leq 1$. If the "environment" consists of only "one spin", then we expect trace(ρ^2) to have "ups" and "downs", and from time to time to have trace$(\rho^2) \sim 1$. In such case we cannot say that the entanglement process is "irreversible". But if the environment consists of many degrees of freedom, as in the case of interaction with "bath", then the loss of "purity" becomes irreversible, and we regard it as a "decoherence process".

To be more specific let us consider the prototype example of interference in Aharonov-Bohm ring geometry. The particle can go from the input lead to the output lead by traveling via either arms of the ring. This leads to interference, which can be tested by measuring the dependence of the transmission on the magnetic flux via the ring. Consider now the situation where there is a spin degree of freedom in one arm [23]. The particle can cause a spin flip if it travels via this arm. In such case interference is lost completely. However, this entanglement process is completely reversible. We can undo the entanglement simply by letting the particle interact with the spin twice the time. Therefore, according to our restrictive definition, this is not a real decoherence process.

Consider now the situation where a particle gets entangled with bath degrees of freedom. If the bath is infinite, then the entanglement process is irreversible, and therefore it constitutes, according to our definition, a decoherence process.

At first sight it seems that for having irreversibility one needs "infinity". This point of view is emphasized in Ref. [60]: Irreversibility can be achieved by having the infinity of the bath (infinitely many oscillators), or of space (a lead that extends up to infinity). In this set of lectures we emphasize a third possibility: Having irreversibility due to the interaction with chaos. Thus we do not need "infinity" in order to have "irreversibility".

18 Interpretation of Decoherence as a Dephasing Process

"Dephasing" is used as a synonym for "decoherence" whenever a semiclassical point of view is adopted. Determining the dephasing (decoherence) time τ_φ for a particle (x, p) that interacts with an environmental degrees of freedom (Q, P) is a central theme in quantum physics. In the absence of such interaction the x motion is coherent, and interference should be taken into account. This means, from semiclassical point of view, that at least two trajectories $x(\tau) = x_{\mathrm{A}}(\tau)$ and $x(\tau) = x_{\mathrm{B}}(\tau)$ have a leading contribution to the probability to travel, say, from $x(0)$ to $x(t)$, as in the prototype example of the two slit experiment.

In the semiclassical framework the probability to travel from one point to some other point is given by an expression that has the schematic form

$$\sum_{A,B} F[x_A, x_B] \exp\left(i \frac{S[x_A] - S[x_B]}{\hbar}\right) , \qquad (35)$$

where $S[x]$ is the classical action, and $F[x_A, x_B]$ is the influence functional. Each pair of trajectories is a "stationary point" of the Feynman-Vernon double path integral. The "diagonal terms" are the so-called classical contribution, while the "off-diagonal terms" are the interference contribution. It should be kept is mind that the validity of the semiclassical approach is a subtle issue [61].

The off-diagonal interference contribution is suppressed due to the interaction with the environment if $|F[x_A, x_B]| \ll 1$. Therefore $|F[x_A, x_B]|$ is called the "dephasing factor". From the definition of the influence functional it is clear that it reflects the probability to "leave a trace" in the environment. Having $|F[x_A, x_B]| = 0$ means that a different "trace" is left in the environment, depending on whether the particle goes via the trajectory $x_A(t)$ or via the trajectory $x_B(t)$. In such case one can regard the interaction with the environment as a "measurement" process. In case of the DLD model (see Section 15) this "trace" can be further interpreted as leaving an excitation along the way. For critical discussion of this point see Appendix C of [54]. In the more general case the notion of "leaving a trace" does not have a simple meaning. All we can say is that decoherence means that the environment is left in different (orthogonal) states depending on the trajectory that is taken by the particle.

The law of "action and reaction" holds also in the world of decoherence studies. Feynman and Vernon have realized that the dephasing factor can be re-interpreted as representing the effect of a c-number noise source (see Section 16). From this point of view the decoherence is due to the "scrambling" of the relative phase by this noise. Hence the reason for using the term "dephasing" as a synonym for "decoherence". The analysis of dephasing using this latter point of view can be found in [54]. See also [63]. At high temperatures it is possible to use a Markovian master equation approach (dynamical semigroups) in order to obtain the (reduced) evolution of the Brownian particle. The Markovian master equation approach is described in other lectures of this school. The master equation in case of the DLD model can be found in Section 3 of [54]. Similar, but not identical master equations are obtained in case of interaction with many body bath [57].

19 Determination of the Dephasing Time

In the above described semiclassical framework, the problem of dephasing reduces to the more restricted problem of studying the dynamics of a time dependent Hamiltonian $\mathcal{H}(Q, P; x(t))$, cf. Fig. 18. Moreover, we see that the Feynman-Vernon dephasing factor is just the absolute value of the fidelity amplitude $F(t)$ that corresponds to Eq. (15). Note however that here we use a more general notion of fidelity: The restricted definition of fidelity (Eq. (17)) is formally obtained if $x_{\mathrm{A}}(\tau)$ and $x_{\mathrm{B}}(\tau)$ are "rectangular pulses".

The dephasing time τ_φ is defined as the time that it takes for $|F(t)|$ to drop significantly from $|F(t)| \sim 1$ to some very small value $|F(t)| \ll 1$. Let us concentrate on the Brownian motion model of Fig. 16a. If the motion of the Brownian particle is characterized by a velocity V, then we have to distinguish between the following possibilities: Having very small "adiabatic" velocities; Having intermediate velocities that allow LRT treatment; And having non-perturbative velocities. In the latter case a semiclassical picture can be justified.

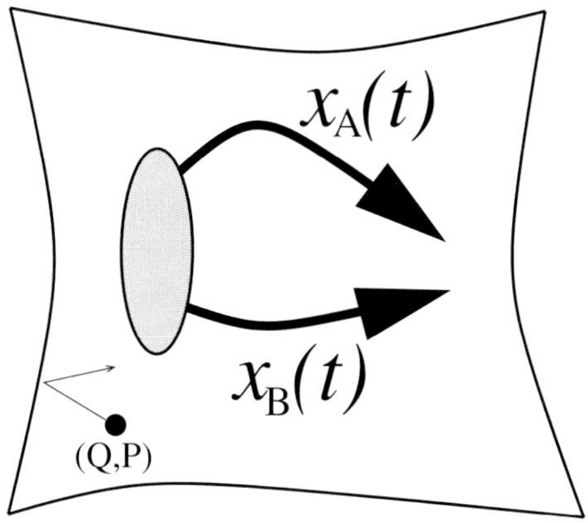

Fig. 18. Schematic illustration of an interference experiment using a semiclassical point of view. The Brownian particle can take either the $x = x_A(t)$ trajectory, or the $x = x_B(t)$ trajectory as in a two slit experiment.

The detailed analysis of the problem can be found in [62]. Here we just quote the final results. In the semiclassical regime

$$\tau_\varphi \;=\; \tau_{\rm col} \;=\; \frac{L_{\rm col}}{v_{\rm E}}\,, \tag{36}$$

where $L_{\rm col}$ is the mean free path between collisions with the Brownian particle. This is the expected naive result. It means that one collision with the Brownian particle is enough in order to "measure" its trajectory. The other extreme case in having extremely small adiabatic velocities. To the extend that we can trust adiabaticity there is no dephasing at all: The gas particle simply "renormalize" the bare potential, which is in fact the Born-Oppenheimer picture. Of course, if we take into account corrections to the adiabatic picture, then we get a finite dephasing time. In the LRT regime of velocities we can estimate the dephasing time as

$$\tau_\varphi \;\approx\; \left(\frac{L_{\rm col}\lambda_{\rm E}^2}{v_{\rm E}V^2}\right)^{\frac{1}{3}}\,. \tag{37}$$

Both results have re-interpretation within the framework of an effective DLD/ZCL model. See [62] for details.

20 Recurrences

Consider ice-cube inside a cup of hot water. After some time it melts and disappears. But if we wait long enough (without time reversal) we have some probability to see the ice-cube re-emerging due to recurrences. The issue of recurrences becomes relevant whenever we consider a closed (un-driven) system. In other words, whenever we do not try to control its evolution from the outside.

There are recurrences both in classical and quantal physics. In the latter case the tendency for recurrences is stronger due to the quasi-periodic nature of the dynamics. However, if the time scale for recurrences is long enough with respect to other relevant time scales, then we can practically ignore these recurrences. Actually it is useful to regard these recurrences as "fluctuations", and to take the standpoint that our interest is only in some "average" or "likely" scenario.

The thermalization process of the particle-environment system is traditionally described as "irreversible". Indeed, if the bath is infinite, then also the time for recurrences of the particle-bath system becomes infinite. On the other hand, if the bath is finite, then we have to consider the recurrences of the particle-bath system. These recurrences can lead back to the initial un-entangled state.

In practice "recurrences" do not constitute a danger for "irreversibility". The time to get un-entangled by recurrences is extremely large (typically larger than the age of the universe). Assuming a chaotic environment, and ignoring issues of level statistics, the time scale for recurrences is at least the Heisenberg time (inverse of the mean level spacing) of the combined particle-environment system. It scales like $\hbar^{-(d+d_0)}$ where d_0 and d are the number of degrees of freedom of the particle and the environment respectively.

It goes without saying that the above issue of recurrences becomes irrelevant if the x motion is treated classically. There is however a twist to this latter statement in the case where the time variation of x is strictly periodic. This is due to dynamical localization effect [6]. Note however that dynamical localization is a very fragile effect: Even in case that it is found, it turns out that it manifests itself only after a time that scales like $\hbar^{-(1+2d)}$, which is much larger than the Heisenberg time of the environment [11].

Acknowledgments

Essential for the promotion of this line of study are the collaborations with Tsampikos Kottos (MPI Gottingen), and with Alex Barnett (Harvard), and lately with Diego Wisniacki (Comision Nacional de Energia Atomica, Argentina). I also thank Shmuel Fishman (Technion) for many useful discussions, and Rick Heller (Harvard), Joe Imry (Weizmann), Bilha Segev (Ben-Gurion) and Uzy Smilansky (Weizmann) for interesting interaction.

References

1. M.C. Gutzwiller, *Chaos in Classical and Quantum Mechanics* (Springer Verlag New York 1990).
2. H.J. Stockmann, "Quantum Chaos: An Introduction" (Cambridge Univ Pr 1999).
3. F. Haake, "Quantum Signatures of Chaos" (Springer 2000).
4. M. Wilkinson, J. Phys. **A21**, 4021 (1988); J. Phys. A **20**, 2415 (1987).

5. M. Wilkinson and E.J. Austin, J. Phys. **A28**, 2277 (1995).
E.J. Austin and M. Wilkinson, Nonlinearity **5**, 1137 (1992).
6. For review and references see [7] and [8].
7. S. Fishman in "Quantum Chaos", *Proceedings of the International School of Physics "Enrico Fermi", Course CXIX*, Ed. G. Casati, I. Guarneri and U. Smilansky (North Holland 1991).
8. M. Raizen in "New directions in quantum chaos", *Proceedings of the International School of Physics "Enrico Fermi", Course CXLIII*, Edited by G. Casati, I. Guarneri and U. Smilansky (IOS Press, Amsterdam 2000).
9. D. Cohen in "New directions in quantum chaos", *Proceedings of the International School of Physics "Enrico Fermi", Course CXLIII*, Edited by G. Casati, I. Guarneri and U. Smilansky, (IOS Press, Amsterdam 2000).
10. D. Cohen, Phys. Rev. Lett. **82**, 4951 (1999).
11. D. Cohen and T. Kottos, Phys. Rev. Lett. **85**, 4839 (2000).
12. D. Cohen, Annals of Physics **283**, 175 (2000).
13. D. Cohen, F.M. Izrailev and T. Kottos, Phys. Rev. Lett. **84**, 2052 (2000).
14. T. Kottos and D. Cohen, Phys. Rev. **E64**, R-065202 (2001).
15. S.W. Doescher and M.H. Rice, Am. J. Phys. **37**, 1246 (1969).
16. A.J. Makowski and S.T. Dembinski, Physics Letters **A154**, 217 (1991).
17. J.V. Jose and R. Cordery, Phys. Rev. Lett. **56**, 290 (1986).
18. J. Blocki, Y. Boneh, J.R. Nix, J. Randrup, M. Robel, A.J. Sierk and W.J. Swiatecki, Ann. Phys. **113**, 330 (1978).
19. S.E. Koonin, R.L. Hatch and J. Randrup, Nuc. Phys. **A283**, 87 (1977).
S.E. Koonin and J. Randrup, Nuc. Phys. **A289**, 475 (1977).
20. D. Cohen and T. Kottos, Phys. Rev. **E63**, 36203 (2001).
21. E. Wigner, Ann. Math. **62** 548 (1955); **65** 203 (1957).
22. G. Casati, B.V. Chirikov, I. Guarneri and F.M. Izrailev, Phys. Rev. **E48**, R1613 (1993); Phys. Lett. **A223**, 430 (1996).
23. Y. Imry, *Introduction to Mesoscopic Physics* (Oxford Univ. Press 1997).
24. Y. Alhassid, Rev. Mod. Phys. **72**, 895 (2000).
25. L.P. Kouwenhoven, C.M. Marcus, P.L. Mceuen, S. Tarucha, R. M. Westervelt and N.S. Wingreen, Proc. of Advanced Study Inst. on Mesoscopic Electron Transport, edited by L.L. Sohn, L.P. Kouwenhoven and G. Schon (Kluwer 1997).
26. M.V. Berry in *Chaos and Quantum Systems*, ed. M.-J. Giannoni, A. Voros, J. Zinn-Justin (Elsevier, Amsterdam, 1991).
27. M. Feingold and A. Peres, Phys. Rev. **A34** 591, (1986);
M. Feingold, D. Leitner and M. Wilkinson, Phys. Rev. Lett. **66**, 986 (1991);
M. Wilkinson, M. Feingold and D. Leitner, J. Phys. **A24**, 175 (1991);
M. Feingold, A. Gioletta, F. M. Izrailev and L. Molinari, Phys. Rev. Lett. **70**, 2936 (1993).
28. H. Attias and Y. Alhassid, Phys. Rev. **E52**, 4776 (1995).
29. T. Guhr, A. Muller-Groeling and H.A. Weidenmuller, Phys. Rep. **299**, 190 (1998).
30. D. Cohen and E.J. Heller, Phys. Rev. Lett. **84**, 2841 (2000).
31. D. Cohen, A. Barnett and E.J. Heller, Phys. Rev. **E63**, 46207 (2001).
32. F. Borgonovi, I. Guarneri and F.M. Izrailev, Phys. Rev. **E57**, 5291 (1998);
L. Benet, F.M. Izrailev, T.H. Seligman and A. Suarez-Moreno, chao-dyn/9912035.
33. E.J. Heller in *Chaos and Quantum Systems*, ed. M.-J. Giannoni, A. Voros, J. Zinn-Justin (Elsevier, Amsterdam, 1991).
34. A. Peres, Phys. Rev. **A30**, 1610 (1984); See also A. Peres, *Quantum Theory: Concepts and Methods* (Dordrecht, 1995).

35. R.A. Jalabert and H.M. Pastawski, Phys. Rev. Lett. **86**, 2490 (2001);
F.M. Cucchietti, H.M. Pastawski and R. Jalabert Physica, **A283**, 285 (2000);
F.M. Cucchietti, H.M. Pastawski and D.A. Wisniacki cond-mat/0102135, to be published as Phys. Rev. E Rapid;
F. Cucchietti et al., nlin.CD/0112015, to be published in Phys. Rev. E.
36. Ph. Jacquod, P.G. Silvestrov and C.W.J. Beenakker, Phys. Rev. **E64**, 055203 (2001).
37. N.R. Cerruti and S. Tomsovic, Phys. Rev. Lett. **88**, 054103 (2002).
38. D.A. Wisniacki and D.Cohen, quant-ph/0111125.
39. T. Prosen, quant-ph/0106149;
T. Prosen and M. Znidaric, J. Phys. **A34**, L681 (2001).
40. T. Kottos and D. Cohen, cond-mat/0201148.
41. See [12] and [9]. The classical theory that is presented in those references integrates ideas that were promoted in previous studies, mainly [42] and [4] and [43].
42. E. Ott, Phys. Rev. Lett. **42**, 1628 (1979);
R. Brown, E. Ott and C. Grebogi, Phys. Rev. Lett, bf 59, 1173 (1987); J. Stat. Phys. **49**, 511 (1987).
43. C. Jarzynski, Phys. Rev. **E48**, 4340 (1993).
44. A clear formulation of the diffusion dissipation relation can be found in [9,12], and also in Appendix A of [45]. It constitutes a refinement/generalization of the dissipation picture which is presented in [4]. The standard textbook formulation of LRT and the fluctuation-dissipation relation can be found in Appendix A of [23].
45. A. Barnett, D. Cohen and E.J. Heller, J. Phys. **A34**, 413 (2001).
46. D. Cohen, unpublished.
47. A. Barnett, D. Cohen and E.J. Heller, Phys. Rev. Lett. **85**, 1412 (2000).
48. J.M. Robbins and M.V. Berry, J. Phys. **A25**, L961 (1992);
M.V. Berry and J.M. Robbins, Proc. R. Soc. Lond. **A442**, 659 (1993);
M.V. Berry and E.C. Sinclair, J. Phys. **A30**, 2853 (1997).
49. O.M. Auslaender and S. Fishman, Phys. Rev. Lett. **84**, 1886 (2000); J. Phys. **A33**, 1957 (2000).
50. D.A. Wisniacki and E. Vergini, Phys. Rev. **E59**, 6579 (1999).
51. C. Jarzynski, Phys. Rev. Lett. **74**, 2937 (1995).
52. D. Cohen, Phys. Rev. **E55**, 1422 (1997).
53. D. Cohen, Phys. Rev. Lett. **78**, 2878 (1997).
54. D. Cohen, J. Phys. **A31**, 8199 (1998).
55. A.O. Caldeira and A.J. Leggett, Physica **A121**, 587 (1983);
A.O. Caldeira and A.J. Leggett, Ann. Phys. (N.Y.) **140**, 374 (1983);
Physica **A121**, 587 (1983).
56. L. Bonig, K. Schonhammer and W. Zwerger, Phys. Rev. **B46**, 855 (1992).
57. B. Vacchini, Phys. Rev. **E63**, 066115 (2001); J. Math. Phys. **42**, 4291 (2001).
58. A. Bulgac, G.D. Dang and D. Kusnezov, Phys. Rev. **E58**, 196 (1998).
59. R.P. Feynman and F.L. Vernon Jr., Ann. Phys. (N.Y.) **24**, 118 (1963).
60. M. Buttiker, cond-mat/0106149.
61. D. Cohen and Y. Imry, Phys. Rev. **B59**, 11143 (1999).
62. D. Cohen, Phys. Rev. **E65**, 026218 (2002).
63. R. Alicki, Phys. Rev. **A65**, 034104 (2002).

Quantum State Control in Cavity QED

T. Wellens and A. Buchleitner

Max-Planck-Institut für Physik komplexer Systeme,
Nöthnitzer Str. 38, D-01187 Dresden

Abstract. We give an outline of two rather complementary approaches to the control of the quantum state of a simple quantum system – the quantized harmonic oscillator. What we will be presenting here in a rather general context, with a minimum of technical machinery, is the implementation of two original control schemes which are hitherto nonstandard when guiding quantum systems into some desired target state. However, we do believe that these novel control schemes bear considerable potential, because of their convergence properties in one case, because of their robustness in the other, and, finally, in the light of recent experimental progress which lowers the barrier between mathematical idealizations and physical reality.

1 Setting the Stage

Before we describe how to prepare an arbitrary quantum state via asymptotic completeness, and how to synchronize the time evolution of a bistable, noisy quantum system with a weak periodic signal, let us first prepare the stage... Why quantum state control? Which are the difficulties, which are the prospects, and which strategies have been devised so far? We do not attempt to be exhaustive here, but shall rather quote few examples of exemplary importance.

One is photochemistry, the attempt to monitor and to control chemical reactions by coherent electromagnetic fields. This topic – which also carries the label "Coherent Control" – is nowadays fully operational, with the advent of femtosecond laser sources during the early nineties of the last century [1], and the availability of attosecond laser pulses to be expected for the near future [2]. It allows us to break molecular bonds in a controlled way [3], to create Rydberg wave packets of arbitrary shape [4–6], and promises progress in cooling molecules [7](so much harder than cooling atoms [8–10]) when combined with spontaneous decay [11]. Even the selective excitation of complex molecules with similar absorption spectra in a solvent has been reported on recently, using a self learning algorithm [12,13]. To achieve such complex tasks, virtually all knobs to tune the control field need to be used – shape and energy of the pulse, the carrier frequency, and its polarization [5,12–14]. In passing, let us note that these self-learning control schemes which essentially rely on nonlinear feedback have been developed in the laboratory [12], and that it remains a largely unexplored but highly intriguing theoretical problem to extract the information on the molecular dynamics which is unavoidably encoded in the optimal control pulse which finally yields the desired experimental result.

The other area where quantum state control is of major importance is a yet emerging field, which is nontheless already very active on the theoretical side: quantum information theory, notably its sub-discipline of quantum information processing. The simplest example – much older than the paradigm it is used here to illustrate – is the dynamics of a spin-$\frac{1}{2}$ particle which is nowadays known under the much more fancy name of a q-bit. Any state of such a system can be represented by a vector $\boldsymbol{r}$ on the Bloch sphere [15,16], according to the equality

$$\rho = \frac{\mathbb{1} + \boldsymbol{r} \cdot \boldsymbol{\sigma}}{2}, \tag{1}$$

where ρ is the density operator which describes the state of the system, $\boldsymbol{\sigma}$ is the vector of the Pauli matrices σ_{x}, σ_{y}, and σ_{z}, and the modulus of $\boldsymbol{r}$ is bounded by unity (the state of the sytem is pure if $|\boldsymbol{r}| = 1$, and mixed if $|\boldsymbol{r}| < 1$). All kinds of quantum information processing essentially are dealing with the manipulation of the state of entangled groups of individual quantum systems – q-bits in the simplest of all thinkable worlds. A group of q-bits makes a quantum register, to be built into future generations of quantum computers, and it is an indispensible prerequisite for running such a beautiful machine to be able to control (and to store) the state of a quantum register and of its individual components.

Hence, we face important fundamental questions, both (and at least) in photochemistry and in quantum information processing:

- How to control our target system?
- Which target states can be reached by which type of control?
- How robust are these control schemes against incoherent processes?

Once again, time, space, and probably also current knowledge are too limited here to give a comprehensive answer to all of them, and we shall discuss only some of the most important aspects inherent in any attempt to find a generally valid strategy.

There are essentially two approaches to control a quantum system, the first of which we call "semiclassical" here. This is the typical scenario of coherent control in photochemistry, and it can be condensed in a time dependent Hamiltonian

$$H(t) = H_0 + \sum_j c_j(t) A_j, \tag{2}$$

with H_0 the target system, which is to be guided into some predefined target state $|\chi_{\mathrm{target}}\rangle$ by suitable choice of the duration and strength of externally applied perturbations (which need not be and in general are not perturbative in nature!). The latter appear as $c_j(t)$ and A_j in (2), where $c_j(t)$ defines the – in general time-dependent – potential, and A_j the coupling to the target. Typically, the $c_j(t)$ represent classical fields (e.g., an electromagnetic field of well-defined frequency and amplitude, perhaps combined with static electric or magnetic fields [5,6]), what justifies the above attribute "semiclassical" – in the sense that the quantum system to be controlled (H_0) is coupled to a classical system through $\sum_j c_j(t) A_j$, and the state of the latter is unaffected by the interaction with the quantum system [5,15]. To maximize the probability that $|\chi_{\text{target}}\rangle$ be prepared within a finite time window, it is crucial

1. to precisely know the initial state of the control system, i.e. the decomposition of its unperturbed initial state $|\chi_0\rangle$ in the H_0 basis;
2. to have optimal control over the time dependence of the $c_j(t)$, such as to optimize diabatic or adiabatic switching at avoided crossings (conical intersections) of the energy level (potential surface) diagram of H [5,14,17].

Note that a considerable part of probability density may get lost in the vicinity of avoided crossings (or conical intersections) if the time scales are not perfectly controlled [14], and that this approach cannot work efficiently anyway if $|\chi_0\rangle$ is not precisely known. The problem cooks down to guiding a Hamiltonian trajectory deterministically from its initial point into the target – for this we have to know where we start and which way we take.

But where can we get with that technique? If we wish to attain any state in the Hilbert space of H_0, this means that we want to induce an arbitrary unitary transformation on $|\chi_0\rangle$, by suitable choice of A_j and $c_j(t)$. Since unitary evolutions are generated by self adjoint operators, we can generate[1] any evolution which is generated by the closed set of operators (i.e. the Lie algebra) resulting from arbitrary commutation operations between the constituents H_0 and A_j of H, i.e.

$$H_0, A_j, [H_0, A_j], [A_j, A_k], \ldots, [A_j, [A_k, A_\ell]], \ldots \ . \tag{3}$$

This set spans the space of self adjoint operators which we can realize to act on the Hilbert space of the system, and therefore generates the corresponding unitary evolutions. It is well known that in this way, indeed, the state of a q-bit can be guided to any point on the Bloch sphere by a classical oscillating field (using π-pulses, $\pi/2$-pulses, cotangent-pulses, etc [15]). However, if we wish, e.g., to control the state of a quantized harmonic oscillator by a classical dipole, through

$$H(t) = \hbar\omega a^\dagger a + d(t)(a^\dagger + a) \tag{4}$$

the Lie algebra is composed of

$$a, a^\dagger, a^\dagger a, a^\dagger \pm a, \mathbb{1} \tag{5}$$

alone, what obviously does not generate all possible unitary evolutions of the harmonic oscillator. Therefore, for an arbitrary target system, nothing guarantees that an arbitrary target state can be unitarily prepared by a small number of simple control potentials A_j [18,19].

Instead of guiding the target system into the target state by application of classical fields, we can alternatively make the target system interact with another quantum system – which we baptize the "handle(-system)". Due to the backaction of the target on the handle, the quantum state of target and handle

[1] As a consequence of $i\hbar\dot{A} = [H, A] + i\hbar\partial A/\partial t$ [15].

will in general be entangled, i.e. the state vector of the total system has the form

$$|\Phi(t)\rangle = \sum_{j,m} c_{j,m}(t)|\psi_j\rangle \otimes |\chi_m\rangle \neq |\psi\rangle \otimes |\chi\rangle, \forall|\chi\rangle, |\psi\rangle \tag{6}$$

which cannot be written as a tensor product on the Hilbert spaces of the handle and target subsystems. Furthermore, in the presence of noise and decoherence, we have to describe the bipartite quantum system by its density operator ρ instead of its state vector. On this level, entanglement prevents us from representing the state as a convex combination of tensor products of density matrices of both subsystems, i.e.

$$\rho(t) \neq \sum_j p_j \rho_j^{|\psi\rangle} \otimes \rho_j^{|\chi\rangle}, \ \sum_j p_j = 1, \ p_j \geq 0 \ . \tag{7}$$

In neither case is the target system in a well-defined state (e.g., if we trace out the handle in (6), the target is left in a mixed state), and we cannot unambiguously infer the state of the target by measuring the state of the handle. This introduces measurement noise in the preparation of the target state, which, in the case of a pure target-handle state, Eq. (6), is entirely of quantum origin.

In such a setting, the aim is to impose a time evolution on the entire system which – at the desired instant – leaves the target component in the desired state ($|\chi_{\mathrm{target}}\rangle$ or $\rho_{\mathrm{target}}^{|\chi\rangle}$). A priori, the time evolution need neither be unitary nor coherent, and projective measurements as well as suitably chosen noise sources – in addition to unitary operations – may be employed to achieve this duty. So far, at least three strategies are discussed in the literature:

1. state preparation by conditional measurement [20,21];
2. coherent evolution into a product state;
3. stochastic synchronization with a weak external signal.

The latter two will be discussed in the sequel, whereas we refer the reader to the original literature [20,21] for the former.

For a specific experimental implementation of the control schemes we will discuss, we consider a sequence of two-level atoms resonantly coupled to a single mode of the quantized radiation field sustained by a high quality microwave cavity. If we allow for weak losses of the cavity to the thermal environment, this system – experimentally realized and almost perfectly well controlled in the micromaser [22,23] – represents a kicked, weakly damped harmonic oscillator (see also [24]), which is described by a combination of the Jaynes-Cummings model [25] (for the coherent atom-field interaction), and of a master equation to account for the damping. Since all the details of the rather straightforward theoretical treatment can be found in the literature, we only summarize those facts which are absolutely essential for our specific purposes.

1.1 Essential Theoretical Tools

Our specific target system will be the quantized harmonic oscillator mentioned above, in *subsequent*, resonant coherent interaction with single two-level atoms, and incoherently coupled to an environment. The Hamiltonian for the entire system atom plus field plus environment can be written as

$$H = \hbar\omega\sigma^\dagger\sigma + \hbar\omega a^\dagger a + \hbar g(a^\dagger\sigma + a\sigma^\dagger) + \hbar\sum \omega_j b_j^\dagger b_j + \hbar\sum (g_j a^\dagger b_j + g_j^* a b_j^\dagger) , \quad (8)$$

where the first term describes an unperturbed two-level atom with an energy spacing $E_u - E_d$ between the upper and lower level $|u\rangle$ and $|d\rangle$, respectively, equal to the frequency ω of the cavity mode. The second term represents the energy of the free field mode, with creation and anihilation operators $a^\dagger$ and a. The third term describes the coherent atom-field interaction in rotating wave approximation [15], with the atomic ladder operators σ and $\sigma^\dagger$, and all these first three terms together are known as the Jaynes-Cummings model [15,25].[2] The fourth term describes the energy of the modes (through their respective creation and annihilation operators, $b_j^\dagger$ and b_j) of the electromagnetic environment (at temperature T), the last term represents the interaction of the cavity field mode with these environment modes, and g and g_j finally are the coupling constants proportional to the coupling matrix elements [26]. Note that describing the environment as a large set of harmonic oscillators of variable frequency ω_j is common practice in this kind of problems [26].

Coherent atom-field time evolution. The Jaynes-Cummings part of the Hamiltonian (8) generates the time evolution of an arbitrary atom-field state

$$|\Phi(t)\rangle = \sum_n [c_{n,u}(t)|u\rangle \otimes |n\rangle + c_{n,d}(t)|d\rangle \otimes |n\rangle], \quad (9)$$

and the eigenstates of the Jaynes-Cummings Hamiltonian are also known as "dressed states" of the atom in the field.[3] The time evolution operator reads, in the interaction picture,

$$U = \exp\left(-i\phi(a^\dagger\sigma + a\sigma^\dagger)\right) , \quad (10)$$

---

[2] Remember the familiar relations $a|n\rangle = \sqrt{n}|n-1\rangle$, $a^\dagger|n\rangle = \sqrt{n+1}|n+1\rangle$, $[a, a^\dagger] = \mathbb{1}$, $\sigma = |d\rangle\langle u| = (\sigma^\dagger)^\dagger$ [15].

[3] In the context of the present lecture, we are dealing with the interaction of atoms and electromagnetic radiation in a specific regime of driving amplitudes and frequencies – the realm of cavity QED [23]. The dressed state picture [26] is particularly adapted for the description of this situation, but is in principle completely general and would also apply to the interaction of atoms with classical electromagnetic fields. However, the latter are essentially described by coherent states of the quantized field, and one often uses the semiclassical limit of the dressed state formalism – the Floquet picture [27,28]. As a matter of fact, the latter is historically older than the former and extremely efficient – as highlighted, e.g., in Zakrzewski's lecture [6] during this school. Only when the radiation field is in a non-classical state – that's the situation we will be facing further down in this lecture – (or when the average occupation number of the mode is comparable to its typical fluctuations) are we obliged to quantize the field degree of freedom of the atom-field system.

where the interaction time is hidden in $\phi = gt_{\text{int}}$, the "Rabi angle". Application of U on a single component of the expansion (9) of $|\Phi(t)\rangle$ gives

$$U[|u\rangle \otimes |n\rangle] = \cos(\phi\sqrt{n+1})|u\rangle \otimes |n\rangle - i\sin(\phi\sqrt{n+1}|d\rangle \otimes |n+1\rangle, \quad (11)$$

$$U[|d\rangle \otimes |n\rangle] = \cos(\phi\sqrt{n})|d\rangle \otimes |n\rangle - i\sin(\phi\sqrt{n}|u\rangle \otimes |n-1\rangle\,, \quad (12)$$

what shows that all the essential dynamics are encoded in the coefficients

$$A_n = \cos(\phi\sqrt{n}),\ B_n = \sin(\phi\sqrt{n})\,. \quad (13)$$

Indeed, these coefficients are the only fundamental ingredients which we need to derive the results of the next chapter.

Some remarks are in order to identify some specific features of the Jaynes-Cummings dynamics:

- If we prepare the atom-field system initially in a state $|u\rangle \otimes |n\rangle$, it will periodically oscillate between $|u\rangle \otimes |n\rangle$ and $|d\rangle \otimes |n+1\rangle$, with the "Rabi frequency"
$$\Omega_n = g\sqrt{n+1}\,, \quad (14)$$
by virtue of (11).
- If the atoms are initially prepared in $|u\rangle$, they can add energy to (i.e., in the quantum optics jargon, "pump") the cavity field – by depositing a photon during the Rabi oscillation – and exit in $|d\rangle$, or leave the occupation of the field mode unaltered – and exit in $|u\rangle$. Hence, if, on exit from the cavity, we detect the atom in either $|d\rangle$ or $|u\rangle$, we can unambiguously conclude on the state of the field, provided we know its initial state on entry of the atom into the cavity. Therefore, the atoms are not only used to pump the field, but also to "probe" it, by a projective measurement (which destroys the atom-field entanglement) after the atom-field interaction.
- If the interaction time t_{int} is chosen such that
$$\phi = \frac{k\pi}{\sqrt{n_t+1}},\ k \in \mathbb{Z} \quad (15)$$
the harmonic oscillator cannot be excited beyond the "trapping state" $|n_t\rangle$, since B_{n_t+1} vanishes in (11).
- The above Hamiltonian (8) describes the interaction of the mode with one single atom. If we want to extract the dynamics of the field under *subsequent* interaction with $N \geq 1$ atoms, we have to multiply the respective single atom time evolution operators, to obtain the complete time evolution operator
$$U_N = U^{(N)}U^{(N-1)}\ldots U^{(1)} \quad (16)$$
where the index discriminates the subsequent atoms interacting with the field.

- If we prepare our system initially in a product state $|\psi_0\rangle \otimes |\chi_0\rangle$, it will evolve into a state with nonvanishing entanglement between the atom and the field, i.e.
$$U_N[|\psi_0\rangle \otimes |\chi_0\rangle] = |\Phi\rangle \neq |\psi_f\rangle \otimes |\chi_f\rangle \ . \tag{17}$$
Note that $|\psi_0\rangle$ and $|\psi_f\rangle$ are themselves – possibly entangled – N-atoms states (acted upon by U_N), independently of the atoms-field entanglement.

Cavity decay. If, on top of the coherent atom-field interaction, we allow for dissipative losses of energy through the cavity walls – described by the environment terms in (8) – we are dealing with what is known as the micromaser dynamics, under the following, additional assumptions on the relevant time scales:

- The temporal separation t_{sep} of two subsequent atoms is large with respect to the atom-field interaction time t_{int}.
- The cavity decay rate $\gamma = 1/t_\gamma$ is much smaller than the pumping rate $r = 1/t_{sep}$ (i.e. the rate at which the atoms enter the cavity).

In toto, this imposes the following separation of time scales

$$t_\gamma \gg t_{sep} \gg t_{int} \ , \tag{18}$$

what implies, in particular, that the cavity damping can be neglected during the coherent interaction between an atom and the field, and that damping is effective only during the time interval when no atom is in the cavity. Therefore, the maser dynamics has a stroboscopic or map-like character: a "kick" (due to coherent atom-field interaction) is followed by a relaxation period (due to damping) – a situation which is reminiscent of the scenario developed by Fannes in his lecture [24].

Due to this separation of time scales, and after coarse graining of the dynamics (remember Alicki's lecture [29], and see also [26]), the complete time evolution of the cavity field is described by the master equation

$$\dot{\rho} = \dot{\rho}|_{\mathrm{at}} + \dot{\rho}|_{\mathrm{env}}, \ t \gg t_{int} \ , \tag{19}$$

where the first term describes the state evolution induced by the coherent atom field interaction, whilst the second describes the damping. Explicitely, the coherent dynamics reads, in the number state basis of the field (i.e., the eigenbasis of $a^\dagger a$):

$$\begin{aligned}\dot{\rho}_{n,m}|_{\mathrm{at}} = & \ r\left[|a|^2 A_{n+1}A_{m+1} + |b|^2 A_n A_m - 1\right]\rho_{n,m} \\ & + r|a|^2 B_n B_m \rho_{n-1,m-1} + r|b|^2 B_{n+1}B_{m+1}\rho_{n+1,m+1} \\ & + ircA_{n+1}B_{m+1}\rho_{n,m+1} - irc^* B_{n+1}A_{m+1}\rho_{n+1,m} \\ & - ircB_n A_m \rho_{n-1,m} + irc^* A_n B_m \rho_{n,m-1} \ ,\end{aligned}$$

for all atoms initially prepared in the state

$$\tau = \begin{pmatrix} |a|^2 & ic \\ -ic & |b|^2 \end{pmatrix} \tag{20}$$

on entry into the cavity.[4] The damping part follows

$$\begin{aligned} \dot{\rho}_{n,m}|_{\rm env} &= -\frac{\gamma}{2}\left[n_b(n+m+2)+(n_b+1)(n+m)\right]\rho_{n,m} \\ &\quad +\gamma n_b\sqrt{n}\sqrt{m}\rho_{n-1,m-1}+\gamma(n_b+1)\sqrt{n+1}\sqrt{m+1}\rho_{n+1,m+1}, \end{aligned} \tag{21}$$

with $n_b = [\exp(\hbar\omega/k_B T) - 1]^{-1}$ the thermal occupation of the mode $\omega_j = \omega$ of the environment. Note from (20,21) that the time evolution induced by the damping (i.e. those terms proportional to γ), or by the incoherent part of the pumping cycle (proportional to r) remain strictly confined to the (sub-)diagonal of the field density matrix they start from. Only the coherence induced part of the dynamics, proportional to the initial atomic coherence c in (20), couples neighbouring (sub-)diagonals. This leads to the build-up of coherence between different number states of the field, even if the latter initially exhibits no off-diagonal occupation.

Obviously, the build-up of coherence as well as the maximum energy which can be deposited by the atoms in the cavity mode is limited by the losses due to damping. If we specialize (19) to the case $|a| = 1$ and $c = b = 0$, i.e. all the atoms are injected into the cavity in the upper state, the maser is pumped incoherently, and a stationary state of the cavity field will be established if losses due to

$$\dot{\rho}_n|_{\rm env} = \gamma\left[n_b n\rho_{n-1} - (n_b(n+1)+(n_b+1)n)\rho_n + (n_b+1)(n+1)\rho_{n+1}\right], \tag{22}$$

proportional to γn for $n_b \ll n$, compensate for the gain given by

$$\dot{\rho}_n|_{\rm at} = r\left[B_n^2\rho_{n-1} - B_{n+1}^2\rho_n\right], \tag{23}$$

proportional to $rB_{n+1}^2 \sim \sin^2(\phi\sqrt{n+1})$. This condition of balanced loss and gain is illustrated in Fig. 1, where the gain and the loss curve are plotted as a function of n. Apparently, there are multiple solutions, and the case depicted in the figure shows two stable (n_1 and n_2) and one unstable (n_3) gain-loss equilibrium. In such a case, the density matrix of the field will exhibit a double peaked occupation, with maxima in the vicinity of the stability points n_1 and n_2. If, furthermore, a finite coherence $c \neq 0$ is fed into the cavity, the field converges into a stationary state as the one plotted in Fig. 2 [30]. Note that the density matrix describes the dynamics of an ensemble average of the maser dynamics, in the sense that it averages over all possible outcomes of a measurement of the state of the cavity field, whereas a real time measurement in an actual experimental run will observe quantum jumps of the cavity field between occupation numbers close to n_1 or n_2 [31], at unpredictable times, such as in Fig. 3. Thanks to the ergodicity property of the maser dynamics [32], the time average over the real time measurement will recover the double peaked stationary state of the master equation (19). It will be the aim of the last chapter of these lectures to devise a way to control these quantum jumps, by synchronizing them with a weak, externally applied periodic signal.

[4] Recall that τ describes a pure state for $|c| = |ab|$, a mixed state for $|c| < |ab|$, and an incoherent mixture for $c = 0$.

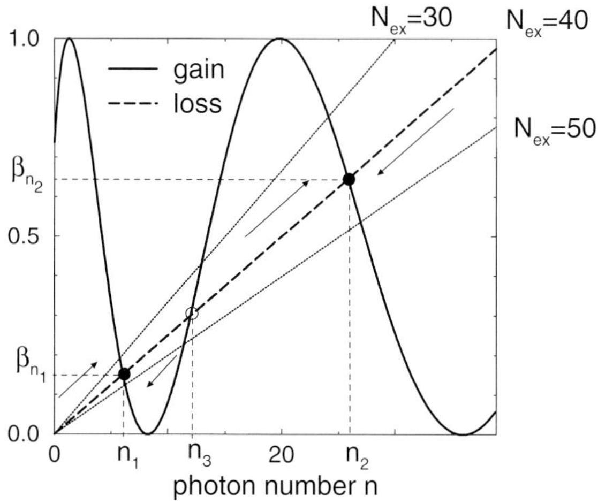

Fig. 1. Gain vs. loss of the maser dynamics, as a function of the photon number n. One unstable (at n_3) and two stable (at n_1 and n_2) equilibria are observed, within a finite interval of the scaled pump rate $N_{\rm ex} = r/\gamma = 30 \ldots 50$. $\beta_n = B_{n+1}^2$, see Eq. (13).

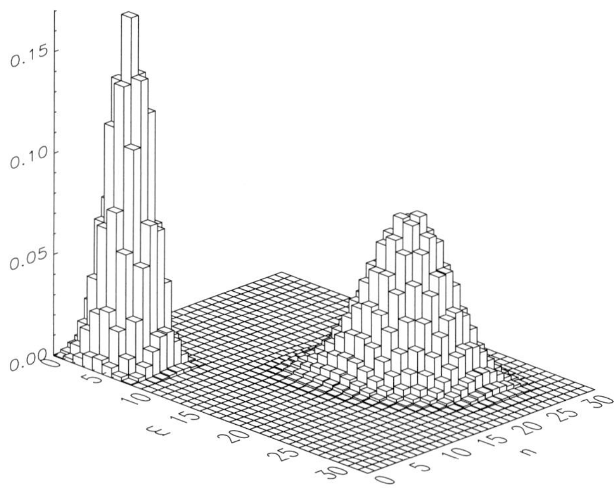

Fig. 2. Stationary state of the photon field density matrix (in the number state representation), with two metastable states centered around $n_1 \simeq 4$ and $n_2 \simeq 21$. Initial atomic superposition, see Eq. (20): $a = \sqrt{0.9}$, $b = \sqrt{0.1}$. $T = 0.5$ K, $\omega = 21.5$ GHz, $\gamma^{-1} = 0.06$ sec, $r = 40\gamma$, $\phi = 1.1$.

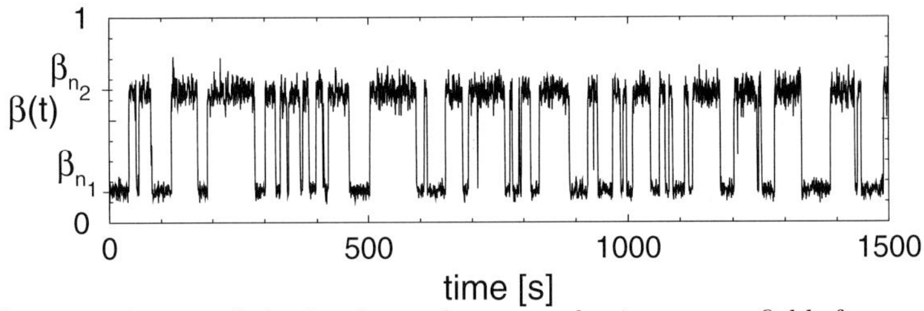

Fig. 3. Quantum jumps of the incoherently pumped micromaser field, for $a = 1$, $T = 0.6$ K, $\omega = 21.5$ GHz, $\gamma^{-1} = 0.06$ sec, $r = 40\gamma$, $\phi = 1.1$. The quantum jumps are monitored by the time dependence of the probability $\beta(t) = \sum_n p_n B_{n+1}^2$ (see Eq. (11); p_n denotes the population distribution of the energy levels of the cavity field) to detect an atom in $|d\rangle$, on exit from the cavity. To smooth the measurement noise, $\beta(t)$ has been averaged over approx. 500 detection events. The typical residence time of the cavity field in either one of the metastable states is of the order of 50 sec.

With the above, we have collected all what we need for our subsequent treatment. None of this material is original, as opposed to what will follow.

2 Quantum State Preparation via Asymptotic Completeness

We now turn to the problem of preparing an arbitrary state of a quantum system, which we chose to be a quantized harmonic oscillator. What we need in order to realize this objective are Eqs. (8-17) from sec. 1.1, a recent mathematical theorem [33] (which, as a matter of fact, has been communicated to us by its authors after they got aware of the work to be presented in the last chapter of these lecture notes), and – to get an idea of the substance of this theorem – an intuitive idea of "asymptotic completeness", a concept from mathematical scattering theory [34].

Let us first restate our problem: We want to prepare an arbitrary, pure state of a single mode of the quantized radiation field, by resonant, coherent interaction with a sequence of two-level atoms. We want to keep the interaction itself as simple as possible, notably unaltered for all atoms, and refrain from any temporal modulation of the parameters which define the atom-field interaction. Indeed, we shall employ the Jaynes-Cummings dynamics with fixed Rabi angle ϕ (i.e. fixed interaction time $t_{\rm int}$). Furthermore, we wish to get around measurement noise, i.e. we aim at preparing the (arbitrary) target state $|\chi_{\rm target}\rangle$ of the field such that it emerges in a *product state* $|\chi_{\rm target}\rangle \otimes |\psi_{\rm final}\rangle$ with the final state of the atoms on exit from the cavity. On a first glance, this seems rather illusive: As we have seen above, the Jaynes-Cummings interaction almost unavoidably entangles the atom and the field, as a consequence of the Rabi cycle. In general, an atom does not leave the cavity in a well defined state on exit from the cavity – it is entangled with the field. The same is of course true for a sequence of atoms.

Since the interaction is fixed for all atoms, the only way to prepare $|\chi_{\mathrm{target}}\rangle$ is through the appropriate preparation of the atoms-field (sic! – we consider a sequence of N atoms injected into the cavity) initial state, which we assume to have product form $|\chi_0\rangle \otimes |\psi_0\rangle$. In particular, we admit entanglement of the initial atomic state, i.e. $|\psi_0\rangle$ has the general form

$$|\psi_0\rangle = \sum_{i_N \ldots i_1} c_{i_N \ldots i_1} |i_N \ldots i_1\rangle \,. \tag{24}$$

With these premises, the asymptotic completeness of the Jaynes-Cummings interaction indeed does insure that there is a solution to our problem, due to the following theorem [33]:

> As $N \to \infty$, each observable A of the field turns into an observable M of the atoms; in symbols:
>
> $$\lim_{N\to\infty} U_N^\dagger (A \otimes \mathbb{1}) U_N = \mathbb{1} \otimes M \,. \tag{25}$$

What are our fellow mathematicians trying to tell us? Note that (25) is formulated in the Heisenberg picture, and that all expectation values therefore have to be evaluated with respect to the initial state $|\chi_0\rangle \otimes |\psi_0\rangle$. If we do that for the specific choice $A = |\chi_{\mathrm{target}}\rangle\langle\chi_{\mathrm{target}}|$, i.e. the projection on the desired field state, (25) turns into

$$\begin{aligned} &\langle\psi_0| \otimes \langle\chi_0| U_N^\dagger (|\chi_{\mathrm{target}}\rangle\langle\chi_{\mathrm{target}}| \otimes \mathbb{1}) U_N |\chi_0\rangle \otimes |\psi_0\rangle \quad \xrightarrow[N\to\infty]{} \\ &\langle\psi_0| \otimes \langle\chi_0| (\mathbb{1} \otimes M) |\chi_0\rangle \otimes |\psi_0\rangle = \langle\chi_0|\chi_0\rangle \langle\psi_0|M|\psi_0\rangle = 1 \,. \end{aligned} \tag{26}$$

The last equality – which is the crucial one – still must be justified: with the specific choice of A as a projector, and given that the time evolution under the action of U_N is unitary, the spectra of $|\chi_{\mathrm{target}}\rangle\langle\chi_{\mathrm{target}}| \otimes \mathbb{1}$ and $\mathbb{1} \otimes M$ are identical, with eigenvalues 0 and 1. Hence, if we choose $|\psi_0\rangle \in \mathrm{range}(M)$, then $M|\psi_0\rangle = |\psi_0\rangle$, justifying the last identity in (26). Therefore, the theorem tells us that

- any field state $|\chi_{\mathrm{target}}\rangle$ can be prepared by choosing an appropriate initial atoms-state of the form (24), *irrespective* of the initial state $|\chi_0\rangle$ of the field. The latter follows from the factorization on the right hand side of (25), since the expectation value of the identity is unity, for any normalized initital state $|\chi_0\rangle$. This is tantamount of saying that
- due to the interaction with many atoms, the field loses the memory of its initial state.

Both these conclusions establish some link to the definition of asymptotic completeness in mathematical scattering theory [34]: Here, a potential or an interaction is said to be asymptotically complete, if the Møller operators Ω_+ and Ω_- (these define the scattering operator $S = \Omega_-^{-1}\Omega_+$, the central object in scattering theory) exist, with $\mathrm{range}(\Omega_+) = \mathrm{range}(\Omega_-)$. As illustrated in Fig. 4,

$$\Omega_\pm = \lim_{t\to\mp\infty} T_{-t} T_t^{(0)} \,, \tag{27}$$

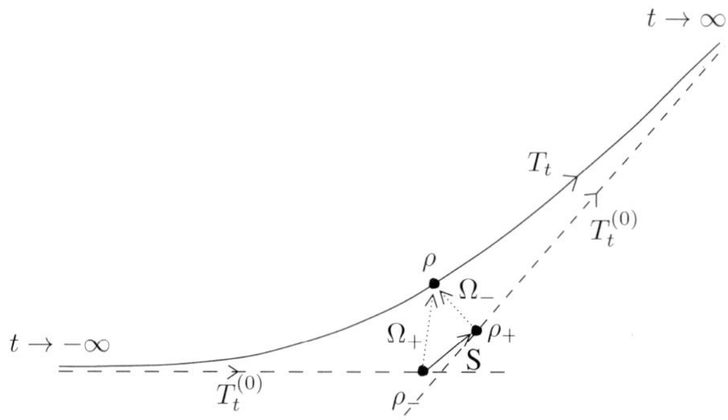

Fig. 4. Pictorial definition of asymptotic completeness, in the usual setting of potential scattering. A particle which was asymptotically free in the distant past will be asymptotically free again in the distant future, provided the Møller operators $\Omega_\pm$ exist and have the same range.

map the free evolution $T_t^{(0)}$ on the interacting evolution $T_{\pm t}$, such as to establish – in the presence of a scattering potential – an unambiguous relation between the asymptotically free states in the distant past and the asymptotically free states in the far future, via

$$\rho_+ = \Omega_-^{-1}\Omega_+\rho_- \,. \tag{28}$$

After contemplating these definitions together with the figure for some time [35], it turns out that asymptotic completeness essentially tells us, in a sloppy phrasing, that every state which is asymptotically free in the distant past will again be asymptotically free in the distant future, and, therefore, in the case of potential scattering of particles, none are swallowed by the target. For our present problem of a sequence of atoms interacting with the quantized field mode, this can be translated [35] into the statement that for any target state of the field (to be prepared after interaction with $N \to \infty$ atoms, i.e. in the distant future) there is a N-atoms, $N \to \infty$, initial state (to be prepared before the interaction with the field, i.e. in the distant past), which is faithfully mapped onto the target state by the atoms-field interaction. Due to the tensor product structure of the Hilbert space supporting the entire dynamics, it equally follows that the initial state of the field (prepared in the distant past) is faithfully mapped on the N-atoms state, $N \to \infty$, *after* the interaction, i.e. in the distant future. Hence, theorem (25) implies that any information which can be stored by a quantized harmonic oscillator can be encoded and equally well decoded by the interaction with an infinite sequence of two-level atoms.

This is, obviously, a very beautiful – mathematical – result. However, as opposed to many mathematical theorems, physics is rarely dealing with asymptotics, but rather with finite numbers. In particular, the theoretical physicist who'll approach an experimentalist with the earth-shaking news that the only thing she has to do to produce an arbitrary quantum state of a harmonic oscillator is entangling $N \to \infty$ two-level atoms is unlikely to be invited for dinner. The controlled entanglement of atoms is so far a formidable experimental task and, at present, controlled experimental entanglement of approx. four [36] or slightly more [37] atoms can be achieved. Therefore, the physically relevant questions here are

- how fast the asymptotic limit of perfect state preparation is reached, and
- whether the state preparation keeps being independent of the initial state of the field, for a finite number $N \ll \infty$ of injected atoms.

Now, for finite N,

$$P_{|\chi_{\text{target}}\rangle} = U_N^\dagger (|\chi_{\text{target}}\rangle\langle\chi_{\text{target}}| \otimes \mathbb{1}) U_N \neq \mathbb{1} \otimes M \,, \tag{29}$$

is no more a projector on the atomic subspace, and the target state cannot be prepared exactly any more. However, we can extract the fidelity F_N of the state preparation as the expectation value of the atomic operator

$$M^{(\rho_0)} = tr_{\text{field}} \left[(\rho_0 \otimes \mathbb{1}) P_{|\chi_{\text{target}}\rangle} \right] , \tag{30}$$

where ρ_0 is the – possibly mixed – initial state of the field, with respect to a suitable N-atoms initial state $|\psi_0\rangle$,

$$F_N = \langle \psi_0 | M^{(\rho_0)} | \psi_0 \rangle \,. \tag{31}$$

Finding the largest achievable fidelity $F_N^{(\max)}$ is thus tantamount to finding the largest eigenvalue of $M^{(\rho_0)}$, together with the associated eigenvector $|\psi_0^{\max}\rangle$, which is the requested N-atoms initial state.

We must remember, though, that $M^{(\rho_0)}$ operates on the 2^N dimensional Hilbert space of N-atoms states, and that diagonalizing this operator rapidly gets a demanding or even unrealistic task – even with current supercomputers much more than $N \simeq 20$ cannot be handled. However, for target states $|\chi_{\text{target}}\rangle = \sum_m c_m |m\rangle$, we can derive an analytic lower bound $F_N = \sum_m |c_m|^2 F_N(m) \leq F_N^{(\max)}$ [35], exclusively given by the Jaynes-Cummings transition amplitudes (13):

$$F_N(m) = 1 - \sum_{k=1}^{m} A_k^{2N} \prod_{k \neq i = 1}^{m} \frac{B_i^2}{B_i^2 - B_k^2} \,, \tag{32}$$

where $|\chi_0\rangle = |0\rangle$ was assumed [35,38]. Note that, for $|\chi_{\text{target}}\rangle$ with a finite expansion in the number state basis, the convergence of (32) with N towards unity will be dominated by the largest value of the A_k^2, for sufficiently large N, leading to *exponential* convergence with the corresponding rate. This is very good news and makes asymptotic completeness a practically relevant property! Fig. 5 illustrates the exponential convergence of $F_N^{(\max)}$ with N towards unity, for a targeted number and phase state, respectively, with maximum photon number $n = 5$, and for the cavity field initially prepared in the vacuum or in the thermal equlibrium state at $T \simeq 1$ K. Besides the exponential convergence, we also observe an initial, rather ineffective phase of the state preparation –

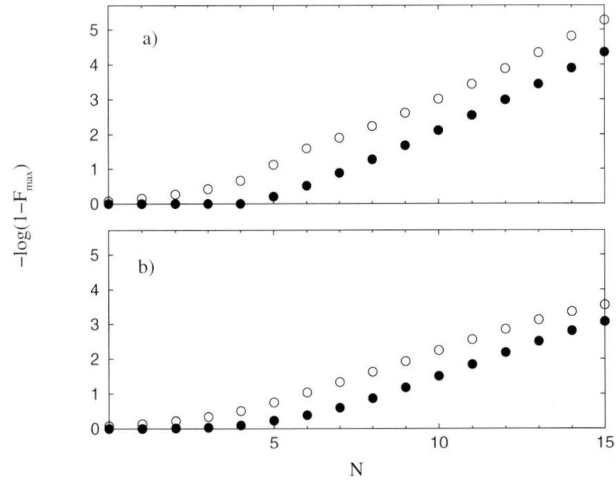

Fig. 5. Maximum fidelity for the preparation of the number state $|5\rangle$ (filled circles), and of the truncated phase state $|\chi_5\rangle = \sum_{i=0}^{5} |i\rangle/\sqrt{6}$ (open circles). $\phi = 0.91$. Initial field state (a) $|0\rangle$, (b) thermal equilibrium state with average photon number $n_b = 0.55$ ($T \simeq 1$ K).

simply due to the fact that we need to inject at least five atoms into the cavity, each of which carries at most one photon, in order to populate number states with $n = 5$. We also note that the rate of convergence is different for the field initially in the vacuum as compared to the thermal initial state. This is due to the nonvanishing population of number states with $n > 5$ in the thermal field initial state, which has to be wiped out during the preparation process [35]. From this plot we can extract yet another useful information: In order to prepare the five photon number state with a fidelity of more than 99%, we need to inject approx. $N = 10$ atoms, what leaves some work to the experimentalist, which is nontheless unlikely to overburden her. As a matter of fact, our numerical results suggest that there is a linear dependence between the maximum photon number in the number state expansion of the target state and the number of atoms which need to be injected to achieve a prechosen fidelity. This is illustrated in Fig. 6. Fig. 7 illustrates the abovementioned independence of the fidelity of the preparation with respect to the initial state of the field, for finite N. We compare $F_{N=10}^{(\max)}$ with the field initially in the thermal state to the fidelity achieved with the optimal state for $\rho_0 = |0\rangle\langle 0|$, for three different target states, and over a finite range of Rabi angles ϕ. The agreement is rather satisfactory, over the entire range of ϕ, with the largest deviations for the coherent target state (a detailed discussion of this issue can be found in [35]).

To conclude, let us address an important question which remained unanswered so far: Which are the conditions for the theorem (25) to hold, and why does the Jaynes-Cummings interaction satisfy them? The mathematical theorem underlying (25) identifies the following sufficient conditions [33]:

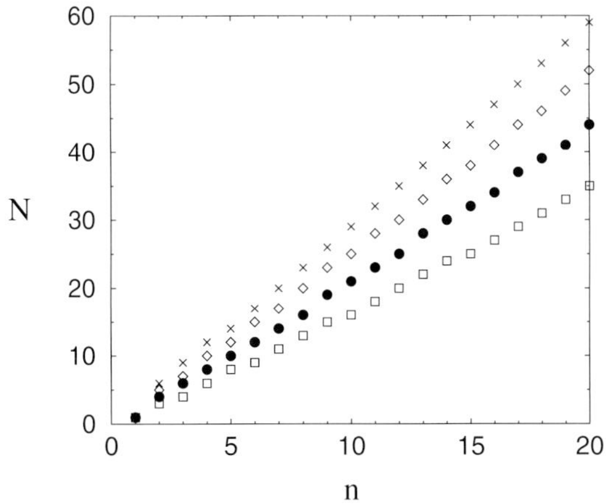

Fig. 6. Minimum number N of atoms needed to prepare the number state $|n\rangle$ starting from the vacuum $|0\rangle$ with fidelity $F \geq 1 - \epsilon$, $\epsilon = 10^{-1}, 10^{-2}, 10^{-3}, 10^{-4}$ (from bottom to top). $\phi = 0.91$.

1. There exists a single-atom state $|\tilde{\psi}\rangle$, such that the cavity field converges into a uniquely defined, pure invariant state $|\tilde{\chi}\rangle$ (independent of the initial field state), upon interaction with an infinite sequence $|\tilde{\psi}\rangle \otimes |\tilde{\psi}\rangle \otimes |\tilde{\psi}\rangle \ldots$.
2. The target-handle interaction is time-reversal symmetric, i.e., given that $|\chi_2\rangle$ can be prepared from $|\chi_1\rangle$, then also the reverse is true.

We can easily verify that for our specific implementation of the theorem:

1. $|\tilde{\psi}\rangle = |d\rangle$, and $|\tilde{\chi}\rangle = |0\rangle$, where trapping states defined by (15) have to be avoided.
2. It is easy to construct a time reversal operator T_N [35], such that

$$U_N^{-1} = (\mathbb{1} \otimes T_N)^{-1} U_N (\mathbb{1} \otimes T_N) . \tag{33}$$

This implies

$$U_N |\chi_1\rangle \otimes |\psi_1\rangle = |\chi_2\rangle \otimes |\psi_2\rangle \tag{34}$$

$$\Rightarrow U_N |\chi_2\rangle \otimes T_N |\psi_2\rangle = |\chi_1\rangle \otimes T_N ||\psi_1\rangle , \tag{35}$$

which is the requested property.

These sufficient conditions for asymptotic completeness suggest the following recipe to "cook" your desired target state $|\chi_{\text{target}}\rangle$:

Step 1: Purify the field by preparation of $|\tilde{\chi}\rangle$ – this is possible due to condition (1) above.

Step 2: Starting from $|\tilde{\chi}\rangle$, prepare the target state – this is possible due to condition (2) above, since condition (1) guarantees that $|\tilde{\chi}\rangle$ can also be prepared starting from $|\chi_{\text{target}}\rangle$.

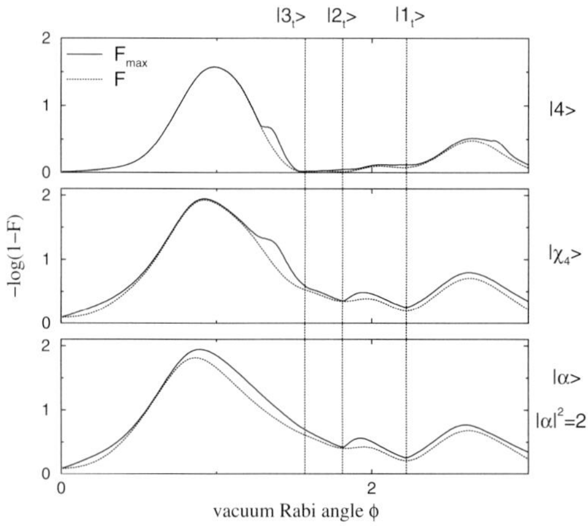

Fig. 7. Approximate independence of the fidelity $F_{N=10}^{(\max)}$ of the state preparation on the initial state $|\chi_0\rangle$ of the field, for different target states (top: four-photon number state; middle: truncated phase state with $n \leq 4$; bottom: coherent state $|\alpha\rangle$ with average photon number $|\alpha|^2 = 2$). Here, the field is prepared in the thermal equilibrium state with $n_b = 1$, and the solid line shows the fidelity obtained for the corresponding optimal atomic initial state $|\psi_0^{\max}\rangle$. The dotted line shows the result to be expected if the atoms are prepared in the optimal atomic initial state corresponding to the vacuum initial state of the field. Apparently, this initial state also does a very good job for the preparation of the various target states out of the thermal equilibrium state. The dotted lines labeled $|n_t\rangle$, $n = 1, 2, 3$, indicate the corresponding trapping state conditions, see Eq. (15).

Note that – since this recipe is deduced from sufficient conditions for asymptotic completeness – nothing guarantees that this is the unique strategy to prepare $|\chi_{\mathrm{target}}\rangle$, but it certainly is a viable one. Indeed, we have strong evidence that the recipe defines the optimal strategy for number state target states (prepared from $|\chi_0\rangle = |0\rangle$), but also found target states which can be reached faster along different paths. Since the definition of "fast" and "path" in the previous statement turns out to be more tricky than one might expect, and since space and time are limited here, we refer the curious reader to [35] for more details.

In summary, we have seen that asymptotic comleteness provides a novel and higly efficient scheme to prepare an arbitrary quantum state of the quantized radiation field, via interaction with a *finite* number of – in general entangled – two-level atoms. The obvious advantages of asymptotic completeness as compared to other schemes of optimal quantum control are

- the independence of the state preparation of the initial state of the target system, i.e. control of the initial state – an indispensible prerequisite, e.g., for traditional coherent control, is *not* required;
- the convergence into a product state of target and handle system, i.e. the exponential suppression of measurement noise with increasing length N of the injected atomic string.

In addition, asymptotic completeness allows highly faithful en- and decoding of (quantum) information in the target system, with exponentially small errors (by virtue of our estimation (32)), and – when focussing on $|\chi_0\rangle$ and $|\psi_f\rangle$ instead of $|\psi_0\rangle$ and $|\chi_{\text{target}}\rangle$ – it can also be employed to create entangled states of the handle system, e.g. from a specific subclass of states of the target system which may be efficiently prepared by other means [39]. It remains an equally challenging and interesting question to characterize the set of entangled handle states which are reachable from certain classes of target states (interacting with initially unentangled handle states), and vice versa. Of course, the efficiency of our preparation scheme has to be payed by the preparation of entangled handle states before the target state preparation, and in some sense this amounts to sweeping the real problem under the rug. However, it must be realized that experimentalists *will* be able to create controlled entanglement of more than just 10 two-level atoms – if not within one year, then certainly within ten years or so [36,37,40,41]. If engineering an arbitrary quantum state of the target (e.g., by coherent control) is more difficult than the entanglement of a string of spins, this experimental effort will merit its price. If, on the other hand, efficient means allow to engineer any state of the target, asymptotic completeness allows to create a class of strongly entangled many-particle states with a dimension comparable to the one of the Hilbert state of the target. Hence, it is unquestionable that asymptotic completeness as a means of quantum state control exhibits large potential for processing quantum information.

In this chapter, we have shown that a string of simple two-level atoms, combined with one of the virtually simplest models for the interaction of radiation with matter, is capable to generate quite intricate phenomena. We did not discuss the influence of decoherence and noise, which certainly affect our above conclusions, however in a surprisingly benign way [35]. Nevertheless, it is clear that in the above scheme of quantum state control – as in arguably most of them – noise and decoherence remain a nuisance, and we have to do our best to isolate target and handle from the environment. On the other hand, it is obvious that a complete isolation from the environment is as utopian an idea as entangling $N \to \infty$ atoms, and we shall therefore approach the problem of quantum state control from a radically different perspective in the next, concluding chapter of this lecture.

3 Control of Quantum Jumps by Stochastic Resonance

Here, our starting hypothesis will be the following: The target system is coupled to the environment, and there is nothing we can do about that. Much as in biological systems living in a typical environment (which they got adapted to in the course of evolution), the latter creates a typical noise level, and the problem is how to induce specific transitions or processes in the target system, minimizing the errors induced by the environment. The solution to this problem is provided by Stochastic Resonance (SR), an extremely robust statistical phenomenon which is implemented in a wide class of natural phenomena, from celestial to microscopic scales. As a matter of fact, SR was introduced [42] as a possible mechanism to explain the apparently near-periodic occurence of ice ages (with a period of approx. 100000 years) in the earth climate, during the last 700000 years. In this particular model, noise from atmospheric dynamics cooperates with a weak periodic signal (with a period of approx. 100000 years) provided from the earths celestial mechanics. Whilst the latter alone is too weak to drive the earth's climate (which is supposed to be bistable) from warm to cold periods (with a difference in the average temperature of approx. 10 K), almost periodic transitions from cold to warm and back are observed when the signal is assisted by the proper amount of noise.

Given the not really satisfactory statistics of climatic changes (in the above example, only seven events define the periodicity), SR remains controversial as the real cause of ice ages (see, however, [43]), though the model of [44] was soon carried over to simple electronic systems with bistability [45]. Next came ring lasers [46], and the field really started to blossom once the relevance of SR for biological systems was realized in the early nineties of the last century. Cray fish predator detection (see the beautiful article by Moss and Wiesenfeld [47], and notably the impressive snap shot of the crayfish getting away from the big fish which is after him – an example for all small fries in delicate circumstances), transport across ion channels, the auditory system, and crickets – all exploit SR such as to detect weak periodic signals in a noisy environment [48]. Also medical applications of SR are under discussion, what promoted the subject twice to The Economist's Science and Technology section [49,50].

What characterizes SR? It is

- a nonlinear effect – most directly illustrated by the noise assisted dynamics of a periodically driven particle in a one dimensional double well potential, with the periodic force not strong enough to carry the particle across the barrier;
- due to the cooperativity of signal and noise, in the sense that the typical time scale (given by Kramers' law [51]) on which the stochastic force drives the particle across the barrier matches half the period of the signal;
- expressed by an optimal signal enhancement at a *nonvanishing* noise level, in the response of the bistable system to the periodic force.

To be a little more specific, consider the bistable potential in Fig. 8, with stable (unstable) equilibria at $x = \pm c$ ($x = 0$), and a potential barrier of height U_0.

If we consider an overdamped particle moving in this potential, additionally subject to a periodic force of strength ϵ and frequency ω_0, and in the presence of a stochastic force $\xi(t)$, with intensity D, the equation of motion reads [52]

$$\dot{x} = -\frac{dU}{dx} + \epsilon \cos(\omega_0 t) + \sqrt{D}\xi(t) , \tag{36}$$

and the potential barrier seen by the particle is modulated with an amplitude $U_1 = \epsilon c$. Provided the modulation of the potential fulfils the adiabaticity condition $\omega_0 \ll U''(\pm c)$, i.e. the time scale of the drive is much larger than the

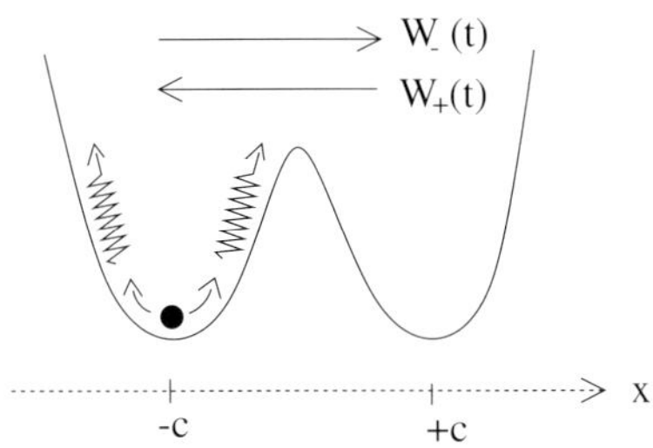

Fig. 8. Typical setting to observe Stochastic Resonance (SR): An overdamped particle moves in a bistable potential, under the combined influence of a weak periodic drive (smooth arrows), and of a stochastic force (ragged arrows). The latter defines transition rates $W_{\pm}$, which are periodically modulated by the periodic signal, see Eq. (37), provided the signal period is much larger than the typical relaxation time scale of the particle within the potential wells.

intrastate relaxation time of the particle, then Kramers' law can be generalized to the following, modulated transition rates

$$W_{\pm} \sim \exp\left(-\frac{2U_0}{D} \pm \frac{2U_1}{D}\cos(\omega_0 t)\right) \tag{37}$$

between the metastable sites. This allows to deduce the following spectral density of the response $x(t)$ of the particle to the forcing, via a suitable Fourier transform [52] of the autocorrelation function $C(\tau) = \langle x(t)x(t+\tau)\rangle$ of the particle:

$$P(\omega) = \text{Noise} + \frac{\frac{\text{const}\times 4\pi\epsilon^2 c^4}{D^2} e^{-\frac{4U_0}{D}}}{\text{const}\times e^{-\frac{4U_0}{D}} + \omega_0^2}\,\delta(\omega-\omega_0)\,. \tag{38}$$

Inspection of the prefactor of the delta function – centered at the frequency of the periodic signal and therefore the signal part of the reponse – shows that the (positive) weight of the signal tends to zero for $D \to 0$ as well as for $D \to \infty$. Hence, besides the Lorentzian noisy background [52] in the spectral density – the first term in (38) – the periodic response of the particle is most pronounced at a *finite* noise level, and this is precisely the signature of SR.

Note, however, that there are different definitions of SR floating around in the literature, and that they do not always come to the same conclusion in as whether a system displays SR or not [53]. The most common are

- the signal to noise ratio (SNR), this is the ratio of signal strength to noise level at $\omega = \omega_0$;
- the signal enhancement – the prefactor of the delta function discussed above;
- and the residence time distribution – the probability density function of the time intervals spent by the particle in a metastable site between entry and exit.

The reason for the fact that different conclusions may be drawn from the application of these different indicators of SR appears to be often the intrawell dynamics of the particle, which may hide [30] the actual noise induced synchronization of the dynamics with the signal – the essence of SR. The current intuition of the authors suggests that the residence time distribution is the most reliable indicator, though possibly requires a larger amount of sampled data, in order to make the synchronization effect emerge from the noisy background.

So far, our outline of SR is completely classical, with Kramers' law and the adiabaticity condition as the essential ingredients. Since SR is so successful in the macroscopic world, from celestial scales to biological and medical applications, it appears somewhat natural to ask what happens when we enter the microcosm. This, however, is the realm of quantum mechanics, and we know that a quantum particle finds other ways to overcome the potential barrier in Fig. 8 than climbing the hill or being kicked over it by a stochastic force. Hence, which is the effect of generic quantum mechanical transition mechanisms such as tunneling or quantum fluctuations? This question was first addressed in the mid-nineties, for various model systems [54,55], and some modifications of the SR mechanism have been described. A first implementation of SR in a fundamental quantum system readily accessible in the laboratory [31] was suggested in [56], with a detailed elaboration of the theory in [30,35,57–59]. The reader will not be too surprised to hear that this fundamental quantum system is, indeed, the micromaser, and that SR is thus a means to synchronize the quantum jumps of the bistable maser field which we mentioned above (see Fig. 3). Since then, various other quantum optical playgrounds for SR have been identified [60–62].

In contrast to the previous chapter, where we controlled the quantum state of the field by a strictly unitary interaction, without any additional random forcing, we now need the full Hamiltonian (8) including the interaction of the field mode with the noisy environment, mimicked by the environment oscillators. If we simulate the maser dynamics generated by this Hamiltonian employing a quantum Monte Carlo technique [23,35,57], we obtain the signal of Fig. 3, which illustrates the time dependence of the probability to detect an atom in $|d\rangle$, *after* the interaction with the cavity mode. Remember that, due to (11,12), the atom and the field are entangled after exit of the atom from the cavity, and only the detection of the atom forces the field into a well defined (metastable) state. However, a single detection event cannot be predicted with certainty, leading to measurement noise because of the entanglement between atoms and field. In the bistable situation depicted in Fig. 2, the detection signal is averaged over approx. 500 detection events, what smoothes the measurement noise, and only leaves the detectable consequence of the jumps of the maser field between the two metastable states which are manifest in the maxima in the occupation of the density matrix of Fig. 2. Indeed, when we *observe* the atom predominantly in its lower state, on exit from the cavity, it is likely to have deposited a photon in the mode, and the cavity field state is therefore likely to be centered around n_2 in Fig. 2. Vice versa, if we observe the exiting atoms predominantly in the upper state, then the state of the mode is centered around n_1 in Fig. 2, due to the quantum mechanical projection postulate. However, whilst the jumps of the maser field occur on a typical time scale of tens of seconds in Fig. 3 (in agreement with the experiments reported in [31]), we cannot predict when a single jump will occur. This is precisely the general setting of SR: The maser field displays bistability, with environment- (vulgo noise-) induced transitions between the metastable states, on time scales much longer than any other characteristic time scale of the maser dynamics, what will allow us to satisfy the adiabaticity requirement of SR. The only difference with respect to the classical examples listed above is the dominantly quantum origin of the noise – measurement noise, on the one hand, and quantum fluctuations of the environment at low temperatures ($T \simeq 0.5$ K), i.e. small values of the thermal occupation $n_b < 1$ of the cavity mode, on the other. Consequently, SR provides a means to reduce our uncertainty about the residence time of the field in either of the metastable states, by synchronizing it with a weak, nondeterministic signal (in the sense that the bistability of the dynamics is not suppressed). Indeed, if we introduce the signal by a simple periodic modulation of the pump rate r of the cavity (experimentally easily achieved through a periodic modulation of the laser which excites the atoms from their ground to the Rydberg level [57]), we still observe random jumps of the cavity field, with a slight periodic modulation of the intrawell dynamics induced by the field, see Fig. 9a. Now we do nothing but to *increase* the noise level by increasing the environment temperature from $T = 0.3$ K to $T = 0.6$ K, and observe ... an *enhanced* periodicity of the quantum jumps of the field, at the period $t_{\mathrm{mod}} = 42$ sec of the signal, see Fig. 9b. Noise and signal cooperate optimally at this temperature, since, in fact, the average residence time of the field in either state is close to $t_{\mathrm{mod}}/2 \simeq 20$ sec at this temperature, in the absence of any modulation [57]. If we further increase the temperature, the noise is too strong, and the stochastic stimulus induces too many transitions per signal period – the signal drowns in the noisy background, see Fig. 9c.

To quantify this observation, we extract the spectral density from the autocorrelation function (or, alternatively, from the master Eq. (19) [57,58]), which is plotted in Fig. 10.

Clearly, there is a pronounced signal peak at the frequency of the coherent drive, and, additionally, at its higher (even and odd) harmonics.

The latter are just an expression of the nonlinear response of the system to the injected signal, and even harmonics do not vanish due to the asymmetry of the transition rates between the metastable states of the cavity field [57]. SR is finally born out when we plot the strength of the signal peak (and equally of the higher harmonics, see [59]) as a function of the environment temperature, see Fig. 11. Clearly, the signal is strongly enhanced, with a maximum at $T = 0.6$ K.

In summary, SR gives us a handle to control the noise induced quantum jumps of the maser field, by tuning only the noise intensity. Since this synchronization effect is essentially due to the matching of the typical time scales of the stochastic forcing and of the periodic drive, we might equally fix the noise level and tune the signal period [57,58]. Quite obviously, the control which we achieve by SR can by no means compete with the almost perfect quantum state control outlined in the preceding chapter.

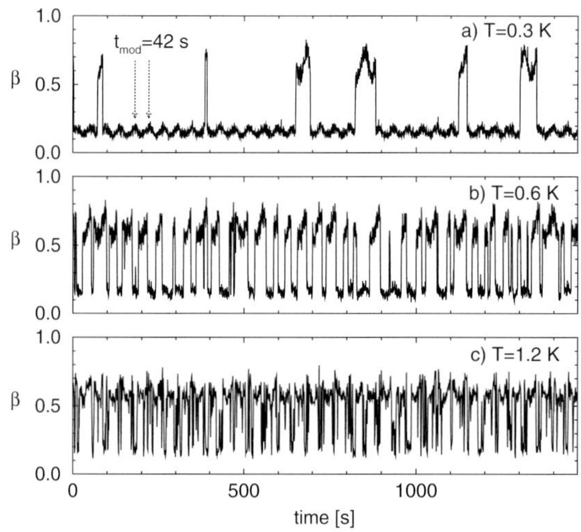

Fig. 9. Time evolution of the probability $\beta = \sum_n p_n B_{n+1}^2$ to detect an atom in $|d\rangle$, with periodically modulated atomic flux $r(t)/\gamma = 40 + 6.9\cos(2\pi t/t_{\mathrm{mod}})$, at different temperatures. Incoherent pumping, $a = 1$, with $\phi = 1.033$. Clearly, optimal synchronization is observed at $T = 0.6$ K.

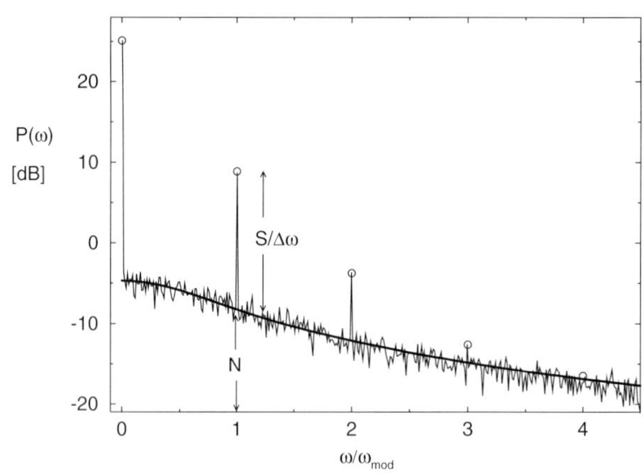

Fig. 10. Power spectral density of the detection signal $\beta(t)$ on output from the maser cavity, for $T = 0.6$ K. Other parameters as in Fig. 9.

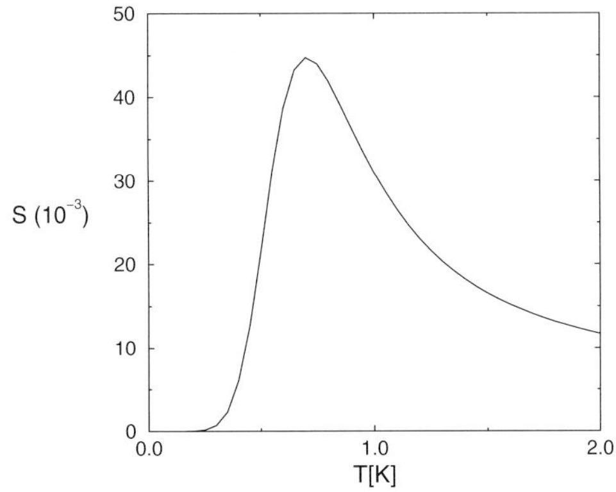

Fig. 11. Temperature dependence of the signal strength in the power spectral density of the detection signal $\beta(t)$. Parameters as in Fig. 9. Optimal synchronization at $T = 0.6$ K is clearly born out by the maximum in the plot.

SR tells us what can be done if we can *only* tune the noise, t_{mod}, or – more realistic in an evolutionary setting, where the species has to adapt to the environment – the potential barrier U_0 and its modulation amplitude U_1. Therefore, when we seek deterministic control over quantum states in low dimensional quantum systems, as illustrated in the preceding chapter, we will certainly try to implement asymptotic completeness or control schemes inspired by a similar vein. However, as we try to control quantum systems of increasing dimension, which will be ever harder to isolate from their noisy environments, SR might bear the larger potential.

References

1. I. R. Sims, M. Gruebele, E. D. Potter and A. H. Zewail, J. Chem. Phys. **97**, 4127 (1992).
2. M. Hentschel et al., Nature **414**, 509 (2001).
3. S. Rice and D. Tannor, Adv. Chem. Phys. **70**, 441 (1988).
4. T. Weinacht, J. Ahn and P. Bucksbaum, Nature **397**, 233 (1999).
5. A. Buchleitner, D. Delande and J. Zakrzewski, Phys. Rep. (2002), in print.
6. J. Zakrzewski, unpublished Lecture (Lądek 2002).
7. H. L. Bethlem et al., Nature **406**, 491 (2000).
8. S. Chu, Rev. Mod. Phys. **70**, 685 (1998).
9. C. Cohen-Tannoudji, Rev. Mod. Phys. **70**, 707 (1998).
10. W. D. Phillips, Rev. Mod. Phys. **70**, 721 (1998).
11. A. Bartana, R. Kosloff and D. J. Tannor, Chem. Phys. **267**, 195 (2001).
12. A. Assion et al., Science **282**, 919 (1998).
13. T. Brixner, N. H. Damrauer, P. Niklaus and G. Gerber, Nature **414**, 57 (2001).
14. J. Zakrzewski and D. Delande, J. Phys. **B30**, L87 (1997).
15. L. Mandel and E. Wolf, *Optical Coherence and Quantum Optics*, Cambridge University Press, Cambridge, 1995.
16. M. A. Nielsen and J. L. Chuang, *Quantum Computation and Information*, Cambridge Univ. Press, Cambridge, 2000.

17. A. Hofmann and R. de Vivie-Riedle, Chem. Phys. Lett. **346**, 299 (2001).
18. S. Lloyd, Phys. Rev. Lett. **75**, 346 (1995).
19. V. Ramakrishna, M. Salapaka, M. Dahleh, H. Rabitz and A. Peirce, Phys. Rev. **A51**, 960 (1995).
20. K. Vogel, V. Akulin and W. Schleich, Phys. Rev. Lett. **71**, 1816 (1993).
21. G. Harel, G. Kurizki, J. McIver and E. Coutsias, Phys. Rev. **A53**, 4534 (1996).
22. D. Meschede, H. Walther and G. Müller, Phys. Rev. Lett. **54**, 551 (1985).
23. S. Haroche, Cavity electrodynamics, in *Fundamental Systems in Quantum Optics, Les Houches, Session LIII*, edited by J. Dalibard, J.-M. Raimond and J. Zinn-Justin, Amsterdam, 1992, North-Holland.
24. M. Fannes, contribution to this volume.
25. E. Jaynes and F. Cummings, Proc. IEEE **51**, 89 (1963).
26. C. Cohen-Tannoudji, J. Dupont-Roc and G. Grynberg, *Atom-Photon Interactions: Basic Processes and Applications*, John Wiley and Sons, New York, 1992.
27. M. G. Floquet, Annales de l'École Normale Supérieure **12**, 47 (1883).
28. J. H. Shirley, Phys. Rev. **138**, B979 (1965).
29. R. Alicki, Quantum stochastic dynamics, in *Dynamical semigroups: dissipation, chaos, quanta.*
30. T. Wellens and A. Buchleitner, Phys. Rev. Lett. **84**, 5118 (2000).
31. O. Benson, G. Raithel and H. Walther, Phys. Rev. Lett. **72**, 3506 (1994).
32. H.-J. Briegel, B.-G. Englert, N. Sterpi and H. Walther, Phys. Rev. **A49**, 2962 (1994).
33. B. Kümmerer and H. Maassen, Infin. Dimens. Anal. Quantum Probab. Relat. Top. **3**, 161 (2000).
34. M. Reed and B. Simon, *Methods of Modern Mathematical Physics, III: Scattering Theory*, Academic Press, San Diego, 1979.
35. T. Wellens, *Entanglement and Control of Quantum States*, PhD thesis, Ludwig-Maximilians-Universität München, 2002.
36. C. Sackett et al., Nature **404**, 256 (2000).
37. A. Rauschenbeutel et al., Science **288**, 2024 (2000).
38. T. Wellens, A. Buchleitner, B. Kümmerer and H. Maassen, Phys. Rev. Lett. **85**, 3361 (2000).
39. S. Brattke, B. T. H. Varcoe and H. Walther, Phys. Rev. Lett. **86**, 3534 (2001).
40. S. Kuhr et al., Science **293**, 278 (2001).
41. G. Guthörlein, M. Keller, K. Hayasaka, W. Lange and H. Walther, Nature **414**, 49 (2001).
42. R. Benzi, A. Sutera and A. Vulpiani, J. Phys. **A14**, L453 (1981).
43. A. Ganopolski and S. Rahmstorf, Phys. Rev. Lett. **88**, 38501 (2002).
44. R. Benzi, G. Parisi, A. Sutera and A. Vulpiani, Tellus **34**, 10 (1982).
45. S. Fauve and F. Heslot, Phys. Lett. **A97**, 5 (1983).
46. G. Vemuri and R. Roy, Phys. Rev. **A39**, 4668 (1988).
47. F. Moss and K. Wiesenfeld, Scientific American (August 1995).
48. K. Wiesenfeld and F. Moss, Nature **373**, 33 (1995).
49. Science and Technology, *Come Feel the Noise*, The Economist, July 29th, 61 (1995).
50. Science and Technology, *Stochastic Resonance: Feel the Noise*, The Economist, September 19th, 99 (1998).
51. H. Kramers, Physica (Utrecht) **7**, 284 (1940).
52. B. McNamara and K. Wiesenfeld, Phys. Rev. **A39**, 4854 (1989).
53. A. Bulsara and L. Gammaitoni, Physics Today **49**, 39 (March 1996).
54. R. Löfstedt and S. Coppersmith, Phys. Rev. **E49**, 4821 (1994).

55. M. Grifoni and P. Hänggi, Phys. Rev. Lett. **76**, 1611 (1996).
56. A. Buchleitner and R. Mantegna, Phys. Rev. Lett. **80**, 3932 (1998).
57. T. Wellens and A. Buchleitner, J. Phys. **A32**, 2895 (1999).
58. T. Wellens, *Stochastische Resonanz im Mikromaser*, Diploma thesis (LMU Munich, 1998).
59. T. Wellens and A. Buchleitner, Chem. Phys. **268**, 131 (2001).
60. L. Viola, E. Fortunato, S. Lloyd, C.-H. Tseng and D. Cory, Phys. Rev. Lett. **84**, 5466 (2000).
61. S. F. Huelga and M. Plenio, Phys. Rev. **A62**, 52111 (2000).
62. L. Sanchez-Palencia, F. R. Carminati, M. Schiavoni, F. Renzoni and G. Grynberg, Phys. Rev. Lett. **88**, 133903 (2002).

Solving Schrödinger's Equation for an Open System and Its Environment

W.T. Strunz

Theoretische Quantendynamik, Fakultät für Physik, Universität Freiburg, Hermann-Herder-Str.3, 79104 Freiburg, Germany

Abstract. The dynamics of an open quantum system is described in terms of a non-Markovian stochastic Schrödinger equation derived straight from a microscopic model for the open system and its environment. We highlight several remarkable features of our approach: its convolutionless formulation, the possibility to derive the corresponding nonlinear version, and the master equation for the ensemble mean. We finally apply it to the standard quantum theory of Brownian motion, here in the Schrödinger picture. The stochastic Schrödinger equation approach to open quantum system dynamics is thus no longer restricted to the special class of Markovian, Lindblad evolution.

1 Introduction

The dynamics of an isolated quantum system is determined by the Schrödinger equation involving the Hamiltonian of the system in question. If one is to describe the dynamics of an open quantum system, several approaches are possible, as described in more detail in the next Sect. 2. In this lecture we include the environment explicitely as part of a huge, isolated total system and apply Schrödinger's equation to both, system and environment [1]. Alternatively, evolution equations for open quantum systems may be derived on axiomatic grounds, demanding density operators to remain density operators during time evolution. This is the approach that leads, for instance, to the Markovian Lindblad class of open system evolution [2].

An understanding of the dynamics of open systems is of importance in most areas of physics, both in the classical and in the quantum domain. The interaction of a distinguished subsystem with its many degree of freedom environment or reservoir induces friction and fluctuations. In the quantum domain, due to entanglement, we encounter an additional phenomenon, known as quantum noise: Even at zero temperature the interaction with an environment leads to a loss of purity of the open system state, since system and environment become entangled. In this case, system and environment together are described by a single state vector yet this ceases to be true for the open system itself which is now in a mixed state.

Theoretical tools to study open system dynamics are well developed (see Sect. 2). Among the most recent developments have been *stochastic Schrödinger equations* for system states ψ_t that are constructed in such a way as to recover the reduced density ρ_t of the open system as an ensemble mean over many realizations, $\rho_t = \mathcal{M}\left[|\psi_t\rangle\langle\psi_t|\right]$. We use the notation $\mathcal{M}[\ldots]$ for ensemble means in order to clearly distinguish them from quantum expectation values, denoted by angular brackets $\langle \ldots \rangle$. Very often, this stochastic approach has numerical advantages since the propagation of the $N \times N$-matrix ρ_t may be replaced by the solution of a Schrödinger equation for the N-dimensional vector ψ_t – the prize to be paid, however, is the need to solve the stochastic equation many times. Sometimes, in particular for large N, and depending on the accuracy required, the stochastic approach proves to be advantageous.

Yet there is more to a stochastic Schrödinger equation description of open system dynamics than merely being a useful numerical tool: the stochastic Schrödinger equation to be developed in this contribution amounts to a Monte-Carlo determination of the total state of system *and* environment, thus providing information about the total state. In our work, the stochastic system states ψ_t correspond to the system part in an expansion of the total state in a particular basis for the environment. Such a representation allows to interprete the stochastic states as *conditional* system states, given the state of the environment; a point of view that clarifies why stochastic Schrödinger equations have emerged naturally in the framework of continuous measurement theories, in particular in quantum optics [3].

So far, most results on stochastic Schrödinger equations have been restricted to the special class of open system evolution of Lindblad type [2] (see also Sect. 2). By contrast, our starting point is a general microscopic model for system and environment and thus covers a far wider range of open system dynamics, generally non-Markovian. We base our approach on the solution of the Schrödinger equation for both, system and environment and determine the corresponding total state in a Monte-Carlo sense. The reduced density operator of the open system is recovered as an ensemble mean without the need to derive any master equation for its time evolution. As we will see, moreover, our stochastic approach turns out to be a useful tool for the derivation of master equations for open system evolution, if that is desired.

This lecture is organized in the following way: first we give a short overview over methods how to describe open system evolution, both classical and quantum, clarifying the aims and methods of this paper. Next, in Sect. 3 we introduce our system and environment model, consisting of the widely used harmonic-oscillator bath, with an arbitrary system Hamiltonian, the interaction between them being linear in the bath part and arbitrary in the system part of the coupling. The corresponding Schrödinger equation is solved in a coherent state representation for the reservoir degrees of freedom [4]. We find a stochastic representation of the generally non-Markovian open system dynamics, the evolution of the current state being dependent on an integral over a memory kernel. For a special memoryless case, namely the standard Markov limit, we recover a well established stochastic representation of Lindblad type evolution, the *quantum state diffusion* equation of Gisin and Percival [5]. A very powerful way how to deal with the memory integral appearing in the non-Markovian case is developed in Sect. 4: for a wide class of interesting open system behavior, a time-local, *convolutionless* description is possible [6]. In this formulation, non-Markovian features of the dynamics are taken into account as time dependent evolution operators that act on the state at the present time only, a considerable simplification. Moreover, the corresponding master equation for the ensemble evolution can easily be derived from the stochastic approach, describing quantum dynamics beyond the class of Lindblad evolution. The linear version of the stochastic equation used so far in the analysis may lead to stability problems in numerical applications, related to the fact that the linear theory does not preserve the norm of the stochastic states. This drawback may be overcome by a nonlinear version of the non-Markovian stochastic Schrödinger equation [6]. The non-linear version corresponds to an expansion of the total state in a physically more appropriate *comoving* environmental coherent state basis. In Sect. 5 we apply our general theory to specific cases and examples: the weak coupling and the near Markov case are treated in more detail. Furthermore, a few exactly soluble models are investigated within our approach – these allow for rigorous results irrespective of coupling strength or specific type of environmental structure. We end our contribution with a special section devoted to the standard quantum Brownian motion model [7] and close with concluding remarks.

2 Open System Dynamics

The interaction of an open system with its environment induces fluctuations and dissipation. Accordingly, the otherwise Hamiltonian evolution is replaced by stochastic and damped motion. Classically, the damped, stochastic motion of a one degree of freedom Brownian particle with Hamiltonian $H_{\mathrm{sys}} = p^2/2m + V(q)$ experiencing some force derived from a potential $V(q)$ may be described by a Fokker-Planck equation [8] for its phase space density $\rho_t(q,p)$,

$$\partial_t \rho_t = \{H_{\mathrm{sys}}, \rho_t\} + \gamma \frac{\partial}{\partial p}(p\rho_t) + m\gamma kT \frac{\partial^2}{\partial p^2}\rho_t \,. \tag{1}$$

The first term on the right hand side indicates the Hamiltonian evolution in terms of a Poisson bracket, the second term represents a friction force with damping constant γ, and finally a diffusion term with diffusion constant $m\gamma kT$ with k Boltzmann's constant and T the temperature.

An alternative dynamical description is based on the stochastic evolution equation of a single Brownian particle, i.e. a Langevin equation

$$\begin{aligned} \dot{q} &= p/m \,, \\ \dot{p} &= -V'(q) - \gamma p + F(t) \,, \end{aligned} \tag{2}$$

or equivalently, $m\ddot{q} + m\gamma\dot{q} + V'(q) = F(t)$, highlighting the friction force $m\gamma\dot{q}$ and the stochastic force $F(t)$ with zero mean and correlations $\mathcal{M}[F(t)F(s)] = 2m\gamma kT\delta(t-s)$ (again, we use $\mathcal{M}[\ldots]$ to denote ensemble means in order to clearly distinguish them from quantum expectation values denoted by angular brackets $\langle\ldots\rangle$).

The two dynamical descriptions (1) and (2) are equivalent in the sense that the phase space distribution ρ_t is recovered from the Langevin equation as the ensemble mean over many realizations of the stochastic motion of a single Brownian particle,

$$\rho_t(q,p) = \mathcal{M}\left[\delta(q-q_t)\delta(p-p_t)\right] . \tag{3}$$

The quantization of such dissipative, diffusive dynamics may be performed in several ways [1]. The microscopic approach, adopted in this paper, takes into account the environment as part of a total Hamiltonian, whereupon Schrödinger's equation is applied to the unitary dynamics of both.

Alternatively, an axiomatic approach to open quantum system dynamics is fruitful, leading for instance in the standard Markov case to *Lindblad* [2] evolution equations for the reduced density operator of an open system, in its simplest form

$$\partial_t \rho = -i[H_{\rm sys}, \rho] + \frac{1}{2}\left([L\rho_t, L^\dagger] + [L, \rho_t L^\dagger]\right) . \tag{4}$$

The single non-unitary contribution in (4) involving an operator L in the Hilbert space of the open system being replaced by a sum or integral over such terms in the general case. For simplicity, we set $\hbar = 1$. Such *master equations* for the evolution of density operators ρ_t can be seen as representing the Schrödinger picture quantum analogue of a diffusion equation like (1).

Sometimes, but not always, master equations derived from a microscopic approach turn out to be of the mathematically desirable Lindblad class (4). Either due to approximations made in their derivation, or through dropping the semigroup property which is requested for (4), it may well happen that useful and even exact master equations are not of Lindblad class - examples will be given below.

This lecture is concerned with a *stochastic* description of open quantum system dynamics. We establish the quantum analogue of the classical Langevin equation (2), in the Schrödinger picture. Such stochastic Schrödinger equations or *quantum trajectories* for system states ψ_t recover the reduced density operator in analogy to (3) as an ensemble mean over stochastic pure states,

$$\rho_t = \mathcal{M}\left[|\psi_t\rangle\langle\psi_t|\right] . \tag{5}$$

Quantum trajectories are intimately related to theories of continuous measurement, or dynamical reduction models. Our work is the non-Markovian generalization of the *quantum state diffusion* equation of Gisin and Percival [5],

$$\partial_t \tilde{\psi}_t = -iH_{\rm sys}\tilde{\psi}_t + (L - \langle L\rangle_t)(z_t^* + \langle L^\dagger\rangle_t)\tilde{\psi}_t - \frac{1}{2}(L^\dagger L - \langle L^\dagger L\rangle_t)\tilde{\psi}_t . \tag{6}$$

We here use its *Stratonovich* version with a complex white noise z_t such that $\mathcal{M}[z_t z_s] = 0$ and $\mathcal{M}[z_t z_s^*] = \delta(t-s)$. The quantum state diffusion stochastic Schrödinger equation (6) is nonlinear due to the appearance of expectation values like $\langle L\rangle_t = \langle\tilde{\psi}_t|L|\tilde{\psi}_t\rangle$ on the right hand side. Importantly, (6) preserves the norm of the stochastic states $\tilde{\psi}_t$.

Dropping all expectation values in (6), we find its linear version,

$$\partial_t \psi_t = -iH_{\text{sys}}\psi_t + Lz_t^* \psi_t - \frac{1}{2}L^\dagger L\psi_t \,, \tag{7}$$

for unnormalized stochastic states ψ_t. Both stochastic equations (6) and (7) recover the reduced density operator according to the crucial property (5). As we will show in the subsequent sections, these stochastic states ψ_t or $\tilde{\psi}_t$ are the system part in an expansion of the total state of system and environment, and thus correspond to (are *relative* to) a certain environmental state. The stochastic states may thus be seen as *conditional* states of the open quantum system: they are the system states given a certain state of the environment.

The central subject of this contribution is to describe a stochastic approach to open quantum systems beyond the class of Lindblad evolution. It is based on a microscopic description of open system dynamics, taking into account the environment explicitely. In fact, by solving Schrödinger's equation for both, system and environment in a particular representation, we are able to derive a stochastic Schrödinger equation description of open quantum system dynamics in a very general setting. It reduces to the *quantum state diffusion* equations (6) and (7) in the limit of Markovian (Lindblad) evolution for the reduced density operator as in (4). Our microscopic approach enables us to deal with general, non-Markovian dynamics.

For completeness, we would also like to mention the traditional stochastic approach to open quantum systems, based on the *Heisenberg* picture [1]. The classical Brownian motion evolution equation in (2) may be considered as a quantum evolution equation for Heisenberg operators, the stochastic force $F(t)$ now being an operator force, however. As a microscopic derivation based on an oscillator model for the environment reveals, the operator force acts in the Hilbert space of the environment, and thus, quantum Langevin equations in the Heisenberg picture are difficult to handle, unless soluble analytically as for linear systems. The crucial feature of the Schrödinger picture stochastic description of open system dynamics developed in this contribution is the replacement of the usual operator noise in quantum Langevin equations by c-number noise driving the stochastic Schrödinger equation. Thus, an efficient numerical solution of the stochastic equation is possible.

3 Microscopic Approach and Non-Markovian Dynamics

Schrödinger's equation applies to closed systems only. If an open quantum system is to be described, a standard approach we follow here is to take into account the environment and solve

$$i\partial_t \Psi_t = H\Psi_t \,, \tag{8}$$

for both, system and environment (we use units such that $\hbar = 1$). The total Hamiltonian

$$H = H_{\text{sys}} + H_{\text{env}} + H_{\text{int}} \,, \tag{9}$$

consists of the Hamiltonian of the open system $H_{\rm sys}$, the Hamiltonian of the environment (alias reservoir or heat bath) $H_{\rm env}$ and, crucially, the interaction between the two, $H_{\rm int}$. For a mixed initial state, (8) should be replaced by the von Neumann equation $\partial_t \rho_{\rm tot} = -i[H, \rho_{\rm tot}]$.

For simplicity, we here assume that system and environment are in a pure joint initial state Ψ_0. Sometimes a more realistic choice is a thermal state for the environment which poses no fundamental difficulty for what we are going to develop, see [6]. However, that choice may lead to somewhat clumsier expressions which is why we stick for the most part of the paper to the pure Ψ_0. Moreover, we simplify further by assuming that system and environment are uncorrelated initially such that the total state is a product of system and environment state, the latter is taken to be the vacuum,

$$|\Psi_0\rangle = |\psi_0\rangle|0_{\rm env}\rangle\,. \tag{10}$$

Due to the generally growing entanglement between system and environment, solving Schrödinger's equation (8) is a formidable task and if one is interested in properties of the open system only, it traditionally appeared appropriate to try to derive an effective evolution equation for the reduced density operator

$$\rho_t = {\rm Tr}_{\rm env}[\rho_{\rm tot}] = {\rm Tr}_{\rm env}\left[|\Psi_t\rangle\langle\Psi_t|\right]\,. \tag{11}$$

If that turns out to be possible, the problem is reduced to a (non-unitary) evolution equation for ρ_t in the Hilbert space of the open system only.

Alternatively, as mentioned in the last Sect. 2, stochastic Schrödinger equations may be employed to describe the dynamics. As soon as the Hilbert space dimension N of the system is very large, this approach is often more efficient since the propagation of an $N\times N$ matrix ρ_t is replaced by the stochastic propagation of an only N-dimensional state ψ_t.

We here develop such a stochastic theory based on the solution of the Schrödinger equation (8). The environment is modeled by the standard collection of harmonic oscillators,

$$H_{\rm env} = \sum_\lambda \omega_\lambda a_\lambda^\dagger a_\lambda \tag{12}$$

and the coupling is taken to be of the form

$$H_{\rm int} = (LB^\dagger + BL^\dagger)\,, \tag{13}$$

with an arbitrary (hermitian or non-hermitian) system operator L and a bath operator B consisting of contributions of all environmental oscillators, $B = \sum_\lambda g_\lambda a_\lambda$, with g_λ a coupling constant to oscillator λ. It turns out to be convenient to change to an interaction representation with respect to the free environment evolution, such that instead of (9) we use

$$H_{\rm tot}(t) = e^{iH_{\rm env}t}H_{\rm tot}e^{-iH_{\rm env}t} = H_{\rm sys} + \left(LB^\dagger(t) + L^\dagger B(t)\right)\,, \tag{14}$$

for the total Hamiltonian with

$$B(t) = e^{iH_{\rm env}t}Be^{-iH_{\rm env}t} = \sum_\lambda g_\lambda a_\lambda e^{-i\omega_\lambda t}\,. \tag{15}$$

The dynamics of the open system is thus influenced by the bath operator $B(t)$, whose statistical properties are captured in its correlation function

$$\alpha(t-s) = \langle B(t)B^\dagger(s)\rangle_{\mathrm{env}} = \sum_\lambda |g_\lambda|^2 e^{-i\omega_\lambda(t-s)} = \int_0^\infty d\omega J(w) e^{-i\omega(t-s)}, \quad (16)$$

here at zero temperature. The last equality in (16) defines the spectral density $J(\omega)$ of bath oscillators.

If the dynamics of system and environment is such that the bath correlation function $\alpha(t-s)$ in (16) may be replaced by a delta function, the reduced dynamics is Markovian and ρ_t evolves according to the Lindblad master equation (4). A general correlation function $\alpha(t-s)$, however, describes memory effects of the environment and matters become exceedingly more difficult. Such *non-Markovian* effects are known to be relevant in many situations, in particular at low temperatures and as soon as narrow energy splittings occur, as in tunneling processes [1]. They are also relevant whenever the environment is structured, as for instance the electro-magnetic vacuum in the presence of a photonic band gap material [9]. Moreover, non-Markovian effects play a role in output coupling dynamics of atoms from a Bose Einstein condensate with the aim to build an atom laser [10]. It may also happen that the dynamics appears to be Markovian, yet the evolution equation is not of the standard Lindblad class (4). This is the case, for instance, in the high-temperature limit of the standard quantum Brownian motion model to be discussed in Sect. 5. In such cases, transient effects (*initial slips*, see also [11]) are important and no contradiction with the axioms for Lindblad evolution occurs.

Rather than trying to derive an evolution equation for the reduced density operator for general bath correlation $\alpha(t-s)$, we here follow a different route to capture non-Markovian effects. We solve the Schrödinger equation (8) for both, system and environment with initial state (10) and total Hamiltonian (14). For that we expand the total state Ψ_t in a coherent state basis for the environmental degrees of freedom.

3.1 Expansion in Coherent States

Coherent states $|z\rangle$ are minimum uncertainty wave packets and remain coherent under harmonic evolution. They are labeled by a complex parameter z, with the defining property $a|z\rangle = z|z\rangle$ for a bosonic annihilation operator a. We here use unnormalized *Bargmann* coherent states [12],

$$|z\rangle \equiv e^{za^\dagger}|0\rangle\,, \quad (17)$$

with $\langle z|z'\rangle = \exp(z^* z')$ such that the resolution of the identity reads

$$\mathbb{1} = \int \frac{d^2z}{\pi} e^{-|z|^2} |z\rangle\langle z|\,. \quad (18)$$

Importantly, Bargmann coherent states $|z\rangle$ are analytical in z.

We now expand the environmental part of the total state Ψ_t of (8) in a fixed Bargmann coherent state basis, using $z = (z_1, \ldots, z_\lambda, \ldots)$ to notate the vector of all environmental coherent state labels, and write

$$|\Psi_t\rangle = \int \frac{d^2 z}{\pi} e^{-|z|^2} |\psi_t(z^*)\rangle |z\rangle\,. \tag{19}$$

Here, $|z^2|$ is to be read as $\sum_\lambda |z_\lambda|^2$, for instance. The system states $|\psi_t(z^*)\rangle = \langle z|\Psi_t\rangle$ relative to $|z\rangle$ are analytical in $z^* = (z_1^*, \ldots, z_\lambda^*, \ldots)$, and they have been assigned all the time dependence of the total state. Note that we refrain from introducing an extra amplitude for the product state $|\psi_t(z^*)\rangle|z\rangle$, that being included in the definition of $\psi_t(z^*)$, which will therefore not remain normalized under time evolution.

We plug (19) into the Schrödinger equation (8) with total Hamiltonian (14) and find

$$i\partial_t \psi_t(z^*) = \left[H_{\rm sys} + L \sum_\lambda g_\lambda^* z_\lambda^* e^{i\omega_\lambda t} + L^\dagger \sum_\lambda g_\lambda \frac{\partial}{\partial_{z_\lambda^*}} e^{-i\omega_\lambda t} \right] \psi_t(z^*)\,. \tag{20}$$

The crucial next step is to identify the combination

$$z_t^* \equiv -i \sum_\lambda g_\lambda^* z_\lambda^* e^{i\omega_\lambda t}\,, \tag{21}$$

appearing in (20) and regard the states $\psi_t(z^*)$ no longer as a function of the vector z^*, but as a functional of z_t^*. Then (20) turns into

$$\partial_t \psi_t(z^*) = -iH_{\rm sys}\psi_t(z^*) + L z_t^* \psi_t(z^*) - L^\dagger \int_0^t ds\, \alpha(t-s) \frac{\delta \psi_t(z^*)}{\delta z_s^*}\,, \tag{22}$$

where $\alpha(t-s)$ is nothing but the (zero temperature) bath correlation function (16). Equation (22) was first derived in [4] and is the central starting point for a stochastic description of open system dynamics beyond the standard Markov Lindblad class. In the next subsection we see that z_t^* can be assigned the role of a stochastic process and thus, (22) turns indeed into a stochastic Schrödinger equation for the states $\psi_t(z^*)$ of the open quantum system.

3.2 Reduced Dynamics

According to expansion (19), the solution of (22) for all vectors $z^* = (z_1^*, \ldots, z_\lambda^*, \ldots)$, or equivalently, for all functions z_t^*, amounts to determining the solution Ψ_t of the Schrödinger equation (8), and thus total knowledge about system and environment. We may, for instance, determine the reduced density operator (11) for the open system and find

$$\rho_t = \int \frac{d^2 z}{\pi} e^{-|z|^2} |\psi_t(z^*)\rangle\langle\psi_t(z^*)|\,. \tag{23}$$

The reduced density operator is the mixture of all states $\psi_t(z^*)$, distributed with the Gaussian probability $e^{-|z|^2}$. We conclude that ρ_t may indeed be determined as an ensemble mean according to (5) through a Monte-Carlo integration in (23), here taken over the Gaussian distribution $\mathcal{M}[\ldots] = \int \frac{d^2z}{\pi} e^{-|z|^2} [\ldots]$ of z-vectors. On these grounds, equation (22) for the system part $\psi_t(z^*)$ of the total state Ψ_t turns indeed into a *non-Markovian stochastic Schrödinger equation* for the determination of the reduced density operator. The *stochastic process* z_t^* from (21) becomes a Gaussian colored process with zero mean $\mathcal{M}[z_t^*] = 0$ and correlations

$$\mathcal{M}[z_t z_s^*] = \alpha(t-s) \quad \text{and} \quad \mathcal{M}[z_t z_s] = 0 \,. \tag{24}$$

Here $\alpha(t-s)$ is again the (zero temperature) bath correlation function (16).

What remains is the task to solve the central equation (22) which is difficult due to the appearance of a functional derivative under the memory integral. We overcome this problem for a wide range of open system dynamics.

4 Convolutionless Formulation of Non-Markovian Dynamics

Memory effects of non-Markovian evolution clearly make themselves felt through the integral over the past in equation (22), involving the bath correlation function and a functional derivative of the current state $\psi_t(z^*)$ with respect to earlier noise z_s^*. In many relevant cases [6], it is possible to replace that functional derivative by some time dependent operator O,

$$\frac{\delta\psi_t(z^*)}{\delta z_s^*} = O(t,s,z^*)\psi_t(z^*) \,, \tag{25}$$

acting in the Hilbert space of the open system on the current state $\psi_t(z^*)$. We indicate that O may depend on the times t and s, and possibly on the (entire history of the) stochastic process z_t^*. Relevant examples of this replacement will be given shortly. One way to determine the operator $O(t,s,z^*)$ in actual applications [6] is to insert the Ansatz (25) in (22) and use the consistency condition

$$\partial_t \frac{\delta\psi_t(z^*)}{\delta z_s^*} = \frac{\delta}{\delta z_s^*}\partial_t\psi_t(z^*) \,. \tag{26}$$

Alternatively, a Heisenberg operator approach to determine $O(t,s,z^*)$ may be useful, as shown in [13].

Once the replacement (25) of the functional derivative by an operator is known – sometimes only approximately – the evolution equation (22) takes the more useful *convolutionless* form [6,14]

$$\partial_t\psi_t(z^*) = \left(-iH_{\text{sys}} + Lz_t^* - L^\dagger\bar{O}(t,z^*)\right)\psi_t(z^*) \,, \tag{27}$$

where we defined the operator $\bar{O}(t,z^*)$ as the integral over the action of the functional derivative over the whole past [15],

$$\bar{O}(t,z^*) = \int_0^t ds\, \alpha(t-s) O(t,s,z^*) \,. \tag{28}$$

The determination of $\psi_t(z^*)$ is now reduced to solving the simple stochastic Schrödinger equation (27). We recall that (27) does not preserve the norm of the states $\psi_t(z^*)$, thus, if only a numerical solution is possible, it is most often more advisable to use its nonlinear, norm-preserving version, which is derived towards the end of this section.

The crucial task remains the determination of the operator $O(t,s,z^*)$ in (25), a problem we now turn our attention to.

Weak coupling. The action of the functional derivative may be expanded in powers of the interaction H_{int}. To lowest order, one finds $\delta\psi_t(z^*)/\delta z_s^* = \left[e^{-iH_{\text{sys}}(t-s)}Le^{iH_{\text{sys}}(t-s)} + \ldots\right]\psi_t(z^*)$ and thus

$$O(t,s) \approx e^{-iH_{\text{sys}}(t-s)}Le^{iH_{\text{sys}}(t-s)}\,, \tag{29}$$

or

$$\bar{O}(t) \approx \int_0^t ds\, \alpha(s) e^{-iH_{\text{sys}}s}Le^{iH_{\text{sys}}s}\,. \tag{30}$$

Note that in this lowest order, $O(t,s)$ and therefore also $\bar{O}(t)$, are independent of the noise z_t^*. A weak coupling stochastic Schrödinger equation equivalent to (27) with the replacement (30) was derived in [16] in a formulation that kept the memory integral over the bath correlation function. The convolutionless formulation (27) is easier to handle, as there is no need to store the state vector $\psi_s(z^*)$ at earlier times $s < t$.

Near Markov. If the bath correlation function $\alpha(t-s)$ falls off rapidly under the memory integral in (22), an expansion of the functional derivative in terms of the time delay $(t-s)$ is sensible [7,15]. With $A_n(t) = \int_0^t ds\; s^n\alpha(s)$ and $n = 0, 1, 2, \ldots$, we find for the relevant integrated operator (28) to first order,

$$\bar{O}(t) \approx A_0(t)L + A_1(t)\left(-i[H,L] + A_0(t)[L,L^\dagger]L\right)\,, \tag{31}$$

neglecting contributions $A_n(t)$ with $n \geq 2$. Again, to the order accepted, the relevant operator replacing the functional derivative turns out to be independent of the noise z_t^*.

Of particular interest is the standard Markov limit. As soon as $\alpha(t-s) = \gamma\delta(t-s)$ with some constant γ, only A_0 is relevant and may be replaced by the constant $\gamma/2$, thus $\bar{O} = \frac{\gamma}{2}L$. In this case, indeed, the stochastic Schrödinger equation (27) (or equivalently, (22)) reduces to the linear version of the Markov quantum state diffusion equation (7). The next order term in expansion (31) will turn out to be relevant for the high temperature limit of the quantum Brownian motion model investigated in Sect. 5.

We remark that very often, Markov and weak coupling approximation are only meaningful in combination, referred to as *Born-Markov* approximation.

Soluble models. For some choices of the system Hamiltonian $H_{\rm sys}$ and the coupling operator L in the interaction Hamiltonian (13), the functional derivative may be replaced by an operator $O(t,s,z^*)$ without any approximation [6]. Two examples are given here, a third, more complicated one follows in Sect. 5.

The first example is an harmonic oscillator with $H_{\rm sys} = \Omega a^\dagger a$, coupled to the environment through a rotating-wave-type coupling, $L = a$ in the total Hamiltonian (9). It turns out [6] that the functional derivative may be replaced by an operator $O(t,s)$ being proportional to the annihilation operator without any approximation,

$$O(t,s) = \frac{c(s)}{c(t)} a\,, \tag{32}$$

where $c(t)$ is a complex function satisfying the equation of motion

$$\dot{c}(t) + i\Omega c(t) + \int_0^t ds\, \alpha(t-s)c(s) = 0\,, \tag{33}$$

involving the bath correlation function, reminiscent of the damped motion of the amplitude of the oscillator. The integrated operator $\bar{O}(t)$ thus becomes

$$\bar{O}(t) = F(t)a\,, \tag{34}$$

without any approximation, where $F(t) = \int_0^t ds\, c(s)\alpha(t-s)/c(t)$.

A similar result may be derived for a damped two-level system with Hamiltonian $H_{\rm sys} = \Omega\sigma_z/2$ and the rotating-wave type coupling $L = \sigma_-$. Similar to the case of the damped harmonic oscillator above, we find

$$O(t,s) = \frac{c(s)}{c(t)} \sigma_- \tag{35}$$

with the very same function $c(t)$ from (33). For this model the relevant integrated operator becomes

$$\bar{O}(t) = F(t)\sigma_- \tag{36}$$

again with $F(t) = \int_0^t ds\, c(s)\alpha(t-s)/c(t)$.

To summerize, in all the cases presented in this section, it is possible to replace the functional derivative by an expression $O(t,s)$ that we know how to handle. The description of the dynamics of the open quantum system is thus reduced to the solution of the simple stochastic Schrödinger equation (27) in the Hilbert space of the open system.

4.1 Ensemble Evolution, Non-Lindblad Master Equations

Our stochastic approach to open quantum systems may be employed to derive the master equation for the ensemble evolution [17]. This is particularly simple and straightforward as soon as the integrated operator $\bar{O}_t$ from (28) in (27) is independent of the noise z_t^* – which happens to be true for all cases mentioned in the last paragraph: weak coupling, near Markov, and those exactly soluble models.

According to our construction, the reduced density operator ρ_t is given by the ensemble mean over the solutions of the stochastic Schrödinger equation (27), which we write here as $\rho_t = \mathcal{M}[P_t]$ with $P_t = |\psi_t(z^*)\rangle\langle\psi_t(z^*)|$. Upon taking the time derivative in (5) and employing (27), we get the expression

$$\dot{\rho}_t = -\,i[H_{\rm sys}, \rho_t] + L\mathcal{M}[z_t^* P_t] - L^\dagger \mathcal{M}[\bar{O}(t, z^*)P_t] \\ + \mathcal{M}[P_t z_t]L^\dagger - \mathcal{M}[P_t \bar{O}^\dagger(t, z^*)]L \,. \qquad (37)$$

This is far from being a closed evolution equation for ρ_t. Using a version of Novikov's theorem [18],

$$\mathcal{M}[P_t z_t] = \mathcal{M}\left[\int_0^t ds\; \alpha(t-s)\frac{\delta}{\delta z_s^*}P_t\right] , \qquad (38)$$

which amounts to a partial integration under the Gaussian probability distribution, we may use the replacement of the functional derivative by the operator O in (25) and write

$$\dot{\rho}_t = -i[H_{\rm sys}, \rho_t] + [L, \mathcal{M}[P_t\bar{O}^\dagger(t, z^*)]] + [\mathcal{M}[\bar{O}(t, z^*)P_t], L^\dagger] \,. \qquad (39)$$

Now we see how a convolutionless evolution equation for the non-Markovian dynamics of the reduced density operator may be derived from the knowledge of the operator O replacing the functional derivative. Crucially, as soon as that replacement is independent of the noise z_t^*, as in all the examples mentioned above, (39) is a convolutionless closed master equation for ρ_t, due to $\mathcal{M}[\bar{O}(t)P_t] = \bar{O}(t)\mathcal{M}[P_t] = \bar{O}(t)\rho_t$, namely

$$\dot{\rho}_t = -i[H_{\rm sys}, \rho_t] + [L, \rho_t\bar{O}^\dagger(t)] + [\bar{O}(t)\rho_t, L^\dagger] \,. \qquad (40)$$

This master equation applies immediately to all the cases mentioned above. In the weak coupling case with $\bar{O}(t)$ from (29) for instance, we recover the so-called Redfield master equation [11,16], including an initial slip captured by the initial time dependence of $\bar{O}(t)$. In the Markov case, keeping only the lowest order in expansion (31), we recover Lindblad's equation (4) for the ensemble evolution. For the two exactly soluble models with $\bar{O}(t)$ from (34) and (36), (40) represents the exact evolution equation, which asymptotically, for large times t, is of Lindblad class. In the last Sect. 5 we discuss yet another exactly soluble case, where (40) turns out not to be of the Lindblad class, asymptotically.

4.2 Nonlinear Non-Markovian Stochastic Schrödinger Equation

The linear non-Markovian convolutionless stochastic Schrödinger equation (27) has the drawback of not preserving the norm of the states ψ_t. This may cause problems in simulations (importance sampling) and is overcome by a time-dependent shift of the probability distribution of the noise z_t^*, leading to the non-linear, non-Markovian stochastic Schrödinger equation

$$\begin{aligned}\partial_t\tilde{\psi}_t = & -iH_{\mathrm{sys}}\tilde{\psi}_t + (L - \langle L\rangle)\tilde{z}_t^*\tilde{\psi}_t \\ & - \left((L^\dagger - \langle L^\dagger\rangle_t)\bar{O}(t,\tilde{z}^*) - \langle (L^\dagger - \langle L^\dagger\rangle_t)\bar{O}(t,\tilde{z}^*)\rangle\right)\tilde{\psi}_t\,, \end{aligned} \tag{41}$$

for normalized states $\tilde{\psi}_t$, as shown in [6]. Here

$$\tilde{z}_t^* = z_t^* + \int_0^t ds\, \alpha^*(t-s)\langle L^\dagger\rangle_s\,, \tag{42}$$

is a shifted noise with the original z_t^* from (24). The crucial property (5) to recover the reduced density operator is preserved for the ensemble of solutions. In the case of Markov (Lindblad) evolution, the above reduces to the non-linear quantum state diffusion equation (6).

5 Quantum Brownian Motion

Of particular interest is the quantized version of Brownian motion (1,2), which may be derived from the general model (9) with the choices $H_{\mathrm{sys}} = p^2/2m + V(q)$ and a coupling $L = q$ [19,20]. In this case the general linear stochastic Schrödinger equation (22) is valid for arbitrary temperature [6,7] by simply extending the zero temperature bath correlation function (16) to finite temperatures,

$$\alpha(t-s) = \int d\omega J(\omega)\left[\coth\left(\frac{\omega}{2kT}\right)\cos\omega(t-s) - i\sin\omega(t-s)\right]. \tag{43}$$

Typical choices for the spectral density $J(\omega)$ is a linear dependence for $\omega \to 0$, and a cutoff at some high frequency Λ. The replacement of the functional derivative by the operator $O(t,s,z^*)$ in (25) which is necessary in order to derive the convolutionless version (27) of the non-Markovian stochastic Schödinger equation (or its non-linear version (41)) cannot be performed in the general case. We mention two relevant cases.

5.1 Classical Limit: High Temperature

In the high temperature limit, the thermal frequency $kT/\hbar$ is assumed the largest frequency involved and the bath correlation function may be replaced by [20]

$$\alpha(t-s) \approx 2m\gamma kT\Delta(t-s) + im\gamma\dot{\Delta}(t-s)\,. \tag{44}$$

Here, $\Delta(t-s)$ is a delta-like function, decaying on the bath correlation time scale Λ^{-1}, where Λ is a cutoff frequency of the spectral density $J(\omega)$ in (43). With

(44), in the limit of high cutoff frequency $\Lambda \to \infty$, the functional derivative may be determined from the expansion (31), including that first order. We find

$$\bar{O}(t) = A_0(t)q - A_1(t)p/m\,, \tag{45}$$

with the functions $A_n(t)$ as in (31). After an initial slip on the inverse cutoff frequency time scale, the asymptotic $\bar{O}$ becomes $\bar{O} = m\gamma kTq + \gamma p/2m$. With (45) we are in the position to determine single runs of equations (27) or its nonlinear version (41) numerically for an arbitrary potential $V(q)$. For applications to a driven, nonlinear system see [7]. As (45) is independent of the noise, the corresponding master equation (40) is valid and turns out to be nothing but the well known high temperature master equation of quantum Brownian motion [20] that reduces to the classical Fokker-Planck equation (1) upon taking the limit $\hbar \to 0$. As shown in [7], single runs of the nonlinear stochastic Schrödinger equation (41) for this case approach minimum-uncertainty wave packets as $\hbar \to 0$, whose centroid follows the classical Langevin equation (2). Thus, (41) may be seen as the quantized version of the classical Langevin equation in the Schrödinger picture, on the level of an individual realization.

5.2 Harmonic Oscillator

Of particular interest is the case of an harmonic oscillator system, $H_{\rm sys} = p^2/2m + m\Omega^2q^2/2$ and standard coupling $L = q$, which allows for a rigorous treatment within this stochastic approach [6,7]. The operator replacement of the functional derivative may be determined exactly and the integrated action (28) of the functional derivative turns out to be

$$\bar{O}(t, z^*) = F(t)q + G(t)p + i\int_0^t ds\; J(t,s)z_s^*\,, \tag{46}$$

and is thus dependent on the entire history of the noise z_s^*. The time dependent functions $F(t), G(t)$, and $J(t,s)$ are complicated, but known (see [6,7]). They incorporate all non-Markovian effects of the dynamics of the central oscillator. With the knowledge of $\bar{O}$, the stochastic Schrödinger equation (27) and its nonlinear variant (41) may be implemented on a computer and the dynamics of the oscillator at any temperature, coupling strength, etc., be determined. Recall that the solution of the stochastic equation (27) amounts to a determination of the total state Ψ_t of this model according to expansion (19).

If one is interested in the corresponding master equation for the evolution of the ensemble mean ρ_t only, the simple argument that led to equ. (40) is no longer applicable here due to the noise dependence of $\bar{O}$. Nevertheless, it is possible to rigorously derive the corresponding convolutionless master equation for non-Markovian evolution (see [21]),

$$\begin{aligned}\dot{\rho}_t = &-i[H_{\rm sys}, \rho_t] \\ &+ \frac{a(t)}{2i}[q^2, \rho_t] + \frac{b(t)}{2i}[q, \{p, \rho_t\}] \\ &+ c(t)[q,[p,\rho_t]] - d(t)[q,[q,\rho_t\}]]\,.\end{aligned} \tag{47}$$

This remarkable finding has been established much earlier using alternative approaches [22]. The terms involving the real time dependent coefficients $a(t)$ and $b(t)$ represent a potential renormalization and damping term, respectively, while the terms accompanied by the real valued $c(t)$ and $d(t)$ represent diffusion. The coefficients are complicated [21,22] and approach their asymptotic constant values on the very short bath correlation time, describing an initial slip. Remarkably, the asymptotic master equation is not of the Lindblad class (4). Again, our general approach to a stochastic representation of open system dynamics allows a rigorous treatment.

6 Conclusions

We establish a framework to describe the dynamics of an open quantum system in terms of pure system states $\psi_t(z^*)$, determined from a non-Markovian *stochastic Schrödinger equation.* The usual reduced density operator is recovered as the ensemble mean over many runs of the stochastic equation. In the standard Lindblad Markov limit, our findings reduce to the quantum state diffusion equation (6). The class of open system dynamics captured by this approach extends far beyond the standard class of Lindblad evolution, as illustrated with many examples. The derivation shows that our approach amounts to a solution of the Schrödinger equation for both, system and environment in a particular representation. For applications, so far it is crucial to find a convolutionless representation for the non-Markovian dynamics, a task we have been able to fulfill for a large variety of open system dynamics. For a numerical solution, the nonlinear, norm-preserving version that we derive is most often superior. Remarkably, our stochastic approach is also useful for the derivation of convolutionless master equations for the ensemble evolution. Applications to the standard quantum Brownian motion model reveal that the stochastic Schrödinger equation thus developed amounts to a quantized version of the classical Langevin equation, here in the Schrödinger picture.

Acknowledgments

It is a pleasure to thank L. Diósi, P. Gaspard, N. Gisin, F. Haake and T. Yu for a wealth of fruitful discussions. Support by the Sonderforschungsbereich 276 “Korrelierte Dynamik hochangeregter atomarer und molekularer Systeme” of the Deutsche Forschungsgemeinschaft is gratefully acknowledged.

References

1. U. Weiss, *Quantum dissipative systems*, (World Scientific, Singapore), 2nd edition, (2000).
2. G. Lindblad, Comm. Math. Phys. **48**, 119 (1976);
   V. Gorini, A. Kossakowski and E.C.G. Sudarshan, J. Math. Phys. **17**, 821 (1976).

3. H. J. Carmichael *An open system approach to quantum optics*, (Springer, Berlin) (1994);
M.B. Plenio and P. L. Knight, Rev. Mod. Phys. **70**, 101 (1998);
C.W. Gardiner and P. Zoller *Quantum Noise*, 2nd ed. (Springer, Berlin) (2000).
4. L. Diósi and W.T. Strunz, Phys. Lett. **A235**, 569 (1997).
5. N. Gisin and I. C. Percival, J. Phys. A: Math. Gen. **25**, 5677 (1992); **26**, 2233 (1993); **26**, 2245 (1993).
6. L. Diósi, N. Gisin and W.T. Strunz, Phys. Rev. **A 61**, 022108 (1998);
W. T. Strunz, L. Diósi and N. Gisin, Phys. Rev. Lett. **82**, 1801 (1999).
7. W.T. Strunz, L. Diosi and N. Gisin, Phys. Rev. Lett. **83**, 4909 (1999);
W.T. Strunz, Chem. Phys. **268** 237 (2001).
8. H. Risken, *The Fokker-Planck equation*, 2nd edition, (Springer, Berlin) (1989).
9. S. John and T. Quang, Phys. Rev. Lett. **74**, 3419 (1995);
B. Garraway, Phys. Rev. **A55**, 4636 (1997);
N. Vats and S. John, Phys. Rev. A **58**, 4168-4185 (1998);
M.W. Jack, M.J. Collett and D.F. Walls, Phys. Rev. **A59**, 2306 (1999);
G.M. Nikolopoulos and P. Lambropoulos, Phys. Rev. **A61** 053812 (2000);
M.W. Jack and J. J. Hope, Phys. Rev. **A63**, 043803 (2001).
10. J.J. Hope, Phys. Rev. **A55**, R2531-R2534 (1997);
H.-P. Breuer, D. Faller, B. Kappeler and F. Petruccione, Phys. Rev. **A60**, 3188-3196 (1999);
J.J. Hope, G.M. Moy, M.J. Collett and C.M. Savage, Phys. Rev. **A61**, 023603 (2000).
11. P. Gaspard and M. Nagaoka, J. Chem. Phys. **111**, 5676 (1999).
12. V. Bargmann, Commun. Pure Appl. Math. **14**, 187 (1961).
13. J. Cresser, Laser Physics **10**, 337 (2000).
14. F. Schwabl and W. Thirring, Ergeb. exakt. Naturwiss. **36**, 219 (1964);
N.G. van Kampen, *Stochastic Processes in Physics and Chemistry*, 2nd ed. (North-Holland, Amsterdam), (1992);
H.-P. Breuer, B. Kappeler and F. Petruccione, Phys. Rev. **A59**, 1633-1643 (1999).
15. T. Yu, L. Diósi, N. Gisin and W. T. Strunz, Phys. Rev. **A60**, 91 (1999).
16. P. Gaspard and M. Nagaoka, J. Chem. Phys. **111**, 5668 (1999).
17. T. Yu, L. Diósi, N. Gisin and W. T. Strunz, Phys. Lett. **A265**, 331 (2000).
18. A. Novikov, Sov. Phys. JETP **20**, 1290 (1965).
19. R.P. Feynman and F.L. Vernon, Ann. Phys. (N.Y.) **24**, 118 (1963).
20. A.O. Caldeira and A.J. Leggett, Physica **A121**, 587 (1983).
21. W.T. Strunz and T. Yu (to be published).
22. F. Haake and R. Reibold, Phys. Rev. **A32**, 2462 (1985);
B.L. Hu, J.P. Paz and Y. Zhang, Phys. Rev. **D45**, 2843 (1992).

Chapter V

Dynamics of Large Systems

Thermodynamic Behavior of Large Dynamical Systems – Quantum 1d Conductor and Classical Multibaker Map –

S. Tasaki

Department of Applied Physics and Advanced Institute for Complex Systems, Waseda University, 3-4-1 Okubo, Shinjuku-ku, Tokyo 169-8555, Japan

Abstract. Two purely dynamical approaches to nonequlibrium statistical mechanics are illustrated using a quantum one-dimensional conductor and an open multibaker map as typical examples. One of them is for infinitely extended systems and the other is based on dynamical systems theory. In both cases, a class of initial states, where reservoirs are in different equilibria, unidirectionally evolve towards the thermodynamically normal steady state for $t \to +\infty$. The two approaches adopt different quantities for the microscopic counterpart of the entropy production. The one is the negative time derivative of the relative entropy between the present and initial states, and the other is the time derivative of the coarse grained entropy. At the steady state, both entropy productions are shown to be nonnegative and consistent with thermodynamics. Moreover, for the open multibaker map, they are shown to be equivalent with each other in a macroscopic scaling limit.

1 Introduction

The understanding of irreversible phenomena including nonequilibrium steady states is a longstanding problem of statistical mechanics. Recently, stimulated by the progress of the dynamical systems theory, a new approach has been developed in classical statistical mechanics. It mainly deals with hyperbolic systems, namely systems where every trajectory is exponentially unstable with respect to the forward and backward time evolutions. So far, several new results have been obtained such as relations among Kolmogorov-Sinai entropy, Lyapunov exponents and transport coefficients (escape-rate formalism [1–3]), simple symmetry of a large deviation distribution of the entropy production in nonequilibrium steady states (fluctuation theorem [4,5]) and fractal distributions describing nonequilibrium steady states [4–7]. This approach is believed to be general enough because of the Gallavotti-Cohen hypothesis [5], which states that the microdynamics of an N-body system for large N can be regarded as a hyperbolic system for the purpose of computing macroscopic properties. The hyperbolic systems are classified into the open conservative systems and dissipative thermostated ones. Here we deal with the former only (for the latter, see [4–6,8] and references therein.)

Statistical mechanics of infinitely extended systems has been investigated rigorously since 1960's [9–11] and many results are obtained such as a precise formulation of phase transitions [11,9] and a proof of ergodicity of an ideal gas with respect to a thermal equilibrium state [10]. Also, nonequilibrium steady states have been investigated for harmonic crystals [13,14], a one-dimensional gas [15], unharmonic chains [16], an isotropic XY-chain [17], a one- dimensional quantum conductor [18] and an interacting fermion-spin system [19]. Entropy production is rigorously studied for a class of large quantum systems [20–22], where the relative entropy between the present and initial states is related to the thermodynamic entropy.

In this article, we illustrate typical features of the statistical mechanics of infinitely extended systems and the dynamical systems approach using representative examples, a quantum one-dimensional lattice conductor and an open multibaker map, respectively. Then, for the open multibaker map, we investigate the entropy production related to the relative entropy. It is compared with the coarse grained entropy production studied by Gaspard [23] and Gilbert and Dorfman [24]. The coarse grained entropy measures the system information with an accuracy determined by the coarse graining cells and its increase is due to information flow from coarser to finer phase space regions. On the other hand, the relative entropy between the present and initial states measures a global difference of the two states and its increase implies that the state at time t continuously deviates from the initial state. Thus, the two entropy productions have different physical meanings. However, as will be shown, they are equivalent in a macroscopic scaling limit.

The rest of this article will be arranged as follows. In Sec. 2, the statistical mechanics of quantum one-dimensional lattice conductor is reviewed as an example of infinitely extended systems. After describing the necessary mathematics and the model, we show that initial states, where two semi-infinite subsystems are in different equilibria, converge to a steady state in the limit of $t \to +\infty$. The transports at the steady state and an entropy production determined from the relative entropy are shown to be consistent with thermodynamics. In Sec. 3, for an open multibaker map with energy coordinate [25,26], we summarize the previous results on the properties of the steady states and the coarse grained entropy production. A multibaker map [7,2,27] is a lattice extension of the conventional baker transformation, which exhibits a deterministic diffusion. And multibaker maps and their generalizations are extensively used to study transport properties of hyperbolic systems. For recent developments, see the review by Tél and Vollmer [27]. In Sec. 4, for the open multibaker map, the relative entropy between the present and initial states is investigated and is shown to have properties consistent with thermodynamics. Moreover, in a macroscopic scaling limit, the relative entropy production and the coarse grained entropy production are shown to be equivalent. The last section is devoted to the summary.

2 Nonequilibrium States of Quantum 1d Lattice Conductor

2.1 1d Lattice Conductor and C* Algebra

Here we consider transports of a one-dimensional conductor consisting of electrons on an infinitely extended chain interacting with a localized potential. Its Hamiltonian is $H = \sum_{j=-\infty}^{+\infty} h_j$. Here the local Hamiltonian h_j given by

$$h_j = \sum_{\sigma=\pm} \left[-\gamma \left\{ c_{j,\sigma}^* c_{j+1,\sigma} + c_{j+1,\sigma}^* c_{j,\sigma} \right\} + \epsilon_j \; c_{j,\sigma}^* c_{j,\sigma} \right] \tag{1}$$

where $\gamma (> 0)$ is the strength of the electron transfer, ϵ_j stands for the potential localized on an interval $[1, L]$ and $c_{j,\sigma}^*$ and $c_{j,\sigma}$ are, respectively, creation and annihilation operators of an electron at site $j (\in \mathbf{Z})$ with spin $\sigma (= \pm)$. They satisfy the anticommutation relations

$$[c_{j,\sigma}, c_{k,\tau}]_+ = [c_{j,\sigma}^*, c_{k,\tau}^*]_+ = \mathbf{0} \qquad [c_{j,\sigma}, c_{k,\tau}^*]_+ = \delta_{jk}\delta_{\sigma\tau}\mathbf{1} \tag{2}$$

where $[A, B]_+ = AB + BA$ is the anticommutator, $\mathbf{0}$ the null operator and $\mathbf{1}$ the unit operator. This setting, however, has a mathematical defect: The series in H does not converge in the sense of operators because $c_{j,\sigma}^*$ and $c_{j,\sigma}$ have norm one: $\|c_{j,\sigma}^*\| = \|c_{j,\sigma}\| = 1$. Thus, approaches without explicitly referring to H are desirable. One of such approaches is the C* algebraic approach (for details, see Ref. [9]).

The key lies in the following observation: Let α_t be an evolution operator acting on dynamical variables: $\alpha_t(A) \equiv e^{iHt} A e^{-iHt}$, then it is defined as an exponential function $\alpha_t = e^{\delta t}$ of the generator $\delta(A) \equiv \frac{d}{dt}\alpha_t(A)|_{t=0}$ because it has a group property $\alpha_{s+t} = \alpha_s \alpha_t$. The generator δ acts, for example, as

$$\delta(c_{j,\sigma}) \equiv \frac{d}{dt}\alpha_t(c_{j,\sigma})|_{t=0} = i[H, c_{j,\sigma}] = -i \left\{ -\gamma \left(c_{j+1,\sigma} + c_{j-1,\sigma} \right) + \epsilon_j c_{j,\sigma} \right\} \tag{3}$$

and the right-hand side is well-defined. Conversely, if one defines the generator δ by the right-hand side of (3), α_t is well defined via $\alpha_t = e^{\delta t}$ without explicitly referring to H.

The C* algebraic approach is formulated as follows: The first ingredient is a collection $\mathcal{A}$ of all dynamical variables which are expressed by $c_{j,\sigma}^*$ and $c_{j,\sigma}$, and possess finite average values. Such dynamical variables are sums of products of the following elements and their limits:

$$B(f, g) \equiv \sum_{\sigma=\pm} \sum_{j=-\infty}^{+\infty} \left\{ f_{j,\sigma} c_{j,\sigma} + g_{j,\sigma} c_{j,\sigma}^* \right\} \tag{4}$$

where $\{f_{j,\sigma}\}$ and $\{g_{j,\sigma}\}$ are square summable. Since sums, products, scalar products, and an adjoint operation (the *- operation) are defined, $\mathcal{A}$ forms a *-algebra. Also, the norm measuring the size of each element is defined and all the limiting elements are included (i. e., $\mathcal{A}$ is complete). Moreover, as $\mathcal{A}$ is an abstraction of a *-algebra of bounded operators on a Hilbert space, its norm satisfies the C* condition: $\|A^*A\| = \|A\|^2$. Such an algebra $\mathcal{A}$ is called a C* algebra.

In statistical mechanics, when a state is specified, the average of any dynamical variable is calculated. Conversely, when average values of all dynamical variables are known, the corresponding state is completely specified. A state is, then, introduced as a map assigning the average value $\omega(A)$ to each dynamical variable A. The map is obviously linear and is called a linear functional. Moreover, $\omega(A^*A) \geq 0$ since A^*A has only positive spectrum, and $\omega(\mathbf{1}) = 1$ since 1 is the only spectrum of the unit $\mathbf{1}$. In addition, $\omega(A)$ is considered to be continuous with respect to A.

The time evolution operator α_t is defined via $\alpha_t = e^{\delta t}$, where the generator δ is given by

$$\delta(A) \equiv \lim_{\Lambda \to +\infty} i[H_\Lambda, A] \ . \tag{5}$$

Here, $H_\Lambda = \sum_{-\Lambda}^{\Lambda} h_j$ is the truncated Hamiltonian. Under certain conditions on h_j which are satisfied by the present model, the limit (5) exists for enough elements $A \in \mathcal{A}$ and, thus, α_t is defined without difficulty.

The evolution of states is defined by $\omega_t(A) \equiv \omega_0(\alpha_t(A))$. Then, the $t \to \pm\infty$ limits of ω_t are naturally defined as

$$\omega_{\pm\infty}(A) \equiv \lim_{t \to \pm\infty} \omega_t(A) \quad (\forall A \in \mathcal{A}) \ . \tag{6}$$

We note that, even when the limits $\lim_{t\to\pm\infty} \alpha_t(A)$ do not exist, the limits $\omega_{\pm\infty}$ in the sense of (6) may exist and that the time reversal symmetric evolution may admit unidirectional state evolution. As will be seen, this is indeed the case.

2.2 Steady States and Transports

We consider a class of initial states as described below: Firstly, the chain is divided into three: $(-\infty, -M]$, $[-M+1, N-1]$ and $[N, +\infty)$ with $M > 0$ and $N > L$. The two semi-infinite segments serve as reservoirs and the finite one as an embedded system. Corresponding to this division, the Hamiltonian H is represented as a sum of a left-reservoir part H_L, a right-reservoir part H_R, an embedded-system part H_S and a reservoir-system interaction V_{int}: $H = H_L + H_R + H_S + V_{int}$. The number operator is decomposed as well: $N = N_L + N_R + N_S$. Then we consider initial states ω_{in} where the left reservoir is in equilibrium with inverse temperature β_L and chemical potential μ_L, and the right reservoir is in equilibrium with β_R and μ_R. Then, we assume the absence of bound states, which guarantees the convergence of initial states to steady states.

Under the conditions just explained, we showed [18] that, for $t \to \pm\infty$, the initial state ω_{in} evolves towards the unique steady state $\omega_{\pm\infty}$ in the sense that for any $B \in \mathcal{A}$, $\lim_{t\to\pm\infty} \omega_{in}(\alpha_t(B)) = \omega_{\pm\infty}(B)$, irrespective to the choice of the separating points M, N and the initial state of the embedded system. The states $\omega_{\pm\infty}$ satisfy Wick's theorem and are fully characterized by the two-point functions:

$$\omega_{+\infty}(c^*_{j\sigma}c_{j'\sigma'}) = \delta_{\sigma\sigma'}\int_0^\pi dq\,\{F_L\left(E_q\right)\psi_q(j)^*\psi_q(j') + F_R\left(E_q\right)\psi_{-q}(j)^*\psi_{-q}(j')\} \tag{7}$$

and $\omega_{-\infty}(c^*_{j\sigma}c_{j'\sigma'}) = \omega_{+\infty}(c^*_{j\sigma}c_{j'\sigma'})^*$. $F_L(E)$ and $F_R(E)$ are the Fermi distribution functions: $F_\alpha(E) = 1/\{\exp\left(\beta_\alpha(E-\mu_\alpha)\right)+1\}$ $(\alpha = L, R)$ of the left and right reservoirs and $E_q = -2\gamma\cos q$.

Eq. (7) gives two-probe Landauer-type formula for the particle flow and the energy flow, which were phenomenologically derived by Sivan and Imry [28]:

$$\langle J^N_{j-1|j}\rangle_{+\infty} \equiv \omega_{+\infty}\left(J^N_{j-1|j}\right) = \int_{-2\gamma}^{2\gamma}\frac{dE}{\pi}|T_{q(E)}|^2\left\{F_L(E) - F_R(E)\right\} \tag{8}$$

$$\langle J^E_{j-1|j}\rangle_{+\infty} \equiv \omega_{+\infty}\left(J^E_{j-1|j}\right) = \int_{-2\gamma}^{2\gamma}\frac{dE}{\pi}E|T_{q(E)}|^2\left\{F_L(E) - F_R(E)\right\} \tag{9}$$

where $\langle\cdots\rangle_{+\infty}$ stands for the average with respect to $\omega_{+\infty}$, $|T_q|^2$ is the transmission coefficient of the potential scattering by ϵ_j, $q(E) \equiv \cos^{-1}\{-E/(2\gamma)\}$, and $J^N_{j-1|j}$ and $J^E_{j-1|j}$ are the particle-flow and energy-flow operators, respectively, from the $(j-1)$th to the jth sites:[1]

$$J^N_{j-1|j} = i\gamma\sum_{\sigma=\pm}\{c^*_{j,\sigma}c_{j-1,\sigma} - c^*_{j-1,\sigma}c_{j,\sigma}\} \tag{10}$$

$$J^E_{j-1|j} = -i\gamma^2\sum_{\sigma=\pm}\{c^*_{j+1,\sigma}c_{j-1,\sigma} - c^*_{j-1,\sigma}c_{j+1,\sigma}\} + \epsilon_j J^N_{j-1|j}\ . \tag{11}$$

As explained in [28] and [18], in the linear response regime, the currents reduce to the well known forms in the linear nonequilibrium thermodynamics.

Note that a noninteracting time-reversible system exhibits irreversible state evolution. This is because of the infinite extension of the system and of the prescription of the long-time limits (6) of the states: An initial state evolves in a way that a domain having steady-state properties is created near the system- reservoir boundaries and expands towards infinity. On the other hand, an arbitrary dynamical variable is well-localized since the generator $B(f,g)$ of the algebra $\mathcal{A}$ is a linear combination of elements decaying at infinity (cf. the sequences f_j and g_j are square summable). Consequently, the contribution to $\omega_t(A)$ from the domain without steady-state properties diminishes as time goes on, and $\omega_t(A)$ tends to the steady-state average of A.

The reason why steady-state transports are consistent with thermodynamics seems to be the preparation of initial states where reservoir states are described by grand canonical ensembles. As an example, let us consider the behavior of electrons with a given energy E in case of $\mu_L = \mu_R$ and $T_L > T_R$. Then the number of electrons with energy E in the left reservoir is greater than that in the

[1] The expression of the energy current operator is different from that in Ref. [18]. This is due to the difference of the definition of the local energy, which is h_j in this article.

right reservoir. Since the flow is proportional to the number of electrons, the flow from the left to the right reservoirs exceeds the opposite flow, and there appears a net flow from the higher-temperature (left) reservoir to the lower- temperature (right) one, as expected from thermodynamics.

2.3 Relative Entropy and Its Change

In the C* algebraic approach, the relative entropy between the present and initial states is considered to be a microscopic counterpart of the thermodynamic entropy. Ojima et al. [20] studied the relative entropy change in various cases such as externally driven equilibrium states and nonequilibrium states due to the interaction among reservoirs in different equilibria, and showed that the time derivative of the relative entropy is related to thermodynamical entropy production. More recently, Jakšić and Pillet [21] found the same result in the case where several reservoirs are prepared initially in different equilibria and showed that the steady-state entropy production agrees with that phenomenologically obtained by Ruelle [22]. Here their results are explained more intuitively for the one-dimensional conductor.

First we consider the behavior of a truncated system which consists of dynamical variables contained in a finite interval $|j| \leq \Lambda$. Then we consider the time evolution generated by the truncated Hamiltonian $H_\Lambda = \sum_{j=-\Lambda}^{\Lambda} h_j$. As in the original case, one has: $H_\Lambda = H_L^\Lambda + H_R^\Lambda + H_S + V_{int}$. The number operators of the left and right reservoirs N_L and N_R are truncated so that N_L^Λ and N_R^Λ commute both with H_L^Λ and H_R^Λ. Since the truncated system is finite dimensional, the initial states ω_{in} is represented by a density operator

$$\rho_{in}^\Lambda = \exp\left(-\beta_L(H_L^\Lambda - \mu_L N_L^\Lambda) - \beta_R(H_R^\Lambda - \mu_R N_R^\Lambda)\right) \hat{\sigma}_S / Z \,,$$

where $\hat{\sigma}_S$ is the density matrix corresponding to the initial state of the embedded system and Z the normalization constant. And the state at time t is given by the density operator $\rho_t^\Lambda = \exp(-iH_\Lambda t)\rho_{in} \exp(+iH_\Lambda t)$.

When $\hat{\sigma}_S$ is invertible, the relative entropy $S(\rho_t^\Lambda|\rho_{in}^\Lambda)$ between ρ_t^Λ and ρ_{in}^Λ is

$$S(\rho_t^\Lambda|\rho_{in}^\Lambda) = -\mathrm{Tr}\left\{\rho_t^\Lambda\left(\ln\rho_t^\Lambda - \ln\rho_{in}^\Lambda\right)\right\} = i\int_0^t ds \langle [H_\Lambda, \ln\rho_{in}^\Lambda]\rangle_s^\Lambda \tag{12}$$

$$= i\int_0^t ds\langle [\beta_L(H_L^\Lambda - \mu_L N_L^\Lambda) + \beta_R(H_R^\Lambda - \mu_R N_R^\Lambda), V]\rangle_s^\Lambda$$

$$-i\int_0^t ds\langle [\ln\hat{\sigma}_S, V]\rangle_s^\Lambda \tag{13}$$

$$= \int_0^t ds\left\{\beta_L\langle J_q^L\rangle_s^\Lambda + \beta_R\langle J_q^R\rangle_s^\Lambda\right\} + \langle \ln\hat{\sigma}_S\rangle_t^\Lambda - \langle \ln\hat{\sigma}_S\rangle_{t=0}^\Lambda \,, \tag{14}$$

where $V \equiv H_S + V_{int}$, $\langle\cdots\rangle_s^\Lambda = \mathrm{Tr}(\rho_s^\Lambda\cdots)$ with ρ_s^Λ the state at time s, operators J_q^L and J_q^R represent heat flows to the embedded system from the left and right reservoirs, respectively,

$$J_q^L \equiv J^E_{-M-2|-M-1} - \mu_L J^N_{-M-1|-M} \qquad J_q^R \equiv -\{J^E_{N|N+1} - \mu_R J^N_{N|N+1}\} \,. \tag{15}$$

By taking the infinite volume limit $\Lambda \to +\infty$, one has

$$S(\omega_t|\omega_{in}) = \int_0^t ds \left\{\beta_L \langle J_q^L \rangle_s + \beta_R \langle J_q^R \rangle_s \right\} + \langle \ln \hat{\sigma}_S \rangle_t - \langle \ln \hat{\sigma}_S \rangle_{t=0} \,, \quad (16)$$

where $\langle \cdots \rangle_t \equiv \omega_t(\cdots)$. Rigorous arguments give the same result. Note that the $\Lambda \to +\infty$ limit of (13) is the Jakšić- Pillet formula (see Theorem 1.1 of Ref [21]).

Since the difference $\ln \rho_t^\Lambda - \ln \rho_{in}^\Lambda$ is the integral of local current operators, the initial state ρ_{in}^Λ is regarded as a local equilibrium approximate to ρ_t^Λ. In this view, the relative entropy $S(\omega_t|\omega_{in})$ is considered to be the difference between the Gibbs and a coarse-grained entropies. Then, as discussed by Dorfman and van Beijeren [29] and by Breymann, Tél and Vollmer [30,27], the negative time derivative of the relative entropy at steady states corresponds to the thermodynamic entropy production σ_R^{st}:

$$\sigma_R^{st} = - \lim_{t \to +\infty} \frac{d}{dt} S(\omega_t|\omega_{in}) \,. \quad (17)$$

This is the result of Ojima et al. [20] and Jakšić- Pillet [21].

As the long time limit of $\frac{d}{dt}\langle \ln \hat{\sigma}_S \rangle_t$ vanishes, the steady-state entropy production is

$$\sigma_R^{st} = -\beta_L \langle J_q^L \rangle_{+\infty} - \beta_R \langle J_q^R \rangle_{+\infty} \quad (18)$$

$$= \int_{-2\gamma}^{2\gamma} \frac{dE}{\pi} |T_{q(E)}|^2 \left\{\beta_L(E-\mu_L) - \beta_R(E-\mu_R)\right\} \left\{F_R(E) - F_L(E)\right\} . \quad (19)$$

where in the last equality the relations (8) and (9) are used. As easily seen from (19), σ_R^{st} is nonnegative and vanishes if and only if $\mu_L = \mu_R$ and $\beta_L = \beta_R$. This is consistent with thermodynamics. In the linear response regime, σ_R^{st} reduces to the quadratic form well known in the linear nonequilibrium thermodynamics. One more interesting relation is: $-\langle J_q^L \rangle_{+\infty} - \langle J_q^R \rangle_{+\infty} = (\mu_L - \mu_R) \langle J_{j-1|j}^N \rangle_{+\infty}$ where the right-hand side corresponds to the Joule heat generated by the electric current and the left-hand side stands for the net heat flow into the reservoirs. This is again consistent with thermodynamics. Note that Ruelle [22] defined the steady-state entropy production by (18) and discussed its positivity.

It is remarkable that a system consisting of noninteracting electrons admits positive entropy production. Presumably, this is because the initial reservoir states are prepared to be in equilibrium. Firstly, by tracing back the derivation of (18), one finds that its expression comes from grand canonical ensembles describing the initial reservoir states. Secondly, when the system is in the steady state, the entropy flows $-\beta_L \langle J_q^L \rangle_{+\infty}$ and $-\beta_R \langle J_q^R \rangle_{+\infty}$ to the left and right reservoirs are not compensated with each other since temperatures and chemical potentials of the reservoirs are different but energy and particle flows are identical. And, we have a nonvanishing entropy production. This argument is independent of the dynamics of the embedded system and could be valid even for systems of noninteracting particles.

3 Nonequilibrium States of Open Multibaker Map

3.1 Open Multibaker Map with Energy Coordinate

The multibaker map with energy coordinate introduced in [25] is a caricature of the first return map of the periodic Lorentz gas. The phase space is a chain $\Gamma = \cup_{n \in \mathbf{Z}} \Gamma_n$ of three-dimensional cells as shown in Fig. 1(a) where

$$\Gamma_n = \left\{(n, x, y, E) | E \in \mathbf{R}^+, \ 0 < x \le a_E, 0 < y \le a_E \right\} \tag{20}$$

$\mathbf{Z}$ and $\mathbf{R}^+$ stand for the sets of integers and of positive real numbers, respectively, and $a_E = ae^E$ with a a positive constant. Hereafter, each subset Γ_n will be referred to as a cell. The variable E corresponds to the kinetic energy of particles in a periodic Lorentz gas and a_E^2 represents the area of a constant "kinetic energy" section of each cell.

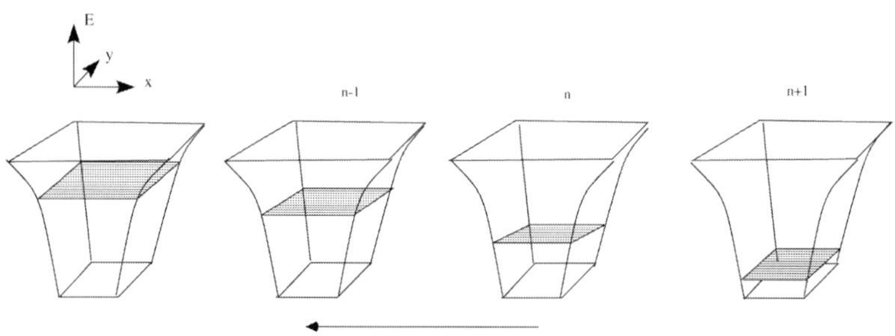

Fig. 1a. Schematic representation of the phase space Γ. The sectional area at "kinetic energy" E depends on E. The arrow represents the applied field and hatched squares corresponds to a constant total energy surface.

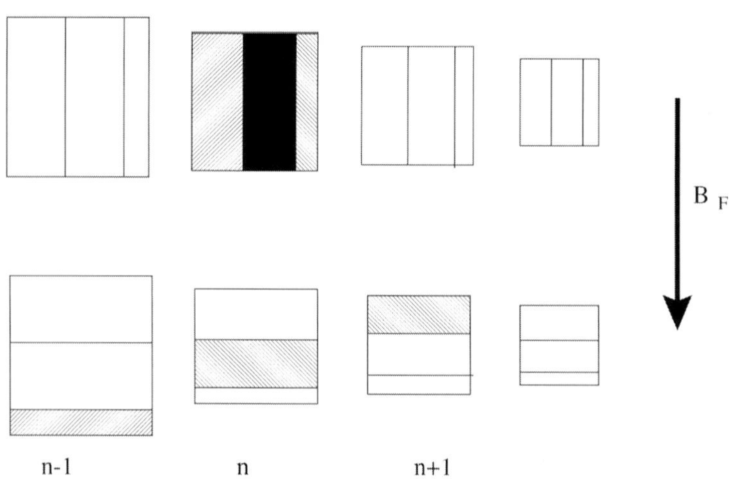

Fig. 1b. Hyperbolic part of the multibaker map B on the constant total energy surface.

The map B is given by [cf. Fig. 1(b)]

$$B(n,x,y,E)=\begin{cases}\left(n-1,\ \dfrac{x}{l^+e^F},\ l^+e^F\, y,\ E+F\right) & \left(0<\dfrac{x}{a_E}\le l^-\right),\\ \left(n,\ \dfrac{x-l^-a_E}{s},\ sy+l^+a_E,\ E\right) & \left(l^-<\dfrac{x}{a_E}\le 1-l^+\right),\\ \left(n+1,\ \dfrac{x-(1-l^+)a_E}{l^-e^{-F}},\right. & \\ \qquad \left.\{l^-y+(1-l^-)a_E\}e^{-F}, E-F\right) & \left(1-l^+<\dfrac{x}{a_E}\le 1\right),\end{cases} \tag{21}$$

where $-F$ is a parameter corresponding to the applied field, $l^{\pm}\equiv 2l/(1+e^{\pm 2F})$ is the transition rate from the nth to the $(n\pm 1)$th cells and $s\equiv 1-2l$ with $0<l\le 1/2$.

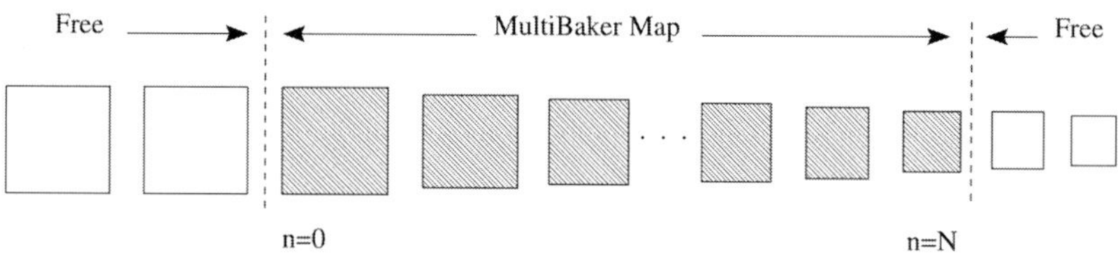

Fig. 2. Overall view of the constant energy surface of the open multibaker map B. On the shaded cells, the map is hyperbolic and, on the rest, the map is a simple shift.

We consider an open system where the multibaker chain of length $N+1$ is embedded between infinitely extended "free chains"(Fig. 2). Then, (21) holds for $n\in[1,N-1]$. The free motion is modeled by a simple shift, i.e.,

$$B(n,x,y,E)=\begin{cases}(n-1,x,y,E)\ , & \left(0<\dfrac{x}{a_E}\le l_f\right),\\ (n,x,y,E)\ , & \left(l_f<\dfrac{x}{a_E}\le 1-l_f\right),\\ (n+1,x,y,E)\ , & \left(1-l_f<\dfrac{x}{a_E}\le 1\right),\end{cases} \tag{22}$$

where $l_f=l^+$ for $n\le -2$ and $l_f=l^-$ for $n\ge N+2$. The sites $n=-1,0,N,N+1$ are joint sites and, as shown in [31], different transformation rules are applied.

As seen from (21), the phase space is stretched along x-axis and contracted along y-axis, or the multibaker part is hyperbolic. And the present model is of chaotic scattering type.

3.2 Phase Space of Infinitely Extended Systems and Poisson Measure

We are interested in dynamics when identical particles are distributed to the whole system with finite density. Conventional approach based on Gibbs' μ-space idea suggests the introduction of an infinite dimensional phase space $\widetilde{X}$ to describe dynamical states, axes of which correspond to individual degrees of freedom of individual particles. Then, one encounters an inconvenience: The phase points in a bounded subset of $\widetilde{X}$ correspond to irrelevant states where infinitely many particles exist in a finite region.

A more convenient description is possible by generalizing Boltzmann's Γ-space approach, where a particle distribution on the one-body phase space was investigated. Analogously, dynamical states are specified by a countable list of particle coordinates in the one-body phase space Γ: $\zeta \equiv \{z_1, z_2, \cdots\}$ with $z_j = (n_j, x_j, y_j, E_j) \in \Gamma$ $(j = 1, 2, \cdots)$. The list is not ordered and repetitions are permitted. Since the particle density is assumed to be finite everywhere, the number of particles included in any bounded closed set K of Γ should be finite, or only configurations ζ satisfying $\#(\zeta \cap K) < +\infty$ are allowed, where $\#$ stands for the number of elements of the subsequent set. The collection of such lists forms the phase space X [23,10]. Then, dynamics is defined as in the case of finite systems. For the open multibaker map, since there is no particle-particle interaction, dynamics of configurations is given by $\hat{B}^t\zeta \equiv \{B^t z_1, B^t z_2, \cdots\}$ $(\zeta \in X)$.

Probability measures on X can be defined by specifying probability of finding m particles in each bounded closed subset $K \subset \Gamma$. When the probability obeys the Poisson distribution, the measure is called the Poisson measure [10]. Namely, let $C_{K,m} = \{\zeta \in X | \#(\zeta \cap K) = m\}$ be a set of configurations with m particles in K, then the Poisson measure P is defined by

$$P(C_{K,m}) = \frac{\nu(K)^m}{m!} e^{-\nu(K)} \tag{23}$$

$$P(C_{K,m} \cap C_{K',m'}) = P(C_{K,m}) P(C_{K',m'}) \quad (\text{if } K \cap K' = \emptyset) \tag{24}$$

where $\nu(K)$ stands for the average number of particles in K and, thus, is a measure on Γ.

When particle-particle interactions are absent and the initial state is described by a Poisson measure P, the subsequent states are also described by Poisson measures. For the open multibaker map, it is given by

$$P_t(C_{K,m}) = P(\hat{B}^{-t} C_{K,m}) = \frac{\nu_t(K)^m}{m!} e^{-\nu_t(K)} \tag{25}$$

where $\nu_t(K) \equiv \nu(B^{-t}K)$ is the evolved measure on Γ. Also equilibrium states of such systems are described by Poisson measures. Note that, even when particle-particle interactions exist, dynamics and probability measures are defined as well (for details, see [10]).

3.3 Steady States and Transports

When the initial measure ν_0 is absolutely continuous with respect to the Lebesgue measure with density $\rho_0(n, x, y, E)$, it is convenient to consider the partially integrated distribution G_t at fixed total energy E, which is defined by

$$G_t(n, x, y, E) \equiv \int_0^y dy' \rho_0\left(B^{-t}(n, x, y', E - nF)\right) . \tag{26}$$

Because of the positivity of 'kinetic energy', when $F > 0$, total energy should satisfy $E \geq (N+1)F$.

The evolution equation of the density $\rho_{t+1} = \rho_t \circ B^{-1}$ gives the evolution equation of G_t. It leads to the equation of continuity as well:

$$\Pi_{t+1}(n,E) - \Pi_t(n,E) = J^N_{n-1|n}(t,E) - J^N_{n-1|n}(t,E)\,, \tag{27}$$

where $\Pi_t(n,E)$ is the average number of particles in the nth cell per total energy

$$\Pi_t(n,E) \equiv \int_0^{a_n(E)} dx G_t(n,x,a_n(E),E)\,, \tag{28}$$

with $a_n(E) = a(E)e^{-nF}$ and $J^N_{n-1|n}(t,E)$ is the particle flows per total energy from the $(n-1)$th to the nth cells.

We consider a class of initial states where particles are distributed on the left and right free chains uniformly with respect to the Lebesgue measure. Let $\rho_\pm(E)$ be the corresponding particle densities per total energy, then one has

$$\rho_0(n,x,y,E) = \rho_-(E) \equiv e^{-\beta_L(E-\mu_L)} \quad (n \leq -1)\,, \tag{29}$$

$$\rho_0(n,x,y,E-(N+2)F) \quad = \rho_+(E) \equiv e^{-\beta_R(E-\mu_R)} \quad (n \geq N+1)\,, \tag{30}$$

where β_L, μ_L and β_R, μ_R are inverse temperatures and chemical potentials of the left and right reservoirs, respectively[2].

If the initial density ρ_0 is piecewise continuously differentiable in x, the partially integrated distribution G_t converges to steady state distributions $G_{\pm\infty}$ in the limit of $t \to \pm\infty$. Particularly, when $0 \leq n \leq N$, the convergence of $G_t(n,x,y,E)$ is exponentially fast. At the state $G_{+\infty}$, the average particle number per total energy in each cell is given by

$$\begin{aligned}\Pi_{+\infty}(n,E) = a_E^2\left(\rho_+(E) - \rho_-(E)\right) \frac{e^{-2(N+1)F}\left(1 - e^{-2(n+1)F}\right)}{1 - e^{-2(N+2)F}} \\ + a_E^2 \rho_-(E) e^{-2nF}\end{aligned} \tag{31}$$

which is an exponential function of site coordinate n if $F \neq 0$ and is a linear function of n if $F = 0$. The steady state carries nonvanishing particle flow per energy

$$\begin{aligned}J^N_{n|n+1}(+\infty,E) &= -l\{\Pi_{+\infty}(n+1,E) - \Pi_{+\infty}(n,E)\} \\ &\quad -l \tanh F\{\Pi_{+\infty}(n+1,E) + \Pi_{+\infty}(n,E)\} \qquad (32)\\ &= -lKa(E)^2\left(\rho_+(E) - \rho_-(E)\right) \qquad (33)\end{aligned}$$

[2] In Ref. [26], reservoir temperatures are not correctly introduced. Eq. (36) of [26] should read as $\pi_\pm(E) = D_\pm(E) p_\pm e^{-E/T_\pm}/z_\pm$ where $D_\pm(E) = a(E)^2 e^{-2n_\pm F} L_\xi$ is the densities of states, $p_\pm$ the mass densities, $T_\pm$ the temperatures and $z_\pm$ the partition functions of the reservoirs. The rest results of [26] are valid except the intermediate formulas (37), (38) and (39).

where K is a positive constant depending on the external field strength F and the system size N. The first and second terms of (32) correspond, respectively, to the diffusive and drift flows. By integrating the flow with respect to the total energy, one finds that the particle flow is again a sum of diffusive and drift terms and that Fick's law of diffusion holds, where the diffusion coefficient corresponds to the parameter l. Fick's law is consistent with thermodynamics. Note that the flow $J^N_{n|n+1}(+\infty, E)$ vanishes if and only if $\rho_+(E) = \rho_-(E)$, or $\beta_L = \beta_R$ and $\mu_L = \mu_R$.

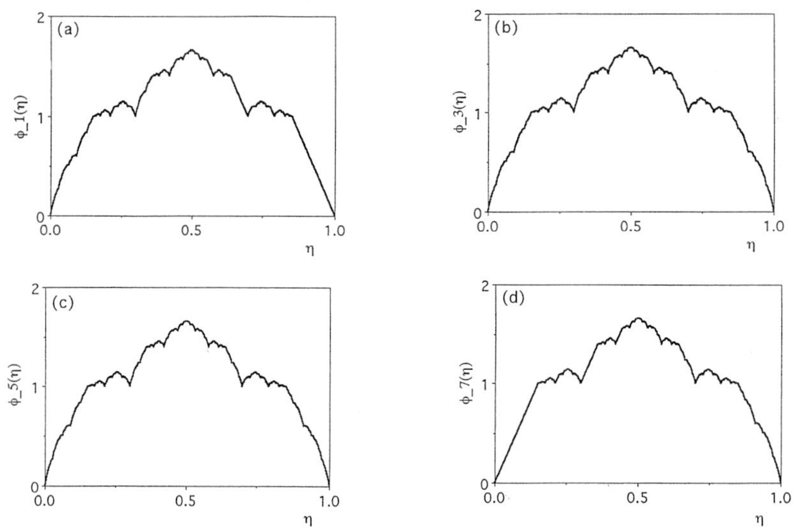

Fig. 3. The fractal part $\varphi_n(\eta)$ of the partially integrated distribution $G_{+\infty}$ vs the rescaled intracell coordinate η for $l^+ = l^- = 0.3$ and $N = 8$. The functions φ_n ($n = 1, 3, 5$ and 7) are shown, respectively, in (a), (b), (c) and (d).

The steady-state distribution $G_{+\infty}$ does not depend on x and is given by

$$G_{+\infty}(n, y, E) = \Pi_{+\infty}(n, E)\, \frac{y}{a^2_{E,n}} - \frac{J^N_{n|n+1}(+\infty, E)}{l a_{E,n}}\, \varphi_n\left(\frac{y}{a_{E,n}}\right) \tag{34}$$

where $\varphi_n(\eta)$ is a fractal function as shown in Fig. 3. Consequently, the intracell distribution of the steady state $G_{+\infty}$ is fractal [7,25]. However, the self-similarity is not complete. Indeed, the steady state measure is absolutely continuous with respect to the Lebesgue measure with a density $\rho_{+\infty}(n, y, E - nF)$. The density takes $\rho_-(E)$ or $\rho_+(E)$ on the complement of the stable manifold of the fractal repeller, which has zero Lebesgue measure [23].

3.4 Coarse Grained Entropy and Its Change

For multibaker maps, Breymann et al. [30] considered that the macroscopic entropy production is given by the change of the difference between a coarse grained and the Gibbs entropy, where individual cells are taken as coarse grained cells. On the other hand, Gaspard [23] claimed that the macroscopic entropy production is given by a coarse grained entropy production calculated via finer coarse graining cells, so that it takes into account the 'information flow' towards finer phase space structure as expressed by the formation of the fractal steady state. Gilbert and Dorfman [24] generalized his approach by introducing the reference Lebesgue measure. The Gaspard-Gilbert-Dorfman entropy production is reviewed by regarding the open multibaker map as a system consisting of infinitely many particles.

Let $\{X_\lambda\}_\lambda$ be a partition of the phase space X (namely $X = \cup_\lambda X_\lambda$ and $X_\lambda \cap X_{\lambda'} = \emptyset$ if $\lambda \neq \lambda'$), then the coarse grained entropy of the state described a measure P is given by

$$S_{cg} = \sum_\lambda P(X_\lambda) \ln \frac{M(X_\lambda)}{P(X_\lambda)} \tag{35}$$

where M is the (unnormalized) Poisson measure associated with the Lebesgue measure. The entropy S_{cg} measures a complexity of the configurations with respect to the partition $\{X_\lambda\}$.

A local coarse grained entropy of a state described by a Poisson measure P_t is introduced as follows: Let $\{D_j\}$ be a partition of the one-body phase space Γ consisting of bounded sets and suppose that the partition satisfies (i) $\Gamma_n = \cup_{D_j \subset \Gamma_n} D_j$, (ii) $D_j \subset \Gamma_n$ implies $BD_j \subset \Gamma_n$ or $\Gamma_{n\pm1}$ and (iii) $BD_j \subset \Gamma_n$ implies $D_j \subset \Gamma_n$ or $\Gamma_{n\pm1}$. Then, for each Γ_n, one has the corresponding phase space partition $X_{\{m_j\}} = \cap_{D_j \subset \Gamma_n} C_{D_j, m_j}$, where $m_j = 0, 1, 2, \cdots$ for every j and $C_{K,m} \equiv \{\zeta \in X | \#(\zeta \cap K) = m\}$. The partition $\{X_{\{m_j\}}\}$ accounts for the configurational variation within Γ_n and, thus, the coarse grained entropy with respect to this partition is regarded as the entropy of the state P_t in Γ_n. Let μ_0 be the Lebesgue measure on Γ, then one has

$$\begin{aligned} P_t(X_{\{m_j\}}) &= \prod_{D_j \subset \Gamma_n} \frac{\nu_t(D_j)^{m_j}}{m_j!} e^{-\nu_t(D_j)} \\ M(X_{\{m_j\}}) &= \prod_{D_j \subset \Gamma_n} \frac{\mu_0(D_j)^{m_j}}{m_j!} \end{aligned} \tag{36}$$

where ν_t is the measure representing the average number of particles and M is taken to be unnormalized. Then one has

$$\begin{aligned} S_{cg}(P_t, \Gamma_n, \{D_j\}) &= \sum_{\{m_j\}} P(X_{\{m_j\}}) \ln \frac{M(X_{\{m_j\}})}{P(X_{\{m_j\}})} \\ &= \sum_{D_j \subset \Gamma_n} \nu_t(D_j) \ln \frac{\mu_0(D_j)}{\nu_t(D_j)} + \nu_t(\Gamma_n) . \end{aligned} \tag{37}$$

Note that $S_{cg}(P_t, \Gamma_n, \{D_j\})$ is extensive with respect to the set Γ_n.

The assumptions (ii) and (iii) give the conservation law:

$$\begin{aligned} S_{cg}(P_{t+1}, \Gamma_n, \{D_j\}) - S_{cg}(P_t, \Gamma_n, \{D_j\}) \\ = J^S_{n-1|n}(t) - J^S_{n|n+1}(t) + \Delta_i S_{cg}(t, \Gamma_n, \{D_j\}) \,, \end{aligned} \tag{38}$$

where $J^S_{n-1|n}(t)$ is the entropy flow and $\Delta_i S_{cg}(t, \Gamma_n, \{D_j\})$ the entropy production given by

$$\begin{aligned} \Delta_i S_{cg}(t, \Gamma_n, \{D_j\}) = \sum_{D_j \subset \Gamma_n} \nu_{t+1}(D_j) \ln \frac{\mu_0(D_j)}{\nu_{t+1}(D_j)} \\ - \sum_{BD_j \subset \Gamma_n} \nu_{t+1}(BD_j) \ln \frac{\mu_0(BD_j)}{\nu_{t+1}(BD_j)} \,. \end{aligned} \tag{39}$$

As discussed in [26] in detail, the coarse grained entropy production $\Delta_i S_{cg}(t, \Gamma_n, \{D_j\})$ is (a) nonnegative, (b) vanishes in the fine-grained limit of the coarse graining cells, and (c) takes the phenomenologically expected expression

$$\Delta_i S_{cg}(+\infty, \Gamma_n, \{D_j\}) \simeq \int_{(N+1)F}^{\infty} dE \frac{J^N_{n|n+1}(+\infty, E)^2}{l \, \Pi_{+\infty}(n, E)} \tag{40}$$

in the middle of the chain and deviates from it near the boundaries. The positivity of the 'kinetic energy' implies $E \geq (N+1)F$ (see after (26)). As pointed out by Gaspard [23] (see also [26]), the fractality of the steady state distribution is considered to be an origin of the coarse grained entropy production. Hence, if the size of the coarse graining cells is smaller than the minimum size of the fractal distribution, the coarse grained entropy cannot 'see' the complexity and is expected to remain constant. This is the case as indicated by the property (b).

Now we summarize the macroscopic scaling limit discussed in Ref. [26]. There, the lattice site n and time t are scaled respectively as $n = L_\xi X$ and $t = L_\tau T$ and the limits $L_\xi \to +\infty$ and $L_\tau \to +\infty$ are taken while keeping the diffusion coefficient $D = l L_\tau / L_\xi^2$ constant. The field strength per unit length $\mathcal{F} = F L_\xi$, the system length $\mathcal{L} = N L_\xi$, the density per energy $\pi_{+\infty} = \Pi_{+\infty} L_\xi$ and the steady-state particle flow per unit time and per unit energy $j = L_\tau J^N_{n|n+1}$ are of order unity. Then, the steady-state entropy production $\sigma^{st}_{cs}(X)$ is

$$\begin{aligned} \sigma^{st}_{cs}(X) &= \lim_{\substack{L_\xi, L_\tau \to +\infty \\ D:\text{fixed}}} L_\tau L_\xi \Delta_i S_{cs}(+\infty, \Gamma_n, \{D_j\}) \\ &= \int_{\mathcal{F}\mathcal{L}}^{\infty} dE \frac{j(E)^2}{D \, \pi_{+\infty}(X, E)} \end{aligned} \tag{41}$$

where the flow $j(E)$ does not depend on X and is given by [26]

$$j(E) = -D \left(2\mathcal{F} \pi_{+\infty}(X, E) + \frac{\partial \pi_{+\infty}(X, E)}{\partial X} \right) \,. \tag{42}$$

4 Relative Entropy Production of Open Multibaker Map

In the previous section, statistical behaviors of the open multibaker map are studied as a typical example of the open hyperbolic systems. And the coarse grained entropy production is shown to be nonnegative. On the other hand, the multibaker map may be regarded as a classical analog to the quantum conductor studied in Sec.2, where the relative entropy between the present and initial states is related to the macroscopic entropy production. So we investigate the relative entropy of the multibaker map and compare it with the coarse grained entropy.

The relative entropy $S(P_t, P_0)$ between the two states P_t and P_0 is given by

$$S(P_t, P_0) = \int_X dP_t(\zeta) \log \left(\frac{dP_0}{dP_t}(\zeta) \right) , \tag{43}$$

where $\frac{dP_0}{dP_t}$ is the density of the initial probability P_0 with respect to the probability P_t. Provided that the initial density ρ_0 is positive everywhere, one has[3]

$$S(P_t, P_0) = \sum_{n=-\infty}^{+\infty} \left\{ \int_{\Gamma_n} dEdxdy \; \rho_t(n,x,y,E) \; \log \frac{\rho_0(n,x,y,E)}{\rho_t(n,x,y,E)} \right.$$
$$\left. + \nu_t(\Gamma_n) - \nu_0(\Gamma_n) \right\} . \tag{44}$$

Since ν_t for finite t is locally different from ν_0, the right-hand side is well defined.

As in the quantum conductor, the negative time derivative of the relative entropy at steady states corresponds to the thermodynamical steady- state entropy production $\Delta_i S_R$:

$$\Delta_i S_R(+\infty) \equiv - \lim_{t \to +\infty} \{ S(P_{t+1}, P_0) - S(P_t, P_0) \}$$
$$= \beta_L J_q^L(+\infty) + \beta_R J_q^R(+\infty) \tag{45}$$
$$= -lK \int_{(N+1)F}^{+\infty} dE a_E^2 \{ \rho_+(E) - \rho_-(E) \} \log \frac{\rho_+(E)}{\rho_-(E)} \tag{46}$$

[3] Eq. (44) can be derived as follows: Let $\{D_j\}$ be a partition of Γ with $\mu_0(D_j) < \epsilon$ and consider a set $X_{\{m_j\}}^{\Lambda\epsilon} = \cap_{D_j \subset \cup_{|n| \le \Lambda} \Gamma_n} C_{m_j, D_j}$. Then, in the limit of $\epsilon \to 0$ and $\Lambda \to +\infty$, the set $X_{\{m_j\}}^{\Lambda\epsilon}$ approaches a set consisting of a single configuration. And $S(P_t, P_0)$ is approximated by

$$\sum_{\{m_j\}} P_t(X_{\{m_j\}}^{\Lambda\epsilon}) \log \frac{P_0(X_{\{m_j\}}^{\Lambda\epsilon})}{P_t(X_{\{m_j\}}^{\Lambda\epsilon})}$$
$$= \sum_{\{m_j\}} P_t(X_{\{m_j\}}^{\Lambda\epsilon}) \sum_{D_j \subset \cup_{|n| \le \Lambda} \Gamma_n} \left\{ m_j \log \frac{\nu_0(D_j)}{\nu_t(D_j)} + \nu_t(D_j) - \nu_0(D_j) \right\} ,$$

which leads to (44) in the limit of $\epsilon \to 0$, $\Lambda \to +\infty$.

where $J_q^L(t)$ and $J_q^R(t)$ are heat flows to the embedded multibaker chain from the left and right reservoirs, respectively:

$$J_q^L(t) = \int_{(N+1)F}^{+\infty} dE(E-\mu_L)J^N_{-1|0}(t,E)$$
$$J_q^R(t) = -\int_{(N+1)F}^{+\infty} dE(E-\mu_R)J^N_{N|N+1}(t,E) \qquad (47)$$

and K is the constant introduced in (33). Eq. (45) is the expression familiar in thermodynamics. As easily seen from (46), $\Delta_i S_R$ is nonnegative and vanishes if and only if $\mu_L = \mu_R$ and $\beta_L = \beta_R$. This is consistent with thermodynamics as well. As before, the sum of the heat flows towards the reservoirs may be regarded as a "Joule heat": $-J_q^L(+\infty) - J_q^R(+\infty) = (\mu_L - \mu_R)J^N(+\infty)$ where $J^N(+\infty)$ is the particle flow at the steady state.

In the scaling limit, one has

$$\begin{aligned}\sigma_R^{st} &= \lim_{\substack{L_\xi\to+\infty, L_\tau\to+\infty \\ D:\text{fixed}}} L_\tau \Delta_i S_R(+\infty) \\ &= \int_{\mathcal{L}\mathcal{F}}^{+\infty} dE\ j(E)\left\{2\mathcal{F}\mathcal{L} + \log\frac{\pi_{+\infty}(\mathcal{L},E)}{\pi_{+\infty}(0,E)}\right\} .\end{aligned} \qquad (48)$$

As easily seen from (41) and (42), the relative entropy production σ_R^{st} is a spatial integral of the coarse grained entropy production density:

$$\int_0^{\mathcal{L}} dX \sigma_{cg}^{st}(X) = \sigma_R^{st} . \qquad (49)$$

Namely, they are equivalent in the macroscopic limit although the physical meanings of the two entropy productions are different.

5 Summary

We illustrated two different approaches to nonequlibrium statistical mechanics, one of which is for infinitely extended systems and the other is based on dynamical systems theory, using a quantum one-dimensional conductor and an open multibaker map as typical examples. In both cases, a class of initial states evolve unidirectionally to a steady state for $t \to +\infty$. And its transport properties are consistent with thermodynamics.

Note that the unidirectional evolution is consistent with dynamical reversibility [18,31]. We illustrate this by the quantum conductor. As shown in Sec.2, as $t \to -\infty$, the state ω_t evolves to another steady state $\omega_{-\infty}$, where transport coefficients have the opposite sign, and which is anti-thermodynamical in this sense. The two steady states are related by the time reversal operation $\mathcal{I}$. Now suppose that a state ω_t is evolving towards $\omega_{+\infty}$ and that the time reversal operation $\mathcal{I}$ is applied at $t = t_R(\gg 1)$. Just before the application of $\mathcal{I}$, the state

ω_t is close to $\omega_{+\infty}$. And since attracting to $\omega_{-\infty}$ for $t \to -\infty$ implies repelling from $\omega_{-\infty}$ for $t \to +\infty$, ω_t is far from $\omega_{-\infty}$. Then, just after the application of $\mathcal{I}$, the state ω_t is close to $\omega_{-\infty}$ and far from $\omega_{+\infty}$. In short, while dynamics causes a continuous state evolution from the 'repeller' $\omega_{-\infty}$ to the 'attractor' $\omega_{+\infty}$, the time reversal operation causes discontinuous change in the opposite direction. In other words, reversibility does not imply the possibility of dynamical evolution in the opposite direction, but ensures the possibility of a nondynamical state-change in the opposite direction. And the dynamical reversibility is compatible with the unidirectional state evolution.

The main difference of the two approaches is the definition of entropy production. For infinitely extended systems, negative time derivative of the relative entropy between the present and initial states is related to the thermodynamic entropy production. The relative entropy production accounts for the unbalance in the heat flows to the reservoirs. On the other hand, for open hyperbolic systems, the coarse grained entropy change is considered to be the macroscopic entropy production. The coarse-grained entropy production measures the information flow towards finer phase-space structure. We have compared the two entropy productions for the open multibaker map and found that they are equivalent in a macroscopic scaling limit, in spite of the different physical meanings. This seems to show that the crucial reason of the positive entropy production is the preparation of equilibrium ensembles for the initial reservoir states. To clarify this point, further investigations are necessary.

Acknowledgments

The author would like to thank Profs. P. Garbaczewski, R. Olkiewicz and organizers of the 38th Winter School of Theoretical Physics, 6-15 February 2002, Ladek Zdrój, for their warm hospitality. He is also grateful to Profs. P. Gaspard, H. van Beijeren, J.R. Dorfman, J. Bellissard and F. Benatti for fruitful discussions. This work is supported by a Grant-in-Aid for Scientific Research (C) from JSPS as well as by Waseda University Grant for Special Research Projects (Individual Research, No.2001A-104) from Waseda University.

References

1. P. Gaspard and G. Nicolis, Phys. Rev. Lett. **65**, 1693 (1990);
   J.R. Dorfman and P. Gaspard, Phys. Rev. **E51**, 28 (1995);
   P. Gaspard and J.R. Dorfman, Phys. Rev. **E52**, 3525 (1995);
   P. Gaspard, Phys. Rev. **E53**, 4379 (1996).
2. P. Gaspard, *Chaos, Scattering and Statistical Mechanics*, (Cambridge Univ. Press, Cambridge, 1998).
3. J.R. Dorfman, *An Introduction to Chaos in Non-Equilibrium Statistical Mechanics*, (Cambridge Univ. Press, Cambridge, 1999).
4. D.J. Evans, E.G.D. Cohen and G.P. Morriss, Phys. Rev. lett. **71** 2401 (1993).
5. G. Gallavotti and E.G.D. Cohen, Phys. Rev. Lett. **74**, 2694 (1995); J. Stat. Phys. **80**, 931 (1995);
   E.G.D. Cohen, Physica A **213**, 293 (1995); Physica **A240**, 43 (1997).

6. N.I. Chernov, G.L. Eyink, J.L. Lebowitz and Ya. G. Sinai, Phys. Rev. Lett. **70**, 2209 (1993); Commun. Math. Phys. **154**, 569 (1993).
7. P. Gaspard, J. Stat. Phys. **68**, 673 (1992);
S. Tasaki and P. Gaspard, J. Stat. Phys. **81**, 935 (1995).
8. C.P. Dettmann, *The Lorentz Gas: A Paradigm for Nonequilibrium Stationary States* in *Hard Ball Systems and the Lorentz Gas*, ed. D. Szász, Encyc. Math. Sci. **101**, (Springer, Berlin, 2000).
9. O. Bratteli and D.W. Robinson, *Operator Algebras and Quantum Statistical Mechanics* vol.1 (Springer, New York, 1987); vol.2, (Springer, New York, 1997).
10. I.P. Cornfeld, S.V. Fomin and Ya.G. Sinai, *Ergodic Theory*, (Springer, New York, 1982);
L.A. Bunimovich et al., *Dynamical Systems, Ergodic Theory and Applications*, Encyclopedia of Mathematical Sciences **100**, (Springer, Berlin, 2000).
11. D. Ruelle, *Statistical Mechanics: Rigorous Results*, (Benjamin, Reading, 1969);
Ya. G. Sinai, *The Theory of Phase Transitions: Rigorous Results*, (Pergamon, Oxford, 1982).
12. G. Klein and I. Prigogine, Physica **19**, 74, 89, 1053 (1953);
R. Brout and I. Prigogine, Physica **22**, 621 (1956).
13. H. Spohn and J.L. Lebowitz, Commun. math. Phys. **54**, 97 (1977) and references therein.
14. J. Bafaluy and J.M. Rubi, Physica **A153**, 129 (1988); *ibid.* **153**, 147 (1988).
15. J. Farmer, S. Goldstein and E.R. Speer, J. Stat. Phys. **34**, 263 (1984).
16. J.-P. Eckmann, C.-A. Pillet and L. Rey-Bellet, Commun. math. Phys. **201**, 657 (1999); J. Stat. Phys. **95**, 305 (1999).
17. T.G. Ho and H. Araki, Proc. Steklov Math. Institute, **228** (2000) 191.
18. S. Tasaki, Chaos, Solitons and Fractals, **12** (2001) 2657; to appear in *Quantum Information V*, T. Hida and K. Saito eds., (World Scientific, Singapore, 2003).
19. V. Jakšić, C.-A. Pillet, Commun. Math. Phys. **226** (2002) 131.
20. I. Ojima, H. Hasegawa and M. Ichiyanagi, J. Stat. Phys. **50** 633 (1988);
I. Ojima, J. Stat. Phys. **56** 203 (1989); *Quantum Aspects of Optical Communications*, (LNP **378**,Springer,1991).
21. V. Jakšić and C.-A. Pillet, Commun. math. Phys. **217**, 285 (2001).
22. D. Ruelle, *Entropy production in quantum spin systems*, math-phys/0006006, 2000.
23. P. Gaspard, J. Stat. Phys. **88**, 1215 (1997); Physica **240A**, 54 (1997).
24. T. Gilbert and J.R. Dorfman, J. Stat. Phys. **96**, 225 (1999).
25. S. Tasaki and P. Gaspard, Theor. Chem. Acc. **102**, 385 (1999);
26. S. Tasaki and P. Gaspard, J. Stat. Phys. **101** 125 (2000).
27. T. Tél and J. Vollmer, *Multibaker Maps and the Lorentz Gas* in *Hard Ball Systems and the Lorentz Gas*, ed. D. Szász, Encyc. Math. Sci. **101**, (Springer, Berlin, 2000).
28. U. Sivan and Y. Imry, Phys. Rev. **B33**, 551 (1986).
29. J.R. Dorfman and H. van Beijeren, Physica **A240**, 12 (1997).
30. W. Breymann, T. Tél and J. Vollmer, Phys. Rev. Lett. **77**, 2945 (1996).
31. S. Tasaki, Adv. Chem. Phys. **122**, 70 (2002).

Coherent and Dissipative Transport in Aperiodic Solids: An Overview

J. Bellissard

Université Paul-Sabatier, 118 route de Narbonne, 31062 Toulouse, France
Institut Universitaire de France, France

Abstract. These lecture notes review the theory of transport in aperiodic solids based on noncommutative geometry. Revisiting transport theory is motivated by the large variety of quantum interference effects that occur at low temperature un such solids. The Mott hopping transport in Anderson's insulators or the strange transport properties of quasicrystals are two important examples. To help the reader not expert in the subject, a heuristic review of the formalism used in irreversible thermodynamics is provided. The first important result given here is the construction of the *noncommutative Brillouin zone*. The second set of results concerns the notion of *spectral and transport exponents* characterizing the coherent part of the transport, namely the part coming uniquely from interference effects. The *Guarneri and BGT inequalities* gives constraints on these exponents. Results on models are also given. The dissipative part of transport is then described within the *relaxation time approximation* (RTA). The main consequence is the *anomalous Drude formula*. Also, occurence of *quantum chaos* is discussed together with its main consequences, namely the existence of quantum diffusion and residual conductivity at low temperature. The case of quasicrystals is emphasized leading to a transition between subdiffusion and quantum chaos when the diffusion exponent satisfies $\overline{\beta}_2 = 1/d$ if d is the dimension of the solid. Kinetic theories available to go beyond the RTA are presented with a model for the Mott hopping transport. A discussion of the Kubo formula for transport cœfficients will end these notes.

1 Introduction

The theory of Solid State Physics started with the seminal paper of P. Drude [1] giving the first microscopic insight on transport properties of metals in terms of the theory of electrons (see [2] for a History of Solid State Physics). It was soon followed by a series of works on phonons including the ones by A. Einstein [3], P. Debye (see [2] Section 1.1) and M. Born & von Karmann [4], before the first diffraction experiment was performed by von Laüe (see [2] Section 1.1). Soon after the creation of Quantum Mechanics, F. Bloch [5] developed the band theory, the main tool available since then for computing the energy spectrum of periodic crystals. Under the influence of Sommerfeld (see [2] Section 2.2), using the Pauli principle for electrons [6], the theory of metal was developed to the extend of understanding what characterized the difference between metals, insulators and semiconductors from a microscopic point of view [7].

It was not until the early seventies that disordered systems were systematically studied under the influence of P.W. Anderson [8] and D. Thouless [9]. It has been recognized that disordered media have entirely new transport properties. Anderson predicted the occurence of a metal-insulator transition depending on how strong is the disorder [10]. As a result, a residual conductivity at low temperature appears in weakly disordered systems due to quantum interferences [11,12]. In strongly disordered ones, the Anderson localization leads to insulating behaviour through Mott's variable range hopping [13,14].

The discovery of quasicrystals in 1984 [15] added a new step in dealing with aperiodic solids. Unlike disordered media, quasicrystals have a long range order even though non periodic. Still their transport properties at low temperature are not yet understood well [16,17]. A striking problem is to understand why quasicrystalline alloys, made of good metals, such as $Al_{62.5}Cu_{25}Fe_{12.5}$ in its icosahedral phase, behave like insulators ? A simple mechanism will be proposed, based on the complexity of quantum interferences due to Bragg reflections, as a possible explanation for such a strange behaviour.

These notes report on a long standing program which started in the early eighties [18], with goal to provide a substitute to Bloch theory for aperiodic solids. The main focus however, will be on the recent developments on transport theory, such that the *anomalous Drude formula*, or various kinetic models within which a complete Kubo formula is established covering most of the situations occuring in aperiodic solids. It will be shown that, due to absence of translation invariance, Bloch theory may be replaced by *Noncommutative Geometry*. The quasimomentum space can then be defined as a *noncommutative manifold* called the Noncommutative Brillouin Zone (NCBZ). A Calculus on such manifold allows to extend most of formulæ valid for crystals to aperiodic solids. Using the full power of functional analysis, it will be possible to investigate the difference between periodic crystals and aperiodic solids.

2 Experimental Aspects

2.1 Why Should One Revisit Transport?

Transport theory goes back to the seminal work by Boltzmann [19]. It came after the first attempt by Clausius [20] to understand why the sound velocity in gas was much smaller than the diffusion one: he got the idea that if gas were made of small particles then this could be easily understood. In his first work, he was supposing that all particles had the same velocity but with random directions. Even though the model was so elementary, it was giving good order of magnitude for the microscopic datas, such as the mean free path, the mean collision time. Moreover he was made to understand the difference between sound and diffusion at a time when the atomic hypothesis was far from being accepted. Soon after, Maxwell [21] realized that the velocity of particle was not unique but given by a distribution depending on the temperature of the gas. This was the starting point of Boltzmann who's goal was to go beyond the very nice argument of Maxwell and to offer a microscopic theory liable to provide formulæ for transport cœfficients such that the viscosity (see the Preface of the French translation of the Boltzmann book by M. Brillouin [22]).

This theory was not immediately recognized as valid. In the early twentieth century, several contributions described transport through using stochastic processes. The Drude model for electrons in metals [1] is the first example. It was not until 1905, however, that Lorentz recognized that it could also be represented by a Boltzmann equation [23]. Later, the study of the Brownian motion by Einstein [24] and then by Langevin [25], gave rise to a different description which eventually led to the foundation of Probability Theory through the notion of Stochastic Processes, thanks to the works by Markov, Levy, Wiener and eventually Kolmogorov in the late thirties. The Boltzmann equation, even though an approximation valid whenever three-body collisions can be neglected, has been used successfully in a large number of situations, including the electron transport in metals [23], the derivation of equations for fluids dynamics, the description of low density self gravitating celestial systems, or the dynamics of atomic nuclei.

These approaches were supplemented in 1931 by the Onsager approach to irreversible thermodynamics [26] and its linear response theory with the reciprocity relations. In 1948, the works by Bogoliubov, Born, Green, Kirkwood and Yvon [27] permitted to supplement the Boltzmann equation in situations for which the density of the system was higher. During the fifties, the contributions by Green [28] and Kubo [29] led to explicit formulæ for the linear response cœfficients in term of the microscopic Hamiltonians. The Green-Kubo formula is actually valid in full generality, for classical and quantum systems as well. One of the striking fact about this theory is that its derivation from first principle is still not well understood today. Since the early seventies, starting with the seminal work by Lanford [30], there has been intensive works with aim to derive the Boltzamnn equations from the first principles. It is not the aim of this work to give a full report on it.

The study of various kind of new conducting materials since the early sixties, including doped semiconductors at low temperature, disordered crystals, such as real metals, amorphous conductors, and more recently, quasicrystals, exhibited new type of transport properties that are not fully understood even today. This is because the electronic transport become sensitive to quantum interferences at low temperature. The large variety of quantum effects that can be produced, depending upon the atomic arrangement or on the chemical bonding, is a source of theoretical difficulties requiring a qualitatively new mathematical approach.

2.2 Known Mechanisms

Several mechanisms have already been identified concerning the temperature behaviour of the electric conductivity in solids.

(i) For metals, the conductivity increases as the temperature decreases. Moreover, the Landau theory of Fermi liquid predicts [31,32]

$$\sigma(T) \stackrel{T\downarrow 0}{\sim} T^{-2},$$

over a large range of temperatures.

(ii) On the opposite side, thermally activated processes, namely whenever a gap arises in the spectrum near the Fermi level, give [14]

$$\sigma(T) \overset{T\downarrow 0}{\sim} e^{-\Delta/T} .$$

(iii) For weakly disordered systems, such as strongly doped semiconductors or normal metals, there is a residual conductivity at low temperature, due to *quantum interferences* [14,11]

$$\sigma(T) \overset{T\downarrow 0}{\sim} \sigma_0 > 0 .$$

(iv) In strongly disordered systems, such as for lightly doped semiconductors, and whenever the Density of States (DOS) is small at the Fermi level, one expects a *Mott variable range hopping* [14] (see Section 2.3), namely in dimension D

$$\sigma(T) \overset{T\downarrow 0}{\sim} \exp\left\{ -\left(\frac{T_0}{T}\right)^{1/(D+1)} \right\} .$$

(v) The case of quasicrystals is special and not yet completely established. Conductivity properties of QC's are rather anomalous when compared to the previous mechanisms. Even though made of good metals, like $Al, Fe, Cu, \cdots$, they behave like insulators at low temperature. Moreover, their conductivity follow a scaling law of the type [16]:

$$\sigma \sim \sigma_0 + AT^\gamma \qquad\qquad 1 < \gamma < 2 ,$$

where σ_0, the residual conductivity, may vanish in some cases. No mechanisms are proposed in conventional books to explain such a behaviour. One of the aim of these notes is to provide a possible explanation [17,33] (see Section 5).

2.3 Mott Variable Hopping Transport

In 1968, Mott [13] proposed the following argument liable to explain the conductivity properties of Anderson insulators at low temperature (see also [14]).

Mott assumes that the solid is a D-dimensional Anderson insulator, namely the electrons are strongly localized. In particular, each energy level of the electron energy spectrum is associated with a position in the solid within a ball of diameter given by the localization length ξ. Moreover, he assumes that the density of states (DOS) n_F at the Fermi level is non vanishing but small enough so as to avoid overlap between electron states. In particular the mean distance between neighbouring electron states is large compare with ξ. The inverse of $n_F \cdot \xi^D$ is a measure of the mean level spacing between states within a ball of diameter ξ. The temperature will be small compare to this scale namely

$$k_B T n_F \xi^D = \frac{T}{T_0} \ll 1 ,$$

if k_B denotes the Boltzmann constant. Then (see Fig. 1), within a small error, all states with energy smaller than the Fermi level E_F are occupied, whereas the ones with energy bigger are empty.

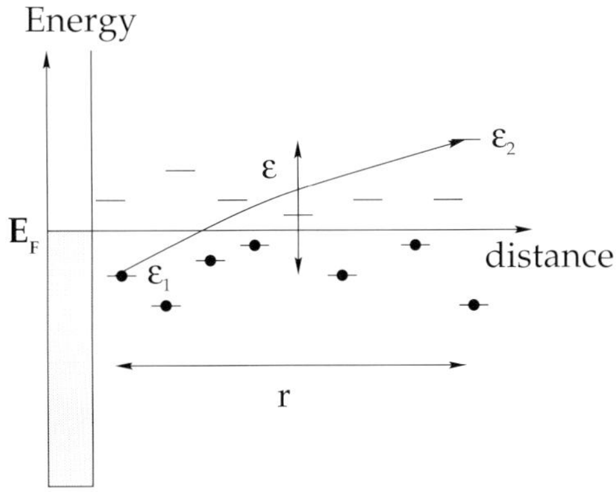

Fig. 1. Mott's Variable Range Hopping

The probability that a phonon of energy ε be produced is proportional to e^{ε/k_BT}. This is true provided $\varepsilon \gg k_BT$. The probability that this phonon be absorbed by an electron of the Fermi sea, with energy $\epsilon_1 < E_F$ to jump in a state of energy $\epsilon_2 > E_F$ is controlled by the tunneling effect forcing such an electron to move in space, since the two states are localized apart from each other. Let then d be the distance between such states, so that the tunneling probability be proportional to $e^{d/\xi}$. Therefore the probability P of transfer of an electron at distance d from its original location is proportional to

$$P \propto \exp - \left\{ \frac{\varepsilon}{k_BT} + \frac{d}{\xi} \right\} .$$

By definition of the DOS, the product $\varepsilon \cdot n_F \cdot d^D$ represents the number of states in an energy interval of width ε localized in a ball of diameter d. This number must be at least 1 so that

$$\varepsilon \cdot n_F \cdot d^D \geq 1 ,$$

the probability of jump P is then optimized provided d assumes its minimum value, which depends on ε. Therefore the conductivity, which is proportional to the sum of contributions of all such jumps, will be controlled by the maximal value of P. Optimizing over ε leads to

$$P \propto \exp - \left\{ \frac{T_0}{T} \right\}^{1/D+1} .$$

Then, the phonon energy optimizing P is given by $\varepsilon_{opt} \approx T^{D/D+1} \gg T$, while the average distance of the jump is given by $d/\xi \approx (T_0/T)^{1/D+1} \gg 1$.

For lightly doped $3D$-semiconductors like silicon (see [14]), the localization length is given by the Bohr radius of the impurity, which is about 100Å. For a concentration of 10^{-9}, the mean distance between impurities is 1000Å which is approximately 10ξ. Then, the typical width of the impurity band is about $1meV$.

Assuming the DOS to be flat on it leads to $k_B T_0 \approx 1eV$ or $T_0 \approx 1.1 \times 10^4 K$! This is huge indeed compare even to the room temperature $k_B T \approx 27meV$. However, this Mott hopping conductivity is hidden as long as the gap between the impurity and the conduction bands is small compare to the temperature. This gap is of the order of $10meV$, requiring a temperature much smaller than $110K$. This example shows that the conductivity is multiplied by a very small factor of order 10^{-4} due to this mechanism! It turns out that this is exactly what happens in the quantum Hall effect (QHE) [34,35]: the Mott variable range hopping controls the fluctuation of the plateaus, leading to the amazing accuracy of this experiment.

2.4 Quasicrystals

The first quasicrystal discovered in 1984 [15] was an alloy of $AlMn$ in a metastable phase. However the transmission electron diffration picture (TEM) was exhibiting a point like spectrum with a 5-fold symmetry, a fact incompatible with the existence of a translation symmetry. It was proposed that such material had a quasiperiodic arrangement of its atoms (see [16] for a review). Since this discovery about 400 alloys have been discovered having a quasicrystalline structure. Most of them have the icosahedral symmetry, a large number a decagonal symmetry. But some other type of symmetries have been observed too. Experiments on the conductivity properties were initiated by the end of the eighties until quite recently. They were performed on high quality sample mainly on $AlFeCu$, $AlPdMn$ and $AlPdRe$ [36], cf. Fig. 2,3. Their main properties are summarized as follows:

QC.1 Their conductivity is low to very low [16].

QC.2 Their conductivity decreases enormously with the temperature, with, in some cases

$$\frac{\sigma_{300K}}{\sigma_{4K}} \geq 300$$

QC.3 The conductivity is not thermally activated.

QC.4 The conductivity decreases with improved sample quality.

QC.5 In the temperature range from $50mK$ to $1000K$, the conductivity follows a scaling law [16].

$$\sigma \sim \sigma_0 + AT^\gamma \qquad\qquad 1 < \gamma < 2\,.$$

QC.6 For most good quality icosahedral compounds, such as $AlCuFe$ or $AlPdMn$, the conductivity flattens below $300K$ to reach a residual conductivity $\sigma(t) \approx \sigma_0 > 0$ (see Fig. 3). However, $\sigma_0 = 0$ for $i - AlPdRe$.

QC.7 In the icosahedral phase of $AlPdRe$, instead, some type of Mott variable range hopping is observed with [38,39]

$$\sigma \propto e^{-(T_0/T)^{1/4}}$$

in a range of temperature which varies from author to author within the interval 0 to $10K$. In [39], $T_0 \sim 6mK$.

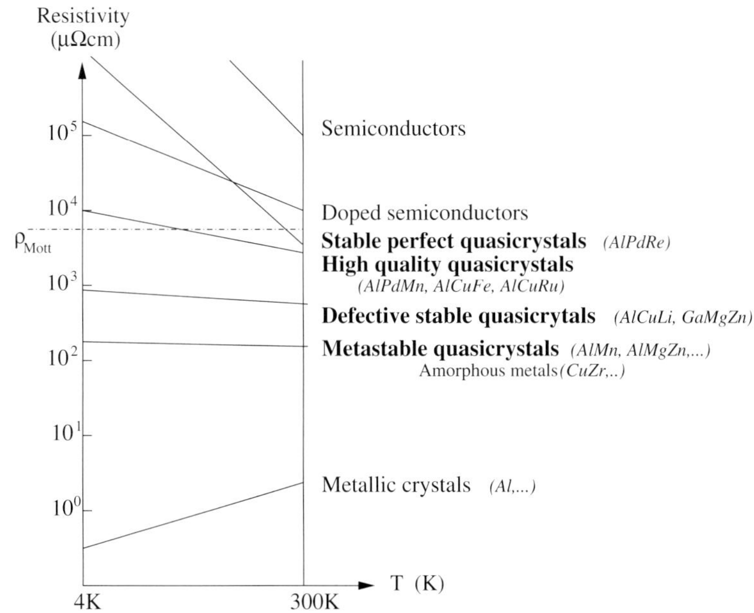

Fig. 2. Typical resistivity of quasicrystals (see C. Berger in [37])

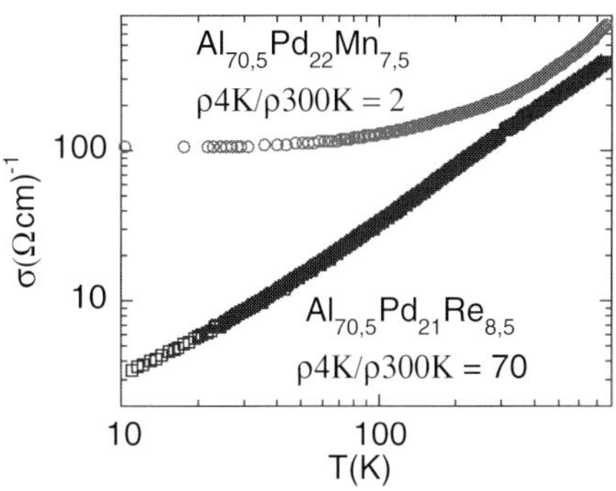

Fig. 3. Conductivity for *AlPdMn* and *AlPdRe* [17]

2.5 Questions

The description given in the last three sections raises several questions:

1. Is it possible to extend the Bloch theory to non periodic solids ?
2. Is there a theory for transport, at least in the one particle approximation, that provides reliable Kubo formulæ for the electric conductivity ?
3. Is there a chance to prove rigorously the Mott estimate for the variable range hopping ?
4. Is it possible to account for the scaling laws that appear in the conductivity of quasicrystals ?

The aim of these notes is to propose a mathematical framework, based upon A. Connes ideas on Noncommutative Geometry [40] which is liable to provide answers to these questions. This theory started in the early eighties [18] and developped over the years [42,35,43]. More recently the theory was extended to coherent [33] and dissipative transport [44,45]. Several of the results shown at the end (see Section 6) are part of a paper in preparation [46].

3 Irreversible Thermodynamics: Heuristic

This Section is a summary of the principle of thermodynamics, whether at equilibrium or out of equilibrium. It can be found in many textbook [47–50]. For a more rigorous approach of the equilibrium statistical mechanics, the reader should look at [51–53]. For reviews on nonequilibrium see [54–57]. This heuristic exposition is given here for the convenience of the reader and to fix the notations and the physical frame within which this work in done.

3.1 Information Theory and Thermodynamics

The main ingredient in the foundation of statistical thermodynamics, is the notion of *information* that was defined and studied in 1948 by Shannon [58] for the purpose of transmission of messages. It turns out that the formalism he developped is very similar to what was in use at the time in Thermodynamics. To be more precise, let $\mathcal{S}$ be a classical system with N particles contained in a box Λ of the physical real space $\mathbb{R}^d$. Its phase space is $\mathfrak{F} = (\Lambda \times \mathbb{R}^d)^N$. In practice, it is not necessary to know with infinite precision the state of the system at all times. So it is convenient to divide the phase space into cells (such as Planck cells if quantum mechanics is considered in a semiclassical approximation). The number of such cells may be at most countable, but it may be convenient to ignore a large number of them on the account that their probability of being visited is very small. For instance the probability for the momentum to reach large values is usually very small. So, one cell can have infinite volume if its probability of visit is small. Let then Ω be the set of such cells. Moreover, the time can also be discretized, as is usually done in dynamical system when using a Poincaré section, and will be labelled by integers. Hence the state of the system at time $n \in \mathbb{Z}$ is a label $q_n \in \Omega$ denoting in which box of the phase space $\mathcal{S}$ is located. In this way, the evolution of the system can be seen as the transmission of a signal through a transducer exactly as Shannon describes it in his paper. Shannon proposed to consider the map $n \mapsto q_n$ as a *stochastic process*. The signal itself is a succession of *digits* namely, in the present situation, a sequence $q_1, q_2, \cdots, q_n$ of box labels through which the orbit of the system is passing. The evolution being deterministic, the evolution equation for the probability distribution of the successive q_n's can be well represented by a short time memory process. Markov processes are only the roughest approximation for such an evolution. Hence at each time $n \in \mathbb{Z}$, the state of the system is rather a *probability distribution* $\mathbb{P}_n$

on Ω. The amount of information such a state contains is measured by the *lack of information* namely its *Shanon entropy* (denoted by H in [58]) given by

$$s(\mathbb{P}) = -\sum_{q\in\Omega} \mathbb{P}(q)\ln\mathbb{P}(q)\,.$$

Quantum system are defined by following a similar chain of arguments. The phase space $\mathfrak{F}$ is then replaced by a Hilbert space $\mathfrak{H}$ as long as the system is contained in a box Λ of finite volume. The stochastic process describing the evolution is now described by a one-parameter group of unitary operators $U(t)$ (here the time may be taken as continuous) while the probability distribution is given by a *density matrix* ρ, that is a positive selfadjoint operator with $\mathrm{Tr}\{\rho\} = 1$. Its entropy is the *von Neumann entropy* given by

$$s(\rho) = -\mathrm{Tr}\{\rho\ln\rho\}\,.$$

An *observable* is either a function $\widehat{F} : q \in \Omega \mapsto \widehat{F}(q) \in \mathbb{R}$, in the classical case, or a selfadjoint operator $\widehat{F}$ on the Hilbert space $\mathfrak{H}$, in the quantum one. Its thermodynamical average is defined as

$$\langle\widehat{F}\rangle = \sum_{q\in\Omega}\mathbb{P}(q)\widehat{F}(q)\,,\ (\text{classical}) \qquad \langle\widehat{F}\rangle = \mathrm{Tr}\{\rho\widehat{F}\}\,,\ (\text{quantum})\,.$$

Classical or quantum systems are subjected to physical constraints, most of them coming from symmetries. The most important one, related to the time translation symmetry, is the conservation of the total energy of the system (*First Law of Thermodynamics*). In the classical case, the energy is a function $\widehat{E} : q \in \Omega \mapsto \widehat{E}(q) \in \mathbb{R}$. In the quantum case it is given by a selfadjoint operator $H = H^*$ acting on $\mathfrak{H}$ ans called the *Hamiltonian*. Because the system is contained in a finite volume, this Hamiltonian has usually a *discrete spectrum*, denoted by $\mathrm{Sp}H$, namely its spectrum is made of a discrete set of eigenvalues with finite multiplicity. In particular the Hamiltonian admits a countable orthonormal basis of eigenvectors. In very much the same way, other conservation laws, such as the total momentum, the total angular momentum, must be taken into account. Also there are other constraints of various origins, geometrical, chemical, *etc.*, such as the conservation of the volume V of the box Λ in which the system evolves, or of the particle number N of each chemical species. Throughout the evolution, these quantities are conserved whenever the system is isolated.

To fix the notation, let $\widehat{X}_1, \cdots, \widehat{X}_K$ be the list of conserved quantities associated with the system. In the classical case, each of the $\widehat{X}_\alpha$'s is a function on Ω which will be supposed to be *real valued* for convenience. If one of the conserved quantities (*e.g.* the momentum) is vectorial or tensorial, then the corresponding coordinates will correspond to some of the $\widehat{X}_\alpha$'s. Therefore, only the labels $q \in \Omega$ with prescribed values X_α of the $\widehat{X}_\alpha$'s are allowed. Let then $\Omega\{X_1, \cdots, X_K\}$ be the subset of such q's. In the quantum case, $\{\widehat{X}_1, \cdots, \widehat{X}_K\}$ is a set of commuting selfadjoint operators. Again because the system is in a finite volume, these operators have usually a discrete spectrum. In particular they have a common orthonormal basis of eigenvectors. Their *joint spectrum* is defined as

$$\mathrm{Sp}\{\widehat{X}_1, \cdots, \widehat{X}_K\} = \{x = (x_1, \cdots, x_K) \in \mathbb{R}^K \,;\, \exists \phi \in \mathfrak{H}, \widehat{X}_\alpha \phi = x_\alpha \phi \, \forall \alpha \,, \phi \neq 0\} .$$

The multiplicity of a joint eigenvalue x is the dimension of the corresponding eigenspace. In most usual cases, this family is *complete* in the sense that its joint spectrum is *simple*, namely each joint eigenvalue has multiplicity one. But this is not necessarily the case. If $I \subset [1, K]$, $\mathfrak{H}\{X_\alpha \,;\, \alpha \in I\}$ will denote the eigensubspace of $\mathfrak{H}$ corresponding to the eigenvalues $x_\alpha = X_\alpha$ for $\alpha \in I$.

In practical situations, no system is totaly isolated. $\mathcal{S}$ may exchange part of the information contained in the conserved quantities with the outside world. However, because these quantities are *extensive*, namely their average value is proportional to the volume V of the box Λ in which the system evolves, the exchange with the outside world can only occur on the boundary $\partial\Lambda$ of this box, so that the relative fluctuation varies like the ratio surface-to-volume S/V of the box, which is very small for large boxes. Therefore, let I_{fluc} be the list of α's corresponding to conserved quantities liable to fluctuate in this manner and let I_{fix} the other labels: the first group of conserved quantities will be called *fluctuating* the other will be *fixed*. It is then equally convenient to consider such a system as defined by the phase space $\Omega\{X_\alpha \,;\, \alpha \in I_{\mathit{fix}}\}$ in the classical case and by the Hilbert space $\mathfrak{H}\{X_\alpha \,;\, \alpha \in I_{\mathit{fix}}\}$ in the quantum one, on which the variables $\widehat{X}_\beta$, $\beta \in I_{\mathit{fluc}}$ are defined. The data of a subset $I_{\mathit{fluc}} \subset [1, K]$ defines a *Gibbs ensemble*. The *microcanonical ensemble* is defined by setting $I_{\mathit{fluc}} = \emptyset$, namely it describes an isolated system.

3.2 Equilibrium Statistical Mechanics

The Second Law of Thermodynamics can be very roughly summarized as follows:

> **Second Law of Thermodynamics:** *During an evolution the total information contained in an isolated system cannot increase*

In other words, the entropy of the system cannot decrease. So the evolution will change the state as long as the entropy increases. But if the state has already reached a maximum of its entropy, its state cannot change anymore. This is precisely the definition of a *Gibbs state* or *equilibrium state*. Its is usual to apply this principle to Gibbs ensembles. This is because, even though they describe partially open systems, such systems are seen as a representative sample of a large isolated system. However, because of the conservation laws, such partially open systems will be subjected to conservation of the variables $\widehat{X}_\alpha$ in the mean whenever $\alpha \in I_{\mathit{fluc}}$

$$\langle \widehat{X}_\alpha \rangle = X_\alpha \,, \qquad \forall \alpha \in I_{\mathit{fluc}} \,. \tag{1}$$

To compute Gibbs states in the classical case, it is therefore enough to find, among all probabilities on $\Omega_{I_{\mathit{fix}}} = \Omega\{X_\alpha \,;\, \alpha \in I_{\mathit{fix}}\}$ satisfying Eq. (1), the one which maximizes the entropy. This a problem of maximizing the functional $s(\mathbb{P})$ with constraints. Because $\mathbb{P}$ is a probability, $\sum_q \mathbb{P}(q) = 1$ gives an extra constraint too. The solution is given in term of *Lagrange multipliers*, namely it is enough to find the critical points of the functional:

$$G(\mathbb{P}) = s(\mathbb{P}) - \lambda_0 \sum_{q \in \Omega_{I_{fix}}} \mathbb{P}(q) - \sum_{\alpha \in I_{fluc}} \lambda_\alpha \langle \widehat{X}_\alpha \rangle \,.$$

Since Ω is a finite set, $\mathbb{P}$ is defined as a vector on $\mathbb{R}^{\Omega_{I_{fix}}}$ and G as a unique maximum because it is strictly concave. This maximum can be computed by solving the equation $\partial G/\partial \mathbb{P}(q) = 0$ which gives:

$$\mathbb{P}(q) = \frac{e^{-\sum_{\alpha \in I_{fluc}} \lambda_\alpha \widehat{X}_\alpha}}{\mathcal{Z}} \,, \qquad \mathcal{Z} = \sum_{q \in \Omega_{I_{fix}}} e^{-\sum_{\alpha \in I_{fluc}} \lambda_\alpha \widehat{X}_\alpha} \,. \tag{2}$$

The normalization constant $\mathcal{Z}$ is called the *partition function*. A very similar result is obtained in the quantum case, provided $\mathbb{P}$ is replaced by ρ and the sum over the q's by the trace over the corresponding Hilbert space.

The other Lagrange multipliers are, in principle, computable through the constraints (1). However, it is usual to proceed differently and to give them a thermodynamical meaning. To do so, the main hypothesis is provided by the *Boltzmann formula*

$$S = k_B \, s(\mathbb{P}) \,, \qquad k_B = 1.38 \times 10^{-23} \, J \cdot K^{-1} \,, \tag{3}$$

where S is the *Clausius entropy*. The Thermodynamics of the 19th century was dealing was infinitesimal changes of equilibrium. In the language defined before, the equilibrium is entirely defined by the data of $\{X_\beta \, ; \, \beta \in I_{fix}\}$ and by the Lagrange multipliers. Equivalently, if the Lagrange multipliers are uniquely defined by (1), the equilibrium is entirely defined by the data of $\{X_1, \cdots, X_K\}$. It is traditional to denote by $U = \langle \widehat{E} \rangle$ the average of the total energy of the system, which will be identified with the variable $\widehat{E} = \widehat{X}_1$ in the sequel. Then $dU, dX_2, \cdots, dX_K$ will denote the infinitesimal variation of the X's under this infinitesimal change of equilibrium. The Clausius entropy was defined as a quantity S, the variation of which being given by

$$T \, dS = dU + \sum_{\alpha=2}^{K} F_\alpha \, dX_\alpha \,, \tag{4}$$

where T is the *temperature* and the F_α's are called the *conjugate quantities*. The Table 1. below gives a list of such quantities.

If the infinitesimal variation is computed directly from the formula (2), it leads to the following relation

$$\lambda_\alpha = \frac{F_\alpha}{k_B T} \,, \qquad \forall \alpha \in I_{fluc} \,. \tag{5}$$

Table 1. Conserved quantities and their conjugate variables in 3D (classical systems)

Name of X	*Notation* $\widehat{X}$	*Average*	*Conjugate Variable F*	*Name of F*
Energy	$\hat{E}(q)$	U	1	T = temperature
Local momentum	$\hat{p}_i(q)$ $i = x, y, z$	$\boldsymbol{p} = (p_x, p_y, p_z)$	$-\boldsymbol{v}$	$\boldsymbol{v}$ = local velocity
Angular momentum	$\hat{L}_i(q)$ $i = x, y, z$	$\boldsymbol{L} = (L_x, L_y, L_z)$	$-\boldsymbol{\omega}$	$\boldsymbol{\omega}$ = local angular velocity
Volume	$\hat{V}(q)$	V	P	P = pressure
Number of particles of species a	$\hat{N}_a(q)$	N_a	$-\mu_a$	μ_a = chemical potential of species a
Magneti-zation	$\hat{M}_i(q)$ $i = x, y, z$	$\boldsymbol{M} = (M_x, M_y, M_z)$	$-\boldsymbol{B}$	$\boldsymbol{B}$ = magnetic field
Electric charge	$\hat{Q}_{el}(q)$	Q_{el}	$-\mathcal{V}$	$\mathcal{V}$ = Electric potential
Deformation tensor *(in solids)*	$\hat{\Sigma}_{i,j}(q)$ $i, j \in \{x, y, z\}$	$\Sigma_{i,j}$	$\Pi_{i,j}$	Π = stress tensor

By convention, the conjugate variable to the energy will be given by $F_1 = 1$. Then the *thermodynamical potentiel* associated with I_{fluc} is defined as

$$A = -k_B T \ln \mathcal{Z} = \sum_{\alpha \in I_{fluc}} F_\alpha X_\alpha - TS \Rightarrow dA = \sum_{\alpha \in I_{fluc}} dF_\alpha X_\alpha - \sum_{\alpha \in I_{fix}} F_\alpha dX_\alpha - SdT\,. \tag{6}$$

3.3 The Local Equilibrium Approximation

If now the system under study is put out of equilibrium, the situation changes because the state describing it will evolve in time. In general, the evolution equation is difficult to describe, but even when this is possible, it is practically too hard to solve and becomes useless. The most common approximation that is used to deal with real systems is called the *Local Equilibrium Approximation*. It is based upon the observation that in most systems, there are various *scales* in space or time. In the simplest possible case there are three main scales called respectively *microscopic, macroscopic* and *mesoscopic*. For example the length and the time scales will be

$$\ell \ll \delta L \ll L, \qquad \tau_{rel} \ll \delta t \ll t. \tag{7}$$

Here ℓ, τ_{rel} are the typical scales at the microscopic level for which the microscopic evolution is computable through, say, perturbation theory. In the case of a gas, for instance, ℓ is of the order of magnitude of the *mean free path*, namely the average length within which no collisions between particles occur, whereas τ_{rel} is the average time between two consecutive collisions experienced by one particle. For electrons in a conductor, ℓ is also the *inelastic mean free path*, namely the average length within which no collisions with phonon occur and τ_{rel} is the average time between two such collisions. The macroscopic scale L is the typical system size whereas t is the observation time. The *mesoscopic length* δL is defined as follows: on the one hand it is large enough so that a cell of size δL can be considered as a Thermodynamical system within Statistical Mechanics; but on the other hand, it is small enough to be considered as infinitesimal compared to L. Similarly the time scales δt is large enough so as

Local Equilibrium Approximation:(LEA) *A cell of mesoscopic size has returned to equilibrium after a mesoscopic period of time.*

On the other hand, δt is small enough to be considered as an infinitesimal if compared to t.

In practice, the typical microscopic time involved in conductors varies between $\tau_{rel} \approx 10^{-12} - 10^{-14}s$ [59]. In the case of copper (Cu), it is $\tau_{rel} = 21 \times 10^{-14}s$([59] p.10). To compute the mean free path, it is sufficient to have an estimate of the conduction electron velocity. In a metal, the conduction electrons have an energy given by the Fermi level E_F. For Cu, the Fermi temperature is $\Theta_F = E_F/k_B = 8.16 \times 10^4 K$ (it varies between $11,000$ and $130,000K$ in metals [59] p.38). An estimate of the velocity is given by $m_* v_F^2/2 = k_B \Theta_F$ where m_* is the effective electron mass. In copper $m_* = 1.3 m_e$, if m_e is the free electron mass. This gives $v_F = 1.7 \times 10^6 m.s^{-1}$ leading to a mean free path $\ell = v_F \tau_{rel} = 3,600\text{Å}$. Thus already in a cell of size 10 times bigger, 1000 collisions occur per collision time. Such a cell can be considered as mesoscopic. In much the same way, if one assumes that the mixing properties of the collision holds, a time 100 times bigger than τ_{rel} should be sufficient to produce a complete loss of memory of the initial condition of a typical electron. This mesoscopic times can be taken as small as $\delta t = 10^{-11}s$ in copper. Hence as long as the measurement process is much slower than δt, the LEA should apply.

The counterpart of the previous arguments is that in modern experiments involving femtosecond lasers, the LEA is clearly violated. This is why such measurement permits to track the details of the particle motion. Morever, using the paradox of quantum mechanics, it is possible to maintain a system in a given state, otherwise unstable, by a constant measurement. This is used today to control coherently some chemical reactions.

Note also that in biological systems, there may be more than three scales. For instance the folding of a protein takes about $1\,s$ to be compared with the typical relaxation time for electronic motion $10^{-14}s$. It turns out that every two orders of magnitude gives rise to a new type of physics, leading to 6 or 7 type of time scales. The spatial structure of such protein is also involved and can be hierarchically described in term of primary, secondary, *etc.* structures [60,61]. Still the (LEA) may give usually an acurate description of biological systems [62].

Thanks to the LEA, the box Λ containing the system $\mathcal{S}$ can be partitioned into cells $\Delta_{\boldsymbol{r}}$ of mesoscopic size centered at points $\boldsymbol{r}$, while the time axis will be also partitioned into intervals $[t, t+\delta t]$ of mesoscopic size. The subsystem $\mathcal{S}(\boldsymbol{r}, t)$ corresponding to the part of $\mathcal{S}$ contained in $\Delta_{\boldsymbol{r}}$ during times in $[t, t+\delta t]$, can be therefore considered at equilibrium. However $\mathcal{S}(\boldsymbol{r}, t)$ is totally opened, namely all $\widehat{X}_\alpha$'s are exchanging information with the neighbouring cells. Therefore such cell requires to be described with the largest possible Gibbs ensemble for which $I_{\mathit{fluc}} = [1, K]$. Thanks to the Section 3.2, the state describing it will be given by a family of conjugate variables, including the temperature, depending on $(\boldsymbol{r}, t)$, leading to the local Gibbs state (see Eq. (2),(5))

$$\mathbb{P}_{(\boldsymbol{r},t)} = \frac{e^{-\beta(\boldsymbol{r},t)\left(\widehat{E}+\sum_{\alpha=2}^{K} F_\alpha(\boldsymbol{r},t)\widehat{X}_\alpha\right)}}{\mathcal{Z}(\boldsymbol{r}, t)}, \qquad \beta(\boldsymbol{r}, t) = \frac{1}{k_B T(\boldsymbol{r}, t)}. \tag{8}$$

A very similar formula holds for quantum systems. In much the same way, the average value of the conserved quantity $\widehat{X}_\alpha$ in the cell will depend on $(\boldsymbol{r}, t)$. But since it is extensive it is proportional to the volume of the cell so that it is itself mesoscopic, hence

$$\delta X_\alpha = \sum_{q\in\Omega(\boldsymbol{r},t)} \mathbb{P}_{(\boldsymbol{r},t)}(q)\,\widehat{X}_\alpha(q),$$

where $\Omega(\boldsymbol{r}, t)$ represents the state space of the cell. The volume $\delta V(\boldsymbol{r}, t)$ is itself liable to fluctuate, however it will be convenient to choose cells of fixed volume δV. Thus the *local density* of $\widehat{X}_\alpha$ is well defined by

$$\rho_\alpha(\boldsymbol{r}, t) = \frac{\delta X_\alpha(\boldsymbol{r}, t)}{\delta V}. \tag{9}$$

3.4 Currents and Transport

Considering the partition into mesoscopic cells as described in Section 3.3, it becomes clear that, on the one hand, the value of the $\widehat{X}_\alpha$'s may vary in time and space, namely from cell to cell. But due to the conservation law, this is possible only if the variation in one cell is compensated by an equal variation of opposite sign outside of this cell. Figure 4 shows an oversimplified situation in which two

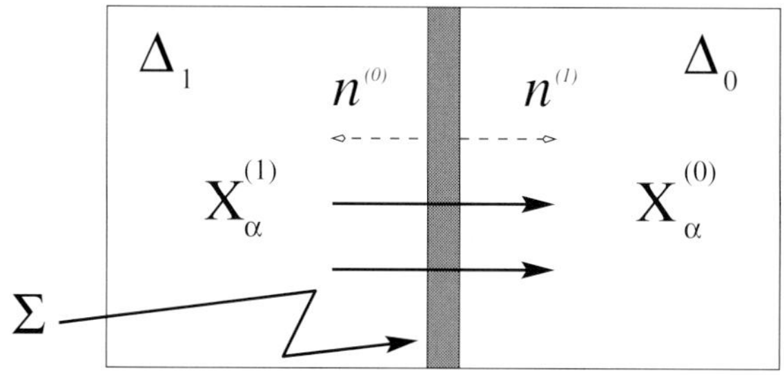

Fig. 4. Transfer, fluxes and currents

cells have been isolated from the rest of the world, but can exchange through an hypothetical membrane separating them.

Considering this example, let $\delta X_\alpha(\boldsymbol{r},t)$ be the variation of $\widehat{X}_\alpha$ in the cell $\Delta^{(1)}$ over a mesoscopic time interval δt. Using the convention of Fig. 4 above, it can be written as

$$\delta X_\alpha(\boldsymbol{r},t) \;=\; -\delta\Phi_\alpha(\boldsymbol{r},t)\delta t \;=\; -\boldsymbol{j}_\alpha(\boldsymbol{r},t)\cdot\boldsymbol{n}^{(1)}\,\delta\Sigma\,\delta t\,,$$

where $\boldsymbol{n}^{(1)}$ is the unit vector normal to the membrane and oriented from the cell $\Delta^{(1)}$ toward the cell $\Delta^{(0)}$, whereas $\delta\Sigma$ is the (mesoscopic) area of the membrane. $\delta\Phi_\alpha(\boldsymbol{r},t)$ represents the *flux* of the variable $\widehat{X}_\alpha$ through the membrane, namely the loss of $\widehat{X}_\alpha$ in the cell $\Delta^{(1)}$ per unit time. The vector $\boldsymbol{j}_\alpha(\boldsymbol{r},t)$ is called the *local current* flowing between the two cells.

Remark 1. *It is important to notice that this definition of the current is* mesoscopic*. There may be contributions to the current that have no microscopic counterpart. An important example in solids is provided by the* phonon drag*: phonons are like waves on the ocean on top of which electrons can surf, producing a charge transfer, thus a contribution to the electric current. Such contribution cannot be defined microscopically.*

Proceeding in much the same way with a more realistic situation where a cell is surrounding by more than one other cell, the current vector can be define accordingly. Summing all the contributions from all faces of the cell leads to the usual *continuity equation*

$$\frac{\partial\rho_\alpha}{\partial t}(\boldsymbol{r},t) + \boldsymbol{\nabla}\cdot\boldsymbol{j}_\alpha(\boldsymbol{r},t) \;=\; 0\,. \tag{10}$$

In much the same way there is a (Clausius) *entropy density* defined as before by

$$s(\boldsymbol{r},t) \;=\; \frac{\delta S(\boldsymbol{r},t)}{\delta V}$$

in each mesoscopic cell. In view of the Eq. (4), thanks to the LEA, its variation in time is given by

$$\frac{\partial s}{\partial t}(\boldsymbol{r},t) \;=\; \sum_{\alpha=1}^{K}\frac{F_\alpha(\boldsymbol{r},t)}{T}\;\frac{\partial\rho_\alpha}{\partial t}(\boldsymbol{r},t)\,.$$

By analogy with the previous definition of a flux, the local entropy current will be defined as

$$\boldsymbol{j}_S(\boldsymbol{r},t) = \sum_{\alpha=1}^{K} \frac{F_\alpha(\boldsymbol{r},t)}{T} \, \boldsymbol{j}_\alpha(\boldsymbol{r},t)\,.$$

Hence the total *entropy production rate* is given by

$$\frac{ds}{dt} = \frac{\partial s}{\partial t} + \boldsymbol{\nabla}\cdot\boldsymbol{j}_S = \sum_{\alpha=1}^{K} \boldsymbol{\nabla}\left(\frac{F_\alpha}{T}\right) \boldsymbol{j}_\alpha \geq 0\,, \tag{11}$$

where the positivity comes from the Second Law of Thermodynamics!

3.5 The Linear Response Theory

The main consequence of the previous Section 3.4 is that currents and fluxes occur because the conjugate variable are not constant in space. Through the LEA, the changes of equilibrium state is slow enough to allow constant return to equilibrium, hence leading to

> **Response theory:** *Each current is a smooth function of the gradient of the conjugate variables.*

Proving such a statement from first principle is still today an open challenge for a Mathematical Physicist. But on the basis of experiment, it is quite acceptable and provides a good frame to interpret the data. Since there should be no current at equilibrium, developing the functionnal relationship between currents and conjugate variables to first order leads to

$$\boldsymbol{j}_\alpha = \sum_{\beta=1}^{K} L_{\alpha,\beta}\, \boldsymbol{\nabla}\left(\frac{F_\beta}{T}\right) + O\left\{\left|\frac{F_\beta}{T}\right|^2\right\}. \tag{12}$$

Linear Response Theory (LRT) consists in neglecting higher order terms in this expansion. In most of the situation this is an excellent approximation.

Example 1. Thermoelectric effects. In the case of a conductor in which only thermal and electric current are considered, the Table 1 gives

$$dQ = T\,dS = dU - \mathcal{V}\,dQ_{el}\,. \tag{13}$$

The left hand side defines the *variation of the heat* produced by the infinitesimal change of equilibrium. But the heat itself *is not* a conserved quantity, while both the energy U and the electric charge Q_{el} are conserved quantities. Here $\mathcal{V}$ is the electric potential. Correspondingly there will be an energy current $\boldsymbol{j}_U$ and an electric current $\boldsymbol{j}_{el}$. Thanks to Eq. (13) it is convenient to define the *heat current* as

$$\boldsymbol{j}_{th} := \boldsymbol{j}_U - \mathcal{V}\,\boldsymbol{j}_{el}\,. \tag{14}$$

On the other hand, the Onsager relations become

$$\begin{aligned} \boldsymbol{j}_U &= L_{u,u}\boldsymbol{\nabla}(\frac{1}{T}) + L_{u,q}\boldsymbol{\nabla}(\frac{-\mathcal{V}}{T}), \\ \boldsymbol{j}_{el} &= L_{q,u}\boldsymbol{\nabla}(\frac{1}{T}) + L_{q,q}\boldsymbol{\nabla}(\frac{-\mathcal{V}}{T}). \end{aligned} \tag{15}$$

For simplicity, let the conductor be isotropic, namely let the Onsager cœfficients above be multiple of the identity matrix. Then, if there is no current, namely if $\boldsymbol{j}_{el} = 0$ so that $\boldsymbol{j}_{th} = \boldsymbol{j}_U$,

$$\boldsymbol{j}_{th} := -\lambda \boldsymbol{\nabla} T, \qquad \lambda = \frac{L_{u,u}L_{q,q} - L_{u,q}L_{q,u}}{T^2 L_{q,q}}, \qquad \textbf{(Fourier's law)}, \tag{16}$$

where λ is called the *thermal conductivity*. Moreover, there is a potential difference created due to the temperature gradient (Seebeck effect)

$$-\boldsymbol{\nabla}\mathcal{V} = \epsilon \boldsymbol{\nabla} T, \qquad \epsilon = \frac{L_{q,u} - \mathcal{V}L_{q,q}}{TL_{q,q}}, \tag{17}$$

where ϵ is called the *thermopower*. At last if the temperature gradient vanishes

$$\boldsymbol{j}_{el} = \sigma \boldsymbol{\mathcal{E}}, \qquad \boldsymbol{\mathcal{E}} = -\boldsymbol{\nabla}\mathcal{V}, \qquad \sigma = \frac{L_{q,q}}{T}, \qquad \textbf{(Ohm's law)} \tag{18}$$

where σ is the *conductivity*. In thermoelectric experiments, the linear response theory is barely violated. It requires the very high electric field found in lasers to start seeing second order effects that are used in nonlinear optics. □

In dimension d, each current takes on values in $\mathbb{R}^d$ so that the $L_{\alpha,\beta}$'s are $d \times d$ real valued matrices, called the *Onsager cœfficients*. The gradient $\boldsymbol{\nabla}(F_\alpha/T)$ plays the rôle of a generalized force and is called an *affinity*. It is quite convenient to consider the Onsager cœfficients as a matrix of blocks, namely as an element of $M_K(\mathbb{R}) \otimes M_d(\mathbb{R})$

$$\mathbb{L} = ((L_{\alpha,\beta}))_{\alpha,\beta=1}^K .$$

It is remarkable that the Second Law of Thermodynamics, as expressed in Eq. (11) implies the following positivity property

$$\textbf{Second Law of Thermodynamics} \qquad \Longrightarrow \qquad \mathbb{L} + \mathbb{L}^t \geq 0. \tag{19}$$

This relation suggests that the only the symmetric part of the matrix $\mathbb{L}$ participates to the dissipation mechanisms responsible for the information loss during the evolution. The antisymmetric part of this matrix contains *nondissipative* response.

In 1931, Onsager [26] published a paper in which he investigated the consequences of the microscopic time reversibility on this cœfficients. The time reversal symmetry is generated by a transformation T_R such that ${T_R}^2 = 1$. There are two irreducible representations corresponding to either the character $\varepsilon = 1$ or to the character $\varepsilon = -1$. In particular, each conserved quantity $\widehat{X}_\alpha$ transforms according to one of this character under T_R denoted by ε_α. Onsager showed that, if the system is submitted to external forces through a set of parameters $\mathcal{P}$, then

$$L_{\beta,\alpha}(\mathcal{P}) = \varepsilon_\alpha\, \varepsilon_\beta\, L_{\alpha,\beta}(T_R\,\mathcal{P})\,, \qquad \textbf{Onsager's reciprocity relations.} \tag{20}$$

The meaning of this formula is that while the affinity of type α can produce a current of type β, conversely the β-affinity produces an α-current with the same Onsager cœfficients up to a sign given by the corresponding characters of the time-reversal symmetry. In the example 1 of thermoelectric effects, it simply means that $L_{u,q} = L_{q,u}$. Then using the positivity (19) leads to $\sigma \geq 0\,, \lambda \geq 0$. Due to the existence of parameters that can change sign under time reversal symmetry, the matrix $\mathbb{L}$ may have non dissipative parts. We infer that these non dissipative contributions are related to toplogical invariants, such as what happens for the Quantum Hall Effect [35,63].

4 Aperiodic Systems

The main problem in transport theory, is to have access to the computation and properties of the transport cœfficients from the microscopic properties of the system. In Solid State physics, most of the theory available in textbooks concern periodic crystal in which the Bloch theory applies. Since the mid-sixties, however, physicists started wondering about what happens for non periodic materials. In this section we propose a formalism developped in various papers since the early eighties [18,64,42,35,43,65], that is a substitute to Bloch theory, whenever Bloch theory fails to apply.

4.1 The Hull

The starting point of the theory consists in considering the set of atomic positions in the ideal case where the atoms are fixed at their equilibrium position at zero temperature. This is a set of points in the ambient space $\mathbb{R}^d$. This set will be considered in the idealized situation for which the solid has infinite volume and is homogeneous in space. This is a convenient approximation that fails only for submicronic samples. Let $\mathcal{L}$ denote this set. It is clear that atoms cannot be too close from each others, due to nucleus repulsion at short distance. Moreover, since the system is idealized at zero temperature, except for special situations, like for zeolites, no lacuna is expected to occur so that there is a maximal size to holes between these points. This can be axiomatized in the following way [66]

Definition 1. *1)- A subset $\mathcal{L}$ of $\mathbb{R}^d$ is* uniformly discrete *if there is $r > 0$ such that every open ball of radius r meets $\mathcal{L}$ at most on one point. Then $\mathcal{L}$ will be called r-discrete.*
2)- A subset $\mathcal{L}$ of $\mathbb{R}^d$ is relatively dense *if there is $R > 0$ such that every closed ball of radius R meets $\mathcal{L}$ at least on one point. Then $\mathcal{L}$ will be called R-dense.*

3)- A subset $\mathcal{L}$ of $\mathbb{R}^d$ is a Delone set *(or also Delaunay) if it is both uniformly discrete and relatively dense. $\mathcal{L}$ will be called (r, R)-Delone if it is r-discrete and R-dense.*
4)- A Delone set $\mathcal{L}$ has finite type whenever $\mathcal{L} - \mathcal{L}$ is locally finite
5)- A Delone set is a Meyer set *whenever $\mathcal{L} - \mathcal{L}$ is itself a Delone set.*

Example 2. 1. A random subset $\mathcal{L}$ of $\mathbb{R}^d$ distributed according to the Poisson distribution with a finite positive density is almost surely discrete, but with probability one, it is neither uniformly discrete nor relatively dense.
2. Let $\mathcal{L}_0$ be a *lattice* in $\mathbb{R}^d$, namely a discrete cocompact subgroup of $\mathbb{R}^d$. Let $\mathcal{L}$ be a random subset of $\mathcal{L}_0$ distributed according to the Bernoulli law on each sites. Then with probability one $\mathcal{L}$ is uniformly discrete (obvious) but not relatively dense. This situation occurs precisely for the distribution of impurity sites in a lightly doped semiconductor [14].
3. Most solids are described, at zero temperature by Delone sets. This is the case for amorphous materials (silicon), glasses, crystals.
4. The cut-and-project method to describe quasicrystals [16] shows that the set of atomic sites of such a material is a Meyer set.

□

In order to represent a uniformly discrete set $\mathcal{L}$, it is convenient to consider its *counting measure* which is the Radon measure on $\mathbb{R}^d$

$$\nu^{(\mathcal{L})} = \sum_{y \in \mathcal{L}} \delta(. - y) . \tag{21}$$

Recall that a Radon measure μ on $\mathbb{R}^d$ is an element of the dual space to the space $\mathcal{C}_c(\mathbb{R}^d)$ of continuous functions with compact support. Since it is not the aim of this paper to give all technical details, the reader should look at [67] to learn more about the natural topology on this space. Thus μ becomes a linear map $\mu : f \in \mathcal{C}_c(\mathbb{R}^d) \mapsto \mu(f) \in \mathbb{C}$ which is continuous. Here $\nu^{(\mathcal{L})}(f) = \sum_{y \in \mathcal{L}} f(y)$ and this sum is finite since f has compact support and since $\mathcal{L}$ is discrete. The measure $\nu^{(\mathcal{L})}$ has the following properties

1. Any ball $B \subset \mathbb{R}^d$ has an integer measure, that is $\nu^{(\mathcal{L})}(B) \in \mathbb{N}$.
2. If B is an open ball of radius less than or equal to r then $\nu^{(\mathcal{L})}(B) \leq 1$, due to the uniform discreteness.
3. If, in addition, $\mathcal{L}$ is relatively dense any closed ball B of radius larger than or equal to R as a measure $\nu^{(\mathcal{L})}(B) \geq 1$.

It is not difficult to check that these three properties characterize $\mathcal{L}$ which is nothing but the support of the measure $\nu^{(\mathcal{L})}$. In other words, there is a one-to-one correspondance between discrete sets and their related measures so that each property of such set can be read on the associated measure and vice versa.

Representing the atomic set by a measure is a convenient tool to describe topologies. The space $\mathfrak{M}(\mathbb{R}^d)$ of Radon measures on $\mathbb{R}^d$ will be endowed with the weak* topology over $\mathcal{C}_c(\mathbb{R}^d)$. This means that a sequence $(\mu_n)_{n \in \mathbb{N}}$ of Radon measures converges to μ if and only if given any continuous function f with compact support on $\mathbb{R}^d$, then $\lim_{n\to\infty} \mu_n(f) = \mu(f)$. Thanks to this language, to say that a sequence $\mathcal{L}_n$ of discrete sets converges to the discrete set $\mathcal{L}$ means that in each open ball B the sets $\mathcal{L}_n \cap B$ converges to $\mathcal{L} \cap B$, say for the Hausdorff distance. Note however, that this convergence needs not being uniform *w.r.t.* B. So it is convenient to denote by $UD_r(\mathbb{R}^d)$ the set of Radon measures on $\mathbb{R}^d$ with support given by an r-discrete set. In very much the same way let $Del_{r,R}(\mathbb{R}^d)$ be the set of Radon measures on $\mathbb{R}^d$ with support given by an (r, R)-Delone sets. Then [43]:

Proposition 1. *1)- The spaces* $UD_r(\mathbb{R}^d)$ *and* $Del_{r,R}(\mathbb{R}^d)$ *are closed and compact in* $\mathfrak{M}(\mathbb{R}^d)$.

2)- The closure $QD(\mathbb{R}^d)$ *of the union* $\bigcup_{r>0} UD_r(\mathbb{R}^d)$ *in* $\mathfrak{M}(\mathbb{R}^d)$, *is nothing but the set of measures with discrete support such that each point in the support has an integer positive mass.*

3)- $UD_r(\mathbb{R}^d)$ *is the closure of the union* $\bigcup_{R>r} Del_{r,R}(\mathbb{R}^d)$.

Remark 2. 1)- means that from any sequence of r-discrete sets it is possible to extract a convergent subsequence that converges to an r-discrete set. In particular the limit of a convergent sequence of r-discrete sets is itself r-discrete.
2)- a measure of $QD(\mathbb{R}^d)$ can be seen as an atomic set in which a finite number of atoms can sit on top of each other.
3)- means that each r-discrete set can be approximated by a sequence of r-discrete Delone sets. □

Remark 3. In much the same way, a Meyer set is an (r, R)-Delone set such that $\mathcal{L} - \mathcal{L}$ is (r', R')-Delone for some $0 < r < R, 0 < r' < R'$. If $Mey_{r,R;r',R'}(\mathbb{R}^d)$ denote such set of measures, it is also compact in the weak* topology. □

Remark 4. The property for $\mathcal{L}$ of having finite type is not preserved under limits. □

Given now $a \in \mathbb{R}^d$, the translation $\mathrm{T}^a : x \in \mathbb{R}^d \mapsto x + a \in \mathbb{R}^d$ acts on $\mathcal{C}_c(\mathbb{R}^d)$ through $\mathrm{T}^a f(x) = f(x - a)$ whenever $f \in \mathcal{C}_c(\mathbb{R}^d)$. Therefore it also acts on the space of Radon measures through $\mathrm{T}^a \mu(f) = \mu(\mathrm{T}^{-a} f)$ whenever $f \in \mathcal{C}_c(\mathbb{R}^d)$ and $\mu \in \mathfrak{M}(\mathbb{R}^d)$. It is elementary to check that these maps are continuous and invertible. Hence the translation group $\mathbb{R}^d$ acts on $\mathfrak{M}(\mathbb{R}^d)$ in a continuous way and

Proposition 2. *The spaces* $QD(\mathbb{R}^d)$, $UD_r(\mathbb{R}^d)$ *and* $Del_{r,R}(\mathbb{R}^d)$ *are* $\mathbb{R}^d$-*invariant.*

Remark 5. In much the same way $Mey_{r,R;r',R'}(\mathbb{R}^d)$ is translation invariant. □

Remark 6. If $\mathcal{L}$ has finite type, then all its translated have finite type. □

This last result allows to define the *Hull* of a uniformly discrete set as follows

Definition 2. *Let $\mathcal{L}$ be a uniformly discrete subset of $\mathbb{R}^d$. Then its Hull is the dynamical system $(\Omega, \mathbb{R}^d, \mathrm{T})$ where Ω is the closure of the $\mathbb{R}^d$-orbit of $\nu^{(\mathcal{L})}$ in $\mathfrak{M}(\mathbb{R}^d)$.*

Remark 7. 1)- Since $\mathcal{L}$ is uniformly discrete there is $r > 0$ such that $\nu^{(\mathcal{L})} \in UD_r(\mathbb{R}^d)$. Hence, by Prop. 2 its orbit is contained in $UD_r(\mathbb{R}^d)$. By Prop. 1 then, Ω is a compact subset of $UD_r(\mathbb{R}^d)$. In particular, any measure $\omega \in \Omega$ defines an r-discrete set $\mathcal{L}_\omega$, namely its support.
2)- The closure of the orbit of any point is obviously translation invariant, so that $\mathbb{R}^d$ acts on Ω through T.
3)- If in addition $\mathcal{L}$ is (r, R)-Delone, the same argument implies that all $\mathcal{L}_\omega$'s are (r, R)-Delone.
4)- If $\mathcal{L}$ has finite type, then so does any elements of the Hull. Actually then $\mathcal{L}_\omega - \mathcal{L}_\omega \subset \mathcal{L} - \mathcal{L}$ for any $\omega \in \Omega$. In particular if $\mathcal{L}$ is Meyer, so does any element of the Hull. □

4.2 Properties of the Hull

Let $\mathcal{L}$ be a uniformly discrete set in $\mathbb{R}^d$ and let Ω its Hull. Then the *canonical transversal* is the subset of $X \subset \Omega$ defined by

$$X = \{\omega \in \Omega \,;\, 0 \in \mathcal{L}_\omega\}\,, \qquad \textbf{(Canonical transversal)}\,. \tag{22}$$

Each orbit meets this transversal on the corresponding atomic set, namely

$$\mathrm{T}^{-x}\omega \in X \qquad \Leftrightarrow \qquad x \in \mathcal{L}_\omega\,.$$

Since $\mathcal{L}$ is uniformly discrete, so does $\mathcal{L}_\omega$, so that there is a minimum distance from one point of X to any other along the orbit. In this sense the orbits are transversal.

Example 3. Let $\mathcal{L}$ be periodic with period group $\mathbb{G}$. If $\mathbb{G}$ is a lattice, namely a discrete subgroup of $\mathbb{R}^d$ that generates $\mathbb{R}^d$ as a vector space, then the Hull is homeomorphic to the torus $\mathbb{R}^d/\mathbb{G}$. Moreover, the transversal is the finite set $X = \mathcal{L}/\mathbb{G}$. □

Example 4. A quasicrystal can be constructed by mean of the *cut-and-project method*. For $N > d$ let $\mathcal{R}$ be a lattice in $\mathbb{R}^N$ and let Δ be a polyhedral fundamental domain. Let then $\mathcal{E}_{\|}$ be a d-dimensional subspace of $\mathbb{R}^N$ meeting $\mathcal{R}$ only at the origin. Then project all points of the strip $\Delta + \mathcal{E}_{\|}$ on $\mathcal{E}_{\|}$ to get $\mathcal{L}$. By identifying $\mathcal{E}_{\|}$ with $\mathbb{R}^d$, $\mathcal{L}$ becomes a Meyer set that is a model for all known quasicrystals [16].

Let $W = \pi_\perp(\Delta)$, where $\pi_\perp$ denote the projection operator on the orthogonal subspace of $\mathcal{E}_{\|}$. Let W be endowed with the coarsest topology such that $W \cap (W + \pi_\perp(a_1)) \cap \cdots \cap (W + \pi_\perp(a_n))$ becomes closed and open for any family $\{a_1, \cdots, a_n\} \subset \mathcal{R}$. Then W is homeomorphic to the transversal [43]. □

In the following it will be convenient to introduce the Hausdorff distance of two sets A, B in $\mathbb{R}^d$, namely $d_H(A,B) = \max\{\sup_{x\in A}\inf_{y\in B}|x-y|\,,\,\sup_{x\in A}\inf_{y\in B}|x-y|\}$.

The first property of such system is given as follows [43]. It is necessary to recall that a dynamical system is *minimal* if every orbit is dense

Proposition 3. *If $\mathcal{L}$ is uniformly discrete but not relatively dense, then Ω admits a fixpoint the orbit of which does not meet X. In particular the Hull is not minimal.*

Consequently, $\mathcal{L}$ must be Delone to have a minimal Hull.

Definition 3. *Let $\mathcal{L}$ be a Delone set in $\mathbb{R}^d$. Then $\mathcal{L}$ is repetitive if for any finite subset $p \subset \mathcal{L}$, and any $\epsilon > 0$, there is $R > 0$ such that any ball of radius R contains a translated of a finite subset p' such that $d_H(p,p') < \epsilon$.*

The following can be found in [66,68,43]

Theorem 1. *Let $\mathcal{L}$ be a Delone set. Then its Hull is minimal if and only if it is repetitive.*

Associated with the transversal X of the Hull, is a *groupoid* $\Gamma(X)$ [69]. This groupoid plays a rôle similar to the notion of *Poincaré map* or *first return map* in the theory of dynamical systems [70,71]. This groupoid is defined as follows. The set of unit $\Gamma^{(0)}$ coincides with X. The set of arrows $\Gamma(X)$ is the set of pairs $(\omega, a) \in X \times \mathbb{R}^d$ such that $\mathrm{T}^{-a}\omega \in X$. Then the range, the source and the composition maps are defined by

$$r(\omega,a) = \omega\,, \qquad s(\omega,a) = \mathrm{T}^{-a}\omega\,, \qquad (\omega,a)\circ(\mathrm{T}^{-a}\omega,b) = (\omega,a+b)\,.$$

The fiber $\Gamma^{(\omega)}$ is $r^{-1}(\{\omega\})$. Endowed with the topology induced by $\Omega \times \mathbb{R}^d$, this is a locally compact groupoid. If $\mathbb{P}$ is an $\mathbb{R}^d$-invariant ergodic measure on Ω, then it induced on $\Gamma(X)$ a *transverse measure* represented by a probability measure $\mathbb{P}_{tr}$ on X [72].

From Def. 1 a uniformly discrete set $\mathcal{L}$ has finite type whenever $\mathcal{L} - \mathcal{L}$ is discrete, thus closed. In such case the following is true

Proposition 4. *A uniformly discrete subset $\mathcal{L}$ of $\mathbb{R}^d$ with finite type admits a Cantorian transversal.*

A characterization of the Hull of a finite type repetitive Delone set has been given in [65]

Theorem 2. *Let $\mathcal{L}$ be a repetitive Delone set with finite type in $\mathbb{R}^d$. Then its Hull is conjugate by homeomorphisms to the projective limit of an inverse sequence of branched oriented flat compact manifolds without boundaries, in which the $\mathbb{R}^d$-action is induced by parallel transport of constant vector fields.*

4.3 Atomic Gibbs Groundstates

In realistic solids, the atomic positions are also determined by their thermodynamical properties. This is because atoms can vibrate around their equilibrium position and can also diffuse through the solid. This motion creates both acoustic waves (phonons) and lacunæ. Nevertheless, the atomic position can still be described by a discrete set, thus an element of $QD(\mathbb{R}^d)$. Hence this last space plays the rôle of the *configuration space* that is needed in Statistical Mechanics. Nevertheless, it is very convenient to describe the atomic motion in solids as a perturbation of the equilibrium position. The acoustic wave are usually treated as phonon degrees of freedom, whereas lacunæ can be seen as impurities. Still, the atomic positions can be seen as typical configurations for the Gibbs measure describing the thermal equilibrium of atoms. Neglecting the atomic motion is equivalent to considering the $T \downarrow 0$ limit $\mathbb{P}$ of the Gibbs measure for the atoms. Then $\mathbb{P}$ can be seen as a *probability measure on* $QD(\mathbb{R}^d)$.

From the mathematical point of view, there is a difficulty. For indeed, $QD(\mathbb{R}^d)$ is not a locally compact space, so that the notion of Radon measure is meaningless. Nevertheless, the weak* topology makes this space a *Polish space* [67]. That is, the topology can be described through a distance for which the space is complete (the choice of such a distance is actually not unique and certainly not canonical). But the beauty of Polish spaces is that there is a genuine theory of probabilities [73] using the Borel approach through σ-additive functions on the σ-algebra of Borel sets. In particular, since the translation group acts on $QD(\mathbb{R}^d)$ in a continuous way, it transforms Borel sets into Borel sets so that it also acts on the space of probabilities by $\mathrm{T}^a\mathbb{P}(A) = \mathbb{P}(\mathrm{T}^{-a}A)$ for A a Borel set. The Prokhorov theorem gives also a very useful criterion for compactness of a family of such probabilities. It turns out that describing the atomic configurations through such probabilities gives rise to several interesting results for physicists [43].

If the solid under consideration is homogeneous, then its Gibbs measure $\mathbb{P}$ is expected to be translation invariant. Moreover, standard results of Statistical Mechanics [51] show that a translation invariant pure phase is described by a Gibbs measure that is ergodic under the translation group. In addition with such consideration, the analysis proposed in Section 4.1 shows that typical configurations of atoms at zero temperature should be at least uniformly discrete, but also Delone unless in very special cases. This is why the following definition can be useful

Definition 4. *An atomic grounstate is a probability measure $\mathbb{P}$ on $QD(\mathbb{R}^d)$ such that*

1. *$\mathbb{P}$ is $\mathbb{R}^d$-invariant;*
2. *$\mathbb{P}$ is $\mathbb{R}^d$-ergodic;*
3. *the space of uniformly discrete sets has $\mathbb{P}$-probability one.*

In addition, $\mathbb{P}$ is called Delone (resp. Meyer) if it gives probability one to the space of Delone (resp. Meyer) sets.

Several results have been obtained in [43] from such a definition.

Theorem 3. *Let $\mathbb{P}$ be an atomic groundstate. Then*

1. *there is $r > 0$ such that $\mathbb{P}$-almost every atomic configuration $\mathcal{L}$ is r-discrete an not r'-discrete for $r' > r$;*
2. *there is a compact subset $\Omega \subset UD_r(\mathbb{R}^d)$ such that for $\mathbb{P}$-almost all atomic configuration $\mathcal{L}$, the Hull of $\mathcal{L}$ is Ω; Ω coincides with the topological support of $\mathbb{P}$;*
3. *if, in addition, $\mathbb{P}$ is Delone (resp. Meyer), there is a unique pair (r, R) (resp. family $(r, R; r', R')$) such that $\mathbb{P}$-almost every configuration is (r, R)-Delone and not $(r", R")$-Delone for $r < r"$ and $R" < R$ (resp. $(r, R; r', R')$-Meyer and not $(r_1, R_1; r'_1, R'_1)$ for $r < r_1$, $R > R_1$, $r' < r'_1$, $R' > R'_1$).*

The next result concerns the notion of *diffraction measure*. Let $\mathcal{L}$ be a point set representing the position of atoms in the solid. Then, the diffraction pattern seen on a screen, in an X-ray diffraction experiment or in a transmission electronic microscope (T.E.M.), can be computed from the Fourier transform of $\mathcal{L}$ restricted to the domain Λ occupied by the sample in $\mathbb{R}^d$. More precisely, the intensity seen on the screen is proportional to

$$I_\Lambda(k) = \frac{1}{|\Lambda|} \left| \sum_{x \in \mathcal{L} \cap \Lambda} e^{\imath \langle k | x \rangle} \right|^2 , \tag{23}$$

where $k \in \mathbb{R}^d$ represents the wave vector of the diffraction beam, the direction of which gives the position on the screen. The Fourier transform of $I_\Lambda(k)$ is given by the following expression: if $f \in \mathcal{C}_c(\mathbb{R}^d)$, with Fourier transform denoted by $\tilde{f}$, then

$$\int_{k \in \mathbb{R}^d} dk \tilde{f}(k) I_\Lambda(k) = \frac{1}{|\Lambda|} \sum_{x,y \in \mathcal{L} \cap \Lambda} f(x - y) = \rho_{\mathcal{L}}^{(\Lambda)}(f) , \tag{24}$$

where $\rho_{\mathcal{L}}^{(\Lambda)}$ will be called the *finite volume diffraction measure*. From Eq. (24), it follows that $\rho_{\mathcal{L}}^{(\Lambda)} \in \mathfrak{M}(\mathbb{R}^d)$ is a positive measure with a Fourier transform being also a positive measure. The main problem is whether such quantity converges as $\Lambda \uparrow \mathbb{R}^d$. The next theorem gives conditions under which convergence holds

Theorem 4. *Let $\mathbb{P}$ be an atomic groundstate. Then:*

(i) For $\mathbb{P}$-almost every $\mathcal{L}$ the family $\rho_{\mathcal{L}}^{(\Lambda)}$ of measures on $\mathbb{R}^d$ converges to a positive measure $\rho_{\mathbb{P}} \in \mathfrak{M}(\mathbb{R}^d)$.
(ii) The distributional Fourier transform of $\rho_{\mathbb{P}}$ is also a positive measure on $\mathbb{R}^d$.

In other words, $\mathbb{P}$ determines in a unique way the diffraction pattern. □

4.4 Bloch Theory

If the solid is a perfect crystal, the set $\mathcal{L}$ is invariant under a *translation group* $\mathbb{G}$. $\mathbb{G}$ is a *lattice* in $\mathbb{R}^d$ namely a discrete subgroup generating $\mathbb{R}^d$ as a vector space. Bloch theory deals with the Schrödinger equation with a $\mathbb{G}$-periodic potential. More precisely, let $\mathcal{H} = L^2(\mathbb{R}^d)$ be the Hilbert space of quantum states. The groups $\mathbb{G}$ is unitarily represented in $\mathcal{H}$ through $U(a)\psi(x) = \psi(x-a)$ whenever $a \in \mathbb{G}$. The Schrödinger operator is a selfadjoint operator $H = H^* = -\Delta + V$ with dense domain, where Δ is the Laplacian on $\mathbb{R}^d$ and V is a locally L^1 $\mathbb{G}$-periodic real valued function. In particular

$$U(a)\, H\, U(a)^{-1} \;=\; H \qquad\qquad \forall a \in \mathbb{G}\,.$$

Therefore H and the $U(a)$'s can be simultaneously diagonalized. Since $\mathbb{G}$ is Abelian, diagonalization of the $U(a)$'s is performed through its character group $\mathbb{G}^*$. Standard results in Pontryagin duality theory imply that $\mathbb{G}^*$ is isomorphic to the quotient $\mathbb{B} = \mathbb{R}^{d\,*}/\mathbb{G}^{\perp}$ of the dual group of $\mathbb{R}^d$ (isomorphic to $\mathbb{R}^d$) by the orthogonal $\mathbb{G}^{\perp}$ of $\mathbb{G}$ in this group. It is a well known fact that $\mathbb{G}^{\perp}$ is a lattice in $\mathbb{R}^d$ (called *the reciprocal lattice* in Solid State Physics [74]) so that $\mathbb{B} = \mathbb{R}^{d\,*}/\mathbb{G}^{\perp}$ is a compact group homeomorphic to a d-torus. $\mathbb{B}$ will be called the *Brillouin zone* (strictly speaking this is slightly different from what crystallographers call Brillouin zone).

The concrete calculation of $\mathbb{B}$ goes as follows: any character of $\mathbb{R}^d$ is represented by an element $k \in \mathbb{R}^{d\,*}$. Since $\mathbb{R}^{d\,*}$ and $\mathbb{R}^d$ can be identified canonically, by using the usual Euclidean structure, k can be seen as a vector $k = (k_1, \cdots, k_d) \in \mathbb{R}^d$. The corresponding character is given by the map

$$\eta_k : x \in \mathbb{R}^d \mapsto e^{\imath \langle k|x\rangle} \in U(1)\,, \qquad\qquad \langle k|x\rangle \;=\; k_1x_1 + \cdots + k_dx_d\,.$$

In particular η_k restricts to a character of $\mathbb{G}$, with the condition that $\eta_k\restriction_{\mathbb{G}} = \eta_{k'}\restriction_{\mathbb{G}}$ if and only if $k - k' \in \mathbb{G}^{\perp}$, where

$$\mathbb{G}^{\perp} \;=\; \{b \in \mathbb{R}^d\,;\; \langle b|a\rangle \in 2\pi\mathbb{Z},\; \forall a \in \mathbb{G}\}\,.$$

Since $\mathbb{B}$ is a compact group, the diagonalization of the $U(a)$'s requires the use of a direct integral decomposition of $\mathcal{H}$ over $\mathbb{B}$, so that

$$\mathcal{H} \;=\; \int_{k\in\mathbb{B}}^{\oplus} d^dk\, \mathcal{H}_k\,, \qquad\qquad H \;=\; \int_{k\in\mathbb{B}}^{\oplus} d^dk\, H_k\,.$$

Here, $\mathcal{H}_k$ is the space of measurable functions ψ on $\mathbb{R}^d$ such that $\psi(x+a) = e^{\imath\langle k|x\rangle}\psi(x)$ for all $a \in \mathbb{G}$ and that $\int_{\mathbb{V}} d^dx|\psi(x)|^2 = \|\psi\|^2_{\mathcal{H}_k} < \infty$, where $\mathbb{V} = \mathbb{R}^d/\mathbb{G}$. H_k is then the partial differential operator formally given by the same expression as H, but with domain $\mathcal{D}_k$ given by the space of elements $\psi \in \mathcal{H}_k$ such that $\partial_i\psi/\partial x_i \in \mathcal{H}_k$, for $1 \leq i \leq d$, and $\Delta_x\psi \in \mathcal{H}_k$. Then H_k is unitarily equivalent to an elliptic operator on the torus $\mathbb{R}^d/\mathbb{G} = \mathbb{V}$. In solid state physics, $\mathbb{V}$ is called the *Wigner-Seitz cell*, whereas it is called the *Voronoi cell* in tiling theory.

Consequently, for each $k \in \mathbb{B}$, the spectrum of H_k is discrete and bounded from below. If $E_i(k)$ denotes the eigenvalues, with a convenient labelling i, the maps $k \in \mathbb{B} \mapsto E_i(k) \in \mathbb{R}$ are called the *band functions*. The spectrum of H is recovered as $\mathrm{Sp}(H) = \bigcup_{i,k\in\mathbb{B}} E_i(k)$ and is called a *band spectrum*. A discrete spectrum is usually liable to be computable by suitable algorithms, since it restricts to diagonalizing large matrices.

This is a short summary of *Bloch theory*. Strutt first realized the existence of band functions [75], but soon after Bloch wrote his important paper [76]. In 1930, Peierls gave a perturbative treatment of the band calculations [77] and Brillouin discussed the $2D$ and $3D$ cases [78]. The reader is invited to look at [74,59] to understand why this theory has been so successful in solid state physics. Let us simply mention that the first explicit calculations of bands in $3D$ were performed in 1933 by Wigner & Seitz [79] on sodium using the cellular method that holds their names. The symmetry properties of the wave function were explicitly used in an important paper by Bouckaert, Smoluchowski & Wigner [80] leading to noticeable simplifications of the band calculation.

4.5 The Noncommutative Brillouin Zone

In Section 4.1 it has been shown that an aperiodic solid is well described by its Hull $(\Omega, \mathbb{R}^d, \mathrm{T})$, namely a dynamical system with group $\mathbb{R}^d$ acting by homeomorphisms on a compact metrizable space Ω. With any dynamical system, there is a canonical C^*-algebranamely the *crossed product* $\mathcal{C}(\Omega) \rtimes \mathbb{R}^d$ [81]. In a similar way, such system can also be described through its canonical transversal X, and its related groupoid $\Gamma(X)$. With any locally compact groupoid Γ, endowed with a *transverse function* [72], and with any *module* δ on Γ, is associated a C^*-algebra$C^*(\Gamma, \delta)$ [69]. In this section, it will be shown that, after a slight modification if the solid is submitted to a magnetic field, $\mathcal{C}(\Omega) \rtimes \mathbb{R}^d$ is the smallest C^*-algebragenerated by the electronic Schrödinger operator and all its translated. Moreover, it will be shown that, $C^*(\Gamma(X))$ is also generated by the matrix of phonon modes that appears in the equations of motion for phonons or by the *effective Hamiltonian* derived form the so-called *tight binding representation* [64,42]. It will also be shown that, for periodic crystals, this algebra is nothing but the set of continuous functions on the Brillouin zone.

Given a uniform magnetic field $B = (B_{\nu\mu})$, namely a real-valued antisymmetric $d \times d$-matrix, the C^*-algebra $C^*(\Omega \times \mathbb{R}^d, B)$ is defined as follows. Let $\mathcal{A}_0$ be the topological vector space $\mathcal{C}_c(\Omega \times \mathbb{R}^d)$ of continuous functions with compact support in $\Omega \times \mathbb{R}^d$. It becomes a topological *-algebra when endowed with the following structure

$$fg\,(\omega, x) = \int_{\mathbb{R}^d} dy\; f(\omega, y)\; g(T^{-y}\omega, x-y)\; e^{i\pi(e/h)\, B\cdot x\wedge y}\,, \tag{25}$$

$$f^*(\omega, x) = \overline{f(T^{-x}\omega, -x)}\,, \tag{26}$$

where $f, g \in C_c(\Omega \times \mathbb{R}^d)$, $B \cdot x \wedge y = \sum B_{\nu\mu} x_\nu y_\mu$ and $\omega \in \Omega$, $x \in \mathbb{R}^d$. Here e is the electric charge of the particle and $h = 2\pi\hbar$ is Planck's constant. This

*-algebra is represented on $L^2(\mathbb{R}^d)$ by the family of representations $\{\pi_\omega;\ \omega \in \Omega\}$ given by

$$\pi_\omega(f)\psi\,(x) \;=\; \int_{\mathbb{R}^d} dy\; f(T^{-x}\omega, y-x)\; e^{-i\pi(e/h)\,B\cdot x\wedge y}\psi(y), \qquad\qquad \psi \in L^2(\mathbb{R}^d)\,, \tag{27}$$

where π_ω is linear, $\pi_\omega(fg) = \pi_\omega(f)\pi_\omega(g)$ and $\pi_\omega(f)^* = \pi_\omega(f^*)$. In addition $\pi_\omega(f)$ is a bounded operator and the representations $(\pi_\omega)_{\omega\in\Omega}$ are related by the covariance condition:

$$U(a)\,\pi_\omega(f)\,U(a)^{-1} \;=\; \pi_{\mathrm{T}^a\omega}(f)\,, \tag{28}$$

where the U's are the so-called *magnetic translations* [82] defined by:

$$U(a)\psi(x) \;=\; \exp\Big\{(ie/\hbar)\int_{[x-a,x]} dy^\mu A_\mu(y)\Big\}\;\psi(x-a)\,, \tag{29}$$

where $\boldsymbol{A} = (A_1, ..., A_d)$ is a vector potential defined by $B_{\mu\nu} = \partial_\mu A_\nu - \partial_\nu A_\mu$, $a \in \mathbb{R}^d$, $\psi \in L^2(\mathbb{R}^d)$ and $[x-a, x]$ is the line segment joining $x-a$ to a in $\mathbb{R}^d$. A C^*-norm on $\mathcal{A}_0$ is defined by

$$\|f\| \;=\; \sup_{\omega\in\Omega} \|\pi_\omega(f)\|\,. \tag{30}$$

Definition 5. *The non-commutative Brillouin zone is the topological manifold associated with the C^*-algebra $\mathcal{A} = C^*(\Omega \rtimes \mathbb{R}^d, B)$ obtained by completion of $\mathcal{A}_0 = \mathcal{C}_c(\Omega\times\mathbb{R}^d)$ under the norm $\|\cdot\|$ defined by Eq. (30).*

For $B = 0$ this construction gives the definition of the C^*-crossed product $\mathcal{C}(\Omega) \rtimes \mathbb{R}^d$ [81,83]. In the case of a perfect crystal (see Section 4.4), with lattice translation group $\mathbb{G}$, the hull $\Omega = \mathbb{R}^d/\mathbb{G}$ is homeomorphic to $\mathbb{T}^d$ (see Example 3) and

Theorem 5. *[84,42] The C^*-algebra$C^*(\mathbb{R}^d/\mathbb{G} \rtimes \mathbb{R}^d, B = 0)$ associated with a perfect crystal with lattice translation group $\mathbb{G}$, is isomorphic to $\mathcal{C}(\mathbb{B})\otimes\mathcal{K}$, where $\mathcal{C}(\mathbb{B})$ is the space of continuous functions over the Brillouin zone and $\mathcal{K}$ the algebra of compact operators.*

Even though the algebra $\mathcal{C}(\mathbb{B}) \otimes \mathcal{K}$ is already non-commutative, its non-commutativity comes from $\mathcal{K}$, the smallest C^*-algebragenerated by finite rank matrices. It describes the possible vector bundles over $\mathbb{B}$. Theorem 5 is the reason to claim that $\mathcal{A}$ generalizes the Brillouin zone for aperiodic systems.

The groupoid C^*-algebraof $\Gamma(X)$ can be defined very similarly. Here $\mathcal{B}_0 = \mathcal{C}_c(\Gamma(X))$ and the structure of *-algebra is given by

$$fg\,(\omega, x) = \sum_{y\in\Gamma^{(\omega)}} f(\omega, y)\; g(T^{-y}\omega, x-y)\; e^{i\pi(e/h)\,B\cdot x\wedge y}\,, \tag{31}$$

$$f^*(\omega, x) \;= \overline{f(T^{-x}\omega, -x)}\,. \tag{32}$$

Setting $\mathfrak{H}_\omega = \ell^2(\Gamma^{(\omega)})$, there is a representation $\widehat{\pi}_\omega$ on $\mathfrak{H}_\omega$ defined by

$$\widehat{\pi}_\omega(f)\psi\,(x) \;=\; \sum_{y\in\Gamma^{(\omega)}} f(T^{-x}\omega, y-x)\; e^{-i\pi(e/h)\,B\cdot x\wedge y}\psi(y), \qquad \psi\in\mathfrak{H}_\omega\,, \quad (33)$$

Giving $\gamma = (\omega, a) \in \Gamma(X)$, there is a unitary operator $U(\gamma) : \mathfrak{H}_{\mathrm{T}^{-a}\omega} \mapsto \mathfrak{H}_\omega$ defined by

$$U(\gamma)\psi(x) \;=\; \exp\left\{(ie/\hbar)\int_{[x-a,x]} dy^\mu A_\mu(y)\right\}\psi(x-a)\,, \qquad \psi\in\mathfrak{H}_{\mathrm{T}^{-a}\omega}\,, \quad (34)$$

such that the covariance condition holds

$$U(\gamma)\,\widehat{\pi}_{\mathrm{T}^{-a}\omega}(f)\,U(\gamma)^{-1} \;=\; \pi_\omega(f)\,. \quad (35)$$

A C^*-norm on $\mathcal{B}_0$ is defined by

$$\|f\| \;=\; \sup_{\omega\in X}\|\widehat{\pi}_\omega(f)\|\,. \quad (36)$$

The C^*-algebra$\mathcal{B} = C^*(\Gamma(X), B)$ is the completion of $\mathcal{B}_0$ under this norm. The main result is the following [84]

Theorem 6. *1)- The C^*-algebras$\mathcal{B} = C^*(\Gamma(X), B)$ and $\mathcal{A} = C^*(\Omega \rtimes \mathbb{R}^d, B)$ are Morita equivalent, namely $\mathcal{A}$ is isomorphic to $\mathcal{B}\otimes\mathcal{K}$.*
2)- For $B = 0$ and for a crystal $\mathcal{L}$ with period group $\mathbb{G}$ and transversal $X = \mathcal{L}/\mathbb{G}$, the C^*-algebra$\mathcal{B}$ is isomorphic to $\mathcal{C}(\mathbb{B})\otimes M_N(\mathbb{C})$ if N is the cardinality of X.

In order to see the Noncommutative Brillouin zone as a manifold, rules of Calculus are required. Integration is provided by using a transverse measure. Let $\mathbb{P}$ be an $\mathbb{R}^d$-invariant ergodic probabilty measure on $\mathbb{R}^d$ and let $\mathbb{P}_{tr}$ be the probability induced on X. Then traces are defined on $\mathcal{A}$ and $\mathcal{B}$ as follows

$$\begin{aligned}\mathcal{T}_{\mathbb{P}}(f) &= \int_\Omega d\mathbb{P}(\omega)\, f(\omega,0)\,, \quad f\in\mathcal{A}_0\,,\\ \hat{\mathcal{T}}_{\mathbb{P}}(f) &= \int_X d\mathbb{P}_{tr}(\omega)\, f(\omega,0)\,, \quad f\in\mathcal{B}_0\,,\end{aligned} \quad (37)$$

$\mathcal{T}_{\mathbb{P}}$ and $\hat{\mathcal{T}}_{\mathbb{P}}$ are traces in the sense that $\mathcal{T}_{\mathbb{P}}(fg) = \mathcal{T}_{\mathbb{P}}(gf)$, and that $\mathcal{T}_{\mathbb{P}}(ff^*) > 0$. Whereas $\mathcal{T}_{\mathbb{P}}$ is not bounded, $\hat{\mathcal{T}}_{\mathbb{P}}$ is actually normalized, namely $\hat{\mathcal{T}}_{\mathbb{P}}(\mathbf{1}) = 1$. Moreover, using the Birkhoff ergodic theorem [70,71], these traces can be seen as *trace per unit volume* in the following sense

$$\mathcal{T}_{\mathbb{P}}(f) \;=\; \lim_{R\uparrow\infty}\frac{1}{|B(0;R)|}\,\mathrm{Tr}\left(\pi_\omega(f)\restriction_{B(0;R)}\right)\,,$$

where $B(x;R)$ denotes the ball of radius R centered at $x\in\mathbb{R}^d$. A similar formula holds for $\hat{\mathcal{T}}_{\mathbb{P}}$. In particular in the crystalline case,

$$\mathcal{T}_{\mathbb{P}}(f) \;=\; \int_{\mathbb{B}} dk\,\mathrm{Tr}\left(\tilde{f}(k)\right)\,,$$

if $\tilde{f}(k)$ denotes the representative of f in $\mathcal{C}(\mathbb{B})\otimes\mathcal{K}$ and dk is the normalized Haar measure on $\mathbb{B} \simeq \mathbb{T}^d$. A similar formula holds for $\hat{\mathcal{T}}_\mathbb{P}$ provided $|B(0;R)|$ is replaced by the number of points in $\mathcal{L}_\omega \cap B(0;R)$. Therefore, these traces appears as the noncommutative analog of the integration over the Brillouin zone.

A positive measure on a topological space defines various spaces of measurable functions, such as the L^p-spaces. In much the same way, a positive trace on a C^*-algebradefines also L^p-spaces [85]. Whenever $1 \leq p < \infty$, $L^p(\mathcal{A}, \mathcal{T}_\mathbb{P})$ is the separation-completion of $\mathcal{A}_0$ with respect to the seminorm $\|A\|_p = \mathcal{T}_\mathbb{P}\{(A^*A)^{p/2}\}^{1/p}$ for $A \in \mathcal{A}_0$. A similar definition is given for $\mathcal{B}$. In particular, $L^2(\mathcal{A}, \mathcal{T}_\mathbb{P})$ coincides with the Hibert space of the GNS-representation π_{GNS} of $\mathcal{A}$ associated with $\mathcal{T}_\mathbb{P}$ [86,87]. Then $L^\infty(\mathcal{A}, \mathcal{T}_\mathbb{P})$ denotes the von Neumann algebra generated by (namely the weak closure of) $\pi_{GNS}(\mathcal{A})$.

Similarly, a differential structure is provided in the following way

$$\nabla f\ (\omega, x) \;=\; \imath \boldsymbol{x} f(\omega, x)\,, \qquad\qquad f \in \mathcal{A}_0 \text{ or} \mathcal{B}_0\,. \tag{38}$$

Here $\boldsymbol{x}$ denotes the vector $x \in \mathbb{R}^d$. It is easy to check that ∇ defines a *-derivation namely it obeys Leibniz rule $\nabla(fg) = \nabla(f)g + f\nabla(g)$ and $\nabla(f)^* = \nabla(f^*)$. Moreover ∇ is the infinitesimal generator of a norm pointwise continuous d-parameter group of automorphisms defined by [81]

$$\eta_{\boldsymbol{k}}(f)\ (\omega, x) \;=\; e^{\imath \boldsymbol{k}\cdot\boldsymbol{x}}\ f(\omega, x)\,, \qquad\qquad f \in \mathcal{A}_0 \text{ or} \mathcal{B}_0\,.$$

Let $\boldsymbol{X}$ be the position operator on $L^2(\mathbb{R}^d)$, (resp. on $\mathfrak{H}_\omega$), defined by $\boldsymbol{X}\psi(x) = \boldsymbol{x}\psi(x)$. Then

$$\pi_\omega\,(\nabla f) \;=\; \imath[\boldsymbol{X}, \pi_\omega(f)]\,,\ \ f \in \mathcal{A}_0\,, \qquad \widehat{\pi}_\omega\,(\nabla f) \;=\; \imath[\boldsymbol{X}, \widehat{\pi}_\omega(f)]\,,\ \ f \in \mathcal{B}_0\,.$$

In the crystalline case, it is easy to show that ∇ coincides with $\partial/\partial k$ in $\mathbb{B}$ [42]. Hence ∇ appears as the noncommutative analog of the derivation in momentum space.

In the philosophy of A.Connes, a noncommutative Geometry for a compact maniflod is given by a spectral triplet $(\mathfrak{A}, \mathcal{H}, D)$ where $\mathfrak{A}$ is a dense subalgebra of a unital C^*-algebra, invariant by holomorphic functional calculus, $\mathcal{H}$ a $\mathbb{Z}_2$-graded Hilbert space on which $\mathfrak{A}$ is represented by degree 0 operators and D is a degree 1 selfadjoint operator with compact resolvent, such that $[D, A] \in \mathcal{B}(\mathcal{H})$ for all $A \in \mathfrak{A}$. Then D plays the rôle of a Dirac operator on the noncommutative space associated with $\mathfrak{A}$, giving both a differential structure and a Riemannien metric corresponding to $ds^2 = D^{-2}$ [40]. Then the dimension spectrum is provided by the set of poles of the ζ-function

$$\zeta_D(s) \;=\; \mathrm{Tr}\left(\frac{1}{|D|^s}\right)\,.$$

If s_0 denote the maximum real pole, then the residue ats_0 provides an integral over the manifold given by

$$\int T \;=\; \lim_{s\downarrow s_0}\,(s - s_0)\,\mathrm{Tr}\left(\frac{1}{|D|^s}T\right) \;=\; \mathrm{Tr}_{\mathrm{Dix}}\left(\frac{1}{|D|^{s_0}}T\right)\,,$$

where $\mathrm{Tr}_{\mathrm{Dix}}$ denotes a *Dixmier trace* on $\mathcal{H}$ [40,41]. Such a structure is also present on the noncommutative Brillouin zone $\mathcal{B}$. For indeed, let $\gamma_1, \cdots, \gamma_d$ be an irreducible representation of the Clifford algebra of $\mathbb{R}^d$ by Dirac matrices satisfying $\gamma_\mu\gamma_\nu + \gamma_\nu\gamma_\mu = 2\delta_{\mu,\nu}$ and $\gamma_\mu^* = \gamma_\mu$ in the finite dimensional Hilbert space Cliff(d). Then Cliff(d) is graded by the matrix $\gamma_0 = \gamma_1\gamma_2\cdots\gamma_d$, so that setting $\mathcal{H} = \mathfrak{H}_\omega \otimes \mathrm{Cliff}(d)$, $G = \mathbf{1}\otimes\gamma_0$, this gives a graded Hilbert space where G defines the graduation. Moreover $\mathfrak{A} = \mathcal{B}_0$ and $D = \sum_{\mu=1}^d \gamma_\mu X_\mu$ give a spectral triplet for which $s_0 = d$. In particular, the Dixmier trace $\mathrm{Tr}_{\mathrm{Dix}}(|D|^{-d})$ exist for $\mathbb{P}_{tr}$-almost every $\omega \in X$ and its common value is given by

$$\mathrm{Tr}_{\mathrm{Dix}}(|D|^{-d}) = \mathrm{Dens}_{\mathbb{P}}(\mathcal{L}_\omega)\,\frac{\pi^d}{\Gamma(1+d/2)}\,,$$

where $\mathrm{Dens}_{\mathbb{P}}(\mathcal{L}_\omega)$ is the density of $\mathcal{L}_\omega$ which exists for $\mathbb{P}$-almost all ω's (see [43] see proof of Theorem 1.12). The numerical factor represents the volume of the unit ball of $\mathbb{R}^d$. Moreover, if $f \in \mathcal{B}_0$ this gives

$$\int f = \mathrm{Tr}_{\mathrm{Dix}}(|D|^{-d}\,\widehat{\pi}_\omega(f)) = \hat{\mathcal{T}}_{\mathbb{P}}(f)\,\frac{\pi^d}{\Gamma(1+d/2)}\,, \qquad \mathbb{P}_{tr} - a.e.\ \omega \in X\,,$$

showing that the Connes integral and the integral over the Brillouin zone coincide modulo normalization.

4.6 Electrons and Phonons

The formalism developped in the previous sections will eventually be useful to describe the quantum motion of electrons and phonons in an aperiodic solid, within the one particle approximation. The interacting case is more involved [88] and will not been considered in these notes.

The quantum motion of an electron in a monoatomic aperiodic solid represented by a uniformly discrete set $\mathcal{L}$ of atomic positions, and submitted to a uniform magnetic field B, is well described by a covariant family of Schrödinger operators of the type

$$H_\omega = \frac{1}{2m}\left(\frac{\hbar}{\imath}\boldsymbol{\nabla} - e\boldsymbol{A}\right)^2 + \sum_{y\in\mathcal{L}_\omega} v(X-y)\,, \qquad \mathrm{on} L^2(\mathbb{R}^d)\,, \quad (39)$$

where $\hbar$ is the Planck constant, m is the electron mass, $\boldsymbol{A}$ is a vector potential and v is an effective atomic potential, that represents the binding forces acting on the valence electron. In practice, an aperiodic solid contains more than one species of atoms, so that $\mathcal{L}$ must be replaced by a family $\mathcal{L}^1, \cdots, \mathcal{L}^r$ of uniformly discrete subsets representing the position of each atomic species, supposing that r species are present. Then each species i acts via an effective potential v_i. Moreover, the number of valence electrons involved in the conduction may be more than one per atom, so that it should be necessary to consider instead an l-body problem if l valence electrons are involved. This latter case will not be considered either, even though its treatment does not represent a major difficulty. In these notes the model described in Eq. (39) will be sufficient.

The Schrödinger operator (39) is not well defined as long as no assumption is given on the nature of the atomic potential v. For the sum over the atomic sites to converge, it is necessary that v decay fast enough at infinity. Also local regularity is necessary. A sufficient condition has been given in [43] (Lemma 2.22), namely let

$$L^1_{K,r}(\mathbb{R}^d) = \{f \in L^1(\mathbb{R}^d); |f(x)| \leq \frac{K}{r^d} \int_{|x-y|<r/2} d^d y\, |f(y)|, \text{ for a.e. x}\}$$

be the set of integrable K-subharmonic functions on $\mathbb{R}^d$. Then

Lemma 1. *Let $v \in L^1_{K,r}(\mathbb{R}^d)$ and $\nu \in UD_r(\mathbb{R}^d)$. Then $\nu * v \in L^\infty_{\mathbb{R}}(\mathbb{R}^d)$ and the map $v \in L^1_{K,r}(\mathbb{R}^d) \mapsto \nu * v \in L^\infty_{\mathbb{R}}(\mathbb{R}^d)$ is continuous.*

The following Proposition is a consequence of [42] Section 2, Theorem 4 and [43] Theorem 2.23

Proposition 5. *1)- For any $\mathcal{L}$ an r-discrete subset of $\mathbb{R}^d$, let $H_{\mathcal{L}}$ be the Schrödinger operator given by Eq. (39). Then, if $v \in L^1_{K,r}(\mathbb{R}^d)$ the map $\mathcal{L} \in UD_r(\mathbb{R}^d) \mapsto H_{\mathcal{L}}$ is strong resolvent continuous.*
2)- If Ω is the Hull of $\mathcal{L}$, the map $\omega \in \Omega \mapsto H_\omega$ is strong resolvent continuous and covariant.

As a consequence, thanks to [42] Theorem 6,

Theorem 7. *Let $\mathcal{L}$ be a uniformly discrete set in $\mathbb{R}^d$ with Hull Ω. Let $\mathcal{A} = C^*(\Omega \rtimes \mathbb{R}^d, B)$ be the C^*-algebraof the corresponding Noncommutative Brillouin zone. Let $\{H_\omega; \omega \in \Omega\}$ be the strong resolvent continuous and covariant family of Schrödinger operators defined by Eq. (39) with $v \in L^1_{K,r}(\mathbb{R}^d)$. Then there is a holomorphic family $z \in \mathbb{C} \setminus \mathbb{R} \mapsto R(z) \in \mathcal{A}$ such that*

$$\pi_\omega(R(z)) = (z - H_\omega)^{-1}, \qquad \forall \omega \in \Omega .$$

As a consequence, it can be said that the Schrödinger oeprator is *affiliated to* the C^*-algebra$\mathcal{A}$: a covariant family (A_ω) of selfadjoint operators is affiliated to $\mathcal{A}$ if, for all $f \in \mathcal{C}_0(\mathbb{R})$, the bounded operator $f(A_\omega)$ can be represented as $\pi_\omega(A_f)$ for some $A_f \in \mathcal{A}$ such that the map $A : f \in \mathcal{C}_0(\mathbb{R}) \mapsto A_f \in \mathcal{A}$ is a bounded $*$-morphism [89–91]. The resolvent map $R(z)$ is then given by A_{r_z} if $r_z : s \in \mathbb{R} \mapsto 1/(z-s) \in \mathbb{C}$ for $z \in \mathbb{C} \setminus \mathbb{R}$. Conversely the resolvent map permits to reconstruct the map A through a contour integral. It has been argued in [42] that the C^*-algebra$\mathcal{A}$ above can be actually reconstructed from the Schrödinger operator $H_{\mathcal{L}}$ itself.

In Solid State Physics the conduction electrons are commonly described through the so-called *tight binding approximation.* This is because only those electrons with energy within $O(k_B T)$ from the Fermi level do contribute to the current. There is no need then to consider the full range of energy to describe these electrons. In particular, the Schrödinger operator can be replaced by its restriction to an energy interval of size $O(k_BT)$ around the Fermi level. However this is not practically accessible in most cases. The method to perform such a reduction goes as follows [64,92,93]. The single atom Schrödinger equation

$$H_1 = \frac{1}{2m}\left(\frac{\hbar}{\imath}\boldsymbol{\nabla} - e\boldsymbol{A}\right)^2 + v(X), \qquad \text{on} L^2(\mathbb{R}^d),$$

has usually a non empty discrete spectrum, described by *quantum numbers*. The valence electron occupy a given eigenstate, denoted by $\phi_v \in L^2(\mathbb{R}^d)$ corresponding to the energy E_v. If the potential v decays fast enough to zero at infinity, ϕ_v decays exponentially fast away from the origin. Then an approximate basis of eigenstate of H_ω is provided by the family $\{U(y)\phi_v;\, y \in \mathcal{L}_\omega\}$ whenever U is the magnetic translation group. Denoting by P_ω the projection on the subspace generated by such family in $L^2(\mathbb{R}^d)$, it is possible to check that $P_\omega = \pi_\omega(P)$ for some projection $P \in \mathcal{A}$. Thanks to an orthonormalization procedure, the family $U(y)\phi_v$ gives rise to an orthonormal basis $\{\psi_y;\, y \in \mathcal{L}_\omega\}$ indexed by the atomic sites, where each ψ_y is exponentially localized near y. By using either the Schur complement (or Feshbach) method [94] or a Grushin method [95] there exists an effective Hamiltonian in the form of a function $z \mapsto H_{eff}(z)$ holomorphic in a neighbourhood of E_v with values in covariant matrices over $\mathcal{L}_\omega$. Moreover, the spectrum of the original problem is given by the implicit equation $E \in \mathrm{Sp}H \Leftrightarrow E \in \mathrm{Sp}H_{eff}(E)$. The advantage of this method is that $H_{eff}(z)$ can be seen as an element of the groupoid C^*-algebra$\mathcal{B}$ [64,42], with matrix elements decreasing exponentially fast away from the main diagonal. In practice only a finite number of such diagonals are kept to compute the spectrum, giving rise to *tight-binding models*. Such a method is actually commonly used also in computer software for the purpose of band calculations in crystals. The various *ab initio* methods of Quantum Chemistry or the functional density calculations (Kohn-Sham method), may lead to an accurate calculation of the parameters involved in such effective Hamiltonian. It is not the purpose of these notes to explain these methods. However, they lead to effective Hamiltonian described by elements of the groupoid C^*-algebra$\mathcal{B}$.

Phonons can also be described in a similar way through $\mathcal{B}$ with zero magnetic field. This is because phonons are the harmonic approximation of the atomic motion around their equilibrium positions. If the atom located at $x \in \mathcal{L}$ is moving around x, let $\boldsymbol{x}+\boldsymbol{u}_x(t)$ be its position at time t. Since x is an equilibrium position for the atom, the potential energy created by its neigbours increases away from x, hence it is expected that $\boldsymbol{u}_x(t)$ stay small at all time. Moreover, the lowest order expansion of the potential energy around x gives rise to a quadratic potential, namely harmonic forces, tending to force the atom back to its position x. Therefore, within this *harmonic approximation*, the classical equation of motion for the atom has the form

$$M\frac{d^2\boldsymbol{u}_x}{dt^2} = \sum_{y\in\mathcal{L}_\omega} \kappa_\omega(x,y)\,(\boldsymbol{u}_y - \boldsymbol{u}_x)\,, \tag{40}$$

where M is the atomic mass and $\kappa_\omega(x,y)$ is the matrix of spring constants between atoms located at y and x. So $\kappa_\omega(x,y) \in M_d(\mathbb{R})$. Phonons are just the quantized degrees of freedom associated with this classical motion. In practice however, the influence of atoms far apart is negligible so that $\kappa_\omega(x,y)$ decays fast enough as $|x-y| \to \infty$. Moreover, the action-reaction principle implies that $\kappa_\omega(x,y) = \kappa_\omega(y,x)^t$. In addition, since it describes an attracting force, $\kappa_\omega(x,y)$ is a positive matrix for any (x,y). Also, the translation invariance of interactions between atoms leads to the covariance condition

$$\kappa_{\mathrm{T}^a\omega}(x+a, y+a) \;=\; \kappa_\omega(x,y)\,.$$

In very much the same way, the translation invariance of interactions between atoms leads to the continuity of $\kappa_\omega(x,y)$ with respect to ω. Hence the map $(\omega,x) \in \Gamma(X) \mapsto \kappa_\omega(0,x) \in M_d(\mathbb{R}^d)$ defines an element κ of $C^*(\Gamma(x)) \otimes M_d$. The equation of motion (40) is usually solved by looking at the stationnary solutions, namely solutions for which

$$\frac{d^2\boldsymbol{u}_x}{dt^2} \;=\; -\tilde{\omega}^2\boldsymbol{u}_x\,,$$

leading to the eigenmode equation

$$\sum_{y\in\mathcal{L}_\omega} \langle x|\widehat{\pi}_\omega(K)y\rangle \boldsymbol{u}_y \;=\; \tilde{\omega}^2\boldsymbol{u}_x\,,$$

$$K(\omega,x) \;=\; \delta_{x,0}\sum_{y\in\mathcal{L}_\omega}\frac{\kappa(\omega,y)}{M} - (1-\delta_{x,0})\frac{\kappa(\omega,x)}{M}\,. \tag{41}$$

It is easy to check that this K defines a positive element of $\mathcal{B}$ so that the eigenmodes are given by the spectrum of $K^{1/2}$. This modes are the plane waves that are allowed to propagate through the crystal. Historically, Einstein [3] was the first to propose a quantized version of these oscillations, supposing that only one mode was propagating. Then he could compute the heat capacity of the solid and shown that it saturates at high temperature leading to the Dulong an Petit law. Soon after, Debye ([2] Section 1.1) introduced a distribution of eigenmodes together with a density of eigenmodes, namely the number of eigenmodes per unit volume and unit of frequency at a given value of $\tilde{\omega}$. He also introduced a cut-off to take into account the uniform discreteness of the crystal, in the form of the *Debye Temperature* Θ_D. In 1912, Born and von Karman [4] performed the first explicit calculation of the eigenmodes in a cubic crystal and discovered the existence of *optical modes*, explaining a small discrepancy between the Debye predictions and the measurements of the heat capacity.

In both cases, whether electrons or phonons, the *density of states* (DOS) is defined in a similar way. In the phonon case, the relevant operator is $K \in \mathcal{B}$, whereas for electrons it is its Hamiltonian H, namely either the Schrödinger operator, in the continuum case, or the effective Hamiltonian in the tight binding representation. Let then H be the name of such an operator. In both cases it is given by a strong resolvent continuous family $(H_\omega)_\omega$ of selfadjoint operators, bounded from below, indexed either by the Hull Ω or by the transversal X. It acts either on $L^2(\mathbb{R}^d)$ or on $\ell^2(\mathcal{L}_\omega)$. In both cases, given a bounded box $\Lambda \subset \mathbb{R}^d$, it is meaningfull to restrict H_ω to Λ provided boundary conditions are prescribed. Let $H_{\omega,\Lambda}$ be this restriction. Then its spectrum is discrete in both cases so that there is only a finite number of eigenvalues (counted with their multiplicities), below $E \in \mathbb{R}$. This allows to define the *integrated density of states* (IDS) as

$$\mathcal{N}(E) = \lim_{\Lambda \uparrow \mathbb{R}^d} \frac{1}{|\Lambda|} \# \{E' \in \mathrm{Sp}(H_{\omega,\Lambda});\, E' \leq E\}.$$

The existence of the limit depends upon the choice of an invariant ergodic probability measure $\mathbb{P}$ on the Hull so that the limit exists only $\mathbb{P}$-almost surely (resp. $\mathbb{P}_{tr}$-almost surely) with respect to ω [64,96–98]. Let then $\mathcal{N}_\mathbb{P}$ denote this common limit. The *Shubin formula* [64,42] expresses it in term of the noncommutative Brillouin zone as

$$\mathcal{N}_\mathbb{P}(E) = \mathcal{T}_\mathbb{P}\left(\chi\{H \leq E\}\right), \qquad \textbf{(Shubin formula)} \qquad (42)$$

with a similar formular in the tight-binding case. Here $\chi\{\cdot \leq E\}$ denotes the characteristic function of the interval $(-\infty, E] \subset \mathbb{R}$. Then $\chi\{H \leq E\}$ is the spectral projection of H seen as an element of the von Neumann algebra $L^\infty(\mathcal{A}, \mathcal{T}_\mathbb{P})$. It is then clear from the definition, that $\mathcal{N}_\mathbb{P}$ is a non decresing non negative function of E. It vanishes below $E_0 \inf \mathrm{Sp}\{H\}$. It is constant on spectral gaps (see Fig 5). In particular, its derivative $d\mathcal{N}_\mathbb{P}/dE$ exists in the sense of a Stieljes-Lebesgue measure and is called the *density of states* (DOS).

The DOS is usually used to expressed thermodynamical quantities such as the *heat capacity*. In the case of phonons, the DOS is given by

$$d\mathcal{N}_\phi(\tilde{\omega}) = \rho_\mathbb{P}\, d\, \hat{\mathcal{T}}_\mathbb{P}(\chi(K \leq \tilde{\omega}^2)),$$

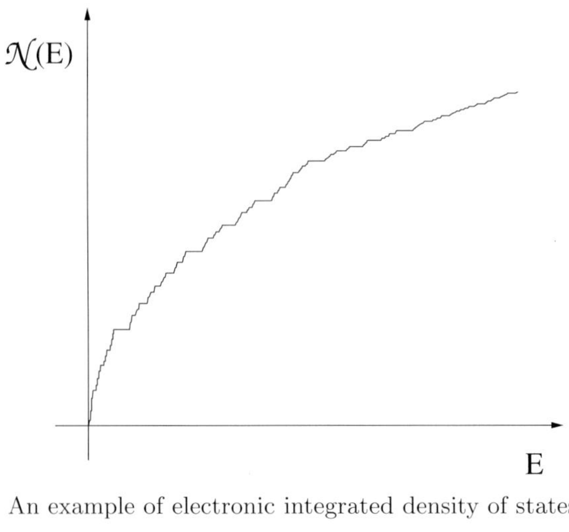

Fig. 5. An example of electronic integrated density of states in 1D

where $\rho_{\mathbb{P}}$ is the atom density (in number of particle per unit volume) in order to get a number of modes per unit volume, and the factor d in front account for the d possible choice of polarization of the acoustic wave. The Debye formula for the phonon contribution to heat capacity per unit volume leads to

$$C_v^{ph} = k_B \int_0^{\tilde{\omega}_D} d\mathcal{N}_\phi(\tilde{\omega}) \left\{ \frac{\hbar\tilde{\omega}/2k_BT}{\sinh(\hbar\tilde{\omega}/2k_BT)} \right\}, \tag{43}$$

where $\tilde{\omega}_D$ is the Debye cut-off frequency of the solid. If $T \uparrow \infty$, Eq. (43) gives the Dulong-Petit law, namely

$$C_V^{ph} \stackrel{T\uparrow\infty}{\longrightarrow} \rho\, k_B\, d\,, \qquad \textbf{(Dulong \& Petit)} \tag{44}$$

As $T \downarrow$, only the low frequencies contribute, so that it is enough to consider the asymptotic of the DOS at $\tilde{\omega} \downarrow 0$. Since low frequencies correspond to large wave lengths, the corresponding phonons do not see the fine structure of the solid and behave like acoustic waves with sound velocity c_α for the polarisation α. In particular $d\mathcal{N}_\phi(\tilde{\omega})$ behaves like $\tilde{\omega}^{d-1}d\tilde{\omega}$, so that

$$C_v^{ph} \propto T^d\,, \qquad \text{as } T \downarrow 0\,.$$

For $d = 3$ this gives the usual T^3 law discovered by Debye.

The electronic contribution to the heat capacity must take the fermionic character of the electrons into account. Then

$$C_v^{el} = T \left(\frac{\partial S}{\partial T} \right)_{V,N},$$

if S is the total entropy in the volume V. The derivative is taken at constant volume and constant electron number. The computation of S is usually performed in the grand canonical ensemble. If $\mathcal{N}_{el}$ denotes the electronic IDS, the chemical potential μ is fixed by the condition

$$2 \int_{\mathbb{R}} d\mathcal{N}_{el}(E)\, \frac{1}{1+e^{-\beta(E-\mu)}} = \frac{N}{V} = 2\mathcal{N}_{el}(E_F)\,.$$

Here N is the total number of electrons in volume V and E_F is called the *Fermi energy*. The factor 2 comes from the spin degrees of freedom. Then the entropy S is given by

$$S = 2k_B V \int_{\mathbb{R}} dE\, \beta^2\,(E-\mu)\, \frac{\mathcal{N}_{el}(E)}{4\cosh^2\{\beta(E-\mu)/2\}}\,. \tag{45}$$

Here $\beta = 1/k_BT$. Setting $x = \beta(E-\mu)/2$ this gives

$$S = 2k_B V \int_0^\infty dx\, \frac{x}{\cosh^2(x)}\, \left(\mathcal{N}_{el}(\mu + 2xk_BT) - \mathcal{N}_{el}(\mu - 2xk_BT)\right)\,. \tag{46}$$

Whenever the IDS is smooth around $E = E_F$, this gives the usual electronic contribution to the heat capacity

$$C_v^{el} \propto T \,, \qquad\qquad \text{as } T \downarrow 0 \,.$$

If, however $\mathcal{N}_{el}$ admits some fractal behaviour near the Fermi energy, then the low temperature behaviour may be modified. It turns out that such a fractal behaviour is hardly seen in practice. This is because the disorder and the interactions as well tend to smooth out the IDS at small energy scales.

4.7 The Master Equation

Within the one particle approximation, it is quite convenient to consider the quantum analog of the Boltzmann approach to kinetic equations. For classical gases Boltzmann was dealing with the one particle density. In the case of electron in an aperiodic solid, this one particle density is provided by a *state* on the C^*-algebra$\mathcal{A}$. Namely a state is a linear map $\phi : \mathcal{A} \mapsto \mathbb{C}$ that is both *bounded* (that is $|\phi(A)| \leq c\|A\|$ for some $c > 0$) and *positive* (that is $\phi(AA^*) \geq 0$ for all $A \in \mathcal{A}$). This state will be normalized so that, by adjoining a unit to $\mathcal{A}$ if necessary, $\phi(\mathbf{1}) = n_{el}$ whenever n_{el} is the electron density (number of electrons per unit volume). This is equivalent to impose $\sup_{\|A\|\leq 1} |\phi(A)| = n_{el}$.

The electrons in a solid are not independent particle. They usually interact with phonons. But the electron-phonon coupling can usually be considered as a small perturbation so that it is convenient to integrate out the phonon degrees of freedom to get an effective evolution for the electrons. This new evolution is no longer unitary, because the electron-phonon interaction produces a loss of information on the electron motion. However, the evolution can always be represented by a one-parameter family of *completely positive maps* $\{\Phi_t; t \geq 0\}$ acting on $\mathcal{A}$ and preserving the unit $\mathbf{1}$ (see a discussion of this point in [99]). A map $\Phi : \mathcal{A} \mapsto \mathcal{A}$ is completely positive if it is linear and if for any $N \in \mathbb{N}$ and any matrix $A = ((A_{i,j}))_{i,j=1}^N$ with matrix elements $A_{i,j} \in \mathcal{A}$, the matrix $((\Phi(A^*A)_{i,j}))_{i,j=1}^N$ defines a positive element in $\mathcal{A} \otimes M_N$. By Stinespring theorem [100], if $\mathcal{A}$ is represented in a Hilbert space $\mathcal{H}$, such map is given by $\Phi(A) = V\pi(A)V^*$ where π is a representation of $\mathcal{A}$ in a different Hilbert space $\mathcal{H}'$ and V is a partial isometry from $\mathcal{H}'$ to $\mathcal{H}$ (Stinespring compression). To get the electron evolution from the interacting theory, it is always necessary to proceed to such a Stinespring compression. If in addition, the electron evolution is considered on a mesoscopic time scale, the evolution becomes *Markovian*. In particular, it can be represented by a *Markov semigroup*, namely of one parameter family of completely positive maps from $\mathcal{A}$ into itself, indexed by $t \in \mathbb{R}_+$ (positive time), and such that

$$\Phi_{t+t'} = \Phi_t \circ \Phi_{t'} \,. \tag{47}$$

To describe this evolution it is convenient to write Eq. (47) as a differential equation in the form

$$\frac{d\Phi}{dt} = \mathfrak{L}\, \Phi_t \,, \tag{48}$$

in which case $\mathfrak{L}$ is called a *Lindbladian.* This is because Lindblad characterized such operators whenever $\mathcal{A}$ is the C^*-algebra of all bounded operators in a Hilbert space

Theorem 8 (Lindblad [99]). *Let Φ_t be a norm continuous Markov semigroup on $\mathcal{B}(\mathcal{H})$ where $\mathcal{H}$ is some separable Hilbert space. Then there are bounded operators H, L_i such that $H = H^*$, $\sum_i L_i^* L_i \in \mathcal{B}(\mathcal{H})$ and that*

$$\frac{d\Phi}{dt} = \mathfrak{L}\,\Phi_t\,, \qquad \text{with} \qquad \mathfrak{L}(A) = \imath[H, A] + \sum_i \left(L_i^* A L_i - \frac{1}{2}\{L_i^* L_i, A\}\right),$$

where $\{A, B\} = AB + BA$. Conversely, any such $\mathfrak{L}$ generates a norm continuous Markov semigroup.

To deal with Statistical Mechanics, it is more convenient to work with states instead of observables. In the case of $\mathcal{B}(\mathcal{H})$ the state space is the set of *density matrices*, namely the positive compact trace class operator with trace 1. Then a Markov semigroup acts on the state space by duality, namely $\rho_t = \rho \circ \Phi_t$. In such a case the Lindbladian acts as follows:

$$\mathfrak{L}^\dagger(\rho) = -\imath[H, \rho] + + \sum_i ([L_i\rho, L_i^*] + [L_i, \rho L_i^*])\,.$$

The evolution equation $d\rho/dt = \mathfrak{L}^\dagger \rho$ is called the *master equation*, which goes back to an early work by Pauli to describe the exchange between matter and radiation (see [54]).

The Lindblad theorem is a noncommutative version of the Levy-Khintchine theorem [101]. In the commutative context, states are nothing but *probability measures* and the Markov semigroups are given by the Chappmann-Kolmogoroff equations describing the evolution of the probability distribution of a stochastic process. If (Ξ, Σ) is a Borel space, where Σ is the corresponding σ-algebra, and if $\mathbb{P}_t$ denotes the probability distribution on Ξ describing the stochastic process at time t, then, for $B \in \Sigma$ and $t > s$

$$\mathbb{P}_t(B) = \int_\Xi P(t-s; B, y)\, \mathbb{P}_s(dy)\,,$$

where the transition probabilities $P(s; \cdot, y)$ satisfy (semigroup property, or Chappmann-Kolmogoroff equations)

$$P(t; B, x) = \int_\Xi P(t-s; B, y)\, P(s; dy, x)\,, \qquad 0 < s < t\,,\ x \in \Xi\,,\ B \in \Sigma\,.$$

In the special case for which $\Xi = \mathbb{R}^d$ and the transition probabilities are given by $P(t; dx, y) = p_t(y - x)dx$, with p_t a probability density, then the Fourier transform of p_t satisfies

$$\int_{\mathbb{R}^d} dx^d\, p_t(x)\, e^{-\imath k\cdot x} = e^{-t\phi(k)}\,,$$

$$\phi(k) = \imath k \cdot x_0 + \int_{\mathbb{R}^d} dG(u) \left(e^{\imath k \cdot u} - 1 - \frac{\imath k \cdot u}{1+u^2} \right) \frac{1+u^2}{u^2} , \qquad (49)$$

where dG is a positive measure of finite mass. If dG admits a mass g at $u = 0$, it contributes for a term $-gk^2/2$ in the *r.h.s.* of (49), namely a gaussian term in p_t.

The Lindblad theorem has practical use in many fields of Physics. It is especially important in Quantum Optics where the explicit computation of the L_i's is related to perturbation theory in QED. Unfortunatley, such a theorem is too restrictive in many situations. First of all, the norm continuity is not natural. In many situations indeed, the generators is not bounded. Thus the norm pointwise continuity, namely $t \in \mathbb{R}_+ \mapsto \Phi_t(A) \in \mathcal{A}$ is continuous for all $A \in \mathcal{A}$, is sufficient in practice. Moreover, the result holds for C^*-algebrasin a Hilbert space and is not necessarily true for a general C^*-algebra.

Example 5 (Fokker-Planck operators). Let $\mathcal{A}$ be the abelian unital C^*-algebraobtained from $\mathcal{C}_0(\mathbb{R})$ by adjoining a unit. Hence $\mathcal{A} = \mathcal{C}(\overline{\mathbb{R}})$ where $\overline{\mathbb{R}}$ is the one point compactification of $\mathbb{R}$. Let $\mathfrak{L}$ be a differential operator. For $\mathfrak{L}$ to be the generator of a norm pointwise continuous Markov semigroup on $\mathcal{A}$ it is necessary that $\mathfrak{L}$ be of the form

$$\mathfrak{L}(f) = \frac{d}{dx} a^2 \frac{d}{dx} + b \frac{d}{dx} - \frac{d}{dx} b + c ,$$

where a, b, c are real continuous functions on $\mathbb{R}$. In particular $\mathfrak{L}$ has no term of degree higher than 2. This is exactly the form of generators of the Fokker-Planck equation describing stochastic processes in the real line. □

In the case studied in Section 4.5 the following example is a generalization of the Levy-Khintchine theorem

Example 6. Let ϕ satisfies Eq. (49). Then, if $\boldsymbol{\nabla}$ denotes the family of *-derivations on $\mathcal{A} = C^*(\Omega \rtimes \mathbb{R}^d, B)$ or on $\mathcal{B} = C^*(\Gamma(X), B)$ (see Section 4.5) defined in Eq. (38)

$$\Phi_t = e^{-t\phi(\boldsymbol{\nabla})}$$

defines a Markov semigroup. □

Example 7 (Model of quantum jumps [45]). Let $\mathcal{H}$ be a separable Hilbert space and let $H = H^*$ be a selfadjoint operator with countable spectrum $S \subset \mathbb{R}$. For $E \in S$ the corresponding eigenprojection is denoted by P_E. Let $\{|j\rangle\,; j \in J\}$ be an orthonormal basis of $\mathcal{H}$, labelled by the set J and made of eigenvectors of H, so that $H|j\rangle = E_j|j\rangle$ where $E_j \in S$ denotes the eigenvalue of H corresponding to the label $j \in J$.

A model of quantum jumps is given by a family $\{\Gamma_{j\to j'}\,; j \neq j' \in S\}$ of non negative numbers representing the probability per unit time of jump from the state j to a state of energy j'. It is assumed that $\sup_j \Gamma_j^- < \infty$ if $\Gamma_j^- = \sum_{j'} \Gamma_{j\to j'}$ is the decay rate of the state j. In much the same way, $\sup_j \Gamma_j^+ < \infty$ if $\Gamma_j^+ = \sum_{j'} \Gamma_{j'\to j}$ is the creating rate of the state j.

In order to describe the jump dynamics, let $W_{j\to j'}$ be the *jump operators* defined by

$$W_{j\to j'}\,|j_0\rangle \;=\; \begin{cases} |j_0\rangle & \text{if} \quad j_0 \neq j\,,\\ |j'\,\rangle & \text{if} \quad j_0 = j\,. \end{cases}$$

Then $W_{j\to j'} = \mathbf{1} + K_{j\to j'}$ where $K_{j\to j'} = (|j'\rangle - |j\rangle)\langle j|$ is an operator of rank 1. In particular $K^*_{j\to j'}K_{j\to j'} = 2|j\rangle\langle j|$.

The Lindbladian associated with such a model is then given by

$$\mathfrak{L}(A) \;=\; \imath[H,A] + \sum_{j\neq j'} \Gamma_{j\to j'}\left(W^*_{j\to j'}AW_{j\to j'} - \frac{1}{2}\{W^*_{j\to j'}W_{j\to j'},A\}\right)\,. \tag{50}$$

It is convenient to express this formula in a basis independent way. The jump probabilities define a bounded operator Γ with matrix elements $\Gamma_{j\to j'}$. Because the decay and creating rates are uniformly bounded, Γ is a bounded operator. $\Gamma^{\pm}$ will denote the diagonal operators with elements given by the decay and creating rates respectively. Then it is convenient to introduce two operations on the space of operators $\mathcal{B}(\mathcal{H})$. The first one is called *averaging*

$$\langle j|\mathbb{E}(A)|j'\rangle \;=\; \delta_{j,j'}\langle j|A|j\rangle\,.$$

Hence $\mathbb{E}$ is a norm 1 projection (*a conditional expectation*). The second operator is given by

$$\langle j|\mathcal{R}(A)|j'\rangle \;=\; \delta_{j,j'}\sum_{j"\in J}\langle j|A|j"\rangle\,.$$

This operator is not defined everywhere in $\mathcal{B}(\mathcal{H})$ and is not necessarily bounded. However it is simple to check that, thanks to the properies of Γ, $\Gamma^- = \mathcal{R}(\Gamma)$ and that, if D is a bounded diagonal operator, $|\mathcal{R}(\Gamma D)| \leq \mathcal{R}(\Gamma|D|) \leq \mathcal{R}(\Gamma)\|D\|$, and $\mathcal{R}(D\Gamma) = D\mathcal{R}(\Gamma)$. An elementary algebra gives

$$\begin{aligned}\mathfrak{L}(A) \;=\; & \imath\left[H + \frac{\Gamma - \Gamma^*}{2\imath}, A\right] - \frac{1}{2}\{\Gamma^-, A\}\\ & - \mathbb{E}\left(\Gamma A + A\Gamma^*\right) + \mathcal{R}\left(\Gamma\mathbb{E}(A) + \mathbb{E}(A)\Gamma\right)\,.\end{aligned} \tag{51}$$

Hence the *r.h.s.* defines the dissipative dynamics in which the Hamiltonian is modified by the term $-\imath(\Gamma - \Gamma^*)/2$. This term corresponds to the *Knight shift* of MNR. Moreover, the second term is a damping term giving a width to any spectral ray of H since Γ^- commutes with H. □

The Model of quantum jumps can be generalized to some examples of C^*-algebrasas shown in the next example.

Example 8 (Variable range hopping on an aperiodic lattice). Here $\mathcal{L}$ denotes a uniformly discrete subset of $\mathbb{R}^d$. Then Ω is its Hull, X the corresponding transversal and $\mathcal{B} = C^*(\Gamma(X))$ the C^*-algebraof the transversal without magnetic field (see Section 4.5). For any $\omega \in X$ let $\mathcal{H}_\omega = \ell^2(\mathcal{L}_\omega)$. Here the

Hamiltonian H_ω is given by multiplication operator $f(\boldsymbol{X})$ if $\boldsymbol{X}$ denotes the position operator acting on $\mathcal{H}_\omega$. In particular, eigenstates are strictly localized. Hence H_ω describes electrons in the extreme case of strongly localized regime. Here H_ω is not necessarily covariant. However the bracket $L_H(A)_\omega = \imath[H_\omega, A_\omega]$ defines a *-derivation of $\mathcal{B}$ since this is the generator of the automorphism group $\eta_t(A)(\omega, x) = e^{\imath t(f(x)-f(0))} A(\omega, x)$ whenever $A \in \mathcal{B}_0$ and $\omega \in X$, $x \in \mathcal{L}_\omega$.

Quantum jumps will be produced if phonons force the electrons to leave their initial states. Thus the jump probability per unit times will be given by a function $\Gamma_{x\to y}(\omega) \geq 0$ with the properties defined in Example 7. In addition, the translation invariance of the dynamics implies that this function be *covariant*, namely $\Gamma_{x\to y}(\omega) = \Gamma_{x-a\to y-a}(\mathrm{T}^{-a}\omega)$ for any $a \in \mathcal{L}_\omega$. Hence $\Gamma \in \mathcal{B}$. Moreover, it is expected that on similar local patches, the transition rates are close, so that $\Gamma_{0\to x}(\omega)$ should be *continuous* on the groupoid $\Gamma(X)$. Thus, $\Gamma^-(\omega) = \sum_{x\in\mathcal{L}_\omega} \Gamma_{0\to x}(\omega)$ is a continuous function on X.

The operations $\mathbb{E}$ and $\mathcal{R}$ are well defined on $\mathcal{B}_0$ since

$$\mathbb{E}(A)(\omega, x) = \delta_{0,x} A(\omega, 0)\,, \qquad \mathcal{R}(A)(\omega, x) = \delta_{0,x} \sum_{y\in\mathcal{L}_\omega} A(\omega, y)\,,$$

following the construction given in Example 7, the corresponding Lindbladian becomes

$$\begin{aligned}\mathfrak{L}(A) &= L_H(A) + \imath\left[\frac{\Gamma - \Gamma^*}{2\imath}, A\right] - \frac{1}{2}\{\Gamma^-, A\} - \mathbb{E}\left(\Gamma A + A\Gamma^*\right) \\ &\quad + \mathcal{R}\left(\Gamma\mathbb{E}(A) + \mathbb{E}(A)\Gamma\right). \end{aligned} \tag{52}$$

This defines a Markov semigroup on $\mathcal{B}$. This semigroup is liable to describe the dissipative evolution of a strongly localized electron due to phonons. It is the basic example for the Mott hopping conductivity. □

5 Coherent Electronic Transport

The coherent part of the electronic transport is the contribution of the quantum interferences to the quantum transport of the electronic wave packet. This is the part of the evolution that occurs before any dissipative mechanism takes place. In practice, this evolution is valid on a time scale τ_{coh} that is the time beyond which collision processes (with phonons say) destroy the phase coherence of the wave packet. This coherence time increases as the temperature decreases, so that the coherent transport eventually dominates at low temperature. In this section it will be assumed that there is no dissipative part in the quantum evolution for the electrons. Moreover, the electrons are considered as independent particles. Should the electrons interact, the description would then be given by a dissipative mechanism beyond a time scale τ_{coll} that represents the average collision times between electrons. In a sense then, the short time dynamics is insensitive to the electron interactions. The Landau theory of Fermi liquid takes care of the interactions and show that indeed the collision time increases like T^{-2} as the temperature T decreases [32]. Therefore again, at low temperature the electron-electron interaction can be neglected.

The main remark that will be developped in this Section is that the quantum wave packet exhibits an anomalous diffusion, namely the mean distance $r(t)$ that it goes after time t follows a scaling law

$$r(t) \overset{t\uparrow\infty}{\approx} t^{\beta}, \qquad 0 \leq \beta \leq 1 .$$

This will have deep consequences for the low temperature behaviour of the transport cœfficients. In particular, the conductivity will behave like a power law in term of the relaxation time (anomalous Drude formula see Section 6.3). This Section is devoted to the definition of these exponents, the relations between them and some of their properties.

5.1 Local and Fractal Exponents of a Borel Measure

In most spectral problems related to the electron dynamics, there are various positive measures, such as the Density of states (DOS) (see Section 4.6), the spectral measure (also called the Local Density of states, LDOS) or the current-current correlation function. In this section scaling exponents are defined for a positive measure on $\mathbb{R}$. Most of the material of this section can be found in [33,102,103].

Let f be a positive measurable function on $(0, b]$. The power decay exponents $\alpha^+, \alpha^-, \alpha^m$ are defined as

$$\alpha^{\pm} = \left\{ \begin{array}{l} \limsup_{\epsilon\downarrow 0} \\ \liminf_{\epsilon\downarrow 0} \end{array} \right\} \frac{\ln f(\epsilon)}{\ln \epsilon}, \qquad \alpha^m = \sup\{\gamma \in \mathbb{R}; \int_0^b d\epsilon \, \frac{f(\epsilon)}{\epsilon^{1+\gamma}} < \infty\} . \tag{53}$$

Therefore $\alpha^- \leq \alpha^m \leq \alpha^+$ and $\alpha^m \geq 0$. Such definition will be shortened by writing

$$f(\epsilon) \overset{\epsilon\downarrow 0}{\sim} \epsilon^{\alpha^\star},$$

where $\star \in \{+, -, m\}$. It can be shown that, whenever f is non decreasing, $\alpha^- = \alpha^m$ [33].

For a positive measurable function on $[b, \infty)$ the power growth exponents $f(t) \overset{t\uparrow\infty}{\sim} t^{\alpha^\star}$ are defined in a similar way.

A Borel measure μ on $\mathbb{R}$ is a positive bounded measure defined as a σ-additive function on the σ-algebra of Borel sets. It can be defined through its primitive $\mathcal{N}_\mu(E) = \int_{-\infty}^{E} d\mu$ which is positive and non decreasing (Stieljes-Lebesgue theory) [104]. The set of Borel measures on $\mathbb{R}$ can also be seen (Riesz theorem) as the space dual to $\mathcal{C}_0(\mathbb{R})$ (space of continuous complex functions on $\mathbb{R}$ vanishing at infinity), with duality defined by $\langle\mu|f\rangle = \int_{\mathbb{R}} d\mu\, f$. It will then be endowed with the *vague topology*: a sequence μ_n of Borel measures converges vaguely to μ if, for all $f \in \mathcal{C}_0(\mathbb{R})$, $\lim_n \int_{\mathbb{R}} d\mu_n f \to \int_{\mathbb{R}} d\mu f$.

A positive Borel measure μ is *absolutely continuous* (*a.c.*) whenever $\mu[E - \epsilon, E + \epsilon] \leq c(E)\, \epsilon$ for almost all E's. Equivalently, the Radon-Nikodym theorem asserts that $\mu(dE) = f(E)dE$ for some Lebesgue integrable non negative function f. The measure μ is *pure-point* (*p.p.*) whenever it can be written as a linear combination of Dirac measures. If μ is neither pure-point nor absolutely continuous, it is called *singular continuous* (*s.c.*). Given two Borel measures μ, ν on $\mathbb{R}$, μ dominates ν whenever for any Borel subset Δ of $\mathbb{R}$, $\mu(\Delta) = 0 \Rightarrow \nu(\Delta) = 0$. They are equivalent if each of them dominates the other. They are orthogonal whenever there is a Borel set Δ such that $\mu(\Delta) = 0$ and $\nu(\mathbb{R} \setminus \Delta) = 0$. The Lebesgue decomposition theorem asserts that any Borel measure on $\mathbb{R}$ can be decomposed in a unique way into the sum of an *a.c.* part, a *p.p.* part and a *s.c.* one, the three contribution being orthogonal to each other.

Given a Borel measure μ on $\mathbb{R}$, its *local dimension* $d^\star_\mu(E)$ is defined by

$$\mu([E-\epsilon, E+\epsilon]) \stackrel{\epsilon\downarrow 0}{\sim} \epsilon^{d^\star_\mu(E)} .$$

Since $\mu([E-\epsilon, E+\epsilon])$ is a non decreasing function of ϵ, $d^m_\mu(E) = d^-_\mu(E)$, so that it is enough to consider $\star \in \{\pm\}$. An important property is the following [33]

Lemma 2. *Let μ, ν be two Borel measures on $\mathbb{R}$. Then*
(i) $d^\star_\nu(E) \leq d^\star_\mu(E)$ for ν-almost every E.
(ii) If μ dominates ν, then $d^\star_\nu(E) = d^\star_\mu(E)$, ν-almost surely.

In particular, using Lebesgue's measure, $0 \leq d^\star_\mu(E) \leq 1$ for μ-almost every E. (Note however that there may be $E \in \mathbb{R}$ for which $d^\star_\mu(E) > 1$, but such E belongs to a set of μ-measure zero.) Moreover equivalent measures have the same local dimension. Hence μ is absolutely continuous then $d^\star_\mu(E) = 1$, μ-almost everywhere. If μ is pure-point then $d^\star_\mu(E) = 0$, μ-almost surely. Thus these exponents are non trivial only for singular continuous measures. Another way to define these exponents is through potential theory [105,106] initially proposed by Frostman [107]

$$d^-_\mu(E) = \sup\{\gamma \in \mathbb{R} ; \int d\mu(E') \frac{1}{|E-E'|^\gamma} < \infty\} .$$

On the other hand [108,109]

Lemma 3. *Let μ be a Borel measure on $\mathbb{R}$ and let $G_\mu(z) = \int_{\mathbb{R}} d\mu(E)/(z-E)$, $\Im m(z) > 0$, its Green's function. Let $\sigma^\star(E)$ be defined as*

$$\Im m\,(G_\mu(E - \imath\epsilon)) \stackrel{\epsilon\downarrow 0}{\sim} \epsilon^{\sigma^\star(E)} .$$

Then $\sigma^-(E) = d^-_\mu(E) - 1$ whenever $d^-_\mu(E) \in [0,2]$.

Given a Borel subset Δ of $\mathbb{R}$, the local dimensions are defined by

$$d^{\star,\pm}_\mu(\Delta) = \left\{ \begin{matrix} \mu - \text{esssup}_{\epsilon\downarrow 0} \\ \mu - \text{essinf}_{\epsilon\downarrow 0} \end{matrix} \right\} d^\star_\mu(E) .$$

Hausdorff and packing dimensions are defined in [110,105,111,106]. Here are the basic definitions. Let F be a subset of $\mathbb{R}$. For $\delta \in (0,1)$, a δ-cover is a covering of F by open sets of diameter less than or equal to δ. Here $d(U)$ denotes the diameter of U. If $s > 0$ let

$$\mathcal{H}^s_\delta(F) \;=\; \inf\{\sum_{i=0}^{\infty} d(U_i)^s \,;\, \{U_i\} \text{ is a } \delta\text{-cover of} F\}\,.$$

Then $\mathcal{H}^s_\delta(F)$ decreases as $\delta \downarrow 0$ so that $\mathcal{H}^s(F) = \lim_{\delta\downarrow 0} \mathcal{H}^s_\delta(F)$ exists and defines a Borel measure called the *s-Hausdorff measure* of F. The Hausdorff dimension $\dim_H(F)$ is the unique value s_0 such that $s < s_0 \Rightarrow \mathcal{H}^s_\delta(F) = \infty$ whereas $s > s_0 \Rightarrow \mathcal{H}^s_\delta(F) = 0$. In much the same way a δ-packing of F is a countable family of disjoint open balls with centers in F and diameter less than or equal to δ. Then

$$\mathcal{P}^s_\delta(F) \;=\; \sup\{\sum_{i=0}^{\infty} d(B_i)^s \,;\, \{B_i\} \text{ is a } \delta\text{-packing of} F\}\,.$$

Again $\mathcal{P}^s_0(F) = \lim_{\delta\downarrow 0} \mathcal{P}^s_\delta(F)$ exists. A Borel measure $\mathcal{P}^s$, the *s-packing measure*, is defined through

$$\mathcal{P}^s(F) \;=\; \inf\{\sum_i \mathcal{P}^s_0(F_i)\,;\, F_i \text{ Borel set }, F \subset \bigcup_i F_i\}\,.$$

The *packing dimension* $\dim_P(F)$ is the unique value s_0 such that $s < s_0 \Rightarrow \mathcal{P}^s_\delta(F) = \infty$ whereas $s > s_0 \Rightarrow \mathcal{P}^s_\delta(F) = 0$.

Whenever μ is a Borel measure on $\mathbb{R}$ and $\Delta \subset \mathbb{R}$ a Borel set, the Hausdorff dimensions $\dim^\pm_H(\mu, \Delta)$ and the packing dimensions $\dim^\pm_P(\mu, \Delta)$ are defined respectively as follows (where $* \in \{H, P\}$)

$$\dim^+_*(\mu, \Delta) \;=\; \inf\{\dim_*(\Delta');\ \Delta' \subset \Delta\,,\ \mu(\Delta') = \mu(\Delta)\}\,,$$

$$\dim^-_*(\mu, \Delta) \;=\; \sup\{\alpha \in \mathbb{R};\ \Delta' \subset \Delta\,,\ \dim_*(\Delta') \le \alpha \Rightarrow \mu(\Delta') = 0\}\,.$$

Then it turns out that (see [102] Appendix):

$$\dim^\pm_H(\mu, \Delta) \;=\; d^{+,\pm}_\mu(\Delta)\,, \qquad\qquad \dim^\pm_P(\mu, \Delta) \;=\; d^{-,\pm}_\mu(\Delta)\,.$$

A finer characterization of the local properties of μ are given by its *fractal exponents* defined by (for $q \ne 1$)

$$\int_\Delta d\mu(E)\; \mu([E-\epsilon, E+\epsilon])^{q-1} \;\stackrel{\epsilon\downarrow 0}{\sim}\; \epsilon^{(q-1)D^\star_{\mu,\Delta}(q)}\,.$$

The function $q \mapsto (q-1)D^\star_\mu(q)$ is concave and $q \mapsto D^\star_\mu(q)$ is non increasing. The fractal exponents are not measure class invariant [33].

Lemma 4 (see [33]). *The local dimensions $d^\star_\mu(E)$ are Borel functions of (μ, E). The local dimensions $d^{\star,\pm}_\mu(\Delta)$ and the fractal exponents $D^\star_{\mu,\Delta}(q)$ are Borel functions of μ.*

The proof of this lemma for d_μ^- can be found in [33]. The same strategy applies to d_μ^+ as well as for $D^\star_{\mu,\Delta}(q)$.

Lemma 5 (see [112], Appendix). *Let $f_\mu(\alpha)$ be the Legendre transform of the function $q \mapsto (q-1)D^\star_\mu(q)$ and let $I_\alpha(\epsilon) = \{E \in \mathbb{R};\ \mu([E-\epsilon, E+\epsilon]) < \epsilon^\alpha\}$. If $D^+_\mu(q) = D^-_\mu(q) = D_\mu(q)$ and $D_\mu(q) < \infty$ for some $q < 0$, then there are positive constants $0 < c < C$ such that*

$$c\epsilon^{\alpha - f_\mu(\alpha)} \;\leq\; \mu(I_\alpha(\epsilon)) \;\leq\; C\epsilon^{\alpha - f_\mu(\alpha)}\,.$$

□

Whenever D^+_μ and D^-_μ do not coincide, there is a weaker inequality due to Barbaroux, Germinet and Tcheremchantsev [113]: if μ has compact support then for any $0 \leq q \leq 1$ there is $\alpha = \alpha(\epsilon, q)$ such that

$$\mu(I_\alpha(\epsilon)) \;\geq\; \epsilon^{(1-q)\alpha}\,\frac{1}{|\ln \epsilon|}\int d\mu(E)\,\mu([E-\epsilon, E+\epsilon])^{q-1}\,, \quad \textbf{(BGT estimate)}\,.$$

5.2 Spectral Exponents

Let H be a selfadjoint operator acting on the separable Hilbert space $\mathcal{H}$. The spectral theory associates with H a $\mathcal{H}$-projection-valued Borel measure Π on $\mathbb{R}$ [114] called the *spectral measure* of H. Furthermore, for any $\phi \in \mathcal{H}$, $\|\phi\| = 1$, let ρ_ϕ be the *spectral measure* of H relative to ϕ, namely for $f \in C_0(\mathbb{R})$,

$$\int d\rho_\phi(E)\, f(E) \;=\; \langle\phi|f(H)|\phi\rangle \;=\; \int \langle\phi|\Pi(dE)|\phi\rangle\, f(E)\,.$$

In physics literature, ρ_ϕ is called the *local density of states* (LDOS).

Definition 6. *Let $E \in \mathbb{R}$ and Δ be a Borel subset of $\mathbb{R}$. The spectral exponents and essential spectral exponents of Π (or H) are defined by*

$$d_\Pi^{\star,+}(\Delta) \;=\; \sup_{\phi\in\mathcal{H}} d_{\rho_\phi}^{\star,+}(\Delta)\,, \qquad d_\Pi^{\star,-}(\Delta) \;=\; \inf_{\phi\in\mathcal{H}} d_{\rho_\phi}^{\star,-}(\Delta)\,.$$

The next result shows that there are *typical* states in $\mathcal{H}$ giving the generic properties of the spectrum

Lemma 6 ([33]). *There exists $\psi \in \mathcal{H}$ with* $d_\Pi^{\star,*}(\Delta) = d_{\rho_\psi}^{\star,*}(\Delta)$.

Let now $\mathcal{L}$ be a uniformly discrete subset of $\mathbb{R}^d$ and let Ω its Hull and X its transversal. Let also $\mathbb{P}$ be an invariant ergodic measure on the Hull and $\mathbb{P}_{tr}$ the induced probability on X. Let then $H = (H_\omega)$ be a covariant strong resolvent continuous family of selfadjoint operators on the Hilbert space $\mathcal{H}_\omega$ corresponding to either $L^2(\mathbb{R}^d)$ (continuous case) or to $\ell^2(\mathcal{L}_\omega)$ (discrete case). Let Π_ω denote the corresponding family of spectral measures. By covariance, Π_ω and $\Pi_{\omega'}$ are equivalent measures whenever ω and ω' belong to the same orbit. Therefore their local dimension coincide. Thanks to Lemma 4 it follows that the maps $\omega \mapsto d_{\Pi_\omega}^{\star,*}(\Delta)$ is Borel and translation invariant, so that, by ergodicity, it is $\mathbb{P}$-almost surely constant ($\mathbb{P}_{tr}$-almost surely constant in the discrete case).

Definition 7. *The LDOS dimensions* $d^{\star;*}_{LDOS}(\Delta)$ *is the common value of the* $d^{\star;*}_{\Pi_\omega}(\Delta)$*'s for* $\mathbb{P}$ *almost all* ω*'s.*

Remark 8. The LDOS dimensions depend on the choice of $\mathbb{P}$ in general. □

In much the same way, the DOS (see Eq. (42)) $d\mathcal{N}_{\mathbb{P}}$ is a Borel measure on $\mathbb{R}$ so that it has its own local dimensions

Definition 8. *The DOS dimensions* $d^{\star;*}_{DOS}(\Delta)$ *are the local dimensions of the density of states* $d\mathcal{N}_{\mathbb{P}}$*.*

The relation between the LDOS and the DOS local dimensions is given by

Lemma 7 ([33]). *If* Δ *is a Borel subset of* $\mathbb{R}$*, then* $d^{-;\pm}_{LDOS}(\Delta) \leq d^{-;\pm}_{DOS}(\Delta)$*.*

In general the inequality is strict. The spectral measure may be pure point with an *a.c.* density of states. This is the case for the Anderson model in one dimension [97,98,96].

5.3 Diffusion Exponents

Let $\mathcal{L}$ be a uniformly discrete subset of $\mathbb{R}^d$ containing the origin and let $\mathcal{H} = \ell^2(\mathcal{L})$. The position operator is defined as in Section 4.5 (see Eq. (38)) by $\boldsymbol{X}\psi(x) = \boldsymbol{x}\psi(x)$, for $\psi \in \mathcal{H}$. Let also $H = H^*$ be a selfadjoint operator on $\mathcal{H}$ such that the corresponding *current operator* $\boldsymbol{J} = \imath[\boldsymbol{X}, H]$ be bounded (actually it is enough that $\boldsymbol{J}$ satisfies $f(H)\boldsymbol{J}$ bounded for all functions f continuous with compact support). For Δ a Borel subset of $\mathbb{R}$, let Π_Δ be the spectral projection of H on Δ. As time goes, the position operator evolves according to $\boldsymbol{X}(t) = e^{\imath tH}\boldsymbol{X}e^{-\imath tH}$. Thus $\boldsymbol{X}(t) - \boldsymbol{X}$ measures the distance that the particle goes between times 0 and t. The average distance the particle goes after time t, starting from an initial state located at $0 \in \mathcal{L}$ and projected on Δ can be measured through the following *transport exponents*

$$\int_{-t}^{+t} \frac{ds}{2t} \langle 0|\Pi_\Delta \left|\boldsymbol{X}(s) - \boldsymbol{X}\right|^q \Pi_\Delta|0\rangle \overset{t\uparrow\infty}{\sim} t^{q\,\beta^\star_q(\Delta)} .$$

Since the current is bounded, $\|\boldsymbol{X}(t) - \boldsymbol{X}\| \leq t\|\boldsymbol{J}\|$. it follows that $0 \leq \beta^\star_q(\Delta) \leq 1$. Even though only few specific values of q matter in practice ($q = 2$ is the most important for transport cœfficients), it may be convenient to take advantage of this varying parameter. The Cesaro mean $\int_{-t}^{+t} ds/2t$ may be replaced by a *gaussian mean* $\int_{\mathbb{R}} ds/t \exp\left(-s^2/t^2\right)$ without changing the exponents. In addition, the constant term $\boldsymbol{X}$ can also be removed without changing the exponents. Whenever $\Delta = \mathbb{R}$, Δ will be omitted.

In much the same way, the previous definition can be extended to the case of a covariant family over the Hull of $\mathcal{L}$. Then the relevant exponents are the *averaged* ones

$$(L_{\Delta,q}(t))^q = \int_{-t}^{+t} \frac{ds}{2t} \int_X d\mathbb{P}_{tr}(\omega) \langle 0|\Pi_{\omega,\Delta} \left|\boldsymbol{X}_\omega(s) - \boldsymbol{X}\right|^q \Pi_{\omega,\Delta}|0\rangle \overset{t\uparrow\infty}{\sim} t^{q\,\overline{\beta}^\star_q(\Delta)} .$$

It is interesting to notice that the last expression admits an interpretation in term of the noncommutative Brillouin zone. Namely

$$\begin{aligned} (L_{\Delta,q}(t))^q &= \int_X d\mathbb{P}_{tr}(\omega)\langle 0|\Pi_{\omega,\Delta}\,|\boldsymbol{X}_\omega(t)-\boldsymbol{X}|^q\,\Pi_{\omega,\Delta}|0\rangle \\ &= \hat{\mathcal{T}}_{\mathbb{P}}\left(|\boldsymbol{\nabla}(e^{-\imath sH})|^q\Pi_\Delta\right) . \end{aligned} \tag{54}$$

Similar definition can be given on the continuum.

The case $q=2$ is special. For indeed the corresponding diffusion exponent enters explicitely in the transport theory. To see this, let the *current-current measure* be defined as follows: let $H=H^*$ be a selfadjoint element affiliated to $\mathcal{A}$ (or to $\mathcal{B}$ in the discrete case; see 4.6). This means that $f(H)\in\mathcal{A}$ for all $f\in\mathcal{C}_0(\mathbb{R})$ [89]. In addition H is such that $\boldsymbol{\nabla}H$ is also affiliated to $\mathcal{A}$ and satisfies $f(H)\boldsymbol{\nabla}Hg(H)\in\mathcal{A}$ for all pair of functions f,g with compact support. Then, thanks to Riesz theorem, there is a Borel measure $dm_{i,j}$ on $\mathbb{R}^2$ defined by

$$\int_{\mathbb{R}^2} dm_{i,j}(E,E')\,f(E)g(E') = \mathcal{T}_{\mathbb{P}}\left(f(H)\partial_i H\, g(H)\partial_j H\right) , \qquad f\; g\in\mathcal{C}_c(\mathbb{R}) . \tag{55}$$

Setting $dm=\sum_{i=1}^d dm_{i,i}$, the result is the following

Proposition 6 ([115,33]). *The diffusion exponents $\overline{\beta}_2^{\pm}$ are given by*

$$m\{(E,E'\in\Delta\times\mathbb{R}\,;\,|E-E'|\leq\epsilon\} \stackrel{\epsilon\downarrow 0}{\sim} \epsilon^{2(1-\overline{\beta}_2^{\star}(\Delta))} .$$

Hence the diffusion exponents β_2 measure the fractal behaviour of the current-current correlation function near the diagonal.

From the definition of the diffusion exponents, if $\overline{\beta}_q^{\star}=0$ then particle is likely to be *localized.* Actually a stronger condition is required. Let the localization length be defined as [116,35]

$$l_q(\Delta)^q = \limsup_{t\to\infty}\int_{-t}^{+t}\frac{ds}{2t}\,\hat{\mathcal{T}}_{\mathbb{P}}\left(|\boldsymbol{\nabla}e^{-\imath Hs}|^q\,\Pi(\Delta)\right) < \infty . \tag{56}$$

Note that this condition is stronger than $\overline{\beta}_q^{\star}(\Delta)=0$, because the integrand may be logarithmically divergent as $t\to\infty$ without changing the diffusion exponent. It turns out that $l_2(\Delta)$ coincides with the notion of localization length used in physics. In several examples (the Anderson model for instance) this condition has been proved to hold.

Theorem 9. *Let the localization condition* (56) *be satisfied for a Borel set $\Delta\subset\mathbb{R}$ and some $q>0$. Then, if $\boldsymbol{\nabla}H\in\mathcal{A}$, the following holds:*
i) [116,35] *$\overline{\beta}_q^{\star}(\Delta)=0$ and $\pi_\omega(H)$ has pure-point spectrum in Δ for $\mathbb{P}_{tr}$-almost every $\omega\in X$.*
ii) [33] *The localization condition* (56) *is invariant by a compact perturbation of H.*

iii) [116,35] *If $q = 2$, there is a $\mathcal{N}$-measurable function l on Δ such that for every Borel subset Δ' of Δ:*

$$l_2(\Delta')^2 = \int_{\Delta'} d\mathcal{N}(E)\, l(E)^2 \,. \tag{57}$$

$l(E)$ is the localization length at the energy E.
iv) [115] *The localization length $l(E)$ is also given by*

$$l_2(\Delta')^2 \;=\; \int_{\Delta\times\mathbb{R}} dm(E,E')\,\frac{1}{|E-E'|^2}\,. \tag{58}$$

Since the current-current measure can also be expressed in term of the average of product of two Green functions, this formula gives the expression of the localization length as computed in physics literature. It justifies the claim that $l(E)$ is the localization length. Similar results are expected for the l_q's.

5.4 The Guarneri and BGT Inequalities

The main result is the so-called *Guarneri inequality* which is supplemented by the BGT one below,

Theorem 10. *Let $\mathcal{L}$ be a uniformly discrete subset of $\mathbb{R}^d$. Let H be a selfadjoint bounded operator on $\ell^2(\mathcal{L})$ such that $\boldsymbol{J} = \imath[\boldsymbol{X}, H]$ be bounded. Then for all $q > 0$*

$$\begin{aligned}
&\text{(i) } \beta_q^-(\Delta)\, d \geq \dim_H^+(\mu,\Delta)\,,\\
&\text{(ii) } \beta_q^+(\Delta)\, d \geq \dim_P^+(\mu,\Delta)\,,\\
&\text{(iii) } \beta_q^{\pm}(\Delta)\, d \geq D^{\pm}_{\mu,\Delta}\Big(\frac{d}{d+q}\Big)\,.
\end{aligned}$$

The inequality (i) appeared in the original Guarneri papers [117,118]. The form it takes above is due to Last [119]. It was extended to the continuous case by Combes [120]. The inequality (ii) first appeared in [102]. The inequality (iii) appeared in [112] with the condition that $D_\mu^+(q) = D_\mu^-(q)$, but the form given above is due to [113] without this condition. The proof is interesting in itself in that it shows where, in energy space, is concentrated the wave packet at large times.
Sketch of the proof: (reproduced from [103]) here only the case $\Delta = \mathbb{R}$ is considered. Let $R > 0$ then

$$\langle 0||\boldsymbol{X}|^q(s)|0\rangle \;\geq\; R^q \sum_{|x|>R} |\langle x|e^{-\imath Hs}|0\rangle|^2 \;=\; R^q\left(1-\sum_{|x|\leq R}|\langle x|e^{-\imath Hs}|0\rangle|^2\right).$$

Let $|0\rangle$ be decomposed as $|\psi\rangle + |\psi^c\rangle$ so that

$$\sum_{|x|\leq R} |\langle x|e^{-\imath Hs}|0\rangle|^2 \;\leq\; 1 - \|\psi\|^2 + \sum_{|x|>R} |\langle x|e^{-\imath Hs}|\psi\rangle|^2 + \text{crossed terms}\,.$$

The crossed term can be treated as the third one on the *r.h.s.*, with a bit of technicalities. Thus

$$\langle 0||\boldsymbol{X}|^q(s)|0\rangle \;\geq\; R^q \left(\|\psi\|^2 - \sum_{|x|>R} |\langle x|e^{-\imath Hs}|\psi\rangle|^2 + \text{crossed terms} \right) .$$

Then ψ is chosen as $\chi(H)|0\rangle$ where χ is the characteristic function of the set $I(t) = I_\alpha(1/t)$ (see Lemma 5) if μ is the spectral measure of H relative to the state $|0\rangle$. Using the large deviation bound, $\|\psi\|^2 = \mu(I_\alpha(1/t)) \geq c t^{f_\mu(\alpha)-\alpha}$. To bound the oscillating terms involving $e^{-\imath Hs}$, it is simpler to use the Gaussian mean instead of the Cesaro mean. Writting $\langle 0|g(H)|x\rangle$ as $\int d\mu(E)g(E)P_x(E)$, with $\int d\mu(E)|P_x(E)|^2 = 1$, and using the Cauchy-Schwarz inequality

$$\int \frac{ds}{t} e^{-s^2/t^2} \, |\langle x|e^{-\imath Hs}|\psi\rangle|^2 \;\leq\; \int_{I(t)} d\mu(E)|P_x(E)|^2 \times \int_{I(t)} d\mu(E')e^{-(E-E')^2t^2} \;\leq\; t^{-\alpha} .$$

This estimate takes care of the remaining terms. Then R is still free. If it is chosen so that $t^{f_\mu(\alpha)-\alpha} \sim R^d t^{-\alpha}$, a little algebra leads to (iii) once α is chosen in the optimal way. □

5.5 A Heuristic Discussion

Coherent transport has been the focus of attention of many numerical studies during the last fifteen years. Existence of models exhibiting anomalous diffusion exponents was pointed out early [121–128]. Thanks to Guarneri's bounds, the following discussion follows. The meaning of the spectral exponents given by various local dimensions, is the following:

1. If the spectrum of H in the Borel set Δ is *absolutely continuous* then $d_{LDOS}(\Delta) = 1$.
2. If the spectrum of H in the Borel set Δ is *pure point* then $d_{LDOS}(\Delta) = 0$.
3. Consequently if $0 < d_{LDOS}(\Delta) < 1$ the electronic spectrum is singular continuous (or *fractal*).

One important problem is whether a singular electronic spectrum can be expected or not. The Anderson model, acting on $\ell^2(\mathbb{Z}^d)$ and defined by

$$(H\psi)(x) \;=\; \sum_{y;|y-x|=1} \psi(y) + \lambda V_x \psi(x) , \qquad \psi \in \ell^2(\mathbb{Z}^d) , \tag{59}$$

where the V_x's are real identically distributed independent random variables, with mean $\langle V_x\rangle = 0$ and covariance $\langle V_x^2\rangle = 1$, gives rise to a pure-point spectrum at large λ or near the band edges [97,98,96,129,130]. Similar result have been obtained for the analog of the Anderson model on $L^2(\mathbb{R}^d)$. Many Schrödinger operators with quasiperiodic potential also give rise to pure point spectrum for large coupling constant [97,98].

One-dimensional Schrödinger operators on quasicrystals or with potentials given by automatic sequences, have singular continuous spectrum [131–135]. A systematic study of operators with singular continuous spectrum has been done by B. Simon *et al.* [109,136–144].

In higher dimension there are not many results available for quasicrystals. The labyrinth model [145–149] is obtained as a cartesian product of one-dimensional models: it shows that as dimension grows absolutely continuous spectrum becomes more and more likely. As for the Anderson model [10], whenever the Schrödinger operator is dominated by the kinetic term, absolutely continuous spectrum becomes more likely as soon as the dimension is greater than or equal to 3. Therefore

Problem 1. *Let $\mathcal{L}$ be a uniformly discrete subset of $\mathbb{R}^d$ with $d \geq 3$. Let Ω be the Hull and $\mathbb{P}$ be an $\mathbb{R}^d$-invariant ergodic probability measure on Ω. Let $v : x \in \mathbb{R}^d \mapsto v(x) \in \mathbb{R}$ be an element of $L^1_{K,r}(\mathbb{R}^d)$ (see Section 4.6 and Lemma 1) for some K, r. Then prove or disprove that there is $E_c \in \mathbb{R}$ such that the part of the spectrum of*

$$H_\omega = -\Delta + \sum_{y \in \mathcal{L}_\omega} v(\cdot - y)$$

contained in $[E_c, \infty)$ is absolutely continuous for $\mathbb{P}$-almost every ω. □

Problem 2. *(i) Let $H = -\Delta + V$ be acting on $L^2(\mathbb{R}^d)$ where V is a bounded uniformly continuous function. Let Ω be the Hull of this operator defined as the strong resolvent closure of its translated [42,64]. Then prove or disprove the same claim as in Problem 1*
(ii) Consider the same problem $H = -\Delta + \lambda V$ be acting on $\ell^2(\mathbb{Z}^d)$ where V is a bounded sequence on $\mathbb{Z}^d$ (including the Hull and the probability $\mathbb{P}$). Prove or disprove that for λ small enough there is an a.c. *spectrum away from the band edges* □

Remark 9. The Anderson transition problem is a special case of Problem 2(ii) whenever V is a random sequence or of Problem 1 if $\mathcal{L}$ is a chosen randomly (with Bernoulli distribution) on a regular lattice. □

Thanks to the last remark, these problems are certainly difficult to handle. Many numerical calculations in the eighties, together with the theory of mesoscopic systems have suggested that at scales of the order of λ^2 near the Fermi level, the effective Hamiltonian of the Anderson model behaves like a Gaussian random matrix [150–154] (see especially [153] for a review). There are many reasons to believe that the same property holds in general for potentials of the kind introduced in Problem 1 and 2, at least if the subset $\mathcal{L}$ is dense enough (say Delone) and aperiodic. In particular, the level spacing distribution of an octagonal quasicrystal belongs to the Wigner-Dyson universality class, suggesting that RMT applies [155]. Then the Random Matrix Theory (RMT) predicts a smooth

DOS, a simple *a.c.* spectrum and also quantum diffusion (see below). If this is the case, the properties of this type of spectrum should be universal close to the Fermi level.

The theory of transport exponents developped in the previous Section 5.3 can also be interpreted as follows (the discussion concentrates on β_2

1. Ballistic motion implies $\beta_q = 1$. This is what happens for periodic Schrödinger operators either on the continuum or on the lattice. From the Guarneri inequality, in dimension $d = 1$, *a.c.* spectrum implies $\beta_q = 1$. Note, however, that there are examples of Hamiltonians with *a.c.* spectrum and subballistic motion [156].
2. Strong localization, on the opposite side, implies $\beta_q = 0$. However there are examples of Hamiltonian in dimension $d = 1$ with $\beta_q = 0$ and no localized states [157]. In dimension $d = \infty$ it is even possible to construct Hamiltonians with *a.c.* spectrum and $\beta_q = 0$ [158].
3. Quantum diffusion corresponds to $\beta_2 = 1/2$. This is what happens for the Anderson Hamiltonian in the Randon Phase Approximation (the so-called *Wegner N-orbital model* [33]; see Section 6.4). This type of behaviour is expected to occur every time RMT applies, thus for the Anderson model in the weak localization regime such as low disorder for $d \geq 3$ or with spin-orbit coupling in $d = 2$ [11]. If the answer to Problem 1 is positive, then $\beta_2 = 1/2$ is expected at low energy scale near the Fermi level on any type of tiling, especially in quasicrytals. As a consequence it should produce a residual conductivity at low temperature (see Section 6.3).
4. Subballistic overdiffusive motion will correspond to $1/2 < \beta_q < 1$. This type of behaviour has been observed in many models of quasicrystals [146,147,159]. They are compatible with *a.c.* spectrum only if $d \geq 2$. However, these models for quasicrystals do not seem to reflect what happens in experiments.
5. Subdiffusive motion corresponds to $0 < \beta_2 < 1/2$. Many examples of models are also known to be subdiffusive [146,147,159]. Note that *a.c.* spectrum implies $\beta_q \geq 1/d$. In [149] examples of labyrinth models with *a.c.* spectrum and β_2 as close as possible from $1/d$ have been constructed. It is believed that in most quasicrystals (either with decagonal or icosahedral symmetry), the electronic Hamiltonian, in the one-particle approximation, should have an *a.c.* spectrum near the Fermi energy with subdiffusive behaviour. It may not be the case for the *AlPdRe* compound reported in Fig. 3 however (see Section 6.5 for a discussion). In any case, it is difficult, to produce non artificial models liable to describe faithfully the transport properties of a realistic quasicrystal.

6 Dissipative Transport in the RTA

Dissipative transport concerns what happen to electrons beyond the coherence time τ_{coh}. Then collisions with other degrees of freedom, mostly with phonons must be taken into account. In principle, the description of this regime requires giving a full account of the interactions between electrons and phonons. This means using the second quantization formalism and an effective Hamiltonian liable to represent the electron-phonon interaction or even the electron-electron interaction. In most cases, at finite volume and within the tight-binding approximation, the most common Hamiltonian that is used in Solid State Physics is the Fröhlich one given below. If Λ is a bounded open subset of $\mathbb{R}^d$, let $\Lambda_\omega = \Lambda \cap \mathcal{L}_\omega$

$$H_{\Lambda,\omega} = \sum_{x,y\in\Lambda_\omega} \left(h(\mathrm{T}^{-x}\omega, y-x)\, f^\dagger_{x,\sigma} f_{y,\sigma} + \kappa_{\alpha,\beta}(\mathrm{T}^{-x}\omega, y-x)\, b^\dagger_{x,\alpha} b_{y,\beta}\right) + \lambda \sum_{x\in\in\Lambda_\omega} (b^\dagger_x + b_x) f^\dagger_x f_x \,, \tag{60}$$

where $f^\dagger_{x,\sigma}, f_{x,\sigma}$ represent the creation-annihilation fermion operators at the site x with spin σ (electrons or holes), $b^\dagger_{x,\alpha}, b_{x,\alpha}$ represent the corresponding creation-annihilation boson operators (phonons) at x with polarization α, $h \in \mathcal{B}$ is the one-fermion Hamiltonian, $\kappa \in \mathcal{B}$ is the matrix of spring constants describing the phonon modes in the harmonic approximation, and λ is the coupling constant between phonons and fermions.

The main difficulty consists in getting a good approximation for the ground-state whenever the density of fermions is fixed and to have an efficient description of the excitation spectrum. This is usually far too complicated in comparison with the efficiency of other approximations when dealing with transport. Actually the electron-phonon coupling is very small so that the collision rate between them is often small enough so as to treat the phonon effect as an external noise. This is the essence of the kinetic models, starting with the Drude model [1], with its more elaborate version using the Boltzmann equations, as was proposed by Lorentz [23]. In this section only the so-called *Relaxation Time Approximation* (RTA) will be considered.

6.1 The Drude Model

In the original Drude approach, electrons in a metal are considered as free classical particles of mass m and charge $-e$. Let n_e be the electron density. The main idea is that while moving, a typical electron experiences collisions at random times with the effect of updating its momentum (see Fig. 6). If no external forces are imposed, such as an electric field, these collisions are supposed to enforce the thermal equilibrium, so that the exchange of energy between the electrons and the metal should be described by the Maxwell-Boltzmann distribution at the metal temperature T. Namely, the momentum p_n just after the nth collision is updated randomly according to the probability distribution

$$f(\boldsymbol{p}) = \frac{1}{(2\pi m k_B T)^{d/2}}\, e^{-p^2/2mk_BT} \,. \tag{61}$$

A more rigorous description of this model can given as follows

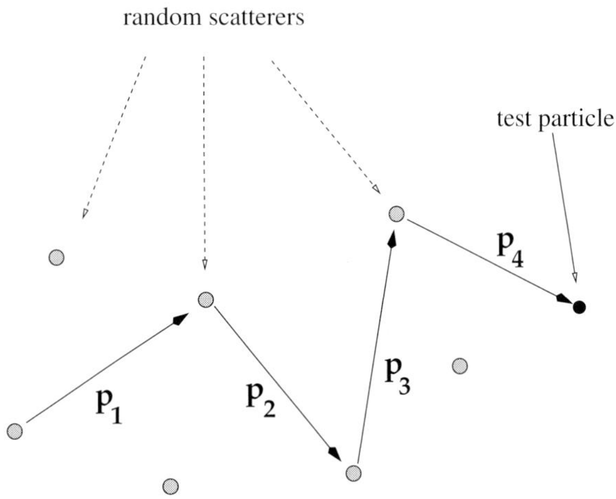

Fig. 6. The Drude model for electrons in metals

1. The collisions times are distributed according to the Poisson law: given a time interval I, the probability that N collisions occur within I is given by $c^N/N!e^{-N}$ where $c = |I|/\tau_{coll}$.
2. The collisions times are labelled by $\cdots < t_1 < 0 \leq t_0 < t_1 < \cdots t_n < t_{n+1} \cdots$. The average value of the time between two consecutive collisions is $\langle t_{n+1} - t_n \rangle = \tau_{coll}$.
3. As a consequence of the Poisson law, the random variables $\tau_n = t_n - t_{n-1}$ are independent identically distributed with common probability density given by the *exponential law*

$$\nu(d\tau) = e^{-\tau/\tau_{coll}} \frac{d\tau}{\tau_{coll}}, \qquad \text{on}[0, \infty) .$$

4. Between two collisions times the electron motion is the free motion of a point particle of mass m and charge $-e$, possibly perturbed by external forces, with initial condition $\boldsymbol{p}(t_n + 0) = \boldsymbol{p}_n$ and position $\boldsymbol{x}_n = \boldsymbol{x}(t_n + 0) = \boldsymbol{x}(t_n - 0)$ (continuous path).
5. The initial conditions $\boldsymbol{p}_n$ are independant random variables distributed according to the Maxwell-Boltzmann distribution (61) at temperature T.

It is an elementary calculation to compute the motion when a small constant and uniform electric field $\boldsymbol{\mathcal{E}}$ is applied to such an electron. For indeed the momentum increases between two consecutive collisions by the amount $\Delta \boldsymbol{p}_n = -\tau_n e \boldsymbol{\mathcal{E}}$. The average current is given by $\boldsymbol{j}_{el} = -en_e \langle \boldsymbol{p}/m \rangle$, so that, since the Maxwell-Boltzmann distribution is isotropic, the average comes entirely from the accelaration due to the electric field to give

$$\boldsymbol{j}_{el} = \sigma \boldsymbol{\mathcal{E}}, \qquad \sigma = \frac{n_e e^2}{m} \tau_{coll}, \qquad \textbf{(Drude)}$$

In much the same way it is also possible to compute the average energy given by the electrons to the metal through the collision process. This energy is transformed into heat. The energy gained by the metal per electron and per collision is given by the average value of the difference $(p(t_n-0)^2-p_{n-1}^2)/2m$. Due to the isotropy of the momentum distribution, the average gain is $\tau_{coll}^2 e^2 \boldsymbol{\mathcal{E}}^2/m$. To get the energy density W gained per unit time, it is enough to multiply this quantity by the electron density and to divise it by the average time between two collisions, giving

$$W = \sigma \boldsymbol{\mathcal{E}}^2 = \boldsymbol{j}_{el} \cdot \boldsymbol{\mathcal{E}}, \qquad \textbf{(Joules law)}$$

It is also possible to compute the heat conductivity supposing that, in a metal, the heat is transported by electrons alone, a reasonnable assumption since the conduction electron have an energy close to the Fermi energy so that their velocity is very high. To compute the heat conductivity it is necessary to assume that there is a small temperature gradient and no electric field. Considering a mesoscopic cylinder with axis parallel to $\boldsymbol{\nabla} T$ and cross-section $d\boldsymbol{\Sigma}$, the heat flux through the middle cross section located at $\boldsymbol{r}$ can be computed by remarking that the electrons arriving from the left experienced their last collision at the point $\boldsymbol{r}-\tau_{coll}\boldsymbol{p}/m$ so that their kinetic energy is given by $d/2k_BT(\boldsymbol{r}-\tau_{coll}\boldsymbol{p}/m)$. In much the same way the electron coming from the right have the energy $d/2k_BT(\boldsymbol{r}+\tau_{coll}\boldsymbol{p}/m)$. In the cylinder the number of such electrons arriving per unit time on each side is the product of the density n_e by the volume $d\boldsymbol{\Sigma}\cdot\boldsymbol{p}/m$ of the cylinder. If the cross section located at $\boldsymbol{r}$ is exactly in the middle of the cylinder, half of such electron come from each side giving a net energy flux

$$\boldsymbol{j}_U \cdot d\boldsymbol{\Sigma} = -k_B\tau_{coll}\frac{n_e \cdot d}{2}\frac{\boldsymbol{p}}{m} \cdot d\boldsymbol{\Sigma}\ (\boldsymbol{\nabla} T(\boldsymbol{r}) \cdot \boldsymbol{p}/m))\ .$$

Averaging over the momentum with respect to the Maxwell-Boltzmann distribution gives

$$\boldsymbol{j}_u = -\lambda \boldsymbol{\nabla} T, \qquad \lambda = \frac{d \cdot \tau_{coll} n_e}{2m} k_B^2 T\,.$$

It is therefore remarkable that the ratio between the electric and heat conductivities be given by a universal constant, leading to the *Wiedemann-Franz law*

$$\frac{\lambda}{\sigma} = \frac{d}{2}\left(\frac{k_B}{e}\right)^2 T, \qquad \textbf{(Wiedemann-Franz law)}\,.$$

The true success of the Drude theory came precisely from the microscopic explanation of this law that had been observed during the 19th century in most metals without explanation. Coming two years after the discovery of the electron by J. J. Thompson, the Drude theory was giving a more solid status to the electron theory. The main criticism, however, is that this theory requires an extra parameter, the collision time τ_{coll}, which is difficult to measure directly. For a long time it was computed through the Drude formula and the measurement of the electric conductivity. For the purpose of the qualitative theory that is presented here, the question of computing τ_{coll} will barely be addressed. But it is only through the method of N-body problem that such a calculation has some chance to be performed.

6.2 The Relaxation Time Approximation

The Drude model can be extended in various ways to take into account that the electronic motion is microscopically not a classical one. In this section the simplest extension will be proposed, namely the so-called *relaxation time approximation* (RTA). Proceeding by analogy, the electron in the aperiodic solid for which the atoms sit on a uniformly discrete set $\mathcal{L}$, with Hull Ω, will be considered as a Drude particle with motion described by the one particle Hamiltonian H given either by an element of $\mathcal{B}$ in the tight-binding approximation, or by a covariant family of selfadjoint operators affiliated to $\mathcal{A}$ in the continuum case. Since these electrons are considered as independent, it is possible to take care of the fermion statistics simply by considering that, at equilibrium, the electron distribution is the Fermi-Dirac density matrix

$$\rho_{\beta,\mu} = \frac{1}{1+e^{\beta(H-\mu)}} ,$$

where $\beta = 1/k_B T$ and μ is the chemical potential defined by the normalization condition

$$\mathcal{T}_{\mathbb{P}}\left(\frac{1}{1+e^{\beta(H-\mu)}}\right) = n_e .$$

In the tight-binding case, the formula is similar provided n_e is replaced by the number of electron per atomic site. The previous formula is meaningful when H is bounded from below. The evolution equation concerns the state of the electron. Since the electron is in the thermal bath created by the atoms (the atomic motion is described by phonons), it cannot be a pure state. Therefore let $\rho(t)$ be the state at time t. Then the motion will be described as follows

1. As in the Drude model, the collisions times are distributed according to the Poisson law with average collision time τ_{coll}.
2. Between two collisions times the evolution of the electron state is given by the Heisenberg equation associated with the one particle Hamiltonian H belonging to the relevant algebra of the Brillouin zone.
3. After each collision the particle state is updated to its equilibrium, namely $\rho(t_n + 0) = \rho_{\beta,\mu}$.

In such a simple model, the computation of the electric conductivity is straightforward. Let $\boldsymbol{\mathcal{E}}$ be a uniform constant electric field be applied to the system (on a mesoscopic scale $\boldsymbol{\mathcal{E}}$ can be considered as uniform indeed). Then the new Hamiltonian is given by $H_{\boldsymbol{\mathcal{E}}} = H - e\boldsymbol{\mathcal{E}} \cdot \boldsymbol{X}$. Unfortunately, the second term has no meaning as an element of the algebra $\mathcal{A}$. However, the equation of motion for the state involves only the commutator with $H_{\boldsymbol{\mathcal{E}}}$. If $\mathcal{L}_H$ denote the *Liouville operator* associated with H, namely $\mathcal{L}_H(A) = \imath[H, A]/\hbar$ for $A \in \mathcal{A}$, where $\hbar$ is the Planck constant, then $\mathcal{L}_{H_{\boldsymbol{\mathcal{E}}}} = \mathcal{L}_H + (e/\hbar)\boldsymbol{\mathcal{E}} \cdot \boldsymbol{\nabla}$. Therefore the Liouville operator defines a genuine $*$-derivation on the C^*-algebraleading to a well defined evolution. Thus, by duality, the evolution of the state is given by

$$\rho(t) = e^{-(t-t_{n-1})\mathcal{L}_{H_{\boldsymbol{\mathcal{E}}}}} \rho_{\beta,\mu} , \qquad t_{n-1} < t < t_n .$$

To compute the electric current, assuming, as before that only electrons contribute, it is enough to consider the current operator $\boldsymbol{J} = -(e/\hbar)\boldsymbol{\nabla} H$. Its average value will then be given by

$$\boldsymbol{j}_{el} = \int_0^\infty \frac{d\tau}{\tau_{coll}} e^{-\tau/\tau_{coll}} \mathcal{T}_{\mathbb{P}} \left(\rho(t_{n-1}+\tau)\boldsymbol{J}\right) .$$

Simple algebra gives, in the limit $\boldsymbol{\mathcal{E}} \downarrow 0$, the conductivity tensor in the form

$$\begin{aligned} \boldsymbol{j}_{el} &= \sigma \boldsymbol{\mathcal{E}} , \\ \sigma_{ij} &= \frac{e^2}{\hbar} \mathcal{T}_{\mathbb{P}} \left(\partial_j \rho_{\beta,\mu} \frac{1}{\hbar/\tau_{coll} - \hbar \mathcal{L}_H} \partial_i H \right) , \qquad \textbf{(Kubo's formula)} \end{aligned} \tag{62}$$

Using the current-current measure (see Eq. 55), the Kubo formula becomes

$$\sigma_{ij} = \frac{e^2}{\hbar} \int_{\mathbb{R}^2} dm(E,E') \frac{f_{\beta,\mu}(E) - f_{\beta,\mu}(E')}{E'-E} \frac{1}{\hbar/\tau_{coll} - \imath(E'-E)} , \tag{63}$$

where $f_{\beta,\mu}$ is the Fermi-Dirac distribution of energy. As the temperature T goes to zero, the collision time τ_{coll} is likely to diverge, so that $\hbar/\tau_{coll}$ decreases to zero, leading to testing more and more of the current-current measure near the diagonal.

If the external electric field is AC with frequency $\tilde{\omega}$, a very similar approach can be performed to compute the linear response cœfficient. It requires, however, some extra technicality, namely the notion of Floquet theory. The treatment can be found in [44] and the result is the following

Lemma 8. *Within the relaxation time approximation the AC (or optical) conductivity is given by the Kubo formula*

$$\sigma_{ij}(\tilde{\omega}) = \frac{e^2}{\hbar} \mathcal{T}_{\mathbb{P}} \left(\partial_j \rho_{\beta,\mu} \frac{1}{\hbar/\tau_{coll} - \hbar \mathcal{L}_H - \imath\tilde{\omega}} \partial_i H \right) , \qquad \textbf{(Kubo's formula)} \tag{64}$$

This approach is actually too simplist. This is because as already noticed in Remark 1, the definition of current should be *mesoscopic*. Here the current was defined microscopically for the electrons. This is possible because electric charges are supported by individual particles. However, this current does not take into account the phonon drag! This is because the latter cannot be defined microscopically. In addition it is not possible either to define the energy current microscopically for the same reason, because the notion of temperature is meaningless microscopically. This will be investigated in Section 7.2.

6.3 The Anomalous Drude Formula

Thanks to Eq.(63) it is now possible to investigate the scaling behaviour of the conductivity at low temperature. For indeed, as $T \downarrow 0$, collisions with phonons become unlikely, because it is more difficult to produce phonons. Thus the relaxation time τ_{coll} should diverge. There are two types of law for the way τ_{coll} scales with T. On the one hand, the Landau theory of Fermi liquid predicts $\tau_{coll} \sim T^{-2}$, due to electron-electron interactions. Actually, this electron-electron interaction is not due to direct Coulomb forces, as explained in [32]. For indeed, the direct Coulomb potential is much too small to explain the values of the low temperature conductivities in metals. Instead electrons interacts through phonons, the same force that leads to the BCS instability. This contribution is much bigger (at least 100 times bigger). On the other hand the Bloch theory for electron-phonon interaction gives $\tau_{coll} \sim T^{-5}$ [59]. The dominant contribution for metal is provided by the shortest time so that the Bloch term can never be seen. As will be seen below, for materials exhibiting subdiffusive coherent transport, the Bloch law will dominate at low temperature.

Using the Lemma 3 and Eq.(63), the behaviour of the DC conductivity as $\tau_{coll} \uparrow \infty$ will be given by the

Proposition 7 ([33]). *Within the relaxation time approximation, as the temperature decreases to zero, the DC conductivity behaves like*

$$\sigma \overset{\tau_{coll}\uparrow\infty}{\sim} \tau_{coll}^{2\beta_F-1}, \qquad \textbf{(anomalous Drude formula)}$$

if $\beta_F = \overline{\beta_2}(E_F)$ *is the diffusion exponent computed at the Fermi energy* E_F.

This formula was proposed on the basis of theoretical arguments by D. Mayou (see a review in [17]) and later supplemented by Sire (see [159]). The rigorous proof was proposed for the first time in [160]. This formula leads to the following discussion

1. For ballistic transport (at Fermi level), $\beta_F = 1$ and the conductivity behaves like in the Drude formula. In presence of several collision times $\tau_{coll}^{(1)}, \cdots, \tau_{coll}^{(n)}$, the resulting Poisson law correspond to a relaxation time given by the *Mathiessen rule* $1/\tau_{rel} = \sum 1/\tau_{coll}^{(i)}$, so that the shortest time dominates.
2. If states near the Fermi level are strongly localized, $\beta_F = 0$ and the previous formula predicts that $\sigma \sim 1/\tau_{coll}$. Such a behaviour can be called *anti-Drude*. This is not observed in practice, because such a formula indicates that the dominant contribution is given by the longest collision time (inverse Mathiessen rule [17]). But because of the Mott hopping transport (see Section 2.3) there is a spectrum of relaxation times with various weights modifying qualitatively this behaviour.
3. For quantum diffusion near E_F, $\beta_F = 1/2$ leading to a *residual conductivity*. This is actually the case for materials for which the Random Matrix Theory applies. This is also the result of the perturbation theory for the Anderson model in the weak localization regime [11]. In particular, if, in quasicrystals, the numerical calculations by [155] suggest that RMT applies to quasicrystals explaining why most of such material exhibit a residual conductivity at low temperature.

4. For subballistic but overdiffusive coherent transport near the Fermi level, the conductivity increases as the temperature decreases leading to a metallic behaviour. It is probably difficult to discriminate between such a behaviour and a normal metal. No example of such material is known.
5. For subdiffusive coherent transport, on the contrary, the conductivity decreases with the temperature, leading to an insulating state. Then Mathiessen rules are inverted [17]: the longest relaxation time dominates. It gives a chance for the Bloch law to show up in such systems.

6.4 Quantum Diffusion and Residual Conductivity

The case of quantum diffusion deserves a special treatment. The main result discussed in this section is the following

Theorem 11. *Let $H = H^*$ be a Hamiltonian affiliated to $\mathcal{A}$ such that its current-current correlation function is absolutely continuous in a neighbourhood $\mathcal{U}$ of the diagonal point (E_F, E_F) where E_F denotes the Fermi level. In addition it is assumed that this density is continuous in $\mathcal{U}$ and denoted by*

$$dm_{i,j}(E,E') = \rho^{(2)}_{i,j}(E,E')\,dE\,dE'\,, \qquad\qquad \rho^{(2)}(E,E') = \sum_{i=1}^{d} \rho^{(2)}_{i,j}(E,E')\,.$$

Then, whenever Δ is contained in a sufficiently small neighbourhood of E_F
(i) The diffusion exponents $\overline{\beta}_2^{\pm}(\Delta)$ are both equal to $1/2$.
(ii) The diffusion cœfficient $D(\Delta) = \lim_{t\uparrow\infty} L_{\Delta,2}(t)^2/t$ is finite and given by

$$D(\Delta) = \pi \int_{\Delta} dE\,\rho^{(2)}(E,E)\,.$$

(ii) If the relaxation time τ_{coll} tends to zero with the temperature, the diagonal part of the AC conductivity tensor admits a finite non zero limit at $T = 0$ given by

$$\sigma_{i,i}(\tilde{\omega}) \restriction_{T\downarrow 0} = \frac{e^2}{\hbar}\,\frac{\pi}{\tilde{\omega}} \int_{E_F-\tilde{\omega}}^{E_F} dE\,\rho^{(2)}_{i,i}(E,E+\tilde{\omega}) + O(\tilde{\omega}) \;\stackrel{\tilde{\omega}\downarrow 0}{\longrightarrow}\; \frac{\pi\,e^2}{\hbar}\,\rho^{(2)}_{i,i}(E_F,E_F)\,.$$

In particular the corresponding material admits a residual conductivity at zero temperature.

Remark 10. The definitions given in this section do not take units into account correctly. Since the trace per unit volume has the unit of L^{-d}, the expression giving $L_{\Delta,q}(t)$ has the unit of $L^{1-d/q}$. To get a length, one possibility is to multiply the expression (54) by the charge carrier density n_e. In much the same way, the definition of the current-current correlation measure (55) gives units $L^{2-d}E^2$ if E is an energy. Consequently the formulæ given in Theorem 11 are not expressed in the correct unit system. □

Sketch of the proof: this result is actually well known from the experts [33]. However it is not clearly published in the literature.

The claim (i) is an elementary consequence of proposition 6.

To prove (ii), the Eq. (54) and the DuHamel formula gives

$$L_{\Delta,2}(t)^2 = \int_{-t}^{+t} \frac{ds}{2t} \int_0^s du \int_0^s du' \, \mathcal{T}_{\mathbb{P}} \left(e^{\imath(u-u')H} \, \nabla H \, e^{-\imath(u-u')H} \, \Pi_\Delta \nabla H \right).$$

Using the current-current correlation measure, the integral over u, u' can be performed easily, leading to

$$= \int_{-t}^{+t} \frac{|s|ds}{2t} \int_\Delta dE_0 \int_{-\infty}^{+\infty} d\xi \, \rho^{(2)}(E_0 + \frac{\xi}{s}, E_0 - \frac{\xi}{s}) \left(\frac{\sin \xi}{\xi} \right)^2,$$

dividing by t and letting $t \uparrow \infty$ gives the expression for $D(\Delta)$.

To prove (iii) the Kubo formula (63) and its AC version (64) will be used. It should be remarked that, as $T \downarrow 0$, then $\beta = 1/k_B T \uparrow \infty$, while the chemical potential converges to the Fermi energy E_F. In addition, the relaxation time $\tau_{coll} \uparrow \infty$. Using the symmetry $\rho^{(2)}_{i,i}(E, E') = \rho^{(2)}_{i,i}(E', E)$, this gives

$$\sigma_{i,i}(\tilde{\omega}) \restriction_{T\downarrow 0} = \frac{e^2}{\hbar} \int_{E<E_F<E'} dE dE' \, \frac{\rho^{(2)}_{i,i}(E, E')}{E' - E} \left(\frac{1}{0^+ + \imath(E' - E - \tilde{\omega})} - \frac{1}{0^+ + \imath(E - E' - \tilde{\omega})} \right).$$

The classical result

$$\frac{1}{x - \imath 0^+} = \mathrm{PP}\frac{1}{x} + \imath\pi\delta(x),$$

where PP denotes the principal part, is the key point in getting the result. □

In [161,44] one example of a model was shown to satisfy the previous properties. This is the so-called N-orbital Wegner model [162], which, in the limit $N \uparrow \infty$ coincides with the coherent phase approximation (CPA) of the Anderson model [163]. Let $\mathcal{L} = \mathbb{Z}^d$ be the regular cubic d-dimensional lattice in $\mathbb{R}^d$. Let H_0 a periodic Hamiltonain of the form

$$H = H_0 + V, \qquad (H_0 \psi)(x) = \sum_{y \in \mathbb{Z}^d \setminus \{x\}} h(y - x) \, \psi(y),$$

$$(V \psi)(x) = V_x \, \psi(x),$$

where h is a sequence of complex numbers such that $\overline{h(x)} = h(-x)$ and $h(x) = O(|x|^{-d-1-\epsilon})$ for some $\epsilon > 0$. Here also the family $(V_x)_{x \in \mathbb{Z}^d}$ is a set of *free variables* with same distributions [164]. Free random variables are a non-commutative extension of independant random variables. A random variable V is a selfadjoint element of a C^*-algebra$\mathcal{A}$ together with a state Φ on $\mathcal{A}$. Its distribution is the probability measure μ_V defined by $\Phi(f(V)) = \int d\mu_V f$ for

$f \in \mathcal{C}_0(\mathbb{R})$. A family $(V_x)_{x\in\mathcal{L}}$ of such random (non commuting) variables is *free* if $\Phi(P_1(V_{x(1)})P_2(V_{x(2)})\cdots P_m(V_{x(m)})) = 0$ for any set of indices (including repeated ones) such that $x(i) \neq x(i+1)$ and every set of polynomials P_i's such that $\Phi(P_i(V_{x(i)})) = 0$ for all $i \in [1, m]$. In Wegner's model, let the C^*-algebragenerated by $f(V_x)$ ($f \in \mathcal{C}_0(\mathbb{R})$) be denoted by $\mathcal{B}_x$. Actually, $\mathcal{B}_x$ is isomorphic to $\mathcal{C}_0(\mathbb{R})$. Then if the V_x's are free, they generate a C^*-algebra$\mathcal{B}$ isomorphic to the infinite free product of the $\mathcal{B}_x$'s. For the Wegner model, $\mathcal{B}$ plays the rôle of the noncommutative Brillouin zone (namely the C^*-algebraof the transversal in the case of a uniform disctrete set). The shift $\mathrm{T}^a(V_x)_{x\in\mathbb{Z}^d} = (V_{x-a})_{x\in\mathbb{Z}^d}$ defines an action of $\mathbb{Z}^d$ by automorphisms. The existence of a common distribution μ to all V_x's defines a unique state Φ_μ on $\mathcal{B}$ such that the V_x's become free random variables [164]. The computation of the average Green function and of the average of a product of two Green functions can be done explicitly to show that such a model satisfies the hypothesis of Theorem 11.

Remark 11. The Anderson model (see Eq. (59)) at small coupling λ, and Gaussian potentials, can be obtained as a special case of the N-orbital model. In such an equivalence, $\mathbb{Z}^d$ is decomposed into blocks of size $O(\lambda^{-2})$ and labelled by $x \in \mathbb{Z}^d$. The Hamiltonian H is restricted to each such box and to an energy interval of size $O(\lambda^2)$. Let V_x the resulting matrix for the box labelled by x. Then, as shown by Magnen *et al.* [151], it is an $N \times N$ Gaussian random matrices where $N = O(\lambda^{-2(d-1)})$ is its dimension. Then the effective Hamiltonian obtained from the Anderson model, in an energy interval of size $O(\lambda^2)$ around the Fermi level, has the form of an N-orbital Wegner model. In the limit $\lambda \downarrow 0$ the V_x's becomes free and it is likely that the results mentionned above apply. This phenomena is the reason why Random Matrix Theory can be used so efficiently in the weak localization regime for disordered systems. □

6.5 Application to Quasicrystals

The main points concerning transport properties of quasicrystals have been given in Section 2.4. The previous discussion may indeed apply to quasicrystal. In a paper by Roche and Fujiwara [165] such discussion has been carried out quantitatively for realistic quasicrystals. Using the Bloch law for the behaviour of τ_{coll}, this gives

$$\sigma \sim T^{5(1-2\beta_F)}.$$

Using an LMTO calculation of the structure, it has been possible to get an estimate of the diffusion exponent for a decagonal $AlCuCo$, giving

$$\beta_F \approx .375 \qquad \Rightarrow \qquad \sigma \sim T^{1.25},$$

giving a scaling law compatible with the experimental data (see Fig. 3). Thus this a mechanism liable to explain the occurence of the scaling laws.

However, this law is not observed all the way down to $T = 0$ in most material. For indeed there is a residual conductivity (see $AlPdMn$ sample in Fig. 3). Several mechanisms may be responsible for such a residual conductivity. For indeed, the DOS of all quasicrystals exhibits a pseudogap exactly at the Fermi energy. At the energy scale of the order of $1eV$ this pseudo gap is created through the Hume-Rothery mechanism namely due to the coincidence of the first (pseudo)-Brillouin zone and the Fermi sphere, the Fermi see decreases its total energy by creating a pseudogap at the Fermi level [17]. However looking at smaller energy scale of the order of $50meV$ instead, tunneling junction experiments [166] show the existence of a pseudogap of the form predict by Altshuler-Ahoronov [167] (see Fig. 7): due to the residual electron-electron Coulomb interaction, the DOS vanishes exactly at the Fermi level like $d\mathcal{N}/dE(E) \sim |E - E_F|^{1/2}$.

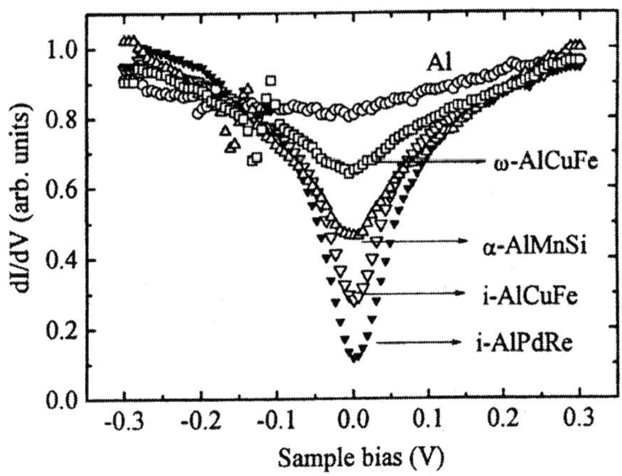

Fig. 7. DOS of quasicrystals measured from tunneling experiments [166]

Experiments show that the pseudogap is actually partially filled, depending on how clean the sample is. This suggests that, like in semiconductors, the low temperature conductivity is dominated by impurities. At very low level of impurities, however, the states in the pseudogap are likely to be localized, which should produce rather a Mott hopping conductivity. But even in very clean samples (with impurity concentration less than 10^{-5} [168]), a residual conductivity is observed.

It seems therefore that another mechanism should be at work. The existence of quantum chaos, seen in the level spacing distribution [155] suggests indeed that the Hamiltonian behaves like a Random Matrix of the Wigner ensemble. Therefore, according to the heuristic discussion in Section 6.3 and the results of Section 6.4, the diffusion exponent should be $\beta_F = 1/2$ at the Fermi level, leading, through the anomalous Drude formula, to a residual conductivity. This way a reasonning has, however, a drawback: as was remarked by Didier Mayou [169], the numerical computation of the level spacing distribution performed by [155] requires the use of a finite sample; it is legitimate to address the question of whether the Wigner-Dyson law is a finite size effect or not then. The answer to Mayou's objection is actually well known and was the purpose of Thouless approach to quantum chaos in mesoscopic systems [170]. In a finite size sample, there are indeed several energy or length scales:

1. At large temperature, namely on large energy scale, the effective Hamiltonian is simply the one-particle one and exhibits quantum diffusion with diffusion exponent β_F. However at low energy scale, the Hamiltonian looks like a random matrix.
2. If the sample size is L, the mean level spacing is given by $\delta E \sim L^{-d}$, since the Hilbert space has dimension $\sim L^d$. The time it takes for the quantum particle to realize that the spectrum is discrete, is the *Heisenberg time* $t_H = \hbar/\delta E \sim L^d$.
3. On the other hand, due to the anomalous quantum diffusion, the time it takes for the wave packet to reach the boundary of the sample, the so-called *Thouless time* t_{Th}, is given by $L \sim t_{Th}^{\beta_F}$, namely $t_{Th} \sim L^{1/\beta_F}$.
4. If the Thouless time is much larger than the Heisenberg time, in the limit of large samples, namely if $\beta_F d > 1$, then the system will never see the anomalous diffusion and the Wigner-Dyson distribution will dominate, leading to a quantum diffusion at low temperature with $\beta_F = 1/2$ and a residual conductivity. On the other hand if $\beta_F d < 1$, then the wave packet will never reach the boundary of the sample fast enough. Thus quantum diffusion dominates all the way to zero temperature.

Its is remarkable that the borderline between *Quantum Chaos* and *Quantum Diffusion* is exactly given by the identity

$$\beta_F d = 1 \qquad \Leftrightarrow \qquad \textbf{diffusion-chaos Thouless borderline}$$

For indeed, due to Guarneri's inequality (see Theorem 10), the quantum diffusion regime requires that the spectrum near the Fermi level, be *singular continuous*! This leads to interpret the difference between the *AlPdMn* and *AlPdRe* samples shown in Fig. 3 as due to the fact that the effective Hamiltonian in the latter has singular continuous spectrum near the Fermi level. Such a claim should be checked. For instance, a numerical *ab initio* calculation to compute an effective Hamiltonian in both sample, may be the way to check that such singular spectrum may occur.

7 Beyond the Relaxation Time Approximation

As pointed out in the previous section, the RTA already gives a good account for many phenomenons concerning the electronic transports in aperiodic media. However, the RTA, especially in the version presented in Section 6, is too limited in many respects. The most important limitation comes from its inability to explain the Mott transport by variable range hopping. For instance, as shown by the discussion in [35] about the accuracy of the plateaus in the Integer Quantum Hall effect, the Mott transport controls the thickness of the plateaus and explains why the measurement of the Hall resistance is so accurate. Besides, the way the Kubo formula was derived in the last Section 6 neglects the mesoscopic part of the current coming from other contributions, such as the phonon drag. In this section an extension of the RTA is proposed along the line drawn by the Drude model. This extension is still phenomenological in that the dissipation is represented by an external noise with many adjustable parameters. But these parameters are actually computable perturbatively from a model of interacting particles. This last computation will not be presented here. The goal here is to derive a compact formula for the transport cœfficients that includes, in principle, all possible terms that can be observed in any materials. The price to be paid for such a generality is the implict character of the parameters entering in the description of the noise terms. The results presented here are taken from [88,45,46].

7.1 The Quantum Jump Models

The Mott argument concerning the hopping conductivity of strongly localized charged particles induced by phonons suggests that instead of considering one type of relaxation time, a better model should include a family of relaxation times associated with various types of jumps. Let the Example 7 be considered again. Here $\mathcal{L}$ will be a uniformly discrete set in $\mathbb{R}^d$. Let H be a selfadjoint operator in the C^*-algebra$\mathcal{B}$ of the transversal. For simplicity H will be chosen in $\mathcal{B}_0$, so that for each $\omega \in X$, its representative $H_\omega = \pi_\omega(H)$ will have only a finite range: there is $R > 0$ such that $\langle x|H_\omega|y\rangle = 0$ for $|x - y| > R$ and all $\omega \in X$. Then $H_{\Lambda,\omega}$ denotes the restriction of H_ω to $\ell^2(\mathcal{L}_\omega \cap \Lambda)$. Let then J a finite set labelling an orthonormal basis of $\ell^2(\mathcal{L}_\omega \cap \Lambda)$ made of eigenvectors of $H_{\Lambda,\omega}$. For each pair (i,j) of elements of J, let $\Gamma_{i\to j}(\omega) \geq 0$ represents the rate of jumps from the state i to the state j. Then the Drude model will be extended as follows

1. Each jump processes occurs at random times $\cdots < t_{i\to j}^{(n)} < t_{i\to j}^{(n+1)} < \cdots$ that are Poisson distributed with an average time $\langle t_{i\to j}^{(n+1)} - t_{i\to j}^{(n)}\rangle = \Gamma_{i\to j}(\omega)^{-1}$.
2. The various jump processes are stochastically independent.
3. At each time $t_{i\to j}^{(n)}$ the collision experienced by the particle results in changing the wave function ψ, representing the instantaneous state of the particle before the collision, into $W_{i\to j}\psi$ (see the definition in Example 7).
4. Between two such collisions, the evolution of the state is provided by $H_{\Lambda,\omega}$.

As discussed in [45] such a model is not physical. For indeed this evolution is not unitary, which is normal, because the jump process being dissipative, cannot preserves the norm of the instantaneous state. But since this model is liable to represents the evolution of an electron (or hole) in the material, the average evolution should conserve the number of particles. Thus the square of the norm of the wave packet $\|\psi\|^2$ should have an average equal to one. This is possible only if the Hamiltonian $H_{\Lambda,\omega}$ is *renormalized* by adding an imaginary part $\imath K_{\Lambda,\omega}$ that will be fixed by demanding that

$$\langle \|\psi(t)\|^2\rangle = 1, \qquad \forall t.$$

It turns out that [45]

Theorem 12. *For the model defined above, the average evolution of the density matrix is given by the Lindbladian (50).*

The next problem is to figure out what happens in the infinite volume limit. This problem was also investigated in [45] in the special case of the variable range hopping (see Example 8). Then the set J is nothing but the set of sites in $\mathcal{L}_\omega \cap \Lambda$. In order to control the infinite volume limit it is necessary to impose some decay to large jumps. Typically such a decay should satisfy

Hypothesis 1. (Transition rates)
Γ.1)- *The transition rate is covariant:* $\Gamma_{x\to y}(\omega) = \Gamma_{x+a\to y+a}(\mathrm{T}^a\omega)$.
Γ.2)- *The map* $(\omega, x) \in \Gamma(X) \mapsto \Gamma_{0\to x}(\omega) \in \mathbb{R}_+$ *is continuous.*
Γ.3)- *The transition rate decays exponentially fast: there are constants* $r > 0$ *and* $C_\Gamma > 0$ *such that uniformly in* ω, $|\Gamma_{x\to y}(\omega)| \leq C_\Gamma \cdot e^{-r|x-y|}$. □

In much the same way, the jump operators should satisfy

Hypothesis 2. (Jump Operators)
W.1)- *The jump operators depend on* ω. *Moreover they are covariant in the following sense:* $T(a)W_{x\to y}(\omega)T(a)^{-1} = W_{x+a\to y+a}(\mathrm{T}^a\omega)$.
W.2)- *The maps* $(\omega, x) \in \Gamma(X) \mapsto \|W_{0\to x}(\omega) - \mathbf{1}\| \in \mathbb{R}_+$ *and* $(\omega, x, y, z) \mapsto \langle y|W_{0\to x}(\omega)|z\rangle \in \mathbb{C}$ *are continuous.*
W.3)- *The jump operators are localized exponentially near the jump from site* x *to site* y, *namely there is a constant* $C_W > 0$ *such that uniformly in* ω

$$|\langle y'|W_{x\to y}(\omega)|x'\rangle - (\delta_{x',y'}(1-\delta_{x,x'}) + \delta_{x,x'}\delta_{y,y'})| \;\leq\; C_W \cdot e^{-r(|x-x'|+|y-y'|)}\,.$$

In particular the $(W_{x\to y}(\omega) - \mathbf{1})$*'s are compact.* □

Note that in condition W.2 above, (ω, x, y, z) belongs to the closed subset of $X \times \mathbb{R}^{d\times 3}$ such that x, y, z all belong to $\mathcal{L}_\omega$. Then under such conditions it was proved in [45] that the instantaneous infinite volume limit evolution exists. However it is not unitary, nor even bounded. Nevertheless the averaged evolution is driven by the following

Proposition 8. *The Lindblad operator defined for* $A \in \mathcal{B}$ *by*

$$\mathfrak{L}_\omega(A) \;=\; \sum_{x,y\in\mathcal{L}_\omega} \Gamma_{x\to y}(\omega)\,\Big\{W_{x\to y}(\omega)^*\,A_\omega\,W_{x\to y}(\omega) - \frac{1}{2}\{W_{x\to y}(\omega)^* W_{x\to y}(\omega), A_\omega\}\Big\}\,, \tag{65}$$

defines a completely dissipative bounded map $\mathfrak{L}$ *on* $\mathcal{B}$. *In particular it generates a Markov semigroup on* $\mathcal{B}$

Sketch of the proof: (i) using the Hypothesis 1 and Hypothesis 2, it is elementary to check that $\mathfrak{L}_\omega(A)$ is covariant and strongly continuous in ω as soon a the series is shown to be convergent. Thus only convergence matters.

(ii) To prove convergence, let $K_{x\to y}(\omega) = (W_{x\to y}(\omega) - \mathbf{1})$. It is a compact operator that decays exponentially fast away from (x, y). Then $\mathfrak{L}_\omega$ can be decomposed into a sum of four terms:

$$\mathfrak{L}_\omega(A) = C_\omega(A) + \frac{1}{2}\{K^{(2)}_\omega, A_\omega\} + \frac{1}{2}([K^*_\omega, A_\omega] + [A_\omega, K_\omega]),$$

where $K, K^{(2)}$ are given by

$$K_\omega = \sum_{x,y\in\mathcal{L}_\omega} \Gamma_{x\to y}(\omega) K_{x\to y}(\omega),$$

$$K^{(2)}_\omega = \sum_{x,y\in\mathcal{L}_\omega} \Gamma_{x\to y}(\omega) K_{x\to y}(\omega)^* K_{x\to y}(\omega),$$

whereas C is defined by

$$C_\omega(A) = \sum_{x,y\in\mathcal{L}_\omega} \Gamma_{x\to y}(\omega) K_{x\to y}(\omega)^* A_\omega K_{x\to y}(\omega).$$

(iii) Clearly C defines a completely positive map, which is completely bounded if and only if $C_\omega(\mathbf{1}) = K^{(2)}_\omega$ is bounded. By covariance, it is enough to show that $K^{(2)}$ defines an element of $\mathcal{B}$ to conclude that C defines a completely positive bounded map from $\mathcal{B}$ into itself.
(iv) That K and $K^{(2)}$ defines elements of $\mathcal{B}$ is proved by proving following estimates

$$\|K\|_{\tilde{L}^{1,\infty}} = \sup_{\omega\in X} \sum_{x'\in\mathcal{L}_\omega} |\langle 0|K_\omega|x'\rangle| < \infty.$$

By the same argument it will then follow that $\|K^*\|_{\tilde{L}^{1,\infty}} < \infty$ showing that $\langle 0|K_\omega|x\rangle$ defines an element of $\mathcal{B}$ since $\|A\|_{\mathcal{B}} \leq \max\{\|A\|_{\tilde{L}^{1,\infty}}, \|A^*\|_{\tilde{L}^{1,\infty}}\}$. A similar estimates holds for $K^{(2)}$. Then, using the Hypothesis 1 and 2 it follows that

$$\|K\|_{\tilde{L}^{1,\infty}} \leq C_\Gamma \cdot C_W \sum_{x,y,x'\in\mathcal{L}_\omega} e^{-r(|x-y|+|y|+|x-x'|)} < \infty,$$

since $\mathcal{L}_\omega$ is uniformly discrete. □

Example 9. In the case of the Mott hopping conductivity (see Example 8)) the following model is relevant:
(i) Let $\mathcal{L}_0$ be a (periodic) lattice in $\mathbb{R}^d$ (for the case of semiconductors, $d = 3$ and $\mathcal{L}_0$ is the diamond lattice).
(ii) Considering a finite alphabet $\mathfrak{A}$ containing the letter 0 and at least another letter, let Ξ be the infinite product $\mathfrak{A}^{\mathcal{L}_0}$. This is a compact completely discontinuous space on which $\mathbb{Z}^d$ acts by the shift. In the case of Silicon say, 0 denotes the silicon sites,whereas the other letters denote the various impurities.
(iii) The probability $\mathbb{P}_\Xi$ is the product measure on Ξ obtained by assigning to each site the letter 0 with probability $(1 - c)$ and the letter $a \in \mathfrak{A} \setminus \{0\}$ the probability $c_a > 0$ such that $\sum_{a\neq 0} c_a = c \ll 1$.

(iv) The set of atomic sites is given by $\mathcal{L}_\xi = \{x \in \mathcal{L}_0; \xi_x \neq 0\}$ if $\xi \in X$.
(v) To each site $x \in \mathcal{L}_\xi$ is assigned a number (the occupation number) $n_x \in \{0,1\}$ that says whether the impurity located at x is ionized or not. Then the transversal X is given by the set of pairs $(\xi, \underline{n})$ where $\underline{n} \in \{0,1\}^{\mathcal{L}_\xi}$ with $\xi \in \Xi$. There is also an obvious action of $\mathbb{Z}^d$. Any $\omega \in X$ has a unique projection ξ on Ξ. So that $\mathcal{L}_\xi$ can actually be denoted by $\mathcal{L}_\omega$ without ambiguity.
(vi) The Hamiltonian describing electrons or holes in the limit of extreme localizations is given by a site potential $V_\omega(x)$ that satisfies the covariance condition. It will be assumed that X is endowed with a probability measure that factorizes through $\mathbb{P}_\Xi$ and makes the $V_\omega(x)$'s independent identically distributed random variables. It is also possible to add hopping terms of the form $\langle x|H_0|y\rangle = t(y-x) = t_0 O(e^{-r|x-y|})$ with t_0 small or r large enough to make this term a small perturbation of the potential term. It is important to remark that the current operator $\boldsymbol{J}$ is entirely coming from H_0, since the position operator commutes with the potential term. In particular there is no coherent transport if $H_0 = 0$. Then the transport will come entirely from the jump processes.
(vii) The jump rates will be given by the Mott argument as

$$\Gamma_{x \to y}(\omega) \;=\; \gamma \cdot e^{-(V_\omega(x) - V_\omega(y))/k_B T - |x-y|/\ell}\, n_x(1-n_y)\,,$$

where ℓ is the localization length of the electrons near the impurities and T the temperature. The term $n_x(1-n_y)$ means that x must contain a particle while the site y must be empty, otherwise, by Pauli's principle, the jump is impossible. If an external electric field is added the local potentials $V_\omega(x)$ are modified into $V_\omega(x) \pm e\boldsymbol{\mathcal{E}} \cdot \boldsymbol{x}$, depending upon the charge carriers are holes or electrons. □

7.2 Kubo's Formula

If there is a family of jump processes leading to the construction of a Lindbladian $\mathfrak{L}$ acting on the algebra $\mathcal{B}$ (or on $\mathcal{A}$) of the Brillouin zone, it is possible to compute the transport cœfficients through the following steps. First of all, the Lindbladian represents the evolution over a mesoscopic scale of time, since it is the result of an averaging over the noise. Then the various currents involve a velocity namely a derivative of the position in time. However, this derivative must be considered as mesoscopic. Therefore the velocity that should be taken into account is given by the following operator called *adiabatic current* [171]

$$\boldsymbol{J}_{ad} \;=\; \frac{d}{dt}\Phi_t(\boldsymbol{X}) \;=\; \frac{\imath}{\hbar}[H, \boldsymbol{X}] + \mathfrak{L}(\boldsymbol{X})\,. \tag{66}$$

In the previous expression, the index ω has been droped. This current is a sum of two terms. The first one $\boldsymbol{J}_{coh} = (\imath/\hbar)[H, \boldsymbol{X}]$ is the *coherent* part, the one that is defined microscopically and produces interference effects for the wave packet. The other $\boldsymbol{J}_{diss} = \mathfrak{L}(\boldsymbol{X})$ is the *dissipative* part and represents the part of the current induced by the jump processes. The phonon drag is contained into this dissipative part. The *electric current* corresponds to $\boldsymbol{J}_{el} = e\boldsymbol{J}_{ad}$ while the *energy current* will be given by $\boldsymbol{J}_U = 1/2\{\boldsymbol{J}_{ad}, H\}$.

Since $\boldsymbol{X}$ is neither bounded nor covariant it is not obvious that the right hand side defines an element of $\mathcal{B}$ or, at least, a selfadjoint operator affiliated to $\mathcal{B}$ (or to $\mathcal{A}$). This has to be checked. However by translation $\boldsymbol{X}$ transform into $\boldsymbol{X} + \boldsymbol{a}\mathbf{1}$. Since any Lindbladian $\mathfrak{L}$ satisfies $\mathfrak{L}(\mathbf{1}) = 0$, it follows that the lack of covariance of $\boldsymbol{X}$ is harmless, so that $\boldsymbol{J}_{ad}$ itself is covariant. In general, however, proving that $\boldsymbol{J}_{ad}$ is affiliated to the algebra will require more details about the model. In the case investigated in Section 7.1 the result is the following

Proposition 9. *Let $\mathfrak{L}$ be defined by (65) where the transition rates and the jump operators obeys to Hypothesis 1 and 2. Let $H = H^*$ be an element of $\mathcal{B}$ such that $\nabla H \in \mathcal{B}$. Then, all components of the adiabatic current $\boldsymbol{J}_{ad}$ are selfadjoint elements of $\mathcal{B}$.*

Proof: By inspection. The coherent part is given by $\boldsymbol{J}_{coh} = \imath[H, \boldsymbol{X}] = \nabla H$ and defines an element of $\mathcal{B}$ by hypothesis. The dissipative part must be estimated as in the proof of Proposition 8. The effect of $\boldsymbol{X}$ is to add terms of the form $|x-y|, |x|, |y|$ or $|x'|$ in front of the exponentials so that the convergences holds anyway. □

The current that is measured in experiment is obtained as follows: the mesoscopic evolution is provided by the Lindbladian in which some external force has been added, such as a (mesoscopic) gradient of the electric potential or of the temperature, in order to force the system out of equilibrium. Here, only the case of time independent external forces is considered. To first order in these external forces, the Lindbladian is given by $\mathfrak{L}_{eq} + \delta\mathfrak{L}$ in which the Hamiltonian part is included. The evolution equation for the density matrix $\rho = \rho_{\beta,\mu} + \delta\rho$ is then given by

$$\frac{d\delta\rho}{dt} = (\mathfrak{L}_{eq}^{\dagger} + \delta\mathfrak{L}^{\dagger})(\rho_{\beta,\mu} + \delta\rho)\,,$$

where $\mathfrak{L}^{\dagger}$ is the dual of $\mathfrak{L}$, namely the action on states instead of observables. A stationnary state is establish if, to first order in the external forces,

$$\delta\rho = -\mathfrak{L}_{eq}^{\dagger\,-1}\delta\mathfrak{L}^{\dagger}(\rho_{\beta,\mu})\,.$$

The thermal average of the current is given by

$$\boldsymbol{j}(t) = \mathcal{T}_{\mathbb{P}}\left(\rho(t)\boldsymbol{J}_{ad}\right)\,.$$

The equilibrium current should vanish and this is possible whenever $\mathfrak{L}_{eq}^{\dagger}\rho_{\beta,\mu} = 0$. Thus ρ can be replaced by $\delta\rho$. In addition the adiabatic current itself admits a part depending on the external forces, namely $\delta\boldsymbol{J} = \delta\mathfrak{L}(\boldsymbol{X})$. Moreover, the measurement is made when a stationary state has been reached, so that

$$\boldsymbol{j} = \mathcal{T}_{\mathbb{P}}\left(\delta\mathfrak{L}(\rho_{\beta,\mu})\,\frac{1}{\mathfrak{L}_{eq}}\,\boldsymbol{J}_{ad}\right) + \mathcal{T}_{\mathbb{P}}\left(\rho_{\beta,\mu}\,\delta\mathfrak{L}(\boldsymbol{X})\right)\,.$$

Developing, this formula exhibit five different terms as follows

$$\boldsymbol{j} = \mathcal{T}_{\mathbb{P}}\left((\mathcal{L}_{\delta H} + \delta\mathfrak{L}_0)(\rho_{\beta,\mu})\,\frac{1}{\mathcal{L}_H + \mathfrak{L}_0}\,(\nabla H + \mathfrak{L}_0(\boldsymbol{X}))\right) + ,\mathcal{T}_{\mathbb{P}}\left(\rho_{\beta,\mu}\,\delta\mathfrak{L}_0(\boldsymbol{X})\right)\,, \qquad \textbf{(Kubo)} \tag{67}$$

where H is the one-particle Hamiltonian, $\mathfrak{L}_0$ is the dissipative part of the Lindbladian at equilibrium. while δH and $\delta\mathfrak{L}_0$ are the linear contributions of the external forces to H and $\mathfrak{L}$.

Since this part of the work has not been completely discussed in the literature yet, no additional comment will be made here, leaving details for future research.

Acknowledgements

These notes are the result of research led since 1993 with various collaborators. I would like to thank especially H. Schulz-Baldes, D. Spehner who were driving forces in this program. I would also like to thank A. Barelli, C. Berger, F. Claro, I. Guarneri, D. Mayou, R. Mosseri, R. Rebolledo, C. Sire, J. Vidal, W. von Waldenfels, X.J. Zhong for their collaborations, suggestions and support all along these years. Let me also thank J. Magnen and V. Rivasseau for their effort to explain me their approach to Anderson's weak localization regime. Let me also thank the Institut Universitaire de France and I.H.É.S. for support while this work was written.

References

1. P. Drude, Ann. Phys., **1**, 566-613, (1900). Ann. Phys., **3**, 369-402, (1900).
2. *Out of the Crystal Maze: Chapters from the History of Solid State Physics*, by Lillian Hoddeson (Editor), Ernest Braun, Jurgen Teichmann, Spencer Weart (Editor), Frederick Seitz, Oxford Univ Press, (1992).
3. A. Einstein, Ann. Phys., **22**, 180-190 (1907).
4. M. Born, T. von Kármán, Phys. Zs., **13**, 297-309 (1912).
5. F. Bloch, Zeit. Phys., **52**, 555-600, (1928).
6. W. Pauli,Zeit. Phys., **31**, 765-783 (1925).
7. A. Wilson, Proc. Roy. Soc., **A133**, 458-491 (1931); Proc. Roy. Soc., **A134**, 277-287 (1931).
8. P. W. Anderson,Phys. Rev., **109**, 1492-1505 (1958).
9. D. J. Thouless, Phys. Rep., **13**, 93-142 (1974).
10. E. Abrahams, P. W. Anderson, D. C. Licciardello and T. V. Ramakrishnan, Phys. Rev. Letters, **42**, 673-676 (1979).
11. G. Bergmann, 107, 1-58 (1984).
12. Y. Imry, *Introduction to Mesoscopic Physics*, (Oxford University Press, 1997).
13. N. F. Mott, *J. Non-Crystal. Solids* **1**, 1 (1968). See also N. F. Mott, *Metal-Insulator Transitions* (Taylor and Francis, London, 1974).
14. B.I. Shklovskii & A.L. Efros, *Electronic Properties of Doped Semiconductors*, Springer, (1984).
15. D. Shechtman, I. Blech, D. Gratias, J. W. Cahn, Phys. Rev. Lett. **51**, 1951-1953 (1984).
16. in *Lectures on Quasicrystals*, F. Hippert & D. Gratias Eds., Editions de Physique, Les Ulis, (1994).
17. S. Roche, D. Mayou & G. Trambly de Laissardière, J. Math. Phys., **38**, 1794-1822 (1997).
18. J. Bellissard, D. Testard, *Quasi Periodic Hamiltonians: a Mathematical Approach*, in *Operator Algebras and Applications*, Coll. AMS, Vol.**2**, pp. 579, (1982).

19. L. Boltzmann, see e.g., *Leçons sur la Théorie des Gaz, Partie I*, Gauthiers-Villars, Paris (1902), reprint by J. Gabay Ed., Paris (1987).
20. R. Clausius, *Théorie Mécanique de la Chaleur*, Librairie Scientifique et Industrielle, Paris, (1868), reprinted by J. Gabay Ed., Paris (1991).
21. J. C. Maxwell, see [19], Preface by M. Brillouin.
22. M. Brillouin, in ref. [19], Preface.
23. H. A. Lorentz, *De beweging der electronen in de metalen*, *Verslagen der Koniglichen Akademie der Wetenschapen Amsterdam*, **13**, 493-508, 565-573, 710-719, (1905). (See also [2] Section 1.3.)
24. A. Einstein, Ann. d. Phyz., **17**, 549, (1905).
25. P. Langevin, C.R. Acad. Sci. (Paris), **146**, 530, (1908).
26. L. Onsager,Phys. Rev., **37**, 405-426, (1931); Phys. Rev., **38**, 2265-2279, (1931).
27. N. N. Bogoliubov, *Lectures on Quantum Statistics* (Gordon and Breach, 1967).
28. M. S. Green, J. Chem. Phys., **22**, 398-413 (1954).
29. R. Kubo,J. Phys. Soc. Japan, **12**, 570-586 (1957).
30. O. Lanford III, *On a derivation of the Boltzmann equation*, in *Nonequilibrium Phenomena 1: The Boltzmann Equation*, eds. J. L. Lebowitz, E. W. Montroll, North-Holland, Amsterdam, (1983).
31. D. Pines, P. Nozières, *The Theory of Quantum Liquids*, Addison-Wesley, (1966).
32. G. Mahan, *Many Particles Physics*, 2nd Printing, Plenum, (1990).
33. H. Schulz-Baldes, J. Bellissard, Rev. Math. Phys., **10**, 1-46, (1998).
34. R. Prange, S. Girvin Eds.,*The Quantum Hall Effect*, Springer,Berlin, (1990).
35. J. Bellissard, H. Schulz-Baldes, A. van Elst, J. Math. Phys., **35**, 5373-5471, (1994).
36. *ICQ7 Conference Proceedings*, Materials Science and Engineering A, **294-296**, (2000).
37. S. Takeuchi, T. Fujiwara Eds., *Quasicrystals*, World Scientific, (1998).
38. C. R. Wang, C. R. Lin, S. T. Lin, Y. Y. Chen, in *Quasicrystals*, pp. 583-586, World Scientific, (1998).
39. J. Delahaye, J. P. Brison, C. Berger, Phys. Rev. Letters, **81**, 4204-4207 (1998).
40. A. Connes, *Noncommutative Geometry*, Acad. Press., San Diego (1994).
41. J. Dixmier, Comptes Rendus Acad. Sc. Paris, **T. 262**, no. 20, 1107-1108 (1966).
42. J. Bellissard, *Gap Labelling Theorems for Schrödinger's Operators*, in *From Number Theory to Physics*, pp. 538-630, Les Houches March 89, Springer, J.M. Luck, P. Moussa & M. Waldschmidt Eds., (1993).
43. J. Bellissard, D. Hermmann, M. Zarrouati, *Hull of Aperiodic Solids and Gap Labelling Theorems*, in *Directions in Mathematical Quasicrystals*, CRM Monograph Series, Volume **13**, (2000), 207-259, M.B. Baake & R.V. Moody Eds., AMS Providence.
44. J. Bellissard, H. Schulz-Baldes, J. Stat. Phys., **91**, 991-1026, (1998).
45. D. Spehner, J. Bellissard,
46. J. Bellissard, R. Rebolledo, D. Spehner, W. von Waldenfels, *The Quantum Flow of Electronic transport I: The finite volume case*, in preparation (2002).
47. H. B. Callen, *Thermodynamics*, John Wiley & Sons, Inc., New-York, London, (1963).
48. B. Diu, C. Guthmann, D. Lederer, B. Roulet, *Physique Statistique*, Hermann, Paris, (1989).
49. H. J. Kreuzer, *Nonequilibrium Thermodynamics and its Statistical Foundations*, Clarendon Press Ed., Oxford, (1981).
50. D. K. Kondepudi, I. Prigogine, *Modern Thermodynamics*, John Wiley & Sons, Part IV, (1998).

51. D. Ruelle, *Statistical Mechanics*, Benjamin, (1969).
52. O. Brattelli & D. Robinson, *Operator Algebras and Quantum Statistical Mechanics*, Vol. I (1979) & Vol. II (1981), Springer Verlag, New York-Heidelberg-Berlin.
53. O. Lanford III, *Entropy and equilibrium states in classical statistical mechanics*, in *Statistical mechanics and mathematical problems*, Lecture Notes in Physics, **20**, 1-113, Springer, Berlin, (1973).
54. Haake F., *Statistical Treatment of Open Systems by Generalized Master Equations*, Springer tracts in modern physics **66**, Springer-Verlag, (1973).
55. H. Spohn, Rev. Mod. Phys., **53**, 569-615, (1980).
56. N. G. van Kampen, *Stochastic Processes in Physics and Chemistry*, second edition, North-Holland, (1992).
57. H. Spohn, *Large Scale Dynamics of Interacting Particles*, **342** Seiten, Texts and Monographs in Physics, Springer-Verlag, Heidelberg und Berlin, (1991).
58. C. E. Shannon,*A Mathematical Theory of Communication*, Bell System Technical Journal, **27**, pp. 379-423 and 623-656, (1948).
See the web site `http://cm.bell-labs.com/cm/ms/what/shannonday/paper.html`
59. N. W. Ashcroft, N. D. Mermin, *Solid State Physics*, Holt, Rinehart and Winston Eds., (1976).
60. C. I. Branden, J. Tooze, *Introduction to Protein Structure*, 2nd Edition, Garland Pub. Inc., New-York, (1998).
61. P. M. Pardalos, D. Shalloway, G. Xue Eds, *Global Minimization of Nonconvex Energy Functions: Molecular Conformation and Protein Folding*, A.M.S. DIMACS Series, **23**, (1995).
62. F. Jülicher, A. Adjari, J. Prost, Rev. Mod. Phys., **69**, 1269-1281 (1997).
63. Y. Avron, *Adiabatic Quantum Transport*, in *Les Houches 1994: Physique quantique mésoscopique*, pp. 741-790, Akkermans E., Montambaux G., Pichard J.-L., Zinn-Justin J. Eds., (Elsevier, 1995).
64. J. Bellissard, *K-Theory of C^*-algebras in Solid State Physics*, in *Statistical Mechanics and Field Theory, Mathematical Aspects*, T.C. Dorlas, M. N. Hugenholtz & M. Winnink Eds., Lecture Notes in Physics, **257**, 99-156, (1986).
65. J. Bellissard, R. Benedetti, J.-M. Gambaudo, *Spaces of Tilings, Finite Telescopic Approximations and Gap-labelling*, math.DS/0109062, submitted to Comm. Math. Phys., (2002).
66. J. C. Lagarias, P. A. B. Pleasants, *Repetitive Delone sets and perfect quasicrystals*, in math.DS/9909033, (1999).
67. J. L. Kelley, *General Topology*, Springer, New-York Heidelberg Berlin, reprint (1975).
68. J. Kellendonk, I. F. Putnam, *Tiling C^*-algebrasand K-theory*, in *Directions in Mathematical Quasicrystals*, CRM Monograph Series, Volume **13**, (2000), 177-206, M.B. Baake & R.V. Moody Eds., AMS Providence.
69. J. Renault, *A Groupoid Approach to C^*-Algebras*, Lecture Notes in Math., **793**, Springer, Berlin (1980).
70. I. Cornfeld, S. Fomin, Ya. G. Sinai, *Ergodic Theory*, Springer-Verlag, New York, (1982).
71. A. Katok, B. Hassenblatt, *Introduction to the Modern Theory of Dynamical Systems*, Cambridge University Press, (1995).
72. A. Connes, *Sur la théorie non commutative de l'intégration*, in *Algèbres d'Opérateurs*, Lecture Notes in Mathematics, **725**, pp. 19-143, Springer, Berlin (1979).
73. K. R. Parthasarathy, *Probability Measures on Metric Spaces*, Academic Press, New York (1967).

74. H. Jones, *The Theory of Brillouin Zones and Electronic States in Crystals*, North Holland, Amsterdam (1960).
75. M. J. O. Strutt, Ann. d. Phys, **84**, 485-506, (1927); Ann. d. Phys, **85**, 129-136, (1928).
76. F. Bloch, Z. Phys., **52**, 555-600, (1928).
77. R. Peierls, Ann. Phys., **4**, 121-148, (1930).
78. L. Brillouin, J. Phys. Radium,**1**, 377-400, (1930).
79. E. Wigner, F. Seitz, Phys. Rev., **43**, 804-810, (1933).
80. L. P. Bouckaert, R. Smoluchowski, E. Wigner, Phys. Rev., **50**, 58-67, (1936).
81. G. K. Pedersen, *C^*-Algebras and their Automorphism Groups*, Academic Press, London (1979).
82. J. Zak, Phys. Rev., **A 134**, 1602-1606, (1964).
83. B. Blackadar, *K-Theory for Operator Algebras*, Springer, New York (1986); 2nd ed., Cambridge University Press,Cambridge (1998).
84. M. Rieffel, (i) *Morita equivalence for operator algebras*, in *Operator algebras and applications*, ed. R. V. Kadison, Amer. Math. Soc., Providence, RI (1982), pp. 285-298; (ii) *Applications of strong Morita equivalence to transformation group C^*-algebras*, ed. R. V. Kadison, Amer. Math. Soc., Providence, RI (1982), pp. 299-310.
85. I. Segal, Ann. of Math., **57**, 401-457, (1953); Ann. of Math., **58**, 595-596, (1953).
86. J. Dixmier, *Les algèbres d'opérateurs dans l'espace hilbertien (Algèbres de von Neumann)*, Gauthier-Villars, Paris (1957); *Les C^*-algèbres et leurs représentations*, Deuxième édition, Gauthier-Villars Éditeur, Paris (1969).
87. M. Takesaki, *Theory of operator algebras. I.*, Springer-Verlag, New York-Heidelberg, (1979).
88. D. Spehner, *Contributions à la théorie du transport électronique dissipatif dans les solides apériodiques*, Ph. D. Thesis, Toulouse, (2000).
89. V. Georgescu, A. Iftimovici, *Crossed Product of C^*-algebrasand Spectral Analysis of Quantum Hamiltonians*, preprint 2002, to be published in *Commun. Math. Phys.*.
90. S. L. Woronowicz, Commun. Math. Phys., **136**, 399-432 (1991);
91. S. L. Woronowicz, K. Napiórkowsky, Rep. Math. Phys., **31**, 353-371 (1992).
92. B. Helffer, J. Sjöstrand, *Opérateurs de Schrödinger avec champs magnétiques faibles et constants*, Séminaire sur les Équations aux Dérivées Partielles, 1988–1989, Exp. No. **XII**, École Polytech., Palaiseau, (1989).
93. C. Gérard, A. Martinez, J. Sjöstrand, Comm. Math. Phys., **142**, 217-244, (1991).
94. H. Feshbach, Ann. Phys., **5**, 357-390 (1958); Ann. Phys., **19**, 287-313 (1962)
95. V. V. Grushin, Math. USSR, Sbornik, **13**, 155-185, (1971).
96. R. Carmona, J. Lacroix, *Random Schrödinger Operators*, Birkhaüser, Basel (1990).
97. H. L. Cycon, R. G. Froese, W. Kirsch, B. Simon, *Schrödinger operators, with application to quantum mechanics and global geometry*, (Springer Study ed.), Texts and Monographs in Physics. Berlin etc., Springer-Verlag, (1987).
98. L. Pastur, A. Figotin, *Spectra of Random and Almost-Periodic Operators*, Springer, (1992).
99. G. Lindblad, Comm. Math. Phys., **48**, 119-130, (1976).
100. W. F. Stinespring, Proc. Amer. Math. Soc., **6**, 211-216, (1955).
101. J. L. Doob, *Stochastic Processes*, John Wiley & Sons Inc., New-York London Sydney, (1953).
102. I. Guarneri, H. Schulz-Baldes, Math. Phys. Elec. J., **5**, # 1 (1999).

103. H. Schulz-Baldes, *Anomalous Transport in Aperiodic Media*, `mp_arc02-03`, Published in XIIIth *International Congress of Mathematical Physics*, pp. 199-205, A. Fokas, A. Grigoryan, T. Kibble, B. Zegarlinski Eds, International Press Boston (2001).
104. W. Rudin, *Real and Complex Analysis*, Third Edition, McGraw-Hill International Editions (1970).
105. K. Falconer, *Fractal Geometry-Mathematical Foundations and Applications*, John Wiley & Sons, (1990).
106. Y. Pesin, *Dimension Theory in Dynamical Systems*, Chicago University Press (1997).
107. O. Frostman, *Potential d'équilibre et Capacités des Ensembles avec quelques Applications à la Théorie des Fonctions*, *Meddel. Lunds univ. Math. Sem.*, **3**, 1-118 (1935).
108. J. X. Zhong, J. Bellissard, R. Mosseri, analysis of energy spectra scaling properties, J. Phys: Condens. Matter, **7**, 3507-3514 (1995).
109. R. del Rio, S. Jitomirskaya, Y. Last, B. Simon, J. d'Analyse Math., **69**, 153-200 (1996).
110. C. A. Rodgers, *Hausdorff Measures*, Cambridge University Press, Cambridge (1970).
111. C. D. Cutler, Suppl. Rend. Circ. Mat. Palermo, **28**, serie II, 319-339, (1992).
112. I. Guarneri, H. Schulz-Baldes, Lett. Math. Phys., **49**, 317-324, (1999).
113. J. M. Barbaroux, F. Germinet, S. Tcheremchantsev,C. R. Acad. Sci. Paris, **330** no 5, 409-414, (2000).
114. M. Reed, B. Simon, *Methods of modern mathematical physics*, Vol. I- IV, Academic Press, London, (1975).
115. J. Bellissard, *Non Commutative Geometry and Quantum Hall Effect*, Published in the *Proceedings of the International Conference of Mathematics, ICM94* (Zürich 94), Birkhäuser (1995).
116. J. Bellissard, *Non Commutative Methods in Semiclassical Analysis*, in *Lecture Notes in Mathematics*, **1589**, edited by S. Graffi, Springer-Verlag, Berlin, (1994).
117. I. Guarneri, Europhys. Letters, **10**, 95-100 (1989).
118. I. Guarneri, Europhys. Letters, **21**, 729-733 (1993).
119. Y. Last, J. Funct. Anal., **142**, 402-445 (1996).
120. J. M. Combes, *Connection between quantum dynamics and spectral properties of time evolution operators*, *Differential Equations with Applications to Mathematical Physics*, Aymes W.F., Harrell E.M., Herold J.V. Eds, Academic Press, Boston (1993).
121. H. Hiramoto, S. Abe, Phys. Rev., **A36**, 5349-5352 (1987).
122. H. Hiramoto, S. Abe, J. Phys. Soc. Japan, **57**, 230-240 (1988); *Dynamics of electrons in a quasiperiodic system. II. Harper's model*, *J. Phys. Soc. Japan*, **57**, 1365-1371 (1988).
123. T. Geisel, R. Ketzmerick, G. Petschel, Phys. Rev. Lett., **66**, 1651 (1991).
124. T. Geisel, R. Ketzmerick, G. Petschel, Phys. Rev. Lett., **67**, 3635 (1991).
125. R. Lima, D. L. Shepelyansky, Model of Quantum Kicked Rotator, Phys. Rev. Lett., **67**, 1377-1380 (1991).
126. T. Geisel, R. Ketzmerick, G. Petschel, *Unbounded quantum diffusion and a new class of level statistics*, in P. Cvitanovic, I. C. Percival, and A. Wirzba, editors, *Quantum Chaos - Theory and Experiment*, page 43. Kluwer Academic Publishers, Dordrecht, (1992).
127. R. Artuso, G. Casati, D. L. Shepelyansky, Phys. Rev. Lett., **68**, 3826-3829 (1992).

128. R. Ketzmerick, K. Kruse, S. Kraut, T. Geisel, Phys. Rev. Letters, **79**, 1959-1962 (1997).
129. M. Aizenman, S. Molchanov, Com. Math. Phys., **157**, 245 (1993).
130. M. Aizenman, Rev. Math. Phys., **6**, 1163 (1994).
131. A. Sütő, Commun. Math. Phys., **111**, 409-415 (1987).
132. A. Sütő, *Singular continuous spectrum on a Cantor set of zero Lebesgue measure for the Fibonacci Hamiltonian*, *J. Stat. Phys.*, **56**, 525-531 (1989).
133. J. Bellissard, B. Iochum, E. Scoppola, D. Testard, Commun. Math. Phys., **125**, 527-543 (1989).
134. J. Bellissard, *Spectral properties of Schrödinger's operator with a Thue-Morse potential*, in *Number theory and physics* (J.M. Luck, P. Moussa, M. Waldschmidt, eds.), Springer Proc. in Phys. **47** (1990).
135. J. Bellissard, A. Bovier, J. M. Ghez, Commun. Math. Phys., **135**, 379-399 (1991).
136. R. del Rio, S. Jitomirskaya, N. Makarov, B. Simon, Bull. Amer. Math. Soc., **31**, 208-212 (1994).
137. R. del Rio, N. Makarov, B. Simon, Commun. Math. Phys., **165**, 59-67 (1994).
138. S. Jitomirskaya, B. Simon, Commun. Math. Phys., **165**, 201-205 (1994).
139. B. Simon, Ann. of Math., **141**, 131-145 (1995).
140. A. Hof, O. Knill, B. Simon, Commun. Math. Phys., **174**, 149-159 (1995).
141. B. Simon, Proc. Amer. Math. Soc., **124**, 1177-1182 (1996).
142. G. Stolz, B. Simon, spectrum, V. Sparse potentials, Proc. Amer. Math. Soc., **124**, 2073-2080 (1996).
143. B. Simon, Commun. Math. Phys., **176**, 713-722 (1996).
144. A. Gordon, S. Jitomirskaya, Y. Last, B. Simon, Acta Math., **178**, 169-183 (1997).
145. C. Sire, Europhys. Lett., **10**, 483-488 (1990).
146. V. G. Benza, C. Sire, Phys. Rev., **B44**, 10343-10345 (1991).
147. B. Passaro, C. Sire, V. G. Benza,Phys. Rev., **B46**, 13751-13755 (1992).
148. J. X. Zhong, R. Mosseri, J. Phys.: Condens. Matter, **7**, 8383-8404 (1995).
149. J. Bellissard, H. Schulz-Baldes, J. Stat. Phys., **99**, 587-594 (2000).
150. B. Altshuler, B. I. Shklovskii, Sov. Phys. JETP, **64**, 127-135 (1986).
151. J. Magnen, G. Poirot, V. Rivasseau, *The Anderson Model as a Matrix Model*, Proceedings of *Advanced Quantum Field Theory* Conference, September 1996, Lalonde les Maures (France), `cond-mat/9611236`.
152. J. Magnen, G. Poirot, V. Rivasseau, *Ward type identities for the 2d Anderson model at weak disorder*, `cond-mat/9801217`.
153. T. Guhr, A. Müller-Groeling, H. A. Weidenmüller, Physics Reports, **299**, 189-425 (1998).
154. A. D. Mirlin, *Statistics of energy levels and eigenfunctions in disordered and chaotic systems: Supersymmetry approach*, Proceedings of the International School of Physics *Enrico Fermi*, Course CXLIII, Eds. G.Casati, I.Guarneri and U.Smilansky, pp.223-298, IOS Press, Amsterdam, (2000), (see `cond-mat/0006421`).
155. M. Schreiber, U. Grimm, R. A. Roemer, J. -X. Zhong, Comp. Phys. Commun., **121-122**, 499-501 (1999).
156. B. Simon, Commun. Math. Phys., **134**, 209-212 (1990).
157. R. del Rio, S. Jitomirskaya, Y. Last, B. Simon, Phys. Rev. Lett., **75**, 117-119 (1995).
158. J. Vidal, R. Mosseri, J. Bellissard, J. Phys. A: Math. Gen., **A32**, 2361-2367 (1999).
159. C. Sire, D. Gratias, Physics, Vol. **319**, p.314, Plenum London (1994).

160. J. Bellissard, H. Schulz-Baldes, *Anomalous transport in Quasicrystals*, in *Proceedings of the 5th International Conference on Quasicrystals*, Editors C. Janot, R. Mosseri, World Scientific, Singapore, (1995).
161. A. Khorunzhy, L. Pastur, Com. Math. Phys., **153**, 605-646 (1993).
162. F. Wegner, Phys. Rev., **B 19**, 783-792 (1979).
163. P. Neu, R. Speicher, J. Stat. Phys. **80**, 1279-1308 (1995).
164. D. Voiculescu, K. Dykema, A. Nica, *Free random variables*, AMS, Providence, Rhode Island, (1992).
165. S. Roche, T. Fujiwara, Phys. Rev., **B58**, 11338-11396 (1998).
166. D. N. Davydov, D. Mayou, C. Berger, C. Gignoux, A. Neumann, A. G. M. Jansen & P. Wyder, Phys. Rev. Letters, **77**, 3173-3176 (1996).
167. B. L. Altshuler, A. G. Aronov, Sov. Phys. JETP **50**, 968 (1979);
see also Altshuler B.L., in *Les Houches 1994: Physique quantique mésoscopique*, Akkermans E., Montambaux G., Pichard J.-L., Zinn-Justin J. Eds., (Elsevier, 1995).
168. C. Berger, private communication (september 2001). I thank Claire Berger for this information about his experiment.
169. D. Mayou, private discussion (1999). I thank Didier Mayou for pointing out this problem.
170. D. J. Thouless, Phys. Rev. Letters, **39**, 1167-1169 (1977).
171. I thank D. Spehner for pointing out this fact.

Scaling Limits of Schrödinger Quantum Mechanics

L. Erdős

School of Mathematics, GeorgiaTech, Atlanta, GA 30332, USA

Abstract. We outline the status of rigorous derivations of certain classical evolution equations as limits of Schrödinger dynamics. We explain two recent results jointly with H.T. Yau in more details. The first one is the derivation of the linear Boltzmann equation as the long time limit of the one-body Schrödinger equation with a random potential. The second one is the mean field limit of high density bosons with Coulomb interaction that leads to the nonlinear Hartree equation.

1 Introduction

The physics of charged particles governs all electric phenomena. From theoretical point of view, these particles can be accurately described using many-body quantum mechanics. In practice, however, the fundamental equation of quantum physics, the Schrödinger equation, is too complicated. The Schrödinger equation is usually replaced with much simpler equations that are computationally more feasible. These equations do not contain all information about the complex system, but they may describe certain quantities of interest with a sufficient precision. The goal is to find the correct approximating equations and to justify rigorously that they are consistent with the Schrödinger theory in certain limiting regimes.

The dynamics of N charged particles is given by the Schrödinger equation,

$$i\partial_t \Psi_t(x_1, \ldots, x_N) = H\Psi_t(x_1, \ldots, x_N) , \tag{1.1}$$

where Ψ_t is the wave function of the particles at positions $x_1, x_2, \ldots, x_N \in \mathbf{R}^d$. The Hamiltonian operator, H, may include interaction among the particles and their interaction with an external field. In general form it is given as

$$H = H_0 + H_{p-p} + H_{p-f} + H_f , \tag{1.2}$$

where

$$H_0 = \sum_{j=1}^{N} -\tfrac{1}{2}\Delta_{x_j} , \tag{1.3}$$

is the (nonrelativistic) kinetic energy;

$$H_{p-p} = \sum_{1\le i<j\le N} \Phi(x_i - x_j) , \tag{1.4}$$

is the two body interaction between the particles; H_{p-f} is the interaction of the particles with the external field and H_f is the field Hamiltonian. H is written in units where $\hbar = m = e = 1$, where m is the mass of the particles and e is their charge. In case of eletrons the unit length is therefore the Bohr radius and the energy is measured in 2 Rydbergs. These atomic units are usually referred to as *microscopic.*

In this lecture note we consider only static random potential field (impurities) and phonon field (lattice vibration). In case of a phonon field the space of wavefunctions must be extended to include also the dynamical phonon variables. Electromagnetic field may also be included by modifying the kinetic energy operator but we do not have results in this direction. We note, however, that it is easy to modify our results to include periodic background potential by changing the dispersion relation $e(k) = \frac{1}{2}k^2$ in (1.3).

The dynamics described above is usually too complex to solve directly. First, in macroscopic material samples the number of particles is big, $N \sim 10^{23}$. Second, the physically interesting time scale may be very long when expressed in atomic (microscopic) units. Finally, if impurities are included, the precise configuration of the environment is usually unknown.

A common difficulty is that memory and correlations are built up along the evolution of the system. Even if the particles are initially independent of each other and the field, the correlation effects appear immediately as the dynamics is turned on and they apparently become stronger as time goes on. However, in certain dynamics and in certain limiting regimes the correlations may disappear or may follow a coherent pattern which renders their description possible. In all these problems the key technical step is to control the correlations.

The most serious complication stems from the large number of particles if a particle-particle interaction is taken into account. If $H_{p-p} \equiv 0$, then the system can (essentially) be described by one particle dynamics. These one-body models will be used to study the long time effects and the influence of the random external fields (static or dynamic). The genuinely many-body models are much harder and they will be considered only without external fields ($H_{p-f} \equiv H_f \equiv 0$) and on microscopic time scale.

We need to assume various idealizations to hope that such complex Schrödinger evolution can be approximated by a simpler dynamics. The following simplifications may become useful:

(i) *Scale separation.* In reality there are often several orders of magnitude between the typical quantum lengthscale and the lengthscale of the measured observable. It is convenient to introduce two sets of space and time coordinates, (x, t), (X, T), called *microscopic and macroscopic variables*, and think of them to describe the same physical space-time in different natural units. In the *kinetic regime*, which is our main interest here, the space and time scales in the same way, hence $(X, T) = (\varepsilon x, \varepsilon t)$ for some dimensionless proportionality constant ε. For example, in case of macroscopic electronic devices of sizes of few centimeters it is natural to measure X in cm and x in Å, hence $\varepsilon = 1\text{Å}/1cm = 10^{-8}$. In order to prove rigorous theorems we will consider the idealized $\varepsilon \to 0$ *scaling limit*.

(ii) *Weakened interactions.* The total effect of interactions upon each particle increases with time and/or with the number of other particles present. Currently we are only able to treat models where the total interaction per particle during the whole time span is comparable with the free kinetic energy. This requires weakening the microscopic interactions usually by a small coupling constant in H_{p-p} and H_{p-f}.

(iii) *Partial information.* A simpler equation cannot represent the evolution of the full complexity of the N-body wavefunction. In most cases, fortunately, the physically interesting quantities can be computed from much simpler objects like the one or two particle density matrices or even from their coarse-grained versions. Mathematically, coarse-graining is described by the weak limit of the scale separation.

(iv) *Randomness* typically helps to reduce correlations. However, we will *not* consider stochastic dynamics as many phenomenological works do. The randomness models the impure medium, but once the random scenery is selected, it is quenched, and the dynamics is Hamiltonian.

In these lectures we will discuss the following models and their limiting equations:

I. One-body models ($H_{p-p} \equiv 0$)

(1) Semiclassical limit $\Rightarrow$ linear Vlasov equation

(2) Weak coupling limit of random fields $\Rightarrow$ linear Boltzmann equation

II. Many-body models ($H_{p-p} \neq 0$)

(1) High density mean field limit

(i) Fermionic particles $\Rightarrow$ nonlinear Vlasov equation

(ii) Bosonic particles $\Rightarrow$ nonlinear Hartree equation

(2) Weak coupling or low density limits $\Rightarrow$ nonlinear Boltzmann equation.

2 One-Body Models

In this section we consider the Schrödinger equation (1.1) for $N = 1$. The kinetic energy is $H_0 = -\frac{1}{2}\Delta$ and $H_{p-p} \equiv 0$.

2.1 Linear Vlasov Equation as Semiclassical Limit

Semiclassical limits are characterized by separating the quantum wavelength from the scale of variation of any other physical data in the problem. For the standard Schrödinger equation this amounts to a potential U and an initial state $\psi_0 = \psi_0^\varepsilon$ that varies on a lengthscale much bigger than the Planck scale (Fig. 1). The initial state may have a phase factor that oscillates on the short scale, but the wavelength is modulated on the large scale (WKB state).

In microscopic coordinates

$$i\partial_t \psi_t^\varepsilon(x) = \Big[-\tfrac{1}{2}\Delta_x + U(\varepsilon x)\Big]\psi_t^\varepsilon(x), \qquad \psi_0^\varepsilon(x) = \varepsilon^{d/2} A(\varepsilon x)\exp\Big(iS(\varepsilon x)/\varepsilon\Big)$$

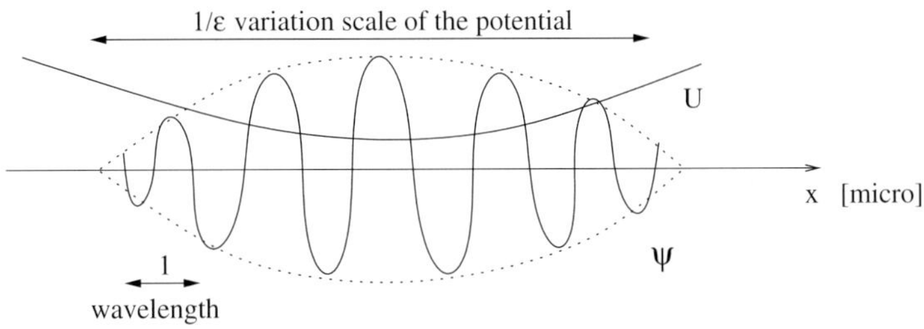

Fig. 1. Wavefunction and potential scales in semiclassics.

or, equivalently, in macroscopic coordinates

$$i\varepsilon\partial_T \Psi_T^\varepsilon(X) = \Big[-\tfrac{\varepsilon^2}{2}\Delta_X + U(X)\Big]\Psi_T^\varepsilon(X), \qquad \Psi_0^\varepsilon(X) = A(X)\exp\Big(iS(X)/\varepsilon\Big), \tag{2.5}$$

under the kinetic scaling $(x\varepsilon, t\varepsilon) = (X, T)$, $\varepsilon \ll 1$. Setting $\varepsilon = \hbar$ in the macroscopic formulation (2.5) gives the standard form of a semiclassical problem.

The key feature of this model is that the WKB form is preserved along the evolution modulo smaller order terms, i.e. $\Psi_T^\varepsilon(X) \approx A_T(X)\exp\Big(iS_T(X)/\varepsilon\Big)$ where the amplitude and the phase functions satisfy the usual Hamilton-Jacobi equations. Here we prefer the slightly more general Wigner formalism that can describe mixed states as well.

We define the *Wigner transform* $W_\psi(x,v)$, $x, v \in \mathbf{R}^d$, of a wavefunction $\psi \in L^2(\mathbf{R}^d)$ as

$$W_\psi(x,v) := \frac{1}{(2\pi)^d}\int_{\mathbf{R}^d} \overline{\psi}\Big(x+\tfrac{z}{2}\Big)\psi\Big(x-\tfrac{z}{2}\Big)e^{iv\cdot z}\mathrm{d}z\,. \tag{2.6}$$

W_ψ is real but in general not positive, hence it cannot be viewed as a genuine phase space density. However, its marginals are the usual configuration and momentum space densities of the wave function ψ

$$\int W_\psi(x,v)\mathrm{d}v = |\psi(x)|^2, \qquad \int W_\psi(x,v)\mathrm{d}x = |\widehat{\psi}(v)|^2.$$

More generally, the integrals of W_ψ on subsets of the phase space represent the phase space content of ψ up to the precision allowed by the uncertainty principle.

The macroscopic Wigner transform is defined by rescaling

$$W_\psi^\varepsilon(X,V) = \varepsilon^{-d}W_\psi(X/\varepsilon, V)\,.$$

Note that in the kinetic scaling the velocity is unscaled. For example, if $\psi = \psi^\varepsilon$ is a WKB state,

$$\psi^\varepsilon(x) = \varepsilon^{d/2}A(\varepsilon x)\,\exp\Big(iS(\varepsilon x)/\varepsilon\Big)\,,$$

then

$$\lim_{\varepsilon\to 0} W_{\psi^\varepsilon}^\varepsilon(X,V) = |A(X)|^2\delta(V - \nabla S(X))\,. \tag{2.7}$$

Theorem 1. *[30,18] Assume that the potential U, the amplitude and the phase functions, A, S, of the initial condition are smooth. Let $\Psi_T^\varepsilon(X)$ be the solution of (2.5). Then the weak limit*

$$W_T(X,V) := \lim_{\varepsilon\to 0} W^\varepsilon_{\Psi^\varepsilon_T}(X,V)\,,$$

of the rescaled Wigner transform of Ψ_T^ε exists for all $T \geq 0$ and it satisfies the linear Vlasov equation

$$(\partial_T + V\cdot\nabla_X)W_T(X,V) = \nabla_X U(X)\cdot\nabla_V W_T(X,V)\,.$$

Remark 1. The theorem is valid for more general initial states beyond the pure WKB states and this is where the Wigner formalism is advantageous. The essential conditions are compactness in phase space and the existence of a limiting macroscopic phase space density of the sequence of initial functions Ψ_0^ε. For the precise conditions see [30,18].

Remark 2. The result shows that the limiting dynamics is fully classical; there is no remnant of quantum mechanics in the limiting equation.

2.2 Linear Boltzmann Equation as Scaling Limit of Quantum Dynamics with Impurities

Impurities in metallic crystals are typically represented by a random potential $U(x)$ whose lengthscale is comparable with the wavelength rather than the lengthscale of the observable. Hence the semiclassical description is not valid any more.

With high concentration of impurities localization occurs [2,3,16,8], i.e., the spectrum of the one-body Hamiltonian $-\frac{1}{2}\Delta_x + U(x)$ is discrete and every initial state remains localized for all time. The basic problem is to study the low concentration regime where conduction occurs ("extended states problem"). In the first approximation, one neglects all interactions among electrons and the question reduces to a one-body Schrödinger equation with a random potential (Anderson model [3]). Except on the Bethe lattice where the space dimension is infinite [1], [23], [24], there are few mathematical results in this direction.

A more modest goal is to reduce the effect of the impurities and describe the dynamics up to the time scale when the total effect of impurities becomes comparable with the kinetic energy. Using the kinetic scaling, $(x\varepsilon, t\varepsilon) = (X,T)$, the effective strength of the impurities must therefore be scaled down by ε if we hope to describe the evolution up to time $t = T/\varepsilon$ for any fixed $T > 0$. We studied two models in their idealized limits: the *low density limit (LDL)* and the *weak coupling limit (WCL)*.

The random potential in the low density case is given by

$$U(x) = \sum_\alpha U_0(x - x_\alpha)\,, \qquad (LDL)\,, \tag{2.8}$$

where U_0 is a smooth localized potential (e.g. compactly supported) and $\{x_\alpha\}$ is an uncorrelated random point process with density $\varrho = \varepsilon$. In the weak coupling case we consider

$$U(x) = \lambda V(x)\,, \qquad (WCL)\,, \tag{2.9}$$

with a coupling constant $\lambda = \sqrt{\varepsilon}$, where V is a centered real Gaussian field, $\langle V(x)\rangle = 0$, with a smooth localized correlation function $G(x-y) := \langle V(x)V(y)\rangle$.

Our method is suitable for other potentials as well that have no long range correlations. For example, one can consider intermediate cases; the key condition is that the total collision effect, $\lambda^2\varrho t$, be ε-independent, i.e. $\lambda^2\varrho \sim \varepsilon$. Notice that in both cases the unscaled potential is given in microscopic variables. Therefore its variational lengthscale is comparable with the wavelength unlike in semiclassical case shown on Fig. 1. The typical LDL and WCL sceneries and the size of the typical quantum wavelength is pictured on Fig. 2.

The analogous models with *classical dynamics* have been studied earlier. In the weak coupling limit the velocity process of the classical dynamics is given by a Brownian motion on the constant energy surface [9,25]. This reflects the scaling limit of a particle subject to many small random "kicks" similarly as the random walk, appropriately rescaled, gives the Brownian motion.

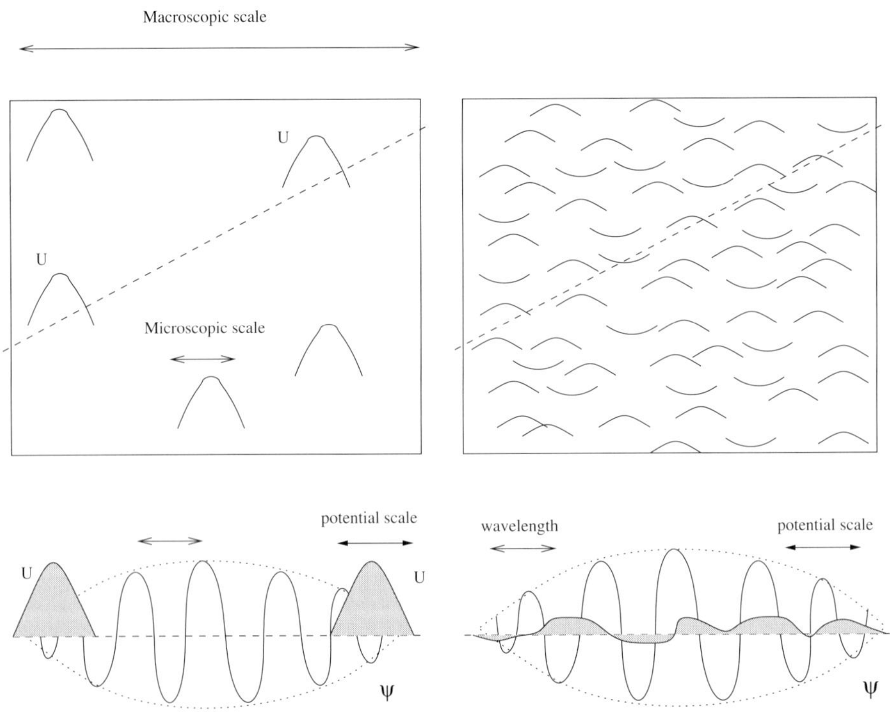

Fig. 2. Typical LDL (left and WCL (right) potential sceneries and a typical wavefunction in a one dimensional cross section along the dashed line.

In the low density limit the classical dynamics converges to the *linear Boltzmann equation* on macroscopic scales [6,17,33],

$$\Big(\partial_T + V\cdot\nabla_X\Big)F_T(X,V) = \int \Big[\sigma(U,V)F_T(X,U) - \sigma(V,U)F_T(X,V)\Big]\mathrm{d}U\,, \quad (2.10)$$

where $F_T(X,V)$ is the time dependent phase space density and $\sigma(U,V)$ is the collision kernel.

The Boltzmann equation is phenomenological. It describes a Markovian process of a freely traveling particle subject to random velocity changes. These collisions are governed by an independent random mechanism without memory and without direct reference to obstacles in configuration space. Unlike the Brownian dynamics, the Boltzmann equation is characterized by describing relatively few but strong "kicks".

For quantum dynamics we have the following result jointly with H.-T. Yau. A short description of the main ideas are found in [10].

Theorem 2. *([13,14]). Let the space dimension $d \geq 2$. Consider the random Schrödinger equation*

$$i\partial_t \psi_t^\varepsilon(x) = \Big[-\tfrac{1}{2}\Delta_x + U(x)\Big]\psi_t^\varepsilon(x)\,, \quad (2.11)$$

with $U(x)$ given in (2.8) or (2.9) with $\lambda = \sqrt{\varepsilon}$. Assume that the initial condition is a normalized WKB state

$$\psi_0^\varepsilon(x) = \varepsilon^{d/2}A(\varepsilon x)\exp\Big(iS(\varepsilon x)/\varepsilon\Big)\,, \quad (2.12)$$

with a smooth amplitude and phase function, in particular it has a limiting macroscopic Wigner distribution (see (2.7))

$$\lim_{\varepsilon\to 0} W_{\psi_0^\varepsilon}^\varepsilon(X,V) = |A(X)|^2\delta(V - \nabla S(X)) =: F_0(X,V)\,. \quad (2.13)$$

In $d=2$ we assume, additionally, that the set $\{(X,V)\ :\ V=0\} \subset \mathbf{R}^d\times\mathbf{R}^d$ has zero measure with respect to the probability measure $F_0(X,V)\mathrm{d}X\mathrm{d}V$. Then for any $T>0$ the statistical average of the limiting macroscopic Wigner transform of the wavefunction $\psi_{T/\varepsilon}^\varepsilon$ at time $t = T/\varepsilon$ exists in a weak sense,

$$F_T(X,V) := \lim_{\varepsilon\to 0}\Big\langle W^\varepsilon_{\psi^\varepsilon_{T/\varepsilon}}(X,V)\Big\rangle\,,$$

and it satisfies the linear Boltzmann equation (2.10) with an initial condition $F_0(X,V)$ and with collision kernel given as follows:

(i) In the LDL case $\sigma(U,V) = 4\pi|T(U,V)|^2\delta(U^2-V^2)$, where $T(U,V)$ is the scattering T-matrix of the single site potential U_0.

(ii) In the WCL case $\sigma(U,V) = 4\pi\widehat{G}(U-V)\delta(U^2-V^2)$.

Remark 1. Certain version of the WCL theorem was proved by Spohn for a sufficiently short time $T \leq T_0$ [34]. The result was extended to higher order correlation functions by Ho, Landau and Wilkins [21].

Remark 2. The theorem is valid for more general initial conditions that have macroscopic phase space density and have a fast uniform decay in momentum space.

Remark 3. The result is formulated for the average of the phase space density, but it is expected to hold for almost all configurations of the random potential.

Remark 4. We have derived an irreversible dynamics from a reversible one. This is, however, not a controversy since information is lost when the weak limit of the rescaled Wigner transform is taken. Microscopic details are neglected at this step, hence the wavefunction cannot be reconstructed from $F_T(X, V)$, in contrast to the semiclassical WKB situation.

Actually, there are many sequences of quantum wavefunctions that yield the same macroscopic phase space density $F_T(X, V)$. The time evolved quantum wavefunction $\psi^\varepsilon_{T/\varepsilon}$ contains delicate correlations with the environment. Therefore our theorem cannot be applied to the time reversed dynamics with initial datum $\widetilde{\psi}^\varepsilon_0 := \psi^\varepsilon_{T/\varepsilon}$ to hope to get ψ^ε_0 back as the backward time evolution, $\widetilde{\psi}^\varepsilon_{-T/\varepsilon} := e^{i(T/\varepsilon)H}\widetilde{\psi}^\varepsilon_0$ of $\widetilde{\psi}^\varepsilon_0$.

Remark 5. Notice that we obtained a classical kinetic equation as a scaling limit of a quantum dynamics. The quantum features of the microscopic models are retained in the collision kernel.

Remark 6. In the low density case both in quantum and in classical mechanics the limiting equation is the linear Boltzmann equation. However, in the weak coupling model the classical and quantum dynamics lead to different equations. In an environment of dense but weak impurities the classical particle suffers many small collisions. In the quantum collision process with a single obstacle, however, the wave function is decomposed into two parts: the free wave and the colliding wave. The free wave passes through as if the obstacle were not present. The colliding wave has a norm of order $O(\lambda^2)$, and its overlap with the free wave is small. The quantum particle rarely "chooses" the colliding wave, but when it does, then it evolves quite differently than the free wave. Therefore, effectively, the quantum particle suffers only a few but strong collisions which leads to the Boltzmann description.

The Boltzmann equation describes the first nontrivial regime beyond free dynamics. If the total effect of the collisions is stronger, or, equivalently, the time scale is longer, then a diffusive limiting dynamics is expected.

One often considers a two step limiting procedure where the kinetic scaling limit is followed by a *second* space/time rescaling of the limiting Markovian evolution (diffusion limit). For example, the long time behavior of the Boltzmann equation is given by a Fokker-Planck equation for the velocities [22]. This approach, however, may lead to an incorrect prediction on the true quantum evolution. Classical memory effects (recollisions) and quantum correlations that are irrelevant on the kinetic time scale may become important on the diffusive scale. The first scaling limit artificially suppresses them. The proper procedure is to consider the full microscopic quantum evolution on the time scale we are interested in, rescale the space appropriately and take only a single scaling limit.

In $d \geq 3$ dimensions we expect that the limiting equation is still diffusive but its diffusion coefficient differs from the one given by the two step limiting argument. In $d = 2$ the diffusive nature is predicted to change to localization on very large scales.

In solid state physics the diffusion coefficient is known to be related to the conductivity via the Einstein relation. Using lowest order perturbation theory in the potential, the conductivity is given by the classical Drude formula [26] as expected from a Markovian kinetic model like the Boltzmann equation. Quantum interferences can, however, substantially modify this picture. The celebrated *scaling theory* of conductance in disordered metals predicts metal-insulator transition in $d = 3$ dimensions and (weak) localization in $d = 2$ (see [31] for review). These arguments are supported by perturbation theory that analyzes higher order Feynman diagrams which have no counterparts in classical mechanics [29]. The so-called "maximally crossed" diagrams express the quantum backscattering effect which is responsible for reducing the classical conductance [31]. In $d = 2$ this effect is believed to lead to localization on exponentially large scales [37]. In $d = 3$ the classical (Boltzmann) conductivity is reduced by the inverse of the mean free path [31].

For definitiveness, we consider the weak coupling case with coupling constant $\lambda = \sqrt{\varepsilon}$. The mean square displacement of the quantum evolution is given by

$$S(t) := \int x^2 \langle |\psi_t(x)|^2 \rangle \mathrm{d}x \,. \tag{2.14}$$

Theorem 2 indicates that $S(T\varepsilon^{-1}) \sim \varepsilon^{-2}$ for any fixed $T > 0$ as $\varepsilon \to 0$, i.e. the Schrödinger dynamics (2.11) is ballistic on time scales $t \sim \varepsilon^{-1}$. We are interested in the behavior of $S(t)$ beyond this scale. The long time evolution of the Boltzmann equation (2.10) is diffusive with a classical Boltzmann diffusion coefficient given by

$$D_B := \lim_{T\to\infty} \frac{1}{T} \int X^2 F_T(X,V) \mathrm{d}X \mathrm{d}V \,.$$

In $d \geq 3$ we expect that the quantum evolution is still diffusive in the kinetic variables. To detect the quantum correction to the diffusion coefficient, we let $\eta > 0$ and we study the evolution up to time $t = \widetilde{T}\varepsilon^{-1-2\eta}$ where $\widetilde{T} > 0$. This corresponds to time $T \sim \varepsilon^{-2\eta}$ on the kinetic scale, hence the wave packet spreads to distances $X \sim \varepsilon^{-\eta}$. On atomic scales it corresponds to $x \sim \varepsilon^{-1-\eta}$. We expect that for any $\widetilde{T} > 0$

$$S(\widetilde{T}\varepsilon^{-1-2\eta}) = \Big(D_B - C_{d,\eta}\varepsilon^{\kappa_d} + o(\varepsilon^{\kappa_d}) \Big) \widetilde{T}\varepsilon^{-2-2\eta} \,, \qquad \varepsilon \to 0 \,. \tag{2.15}$$

The second term on the right is the correction to the classical diffusion with some $C_{d,\eta} > 0$ and $\kappa_d \geq 0$. Explicit computations of several Feynman diagrams show [31] that in $d \geq 3$ dimensions $\kappa_d = d-2$ and $C_{d,\eta}$ is independent of η. Thus the quantum diffusion coefficient is still given by D_B with corrections. In $d = 2$ one expects $\kappa_d = 0$, $C_{d,\eta} > 0$, i.e. the Boltzmann diffusion coefficient does not predict the quantum dynamics correctly beyond the kinetic time scale $t \gg \varepsilon^{-1}$. The diagrams yielding $C_{d,\eta}$ diverge as $\eta \to \infty$. In the physics literature this breakdown of the perturbation theory is commonly interpreted as an indication for localization in $d = 2$.

2.3 Linear Boltzmann Equation as Scaling Limit of Quantum Dynamics with Weakly Coupled Phonons

Let $c_k^\dagger, c_k$ be bosonic creation and annihilation operators representing phonons in mode $k \in \mathbf{R}^d$. The phonon dispersion relation is denoted by $\omega(k)$ and we assume that it is sufficiently small and smooth. The Hamiltonian of the phonon field is given by

$$H_f = \int \omega(k) c_k^\dagger c_k \mathrm{d}k \,,$$

acting on the bosonic Fock space $\mathcal{H}_f$.

We assume that the phonon field is weakly coupled to the charged particle. The interaction Hamiltonian is therefore given by

$$H_{p-f} = i\lambda \int Q(k) \Big[e^{-ikx} c_k^\dagger - e^{ikx} c_k \Big] \mathrm{d}k \,,$$

as a multiplication operator on the Hilbert space, $\mathcal{H}_p = L^2(\mathbf{R}_x^d)$, of the particle at x. The function $Q(k)$ models the details of the interaction and we assume that it is real, symmetric and decays sufficiently for large momenta.

The state of the system at time t is represented by a density matrix Γ_t acting on the joint Hilbert space of the phonons and the particle, $\mathcal{H}_f \otimes \mathcal{H}_p$. The time evolution is given by the Schrödinger equation for density matrix

$$i\partial_t \Gamma_t = [H, \Gamma_t] \,, \tag{2.16}$$

where $H = H_0 + H_f + H_{p-f}$ (see (1.2)) and $[A, B] = AB - BA$ is the commutator.

We assume that the phonons are initially in equilibrium at inverse temperature β and chemical potential μ, i.e., their state is given by

$$\gamma_f := Z^{-1} \exp\Big(-\beta H_f + \beta\mu \int c_k^\dagger c_k \mathrm{d}k \Big) .$$

Here Z is the normalization such that $\mathrm{Tr}_{\mathcal{H}_f}\, \gamma_f = 1$. If we start from a pure state ψ_0 for the particle, then its initial density matrix is $\gamma_p = |\psi_0\rangle\langle\psi_0|$, the projection matrix onto ψ_0 in $\mathcal{H}_p$. The total initial density matrix is $\Gamma_0 = \gamma_p \otimes \gamma_f$. Finally we are interested only in the evolution of the particle, therefore we integrate out the phonon variables and consider the partial trace $\gamma_t := \mathrm{Tr}_{\mathcal{H}_f} \Gamma_t$.

We again consider the kinetic scaling limit, $(X, T) = (x\varepsilon, t\varepsilon)$ with $\varepsilon = \lambda^2$ and we are interested in the macroscopic phase space profile of the reduced density matrix γ_t. The Wigner transform (2.6) conveniently generalizes to one particle density matrices γ as

$$W_\gamma(x, v) = \frac{1}{(2\pi)^d} \int_{\mathbf{R}^d} \gamma\Big(x + \tfrac{z}{2}, x - \tfrac{z}{2} \Big) e^{iv\cdot z} \mathrm{d}z \,,$$

and we again consider its rescaling

$$W^\varepsilon_\gamma(X, V) = \varepsilon^{-d} W_\gamma(X/\varepsilon, V) \,.$$

The statement of the theorem is analogous to Theorem 2. A short description of the proof is found in [11].

Theorem 3. *([12]) Let $d \geq 3$ and $\lambda = \sqrt{\varepsilon}$. Let Γ^ε_t solve the Schrödinger equation (2.16) with initial condition $\Gamma_0 = \gamma_p \otimes \gamma_f$, where the initial state $\psi_0 = \psi^\varepsilon_0$ of the particle is a WKB state (2.12). Then for any $T > 0$ the weak limit of the rescaled Wigner transform of the reduced density matrix, $\gamma_t = Tr_{\mathcal{H}_f} \Gamma_t$, at time $t = T/\varepsilon$ exists,*

$$F_T(X, V) := \lim_{\varepsilon \to 0} W^\varepsilon_{\gamma_{T/\varepsilon}}(X, V) \,,$$

and it satisfies the linear Boltzmann equation (2.10) with initial condition $F_0(X, V)$ (2.13) and collision kernel

$$\begin{aligned} \sigma(V, V) : = 2\pi |Q(V - V)|^2 \Big\{ & \Big(\mathcal{N}(V - V) + 1\Big) \delta\Big(\tfrac{1}{2}V^2 - \tfrac{1}{2}V^2 + \omega(V - V)\Big) \\ & + \mathcal{N}(V - V) \delta\Big(\tfrac{1}{2}V^2 - \tfrac{1}{2}V^2 - \omega(V - V)\Big) \Big\} \,. \end{aligned} \tag{2.17}$$

Here

$$\mathcal{N}(k) := Tr_{\mathcal{H}_f} \gamma_f c^\dagger_k c_k = \frac{1}{e^{\beta(\omega(k) - \mu)} - 1} \,,$$

is the expected number of phonons in mode k. The two terms in (2.17) correspond to phonon emission and absorption, respectively.

Remark 1. The evolution problem for the electron density matrix, $\gamma_t = \mathrm{Tr}_{\mathcal{H}_f} \Gamma_t$, is formally equivalent to the Schrödinger equation

$$i\partial_t \gamma_t = \Big[H_0 + \lambda V, \gamma_t\Big] \,,$$

with a time dependent Gaussian random potential $V(x, t)$ with covariance (in Fourier space)

$$\Big\langle \overline{\widehat{V}(p, t)} \widehat{V}(q, s) \Big\rangle = \delta(p - q) |Q(p)|^2 \Big[\Big(\mathcal{N}(p) + 1\Big) e^{i(t-s)\omega(p)} + \mathcal{N}(p) e^{-i(t-s)\omega(p)}\Big] \,. \tag{2.18}$$

This means that the formal perturbation expansions of these two problems coincide term by term. However, this connection is only formal and in our rigorous proof we cannot and do not make use of it.

Remark 2. Long time evolution of a *microscopically localized* electron weakly coupled to a phonon bath was studied in [7] in the dipole approximation (see also references therein). The limiting equation is diffusive (Fokker-Planck) already on the first nontrivial time scale (in the van Hove limit). In this case there is no Boltzmann equation before diffusion emerges. The diffusion mechanism is quite different; it is a resonance effect between certain phonon modes and the eigenfrequencies of the confinement.

3 Many-Body Models

3.1 Mean Field Limit at High Density

We consider N interacting quantum particles in a microscopic region of $\mathbf{R}^3$. We will consider the $N \to \infty$ limit. Since there is no scale separation in this model, the system is at high density. We neglect external fields, i.e. we consider only the kinetic energy term (H_0) and the pair interaction term (H_{p-p}) in the Hamiltonian (1.2). The Schrödinger equation is given by

$$i\partial_t \Psi_t(x_1, \cdots, x_N) = H_N \Psi_t(x_1, \cdots, x_N) , \tag{3.19}$$

with

$$H_N = \sum_{j=1}^{N} -\tfrac{1}{2}\Delta_{x_j} + \lambda \sum_{i<j}^{N} \Phi(x_i - x_j) . \tag{3.20}$$

The lengthscale of the potential is comparable with the size of the system, therefore every particle interacts with every other particle. There are $O(N^2)$ effective interaction terms in the Hamiltonian.

A weak coupling parameter λ is introduced in front of the interaction. Its size is chosen in such a way that the total kinetic and potential energies be of the same order of magnitude. This is the first nontrivial regime where the interaction influences the evolution.

The kinetic energy in the high density regime depends on the statistics of the particles. For *bosons* the total kinetic energy scales like $O(N)$ hence $\lambda = \frac{1}{N}$ is the correct coupling. For *fermions* the total kinetic energy is $O(N^{5/3})$ hence $\lambda = N^{-1/3}$ is chosen.

In both cases we assume that the one particle density matrix of the sequence of initial states $\Psi_0 = \Psi_0^{(N)}$ has a limiting phase space density $f_0(x, v)$ as $N \to \infty$, perhaps after appropriate rescaling. Let $\varrho_0(x) = \int_{\mathbf{R}^3} f_0(x, v) dv$ be the configuration space density. One expects that a limiting profiles, $f_t(x, v)$ and $\varrho_t(x)$, exist at later times as well. Each particle is then subject to the cumulative potential generated by all the other particles and this potential can therefore be approximated by $\Phi * \varrho_t$ at time t. The evolution of ϱ_t is given self-consistently as a one particle evolution in a self-generated mean field potential $\Phi * \varrho_t$.

This heuristics neglects possible correlations. The true wave function describes an N-body state. Even if the state was uncorrelated initially (i.e. product state for bosons or Slater determinant for fermions), it does not remain so at later times. The key technical difficulty is to show that all correlation effects are negligible in the $N \to \infty$ limit.

Nonlinear Vlasov equation as a high density limit of fermions. We consider the Schrödinger equation (3.19), (3.20) with $\lambda = N^{-1/3}$ acting on the fermionic wave functions, $\Psi \in \bigwedge_{j=1}^{N} L^2(\mathbf{R}^3)$. The kinetic energy per particle is $O(N^{2/3})$ hence the velocity is $O(N^{1/3})$ and the typical wavelength is $O(N^{-1/3})$.

Time and velocity must be rescaled to expect a limiting equation, space is unscaled. We define the weak limit of the rescaled Wigner transform of the one particle density matrix γ_t:

$$f_T(X,V) := \lim_{N\to\infty} W_{\gamma_{T/N^{1/3}}}(X, VN^{1/3}), \tag{3.21}$$

and let $\varrho_T(X) := \int f_T(X,V)\mathrm{d}V$ be the configuration space density.

The spatial variation of the effective mean field potential $\Phi_{eff} := \Phi * \varrho_T$ is $O(1)$. Since the wavelength is much shorter than the potential scale, this is effectively the semiclassical regime and the limiting equation is the nonlinear self-consistent Vlasov equation

$$(\partial_T + V\cdot\nabla_X) f_T(X,V) = \nabla\Phi_{eff}(X)\cdot\nabla_V f_T(X,V)\,, \qquad \Phi_{eff} := \Phi * \varrho_T\,. \tag{3.22}$$

The existence of the limit (3.21) and equation (3.22) have been shown by Narnhofer and Sewell [32] for analytic interaction potential Φ and localized initial data with finite kinetic energy and without correlation. Spohn [35] has generalized this result for potentials satisfying $\int |\widehat{\Phi}(k)|k^2 dk < \infty$ using a different method. There is no rigorous result for the Coulomb potential $\Phi(x) = |x|^{-1}$ yet.

In this model the equation is fully classical, but the fermionic quantum statistics dictates the proper rescaling of time and velocity.

Nonlinear Hartree equation as a high density limit of bosons. We consider the Schrödinger equation (3.19), (3.20) with $\lambda = N^{-1}$ acting on symmetric (bosonic) wave functions $\Psi \in \bigotimes_{j=1}^N L^2(\mathbf{R}^3)$. The kinetic energy per particle is $O(1)$, hence no time or velocity rescaling is needed. The equation remains fully quantum, but the many-body potential can be replaced by a self generated one-body potential if the particles have negligible correlations.

We assume that the initial state is a product function,

$$\Psi_0(x_1,\ldots,x_N) = \prod_{j=1}^N \psi_0(x_j)\,,$$

and we expect that such factorization remains approximatively true at later times as well,

$$\Psi_t(x_1,\ldots,x_N) \approx \prod_{j=1}^N \psi_t(x_j)\,. \tag{3.23}$$

In case of factorized states, the potential at any point $x \in \mathbf{R}^3$ is given by

$$x \mapsto \frac{1}{N}\sum_{j=1}^N \Phi(x - x_j)|\psi(x_j)|^2 \approx \Phi_{eff}(x) := (\Phi * |\psi|^2)(x)\,,$$

by the law of large numbers in the limit $N \to \infty$. Therefore the effective one-body equation governing the evolution of ψ_t is the *nonlinear Hartree equation*

$$i\partial_t\psi_t = -\tfrac{1}{2}\Delta\psi_t + \left(\Phi * |\psi_t|^2\right)\psi_t\,, \tag{3.24}$$

which is special type of nonlinear Schrödinger equations.

The key difficulty is the proof of the approximation (3.23). It is a fundamental question of many body dynamics to find an effective way to compare two normalized N-body wavefunctions, Ψ and Ψ'. Typically, the natural $L^2(\mathbf{R}^{3N})$ norm, $\|\Psi - \Psi'\|_2$, is too strong. For example, let f, f' and g be orthogonal normalized one-body states and $\Psi := \mathcal{S}\Big[f \otimes \bigotimes_{j=2}^N g\Big]$, $\Psi' := \mathcal{S}\Big[f' \otimes \bigotimes_{j=2}^N g\Big]$ where $\mathcal{S}$ is the symmetrization operator. Then all but one electrons are in the same state, nevertheless $\|\Psi - \Psi'\|_2 = \|f - f'\|_2 = \sqrt{2}$. However the normalized (Tr $\gamma = 1$) one particle density matrices,

$$\gamma_\Psi^{(1)} = \frac{1}{N}\,|f\rangle\langle f| + \frac{N-1}{N}\,|g\rangle\langle g|\,, \qquad \gamma_{\Psi'}^{(1)} = \frac{1}{N}\,|f'\rangle\langle f'| + \frac{N-1}{N}\,|g\rangle\langle g|\,,$$

are close in trace norm; $\mathrm{Tr}\Big|\gamma_\Psi^{(1)} - \gamma_{\Psi'}^{(1)}\Big| = O(N^{-1})$.

Unfortunately, there is no closed equation for the evolution of the one particle density matrix of an interacting N-body system. Therefore we have to introduce a whole hierarchy of density matrices and a system of coupled equations for their evolution, called the BBGKY-hierarchy (see (3.25) below).

We define the k-particle density matrix of the N-body wave function Ψ_t at time t as

$$\gamma_{N,t}^{(k)}(X_k; X_k') := \int \Psi_{N,t}(X_k, Y_{N-k})\overline{\Psi}_{N,t}(X_k', Y_{N-k})\mathrm{d}Y_{N-k}\,, \qquad 1 \le k \le N\,,$$

where $X_k = (x_1, x_2, \ldots, x_k)$ and $Y_{N-k} = (x_{k+1}, \ldots, x_N)$. Notice that $\gamma_{N,t}^{(N)} = |\Psi_{N,t}\rangle\langle\Psi_{N,t}|$ is the full density matrix for $k = N$. Moreover, for the initial product function $\Psi_0 = \prod_{j=1}^N \psi_0(x_j)$ the k-particle density matrices also factorize

$$\gamma_{N,0}^{(k)} = \bigotimes_{j=1}^{k} \gamma_{\psi_0}\,,$$

with $\gamma_{\psi_0} := |\psi_0\rangle\langle\psi_0|$.

The *N-body BBGKY-hierarchy* is the following system of equations for the sequence of density matrices $\Big(\gamma_{N,t}^{(1)}, \gamma_{N,t}^{(2)}, \ldots, \gamma_{N,t}^{(N)}\Big)$

$$i\partial_t \gamma_{N,t}^{(k)} = \Big[H_N^{(k)}, \gamma_{N,t}^{(k)}\Big] + \frac{N-k}{N} B_k \gamma_{N,t}^{(k+1)}\,, \qquad k = 1, 2, \ldots N\,, \tag{3.25}$$

where

$$H_N^{(k)} := \sum_{j=1}^{k} -\tfrac{1}{2}\Delta_{x_j} + \frac{1}{N}\sum_{i<j}^{k} \Phi(x_i - x_j)\,,$$

is the evolution of k particles and B_k is the collision operator

$$(B_k\gamma^{(k+1)})(X_k; X_k') := \sum_{j=1}^{k} \int \Big[\Phi(x_j - z) - \Phi(x_j' - z)\Big]\gamma^{(k+1)}(X_k, z; X_k', z)\mathrm{d}z\,,$$

acting on a $(k+1)$-particle density matrix $\gamma^{(k+1)}$.

These equations are obtained easily from the full N-body Schrödinger equation

$$i\partial_t \gamma_{N,t}^{(N)} = \Big[H_N, \gamma_{N,t}^{(N)} \Big] ,$$

by integrating out $N-k$ variables. In fact, the BBGKY-hierarchy (3.25) looks like a tautology, since all equations for $k < N$ easily follow from the last equation for $k = N$. Nevertheless it enables us to pass to the $N \to \infty$ limit, $\gamma_{N,t}^{(k)} \to \gamma_{\infty,t}^{(k)}$, and to obtain an *infinite* BBGKY-hierarchy, at least formally

$$i\partial_t \gamma_{\infty,t}^{(k)} = \Big[\sum_{j=1}^{k} -\tfrac{1}{2}\Delta_{x_j}, \gamma_{\infty,t}^{(k)} \Big] + B_k \gamma_{\infty,t}^{(k+1)} , \qquad k = 1, 2, \ldots . \tag{3.26}$$

The solution can be written in integral form as

$$\gamma_{\infty,t}^{(k)} = U(t)\gamma_{\infty,0}^{(k)} - i \int_0^t U(t-s) B_k \gamma_{\infty,s}^{(k+1)} \mathrm{d}s \qquad k = 1, 2, \ldots , \tag{3.27}$$

where $U(t)\gamma := e^{-itH_0} \gamma e^{itH_0}$ is the free evolution on any density matrix γ. Notice that there is no "last" equation in this hierarchy, but it still may give nontrivial information about the evolution of k-particle density matrices. In particular, it is easy to check that if ψ_t satisfies the nonlinear Hartree equation (3.24), then the product k-particle density matrices, $\bigotimes_1^k \gamma_{\psi_t}$ satisfy the infinite hierarchy (3.26).

Now we are ready to state the main theorem joint with H.T. Yau.

Theorem 4. *[15] Let $\Phi(x) := \pm |x|^{-1}$ be the attractive or repulsive Coulomb potential. Assume that the initial state is factorized, $\Psi_0 = \prod_{j=1}^N \psi_0(x_j)$ with $\psi_0 \in H^2(\mathbf{R}^3)$ and let ψ_t be the solution of the nonlinear Hartree equation (3.24) with initial condition $\psi_{t=0} := \psi_0$. Let $\Psi_{N,t}$ be the solution of the Schrödinger equation (3.19) with $\lambda = N^{-1}$ and let $\gamma_{N,t}^{(k)}$ be the k-particle density matrix at time t for $k = 1, 2, \ldots, N$. Then for each fixed k*

$$\lim_{N\to\infty} \gamma_{N,t}^{(k)} = \bigotimes_{j=1}^{k} \gamma_{\psi_t} , \tag{3.28}$$

in the weak topology of $L^\infty(\mathbf{R}_+, \langle t\rangle^{-1}\mathrm{d}t, \mathcal{L}^1(H^{\otimes k}))$, where $\langle t \rangle = (1+t^2)^{1/2}$. In particular this means that for any compact operator K acting on the k-particle space $\bigotimes_1^k L^2(\mathbf{R}^3)$ we have*

$$\lim_{N\to\infty} Tr\Big[K\gamma_{N,t}^{(k)} \Big] = Tr\Big[K\Big(\bigotimes_{j=1}^{k} \gamma_{\psi_t} \Big) \Big] .$$

The first result in this direction was proven by Hepp [20]. This work formulated an analogous result for field operators and the interacting potential Φ needed to be differentiable. Spohn [36] generalized the result for any bounded potential by iterating the infinite BBGKY hierarchy. The basic idea was to control this expansion in trace norm. Ginibre and Velo [19] have greatly extended these results to allow singular potentials (including Coulomb). However, they worked in a second quantized framework and they used a special quasi-free initial state that describes certain excitations around the product state. In particular, this method does not apply to wave functions with a fixed particle number.

A different route was followed in [4]. They showed that any w^*-limit point of the solutions to the N-body hierarchy (3.25) satisfies the infinite hierarchy (3.26). This is valid under very general conditions on the potential. However, to conclude the limiting statement (3.28) one needs to prove that the infinite hierarchy (3.26) has a unique solution. This would guarantee that any limit point must be equal to the known product solution to the infinite hierarchy, in particular the limit exists. This was proven only for bounded potential in [4] by controlling the expansion in trace norm, similarly to Spohn's method [36]. To cover the case of the Coulomb potential, one needs a stronger *a priori* control and this is the content of [15] (see also the expository paper [5]).

Basic idea of the proof. The basic problem is to find the appropriate normed space of density matrices such that

(i) $\lim_{N\to\infty}\gamma_{N,t}^{(k)}$ exists in this norm for each fixed $k = 1, 2, \ldots$ and the limit satisfies the infinite BBGKY-hierarchy (3.26);

(ii) the infinite BBGKY-hierarchy (3.26) has a unique solution in this space.

From (i) and (ii) the statement of the Theorem easily follows. Note that these two questions are related: existence of the limit is easier in a weaker norm, while uniqueness is more likely to hold in a stronger norm. Weak limit via a compactness argument requires an *a priori* bound on $\gamma_{N,t}^{(k)}$ uniformly in N.

For example, the trace norm, $\text{Tr}|\gamma_{N,t}^{(k)}| = 1$, is preserved but the collision term B_k is not bounded in this norm unless Φ is bounded (see Spohn's proof [36]). Without such control one cannot prove stability and uniqueness of the solution to the infinite hierarchy (3.26). On the other hand, it is easy to see that the solution to (3.26) is unique in the C^∞-class. But this norm is too strong to prove *a priori* estimates on $\gamma_{N,t}^{(k)}$ since already the helium wavefunction $N = 2$ is not smooth at $x_1 = x_2$.

We define the following *mixed Sobolev norm* for k-particle density matrices

$$\|\gamma^{(k)}\|_{H^{1,(k)}} := \text{Tr}\Big|S_1 S_2 \ldots S_k\, \gamma^{(k)} S_k \ldots S_2 S_1\Big| , \tag{3.29}$$

with $S_j := (I - \Delta_{x_j})^{1/2}$. This norm controls one derivative in each variable of the kernel of the density matrix $\gamma^{(k)}(X_k, X_k')$. Such a control is sufficient to balance even an $|x|^{-2}$ singularity by Hardy's inequality

$$\frac{1}{4|x|^2} \le -\Delta_x \qquad x \in \mathbf{R}^3 . \tag{3.30}$$

It is therefore possible to show that the collision operator is bounded in the norm (3.29),

$$\|B_k \gamma^{(k+1)}\|_{H^{1,(k)}} \le (const.) k \|\gamma^{(k+1)}\|_{H^{1,(k+1)}} ,$$

and this guarantees that the solution of the integral equation (3.27) is unique by iteration. This completes step (ii) above.

As to step (i), we want to show that

$$\|\gamma_{N,t}^{(k)}\|_{H^{1,(k)}} \le C^k ,$$

uniformly in N with some constant C depending only on the initial condition. The N-body dynamics conserves any function of the energy, in particular the k-th power of the Hamiltonian, $\langle \Psi_{N,t}, H_N^k \Psi_{N,t}\rangle$, is preserved for any t. Using Hardy-type inequalities, one expects that H_N^k is comparable with the k-th power of the kinetic energy alone, i.e.

$$C_1^k H_N^k \le \Big(\sum_{j=1}^N -\tfrac{1}{2}\Delta_{x_j} \Big)^k \le C_2^k H_N^k , \tag{3.31}$$

with appropriate positive constants C_1, C_2. This would pass the *a priori* control from the powers of the energy, $\langle \Psi_{N,t}, H_N^k \Psi_{N,t}\rangle$, onto the Sobolev norm $\|\gamma_{N,t}^{(k)}\|_{H^{1,(k)}}$.

The proof of (3.31) is nontrivial. For $k = 1$ it easily follows from Hardy's inequality (3.30), but operator inequalities cannot be raised to higher powers. The inequality $0 \le A \le B$ for operators does not imply $A^2 \le B^2$. The actual proof of (3.31) uses commutator estimates and controlling $\nabla \Phi$ by Hardy's inequality.

However, there is one more complication in the proof of (3.31). High powers of the Coulomb potential $\Phi(x_i - x_j)$ appear in H_N^k. We note that already the third power of $\Phi(x) = |x|^{-1}$ is not locally integrable, hence H_N^k is infinite even on C_0^∞ functions. To overcome this difficulty, we cutoff the Coulomb potential at scale $O(N^{-1/2})$. For the cutoff Hamiltonian the argument outlined above is correct. Then we use L^2-norm to control the effect of this small cutoff. As an additional technicality, we also need to regularize the initial data $\psi_0 \in H^2$ and control the effect of the regularization. For details we refer to [15].

3.2 Low Density and Weak Coupling Limits of Interacting Systems

Finally, we briefly discuss the many-body version of the models in Section 2. We again consider the interacting many-body Schrödinger equation (3.19), (3.20) with a localized smooth potential Φ, but now we are interested in the long time evolution of the system at low densities or at moderate densities with weak coupling. We therefore consider the kinetic scaling limit $(\varepsilon x, \varepsilon t) = (X, T)$.

Let Vol be the total volume occupied by the system. We set the density $\varrho := N/Vol$ and the coupling constant λ in such a way that the total effect of interactions per particle up to time $t = T/\varepsilon$ is $O(1)$. If the particle distribution is more or less uniform and it remains so for later times as well, then this requirement again sets $\lambda^2 \varrho t = O(1)$. The two extreme cases are the low density ($\varrho = O(\varepsilon)$, $\lambda = O(1)$) and the weak coupling ($\varrho = O(1)$, $\lambda = \sqrt{\varepsilon}$) regimes.

Each particle feels a potential field generated by the other particles. If their distribution is random, then it is effectively the one-body LDL or WCL model discussed in Section 2. Therefore it is expected that the evolution of the one particle marginals is described by a self-consistent *nonlinear Boltzmann equation* in the scaling limit. More precisely, the weak limit of the rescaled Wigner transform of the one particle density matrix $\gamma_{N,t}^{(1)}$ converges at least for short macroscopic times, $T \le T_0$,

$$F_T(X,V) := \lim_{\varepsilon \to \infty} W^{\varepsilon}_{\gamma^{(1)}_{N,T/\varepsilon}} ,$$

and the limit satisfies the equation

$$\Big(\partial_T + V \cdot \nabla_X\Big) F_T(X,V) = I[F](X,V) \, .$$

The nonlinear collision mechanism $I[F](X,V)$ is given by

$$\int \sigma(U,V;U',V')\Big\{F(U')F(V')[1-F(U)][1-F(V)] \\ - F(U)F(V)[1-F(U')][1-F(V')]\Big\} \mathrm{d}U \mathrm{d}U' \mathrm{d}V' \, ,$$

(omitting the X and T variables) with a collision kernel

$$\sigma(U,V;U',V') := \varrho_0 \sigma_{sc}(U-U')\delta\Big(U+U'-V-V'\Big)\delta\Big(U^2+(U')^2-V^2-(V')^2\Big) \, ,$$

where σ_{sc} is the one-body scattering kernel of the potential Φ. For the classical dynamics this result has been proven by O. Lanford [27], [28] in the low density case with hard core interaction.

There is no rigorous derivation of the nonlinear Boltzmann equation from many-body quantum dynamics. The major difficulty is that correlations are created along the dynamics and one cannot separate the evolution of a selected particle from the rest. It seems very hard to justify the heuristic argument that the remaining $N-1$ particles behave as a random background for the selected "average" electron.

The analogous problem of accumulating correlations in the high density cases discussed in Section 3.1 is much easier. The particles were either in the semiclassical regime (fermions) or they behaved in a coherent way (bosons). Moreover, the mean field coupling constant was so small that the total potential energy was bounded. In the Boltzmann problem typically only nearby particles interact and the coupling constant is set accordingly. The total potential energy is uncontrollable if many particles happen to cluster. One expects that such concentration cannot occur at least within short macroscopic time if the initial state is sufficiently uncorrelated. Whatever "obvious" this statement sounds, the mathematical technique is lacking for a rigorous proof.

References

1. M. Aizenman, Rev. Math. Phys. **6**, 1163–1182 (1994)
2. M. Aizenman and S. Molchanov, Commun. Math. Phys. **157**, 245–278 (1993)

3. P. Anderson,Phys. Rev. **109**, 1492–1505 (1958)
4. C. Bardos, F. Golse and N. Mauser, Methods Appl. Anal. **2**, 275-293 (2000)
5. C. Bardos, L. Erdős, F. Golse, N. Mauser and H.-T. Yau, *Derivation of the Schrödinger-Poisson equation from the quantum N-body problem.* Submitted to C. R. Acad. Sci. (2001)
6. C. Boldrighini, L. Bunimovich and Y.Sinai, J. Stat. Phys. **32**, 477–501, (1983)
7. F. Castella, L. Erdős, F. Frommlet and P. A. Markowich, J. Stat. Phys. **100** no. (3/4) 543-601, (2000)
8. H. von Dreifus and A. Klein, Commun. Math. Phys. **140**, 133–147 (1991)
9. D. Durr, S. Goldstein and J. Lebowitz, Commun. Math. Phys. **113**, 209–230 (1987)
10. L. Erdős, *Linear Boltzmann equation as the weak coupling limit of the random Schrödinger equation* Operator Theory Advances and Applications, Vol. **108**, 233-242. Eds. J. Dittrich, P. Exner and M. Tater, Birkhäuser (1999)
11. L. Erdős, *Long time dynamics of an electron in a weakly coupled phonon field.* Proceedings of the XIII. International Congress on Mathematical Physics, 273-279, International Press, 2001.
12. L. Erdős, *Linear Boltzmann equation as the long time dynamics of an electron weakly coupled to a phonon field.* To appear in J. Stat. Phys., 2002.
13. L. Erdős, H.-T. Yau, *Linear Boltzmann equation as scaling limit of quantum Lorenz gas*, Advances in Differential Equations and Mathematical Physics. Contemporary Math. **217**, 137–155 (1998)
14. L. Erdős, H.-T. Yau, Commun. Pure Appl. Math. **LIII**, 667-735 (2000)
15. L. Erdős, H.-T. Yau, *Derivation of the nonlinear Schrödinger equation from a many-body Coulomb system.* Accepted for Adv. Theor. Math. Phys. (2001)
16. J. Fröhlich and T. Spencer, Commun. Math. Phys. **88**, 151–184 (1983)
17. G. Gallavotti, *Rigorous theory of the Boltzmann equation in the Lorentz gas*, Nota interna n. 358, Univ. di Roma (1970)
18. P. Gérard, P. A. Markowich, N. J. Mauser and F. Poupaud, Comm. Pure Appl. Math. Vol. L., 323-379 (1997); Erratum: Comm. Pure Appl. Math. Vol. LIII., 280-281 (2000)
19. J. Ginibre and G. Velo, Commun. Math. Phys. **66**, 37-76 (1979) and **68**, 45-68 (1979)
20. K. Hepp, Commun. Math. Phys. **35**, 265-277 (1974)
21. T. G. Ho, L. J. Landau and A. J. Wilkins, Rev. Math. Phys. **5**, 209–298 (1992)
22. A. M. Il'in and R. Z. Khas'minskii, Theory Probab. Its Appl. (USSR) erdos**9**, 421–444 (1964)
23. A. Klein, Math. Res. Lett. **1**, 399–407 (1994)
24. A. Klein, Commun. Math. Phys. **177**, 755–773 (1996)
25. H. Kesten and G. Papanicolaou, Commun. Math. Phys. **78**, 19–63 (1980)
26. W. Kohn, J. M. Luttinger, Phys. Rev. (2) **108**, 590-611 (1957)
27. O. E. Lanford, *Time evolution of large classical systems*, ed. J. Moser, Lecture Notes in Physics Vol. **38**, 1–111 (1975)
28. O. E. Lanford, Astérisque **40**, 117–137 (1976)
29. J. S. Langer, T. Neal, Phys. Rev. Lett. **16**, 984-986 (1966)
30. P.L. Lions and T. Paul, Rev. Mat. Iberoamericana, **9**, 553-618 (1993)
31. P. A. Lee, T. V. Ramakrishnan, Rev. Mod. Phys. **57**, 287-337 (1985)
32. H. Narnhofer and G. L. Sewell, Comm. Math. Phys. **81**, 9-24 (1981)
33. H. Spohn, Commun. Math. Phys. **60**, 277–290 (1978)
34. H. Spohn, J. Stat. Phys., **17**, 385-412 (1977)

35. H. Spohn, Math. Meth. in Appl. Sci. **3**, 445-455 (1981)
36. H. Spohn, Rev. Mod. Phys. **52** no. 3, 569–615 (1980)
37. D. Vollhardt, P. Wölfle, Phys. Rev. B **22**, 4666-4679 (1980)

Subject Index

LIST OF KARPACZ SCHOOLS OF THEORETICAL PHYSICS
Bibliography: 1964 – 2002

I – 1964: R.S. Ingarden (unpubl.), *Particles, Fields and Superconductivity,*

II – 1965: W. Ziętek (ed.), *Symmetries in Particle and Condensed Matter Physics,* Mimeographed Lecture Notes of the University of Wrocław

III – 1966: Z. Galasiewicz (ed.), *Statistical Physics of Condensed Matter,* Acta Universitatis Wratislaviensis, No. 80

IV – 1967: J. Rzewuski (ed.), *Functional Methods in Quantum Field Theory and Statistical Mechanics,* Acta Universitatis Wratislaviensis, No. 88, 89, 90

V – 1968: J. Łopuszański (ed.), *Axiomatic Approach to Quantum Field Theory and Many Body Problem,* Acta Universitatis Wratislaviensis, No. 98, 99, 113

VI – 1969: J. Przystawa (ed.), *Group Theory and Statistical Physics of Condensed Matter,* Mimeographed Lecture Notes of the University of Wrocław

VII – 1970: Z. Galasiewicz (ed.), *Liquid Helium and Many Body Problems,* Acta Universitatis Wratislaviensis, No. 141

VIII – 1971: J. Lukierski (ed.), *New Developments in Relativistic Quantum Field Theory and its Applications,* Acta Universitatis Wratislaviensis, No. 164, Vol. 1 and 2

IX – 1972: K. Wojciechowski (ed.), *Theory of Metals and the Many Body Problem,* Acta Universitatis Wratislaviensis, No. 181, Vol. 1 and 2

X – 1973: J. Lukierski (ed.), *New Developments in Relativistic Quantum Field Theory,* Acta Universitatis Wratislaviensis, No. 207

XI – 1974: J. Łopuszański, A. Pękalski, J. Przystawa (eds.), *Magnetism in Metals and Metallic Compounds,* Plenum Press, London

XII – 1975: W. Garczyński (ed.), *Functional and Probabilistic Methods in Quantum Field Theory,* Acta Universitatis Wratislaviensis, No. 368

XIII – 1976: J. Lukierski (ed.), *Recent Development in Relativistic Quantum Field Theory and Its Application,* Acta Universitatis Wratislaviensis, No. 389

XIV – 1977: Z. Galasiewicz (ed.), *Collective Effects in Condensed Media,* Acta Universitatis Wratislaviensis, No. 436

XV – 1978: W. Karwowski (ed.), *Mathematical Aspects of Quantum Field Theory,* Acta Universitatis Wratislaviensis, No. 519

XVI – 1979: A. Pękalski, J. Przystawa (eds.), *Modern Trends in the Theory of Condensed Matter,* Lecture Notes in Physics No. 115, Springer–Verlag, Berlin

XVII – 1980: L. Turko, A. Pękalski (eds.), *Developments in the Theory of Fundamental Interactions,* Studies in High Energy Physics, Vol. 3, Harwood Acad. Publ., Chur

XVIII – 1981: W. Garczyński (ed.), *Gauge Field Theories: Theoretical Studies and Computer Simulations,* Studies in High Energy Physics, Vol. 4, Harwood Acad. Publ., Chur

1982: *The School did not take place because of the Martial Law in Poland,*

XIX – 1983: B. Milewski (ed.), *Supersymmetry and Supergravity,* World Scientific, Singapore

XX – 1984: A. Pękalski, J. Sznajd (eds.), *Static Critical Phenomena in Inhomogeneous Systems,* Lecture Notes in Physics No. 206, Springer–Verlag, Berlin

XXI – 1985: L. Michel, J. Mozrzymas, A. Pękalski (eds.), *Spontaneous Symmetry Breakdown and Related Subjects,* World Scientific, Singapore

XXII – 1986: A. Jadczyk (ed.), *Fields and Geometry*, World Scientific, Singapore

XXIII – 1987: T. Paszkiewicz (ed.), *Physics of Phonons*, Lecture Notes in Physics No. 285, Springer–Verlag, Berlin

XXIV – 1988: R. Gielerak, W. Karwowski (eds.), *Stochastic Methods in Mathematical Physics*, World Scientific, Singapore

XXV – 1989: Z. Haba, J. Sobczyk (eds.), *Functional Integration, Geometry and Strings*, Birkhaeuser Verlag, Basel

XXVI – 1990: Z. Galasiewicz, A. Pękalski (eds.), *Ordering Phenomena in Condensed Matter Physics*, World Scientific, Singapore

XXVII – 1991: P. Garbaczewski, Z. Popowicz (eds.), *Nonlinear Fields: Classical, Random, Semicassical*, World Scientific, Singapore

XXVIII – 1992: R. Gielerak (ed.), *Infinite Dimensional Geometry in Physics*, Elsevier–North Holland, Amsterdam

XXIX – 1993: T. Paszkiewicz, K. Rapcewicz (eds.), *Die Kunst of Phonons*, Plenum Press, London

XXX – 1994: J. Lukierski, Z. Popowicz, J. Sobczyk (eds.), *Quantum Groups. Formalism and Applications*, Polish Scientific Publishers PWN, Warsaw

XXXI – 1995: P. Garbaczewski, M. Wolf, A. Weron (eds.), *Chaos – The Interplay Between Stochastic and Deterministic Behaviour*, Lecture Notes in Physics No. 457 (1995) Springer-Verlag, Berlin

XXXII – 1996: J. Przystawa, Z. Petru K. Rapcewicz (eds.), *From Quantum Mechanics to Technology*, Lecture Notes in Physics No. 477 (1996) Springer-Verlag, Berlin

XXXIII – 1997: Z. Hasiewicz, J. Jaskólski, J. Sobczyk (eds.), *Duality - String and Fields*, Nucl. Phys.B (Proc. Suppl.) vol. 61A, pp.188 (1998).

XXXIV – 1998: Z. Petru, *From Quantum Mechanics to Quantum Technology*, (unpublished)

XXXV – 1999: J. Kowalski-Glikman (ed.), *Towards Quantum Gravity*, Lecture Notes in Physics (2000) No. 541, pp. 376, Springer-Verlag, Berlin

XXXVI – 2000: A. Pekalski, *Exotic Statistical Physics*, Physica A vol. 285, Nos. 1-2, pp. 238 (2000)

XXXVII – 2001: J. Lukierski, J. Rembielinski (eds.), *New Developments in Fundamental Interaction Theories*, AIP Conference Proceedings vol. 589, pp. 450 (2001)

XXXVIII – 2002: P. Garbaczewski, R. Olkiewicz (eds.), *Dynamical Semigroups: Dissipation, Chaos, Quanta*, (This volume)

Druck: Strauss Offsetdruck, Mörlenbach
Verarbeitung: Schäffer, Grünstadt

Lecture Notes in Physics

For information about Vols. 1–560
please contact your bookseller or Springer-Verlag

Vol. 561: H. J. Carmichael, R. J. Glauber, M. O. Scully (Eds.), Directions in Quantum Optics. XVII, 369 pages. 2001.

Vol. 562: C. Lämmerzahl, C. W. F. Everitt, F. W. Hehl (Eds.), Gyros, Clocks, Interferometers...: Testing Relativistic Gravity in Space. XVII,507 pages. 2001.

Vol. 563: F. C. Lázaro, M. J. Arévalo (Eds.), Binary Stars. Selected Topics on Observations and Physical Processes. 1999.IX, 327 pages. 2001.

Vol. 564: T. Pöschel, S. Luding (Eds.), Granular Gases. VIII, 457 pages. 2001.

Vol. 565: E. Beaurepaire, F. Scheurer, G. Krill, J.-P. Kappler (Eds.), Magnetism and Synchrotron Radiation. XIV, 388 pages. 2001.

Vol. 566: J. L. Lumley (Ed.), Fluid Mechanics and the Environment: Dynamical Approaches. VIII, 412 pages. 2001.

Vol. 567: D. Reguera, L. L. Bonilla, J. M. Rubí (Eds.), Coherent Structures in Complex Systems. IX, 465 pages. 2001.

Vol. 568: P. A. Vermeer, S. Diebels, W. Ehlers, H. J. Herrmann, S. Luding, E. Ramm (Eds.), Continuous and Discontinuous Modelling of Cohesive-Frictional Materials. XIV, 307 pages. 2001.

Vol. 569: M. Ziese, M. J. Thornton (Eds.), Spin Electronics. XVII, 493 pages. 2001.

Vol. 570: S. G. Karshenboim, F. S. Pavone, F. Bassani, M. Inguscio, T. W. Hänsch (Eds.), The Hydrogen Atom: Precision Physics of Simple Atomic Systems. XXIII, 293 pages. 2001.

Vol. 571: C. F. Barenghi, R. J. Donnelly, W. F. Vinen (Eds.), Quantized Vortex Dynamics and Superfluid Turbulence. XXII, 455 pages. 2001.

Vol. 572: H. Latal, W. Schweiger (Eds.), Methods of Quantization. XI, 224 pages. 2001.

Vol. 573: H. M. J. Boffin, D. Steeghs, J. Cuypers (Eds.), Astrotomography. XX, 434 pages. 2001.

Vol. 574: J. Bricmont, D. Dürr, M. C. Galavotti, G. Ghirardi, F. Petruccione, N. Zanghi (Eds.), Chance in Physics. XI, 288 pages. 2001.

Vol. 575: M. Orszag, J. C. Retamal (Eds.), Modern Challenges in Quantum Optics. XXIII, 405 pages. 2001.

Vol. 576: M. Lemoine, G. Sigl (Eds.), Physics and Astrophysics of Ultra-High-Energy Cosmic Rays. X, 327 pages. 2001.

Vol. 577: I. P. Williams, N. Thomas (Eds.), Solar and Extra-Solar Planetary Systems. XVIII, 255 pages. 2001.

Vol. 578: D. Blaschke, N. K. Glendenning, A. Sedrakian (Eds.), Physics of Neutron Star Interiors. XI, 509 pages. 2001.

Vol. 579: R. Haug, H. Schoeller (Eds.), Interacting Electrons in Nanostructures. X, 227 pages. 2001.

Vol. 580: K. Baberschke, M. Donath, W. Nolting (Eds.), Band-Ferromagnetism: Ground-State and Finite-Temperature Phenomena. IX, 394 pages. 2001.

Vol.581: J. M. Arias, M. Lozano (Eds.), An Advanced Course in Modern Nuclear Physics. XI, 346 pages. 2001.

Vol.582: N. J. Balmforth, A. Provenzale (Eds.), Geomorphological Fluid Mechanics. X, 579 pages. 2001.

Vol.583: W. Plessas, L. Mathelitsch (Eds.), Lectures on Quark Matter, XIII, 334 pages. 2002.

Vol.584: W. Köhler, S. Wiegand (Eds.), Thermal Nonequilibrium Phenomena in Fluid Mixtures. XVII, 470 pages. 2002.

Vol.585: M. Lässig, A. Valleriani (Eds.), Biological Evolution and Statistical Physics. XI, 337 pages. 2002.

Vol.586: Y. Auregan, A. Maurel, V. Pagneux, J.-F. Pinton (Eds.), Sound–Flow Interactions. XVI, 286 pages. 2002

Vol.587: D. Heiss (Ed.), Fundamentals of Quantum Information. Quantum Computation, Communication, Decoherence and All That. XIII, 265 pages. 2002.

Vol.588: Y. Watanabe, S. Heun, G. Salviati, N. Yamamoto (Eds.), Nanoscale Spectroscopy and Its Applications to Semiconductor Research. XV, 306 pages. 2002.

Vol.589: A. W. Guthmann, M. Georganopoulos, A. Marcowith, K. Manolakou (Eds.), Relativistic Flows in Astrophysics. XII, 241 pages. 2002

Vol.590: D. Benest, C. Froeschlé (Eds.), Singularities in Gravitational Systems. Applications to Chaotic Transport in the Solar System. XI, 215 pages. 2002

Vol.591: M. Beyer (Ed.), CP Violation in Particle, Nuclear and Astrophysics. XI, 334 pages. 2002

Vol.592: S. Cotsakis, L. Papantonopoulos (Eds.), Cosmological Crossroads. An Advanced Course in Mathematical, Physical and String Cosmology. XVI, 477 pages. 2002

Vol.593: D. Shi, B. Aktaş, L. Pust, F. Mikhailov (Eds.), Nanostructured Magnetic Materials and Their Applications. XII, 289 pages. 2002

Vol.594: S. Odenbach (Ed.),Ferrofluids. Magnetical Controllable Fluids and Their Applications. X, 257 pages. 2002

Vol.595: C. Berthier, L. P. Lévy, G. Martinez (Eds.), High Magnetic Fields. Applications in Condensed Matter Physics and Spectroscopy. X, 493 pages. 2002

Vol.596: F. Scheck, H. Upmeier, W. Werner (Eds.), Noncommutative Geometry and the Standard Model of Elememtary Particle Physics. XIV, 415 pages. 2002

Vol.597: P. Garbaczewski, R. Olkiewicz (Eds.), Dynamics of Dissipation. X, 512 pages. 2002

Monographs

For information about Vols. 1–29 please contact your bookseller or Springer-Verlag

Vol. m 30: A. J. Greer, W. J. Kossler, Low Magnetic Fields in Anisotropic Superconductors. VII, 161 pages. 1995.

Vol. m 31 (Corr. Second Printing): P. Busch, M. Grabowski, P.J. Lahti, Operational Quantum Physics. XII, 230 pages. 1997.

Vol. m 32: L. de Broglie, Diverses questions de mécanique et de thermodynamique classiques et relativistes. XII, 198 pages. 1995.

Vol. m 33: R. Alkofer, H. Reinhardt, Chiral Quark Dynamics. VIII, 115 pages. 1995.

Vol. m 34: R. Jost, Das Märchen vom Elfenbeinernen Turm. VIII, 286 pages. 1995.

Vol. m 35: E. Elizalde, Ten Physical Applications of Spectral Zeta Functions. XIV, 224 pages. 1995.

Vol. m 36: G. Dunne, Self-Dual Chern-Simons Theories. X, 217 pages. 1995.

Vol. m 37: S. Childress, A.D. Gilbert, Stretch, Twist, Fold: The Fast Dynamo. XI, 406 pages. 1995.

Vol. m 38: J. González, M. A. Martín-Delgado, G. Sierra, A. H. Vozmediano, Quantum Electron Liquids and High-Tc Superconductivity. X, 299 pages. 1995.

Vol. m 39: L. Pittner, Algebraic Foundations of Non-Com-mutative Differential Geometry and Quantum Groups. XII, 469 pages. 1996.

Vol. m 40: H.-J. Borchers, Translation Group and Particle Representations in Quantum Field Theory. VII, 131 pages. 1996.

Vol. m 41: B. K. Chakrabarti, A. Dutta, P. Sen, Quantum Ising Phases and Transitions in Transverse Ising Models. X, 204 pages. 1996.

Vol. m 42: P. Bouwknegt, J. McCarthy, K. Pilch, The W3 Algebra. Modules, Semi-infinite Cohomology and BV Algebras. XI, 204 pages. 1996.

Vol. m 43: M. Schottenloher, A Mathematical Introduction to Conformal Field Theory. VIII, 142 pages. 1997.

Vol. m 44: A. Bach, Indistinguishable Classical Particles. VIII, 157 pages. 1997.

Vol. m 45: M. Ferrari, V. T. Granik, A. Imam, J. C. Nadeau (Eds.), Advances in Doublet Mechanics. XVI, 214 pages. 1997.

Vol. m 46: M. Camenzind, Les noyaux actifs de galaxies. XVIII, 218 pages. 1997.

Vol. m 47: L. M. Zubov, Nonlinear Theory of Dislocations and Disclinations in Elastic Body. VI, 205 pages. 1997.

Vol. m 48: P. Kopietz, Bosonization of Interacting Fermions in Arbitrary Dimensions. XII, 259 pages. 1997.

Vol. m 49: M. Zak, J. B. Zbilut, R. E. Meyers, From Instability to Intelligence. Complexity and Predictability in Nonlinear Dynamics. XIV, 552 pages. 1997.

Vol. m 50: J. Ambjørn, M. Carfora, A. Marzuoli, The Geometry of Dynamical Triangulations. VI, 197 pages. 1997.

Vol. m 51: G. Landi, An Introduction to Noncommutative Spaces and Their Geometries. XI, 200 pages. 1997.

Vol. m 52: M. Hénon, Generating Families in the Restricted Three-Body Problem. XI, 278 pages. 1997.

Vol. m 53: M. Gad-el-Hak, A. Pollard, J.-P. Bonnet (Eds.), Flow Control. Fundamentals and Practices. XII, 527 pages. 1998.

Vol. m 54: Y. Suzuki, K. Varga, Stochastic Variational Approach to Quantum-Mechanical Few-Body Problems. XIV, 324 pages. 1998.

Vol. m 55: F. Busse, S. C. Müller, Evolution of Spontaneous Structures in Dissipative Continuous Systems. X, 559 pages. 1998.

Vol. m 56: R. Haussmann, Self-consistent Quantum Field Theory and Bosonization for Strongly Correlated Electron Systems. VIII, 173 pages. 1999.

Vol. m 57: G. Cicogna, G. Gaeta, Symmetry and Perturbation Theory in Nonlinear Dynamics. XI, 208 pages. 1999.

Vol. m 58: J. Daillant, A. Gibaud (Eds.), X-Ray and Neutron Reflectivity: Principles and Applications. XVIII, 331 pages. 1999.

Vol. m 59: M. Kriele, Spacetime. Foundations of General Relativity and Differential Geometry. XV, 432 pages. 1999.

Vol. m 60: J. T. Londergan, J. P. Carini, D. P. Murdock, Binding and Scattering in Two-Dimensional Systems. Applications to Quantum Wires, Waveguides and Photonic Crystals. X, 222 pages. 1999.

Vol. m 61: V. Perlick, Ray Optics, Fermat's Principle, and Applications to General Relativity. X, 220 pages. 2000.

Vol. m 62: J. Berger, J. Rubinstein, Connectivity and Superconductivity. XI, 246 pages. 2000.

Vol. m 63: R. J. Szabo, Ray Optics, Equivariant Cohomology and Localization of Path Integrals. XII, 315 pages. 2000.

Vol. m 64: I. G. Avramidi, Heat Kernel and Quantum Gravity. X, 143 pages. 2000.

Vol. m 65: M. Hénon, Generating Families in the Restricted Three-Body Problem. Quantitative Study of Bifurcations. XII, 301 pages. 2001.

Vol. m 66: F. Calogero, Classical Many-Body Problems Amenable to Exact Treatments. XIX, 749 pages. 2001.

Vol. m 67: A. S. Holevo, Statistical Structure of Quantum Theory. IX, 159 pages. 2001.

Vol. m 68: N. Polonsky, Supersymmetry: Structure and Phenomena. Extensions of the Standard Model. XV, 169 pages. 2001.

Vol. m 69: W. Staude, Laser-Strophometry. High-Resolution Techniques for Velocity Gradient Measurements in Fluid Flows. XV, 178 pages. 2001.

Vol. m 70: P. T. Chruściel, J. Jezierski, J. Kijowski, Hamiltonian Field Theory in the Radiating Regime. VI, 172 pages. 2002.

Vol. m 71: S. Odenbach, Magnetoviscous Effects in Ferrofluids. X, 151 pages. 2002.

Vol. m 72: J. G. Muga, R. Sala Mayato, I. L. Egusquiza (Eds.), Time in Quantum Mechanics. XII, 419 pages. 2002.